THE PHYSICS OF WARM NUCLEI

OXFORD STUDIES IN NUCLEAR PHYSICS

General editor: P. E. Hodgson

1. J. L. Emmerson: *Symmetry principles in particle physics*
2. J. M. Irvine: *Heavy nuclei, superheavy nuclei, and neutron stars*
3. I. S. Towner: *A shell-model description of light nuclei*
4. P. E. Hodgson: *Nuclear heavy ion reactions*
5. R. D. Lawson: *Theory of nuclear shell model*
6. W. E. Frahn: *Diffractive process in nuclear physics*
7. S. S. M. Wong: *Nuclear statistical spectroscopy*
8. N. Ullah: *Matrix ensembles in the many-nucleon problem*
9. A. N. Antonov, P. E. Hodgson, I. Zh. Petkov: *Nucleon momentum and density distributions in nuclei*
10. D. Bonatsos: *Interacting boson models of nuclear structure*
11. H. Ejiri, M. J. A. deVoigt: *Gamma-ray and electron spectroscopy in nuclear physics*
12. B. Castel, I. S. Towner: *Modern theories of nuclear physics moments*
13. R. F. Casten: *Nuclear structure from a simple perspective*
14. I. Zh. Petkov, M. V. Stitsov: *Nuclear density function theory*
15. I. Gadioli, P. E. Hodgson: *Pre-equilibrium nuclear reactions*
16. J. D. Walecka: *Theoretical nuclear and subnuclear physics*
17. D. N. Poenaru, W. Greiner: *Handbook of nuclear properties*
18. P. Fröbrich, R. Lipperheide: *Theory of nuclear properties*
19. F. Stancu: *Group theory in subnuclear physics*
20. S. Boffi, C. Giusti, F. D. Pacati, M. Radici: *Electromagnetic response of atomic nuclei*
21. G. Ripka: *Quarks bound by chiral fields*
22. H. Ejiri, H. Toki (eds): *Nucleon–hadron many-body systems*
23. R. F. Casten: *Nuclear structure from a simple perspective, second edition*
24. M. N. Harakeh, A. van der Woude: *Giant resonances: fundamental high-frequency modes of nuclear excitation*
25. H. Hofmann: *The physics of warm nuclei: with analogies to mesoscopic systems*

THE PHYSICS OF WARM NUCLEI
with analogies to mesoscopic systems

HELMUT HOFMANN

Department of Physics
Technical University of Munich

OXFORD
UNIVERSITY PRESS

Great Clarendon Street, Oxford OX2 6DP

Oxford University Press is a department of the University of Oxford.
It furthers the University's objective of excellence in research, scholarship,
and education by publishing worldwide in

Oxford New York

Auckland Cape Town Dar es Salaam Hong Kong Karachi
Kuala Lumpur Madrid Melbourne Mexico City Nairobi
New Delhi Shanghai Taipei Toronto

With offices in

Argentina Austria Brazil Chile Czech Republic France Greece
Guatemala Hungary Italy Japan Poland Portugal Singapore
South Korea Switzerland Thailand Turkey Ukraine Vietnam

Oxford is a registered trade mark of Oxford University Press
in the UK and in certain other countries

Published in the United States
by Oxford University Press Inc., New York

© Helmut Hofmann 2008

The moral rights of the author have been asserted
Database right Oxford University Press (maker)

First published 2008

All rights reserved. No part of this publication may be reproduced,
stored in a retrieval system, or transmitted, in any form or by any means,
without the prior permission in writing of Oxford University Press,
or as expressly permitted by law, or under terms agreed with the appropriate
reprographics rights organization. Enquiries concerning reproduction
outside the scope of the above should be sent to the Rights Department,
Oxford University Press, at the address above

You must not circulate this book in any other binding or cover
and you must impose the same condition on any acquirer

British Library Cataloguing in Publication Data
Data available

Library of Congress Cataloging in Publication Data
Data available

Printed in Great Britain
on acid-free paper by
Biddles Ltd., www.biddles.co.uk

ISBN 978–0–19–850401–6

1 3 5 7 9 10 8 6 4 2

To Helga

PREFACE

Ein Buch, das nicht wert ist, wenigstens zweimal gelesen zu werden, ist auch nicht wert, dass man es einmal liest.[1]

Von allen Unmöglichkeiten ist es die größte, ein Buch zu schreiben, das allen gefällt.[2]

Karl Julius Weber

A full comprehension of present-day nuclear physics is no longer possible solely on the basis of a knowledge of quantum mechanics. In the past decades nuclear physics has developed into a field which borrows methods and concepts from diverse areas in physics. For an understanding of the nuclear force, for instance, formulations in terms of quantum chromodynamics are in sight. To follow this line requires the application of quantum relativistic field theory, together with elements of elementary particle physics. In typical accelerator experiments with heavy ions, on the other hand, nuclear compounds are created with large excitation energies for which methods of statistical physics are indispensable. It is true that a similar situation has already been faced in conventional reaction theory. Different methods are necessary, however, to cope with the problem of collective motion. In many cases the latter is of a timescale much larger than that of the residual degrees of freedom. This is the realm of non-equilibrium statistical mechanics with transport equations at one's disposal.

In this book we concentrate on examples which fall in this latter category. The notion "warm nuclei" is meant to indicate that only very moderate thermal excitations will be considered. A "nuclear temperature" should not only be much smaller than the Fermi energy, but will be assumed small enough so that some of those many-body features are still present, which in a sense represent traditional nuclear physics. Take shell effects as one of the most prominent examples, known to disappear at $k_B T \simeq 3$ MeV. The study of this transition will play a major role in our discussion, in particular because of its intriguing implications for transport properties. We will, however, refrain from describing any effects which go beyond this regime. In fact for hotter systems the nucleons' dynamics may no longer be associated with that in a deformed mean field. At still larger energies subnucleonic degrees of freedom may be excited, for which field theoretical concepts are involved, and which are beyond the scope of this book.

It should be evident that concepts of statistical mechanics cannot be taken over to nuclear physics without serious forethought. This holds true both for

---

[1] A book not worthy of being read twice is not worthy of being read once.
[2] The greatest of all impossibilities is to write a book which pleases everyone.

static and dynamic properties, as well as for those involving irreversible features. Most of the problems are related to the fact that the nucleus is a small, self-bound and thermally isolated system. For example, a temperature can be defined only within some inherent uncertainty, and will not stay constant in any dynamic process. This is so even in the reversible case as required by the constancy of entropy. The latter feature is intimately related to the extension of the quantal adiabaticity theorem to statistical mechanics, and thus to the difference of the susceptibilities involved. Another problem is that collective degrees of freedom are not given per se; they must be introduced obeying conditions of self-consistency with those of the nucleons. This feature is already known for the conventional case of zero thermal excitation, but it still plays an important role at finite excitations. It is predominantly here that irreversible features come into play, which for nuclear physics are not unexpected. Indeed, they have already shown up in traditional theories of nuclear reactions, like those for overlapping resonances and in relation to the nuclear compound model.

A large part of the book is devoted to explaining and developing such aspects at the borders where microscopic and macroscopic physics meet. Some of them may be applied to other finite Fermi systems, like metal clusters and quantum dots or wires. In some cases one may simply dwell on analogies, like the appearance of nuclear dissipation and the absorption of electromagnetic waves in metal clusters. However, as is known, many methods have been applied outside of nuclear physics which were originally developed solely for that domain.

To cover such a wide field is not an easy task. The general reader must not only be introduced to the basics of nuclear physics (and of mesoscopic systems), but also to elementary topics of the theoretical tools involved. For this reason the book is split into different parts. The first part is meant to acquaint the reader with methods aimed at an understanding of the essentials of collective motion and transport problems used in nuclear physics. In the second and third parts applications to nuclear reactions and mesoscopic systems are given. The fourth part contains a thorough discussion of several theoretical methods and is kept at a more general level, but with emphasis on the specific problems discussed in the other parts. The overall level of the presentation is oriented at a general, alert reader, not necessarily a specialist in nuclear physics. In fact, parts 1 and 4 are meant to serve as self-contained introductions to the fields presented there. Hardly more than a basic knowledge of quantum mechanics is required for an easy understanding of such diverse fields as the theories of reaction, transport or linear response, as well as of the methods of Wigner transformations or functional integrals.

ACKNOWLEDGEMENTS

It is a great pleasure to acknowledge the credit due to the many scientists and colleagues with whom I had close contact over the years in which my understanding of the topics of this book has grown steadily albeit sometimes very slowly. First I should like to acknowledge the important contributions from my former collaborators R. Alkofer, G.-L. Ingold, F.A. Ivanyuk, A.S. Jensen, D. Kiderlen, A.G. Magner, R. Nix, C. Ngo, C. Rummel, R. Samhammer, P.J. Siemens, R. Sollacher and S. Yamaji for the fruits of joint publications which have been incorporated within these pages. I should also like to express my deep obligation and the great benefit I have had from discussions with Professors A. Bohr, R. Balian, M. Brack, W. Brenig, K. Dietrich, R.J. Glauber, W. Götze, D. Kusnezov, B.R. Mottelson, D. Pines, P. Ring, V.M. Strutinsky, N.G. van Kampen, M. Vénéroni, H.A. Weidenmüuller, V.G. Zelevinsky and S. Aaberg, in which I have been informed about pertinent methods and ideas that have often prevented me from choosing inappropriate paths. My final thanks go to Professor J.A. Wheeler who was the first to suggest that I take up the endeavor of writing this book, to Professor R.H. Lemmer who gave the final push to this project, to Professor W.U. Schröder for his continued interest during the preparation period, and to Dr. F.A. Ivanyuk for his invaluable help in the preparation of figures, as well as Dr. R.R. Hilton for much advice not only about questions of proper English style.

COPYRIGHT PERMISSIONS

The author gratefully acknowledges permission to reprint figures previously published in various journals. The references to these figures can be found in the figure captions. Thanks are due both to the authors as well as to the publishers. This concerns the following entries and publishers or journals:

- Elsevier for Figs. 3.2, 5.1, 6.1-6.11, 7.1, 7.4, 11.2, 13.1, 13.3, 17.2.
- American Physical Society for Figs. 3.2, 4.2, 6.4, 12.1-12.8
- Springer Science and Business Media for Figs. 4.1, 4.3 and 6.7
- Progress of Theoretical Physics for Figs. 5.5 and 6.6.
- Acta Physica Polonica B for Fig. 17.1
- Nature for Fig. 5.4
- World Scientific (IJMPE) Fig. 7.4

CONTENTS

I BASIC ELEMENTS AND MODELS

1 Elementary concepts of nuclear physics — 3
- 1.1 The force between two nucleons — 3
  - 1.1.1 Possible forms of the interaction — 4
  - 1.1.2 The radial dependence of the interaction — 5
  - 1.1.3 The role of sub-nuclear degrees of freedom — 6
- 1.2 The model of the Fermi gas — 7
  - 1.2.1 Many-body properties in the ground state — 9
  - 1.2.2 Two-body correlations in a homogeneous system — 11
- 1.3 Basic properties of finite nuclei — 14
  - 1.3.1 The interaction of nucleons with nuclei — 14
  - 1.3.2 The optical model — 19
  - 1.3.3 The liquid drop model — 23

2 Nuclear matter as a Fermi liquid — 30
- 2.1 A first, qualitative survey — 30
  - 2.1.1 The inadequacy of Hartree–Fock with bare interactions — 30
  - 2.1.2 Short-range correlations — 32
  - 2.1.3 Properties of nuclear matter in chiral dynamics — 34
  - 2.1.4 How dense is nuclear matter? — 35
- 2.2 The independent pair approximation — 36
  - 2.2.1 The equation for the one-body wave functions — 36
  - 2.2.2 The total energy in terms of two-body wave functions — 37
  - 2.2.3 The Bethe–Goldstone equation — 38
  - 2.2.4 The G-matrix — 42
- 2.3 Brueckner–Hartree–Fock approximation (BHF) — 44
  - 2.3.1 BHF at finite temperature — 48
- 2.4 A variational approach based on generalized Jastrow functions — 48
  - 2.4.1 Extension to finite temperature — 50
- 2.5 Effective interactions of Skyrme type — 52
  - 2.5.1 Expansion to small relative momenta — 52
- 2.6 The nuclear equation of state (EOS) — 55
  - 2.6.1 An energy functional with three-body forces — 55
  - 2.6.2 The EOS with Skyrme interactions — 56
  - 2.6.3 Applications in astrophysics — 59
- 2.7 Transport phenomena in the Fermi liquid — 61
  - 2.7.1 Semi-classical transport equations — 62

3 Independent particles and quasiparticles in finite nuclei — 67
- 3.1 Hartree–Fock with effective forces — 67
  - 3.1.1 H–F with the Skyrme interaction — 67
  - 3.1.2 Constrained Hartree–Fock — 68
  - 3.1.3 Other effective interactions — 69
- 3.2 Phenomenological single particle potentials — 70
  - 3.2.1 The spherical case — 70
  - 3.2.2 The deformed single particle model — 73
- 3.3 Excitations of the many-body system — 78
  - 3.3.1 The concept of particle-hole excitations — 78
  - 3.3.2 Pair correlations — 79

4 From the shell model to the compound nucleus — 85
- 4.1 Shell model with residual interactions — 85
  - 4.1.1 Nearest level spacing — 86
- 4.2 Random Matrix Model — 87
  - 4.2.1 Gaussian ensembles of real symmetric matrices — 87
  - 4.2.2 Eigenvalues, level spacings and eigenvectors — 89
  - 4.2.3 Comments on the RMM — 91
- 4.3 The spreading of states into more complicated configurations — 93
  - 4.3.1 A schematic model — 94
  - 4.3.2 Strength functions — 95
  - 4.3.3 Time-dependent description — 98
  - 4.3.4 Spectral functions for single particle motion — 99

5 Shell effects and Strutinsky renormalization — 104
- 5.1 Physical background — 106
  - 5.1.1 The independent particle picture, once more — 107
  - 5.1.2 The Strutinsky energy theorem — 108
- 5.2 The Strutinsky procedure — 109
  - 5.2.1 Formal aspects of smoothing — 109
  - 5.2.2 Shell correction to level density and ground state energy — 110
  - 5.2.3 Further averaging procedures — 113
- 5.3 The static energy of finite nuclei — 115
- 5.4 An excursion into periodic-orbit theory (POT) — 119
- 5.5 The total energy at finite temperature — 122
  - 5.5.1 The smooth part of the energy at small excitations — 123
  - 5.5.2 Contributions from the oscillating level density — 124

6 Average collective motion of small amplitude — 128
- 6.1 Equation of motion from energy conservation — 129
  - 6.1.1 Induced forces for harmonic motion — 129
  - 6.1.2 Equation of motion — 131
  - 6.1.3 One-particle one-hole excitations — 133

6.2	The collective response function		134
	6.2.1	Collective response and sum rules for stable systems	137
	6.2.2	Generalization to several dimensions	139
	6.2.3	Mean field approximation for an effective two-body interaction	141
	6.2.4	Isovector modes	143
6.3	Rotations as degenerate vibrations		143
6.4	Microscopic origin of macroscopic damping		145
	6.4.1	Irreversibility through energy smearing	146
	6.4.2	Relaxation in a Random Matrix Model	150
	6.4.3	The effects of "collisions" on nucleonic motion	150
6.5	Damped collective motion at thermal excitations		154
	6.5.1	The equation of motion at finite thermal excitations	154
	6.5.2	The strict Markov limit	157
	6.5.3	The collective response for quasi-static processes	160
	6.5.4	An analytically solvable model	164
6.6	Temperature dependence of nuclear transport		166
	6.6.1	The collective strength distribution at finite T	166
	6.6.2	Diabatic models	171
	6.6.3	T-dependence of transport coefficients	177
6.7	Rotations at finite thermal excitations		185

7 Transport theory of nuclear collective motion   190

7.1	The locally harmonic approximation		191
7.2	Equilibrium fluctuations of the local oscillator		194
7.3	Fluctuations of the local propagators		196
	7.3.1	Quantal diffusion coefficients from the FDT	200
7.4	Fokker–Planck equations for the damped harmonic oscillator		203
	7.4.1	Stationary solutions for oscillators	203
	7.4.2	Dynamics of fluctuations for stable modes	206
	7.4.3	The time-dependent solutions for unstable modes and their physical interpretation	207
7.5	Quantum features of collective transport from the microscopic point of view		209
	7.5.1	Quantized Hamiltonians for collective motion	210
	7.5.2	A non-perturbative Nakajima–Zwanzig approach	217

II COMPLEX NUCLEAR SYSTEMS

8 The statistical model for the decay of excited nuclei   225

8.1	Decay of the compound nucleus by particle emission		225
	8.1.1	Transition rates	225
	8.1.2	Evaporation rates for light particles	228
8.2	Fission		229
	8.2.1	The Bohr–Wheeler formula	229

	8.2.2 Stability conditions in the macroscopic limit	232
9	**Pre-equilibrium reactions**	**235**
	9.1 An illustrative, realistic prototype	236
	9.2 A sketch of existing theories	242
	9.2.1 Comments	244
10	**Level densities and nuclear thermometry**	**246**
	10.1 Darwin–Fowler approach for theoretical models	246
	10.1.1 Level densities and Strutinsky renormalization	247
	10.1.2 Dependence on angular momentum	251
	10.1.3 Microscopic models with residual interactions	252
	10.2 Empirical level densities	254
	10.3 Nuclear thermometry	257
11	**Large-scale collective motion at finite thermal excitations**	**262**
	11.1 Global transport equations	262
	11.1.1 Fokker–Planck equations	262
	11.1.2 Over-damped motion	265
	11.1.3 Langevin equations	267
	11.1.4 Probability distribution for collective variables	269
	11.2 Transport coefficients for large-scale motion	270
	11.2.1 The LHA at level crossings and avoided crossings	275
	11.2.2 Thermal aspects of global motion	278
12	**Dynamics of fission at finite temperature**	**280**
	12.1 Transitions between potential wells	280
	12.1.1 Transition rate for over-damped motion	281
	12.2 The rate formulas of Kramers and Langer	283
	12.3 Escape time for strongly damped motion	288
	12.4 A critical discussion of timescales	291
	12.4.1 Transient- and saddle-scission times	293
	12.4.2 Implications from the concept of the MFPT	296
	12.5 Inclusion of quantum effects	299
	12.5.1 Quantum decay rates within the LHA	300
	12.5.2 Rate formulas for motion treated self-consistently	302
	12.5.3 Quantum effects in collective transport, a true challenge	307
13	**Heavy-ion collisions at low energies**	**308**
	13.1 Transport models for heavy-ion collisions	309
	13.1.1 Commonly used inputs for transport equations	313
	13.2 Differential cross sections	317
	13.3 Fusion reactions	319
	13.3.1 Micro- and macroscopic formation probabilities	321

	13.4 Critical remarks on theoretical approaches and their assumptions	326
14	**Giant dipole excitations**	**330**
	14.1 Absorption and radiation of the classical dipole	330
	14.2 Nuclear dipole modes	332
	14.2.1 Extension to quantum mechanics	333
	14.2.2 Damping of giant dipole modes	333

III MESOSCOPIC SYSTEMS

15	**Metals and quantum wires**	**341**
	15.1 Electronic transport in metals	341
	15.1.1 The Drude model and basic definitions	341
	15.1.2 The transport equation and electronic conductance	342
	15.2 Quantum wires	344
	15.2.1 Mesoscopic systems in semiconductor heterostructures	344
	15.2.2 Two-dimensional electron gas	345
	15.2.3 Quantization of conductivity for ballistic transport	346
	15.2.4 Physical interpretation and discussion	348
16	**Metal clusters**	**350**
	16.1 Structure of metal clusters	350
	16.2 Optical properties	351
	16.2.1 Cross sections for scattering of light	353
	16.2.2 Optical properties for the jellium model	354
	16.2.3 The infinitely deep square well	355
17	**Energy transfer to a system of independent fermions**	**361**
	17.1 Forced energy transfer within the wall picture	361
	17.1.1 Energy transfer at finite frequency	363
	17.1.2 Fermions inside billiards	365
	17.2 Wall friction by Strutinsky smoothing	366

IV THEORETICAL TOOLS

18	**Elements of reaction theory**	**373**
	18.1 Potential scattering	373
	18.1.1 The T-matrix	373
	18.1.2 Phase shifts for central potentials	379
	18.1.3 Inelastic processes	380
	18.2 Generalization to nuclear reactions	382
	18.2.1 Reaction channels	382
	18.2.2 Cross section	383
	18.2.3 The T-matrix for nuclear reactions	384
	18.2.4 Isolated resonances	385

	18.2.5 Overlapping resonances	389
	18.2.6 T-matrix with angular momentum coupling	389
18.3	Energy averaged amplitudes	390
	18.3.1 The optical model	390
	18.3.2 Intermediate structure through doorway resonances	392
18.4	Statistical theory	395
	18.4.1 Porter–Thomas distribution for widths	396
	18.4.2 Smooth and fluctuating parts of the cross section	396
	18.4.3 Hauser–Feshbach theory	401
	18.4.4 Critique of the statistical model	403

19 Density operators and Wigner functions — 406
- 19.1 The many-body system — 406
  - 19.1.1 Hilbert states of the many-body system — 406
  - 19.1.2 Density operators and matrices — 406
  - 19.1.3 Reduction to one- and two-body densities — 408
- 19.2 Many-body functions from one-body functions — 410
  - 19.2.1 One- and two-body densities — 411
- 19.3 The Wigner transformation — 412
  - 19.3.1 The Wigner transform in three dimensions — 412
  - 19.3.2 Many-body systems of indistinguishable particles — 414
  - 19.3.3 Propagation of wave packets — 415
  - 19.3.4 Correspondence rules — 416
  - 19.3.5 The equilibrium distribution of the oscillator — 417

20 The Hartree–Fock approximation — 420
- 20.1 Hartree–Fock with density operators — 420
  - 20.1.1 The Hartree–Fock equations — 422
  - 20.1.2 The ground state energy in HF-approximation — 423
- 20.2 Hartree–Fock at finite temperature — 424
  - 20.2.1 TDHF at finite T — 425

21 Transport equations for the one-body density — 426
- 21.1 The Wigner transform of the von Neumann equation — 426
- 21.2 Collision terms in semi-classical approximations — 428
  - 21.2.1 The collision term in the Born approximation — 430
  - 21.2.2 The BUU and the Landau–Vlasov equation — 431
- 21.3 Relaxation to equilibrium — 432
  - 21.3.1 Relaxation time approximation to the collision term — 434
  - 21.3.2 A few remarks on the concept of self-energies — 435

22 Nuclear thermostatics — 437
- 22.1 Elements of statistical mechanics — 437
  - 22.1.1 Thermostatics for deformed nuclei — 438
  - 22.1.2 Generalized ensembles — 443
  - 22.1.3 Extremal properties — 448

22.2	Level densities and energy distributions	450
	22.2.1 Composite systems	452
	22.2.2 A Gaussian approximation	454
	22.2.3 Darwin–Fowler method for the level density	457
22.3	Uncertainty of temperature for isolated systems	461
	22.3.1 The physical background	461
	22.3.2 The thermal uncertainty relation	463
22.4	The lack of extensivity and negative specific heats	464
22.5	Thermostatics of independent particles	466
	22.5.1 Sommerfeld expansion for smooth level densities	469
	22.5.2 Thermostatics for oscillating level densities	470
	22.5.3 Influence of angular momentum	472

23 Linear response theory 475

23.1	The model of the damped oscillator	475
23.2	A brief reminder of perturbation theory	478
	23.2.1 Transition rate in lowest order	480
23.3	General properties of response functions	482
	23.3.1 Basic definitions	482
	23.3.2 Basic properties	484
	23.3.3 Dissipation of energy	486
	23.3.4 Spectral representations	487
23.4	Correlation functions and the fluctuation dissipation theorem	489
	23.4.1 Basic definitions	489
	23.4.2 The fluctuation dissipation theorem	491
	23.4.3 Strength functions for periodic perturbations	492
	23.4.4 Linear response for a Random Matrix Model	493
23.5	Linear response at complex frequencies	496
	23.5.1 Relation to thermal Green functions	497
	23.5.2 Response functions for unstable modes	498
	23.5.3 Equilibrium fluctuations of the oscillator	499
23.6	Susceptibilities and the static response	501
	23.6.1 Static perturbations of the local equilibrium	501
	23.6.2 Isothermal and adiabatic susceptibilities	504
	23.6.3 Relations to the static response	505
23.7	Linear irreversible processes	507
	23.7.1 Relaxation functions	507
	23.7.2 Variation of entropy in time	510
	23.7.3 Time variation of the density operator	512
	23.7.4 Onsager relations for macroscopic motion	515
23.8	Kubo formula for transport coefficients	518

24 Functional integrals 522

24.1	Path integrals in quantum mechanics	522
	24.1.1 Time propagation in quantum mechanics	522

		24.1.2 Semi-classical approximation to the propagator	525

- 24.1.2 Semi-classical approximation to the propagator — 525
- 24.1.3 The path integral as a functional — 531
- 24.2 Path integrals for statistical mechanics — 532
  - 24.2.1 The classical limit of statistical mechanics — 535
  - 24.2.2 Quantum corrections — 536
- 24.3 Green functions and level densities — 539
  - 24.3.1 Periodic orbit theory — 540
  - 24.3.2 The level density for regular and chaotic motion — 542
- 24.4 Functional integrals for many-body systems — 543
  - 24.4.1 The Hubbard–Stratonovich transformation — 543
  - 24.4.2 The high temperature limit and quantum corrections — 546
  - 24.4.3 The perturbed static path approximation (PSPA) — 548

25 Properties of Langevin and Fokker–Planck equations — 554

- 25.1 The Brownian particle, a heuristic approach — 554
  - 25.1.1 Langevin equation — 554
  - 25.1.2 Fokker–Planck equations — 557
  - 25.1.3 Cumulant expansion and Gaussian distributions — 559
- 25.2 General properties of stochastic processes — 560
  - 25.2.1 Basic concepts — 561
  - 25.2.2 Markov processes and the Chapman–Kolmogorov equation — 562
  - 25.2.3 Fokker–Planck equations from the Kramers–Moyal expansion — 564
  - 25.2.4 The master equation — 568
- 25.3 Non-linear equations in one dimension — 570
  - 25.3.1 Transport equations for multiplicative noise — 570
  - 25.3.2 Properties of the general Fokker–Planck equation — 572
- 25.4 The mean first passage time — 573
  - 25.4.1 Differential equation for the MFPT — 574
- 25.5 The multidimensional Kramers equation — 575
  - 25.5.1 Gaussian solutions in curvilinear coordinates — 576
  - 25.5.2 Time dependence of first and second moments for the harmonic oscillator — 578
- 25.6 Microscopic approach to transport problems — 580
  - 25.6.1 The Nakajima–Zwanzig projection technique — 580
  - 25.6.2 Perturbative approach for factorized coupling — 582

V AUXILIARY INFORMATION

26 Formal means — 589

- 26.1 Gaussian integrals — 589
- 26.2 Stationary phase and steepest decent — 590
- 26.3 The δ-function — 591
- 26.4 Fourier and Laplace transformations — 591

26.5	Derivative of exponential operators	592
26.6	The Mori product	592
26.7	Spin and isospin	593
26.8	Second quantization for fermions	594

27 Natural units in nuclear physics   596

References   597

Index   615

PART I

BASIC ELEMENTS AND MODELS

1

ELEMENTARY CONCEPTS OF NUCLEAR PHYSICS

1.1 The force between two nucleons

Nucleons interact with each other mainly through the strong interaction which obeys fundamental symmetries. Discarding at first a possible dependence on the nucleon's spins and momenta the Hamiltonian for two assumedly point-like particles can be written as

$$\hat{H} = \frac{\hat{\mathbf{p}}_1^2}{2m_1} + \frac{\hat{\mathbf{p}}_2^2}{2m_2} + \hat{V}(\mathbf{r}) = \frac{\hat{\mathbf{p}}^2}{2\mu} + \hat{V}(\mathbf{r}) + \frac{\hat{\mathbf{P}}^2}{2M} \qquad (1.1)$$

with $M = m_1 + m_2$ and $\mu = m_1 m_2 / M$ being the total and the reduced mass, respectively. The second form on the right follows directly from the definition of the vectors of relative motion for coordinates and momenta,

$$\mathbf{r} = \mathbf{r}_1 - \mathbf{r}_2 \qquad \mathbf{p} = (m_2 \mathbf{p}_1 - m_1 \mathbf{p}_2)/M, \qquad (1.2)$$

and the corresponding ones for the center of mass,

$$\mathbf{R} = (m_1 \mathbf{r}_1 + m_2 \mathbf{r}_2)/M \qquad \mathbf{P} = \mathbf{p}_1 + \mathbf{p}_2. \qquad (1.3)$$

Translational invariance is reflected by the property of $\hat{H}$ to commute with the total momentum $\mathbf{P}$. As the latter is a *conserved quantity*, the center of mass motion of the two particles can be split off such that only the Hamiltonian of relative motion $\hat{H}_{\rm rel} = \hat{\mathbf{p}}^2/2\mu + \hat{V}$ needs to be considered.

It turns out that the strong interaction $\hat{V}$ is practically the same no matter which combinations of the two nucleons one takes. In accord with this feature, neutrons and protons may be treated as two realizations of one object, the nucleon. Formally, this can be achieved by introducing the concept of isospin. It is described by a vector operator $\hat{\mathbf{t}}$ whose components fulfill the same commutation rules as angular momentum or spin. Hence, one may apply the same representation as for a spin 1/2 particle: The states $|p\rangle$ and $|n\rangle$ are the two components of one spinor and represent a proton or a neutron, respectively. By convention we will choose the $|p\rangle$ to belong to the eigenvalue 1/2 of the z-component $\hat{t}_z$ and $|n\rangle$ to $-1/2$. Either state can be changed into the other by introducing the common ladder operators: $\hat{t}_+|n\rangle = |p\rangle$ and $\hat{t}_-|p\rangle = |n\rangle$. For a two-particle system the total isospin can be calculated according to the rules of combining angular momenta, namely $\hat{\mathbf{T}} = \hat{\mathbf{t}}_1 + \hat{\mathbf{t}}_2$. The two-particle state may then be symmetric or antisymmetric with respect to an exchange of the partners, depending on the quantum number T, formal details details are found in Section 26.7. The fact of the NN interaction being charge independent may simply be written in the form $[\hat{H}, \hat{\mathbf{T}}] = [\hat{V}, \hat{\mathbf{T}}] = 0$. For a more elaborate discussion of these features we refer to Section I.6 of (de Shalit and Feshbach, 1974).

Besides conservation of isospin and total momentum, which reflect symmetries with respect to charge exchange and translation in space, the strong interaction obeys additional conservation laws. First of all there is the rotational invariance. Again, this symmetry goes along with a vanishing commutator of the interaction $\hat{V}$ with the corresponding generator for rotations, which here is the total angular momentum $\hat{\mathbf{J}} = \hat{\mathbf{L}} + \hat{\mathbf{S}}$, involving both orbital part as well as spin, $[\hat{H}, \hat{\mathbf{J}}] = [\hat{V}, \hat{\mathbf{J}}] = 0$. In the center of mass system the total orbital angular momentum is given by

$$\hat{\mathbf{L}} = \hat{\mathbf{l}}_1 + \hat{\mathbf{l}}_2 = \hat{\mathbf{l}} \quad \text{where} \quad \hat{\mathbf{L}}_{c.m.} = 0 \quad \text{(c.m.-system)}, \quad (1.4)$$

with $\hat{\mathbf{l}}$ being the angular momentum of relative motion, $\hat{\mathbf{l}} = [\mathbf{r} \times \mathbf{p}]$. For the combination of spin and angular momenta see Section 26.7.

In addition there are the symmetries for time reversal ($t \to -t$) and parity or space reflection ($\mathbf{r} \to -\mathbf{r}$). With respect to the latter it must be noted that *parity is not conserved in weak interactions*. The latter may play some role in nuclear systems, both in the two-body problem (see e.g. (Ericson and Weise, 1988)) as well as in heavier nuclei, for instance in β-decay. Such problems will not be touched on in this book. Likewise, there is no room to discuss nuclear properties which involve antiparticles. For the latter a further invariance of the strong interaction comes into play, that with respect to charge conjugation turning particles into antiparticles.

1.1.1 *Possible forms of the interaction*

The fundamental symmetry properties enable one to constrain possible forms of phenomenogical NN interactions. There are good reasons to assume that an almost *static* approximation will do, meaning that the interaction should involve only low powers in the relative momentum of the NN system. This is suggested both from experimental evidence as well as from considerations which engage sub-nuclear degrees of freedom. Since one expects both spin and angular momentum to play an important role an ansatz like the following one is suggested:

$$\hat{V} = \hat{V}_c(r) + \hat{V}_T(r)\hat{S}_{12} + \hat{V}_{LS}(r)\hat{\mathbf{L}} \cdot \hat{\mathbf{S}} + \hat{V}_{LL}(r)\hat{L}_{12} \quad (1.5)$$

with the *central* interaction being given by

$$\hat{V}_c = \hat{V}_0(r) + \hat{V}_\sigma(r)(\hat{\boldsymbol{\sigma}}_1 \cdot \hat{\boldsymbol{\sigma}}_2) + \hat{V}_\tau(r)(\hat{\boldsymbol{\tau}}_1 \cdot \hat{\boldsymbol{\tau}}_2) + \hat{V}_{\sigma\tau}(r)(\hat{\boldsymbol{\sigma}}_1 \cdot \hat{\boldsymbol{\sigma}}_2)(\hat{\boldsymbol{\tau}}_1 \cdot \hat{\boldsymbol{\tau}}_2) \quad (1.6)$$

and with similar expressions for $\hat{V}_T, \hat{V}_{LS}$ and $\hat{V}_{LL}$. In (1.5) the so-called *tensor operator* S_{12} appears:

$$\hat{S}_{12} = \frac{12}{r^2}(\hat{\mathbf{s}}_1 \cdot \hat{\mathbf{r}})(\hat{\mathbf{s}}_2 \cdot \hat{\mathbf{r}}) - 4(\hat{\mathbf{s}}_1 \cdot \hat{\mathbf{s}}_2) = \frac{6}{r^2}(\hat{\mathbf{S}} \cdot \hat{\mathbf{r}})^2 - 2\hat{\mathbf{S}}^2. \quad (1.7)$$

Clearly, this tensor force is not spherically symmetric in coordinate space. Such a component is necessary, however, to explain the measured asymmetry of the charge distribution of the deuteron. In (1.5) the first two terms represent *purely static* (velocity independent) forces. The third term is a typical *spin-orbit* interaction; it is linear in the relative velocity or momentum. Finally, the $\hat{L}_{12}$ of the last term of (1.5) is given by

$$\hat{L}_{12} = \left(\delta_{LJ} + (\hat{\boldsymbol{\sigma}}_1 \cdot \hat{\boldsymbol{\sigma}}_2)\right) \widehat{\mathbf{L}}^2 - \left(\widehat{\mathbf{L}} \cdot \widehat{\mathbf{S}}\right)^2 \tag{1.8}$$

and in a sense acts as a spin-orbit force of second order, see also sections 1-4 and 2-5 of (Bohr and Mottelson, 1969). It is not difficult to prove that (1.5) fulfills the basic symmetry relations mentioned previously.

1.1.2 *The radial dependence of the interaction*

In the general expressions for the interaction as in (1.5) there appear functions of r. A moment's thought tells one that these forces cannot only have attractive parts. Otherwise, it would be very hard to imagine that nuclei could exist as stable objects. The same would hold true for nuclear matter or neutron matter which supposedly exists in one way or another in stars. One needs some mechanism to prevent the nuclear matter from collapsing. The most obvious possibility, at least within the non-relativistic approach we are looking at, is offered by attributing to the interaction a *repulsive core*. The latter will not allow the nucleons to come too close. How in the end this leads to stable configurations at values of the density close to the observed one is the subject of quite involved theories; some of which will be described in Chapter 2. The existence of a repulsive component is confirmed in scattering experiments of nucleons on nucleons at higher energies, above about 200 MeV. This may be attributed to a more or less "hard" core with a radius of $r_c \simeq 0.4 - 0.5$ fm. Such a core does not influence the cross section very much at lower energies.

Knowing such elementary features the following scheme may then be set to find acceptable functions for all the form factors $V_i(r)$ necessary for an ansatz like that given by (1.5). These forms need to have both attractive as well as repulsive parts like the one shown in Fig. 1.1, adapted from the so-called "Reid soft core potential" (Reid, 1968) in the $L = S = 0$ channel. Once these functions are specified by a set of parameters the latter may be fixed by calculating differential cross sections and comparing them with experimental results. In this way one may hope to determine the interaction $\hat{V}$. Unfortunately, for several reasons this method breaks down at energies above several hundred MeV, one being that in this regime meson production starts to play an important role. Nevertheless, it allows one to obtain interactions which may be applied to procedures of many-body physics such that in the end one may deduce properties of nuclear matter or even that of finite nuclei.

To build on this more conventional way of picturing the NN interaction is in line with the general scheme of the book, in which we like to concentrate on the typical many-body effects which have been developed to understand the physics of heavier nuclei. For this reason we want to touch upon sub-nuclear degrees of freedom only marginally.

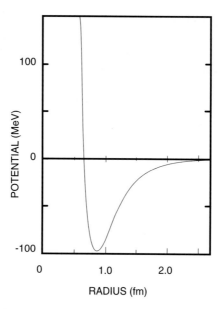

FIG. 1.1. A typical form of the radial dependence of the nucleon–nucleon interaction, adapted from (Reid, 1968).

1.1.3 *The role of sub-nuclear degrees of freedom*

In the past decades much efforts have been undertaken to understand the *physics of the nucleon* in terms of quantum chromodynamics (QCD), which suggests that the nucleon is built out of quarks and gluons. There is no doubt that considerable progress has been achieved. Unfortunately, this is less so for the NN interaction, at least on the more quantitative level. It is probably fair to say that to date still no satisfactory answer to this problem has been found which could be used for the description of heavier nuclei or even for the idealized case of nuclear matter at zero temperature.

This situation is very different as soon as one keeps the nucleon as an elementary particle and takes into account in explicit fashion mesonic degrees of freedom. As has been well known since the pioneering work of H. Yukawa (1935), the exchange of mesons between nucleons leads to a mutual force, by analogy with the Coulomb force which in a field theoretical approach comes about as an exchange of photons. Again, we would not like to go into any details. The interested reader is referred to (Siemens and Jensen, 1987) for an introductory discussion on the construction of an interaction from meson exchange. An exhaustive presentation is found in (Ericson and Weise, 1988). On the very qualitative level, the essential elements can be summarized as follows.

The attractive, long-range part of the interaction can be explained by the exchange of π-mesons, even on a quantitative basis. In this way one also understands the appearance of the tensor force. Important information on the role of the pion in the NN interaction can be deduced from studies of typical properties of the deuteron (Ericson and Weise, 1988). Furthermore, the exchange of ρ and ω mesons allows one to understand on a qualitative level the existence of a repulsive part. Typically, the range of the forces are given by the Compton wavelength of the associated meson, which for a pion is $\hbar/(m_\pi c) \simeq 1.4$ fm and for the ρ meson $\simeq 0.3$ fm (or the ω which has about the same mass as the ρ meson). It must be said, however, that for the shorter distances the exchange model does not work well. Partly this is due to the fact that the smaller the distances being probed in the reaction, the higher must be the wave number and thus the energy. But at higher energies one needs to consider multiple meson exchange, which is difficult to handle in detail. Moreover, at the smaller inter-nuclear distances it becomes somewhat difficult to visualize the two nucleons as independent particles, together with the "existence of a meson in between". Considering such features it appears astonishing that by meson exchange one may nevertheless explain many properties at intermediate distances correctly (Ericson and Weise, 1988). It may thus be no surprise that this picture can be used for parameterizations of the NN interactions. The problem at short distances is coped with by introducing some free parameters again. These interactions have a much more complicated structure than the Reid potential mentioned previously. Amongst other things, they do not only depend on spin, angular momentum and the distance of relative motion, but on other combinations of the particle's momenta as well (in some low power of it). Moreover, the excitation of the Δ resonance of the nucleon can be seen to deliver an effective three-body force in the nuclear medium. As we shall see later, such three-body forces turn out to be necessary for obtaining correct values for nuclear binding, both for nuclear matter as well as for finite nuclei. For recent reviews on these features we recommend (Baldo, 1999), (Brockmann and Machleidt, 1999), (Bombaci, 1999),(Brockmann and Machleidt, 1999), (Heiselberg and Pandharipande, 2000). Likewise we should like to mention the newer development applying effective field theories; for a review see e.g. (Beane *et al.*, 2001). In the past few years great success has been achieved to connect such models to low-energy QCD. One is guided by the chiral symmetry and accounts for the manifestation of its spontaneous breaking by the very fact of a finite pion mass. This opens the way for systematic and consistent perturbation expansions which constrain possible forms of effective Lagrangians, for a recent and concise report see (Weise, 2005). In Section 2.1.3 applications of this approach to nuclear matter will briefly be described.

1.2 The model of the Fermi gas

The reader may already be familiar with this simple model from quantum mechanics or quantum statistics. We will at first restrict ourselves to zero temperature postponing the generalization to later sections. One assumes the fermions to be free particles the motion of which is governed by the Schrödinger equation

$$-\frac{\hbar^2 \Delta}{2m}\psi_{\mathbf{k}}(\mathbf{x}) = \epsilon(\mathbf{k})\psi_{\mathbf{k}}(\mathbf{x}) \,. \tag{1.9}$$

For a homogeneous system with translational invariance the single particle wave function $\psi_\mathbf{k}(\mathbf{x})$ in space will be a plane wave. Discarding for the moment any additional degrees of freedom and assuming motion in three dimensions, it is thus given by

$$\psi_\mathbf{k}(\mathbf{x}) = \mathcal{V}^{-1/2} \exp(i\mathbf{k}\mathbf{x}) \equiv \langle \mathbf{x} | \mathbf{k} \rangle. \tag{1.10}$$

It is convenient to assume periodic boundary conditions,

$$\psi_\mathbf{k}(x, y, z) = \psi_\mathbf{k}(x + L_x, y + L_y, z + L_z), \tag{1.11}$$

with the L_i being the lengths of a box of volume $\mathcal{V} = L_x L_y L_z$. They pose the following restrictions on the wave numbers

$$k_i = \frac{2\pi}{L_i} n_i \qquad n_i = 0, \pm 1, \pm 2 \ldots \qquad \text{for} \qquad i = x, y, z. \tag{1.12}$$

The normalization of the wave function may thus be written as

$$\int_0^{L_x} dx \int_0^{L_y} dy \int_0^{L_z} dz \, \psi_\mathbf{k}^*(x, y, z) \psi_{\mathbf{k}'}(x, y, z) = \delta_{k_1 k_1'} \delta_{k_2 k_2'} \delta_{k_3 k_3'} \equiv \delta_{\mathbf{k}\mathbf{k}'} \tag{1.13}$$

where on the right-hand side Kronecker symbols appear, rather than δ-functions as would be given for plane waves in an infinitely large box, for which the k_i would be continuous variables. Indeed, it is often of advantage to work with such a continuum of states. Then the summation over the discrete indices appearing in (1.12) can be replaced by the following integration in (three-dimensional) $\mathbf{k}$ space

$$\sum_\mathbf{k} f(\mathbf{k}) \equiv \sum_{n_x, n_y, n_z} f(n_x, n_y, n_z) \quad \longrightarrow \quad \mathcal{V} \int \frac{d^3 k}{(2\pi)^3} f(\mathbf{k}), \tag{1.14}$$

which is easier to handle. Such an approximation should not matter too much as long as one is interested only in average quantities, or to get a first orientation on the order of magnitudes of various quantities. If spin and (and isospin) degrees of freedom are to be included, represented by spinors like $\chi(m_s, m_\tau)$, the form (1.10) generalizes to

$$\varphi_\alpha(\mathbf{x}) = \psi_\mathbf{k}(\mathbf{x}) \chi(m_s, m_\tau). \tag{1.15}$$

For the level density this implies including a spin-isospin degeneracy factor $g_{\text{s.i.}}$, which equals 4 if one counts neutron and proton as the same particle. In most cases only the absolute value of the wave vector matters. Calling $dn(k)$ the number of levels in the intervals dn_i ($i = 1 - 3$) one gets

$$dn(k) = g_{\text{s.i.}} \frac{dk_x dk_y dk_z}{(2\pi)^3} \mathcal{V} \equiv g_{\text{s.i.}} \frac{\mathcal{V}}{(2\pi)^3} 4\pi k^2 \, dk. \tag{1.16}$$

The density of levels $g(e) = dn_e/de$ in energy space can be obtained whenever the relation $e = e(\mathbf{k})$ is known. For the Fermi gas this is simply given by the kinetic energy, which is to say

$$e(k) = \frac{\hbar^2}{2m} k^2 \quad \Longleftrightarrow \quad 2k\,dk = \frac{2m}{\hbar^2} de. \tag{1.17}$$

Notice that this is still correct even if the particles move in a medium–provided the effects of the latter can be accounted for by an effective mass. Often this is a good approximation for the conduction electrons in metals or semiconductors. Also for the motion of nucleons in a nuclear medium, like for the model of infinite nuclear matter, there may exist such a relation $e = e(\mathbf{k})$, although in the general case it will not be given just by the kinetic energy. From (1.16) one gets for dn_e

$$dn_e(e) = g_{\text{s.i.}} \frac{\mathcal{V}}{(2\pi)^2} \left(\frac{2m}{\hbar^2}\right)^{3/2} \sqrt{e}\,de. \tag{1.18}$$

Here a small detour is in order to systems of dimensions $d = 2$ or 1, which may be realized for electrons in semiconductors, for instance. The equations from before are easily generalized to any d. Eq. (1.16) turns into

$$dn = g_{\text{s.i.}} \left(\frac{1}{2\pi}\right)^d \mathcal{V}_\text{d}\,d^d k, \tag{1.19}$$

if $\mathcal{V}_\text{d}$ and $d^d k$ represent the volumes in d-dimensional space and $\mathbf{k}$-space, respectively. For the electron gas the $g_{\text{s.i.}}$ would have to be chosen equal to 2. For the level density dn_e/de one gets the following relations:

$$dn_e(e) = g_{\text{s.i.}} \frac{S}{2\pi} \frac{m}{\hbar^2} de, \quad \text{for} \quad d = 2, \tag{1.20}$$

with $\mathcal{V}_\text{d} = S = L_x L_y$ and $d^2 k = 2\pi k\,dk = 2\pi(m/\hbar^2)de$, and

$$dn_e(e) = g_{\text{s.i.}} \frac{L}{2\pi} \sqrt{\frac{m}{2\hbar^2}} \frac{de}{\sqrt{e}} \quad \text{for} \quad d = 1, \tag{1.21}$$

with $\mathcal{V} = L$ and dk for the volume element in k-space. Notice that for $d = 2$ the dn_e is *independent* of energy.

1.2.1 Many-body properties in the ground state

For the many-body ground state all single particle states are filled up to the so-called Fermi energy, or the *Fermi momentum* k_F. Suppose there are A particles in the box of volume $\mathcal{V}$.[3] This number must be given by

$$A \equiv n(k_\text{F}) = \int_0^{k_\text{F}} dn(k) = \frac{g_{\text{s.i.}}}{6\pi^2} \mathcal{V} k_\text{F}^3 \tag{1.22}$$

---

[3]Similar results may also be derived for a system in which the particles are put inside a cubic box of length a with infinitely deep walls. As in this case the density of nucleons vanishes outside the box one may in addition study surface effects both on the density of states as well as on the total kinetic energy, for details see e.g. (de Shalit and Feshbach, 1974)

such that the density of particles becomes

$$\rho_0 = \frac{A}{V} = g_{\text{s.i.}} \frac{1}{6\pi^2} k_F^3 \,. \tag{1.23}$$

For the total (kinetic) energy per particle one gets

$$\frac{E_{\text{kin}}}{A} = \frac{\int_0^{k_F} e(k) \frac{dn(k)}{dk} dk}{\int_0^{k_F} dn} = \frac{\hbar^2}{2m} \frac{\int_0^{k_F} k^4 dk}{\int_0^{k_F} k^2 dk} = \frac{3}{5} \frac{\hbar^2 k_F^2}{2m} \equiv \frac{3}{5} e_F \,. \tag{1.24}$$

On the very right the Fermi energy e_F has been introduced with the help of which the level density $g(e)$ can be evaluated to (recalling (1.18))

$$g(e) = \frac{dn_e}{de} = \frac{3}{2} \sqrt{\frac{e}{e_F}} \frac{A}{e_F} \,, \tag{1.25}$$

a form which is independent of $g_{\text{s.i.}}$ and thus valid for any Fermi gas of A particles in three dimensions.

Let us apply these expressions to a simple picture of a heavy nucleus. It turns out that the density in its interior takes on the value $\rho_0 \simeq 0.172$ fm^{-3}. Using a degeneracy factor of $g_{\text{s.i.}} = 4$ this implies the following values $k_F \simeq 1.36$ fm^{-1}, $\epsilon_F \simeq 38$ MeV and $E_{\text{kin}}/A \simeq 23$ MeV for Fermi momentum, Fermi energy and the total energy per particle, respectively; for the natural units of nuclear physics see Chapter 27. As it should be, the kinetic energy per particle is a positive quantity. The total energy E/A, on the other hand, must be negative; otherwise the system would not be bound. As we shall see later the E/A is known to be about -16 MeV. The difference has to come from the particle's mutual interaction which in the present formulation of the Fermi *gas* model was discarded entirely. Suppose this interaction may be treated on average by way of a mean field (or potential) U. The latter then has to be negative and might be accounted for by simply adding U to the kinetic energy E_{kin} (subtracting the absolute value $|U|$) to get $E = E_{\text{kin}} + U \simeq -16A$ MeV. Thus the potential energy per particle must be quite large, namely of the order of $U/A = -E_{\text{kin}}/A - 16 \simeq -39$ MeV. Later on we will address the problem of calculating U in more quantitative fashion.

Symmetry energy For most nuclei the neutron number differs from the proton number. On the level of the Fermi gas model this may be accounted for by studying two independent gases, in which case the total energy would be given by

$$E_{\text{kin}} = \frac{3}{5} e_F \left(e_F^{(n)} N + e_F^{(p)} Z \right) , \tag{1.26}$$

as follows directly from (1.24). The two different Fermi energies can be expressed by the one associated with the total particle number $A = N + Z$

$$e_F^{(n)} = \left(\frac{2N}{A} \right)^{2/3} e_F \quad \text{and} \quad e_F^{(p)} = \left(\frac{2Z}{A} \right)^{2/3} e_F \,. \tag{1.27}$$

One only needs to calculate the corresponding Fermi momenta from (1.22) using $g_{\text{s.i.}} = 4$ for particle number A and $g_{\text{s.i.}} = 2$ for the N neutrons and the Z protons. Thus the total energy may be written as

$$E_{\text{kin}} = \frac{3}{5}\left(\left(\frac{2N}{A}\right)^{2/3} N + \left(\frac{2Z}{A}\right)^{2/3} Z\right). \tag{1.28}$$

This form allows for an expansion into a power series with respect to the neutron excess $\lambda = (N-Z)/A$. A short calculation shows that to order two this becomes

$$E_{\text{kin}} \simeq \frac{3}{5} e_F A + \frac{1}{3} e_F \frac{(N-Z)^2}{A}. \tag{1.29}$$

Taking the value given above for the Fermi energy the coefficient of the symmetry term takes on the value 13, which, as we shall see later in Section 1.3.3.2, is about half the realistic one.

1.2.2 Two-body correlations in a homogeneous system

Suppose we are given a system of $2n$ neutrons and protons, such that the total number of nucleons becomes $A = 4n = 2N = 2Z$, with n being an integer. The ground state is given by the Slater determinant $|\Phi^{\text{SL}}_{\alpha_1 \ldots \alpha_A}(1,2,\ldots A))\rangle$ built out of single particle states like those given in (1.15). Let us enumerate them by associating the index $l = 1 \ldots A$ with the set of quantum numbers α_l. The one-body density operator

$$\hat{\rho}^{(1)}_{\text{FG}} = A \,\text{tr}|_{2\ldots A}\, |\Phi^{\text{SL}}_{\alpha_1\ldots\alpha_A}(1,\cdots,A)\rangle\langle\Phi^{\text{SL}}_{\alpha_1\ldots\alpha_A}(1,\cdots,A)| \tag{1.30}$$

can then be seen to become

$$\hat{\rho}^{(1)}_{\text{FG}} = \sum_{l=1}^{A}|\alpha_l\rangle\langle\alpha_l| = \sum_{i=1}^{n}|\mathbf{k}_i\rangle\langle\mathbf{k}_i|\sum_{\mu=1}^{4}|\chi_\mu\rangle\langle\chi_\mu|. \tag{1.31}$$

For formal details of the derivation we may refer to Chapter 19 (see Section 19.2.1 in particular), together with Section 26.7 for the treatment of the spin-isospin degrees of freedom. In (1.31) the summation over the (4)-spinors just leads to the unit $\hat{1}_4$ matrix in four dimensions. Therefore the density matrix becomes

$$\hat{\rho}^{(1)}_{\text{FG}}(\mathbf{x};\mathbf{x}') = \langle\mathbf{x}|\hat{\rho}^{(1)}_{\text{FG}}|\mathbf{x}'\rangle = \frac{\hat{1}_4}{\mathcal{V}}\sum_{i=1}^{n} e^{i\mathbf{k}_i(\mathbf{x}-\mathbf{x}')} \tag{1.32}$$

This is still an operator in spin and isospin space. Averaging over these degrees one gets $\overline{\rho}^{(1)}_{\text{FG}}(\mathbf{x};\mathbf{x}') = \text{tr}|_{\sigma\tau}\hat{\rho}^{(1)}_{\text{FG}}(\mathbf{x};\mathbf{x}') = (4/\mathcal{V})\sum_{i=1}^{n} e^{i\mathbf{k}_i(\mathbf{x}-\mathbf{x}')}$, the diagonal matrix elements of which reduce to the space density $\overline{\rho}^{(1)}_{\text{FG}}(\mathbf{x};\mathbf{x}) = A/\mathcal{V} = \rho_0$.

The two-body density gives an interesting insight into those space which are generated by the Pauli principle. Let us just sketch the calculation for the simple case when spin and isospin degrees of freedoms are discarded. The two-body density matrix is given by (19.42). Inserting there the space part of the one-body density of (1.32), the diagonal elements become

$$\rho^{(2)}_{FG}(\mathbf{x}_1, \mathbf{x}_2; \mathbf{x}_1, \mathbf{x}_2) = \frac{1}{\mathcal{V}^2} \sum_{k,l=1}^{n} \left(1 - \cos(2\mathbf{k}_{kl}\mathbf{r})\right), \qquad (1.33)$$

with the vectors $\mathbf{r} = \mathbf{x}_1 - \mathbf{x}_2$ of relative distance and $\mathbf{k}_{kl} = (\mathbf{k}_1 - \mathbf{k}_2)/2$ of relative momentum (in units of $\hbar$). Obviously, the magnitude of $\mathbf{k}_{kl}$ is bound by the Fermi momentum, $|\mathbf{k}_{kl}| = |(\mathbf{k}_k - \mathbf{k}_l)|/2 \leq k_F$. To proceed further we again replace summations over discrete indices by integrals over wave numbers making use of (1.14). Writing $\cos(\mathbf{k}_k\mathbf{r} - \mathbf{k}_l\mathbf{r}) = \cos\mathbf{k}_k\mathbf{r}\cos\mathbf{k}_l\mathbf{r} + \sin\mathbf{k}_k\mathbf{r}\sin\mathbf{k}_l\mathbf{r}$ integrals of the type

$$\int \frac{d^3k}{(2\pi)^3} \cos(\mathbf{k}\mathbf{r}) = \frac{1}{(2\pi)^2} \int k^2 dk d(\cos\theta) \cos(kr\cos\theta)$$
$$= \frac{1}{2\pi^2 r^2} \left[\frac{\sin(k_F r)}{r} - k_F \cos(k_F r) \right] \qquad (1.34)$$

appear. Since the ones over $\sin\mathbf{k}_k\mathbf{r}$ vanish, the two-body density becomes

$$\rho^{(2)}_{FG}(\mathbf{x}_1, \mathbf{x}_2; \mathbf{x}_1, \mathbf{x}_2) = \left(\rho_0^2\right) g_-(k_F r) \qquad \text{for} \qquad P^x_{12} = -1. \qquad (1.35)$$

The function $g_-(u)$ is defined as

$$g_-(u) = 1 - (C(u))^2 \qquad \text{with} \qquad C(u) = \frac{3}{u^2}\left(\frac{\sin u}{u} - \cos u\right). \qquad (1.36)$$

The P^x_{12} introduced in (1.35) represents the permutation operator in coordinate space. As the two particle state must be totally antisymmetric its eigenvalue must be -1 if spin and isospin are discarded.

In the general case one must distinguish between the three different possibilities of permutations given by the space part, the spin and the isospin degrees of freedom. Calling the corresponding operators for permutations P^x_{12}, P^σ_{12} and P^τ_{12}, respectively one needs to have

$$P^x_{12} P^\sigma_{12} P^\tau_{12} = -1. \qquad (1.37)$$

For a given symmetry of the space part the product of the two other permutations may thus be expressed as

$$P^\sigma_{12} P^\tau_{12} = \begin{cases} -1 & \text{if } P^x_{12} = +1 \\ +1 & \text{if } P^x_{12} = -1. \end{cases} \qquad (1.38)$$

The second case has just been treated. A totally symmetric space part can be handled analogously. One may convince oneself by direct calculation that the diagonal element of the density matrix is given by an expression like (1.33) but where the signs in front of the cosines turn into a plus. Likewise, the final result can be expressed like before in (1.35) with the $g_+(u)$ replaced by $g_+(u) = 1 + (C(u))^2$. After averaging over spin and isospin one obtains

$$\bar{\rho}_{FG}^{(2)} = \rho_0^2 \left(1 - \frac{1}{4}(C(k_F r))^2\right), \qquad (1.39)$$

with the ρ_0 being the actual nuclear density $A/\mathcal{V}$. The factor $1/4$ comes about in the following way. The average over spin-isospin implies that the functions g_- and g_+ get weights of $10/16$ and $6/16$, respectively. There are ten possibilities with $P_{12}^\sigma P_{12}^\tau = 1$ and six with $P_{12}^\sigma P_{12}^\tau = -1$: In the former case they consist of combinations of the three triplets for both spin and isospin plus the one of the spin and isospin singlets; a spin-isospin antisymmetric state requires the combination of singlets with triplets. It is then easy to see that the appropriate synthesis of both cases leads to (1.39), indeed. As indicated earlier, the functional form (1.39) represents those two-body correlations in the system of independent nucleons which are there simply because of the Pauli principle. They have nothing to do with correlations which may come from the two-body interaction. As the $C(k_F r))$ approaches 1 for $r \to 0$ the correlation reduces to the value of $1/4$ at zero (relative) distance (it vanishes for the totally antisymmetric space part). This implies that on average the nucleons repel each other leaving a dip in the correlation function the extension of which is of order $r \simeq (2-3)/k_F$ (where the contribution from C^2 becomes negligibly small). It is instructive to calculate the average deviation $1/\rho_0 \int d^3 x (\bar{\rho}_{FG}^{(2)} - \rho_0^2)$ of the correlation in the two-body density from the uncorrelated one, which is easily seen to become $-(\rho_0/4) \int d^3 x (C(k_F r))^2 = -1$. This result may be interpreted by saying that in the neighborhood of each nucleon one nucleon less is found than in the uncorrelated system. For further discussions we like to recommend the textbooks by A. de Shalit and H. Feshbach (1974) and A. Bohr and B.R. Mottelson (1969).

In momentum space the two particles fill two Fermi spheres to the Fermi momentum k_F each. Of interest is the absolute value $|\mathbf{k}_{kl}|$ of their relative momentum, which is bound by k_F. The corresponding probability distribution $P(k)$ may be defined as

$$P(k) = \frac{\int d^3 k_1\, d^3 k_2\, \delta\left(\left|\frac{1}{2}(\mathbf{k}_k - \mathbf{k}_l)\right| - k\right)}{\int d^3 k_1\, d^3 k_2}. \qquad (1.40)$$

By a straightforward calculation it is readily shown to become

$$P(k)dk = \frac{6}{\pi k_F^3}\left[1 - \frac{3}{2}\frac{k}{k_F} + \frac{1}{2}\left(\frac{k}{k_F}\right)^3\right] 4\pi k^2\, dk \quad \text{for} \quad k \leq k_F. \qquad (1.41)$$

The $P(k)$ tends to zero both for $k \to 0$ as well as for $k \to k_F$ and it reaches a maximal value at $k \simeq 0.6 k_F$. The most probable kinetic energy of relative motion therefore is (with the reduced mass $\mu = m_N/2$) $e_r = 2 \cdot (0.6)^2 \cdot e_F \simeq$ 27 MeV. This is the energy the two particles have in their center of mass system (defined such that their total momentum is zero). In the lab frame this would correspond to a scattering of a nucleon of twice the energy of a nucleon at rest. The maximal relative kinetic energy, which corresponds to the maximal relative momentum k_F, is twice the Fermi energy of one particle, viz of the order of 80 MeV. For a nucleon–nucleon reaction in the lab this corresponds to an energy of about 160 MeV. Suppose nucleonic dynamics inside a nucleus were to show no correlations beyond those of the Fermi gas. Then it would suffice to perform nucleon–nucleon scattering experiments in the lab up to projectile energies of that order to get the required information on the (bare) two-body interaction. However, it turns out that correlations of shorter range are important implying that the nucleon–nucleon interaction must also be tested at higher energies.

1.3 Basic properties of finite nuclei

An important means of measuring properties of nuclei is in reactions with other particles. Such measurements are undertaken in accelerator experiments, which may involve elementary particles or nuclear ions, but one may also simply expose nuclei to the gamma radiation and neutrons. Over the years one has been able to accelerate heavier and heavier ions. In the present chapter we will look at the interaction of nucleons with nuclei, mainly to get a first orientation of how nucleons propagate in a nuclear medium. Later in Chapter 13 we will return to heavy-ion collisions to learn about essential features of such more complex reactions.

1.3.1 *The interaction of nucleons with nuclei*

One way of treating the collision of a nucleon N with a nucleus $\mathcal{N}$ is to apply potential scattering. One only has to find appropriate forms of the potential $U(r)$.

1.3.1.1 *The approximation of a real mean field for nucleonic motion* It is an established fact that in N-$\mathcal{N}$ reactions at low energies events occur with high probability which may indeed be described simply as a propagation of the nucleon in the mean field of all the others–although historically it did not at all look like that for quite some time. Indeed, already in the 1930s description of the interaction of a nucleon with a nucleus had been attempted by means of a *real potential* $U(r)$. The problem was that in this way it was *impossible* to get the narrow resonances on the scale of electronvolts seen in neutron scattering, to which we will come to shortly. One may convince oneself (see exercise 1) that the widths of resonances in a potential of realistic size are in the range of MeV. For this and other reasons, the model of *independent particles* moving in a mean field $U(r)$ had almost be given up. However, in 1949 and thereafter, new evidence about nuclear structure appeared in favor of this picture. In that year M. Goeppert-Mayer and J.H.D. Jensen (see (1955)) suggested the inclusion of

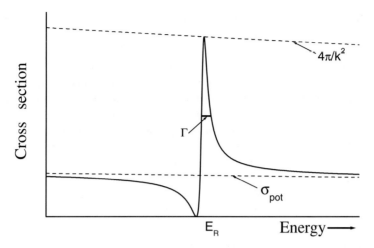

FIG. 1.2. The behavior of the cross section $\sigma(E)$ of eqn(1.43) in the neighborhood of the resonance. The cross section $\sigma_{\text{pot}}(E)$ for pure potential scattering is obtained for $\Gamma = 0$.

a strong spin-orbit interaction $U_{ls}(r)$. With such an interaction it was possible to explain the so-called *magic numbers*. These refer to the particle number for which nuclei were known to be particularly stable. The breakthrough came by choosing the sign of the spin-orbit interaction opposite to that which one knew from atomic physics. This observation led to the shell model. In its simplest version the nucleons move *independently* of each other, which is to say that no interaction is taken into account beyond that of the mean field $U(r) + U_{ls}(r)$. However, as *there are interactions V_{res} residual to the mean field*, this model must be considered a first approximation only.

1.3.1.2 *Compound reactions* The presence of the V_{res} implies that in reality the reaction of N with $\mathcal{N}$ is a more complex process. The physical reason for this behavior is not difficult to understand. When a N hits a $\mathcal{N}$ the *substructure* of the latter becomes relevant at practically all energies. This is an essential, qualitative difference from the NN scattering where the sub-N degrees of freedom, which actually consist of different types of particles, manifest themselves only at larger energies. For N-$\mathcal{N}$ reactions, on the other hand, the impinging nucleon "sees" partners of its own type even at low energies, with which it may perform individual NN collisions, before it will leave the nucleus again.

Typically, in such experiments one uses slow neutrons (or protons) hitting a nucleus of neutron number N and proton number Z. Here "slow" means that their initial kinetic energy is not larger than a few eV. When a neutron enters the nucleus the total available excitation energy is given by this kinetic energy plus its binding energy. The latter is determined by the *separation energy* $S_n(N+1, Z)$ of the least bound neutron of the compound nucleus with mass number $A + 1$, and may thus be calculated from the difference of the corresponding binding energies as $S_n(N+1,Z) = B(N+1,Z) - B(N,Z)$. This difference typically ranges between 5 and 10 MeV.

A picture of the angle integrated cross section as function of the incident kinetic energy is shown in Fig. 1.2 for one resonance. It can be said to be generic for resonances produces in reactions with slow neutron, see e.g. Fig. 2-8 of (Bohr and Mottelson, 1969). The widths of these resonances are smaller by orders of magnitude when compared to the several MeV wide resonances of potential scattering. Qualitatively and physically correct, this can be understood from the time-energy uncertainty relation

$$\delta E\, \tau_{\text{delay}} \simeq \Gamma\, \tau_{\text{delay}} \simeq \hbar. \tag{1.42}$$

Estimating the δE as the width of the resonance, it is seen that the nucleon is re-emitted only after the time delay τ_{delay}. The latter is much larger than the time Δt it takes (see exercise 1) for a slow neutron to cross a potential well typical for the mean field of the nucleus. Actually, this Δt is in the same relation to the width ΔE of the shape resonances as τ_{delay} is to δE. As the δE is in the range of a few eV, the corresponding time delay τ_{delay} is larger by several orders of magnitude. This feature, together with the fact of the interaction between the nucleons being strong, led N. Bohr to describe the N-$\mathcal{N}$ reaction as a sequence of individual N–N collisions rather than a motion of the incoming nucleon in the mean field of the others. He compared the situation to that of an ensemble of billiard balls put into a shallow basin. The additional ball brought into motion would initially collide with all, or many, of the other balls (nucleons) before one is again leaving the ensemble with some finite probability. Put in pictorial terms this is the essence of the model of the *compound nucleus* presented in a talk given at the Copenhagen Academy by N. Bohr (1936). As the basic hypothesis, N. Bohr assumed that the available energy is *equally shared among all nucleons*, as it would be in a microcanonical ensemble of statistical mechanics. As the system passes through such a state before it again decays the latter process should be independent of the way the intermediate state was produced. In the literature this statement is referred to as *N. Bohr's independence hypothesis: The "exit channel" is independent of the "entrance channel"*, save for those quantities which are conserved because of underlying symmetries, like total energy, total angular momentum, parity etc. N. Bohr's independence hypothesis is critically examined in Section 18.4.4 after the cross section has been derived thoroughly within nuclear reaction theory in Chapter 18. In Section 18.2.4.1 the following formula for the cross section is derived

$$\sigma(E) = \frac{4\pi}{k^2}\left|\sin\delta - e^{i\delta}\frac{\Gamma/2}{E - E_R + i\Gamma/2}\right|^2, \tag{1.43}$$

with k being the wave number for relative motion of energy $E = \hbar^2 k^2/2m$. This formula determines the Breit–Wigner type resonance at $E = E_R$ of width

Γ, together with its interference with the "background". The latter is due to potential scattering, which itself would occur with cross section $(4\pi/k^2)\sin^2\delta$. In Fig. 1.2 the implication of the interference is clearly visible: the cross section first decreases below that of the background before it starts to rise sharply.

A closer look exhibits that for resonance scattering the Bohr hypothesis is in a sense naturally fulfilled without having to rely on statistical concepts; for details see the discussion in Section 18.4. However, with increasing energies the resonances lie closer and closer, reflecting the fact that the density of levels increases exponentially. It soon becomes infeasible to treat resonances in explicit fashion, as it is impossible to describe the scattering process by solving Schrödinger equations for the correct many-body wave functions. This feature is intimately connected to the picture of the compound nucleus introduced above. The latter necessarily must include quantum states which are much more complex than those of the independent particle model. Whereas the latter are Slater determinants, the more complicated states involve correlations beyond those implied by the Pauli principle. Like the example of the billiard balls, these correlations come up in individual nucleon–nucleon collisions but manifest themselves only after times much longer than the transit time through the nucleus. Within such a picture one may distinguish between prompt events and the delayed ones of the compound reactions. In the *statistical model* of nuclear reactions, developed in greater detail in Section 18.4, one then argues that no phase correlations in the amplitudes (or T-matrices) exist which are associated with the prompt or delayed processes. As a consequence the cross section is just a sum of probabilities

$$\overline{\sigma} = \sigma^{(P)} + \overline{\sigma^{(FL)}}, \tag{18.133}$$

without interferences between the two types of reactions. Here, "P" stands for "prompt" and the suffix "FL" for "fluctuating" is to indicate that the cross section of compound reactions varies on a much shorter energy scale than the $\sigma^{(P)}$ of the shape resonances. The bar put over the σs in (18.133) is to reflect that we are dealing with averaged quantities. For the statistical model this average is to be understood as an ensemble average carried out in practice as an *energy average*. In the spirit of statistical mechanics this implies an *ergodic hypothesis*. The necessity of averaging over energies is already required by experimental demands: With increasing energy it becomes less and less feasible to measure the structure of individual resonances in detail. This is possible only around and slightly above neutron threshold, i.e. when the excitation energy becomes equal to or a bit larger than the separation energy S_n. As the average spacing D of levels decreases exponentially with E one soon reaches the regime where the widths Γ become larger or even much larger than D, which is to say the resonances begin to *overlap* more or less strongly.

As indicated earlier the independence hypothesis of N. Bohr implies that the compound reaction takes place in two steps, the production of the compound nucleus and its subsequent decay, and that during the first stage a statistical equilibrium is reached in the sense of a microcanonical ensemble. Of course, this may happen *only for sufficiently fast relaxation* to equilibrium. Indeed, there is a whole class of reactions for which this condition is *not* fulfilled. They are called *pre-compound or pre-equilibrium* processes and will be discussed in Chapter 9.

1.3.1.3 *Gross and intermediate structure* As indicated previously, energy averages may be advantageous for experimental reasons, eventually even at lower energies. Interesting features arise when such averages are performed on amplitudes, which in the end imply loss of probability flux. The *total cross section* is no longer determined by that of elastic scattering, called σ_el in the sequel. Rather one has to account for absorption to get

$$\sigma_\text{tot} = \sigma_\text{abs} + \sigma_\text{el}. \tag{1.44}$$

One source of absorption is found in *inelastic processes* where part of the incoming flux may flow into other open channels. After the reaction, the two fragments may differ from the initial ones in respect to e.g. particle number or excitation energy. However, loss of probability may not only come from genuine reactions. If *compound states* are excited they *take away flux from the prompt* processes, simply because of the time delay involved for re-emission. This part of σ_abs is obtained by energy averaging the wave functions for compound elastic scattering. Altogether the absorption cross section is then given by the sum

$$\sigma_\text{abs}^\text{opt} = \overline{\sigma_\text{rea}} + \overline{\sigma_\text{el}^{(C)}}. \tag{1.45}$$

Here, the averaging interval Δ must be large enough to totally wipe out the detailed structure of the compound resonances. This ensures that no delayed processes are counted anymore. For elastic scattering, which is to say for $\sigma_\text{rea} = 0$, averaging the (compound) wave function delivers the only source of absorption. Due to the energy average, the Schrödinger equation for the prompt channel acquires an imaginary part for the potential, which is shown explicitly in Section 18.3.1 and demonstrated further in Section 18.4.3. Due to analogies with propagation of waves in dispersive media this approximation goes along with the notion of the *optical model*, details of which will be discussed below. For the inclusion of inelastic processes we refer to the literature, like e.g. (Feshbach, 1992) or (Hodgson, 1994). By analogy with (1.44) the total cross section is then written as $\sigma_\text{tot}^\text{opt} = \sigma_\text{rea}^\text{opt} + \sigma_\text{el}^\text{opt}$. In Section 18.4.3.1 a simple but instructive model is used to demonstrate the connection of the optical to the compound model.

So far variations in the cross section have been discussed which may be associated with widths of resonances of very different size, the Γ_GS for the *gross structure* seen in the optical model, and the very small Γ_CN for compound scattering. Structure of an *intermediate* nature is also seen where the typical widths Γ_d lie in between the former two,

$$\Gamma_\text{GS} \gg \Gamma_\text{d} \gg \Gamma_\text{CN}. \tag{1.46}$$

This happens when the so-called *doorway mechanism* is involved (for which reason the index "d" is used). In this case the entrance channel couples to the "doorway state" but not directly to the more complicated compound states. The latter are only reached through doorway states. Both couplings, the one from the entrance- to the doorway state and from there to the compound states, may be parameterized in terms of a $\Gamma_d^\uparrow$ and a $\Gamma_d^\downarrow$, respectively. For the sake of simplicity let us assume that just one open channel exists, the general case being discussed in Section 18.3.2. For the elastic cross section the generalization of (1.43) then reads

$$\bar{\sigma}(E) = \frac{4\pi}{k^2} \left| \sin\delta - e^{i\delta} \frac{\Gamma_d^\uparrow/2}{E - E_d + \frac{i}{2}\left(\Gamma_d^\uparrow + \Gamma_d^\downarrow\right)} \right|^2, \qquad (1.47)$$

The denominator in the resonance term is still determined by the *full width* $\Gamma_d = \Gamma_d^\uparrow + \Gamma_d^\downarrow$. In the numerator, however, only the width $\Gamma_d^\uparrow$ for the coupling to the exit channel appears. This feature implies loss of probability, as may be seen from the (in this case) diagonal element of the S-matrix, which is given by

$$\bar{S} = e^{2i\delta} \frac{E - E_d + \frac{i}{2}\left(\Gamma_d^\uparrow - \Gamma_d^\downarrow\right)}{E - E_d + \frac{i}{2}\left(\Gamma_d^\uparrow + \Gamma_d^\downarrow\right)}. \qquad (1.48)$$

This is easily verified from (1.47) knowing the general relation $S = 1 - 2\pi i T$ to the T-matrix and that (for appropriate normalization of the wave functions) the cross section is given by $\sigma(E) = (4\pi^2/k^2)|T|^2$. Evidently, the absolute value of $\bar{S}$ is less than unity, and, hence, flux is not conserved; for a detailed discussion of these properties see Section 18.1. A more direct and useful way of exhibiting this "violation of unitarity" is given by looking at the so-called transmission coefficient $\mathcal{T} = 1 - \left|\bar{S}\right|^2$, for which (1.48) leads to

$$\mathcal{T}(E) = \frac{\Gamma_d^\uparrow \Gamma_d^\downarrow}{(E - E_d)^2 + \frac{1}{4}\left(\Gamma_d^\uparrow + \Gamma_d^\downarrow\right)^2}. \qquad (1.49)$$

1.3.2 The optical model

Above it has been argued that the optical model is introduced to describe the prompt part of the scattering of single particles at nuclei. In a microscopic sense this requires one to perform energy averages over an interval Δ which is larger than both the widths Γ_{CN} of the compound resonances (of the order of fractions of an eV to a few hundred of keV), as well as their average spacing D. On the other hand, the Δ must be small enough to preserve the typical structure in the single particle motion, which varies over ranges of several MeV. As mentioned earlier, averaging implies that the single particle potential gains an imaginary part. In general the effective, *non-Hermitian* Hamiltonian will be of non-local form. In practical applications this is approximated by a local part of the type (see e.g. (Feshbach, 1992), (Hodgson, 1994) or (Mahaux and Sartor, 1991))

$$\hat{h}_{\rm OM} = \frac{\hat{\mathbf{p}}^2}{2m_N} + \hat{U} - i\hat{W} \tag{1.50}$$

(with both $\hat{U} = \hat{U}^\dagger$ and $\hat{W} = \hat{W}^\dagger$). Such a model was introduced into nuclear physics by H. Bethe in (1940). Important progress in the microscopic derivation of the general optical model was achieved by H. Feshbach, C.E. Porter and V.F. Weisskopf (1954) in the early 1950s. In Section 18.3 a possible formal derivation is described using Feshbach's projection technique. This leaves open the question of how the microscopic input for the Hamiltonian is to be obtained upon which the projection technique is based. This latter question is closely related to the fundamental problem of how the nuclear many-body system is to be treated starting from the nucleon–nucleon interacting. This topic will be addressed in the next chapter.

1.3.2.1 *The optical model and the loss of probability* The imaginary part of the potential is associated with loss of probability, as measured by the density $\rho(\mathbf{r},t) = \psi^*(\mathbf{r},t)\psi(\mathbf{r},t)$ together with the continuity equation. For a $\hat{H}$ of the form of (1.50) the latter reads

$$\nabla \cdot \mathbf{j}(\mathbf{r},t) + \frac{\partial}{\partial t}\rho(\mathbf{r},t) = -\frac{2}{\hbar}W(\mathbf{r})\rho(\mathbf{r},t), \tag{1.51}$$

where $\mathbf{j}(\mathbf{r},t) = (\hbar/2im_N)(\psi^*(\mathbf{r},t)\nabla\psi(\mathbf{r},t) - \psi(\mathbf{r},t)\nabla\psi^*(\mathbf{r},t))$ is the current density. The $\psi(\mathbf{r},t)$ is the solution of the time-dependent Schrödinger equation in coordinate representation,

$$\frac{\hbar}{i}\frac{\partial}{\partial t}\psi(\mathbf{r},t) = \hat{h}_{\rm OM}\psi(\mathbf{r},t), \tag{1.52}$$

with the adjoint equation for $\psi^*(\mathbf{r},t)$ obtained by complex conjugation, in which way the $\hat{h}_{\rm OM}^\dagger$ appears.

It is helpful to look at a schematic one-dimensional case for which the U and W are assumed constant. The stationary Schrödinger equation is then

$$\hat{h}_{\rm OM}\phi(x) \equiv \left(-\frac{\hbar^2}{2m_N}\frac{\partial^2}{\partial x^2} + \hat{U} - i\hat{W}\right)\phi(x) = E\phi(x) \quad \text{(with} \quad E = E^*) \tag{1.53}$$

and is satisfied by the ansatz $\phi \sim \exp(iKx) \equiv \exp((ik-\kappa)x)$, where k is supposed to be positive, to ensure a wave propagating from left to right. Its value, as well as that of κ, the real and imaginary parts of K, can be determined from the two equations $\hbar^2(k^2 - \kappa^2) = 2m_N(E - U)$ and $\hbar^2 k\kappa = m_N W$. It is seen that for positive W the κ is positive, too. This means that a wave travelling in the x direction experiences an attenuation according to $|\phi(x)|^2 \sim \exp(-2\kappa x)$: The imaginary part of the potential describes a decrease in intensity due to the "absorption" of particles of given energy. One may thus define a *mean free path*

$$\lambda = \frac{1}{2\kappa} = \frac{\hbar^2}{2m_N}\frac{k}{W}. \tag{1.54}$$

Notice that the eigenvalues E of the Hamiltonian $\hat{h}_{\text{OM}}$ are *real* reflecting the fact that we are dealing with *elastic scattering*.

At this stage it is worthwhile to briefly mention the close analogy with electromagnetic waves. Optical properties of dielectric materials are specified by the dielectric function $\varepsilon(\omega)$ which may be split into its real and imaginary parts by writing $\varepsilon(\omega) = \varepsilon'(\omega) + i\varepsilon''(\omega)$. Any finite $\varepsilon''(\omega)$ implies loss of energy of the electromagnetic wave to the medium. In such a case the index of refraction $n = \sqrt{\varepsilon}$ becomes complex (assuming the magnetic susceptibility μ to be close to unity). As in the case discussed before, the attenuation of the electromagnetic wave can be described by introducing a complex wave vector.

1.3.2.2 *The realistic three-dimensional case* As indicated above, in the past one has been able to develop methods that allow one to understand nuclear dynamics on a microscopic basis. Unfortunately, this knowledge does not always suffice to actually calculate basic quantities on the same level. Rather, one has to rely on phenomenological approaches. One needs to invent a plausible, physically reasonable *form for the average potential*. This involves a set of parameters which may then be found by optimizing the evaluated cross section to fit experimental results best. In this way one gets important information about details of such potentials, which one may then try to understand from a "microscopic" point of view. Equally important is the prediction of the outcome of other experiments, which is possible because the number of parameters is much smaller than the available data. To calculate the cross sections for a given potential one needs to solve numerically the Schrödinger equation to find the phase shifts and from them the scattering amplitude and thus the cross section. As explained in Section 18.1.3, a partial wave analysis can also be applied to absorptive processes. For an instructive example the reader is referred to the case of scattering at a black sphere discussed in an exercise in Section 18.1.3.

Because of the importance of the spin-orbit interaction a corresponding term has to be added to the Hamiltonian (1.50) to get

$$\hat{h}_{\text{OM}} = \frac{\hat{\mathbf{p}}^2}{2m_N} + U(r) - iW(r) + U_{ls}(r)\,\hat{\mathbf{l}}\cdot\hat{\mathbf{s}}. \tag{1.55}$$

The $r = |\vec{r}|$ measures the distance of the nucleon whose motion we study to the center of the heavy nucleus. Here, $\hat{\mathbf{l}} = [\hat{\mathbf{r}} \times \hat{\mathbf{p}}]$ and $\hat{\mathbf{s}}$ are the operators for its orbital angular momentum and spin, respectively. For the three functions of r appearing in (1.55) the following forms have turned out to be useful (see e.g. (Mahaux and Sartor, 1991), (Gadioli and Hodgson, 1992), (Hodgson, 1994))

$$U(r) = U_u(E) f\left(\frac{r - R_u}{a_u}\right) \tag{1.56}$$

$$W(r) = W_v(E) f\left(\frac{r - R_v}{a_v}\right) - 4W_s(E) a_s \frac{\partial}{\partial r} f\left(\frac{r - R_s}{a_s}\right) \tag{1.57}$$

$$U_{ls}(r) = U_{ls} \frac{1}{r} \frac{\partial}{\partial r} f\left(\frac{r - R_{ls}}{a_{ls}}\right)$$

If a proton interacts with the nucleus one needs to add to $U(r)$ its Coulomb potential $U_C(r)$ in the field of the other Z protons, which are usually assumed to be distributed homogeneously. The $f(x)$ is the Fermi function

$$f(x_i) = \frac{1}{1 + \exp(x_i)} \quad \text{with} \quad x_i = (r - R_i)/a_i. \tag{1.58}$$

The values for the radii orient themselves at the space distribution of the density to which we will come to below. For the real part of the mean field of A particles one has, for instance,

$$R_u(A) = r_u A^{1/3} \quad \text{with} \quad r_u = 1.25 \text{ fm} \tag{1.59}$$

This value of r_u is slightly larger than the one for the density given below in (1.61). In this difference the finite range of the two-body interaction is hidden. Like the density, the potential has a diffuse surface determined by the parameter a_u, whose value is about $a_u = 0.65$ fm. The potential is uniform in the interior and has a surface of thickness $4.4a$ (measuring the range where the potential falls from 0.9 to 0.1 of its value in the interior). In nuclear physics a potential of this shape is called a Woods–Saxon potential. Its depth parameter U_u and the strength of the spin-orbit interaction are of order -50 MeV and 30 MeV fm$^2/\hbar^2$, respectively, but they may vary with A in a different manner. Moreover, they depend on isospin, which is to say on $N - Z$, and on energy. It must be said that these features only describe global properties. The determination of more details of the optical potential is indeed a subtle problem in the way microscopic aspects are to be combined with phenomenological trial and error. The interested reader may wish to look into the review article by C. Mahaux and R. Sartor (1991). For a recent exhaustive study of this type we may refer to (Bauge et al., 1998).

For the present purpose it may suffice to consider briefly the following two properties of the imaginary part. As seen from (1.57) the W has a *volume part* W_v and a *surface part* W_s. Whereas the W_v is larger in the interior of the nucleus, the W_s is peaked at the surface: The derivative of the Fermi function is maximal at R_s and drops on both sides over a distance determined by the a_s. A further remark may be made on the energy dependence of the functions $W_v(E)$ and $W_s(E)$: At low energies the $W_v(E)$ is known to increase roughly linearly with E whereas the $W_s(E)$ decreases; a reasonable estimate for the $W_v(E)$ is $W_v(E) \simeq 0.22E$ for $E \gtrsim 10$ MeV. This reflects the general trend that at larger energies the volume part becomes more important. Later on we will discuss on a more microscopic level why the single particle motion becomes increasingly damped the more its energy deviates from the Fermi energy. In fact, along with this behavior we will encounter a similar variation with the temperature of heated systems.

1.3.3 The liquid drop model

An important characteristic for the nature of nuclear dynamics is the mean free path λ of nucleons moving in the nuclear medium. If the λ were much smaller than the nuclear diameter one would be inclined to conclude that the nucleus as a whole behaves as a drop of liquid. Indeed, this was the conjecture of N. Bohr and F. Kalckar (1937) drawn in analogy with the compound model. We will return to this question below after having examined whether or not such a hypothesis still holds true in the light of the experience one has gained in the meantime.

1.3.3.1 The puzzle of the mean free path

In (Siemens and Jensen, 1987) the following example is used to obtain some reasonable orientation for a typical case at zero thermal excitations. For the interior of the nucleus the mean free path may be estimated on the basis of our considerations in Section 1.3.2.1. Using the estimates for W_v one finds the nucleonic mean free path to be of the order of the size of a heavier nucleus. For the example of a neutron of 40 MeV one gets $\lambda \simeq 5$ fm $\simeq R$. This is an astonishingly *large* number on account of the large value of the angle integrated cross section σ for the NN scattering. For the same value for the relative kinetic energy one would get $\sigma \simeq 17$ fm^2. Estimating the mean free path in the way one would do if the nucleons were to behave like a system of classical particles one would find

$$\lambda_{\text{class}} \approx \frac{1}{\rho\sigma} \approx 0.4 \text{ fm} \ll \lambda. \tag{1.60}$$

In such a case the nucleons would behave like molecules or atoms of a classical liquid–or like the ensemble of billiard balls mentioned above. The form (1.60) assumes the subsequent collisions of the nucleons to be independent of each other, which is actually not the case as we shall argue below. Within such a picture, formula (1.60) is immediately understood if one realizes that $\lambda_{\text{class}}\sigma$ defines that volume in which on average a moving nucleon meets one scattering partner, for which reason one must have $\rho\sigma\lambda_{\text{class}} \approx 1$. To get the number shown in (1.60) the density has been taken to be 0.17 nucleons per fm^3, see below.

In fact, even in the book by J.M. Blatt and V.F. Weisskopf (1952) it was believed that the mean free path could be estimated as in (1.60). As just stated, now we know better. However, that does not mean that the concept of the compound nucleus must be totally ruled out. It only has to be modified and put into a more correct or appropriate quantum basis. After all, sharp resonances *are* seen in experiments. As indicated in the previous section, they can be explained only by accounting for a strong interplay of the wave function of the projectile with an adequate many-body function of the target nucleus.

1.3.3.2 The picture of a liquid drop

Many properties of finite nuclei exhibit a smooth behavior when looked at as a function of particle number or the total energy. In this sense the nucleus is said to exhibit "macroscopic" behavior. The truth is that many of these features show similarities to those of a liquid drop. Let us discuss a few of them, beginning with the density distribution in a three-dimensional space.

The density distribution: It can be measured in scattering experiments, in particular using electrons which do not underlie the strong interaction. This is of some advantage, as for example there will be no absorption. Using electrons of sufficiently high energies, their scattering can be described in Born approximation. This allows for an immediate deduction of the form factor, which in turn is a direct measure of the distribution of the protons. Such charge densities follow largely the functional form $\rho(r) = \rho_0[1 + \exp(r - R/a)]^{-1}$ with the three parameters, the density ρ_0 in the interior, the radius R and the surface diffuseness a. One finds

$$R(A) = r_0 A^{1/3}, \quad \text{with} \quad r_0 = 1.1 - 1.15 \text{ fm} \quad \text{and} \quad a \simeq 0.55 \text{ fm}. \quad (1.61)$$

From the radius the average density can be estimated to be

$$\rho_0 = \frac{A}{V} = \frac{3}{4\pi r_0^3} \simeq 0.16 - 0.17 \quad \text{nucleons fm}^{-3}. \quad (1.62)$$

Its value corresponds quite closely to the central density of a heavy nucleus. Combining (1.23) with (1.62), namely $\rho_0 = 3/(4\pi r_0^3)$, one gets for the dimensionless quantity $r_0 k_F$ the following simple relation:

$$r_0 k_F = \left(\frac{9\pi}{8}\right)^{1/3} \simeq 1.523 \quad (1.63)$$

The fact that ρ_0 is largely *independent* of the size of the nucleus is related to the short range nature of the strong interaction. Inside the nuclear medium the individual nucleon encounters other nucleons only in its immediate neighborhood. In this sense one speaks of "nuclear saturation". Unlike the electrons in an atom, on average the nucleons do not come closer when additional particles are added. It is only that the radius changes according to (1.61). This behavior is analogous to the one of molecules in a liquid, which may not be too much of a surprise looking at the form of the average two-body interaction. The latter shows similarities to the van der Waals force, being repulsive at short distances and attractive at larger ones.

The static energy as function of particle number: Let us next look at the binding energy $B(N, Z)$ of a nucleus of N neutrons and Z protons. It is determined by the mass decrement of the $N + Z$ nucleons to the $M(N, Z)$ of the combined system,

$$B(N, Z) = N m_n c^2 + Z m_p^2 - M(N, Z) c^2 \equiv -E(N, Z). \quad (1.64)$$

By definition the $B(N, Z)$ is a positive quantity whose negative value we associate with the energy $E(N, Z)$. As a most striking feature, the variation of $B(N, Z)$

(or $E(N,Z)$) with mass number $A = N+Z$ shows oscillations around an average trend of very small amplitude. Their magnitude is of the order of less than 1%. The average trend is well represented by formulas of the type

$$E_{\text{LD}}(N,Z) = E_{\text{LD}}^{\text{vol}} + E_{\text{LD}}^{\text{surf}} + E_{\text{LD}}^{\text{Coul}} \qquad (1.65)$$

with

$$E_{\text{LD}}^{\text{vol}} = -a_{\text{vol}}\left(1 - \kappa_{\text{vol}}\frac{(N-Z)^2}{A^2}\right) A \qquad (1.66)$$

$$E_{\text{LD}}^{\text{surf}} = a_{\text{surf}}\left(1 - \kappa_{\text{surf}}\frac{(N-Z)^2}{A^2}\right) A^{2/3} \qquad (1.67)$$

$$E_{\text{LD}}^{\text{Coul}} = \frac{3}{5}\frac{Z^2 e^2}{R_c} \qquad (1.68)$$

and the following values of the parameters involved: $a_{\text{vol}} \simeq 16$ MeV, $a_{\text{surf}} \simeq 21$ MeV, $\kappa_{\text{vol}} \simeq 1.9$, $\kappa_{\text{surf}} \simeq 2.3$ and $R_c \simeq 1.16\, A^{1/3}$ fm. In the literature such formulas are commonly associated with the names of H.A. Bethe and C.F. von Weizsäcker. Here, we confine ourselves to a simple version just to exhibit the essential features. In the literature much more involved forms are in use, in particular by the Berkeley–Los Alamos collaboration (see e.g. (Möller et al., 1995)). Such refinements are necessary for truly quantitative descriptions of nuclear masses, which can be obtained after considering contributions from the quantal single particle motion; these problems will be addressed in Chapter 5. Whereas the liquid drop model treats nuclear matter as incompressible, this assumption is eased in the "droplet model" developed by W.D. Myers and W.J. Swiatecki (1969). Moreover, neutrons and protons are no longer assumed to have a common surface. Altogether this leads to terms additional to those shown in (1.65)–(1.68) by which the series in terms of $A^{-1/3}$ continues down to $1/A$ (for details and further references see (Möller et al., 1995)). Another improved formulation of the traditional liquid drop formula can be found in (Pomorski and Dudek, 2003).

Whereas the B itself takes on values of the order of up to about 1800 MeV for the heaviest nuclei, the deviations from the form (1.65) are less than about 10 MeV! (Commonly one plots the quantity B/A which is known to increase from about 5 MeV for the lightest nuclei (not counting the special cases of the deuteron and the α particle) to about 8.8 MeV in the iron region, to drop almost linearly afterwards down to a value of about 7.4 MeV.

The expansion (1.65) is typical for a so-called *leptodermous* system, dominated by a large volume and a relatively thin skin. Often one approximates the smooth space density by a sharp one with a vanishing diffuseness parameter a, but the more detailed forms of (Möller et al., 1995) take into account a diffuse surface. The leading term in (1.65) is given by the so-called "volume energy" which would be there also for an infinitely large system. The main correction is proportional to the size of the surface, hence the name "surface energy". Both terms contain (comparatively small) isospin corrections. Finally, there is the energy corresponding to a homogeneously charged sphere. The whole picture resembles that of a charged liquid drop with a more or less pronounced sharp surface, for which reason we have put the index LD on the binding energy $B_{\rm LD}$ of formula (1.65). The expansion (1.65) may be considered a prime example of *macroscopic behavior* which can be seen in many nuclear properties. Often such quantities behave smoothly in respect to their variation in particle number, or in other quantities by which such properties may be parameterized such as the energy or deformation of the nuclear shape. Quantum effects which accompany the smallness of a system only show up in the fluctuations around the average trend. Although such deviations may be small, they may be decisive in understanding essential nuclear properties.

The deformed liquid drop model: According to (1.65) the liquid drop energy is governed by the volume energy, the surface energy and the Coulomb energy, calculated for spherical shapes, discarding again the other terms mentioned before. It is an established fact that the density distribution of nuclei may be deformed. The degree of deformation may be characterized by some parameter Q_μ. One is mostly concerned with deformations which maintain the volume constant. It is, therefore, only the surface energy and the Coulomb energy which vary with the Q_μ. The coordinate dependence of the total energy may thus be simply expressed by multiplying the corresponding terms of the Bethe–Weizsäcker formula by appropriate factors, namely

$$E_{\rm LD}^{\rm surf} = a_{\rm surf} A^{2/3} S_0 \qquad E_{\rm LD}^{\rm coul} = \frac{3}{5}\frac{Z^2 e^2}{R_c} C_0 \,. \tag{1.69}$$

The one for the surface is simply given by $S_0(Q) = S(Q)/(4\pi R^2(A))$. One assumes that the density which is constant in the interior, has the value given in (1.62) and drops to zero sharply at the radius $R(A)$ given by (1.61). Evidently, the $S_0(Q)$ is larger than 1 for any shape Q which differs from that of a sphere. Conversely, the C_0 for the Coulomb energy must be smaller than unity, because for a deformed system the charges lie farther apart from one another.

For small deviations from a sphere the deformation parameters may be introduced as follows. One parameterizes the shape as $\rho(r,\vartheta,\phi) = \rho_0\,\Theta(R(\vartheta,\phi) - r)$, where $R(\vartheta,\phi)$ is the absolute value of the radius vector which starts at the center of mass and which points in the direction (ϑ,ϕ). The function $R(\vartheta,\phi)$ may be expanded in spherical harmonics in the form

$$R(\vartheta,\phi) = R_0\left(1 + \sum_{\lambda\mu}\alpha_{\lambda\mu}Y^*_{\lambda\mu}(\vartheta,\phi)\right) \tag{1.70}$$

with

$$\alpha_{\lambda\mu} = (-1)^\mu \alpha^*_{\lambda-\mu} = R_0^{-1}\int d\Omega\, R(\vartheta,\phi) Y_{\lambda\mu}(\vartheta,\phi)\,. \tag{1.71}$$

The R_0 is the radius of the sphere (of identical volume) which in the nuclear case is related to the particle number A by $R_0 = r_0 A^{1/3}$; as indicated in (1.68), for the charge density $r_0 \approx 1.24$. Volume conservation requires the condition $\alpha_0 = -\frac{1}{4\pi}\sum_{\lambda\mu}|\alpha_{\lambda\mu}|^2$. Another one is to be put for the $\alpha_{1\mu}$ to ensure that the center of mass remains at rest; proofs of these relations can be found in Appendix 6A of (Bohr and Mottelson, 1975).

Denoting by X either the surface or the Coulomb energy, to order $|\alpha_{\lambda\mu}|^2$ the latter can be written in the form

$$E^X_{\mathrm{LD}}(\alpha_{\lambda\mu}) = E^X_{\mathrm{LD}} + \frac{1}{2}\sum_{\lambda\mu}(C^{\mathrm{LD}}_\lambda)_X |\alpha_{\lambda\mu}|^2, \qquad (1.72)$$

with the spherical parts E^X_{LD} being given by (1.67) and (1.68). The stiffness for the surface energy turns out to be

$$(C^{\mathrm{LD}}_\lambda)_{\mathrm{surf}} = \frac{1}{4\pi}(\lambda-1)(\lambda+2)A^{2/3}a_{\mathrm{surf}}, \qquad (1.73)$$

and the one for the Coulomb energy

$$(C^{\mathrm{LD}}_\lambda)_{\mathrm{Coul}} = -\frac{3}{2\pi}\frac{\lambda-1}{2\lambda+1}\frac{e^2}{r_0}Z^2 A^{-1/3}. \qquad (1.74)$$

Summing up both, it is seen that for quadrupole distortions ($\lambda = 2$) and for

$$\frac{Z^2}{A} > \left(\frac{Z^2}{A}\right)_{\mathrm{crit}} = \frac{10}{3}\left(\frac{r_0 a_{\mathrm{surf}}}{e^2}\right) \approx 49 \qquad (1.75)$$

the overall stiffness becomes negative, such that the nucleus is *unstable*: it may be driven apart by the Coulomb force. This feature may be expressed through the *fissility parameter* x

$$x = \frac{Z^2}{A}\left(\frac{Z^2}{A}\right)^{-1}_{\mathrm{crit}} = \frac{E^{\mathrm{Coul}}_{\mathrm{LD}}}{2E^{\mathrm{surf}}_{\mathrm{LD}}} \approx 0.0205\frac{Z^2}{A}. \qquad (1.76)$$

The instability occurs when x becomes larger than unity. Its expression in terms of the Coulomb and surface energy of the spherical drop is easily recognized from (1.67) and (1.68). These estimates go back to the famous paper by N. Bohr and J.A. Wheeler (1939). Later in Section 8.2.2 we are will study refinements of these properties due to the possibility of nuclear rotations.

For quadrupole deformations (at $\lambda = 2$) it is customary to transform to the body fixed frame by writing

$$\alpha_{2\mu} = \sum_\nu a_{2\mu}\mathcal{D}^2_{\mu\nu}(\phi,\theta,\psi). \qquad (1.77)$$

Here, ϕ, θ and ψ are the three Euler angles which specify the body fixed frame with respect to the lab frame and $\mathcal{D}^2_{\mu\nu}(\phi,\theta,\psi)$ is the transformation matrix. The $a_{2\mu}$ are subject to the conditions $a_{2-1} = a_{21} = 0$ and $a_{2-2} = a_{22}$. For the two parameters which besides the (ϕ, θ, ψ) are independent it is convenient to introduce the β and γ defined through $a_{20} = \beta \cos\gamma$ and $a_{2-2} = a_{22} = \sqrt{1/2}\,\beta\sin\gamma$. The β specifies the size of the deformation and γ may be viewed as a kind of polar angle. To linear order in β the shape represents an ellipsoid with the increment of the three principal axes given by

$$\delta R_k = \left(\frac{5}{4\pi}\right)^{1/2} \beta\, R_0 \cos\left(\gamma - k\frac{2\pi}{3}\right) \qquad k = 1, 2, 3. \qquad (1.78)$$

This relation allows one to understand the geometrical meaning of the parameters; for a detailed description, also of the underlying symmetries we may refer to App.6B-1 of (Bohr and Mottelson, 1975).

1.3.3.3 *The liquid drop model and the compound nucleus* Already in their pioneering paper (1937) N. Bohr and F. Kalckar had suggested generalizing the *liquid drop model to include nuclear dynamics*. Drawing on analogies to the dynamics of a drop of water it became possible to envisage that a nucleus may perform collective motion, such as vibrating about a spherical configuration. Another and less trivial example is nuclear fission, which was discovered in the following year. The liquid drop model delivers the simplest way to understand the most basic physics of this process. The timely sequence to the advent of the compound model in 1936 is no accidence. Indeed, in this paper the authors speak of the LDM as being the *classical analog* to the compound picture. It has been said above that the compound model assumes the existence of strong correlations among the nucleons. Moreover, in 1937 the mean free path of nucleons inside the nucleus was still meant to be strictly smaller than the nuclear diameter. As mentioned before, in these days the λ was estimated on the basis of (1.60). In a classical interpretation such features are characteristic of a liquid.

We shall see later that nuclear matter behaves like a quantum liquid, which at $T = 0$ is essentially *non-viscous*. (It is interesting to note that Bohr and Kalckar also left out the complications of viscosity, although they were aware that this might be too drastic an assumption). For higher temperatures, where the impact of the Pauli principle diminishes, one may expect the quantum liquid to resemble a classical one more and more. One way to see this feature is by studying the temperature dependence of nuclear viscosity. Unfortunately, this is such a hard task that to date there is still no convincing answer; we will return to this problem later.

For the static interpretation of the liquid drop model, on the other hand, there is evidence that its validity is warranted at higher temperatures. The smooth part of the total energy, as represented by the Bethe–Weizsäcker formula, corresponds to the free energy when calculated on a quantal basis for temperatures of $T \approx 2$ MeV. As we shall see later, at such excitations those quantum effects which are responsible for the fluctuations of nuclear masses or their binding energies have already disappeared.

1.3.3.4 First observations about nuclear transport

It has been mentioned that the compound model of nuclear reactions assumes that in an intermediate step the excitation brought into the nuclear system is equally shared among all degrees of freedom, very much in the spirit of a statistical equilibrium. It has also been said that these conditions are not always met. There is a whole class of types of reactions which go along with the concept of *pre-equilibrium*. This notion indicates that one may wish to describe nuclear dynamics as a transport process. This turns out necessary if processes are involved in which many nucleons participate "collectively", and which already for this reason occur on long timescales. One may then introduce the concept of *partial equilibria* for the "fast" degrees of freedom. That nuclei exhibit collective motion has known been at least since the discovery of fission and has become manifest in the 1950s for other degrees of freedom as well. Notwithstanding, it was not before the late 1970s that transport models became fashionable, *again* one might add. Indeed, after the liquid drop model of Bohr and Kalckar, it was H.A. Kramers who already in (1940) had set up a transport model to describe the dynamics of nuclear fission. He implied that collective dynamics was irreversible, tracing dissipation back to viscosity of the nuclear fluid. This hypothesis were fully justified if the mean free path λ would indeed be short compared to the overall size of the nucleus. It was only after the development of the shell model by M. Goeppert-Mayer and J.H.D. Jensen that one realized the opposite were true, at least at zero or small excitations.

This short discussion shows the relevance of the microscopic understanding of the damping of single particle motion–as reflected in the imaginary part of the mean field, for instance–for the understanding of the nature of damping of collective motion. It has already been indicated that λ can be expected to become shorter when the nucleus gets heated up. For the moment let us take this for granted. Evidently, when λ becomes shorter than the dimension of the nucleus, one may expect that the nuclear fluid becomes genuinely viscous, in the sense of being collision dominated. Even before that happens, one may anticipate that the particle dynamics follows transport equations similar to that of the Boltzmann equation, which is known to describe the dynamics of gases (and of liquids to some extent) away from equilibrium. Indeed, in nuclear physics such kind of transport equations find application as well, not only the one of Kramers.

Exercise

1. Potential scattering of s-waves: (i) Calculate the width of a resonance applying time independent scattering theory.
   (ii) Discuss the extension to the scattering of wave packets and the association of the resonance width to a time-delay.

2

NUCLEAR MATTER AS A FERMI LIQUID

In this chapter we will try to get some insight into the question of how the *bare* nucleon–nucleon interaction V_{ij} may lead to saturation of nuclear matter. Here, bare simply means that we look at the collision of nucleon i with nucleon j as if it were not influenced by the presence of any others. In this spirit the V_{ij} is to be chosen as outlined in Section 1.1. For the A-particle system the Hamiltonian may then look like

$$\hat{H} = \hat{T} + \hat{V} = \sum_i^A \hat{t}_i + \frac{1}{2}\sum_{i\neq j} \hat{V}_{ij} \,. \tag{2.1}$$

One of the simplest approaches possible to account for the interaction is Hartree–Fock. We shall begin by examining the applicability of such an approach but will soon see that no stable system is found in this way. Rather, correlations beyond those of Hartree–Fock for which fancier methods are required turn out important (these will be discussed later). As the NN interaction only depends on the vector **r** of relative motion we will frequently make use of the transformation given by (1.2) and (1.3) discarding the mass difference of neutrons and protons.

In Section 2.1.3 a short excursion will be performed to chiral perturbation theory, which delivers another mechanism for the saturation of nuclear matter. Otherwise, throughout the book we will stick to the more traditional point of view which is based on potential models. This should be understand not only with respect to the two-body interaction but also to the nucleons' mean field, which we will construct later on. Indeed, generalizing the single particle potential to the one of the deformed shell model will deliver the microscopic basis for generating collective dynamics.

2.1 A first, qualitative survey

2.1.1 The inadequacy of Hartree–Fock with bare interactions

Details about the Hartree–Fock approximation formulated with density matrices can be found in Chapter 20. The essential assumption is to assume the *motion of the particles to be independent of each other*. As in the Fermi gas, the two-body density operator $\hat{\rho}^{(2)}_{\rm HF}$ can thus be expressed by the one-body density operator $\hat{\rho}^{(1)}_{\rm HF}$, for which we will use the symbol $\hat{\rho}$ in the following. With respect to any single particle basis $|\alpha_l\rangle \equiv |l\rangle$ one may write $\langle mn|\hat{\rho}^{(2)}|kl\rangle = \rho_{mk}\rho_{nl} - \rho_{ml}\rho_{kn}$ with the two-body function being just given by the product of one-body functions, $|kl\rangle = |k\rangle|l\rangle$, see eqn(19.44). If the particles interact with each other by a two-body interaction, the energy functional $E[\Psi^{\rm HF}]$ can be expressed by the one-body density as $E[\hat{\rho}] = \sum_{kl} \langle k|\hat{t}|l\rangle \rho_{lk} + (1/2) \sum_{klmn} \langle kl|\hat{V}_{12}|mn\rangle (\rho_{mk}\rho_{nl} - \rho_{ml}\rho_{nn})$. This functional is to be used for the variational principle to find the optimal one-body density. Here we want to apply it to a homogeneous system. To simplify our considerations we will neglect the exchange term, for which reason it is more correct to speak of Hartree rather than of Hartree–Fock. For the crude qualitative understanding we are aiming at here, this exchange term is not too important although its influence cannot be discarded for quantitative studies (de Shalit and Feshbach, 1974). For an explicit calculation including this effect, see e.g. the book by A.L. Fetter and J.D. Walecka (Fetter and Walecka, 1971). Discarding furthermore any dependence of the interaction on isospin, spin and angular momentum (a discussion including those may be found in Section II.7 of (de Shalit and Feshbach, 1974)) the contribution of the interaction to the total energy turns out to be (see (20.20))

$$E_{\text{pot}}[\rho] = A\left(\frac{1}{2}\int d^3r V(\mathbf{r})\right) \rho. \qquad (2.2)$$

Here, the relative distance $\mathbf{r}$ between the particles was introduced. The vector of their center of mass was integrated over the volume $\mathcal{V}$, which together with the constant one-body density of homogeneous nuclear matter determines the particle number $A = \mathcal{V}\rho$. The contribution from the potential energy to the energy per particle $\mathsf{E} = E/A$ has turned out linear in ρ; introducing the equilibrium density we may write: $\mathsf{E}_{\text{pot}}[\rho] = E_{\text{pot}}[\rho]/A = -\chi(\rho/\rho_0)$. Later on in this chapter this feature will be seen to hold true also for more refined treatments of the interaction, with a different value of the coefficient χ, of course. Next we need to calculate the functional $\mathsf{E}_{\text{kin}}[\rho]$ for the kinetic energy. Since the single particle wave functions will still be given by plane waves we know the result from the discussion of the Fermi gas model for which the kinetic energy was seen to be $E_{\text{kin}} = (3/5 e_F)A$, recall (1.24). Expressing e_F through the Fermi momentum and using (1.23) the dependence on the density becomes proportional to $\rho^{2/3}$. Adding up both contributions the total energy per particle becomes

$$\mathsf{E}[\rho] = \mathsf{E}(\rho/\rho_0) = \lambda \left(\rho/\rho_0\right)^{2/3} - \chi\left(\rho/\rho_0\right) \qquad \text{with} \qquad (2.3)$$

$$\lambda = \rho_0^{2/3} \frac{3}{10} \left(\frac{3\pi^2}{2}\right)^{2/3} \frac{\hbar^2}{m} \simeq 74 \rho_0^{2/3} \text{ MeV} \simeq 23 \text{ MeV}, \qquad (2.4)$$

$$\chi \approx \rho_0 \frac{2}{3} \pi V_0 b^3 \simeq 320 \rho_0 \text{ MeV} \simeq 55 \text{ MeV}. \qquad (2.5)$$

For the potential energy the coefficient χ was estimated for a schematic interaction of the type $V(r) = -V_0 \Theta(b-r)$ where the repulsive core has been discarded. The parameter b for the range of the attraction is known to be approximately given by the pion's Compton wavelength $b \approx \hbar/(\mu_\pi c) \simeq 1.45$ fm, with μ_π being the pion mass. Typically, the potential depth is of the order of $V_0 \simeq 50$ MeV.

A closer inspection of the function $\mathsf{E}(\rho)$ reveals that it does *not exhibit a minimum*. There is a stationary point where $\partial \mathsf{E}/\partial \rho$ vanishes but this corresponds to a *maximum* of the energy; it is found at $\rho_1 = (2\lambda/3\chi)^3$ fm$^{-3} \simeq 0.004$ fm$^{-3} \ll \rho_0 = 0.172$ fm^{-3}. Hence, this simple model does not lead to stability or saturation. The reader may argue that we have left out an important piece of the two-body interaction, the repulsive core at small distances. Indeed, this force should have important consequences, as one may already guess from the example of the van der Waals gas. There, the interaction is similar, namely attractive at large distances but repulsive at small ones, and this model is capable of treating the phase transition to the liquid. The point is, however, that this feature *cannot adequately* be treated in Hartree approximation. Indeed, introducing a finite repulsive core (within our model of ignoring spins etc.) the potential term would still be linear in ρ, as is easily recognized from the expression given in (2.2). Eventually, depending on the strength of this repulsive part the integral may even attain a positive value. If one wants (or needs) to simulate the repulsive part by a *hard core* the integral in (2.2) would diverge! But as we shall soon see, the introduction of the repulsive core helps to solve the problem, one only needs to treat it with methods more advanced than Hartree–Fock.

2.1.2 Short-range correlations

As we know from Section 1.2.2, the Pauli principle prohibits two particles to be at the same place, which means at $r = 0$. A repulsive core extends this forbidden zone to a finite volume. Let us draw on this picture idealizing the core by a hard one. This example has played an important role in many-body physics for the understanding of short-range correlations. Sometimes this idealization is carried further by neglecting any attractive part of the potential. This has been so in the past (see e.g. (Fetter and Walecka, 1971) or other books which address nuclear physics in particular, such as (Bohr and Mottelson, 1969), (de Shalit and Feshbach, 1974) and (Ericson and Weise, 1988), with references given therein), but still remains to be of interest for present studies, for instance when the *hard sphere gas* as a schematic many-body system is treated within "molecular dynamics".

The effects of a hard core can be understood on the semi-quantitative level by the following plausible consideration. Let us first rewrite the kinetic energy per particle in terms of the radius r of that volume which on average is occupied by one nucleon, and which thus corresponds to the density $\rho = 3/(4\pi r^3)$ (recall 1.62). The kinetic energy per particle has been calculated in (1.24) to $3\epsilon_F/5$, and the relation of k_F to the radius was seen in (1.63) to be given by $rk_F = (9\pi/8)^{1/3}$. Notice that the index "0" has been dropped here, simply because the density (and thus the r) is to be considered a free parameter. The expression we aim at is given by:

$$\mathsf{E}_{\text{kin}}[\rho] = \frac{3}{5} \left(\frac{9\pi}{8} \right)^{2/3} \frac{\hbar^2}{2mr^2}. \tag{2.6}$$

At this place we need to invoke Heisenberg's uncertainty relation. Evidently, the form (2.6) is in accord with $\Delta p \sim \hbar/r$. The kinetic energy is just made up by those momenta which are allowed for a situation where one particle is confined to a volume of radius r. However, the hard core reduces this volume, for which reason the kinetic energy effectively must become larger. This fact may be accounted for in (2.6) by replacing r in the denominator by $r - r_c$, or by $r - 0.8 r_c$ as a more careful analysis shows. As a matter of fact for low densities it may be shown (Bohr and Mottelson, 1969; de Shalit and Feshbach, 1974) to represent the expansion

$$\mathsf{E}_{\text{kin}}[\rho] = e_{\text{F}} \left[\frac{3}{5} + \frac{2}{\pi} k_{\text{F}} a + \frac{12}{35\pi^2}(11 - 2\ln 2)(k_{\text{F}} a)^2 + 0.78(k_{\text{F}} a)^3 + \ldots \right] \quad (2.7)$$

in powers of $k_{\text{F}} a$, with a being the scattering length which for a hard core potential is identical to the core radius $|a| = r_c$ (see textbooks on quantum mechanics, e.g. (Schwabl, 1995)). As $r_c \simeq 0.4 - 0.5$ fm, from (1.23) at saturation density of nuclear matter one gets $k_{\text{F}} a \simeq 0.5 - 0.7$. Actually, the first order term corresponds to the Hartree–Fock energy of a repulsive, but finite core. The higher order contributions can be explained (see (Fetter and Walecka, 1971) with references to the original literature) on the basis of the so-called Bethe–Goldstone equation, which we will come to below.

For the moment such details are immaterial; we will be satisfied with the form (2.6) but r replaced by $r - 0.8 r_c$. Introducing again the density by writing $0.8 r_c / r = 1.6 f_c^{1/3} (\rho / \rho_0)^{1/3}$ with the parameter $f_c = (r_c / 2 r_0)^3$ one gets

$$\mathsf{E}(x) = \lambda \frac{x^{2/3}}{\left(1 - 1.6 f_c^{1/3} x^{1/3} \right)^2} - \chi x \quad \text{with} \quad x = \rho / \rho_0 \quad (2.8)$$

and the coefficients λ and χ given in (2.5) and (2.4). As compared to the mere Fermi gas, the "modified" kinetic energy, more precisely the kinetic energy plus the contribution from the hard core, first rises more strongly with x to saturate at larger x. As a consequence, the $\mathsf{E}(x)$ now develops a minimum. At which density this is found and how big the corresponding E is depends sensitively on the parameters one uses, in particular on the f_c. The latter may be expected to range between $f_c \simeq 0.005$ for $r_c = 0.4$, $r_0 = 1.15$ fm to $f_c \simeq 0.0117$ for $r_c = 0.5$, $r_0 = 1.1$ fm. For a $f_c \simeq 0.01$ the minimum comes up at the right position, but the value for $\mathsf{E}(x_{min})$ is quite wrong. Enlarging slightly the coefficient for the potential energy the binding gets stronger, but the value for the equilibrium density is wrong. Of course, one might to try to get both the equilibrium density as well as the energy right by varying the parameters involved. However, in this way the physics behind the manipulation with the kinetic energy would remain obscure. Indeed, a microscopic theory which is based on the potential model (for the two-body interaction) requires more involved methods to treat the "scattering" of the nuclei in the nuclear medium adequately. Before we address this problem in the next section, we should like to briefly discuss a very new development based on chiral perturbation theory. Amongst others, it makes use of the observation that an expansion of the energy per particle E in low powers is much more promising if it is done not with respect to density but to the Fermi momentum k_F.

2.1.3 Properties of nuclear matter in chiral dynamics

Suppose the energy per particle is written as the following expansion in powers of the Fermi momentum (Lutz et al., 2000)

$$\mathsf{E}(k_F) = \frac{3k_F^2}{10m} + \sum_{n \geq 3} \mathcal{F}_n k_F^n \qquad (\hbar = 1) \tag{2.9}$$

Restricting the sum to the first two terms the parameters $\mathcal{F}_3$ and $\mathcal{F}_4$ may be fitted to the equilibrium values of the density and the energy. As a result one gets a negative $\mathcal{F}_3$ and a positive $\mathcal{F}_4$, representing binding and repulsion, respectively. From the functional form $\mathsf{E} = \mathsf{E}(k_F)$ one may evaluate the stiffness around the minimum, which determines the incompressibility coefficient $\kappa = k_F^2 \partial^2 \mathsf{E}/\partial k_F^2 \big|_{\rho_0}$, see Section 2.6.2. The number one gets in this way is in good agreement with empirical values; choosing $\rho_0 = 0.16$ fm^{-3} (or $k_F = 1.33$ fm^{-1}) and $\mathsf{E}_0 = -16$ MeV one gets $\kappa = 236$ MeV.

This nice agreement is a remarkable result in itself. However, it even turns out that the expansion (2.9) can be justified microscopically within chiral perturbation theory (ChPT) and thus connected to low energy QCD (Kaiser et al., 2002). Because of the apparent success of this approach we like to summarize its main arguments; a concise review can be found in (Weise, 2005). ChPT benefits from the fact that for low energy nuclear phenomena there exists a clear separation between the masses of the pions, on the one hand, and those of the nucleons themselves, the actual constituents of nuclei, as well as the slightly heavier Δ-resonance. The pions are interpreted as Goldstone bosons, with their very existence manifesting the spontaneous breaking of the chiral symmetry. Between these two sorts of hadrons there exists a mass gap which is attributed to the scale $4\pi f_\pi \simeq 1$ GeV, where f_π is the pion decay constant. In a finite nucleus, with its low density the associated small Fermi momentum delivers another small scale. From the discussion in Section 1.2.2 we know that the distribution of the relative momenta, which cannot be larger than $2k_F$ anyhow, has its maximum at about $0.6k_F$. This relative momentum of two nucleons equals the momentum the pion can have being exchanged between them. Treating the pion in a relativistic formalism one needs to compare its mass with its momentum. Taking for the latter the possible values one realizes that the latter are of the same order as the mass, indeed $k_F \hbar c \simeq 260$ MeV $\simeq 2m_\pi$. Such features are understood as a strong hint that pions must be taken into account explicitly whenever low energy properties are to be described.

Within ChPT the binding energy of nuclear matter may be calculated, for instance, by summing up contributions from all terms up to the second order in pion exchange. The only constants which enter are the masses of the pions and of the nucleons as well as the pion–nucleon coupling constant, which are all known. In addition there is one adjustable parameter Λ which can be interpreted as representing the contribution from short-range interactions. Actually, this Λ is introduced to regularize diverging loop integrals in momentum space which come up in graphs for iterated pion exchange. It turns out that their contribution is proportional to $\Lambda\rho$ and thus resembles that of a nucleon–nucleon interaction of contact form (which is to say an interaction of very short-range, which in different language might be mediated by the exchange of heavy mesons). Whereas this contributes to the negative, attractive $\mathcal{F}_3$ term, the $\mathcal{F}_4$ term is solely determined by the two-pion exchange. It becomes repulsive because of Pauli blocking in intermediate nucleon states. The value found for $\mathcal{F}_4$ is within 10% of the empirical one obtained in the fit discussed above. A similar statement can be made for the $\mathcal{F}_3$, if only for the cutoff parameter the (very reasonable) value of $\Lambda \simeq 0.65$ GeV is chosen. Later in Section 2.6.2 we will briefly report on applications to finite temperature in which terms up to $\mathcal{F}_6$ have been considered together with an explicit inclusion of the Δ-resonance.

2.1.4 How dense is nuclear matter?

If counted just as mass per volume, there is no doubt that the interior of nuclei represent the highest density of any stable matter we have on earth. In heavy ion experiments one may find configurations where this value may be exceeded a few times, but only for not more than about 10^{-22} sec. Likewise, in the universe there may be stellar objects with a few times nuclear matter density, like neutron stars and such–if we ignore the more exotic cases of black holes etc. However, this number does not tell us which assumptions we ought to make as a starting point for a theoretical description. A much more interesting and relevant quantity is the ratio f of the volume employed by the interaction to that which on average is available to one particle. For an interaction of range b between *two* (point-like) particles, whose average spacing is $2r_0$, one may define the f to be

$$f = \left(\frac{b}{2r_0}\right)^3 = \frac{\pi}{6}\rho b^3. \qquad (2.10)$$

The form given on the right is obtained from $\rho = 3/(4\pi r_0^3)$.

How should we estimate the range b which is to appear in this formula? Let's first make a "conservative" guess arguing that $b \approx r_0 \approx 1$ fm, corresponding roughly to the minimum of the potential, say as given in Fig. 1.1. Even in this case we would get the small number of $f \approx 1/8$ saying that nuclear matter is *dilute rather than dense*! This number becomes much smaller if we identify b by the hard core radius r_c, as done in (2.8) for the estimate of the *quantal* kinetic energy. On the other hand, on the qualitative level we are interested in here, the model of hard spheres allows us to compare nuclear matter with ordinary classical liquids, see e.g. (Chaikin and Lubensky, 1995) or (Curtin and Ashcroft, 1985). Characteristic features of the latter are determined by the f of (2.10), indeed. Taking for r_0 the 1.1 fm of (1.61) and $r_c \simeq 0.4 - 0.5$ fm for the hard core radius, one obtains the very small number $f \simeq 10^{-2}$. In this sense nuclear matter is much much more dilute than ordinary liquids, and even more than solids, of course. If treated as classical objects, *the liquid of hard spheres freezes* at $f = 0.495$, becomes "glassy" at $f = 0.52$ and solid at $f = 0.545$; in between $f = 0.495$ and $f = 0.545$ the liquid and the solid coexist. At still higher densities the solid state takes on other forms: At $f = 0.638$ the system becomes "amorphous" in the sense of having *"random close packed spheres"* and at $f = 0.7405$ *periodic close packing* occurs (Curtin and Ashcroft, 1985). These values are obtained by searching for the extrema of entropy of three-dimensional systems, in the thermostatic limit and as a function of density. (In this context, the latter concept is to be understood in the same way as we define nuclear matter: Both volume and particle number become infinitely large with the ratio kept fixed.)

2.2 The independent pair approximation

The independent pair approximation is an important tool in describing many-fermion systems. It has challenged physicists since the 1950s. Often the theory behind it is simply referred to as "Brueckner theory" (after K.A. Brueckner), but as we shall see it has many facets the clarification of which are associated with other names as well, in particular those of H.A. Bethe, J. Goldstone, C. Mahaux and others. This approximation is coined to handle cases where the two-body interaction between the particles has a strong repulsive core. Below we shall try to give an elementary outline of the theory discussing only the most important issues, following largely the exposition in (de Shalit and Feshbach, 1974). We shall skip most of the formal details trying to keep the discussion as transparent as possible. Naturally, we will not be able to touch more subtle features. For this it would be necessary to exploit Green functions, or field theoretical methods, rather, which is way beyond the scope of this book; for further reading we recommend the accounts by A.L. Fetter and J.D. Walecka (1971) or J.W. Negele and H. Orland (1987).

2.2.1 The equation for the one-body wave functions

One great advantage of studying nuclear matter is found in the fact that one is dealing with an infinite system. Because of translational invariance, and after posing the same periodic boundary conditions as introduced in (1.11) for the Fermi gas, the space part of the one-body wave functions is still given by the plane wave $\psi_{\mathbf{k}_j}(\mathbf{x}) = \langle \mathbf{x} | \mathbf{k}_j \rangle$ of (1.10), a feature which greatly facilitates the calculations. This property remains unchanged in the presence of interactions. In its most general form the Schrödinger equation of the one-body problem in coordinate representation may be written as

$$-\frac{\hbar^2 \Delta}{2m}\psi_\alpha(\mathbf{x}) + \int d^3x' U(\mathbf{x}-\mathbf{x}')\psi_\alpha(\mathbf{x}') = e_\alpha \psi_\alpha(\mathbf{x}), \qquad (2.11)$$

with a non-local operator for the one-body potential U (recall the exchange term in Hartree–Fock as an example). That U only depends on the relative distance is due to the system's translational invariance: U must not depend on any fixed position, which is to say that it remains unchanged if the origin is shifted by any finite amount, viz $U(\mathbf{x} + \Delta\mathbf{x}, \mathbf{x}' + \Delta\mathbf{x}) = U(\mathbf{x}, \mathbf{x}')$. Consequently, its influence only shows up in the dispersion relation between momentum and energy, as becomes obvious after Fourier-transforming (2.11). Suppose the wave function turns into $\psi_\alpha(\mathbf{k}) = \int d^3x e^{-i\mathbf{k}\mathbf{x}} \psi_\alpha(\mathbf{x})$ then the Schrödinger equation leads to

$$(\hbar \mathbf{k}_\alpha)^2/2m + U(\mathbf{k}_\alpha) = e_\alpha(\mathbf{k}_\alpha). \qquad (2.12)$$

This follows from the folding theorem of Fourier transforms and after factorizing out the wave function.

We may use (2.12) to briefly introduce the concept of the *effective mass* on perhaps the simplest possible level; later on we will learn more advanced definitions and applications. For small momenta it may be possible to approximate the $U(\mathbf{k}_\alpha)$ by an expansion to low powers in $|\mathbf{k}_\alpha|$. Because of $U(-\mathbf{x}) = U(\mathbf{x})$, $U(\mathbf{k})$ must be an even function, too. Thus no term linear in k_α survives, such that one may write

$$e_\alpha(\mathbf{k}_\alpha) = \frac{\hbar^2 (\mathbf{k}_\alpha)^2}{2m} + U_0 + \frac{\hbar^2 (\mathbf{k}_\alpha)^2}{2m} U_1 + \ldots \equiv U_0 + \frac{\hbar^2 (\mathbf{k}_\alpha)^2}{2m^*} \qquad (2.13)$$

with the effective mass m^* being given by the relation $m/m^* = 1 + U_1$. In general, this concept has proven very successful. It may be generalized to portray properties of the single particle potential not just at $k = 0$ but in other regions, for instance around the Fermi momentum k_F. There, also expansions with respect to energy are in use.

2.2.2 The total energy in terms of two-body wave functions

In the following we want to assume that there exists a Schrödinger equation for the (antisymmetrized) A-particle wave function,

$$\left(\sum_i^A \hat{t}_i + \frac{1}{2} \sum_{i \neq j}^A \hat{V}(\mathbf{x}_i - \mathbf{x}_j) \right) \Psi_\alpha(1, \ldots, A) = E_\alpha \Psi_\alpha(1, \ldots, A). \qquad (2.14)$$

Later on we may at times concentrate on the ground state, but at present it is sufficient to assume the energy to be small enough such that we may neglect the possibility of decay, say by particle emission. Thus we will be dealing with the common bound state problem for which the energy will be a real quantity.

Since the Hamiltonian consists only of a sum of one- and two-body operators it can be expected that the total energy can be expressed by one and two-body wave functions (see Section 19.1). This conjecture can be proven rigorously by deriving from (2.14) the following expression for the total energy E_α:

$$E_\alpha = \sum_i^A e_{\alpha_i} + \sum_{i<j}^A \langle \alpha_i \alpha_j | \hat{V}(\mathbf{x}_i - \mathbf{x}_j) | [\psi_{\alpha_i, \alpha_j}(i,j)] \rangle. \qquad (2.15)$$

The first term accounts for the single particle content of E_α with the single particle energy e_{α_i} determined through $\hat{t}_i | \alpha_i \rangle \equiv \hat{t}_i | \varphi_{\alpha_i} \rangle = e_{\alpha_i} | \varphi_{\alpha_i} \rangle$. In the present choice it is simply the kinetic energy but one might introduce the contribution from a mean field $\hat{U} = \sum_i \hat{U}(i)$ by adding this $\hat{U}$ to the total kinetic energy of the Hamiltonian in (2.14) and subtracting it in the term which involves the two-body interaction. The (yet unknown) two-body functions $|[\psi_{\alpha_i,\alpha_j}(i,j)]\rangle$ are defined as

$$[\psi_{\alpha_k,\alpha_l}(k,l)] = \int d(1)\ldots d(k-1)d(k+1)\ldots d(l-1)d(l+1)\ldots d(A)\, \Phi^*_{\alpha,kl}\Psi_\alpha. \tag{2.16}$$

The $\Phi_{\alpha,kl}$ is obtained from the auxiliary function $\Phi_\alpha = \varphi_{\alpha_1}(1)\varphi_{\alpha_2}(2)\ldots\varphi_{\alpha_A}(A)$ by leaving out the states associated with α_k and α_l,

$$\Phi_{\alpha,kl} = \prod_{i=1(i\neq k,l)}^{A} \varphi_{\alpha_i}(i) \qquad \alpha_k < \alpha_l. \tag{2.17}$$

(In the Φ_α all quantum numbers are meant to differ from each other, a condition which may be put as $\alpha_1 < \alpha_2 \ldots < \alpha_A$). To obtain the form (2.15) for the energy the solution of (2.14) had to be normalized like $\int d(1)\ldots d(A)\Phi^*_\alpha \Psi_\alpha(1,\ldots,A) = 1$. Notice that the $[\psi_{\alpha_k,\alpha_l}(k,l)]$ is antisymmetric, and it can be shown to have no components in the occupied states except for $\langle[\varphi_{\alpha_k\alpha_l}]|[\psi_{\alpha_k,\alpha_l}(k,l)]\rangle$ (see Chapter III of (de Shalit and Feshbach, 1974)).

In many respects the result (2.15) is very interesting, indeed. First of all, it says that for the calculation of the total energies of the system it is sufficient to know the two-body part of the wave function as given by (2.16). In practice this may not be easy to find, as one would have to know the exact many-body function. But this property shows the lines along which one may proceed and which will be the topic of the next subsection: One needs to invent a model or an approximation which allows one to calculate the $[\psi_{\alpha_k,\alpha_l}(k,l)]$ *without* solving the many-body problem. Interestingly enough, besides the $[\psi_{\alpha_k,\alpha_l}(k,l)]$ it is only the wave functions and eigenvalues of the kinetic energy which one needs to know for evaluating the E_α, which are trivial, of course. However, as indicated previously, it will turn out useful later on to introduce an average one-body potential.

2.2.3 The Bethe–Goldstone equation

The aim of this section is to derive a possible equation for the two-body function introduced in the last subsection. Let us begin looking at the scattering of the two nucleons which interact with each other through $\hat{V}$. If this scattering would happen in free space one would have to find the two-body wave function from the Schrödinger equation

$$\left(\hat{t}_1 + \hat{t}_2 + \hat{V}_{12}\right)|\varphi(12)\rangle = E_{12}|\varphi(12)\rangle, \tag{2.18}$$

where the $\hat{t}_{1/2}$ are the operators for the kinetic energy of particle 1 or 2.

Before we discuss an important modification of (2.18), a few remarks on our goal seem necessary as well as on the consequences to be deduced from it. Most important of all, the scattering of the two nucleons takes place inside nuclear matter. Therefore, the energy appearing on the right-hand side cannot be a free parameter, as it would be the case for genuine scattering in free space.[4] From the discussion of the last subsection we expect it to appear in an expression like (2.15) for the total energy, which corresponds to a bound state of the many-body system. In this case the energy E_{12} will have to be found from the solution of an adequate equation which determines the "scattering in the medium", say from the states $|\alpha\rangle|\beta\rangle$ to *final states* $|\alpha'\rangle|\beta'\rangle$. However, inside the medium such a scattering underlies certain restrictions. If the $|\alpha'\rangle|\beta'\rangle$ are already *occupied* the scattering is *prohibited*. Expanding the two particle wave function into a series of unperturbed functions this fact shows up in the following property

$$|\varphi_{\alpha\beta}\rangle = |\alpha\beta\rangle + \sum_{k'_\alpha, k'_\beta > k_F} C^{\alpha\beta}_{\alpha',\beta'} |\alpha'\beta'\rangle. \tag{2.19}$$

The second part on the right represents the "scattered wave". It cannot have components inside the "Fermi sea", simply because at zero temperature all states with $k'_\alpha, k'_\beta \leq k_F$ are fully occupied. Such features may be accounted for by changing (2.18) to

$$\left(\hat{t}_1 + \hat{t}_2 + \hat{Q}_{\alpha\beta}\hat{V}_{12}\right)|\varphi_{\alpha\beta}\rangle = E_{\alpha\beta}|\varphi_{\alpha\beta}\rangle, \tag{2.20}$$

the so-called *Bethe–Goldstone* equation (1957). Here, the projector

$$\hat{Q}_{\alpha\beta} = |\alpha\beta\rangle\langle\alpha\beta| + \sum_{k'_\alpha, k'_\beta > k_F} |\alpha'\beta'\rangle\langle\alpha'\beta'| = \hat{1} - \hat{P}_{\alpha\beta} \tag{2.21}$$

is used. It projects on all states which, in accord with the Pauli principle, may be reached by the coupling $\hat{V}_{12}$. Backscattering into the initial ones is allowed, of course. Conversely, the $\hat{P}_{\alpha\beta}$ projects on the complementary part of the Hilbert space of the two particle wave functions. Remember that the ones used here are not antisymmetrized, but equation (2.20) can as well be formulated for $|[\varphi_{\alpha\beta}]\rangle = |\varphi_{\alpha\beta}\rangle - |\varphi_{\beta\alpha}\rangle$.

We are now going to discuss some general features, some of which come up after expressing the solution in terms of coordinates and momenta of relative and center of mass motion, **r** and **R**, respectively.

---

[4]It should be kept in mind that we are not seeking that state in the two-body system which would be bound by the presence of the two-body interaction alone. In "free space" this would be the one for the loosely bound deuteron. Its binding is too weak not to be broken up by the interaction with the other nucleons: At about normal density, nuclear matter does not consist of deuterons, nor of any other clusters, but simply of an assembly of nucleons! However, at much lower densities homogeneous matter becomes unstable, with the more stable configuration consisting of smaller clusters. This feature is not restricted to nuclear matter, in fact the formation of galaxies happens by a similar mechanism. In nuclear physics this phenomenon has been seen in heavy-ion collisions, where after a compression phase matter expands. It has been speculated whether or not the process of cluster formation may be understood as a phase transition from the nuclear liquid to nuclear fog.

(1) *Behavior at small distances:* Because of the repulsive core of the interaction the $\varphi_{\alpha\beta}$ must vanish when $r \to 0$. Let us take the model case of a hard core potential, which in the development of the Bethe–Goldstone equation and Brueckner theory has played an important role. Certainly, the matrix element $\langle \alpha\beta | \hat{V}_{12} | [\varphi_{\alpha\beta}] \rangle$ must be finite, which is possible only if $\langle \mathbf{rR} | \varphi_{\alpha\beta} \rangle = 0$ for $r < r_c$. This implies that the first derivative (with respect to r) will vanish inside the core, which in turn means this derivative to be discontinuous at $r = r_c$ (as it also would be the case for the Schrödinger equation).

(2) *Behavior at large distances:* Since the interaction is of short range, at larger distances the "correlated" two-body functions must turn into a wave function which solves eqn(2.20) but with $V = 0$, which is of course nothing but the Schrödinger equation for free particles. Expressed in terms of the expansion (2.19) this means that the scattered part of the wave function must tend to zero at large distances: $C^{\alpha\beta}_{\alpha',\beta'} \langle \mathbf{rR} | \alpha'\beta' \rangle \to 0$ for $r \to \infty$. This feature is in accord with the fact that energy is conserved for elastic scattering. In terms of the expansion (2.19), the only component which has the same energy as the initial (two-body) state $|\alpha, \beta\rangle$ is this one itself. Hence, the correlated wave function must behave as $\langle \mathbf{rR} | \varphi_{\alpha\beta} \rangle \longrightarrow \langle \mathbf{rR} | \alpha\beta \rangle$ for $r \gg r_h$. Here, it was assumed that the transition back to the initial state happens beyond some distance r_h, which in the literature is often called "healing distance". Likewise, one speaks of the "wound" which the interaction produces in the unperturbed two-body function $\langle \mathbf{rR} | \alpha\beta \rangle$.

It is worth reflecting a bit on this important property. On the one hand, the situation is similar to that of a scattering in free space: Beyond the interaction the scattered wave must satisfy the free Schrödinger equation. However, there is an essential difference. In free space the scattered wave may occupy states with different relative momentum $\mathbf{k}$ if only the energy is unchanged: The asymptotic form of the wave has a forward part *plus the outgoing spherical wave*. In this sense the "wound" never heals, which is to say the healing distance is infinitely large.

(3) *Normalization:* The behavior of the wave function at large distances has important consequences for the normalization. In free space this requires special precautions for the normalization of a scattering state, essentially different from that of a bound state. Here, no such problem arises, which is not too surprising, after all. Remember that the two-body state we are looking for will in the end be used to construct the bound state for the many-body system. Moreover, because of the healing property, the norm of the ket $|\varphi_{\alpha\beta}\rangle$ may be related to that of the unperturbed two-body state. The following convention turns out convenient: $\langle \alpha\beta | [\varphi_{\alpha\beta}] \rangle = 1$.

(4) *Relation to the total energy:* We may now start establishing relations to the energy of the many-body system given in (2.15). Finally, we will argue to replace in this form the wave function $|[\psi_{\alpha_k,\alpha_l}(k,l)]\rangle$ by the solution of the Bethe–Goldstone equation. To this end it will be convenient to introduce a shorthand notation for the relevant two-body matrix element by writing

$$e_{\alpha\beta} \equiv \langle\alpha\beta|\hat{V}_{ij}|[\varphi_{\alpha\beta}]\rangle = \langle\alpha\beta|\hat{Q}_{\alpha\beta}\hat{V}_{ij}|[\varphi_{\alpha\beta}]\rangle, \qquad (2.22)$$

After multiplying (2.20) from left by $\langle\alpha\beta|$ and by observing the normalization the correlated two-body energy $E_{\alpha\beta}$ becomes $E_{\alpha\beta} = e_\alpha + e_\beta + e_{\alpha\beta}$. The total energy of the system can be expressed as

$$E_\alpha = \sum_i e_{\alpha_i} + \sum_{i<j} e_{\alpha_i\beta_j}. \qquad (2.23)$$

The summation is understood to be done over those states which are occupied by the A particles. For the ground state, the case we mainly have in mind here, it would be the states corresponding to the lowest possible single particle energies. The second term in (2.23) turns out to be proportional to the density, $\sum_{i<j} e_{\alpha_i\beta_j} \propto A\rho$ (see exercise 1). In this sense the contribution of the interaction to the total energy E_α has a similar behavior as the one found in Section 2.1.1 for our simple model.

(5) *A few remarks on the justification of the Bethe–Goldstone equation.* So far this equation has only been postulated. One possible justification could be found in constructing the many-body wave function we started off in Section 2.2.2. This problem is discussed in section III.4 of (de Shalit and Feshbach, 1974); here, we only like to mention a few elementary facts. It should be obvious that, starting from Hartree–Fock, the approximation which may lead to an improved equation ought to take into account the next hierarchy of possible correlations. In Hartree–Fock only the Pauli principle is taken into account, which implies appropriate antisymmetrization of a product of one-body functions. The next step is to consider three-body functions constructed in terms of correlated two-body functions. This is to say one should put

$$|[\varphi_{\alpha\beta\gamma}(123)]\rangle = |[\varphi_{\alpha\beta}(12)]\rangle|\varphi_\gamma(3)\rangle - |[\varphi_{\alpha\gamma}(12)]\rangle|\varphi_\beta(3)\rangle - |[\varphi_{\gamma\beta}(12)]\rangle|\varphi_\alpha(3)\rangle \qquad (2.24)$$

(mind the signs with respect to the permutations of the initial order $\alpha\beta\gamma$) and ought to proceed similarly with the four and five particle functions and so on. As shown by W. Brenig in (1957) the $|[\varphi_{\alpha\beta}(12)\rangle]$ appearing here must satisfy (2.20). This feature reflects very clearly why the physics behind the Bethe–Goldstone equation may be associated with the name *independent pair approximation*: Only if genuine three-body correlations may be ruled out one may hope that this procedure (alone) may lead to an adequate approximation to the complicated many-body wave function. Such a situation may be given at low density. One crucial criterion is the smallness of the healing distance to the average spacing between two particles. By analogy with the f introduced in (2.10) one could write $f_h \equiv (r_h/2r_0)^3 \ll 1$. This implies that the chance of having three particles together at the same point in space–within a volume in which the two particle wave functions differ from their unperturbed form–is very small. Evidently the r_h must be larger than the hard core radius; how much larger depends on the attractive part of the interaction. Denoting the range of the latter by b one would expect r_h to be of this order. Unfortunately, it has turned out that in practice these conditions are not fulfilled as well as one would wish. Three particle correlations cannot be neglected entirely. Please note that we are talking about three-body correlations which may exist for two-body interactions; this is to be clearly distinguished from the three-body forces which are taken into account in the Jastrow method, to which we will return below.

2.2.4 The G-matrix

In our previous discussion, the Bethe–Goldstone equation was based on the single particle wave functions of the Fermi gas, which means that the corresponding single particle energies were just those of the kinetic energy. If we were only interested in getting the total energy of the system this would do no harm, provided we were able (i) to solve the Bethe–Goldstone equation satisfactorily, (ii) to calculate the correlated pair energies $e_{\alpha\beta}$, and (iii) to get from such results proper values of the equilibrium density and the corresponding energy per particle. However, it turns out that these constraints are not met. To actually help solving all these problems it is useful to introduce an average single potential. Moreover, this step is suggested for physical reasons. After all we know from experience that many properties of nuclei can be described by the shell model, where the nucleons move in a mean field. Therefore, one would expect that this concept should be helpful also for nuclear matter. Fortunately, it is not difficult to change the (2.20) to

$$\left(\hat{h}_1 + \hat{h}_2 + \hat{Q}_{\alpha\beta}\hat{V}_{12}\right)|\varphi_{\alpha\beta}\rangle = E_{\alpha\beta}|\varphi_{\alpha\beta}\rangle, \tag{2.25}$$

where now the single particle wave functions satisfy the equation $\hat{h}|\alpha\rangle = e_\alpha|\alpha\rangle$ with a Hamiltonian $\hat{h} = \hat{t} + \hat{U}$. From a formal point of view the single particle potential $\hat{U}$ may be introduced into the original Hamiltonian (2.1) by the trivial manipulation $\hat{H} = \hat{T} + \hat{V} = \left(\hat{T} + \hat{U}\right) + \left(\hat{V} - \hat{U}\right) = \hat{H}_0 + \hat{H}_1$. Evidently, by a proper choice of $\hat{U}$ this "renormalization" of unperturbed and interacting parts may help to apply perturbation theory in treating the influence of $\hat{H}_1$. As it will turn out, the $\hat{U}$ may largely be determined in self-consistent fashion, for which reason equation (2.25) is called the *self-consistent Bethe–Goldstone equation*. Indeed, for states below the Fermi surface *the $\hat{U}$ can be defined by analogy with Hartree–Fock, with the only (but essential) difference that it is not the bare interaction which appears but the one which is defined through an integral equation based on the Bethe–Goldstone equation*. The procedure to find this new operator is very similar to the derivation of the equation for the "T-matrix" in scattering theory. Here and in the sequel it will turn out convenient to introduce the *Pauli operator* $\hat{Q}_F$. It is defined as that part of the $\hat{Q}_{\alpha\beta}$ which projects *outside* the Fermi sea; this means writing

$$\hat{Q}_{\alpha\beta} = |\alpha\beta\rangle\langle\alpha\beta| + \sum_{k'_\alpha, k'_\beta > k_F} |\alpha'\beta'\rangle\langle\alpha'\beta'| = |\alpha\beta\rangle\langle\alpha\beta| + \hat{Q}_F. \tag{2.26}$$

Next we rewrite (2.25) separating onto different sides the c-number term and the one-body parts, on the one hand, and the term involving the interaction, on the other hand. Furthermore, we may make use of the completeness relation $\sum_{\mu\nu} |\mu\nu\rangle\langle\mu\nu| = 1$ as well as of the definition (2.26) of $\hat{Q}_F$ to get

$$\left(E_{\alpha\beta} - \hat{h}_1 - \hat{h}_2\right)|\varphi_{\alpha\beta}\rangle = \hat{Q}_{\alpha\beta}\hat{V}|\varphi_{\alpha\beta}\rangle = \sum_{\mu\nu}|\mu\nu\rangle\langle\mu\nu|\hat{Q}_{\alpha\beta}\hat{V}|\varphi_{\alpha\beta}\rangle$$
$$= \left(E_{\alpha\beta} - \hat{h}_1 - \hat{h}_2\right)|\alpha\beta\rangle + \sum_{\mu\nu}|\mu\nu\rangle\langle\mu\nu|\hat{Q}_F\hat{V}|\varphi_{\alpha\beta}\rangle. \tag{2.27}$$

To simplify the notation, here and in the remaining parts of this section the interaction $\hat{V}$ between two nucleons will be written without indices. The second line is obtained by taking into account that $\langle\alpha\beta|V|\varphi_{\alpha\beta}\rangle = E_{\alpha\beta} - e_\alpha - e_\beta$, which is identical to (2.22) with the exception of using $|\varphi_{\alpha\beta}\rangle$ rather than $|[\varphi_{\alpha\beta}]\rangle$. Notice that only states with $e_\mu, e_\nu > e_F$ contribute to the remaining sum, which is automatically taken care of by the projector $\hat{Q}_F$.

As the final step we want to invert the operator which stands on the left-hand side of (2.27). This turns out to be possible with the following result:

$$|\varphi_{\alpha\beta}\rangle = |\alpha\beta\rangle + \sum_{\mu\nu}|\mu\nu\rangle\langle\mu\nu|\frac{\hat{Q}_F}{E_{\alpha\beta} - \hat{h}_1 - \hat{h}_2}\hat{V}|\varphi_{\alpha\beta}\rangle. \tag{2.28}$$

The essential point is that because of $E_{\alpha\beta} - e_\mu - e_\nu \neq 0$, even for $\alpha = \mu$ and $\beta = \nu$, the inverse of $E_{\alpha\beta} - \hat{h}_1 - \hat{h}_2$ exists (and is identical to the inverse of $E_{\alpha\beta} - e_\mu - e_\nu$). Equation (2.28) is an "integral" equation for the two-body function $|\varphi_{\alpha\beta}\rangle$, which is equivalent to the form (2.25). Actually, in the way this equation is written here the "integrals" are just sums over $\mu\nu$. In the end this is a consequence, or a benefit rather, of our use of periodic boundary conditions. Given equation (2.28) one may define an operator $\hat{G}$ whose matrix elements in the basis of uncorrelated two particle states are given by

$$\langle\gamma\delta|\hat{G}|\alpha\beta\rangle \equiv \langle\gamma\delta|\hat{V}|\varphi_{\alpha\beta}\rangle. \tag{2.29}$$

One easily verifies that this matrix satisfies the equation

$$\langle\gamma\delta|\hat{G}|\alpha\beta\rangle = \langle\gamma\delta|\hat{V}|\alpha\beta\rangle + \sum_{\mu\nu}\langle\gamma\delta|\hat{V}|\mu\nu\rangle\frac{\langle\mu\nu|\hat{Q}_F|\mu\nu\rangle}{E_{\alpha\beta} - e_\mu - e_\nu}\langle\mu\nu|\hat{G}|\alpha\beta\rangle \tag{2.30}$$

for which reason one may write for the operator itself

$$\hat{G}[E_{\alpha\beta}] = \hat{V} + \hat{V}\frac{\hat{Q}_F}{E_{\alpha\beta} - \hat{h}_1 - \hat{h}_2}\hat{G}[E_{\alpha\beta}]. \tag{2.31}$$

Here, the dependence of the $\hat{G}$ on the quantity $E_{\alpha\beta}$ is indicated explicitly. Its matrix is commonly called the G-matrix, sometimes but less often the K-matrix.

As indicated previously, the definition of the G-matrix through (2.29) has many similarities to that of the T-matrix in scattering theory for collisions in free space. For the reaction of two partners by the mutual interaction $\hat{V}$ the integral equation for the T-matrix reads

$$\hat{T}[E] = \hat{V} + \hat{V} \frac{1}{E - \hat{h}_1 - \hat{h}_2 + i\epsilon} \hat{T}. \tag{2.32}$$

Here, E is a free and continuous parameter which represents the energy available to the two particles as determined by the accelerator. The unperturbed Hamiltonians $\hat{h}_1$ and $\hat{h}_2$ for the two particles have real eigenvalues. Therefore, the "Green function" would diverge without introducing the $+i\epsilon$ (with the plus sign being needed to warrant *outgoing* spherical waves). Likewise, the momenta are continuous variables such that the corresponding matrix equation becomes a genuine integral equation. Moreover, as the states into which the particles scatter are empty, no projection operator is involved.

Actually, it is desirable to generalize the equation (2.31) for the G-matrix to

$$\hat{G}[w] = \hat{V} + \hat{V} \frac{\hat{Q}_F}{w - \hat{h}_1 - \hat{h}_2 + i\epsilon} \hat{G}[w]. \tag{2.33}$$

Now the energy w is not fixed like the $E_{\alpha\beta}$ from before. Rather, it may be considered to represent an independent variable. A proper definition of the energy denominator is ensured by the presence of the $+i\epsilon$, with the plus sign chosen to warrant causality. Notice that this modification is of no influence whenever $w - \hat{h}_1 - \hat{h}_2 \equiv w - e_\mu - e_\nu$ cannot vanish.

2.3 Brueckner–Hartree–Fock approximation (BHF)

So far the mean field U has not been defined yet. This may be done by analogy with Hartree–Fock, where according to Section 20.1.2 the total potential energy may be written as

$$E_{\text{pot}}^{\text{HF}} = \frac{1}{2} \sum_{i,j=1}^{A} \langle \alpha_i \beta_j | \hat{V} \Big(|\alpha_i \beta_j\rangle - |\beta_j \alpha_i\rangle \Big) \equiv \frac{1}{2} \sum_{i}^{A} U^{\text{HF}}(\alpha_i). \tag{2.34}$$

Exploiting the very definition (2.29) of the G-matrix and in accord with (2.22) and (2.23) we may replace here $\hat{V}$ by a $\hat{G}[w]$ to get

$$E_{\text{pot}}^{\text{BHF}} = \frac{1}{2} \sum_{i,j=1}^{A} \langle \alpha_i \beta_j | \hat{G}[w] \Big(|\alpha_i \beta_j\rangle - |\beta_j \alpha_i\rangle \Big). \tag{2.35}$$

Recalling the fundamental feature of Hartree–Fock to assume the particles to be independent of each other, it would not be consistent to evaluate the G-matrix at the correlated energy $E_{\alpha_i \beta_j}$. Rather, the w here is to be chosen as $w = e_{\alpha_i} + e_{\beta_j}$, implying for $U = U^{\mathrm{BHF}}$ the following definition:

$$U(\alpha_i) = \sum_{j=1}^{A} \langle \alpha_i \beta_j | \hat{G}[e_{\alpha_i} + e_{\beta_j}] \Big(|\alpha_i \beta_j\rangle - |\beta_j \alpha_i\rangle \Big). \tag{2.36}$$

The total energy of the ground state is given by

$$E_0^{\mathrm{BHF}} = E_{\mathrm{kin}}^{\mathrm{BHF}} + E_{\mathrm{pot}}^{\mathrm{BHF}} = \sum_{i=1}^{A} \left(t_{\alpha_i} + \frac{1}{2} U(\alpha_i) \right) = \sum_{i=1}^{A} \left(e_{\alpha_i} - \frac{1}{2} U(\alpha_i) \right), \tag{2.37}$$

where the summation is understood to be one over the A lowest possible single particle states.

Please notice that by this procedure one has taken advantage of Hartree–Fock, on the one hand, and benefits, otherwise, from the fact that by introducing the G-matrix short-range correlations can be taken into account. It is only by the latter feature that matrix elements of the "effective interaction" $\hat{G}$ become finite on the level of independent particle motion. Evidently, within this Brueckner–Hartree–Fock approximation the self-consistency procedure is much more complicated than it is for an ordinary interaction $\hat{V}$ (provided that latter can be treated on the HF level). Here the G-matrix itself is a functional of the energies and thus of the mean field. This new self-consistency problem has first been suggested by K.A. Brueckner (1954).

Unfortunately, there is still an ambiguity in specifying the mean field. Inspecting the definition of the potential through (2.36) it becomes clear that this $U(\alpha_i)$ is fixed only for the occupied states $|\alpha_i\rangle$, i.e. those with momenta k_{α_i} below k_{F}, the so-called *hole states*. However, for the integral equation (2.30) knowledge of matrix elements with states above the Fermi surface is required. For those various options of the mean field $U(\alpha_i)$ have been in use. We like to mention only the following two, the *standard choice* (see the review articles by H.A. Bethe (1971a) and B. Day (1978)) and the *continuous choice* (Jeukenne et al., 1976):

$$U(\alpha_i; k_{\alpha_i} > k_{\mathrm{F}}) = \begin{cases} 0 & \text{standard} \\ \mathrm{Re} \sum_{j=1}^{A} \langle \alpha_i \beta_j | \hat{G}[e_{\alpha_i} + e_{\beta_j}] \Big(|\alpha_i \beta_j\rangle - |\beta_j \alpha_i\rangle \Big) & \text{continous} \end{cases} \tag{2.38}$$

In the second case the real part of the energy denominator in (2.33) may vanish (simply because the e_{α_i} itself is larger than e_{F}) such that here the precaution of introducing an imaginary part pays off. As an immediate consequence of this feature, the mean field will become complex. For the BHF procedure only the real part is to be considered. The implications on the density dependence of the ground state energy $E(\rho)$ will be addressed in the next subsection.

As we know from the discussion in Section 1.3 a complex mean field has a real *physical meaning*, in the sense of implying measurable effects. Already for this reason the continuous choice is of great advantage as only this definition of the potential U allows for generalizing the BHF-approximation to calculate empirical quantities other than the ground state energy. Much work in this field has been done by C. Mahaux and his group. For an earlier summary see the review articles by J.P. Jeukenne, A. Lejeune and C. Mahaux (1976) and for more modern ones those by R Sartor (1999) and by W.H. Dickhoff (1999). To elaborate further would be beyond the scope of the present discussion, but in Section 4.3 we will return to the question of the physical, microscopic origin of the appearance of damping in single particle motion. It may be said that in general this one-body field $U(e, k)$ will depend on energy e and the wave number k separately. Actually, features of this type are common to all many-body systems: In equations for quantities representing one-body properties, say the one-body Green function, the presence of a complex $U(e, k)$ implies complex self-energies.

Numerical results for nuclear matter: One major goal of Brueckner theory has always been to understand the empirical saturation properties of nuclear matter, which are reflected in the equilibrium values $\mathsf{E}_0 \approx -16$ MeV and $\rho_0 \approx 0.165$ fm^{-3} for the binding energy per particle and the density, respectively. Truth is, however, that these values "cannot be reproduced by a model in which structureless nucleons interact via realistic two-body interactions", to quote from the review article (1989) by C. Mahaux and R. Sartor. This statement refers to the non-relativistic version of BHF calculations, to which we have restricted ourselves to above. Indeed, extensions to relativistic treatments seem to improve the situation (see (Brockmann and Machleidt, 1984), (ter Haar and Malfliet, 1987) and (Brockmann and Machleidt, 1999), with many references to earlier work). To some extent, this is due to the fact that relativistic many-body effects lead to a different density dependence, which in turn provides an additional saturation mechanism. However, it is perhaps fair to say that less experience has been gained in this newer field. For a discussion of some open problems see (Brockmann and Machleidt, 1999). In the following we concentrate on the non-relativistic approach.

The problem one faces can be explained comparing computations of the energy per particle $\mathsf{E}(\rho)$ as function of the density ρ, which are done for the same two-body interaction as input. Indeed, these curves exhibit a minimum, indicating that saturation can be explained in principle. However, this minimum is not found at the right place, neither with respect to E_0 nor ρ_0. The deviations are much larger than the uncertainties in the empirical values themselves. The "continuous choice" (cBHF) of the mean field turns out to be superior to the "standard choice" (sBHF), both being specified in (2.38). Whereas for the sBHF the E_0 was about -10 MeV large, the one for the cBHF had the right magnitude, albeit with the density ρ_0 being too large by about 60%. Moreover, the results no longer lie above those obtained applying the variational principle to wave functions of Jastrow type, which will be described in Section 2.4, and which can be understood to deliver an "upper bound". In addition the cBHF-curves are close those where correlations beyond the independent pair approximation were taken into account by B. Day (1978), an approach which is based on earlier work by H.A. Bethe and collaborators (see the review (Bethe, 1971a)). Speaking in terms of Feynman or Goldstone diagrams those which involve "3 holes and rings" are taken into account. Such contributions to the total energy become important if the actual density is not as low as needed in Brueckner theory. Indeed, each "hole line" adds contributions which scale linearly in the density ρ. The BHF theory outlined in our discussion above involves just two such lines for which the associated momenta run up to the Fermi momentum. The crucial parameter in such an expansion is related to the one expressed before by the healing distance, $(r_h/2r_0)^3$. A recent critical discussion of such an expansion can be found in (Sartor, 1999).

Let us return to the statement from before that the saturation properties of nuclear matter cannot be reproduced on the basis of two-body forces. It turns out that all results obtained with Brueckner type calculations lie on the so-called "Coester band" (Coester et al., 1970). Depending on the strength of the tensor force, as measured by the content of the d-wave contribution in the deuteron, binding is underestimated for lower densities, the more the larger the strength. Binding increases with the density but definitely lies to the right of the empirical values. All these calculations have been done with two-body forces which are realistic in the sense of describing correctly the scattering of two nucleons as well as the properties of the deuteron, as the only bound two-body system.

One possible way to the problem is to introduce three-body forces, as done also in the Jastrow approach discussed in Section 2.4. A method of how to incorporate a three-body force into Brueckner theory (Baldo et al., 1997) is described in the review articles by M. Baldo (1999) and I. Bombaci (1999). From the original three-body force with two unknown parameters an effective two-body interaction is derived by averaging over the position of the third particle, accounting for the two-body correlations in a Fermi system. This two-body interaction is then used for a BHF calculation. The two unknown parameters are fixed at the correct saturation point. In this way an EOS is obtained which may be used for other densities, for instance in astrophysical applications. It can be said that this inclusion of three-body forces simulates to some extent effects included in the relativistic Dirac-BHF mention above (see Section 3.3 of (Bombaci, 1999)).

T.T.S. Kuo and his group have moved in another direction to improve on the BHF approach. The latter does not account for density fluctuations around the stable minimum. Quite generally, such fluctuations are known to reduce the ground state energy, even on the RPA level. This latter feature will be discussed below in Chapter 6. This effect is taken into account for nuclear matter in Kuo's work. A nice pedagogical review is given in (Kuo et al., 1999), with many references to original work. In this way the minimum of the EOS comes closer to the empirical regime: one gets enhanced binding at smaller densities.

2.3.1 BHF at finite temperature

In this subsection we briefly want to describe how Brueckner–Hartree–Fock may be extended to finite thermal excitations if the latter are parameterized by a grand canonical ensemble; see also (Baldo, 1999), (Bombaci, 1999). To begin with one needs to generalize the Bethe–Goldstone equation in the form (2.31) to

$$\hat{G}[w,T] = \hat{V} + \hat{V} \sum_{i,j} \frac{|\alpha_i\beta_j\rangle Q_F^T \langle\alpha_i\beta_j|}{w - e_{\alpha_i} - e_{\alpha_j}} \hat{G}[w,T] . \qquad (2.39)$$

Here, the Q_F^T is the extension of the Pauli operator Q_F defined in (2.26) to finite temperature. In the single particle representation used in (2.39) it may be expressed by the Fermi occupation numbers $n(e_i, T) = [1 + \exp((e_i - \mu)/T)]^{-1}$ as

$$Q_F^T = (1 - n(e_i, T))(1 - n(e_i, T)) . \qquad (2.40)$$

Indeed, this Q_F^T defines the probability that the states $|\alpha_i\rangle$ and $|\beta_j\rangle$ are unoccupied. The single particle potential has to be changed from (2.36) to

$$U(\alpha_i, T) = \sum_j n(e_j, T) \langle\alpha_i\beta_j| \hat{G}[e_{\alpha_i} + e_{\beta_j}; T] \Big(|\alpha_i\beta_j\rangle - |\beta_j\alpha_i\rangle\Big). \qquad (2.41)$$

The contribution of this potential to the internal energy becomes

$$\begin{aligned}\mathcal{E}_{\text{pot}}^{\text{BHF}}(T) &= \frac{1}{2} \sum_i n(e_i, T) U(\alpha_i, T) \\ &= \frac{1}{2} \sum_{i,j} n(e_i, T) n(e_j, T) \langle\alpha_i\beta_j| \hat{G}[e_{\alpha_i} + e_{\beta_j}; T] \Big(|\alpha_i\beta_j\rangle - |\beta_j\alpha_i\rangle\Big) .\end{aligned} \qquad (2.42)$$

Of course, the average occupation numbers must also be considered for the total kinetic energy, $\mathcal{E}_{\text{kin}}^{\text{BHF}}(T) = \sum_i n(e_i, T) t_{\alpha_i}$. The single particle energy $e_i(T)$ is to be calculated as $e_i(T) = t_{\alpha_i} + U(\alpha_i, T)$ and, hence, becomes temperature dependent. For any given value of T this set of equations has to be solved self-consistently. Applications and problems with this approach are reported in (Baldo, 1999).

2.4 A variational approach based on generalized Jastrow functions

A conceptually simple way of accounting for the short-range correlations is given by the Jastrow method. To get ground state properties one exploits a variational procedure to $E[\Psi_0^J] = \langle\Psi_0^J|\hat{H}|\Psi_0^J\rangle / \langle\Psi_0^J|\Psi_0^J\rangle$, but for an improved ansatz for the many-body wave functions $|\Psi_0^J\rangle$. Different to the Slater determinants used for Hartree–Fock, these new model functions should account for the fact that the repulsive core prevents nucleons to come closer than simply determined by the Pauli principle. Unfortunately, practical applications are not at all easy. They require Monte Carlo methods to evaluate the complicated integrals which then appear in the relevant matrix elements.

Let us begin demonstrating essential features in constructing the wave functions using for the interaction the simplifying assumption of having a hard core. Of primary interest are the two-body wave functions. In Hartree–Fock they would be given by $|[\alpha\beta]\rangle = (|\alpha\beta\rangle - |\beta\alpha\rangle)/2$, built from the factorized states $|\alpha\beta\rangle = |\alpha\rangle|\beta\rangle$. Whereas the latter do not vanish at the origin the $|[\alpha\beta]\rangle$ certainly do. Actually, as we know from the discussion of the Fermi gas in Section 1.2.2, the wave functions $\langle \mathbf{rR}|[\alpha\beta]\rangle$ are small for some finite r around $r = 0$. However, this is not enough. Rather, one needs to have the stronger condition $\langle \mathbf{rR}|\varphi_{\alpha\beta}\rangle = 0$ for $r \leq r_c$. Evidently, with wave functions of this property the matrix elements of hard core interactions may be well defined. Such properties are taken care of by generalizing the variational ansatz to

$$\Psi^{\mathrm{J}}(\mathbf{x}_i \ldots \mathbf{x}_A) = \left(\prod_{i<j} f^c(r_{ij})\right) \Phi^{\mathrm{SL}}(\mathbf{x}_i \ldots \mathbf{x}_A) \quad \text{with} \quad f^c(u) = 0 \quad u < u_{\min}, \tag{2.43}$$

with r_{ij} being the relative distance between particle i and j. Spin and isospin degrees of freedom are meant to be included both in the arguments of the Slater determinant Φ^{SL} as well as in the correlated Jastrow function Ψ^{J}. The presence of the factors $f^c(u)$ warrants the total wave function to vanish whenever the two nucleons i, j come too close. The functions $f^c = f^c(u)$ are then subject to variation which allows one to find optimal choices within the form one likes to work with.

For realistic calculations the $f^c(r_{ij})$ are generalized to include spin and isopin degrees of freedom themselves; for their form one orients oneself at those combinations which have been discussed in Section 1.1 for the two-body interaction. The Ψ^{J} are then changed to

$$\Psi^V = \left(\mathcal{S}\prod_{i<j}\hat{F}_{ij}\right)\Phi^{\mathrm{SL}} \quad \text{with} \quad \hat{F}_{ij} = \sum_{p=1}^{8} f^p(r_{ij})\hat{O}_{ij}^p. \tag{2.44}$$

The set of operators $\hat{O}_{ij}^s$ is defined such that for $s = 1$–4 they are $\hat{1}$, $(\hat{\boldsymbol{\sigma}}_i\cdot\hat{\boldsymbol{\sigma}}_j)$, $\widehat{\mathbf{L}\cdot\mathbf{S}}$, $\hat{S}_{ij}$. The remaining four are given by combinations with the isospin operators $\hat{O}_{ij}^{s+4} = \hat{O}_{ij}^s (\hat{\boldsymbol{\tau}}_i \cdot \hat{\boldsymbol{\tau}}_j)$, where again $s = 1\ldots 4$. The *symmetry operator* $\mathcal{S}$ is necessary because the correlation operators $\hat{F}_{ij}$ do not commute with each other. For the nuclear case the presence of the tensor force $\hat{S}_{ij}$ (see eqn(1.5) and (1.7)) is important to understand nuclear binding quantitatively.

For the sake of simplicity in the ansatz (2.44) three-body correlations have been left out, which have mainly been considered in application to atomic helium liquids and to few body systems in nuclear physics, see (Pandharipande and Ravenhall, 1989) and (Heiselberg and Pandharipande, 2000). For ^{16}O, for instance, three-body correlations are known to reduce the energy by approximately 1 MeV/nucleon (Heiselberg and Pandharipande, 2000). In fact, such three-body correlations may be induced both by two as well as by three-body forces (see Section 3.5 of (Heiselberg and Pandharipande, 2000)).

As mentioned above, nuclear binding energies cannot be explained on the level of two-body forces alone. This is correct both for nuclear matter, as discussed in Section 2.3 in connection to Brueckner theory, as well as for light nuclei (Heiselberg and Pandharipande, 2000). Corrections to the total binding energies of the order of less than 10% are needed. Following the notation of (Heiselberg and Pandharipande, 2000) the ansatz for the Hamiltonian can be written as

$$\hat{H} = \sum_i \frac{\hat{p}_i^2}{2m} + \sum_{i<j} \hat{V}(i,j) + \sum_{i<j<k} \hat{V}(i,j,k). \qquad (2.45)$$

Two-body interaction $\hat{V}(i,j)$ is of the type discussed in Section 1.1. They may be fixed by studying nucleon–nucleon scattering. Unfortunately, the three-body forces $\hat{V}(i,j,k)$ can hardly be measured. Their very existence, on the other hand, already follows from looking at pion exchange: Say nucleon i exchanges a pion with nucleon j exciting the so-called Δ resonance in j which in turn decays by mediating another pion to nucleon k. This process effectively results in some attractive interaction $\hat{V}^{2\pi}(i,j,k)$. However, to get reasonable binding energies one needs to have an additive repulsive part, such one may write

$$\hat{V}(i,j,k) = \hat{V}^{2\pi}(i,j,k) + \hat{V}^R(i,j,k). \qquad (2.46)$$

Both terms contain parameters which may be adjusted to obtain the experimental binding energies of ^3He and ^4He as well as well as reasonable nuclear matter properties. In this way one is able to calculate binding energies of light nuclei to an accuracy of the order of 1% (by the so-called "Greens Function Monte Carlo" approach).

It should be noted that in recent applications an additional relativistic boost correction to the interaction has been considered (Akmal et al., 1998), see also (Heiselberg and Pandharipande, 2000). For the two-body part this implies to replace the $\hat{V}(i,j)$ by

$$\hat{V}(i,j;\mathbf{P}_{ij}) = \hat{V}(i,j) + \delta\hat{V}(i,j;\mathbf{P}_{ij}). \qquad (2.47)$$

Here, $\hat{V}(i,j)$ is the interaction as it comes out from nucleon–nucleon collisions done in the center of mass system with $\mathbf{P}_{ij} = 0$. The correction $\delta\hat{V}(i,j;\mathbf{P}_{ij})$ is calculated to second order in $\mathbf{P}_{ij}$, by exploiting features of relativistic mechanics and field theory. The contribution of the $\delta\hat{V}(i,j;\mathbf{P}_{ij})$ to the energy is repulsive and thus corrects the influence of the repulsive three-body force $\hat{V}^R(i,j,k)$.

2.4.1 Extension to finite temperature

At finite thermal excitations, the variational method basis on the formula

$$\mathcal{F} \leq \mathcal{E}_V - T\mathcal{S}_V. \qquad (2.48)$$

derived in Section 22.1.3. The expression on the right-hand side combines the average internal energy $\mathcal{E}_V$ and the "heat" expressed by the entropy $\mathcal{S}_V$,

$$\mathcal{E}_V = \operatorname{tr} \hat{\rho}_V \hat{H} \quad \text{and} \quad \mathcal{S}_V = -\operatorname{tr} \hat{\rho}_V \ln \hat{\rho}_V . \qquad (2.49)$$

The density operator $\hat{\rho}_V$ is assumed to be that of the canonical ensemble at fixed temperature T. The approximation comes in by taking some guess for the Hamiltonian $\hat{H}^V$ which specifies the density to $\hat{\rho}^V = \exp(-\hat{H}^V/T)/Z^V$. As $\mathcal{E}_V - T\mathcal{S}_V$ is bound from below, this $\hat{H}^V$ may be found through a variational procedure, for instance by exploiting for the latter Jastrow type wave functions. This method was first suggested and worked out for applications in nuclear physics by K.E. Schmidt and V.R. Pandharipande (1979). These authors assumed the ansatz given in (2.44) to be valid also for the excited states Ψ_n^V, with the same $\hat{F}_{ij}$ as found from the ground state. As the $\hat{F}_{ij}$ are not unitary operators the Ψ_n^V are not necessarily orthogonal, although the underlying Slater determinants Φ_n^{SL} are. It is argued that this should not be too big a problem at smaller temperatures. The Hamiltonian $\hat{H}^V$ is assumed to be of one-body nature with its eigenvalues thus being given by

$$E_n^V = \sum_{\mathbf{k}} e(\mathbf{k}, T, \rho) \, n_{\mathbf{k}}^{(m)}, \qquad (2.50)$$

satisfying the many-body Schrödinger equation

$$\hat{H}^V \Psi_n^V = E_n^V \Psi_n^V . \qquad (2.51)$$

The $n_{\mathbf{k}}^{(m)}$ are the occupation numbers (being 0 or 1) which specify the many-body state Φ_m^{SL}. The ρ in the single particle energy $e(\mathbf{k}, T, \rho)$ represents the space density of nuclear matter. Within the single particle approximation for the $\hat{H}^V$ the entropy $\mathcal{S}_V$ will simply be given by

$$\begin{aligned}\mathcal{S}_V(T, \rho) = -\sum_{\mathbf{k}} \Big\{ &[1 - n(e(\mathbf{k}), T, \rho)] \ln [1 - n(e(\mathbf{k}), T, \rho)] \\ &+ n(e(\mathbf{k}), T, \rho) \ln n(e(\mathbf{k}), T, \rho) \Big\} ,\end{aligned} \qquad (2.52)$$

see Section 22.5. The internal energy $\mathcal{E}_V = \operatorname{tr} \hat{\rho}^V \hat{H}$ involves the correct Hamiltonian $\hat{H}$ and has to be calculated with Monte Carlo techniques similar to the ground state energy. It will be a functional of the single particle energies $e(\mathbf{k}, T, \rho)$. The latter are parameterized as

$$e(\mathbf{k}, T, \rho) = \frac{\hbar^2 \mathbf{k}^2}{2m} \left[1 + A(T, \rho) \exp\left(-B(t, \rho) k^2\right) \right] \qquad (2.53)$$

with the two parameters A and B being subject to variation to minimize the approximate free energy $\mathcal{F}^V$ as given on the right of (2.48). Once this is known for given T and ρ all thermodynamic quantities may be calculated by appropriate differentiation (see Chapter 22). Applications will be discussed below in Section 2.6.

2.5 Effective interactions of Skyrme type

In the previous sections two approaches were presented which allow one to understand the phenomenon of nuclear saturation on the basis of the bare interaction between two nucleons. It was also seen, however, that neither one is able to produce quantitatively correct results. This is possible only after introducing *effective interactions*, which in the Jastrow method contains three-body parts. A similar extension may turn out necessary when one starts from Brueckner theory, and we will return to this feature below. At first we would like to remain on the level of two-body forces and develop a scheme which allows one to actually perform Hartree–Fock calculations without being forced to go through the tedious Brueckner–Hartree–Fock procedure. This goal can be reached by deriving an effective interaction from general properties of the G-matrix. In this way it will become possible to perform calculations within the Hartree–Fock approach not only to infinite matter, to get the equation of state, for instance, but even to real, finite nuclei. At some intermediate step it will become necessary to make some ansatz, in the sense of introducing a certain functional form with some open parameters. The latter may then be fixed by looking at certain properties or quantities, comparing their theoretical values with those known from experimental evidence. For instance, one may take information about the total static energy of some prominent nuclei, like their binding energies etc. Such a phenomenological approach is acceptable if in this way it becomes possible to predict many other features and properties.

The guiding principle for finding the effective interaction is to observe the proper dependence of the G-matrix on density and momentum. There are essentially two features which one exploits, namely the local density approximation an expansion to small relative momenta. For finite nuclei the space density varies quite smoothly with the distance from the center. The interior is almost flat anyway, a fact which originally motivated us to introduce the concept of nuclear matter at all. Let us just extrapolate this concept over the whole nucleus. This means taking the calculated energy density functional $E[\rho]$ and applying it everywhere, not just in the interior but to the surface as well. In this sense the E becomes a function of space, through the coordinate dependence of the density: $E(\mathbf{r}) = E[\rho(\mathbf{r})]$. As for the momentum dependence, we may recall that inside a finite nucleus the relative momenta take on moderately large values only. For the Fermi gas we know from (1.41) that the distribution of relative momenta has its maximum at about $0.6 k_{\rm F}$ corresponding to an energy of relative motion of about 27 MeV. Looking at the G-matrix in momentum space $\langle \mathbf{k'}_1 \mathbf{k'}_2 | \hat{G} | \mathbf{k}_1 \mathbf{k}_2 \rangle$, we may then argue that an expansion to low order in the relative momenta may do.

2.5.1 *Expansion to small relative momenta*

Evidently, this expansion will again be governed by *symmetry rules*. Indeed, as suggested by the relation (2.31) between $\hat{G}$ and $\hat{V}$, in general these rules are (almost) the same as those for $\hat{V}$ itself. All expressions have to be scalars with respect to rotations and must warrant conservation of total linear momentum of the two particles involved in this matrix element. However, since matrix elements depend on the states of the particles, new features might come into play. Time reversal invariance, for instance, may be broken in the wave function of an odd nucleus. Such subtleties will be put aside in the following. We only want to discuss the most basic steps necessary to understand how in principle one may get the matrix elements $\langle\alpha\beta|\hat{G}|\gamma\delta\rangle$ of $\hat{G}$ for the two-body states $|\gamma\delta\rangle$ needed in Hartree–Fock, say for an even even nucleus. To simplify the discussion we will save writing down all the terms. The full form used in present-day calculations will be given later when we will address finite nuclei.

To second order in the relative momenta, the G-matrix in momentum space will look like

$$\langle \mathbf{k}'_1 \mathbf{k}'_2 | \hat{G} | \mathbf{k}_1 \mathbf{k}_2 \rangle = (2\pi)^3 \delta(\mathbf{k}_1 + \mathbf{k}_2 - \mathbf{k}'_1 - \mathbf{k}'_2) \left\{ t_0^s(\rho) + \frac{t_1^s(\rho)}{8} \left[(\mathbf{k}_1 - \mathbf{k}_2)^2 + (\mathbf{k}'_1 - \mathbf{k}'_2)^2 \right] + \frac{t_2^s(\rho)}{4} \left[(\mathbf{k}_1 - \mathbf{k}_2)(\mathbf{k}'_1 - \mathbf{k}'_2) \right] + \ldots \right\}. \quad (2.54)$$

The δ-function warrants conservation of total momentum. The coefficients in the expansion will depend on spin through the spin-exchange operator $\hat{P}_\sigma = (1 + \hat{\boldsymbol{\sigma}}_1 \hat{\boldsymbol{\sigma}}_2)/2$, for instance like $t_\nu^s(\rho) = t_\nu(\rho)(1 + x_\nu \hat{P}_\sigma)$. In practical applications this density dependence is mostly neglected, with one exception to which we will come to below. Usually neutrons and protons are treated separately, for which reason no isospin operators appear in this formula.

To continue from here on some straightforward calculations are necessary which will be left to the reader. It is convenient to work with Fourier transforms in infinite space. In this case the plane waves will just be given by $\langle \mathbf{r} | k \rangle = \exp(i\mathbf{k}\mathbf{r})$, with the $1/\sqrt{\mathcal{V}}$ being absorbed in the integral which replaces the sum over momenta. Transforming to relative and center of mass coordinates the expressions given in (2.54) turn into

$$\langle \mathbf{K}' \mathbf{k}' | \hat{G} | \mathbf{K} \mathbf{k} \rangle = (2\pi)^3 \delta(\mathbf{K} - \mathbf{K}') \left\{ t_0^s(\rho) + \frac{t_1^s(\rho)}{2} \left[\mathbf{k}^2 + (\mathbf{k}')^2 \right] + t_2^s(\rho) [\mathbf{k}\mathbf{k}'] \right\}. \quad (2.55)$$

The G-matrix in coordinate space splits into two factors, $\langle \mathbf{r}'\mathbf{R}' | \hat{G} | \mathbf{r}\mathbf{R} \rangle = \delta(\mathbf{R} - \mathbf{R}')\langle \mathbf{r}' | \hat{G} | \mathbf{r} \rangle$, with the non-trivial one for relative motion is given by

$$\langle \mathbf{r}' | \hat{G} | \mathbf{r} \rangle = \int \frac{d^3k\, d^3k'}{(2\pi)^6} \exp\bigl(i(\mathbf{k}'\mathbf{r}' - \mathbf{k}\mathbf{r})\bigr) \times \left\{ t_0^s(\rho) + \frac{t_1^s(\rho)}{2} \left[\mathbf{k}^2 + (\mathbf{k}')^2 \right] + t_2^s(\rho) [\mathbf{k}\mathbf{k}'] + \ldots \right\}. \quad (2.56)$$

Exploiting a few elementary properties of Fourier transforms one finally gets

$$\langle \mathbf{r}'|\hat{G}|\mathbf{r}\rangle = t_0^s(\rho)\delta(\mathbf{r}')\delta(\mathbf{r}) + \frac{t_1^s(\rho)}{2}\Big[\delta(\mathbf{r}')\left(\Delta_{\mathbf{r}}\delta(\mathbf{r})\right) + \left(\Delta_{\mathbf{r}'}\delta(\mathbf{r}')\right)\delta(\mathbf{r})\Big]$$
$$+ t_2^s(\rho)\Big[\left(\nabla_{\mathbf{r}'}\delta(\mathbf{r}')\right)\left(\nabla_{\mathbf{r}}\delta(\mathbf{r})\right)\Big] + \ldots . \tag{2.57}$$

In the end we need to calculate matrix elements like $\langle\alpha\beta|\hat{G}|\gamma\delta\rangle$ for two particle states. Transforming the latter to coordinate space once more and after exploiting (2.57) one finally obtains

$$\langle\alpha\beta|\hat{G}|\gamma\delta\rangle = \int d^3r d^3R\, \langle\alpha\beta|\mathbf{r}\mathbf{R}\rangle\, \Big\{ t_0^s(\rho)\delta(\mathbf{r})$$
$$- \frac{t_1^s(\rho)}{2}\Big[\delta(\mathbf{r})\overrightarrow{\Delta_{\mathbf{r}}} + \overleftarrow{\Delta_{\mathbf{r}}}\delta(\mathbf{r})\Big] - t_2^s(\rho)\Big[\overleftarrow{\nabla_{\mathbf{r}}}\delta(\mathbf{r})\overrightarrow{\nabla_{\mathbf{r}}}\Big] + \ldots \Big\}\langle\mathbf{r}\mathbf{R}|\gamma\delta\rangle . \tag{2.58}$$

The arrows above the differential operators are meant to indicate onto which wave functions they should be applied. This formula says that "effectively" we are calculating two-body matrix elements of an "interaction" which is given by the sum of terms put inside the curly brackets:

$$V_{\text{sky}}(\mathbf{r}) = \Big\{ t_0^s(\rho)\delta(\mathbf{r}) + \frac{t_1^s(\rho)}{2}\Big[\delta(\mathbf{r})\overrightarrow{\Delta_{\mathbf{r}}} + \overleftarrow{\Delta_{\mathbf{r}}}\delta(\mathbf{r})\Big] + t_2^s(\rho)\Big[\overleftarrow{\nabla_{\mathbf{r}}}\delta(\mathbf{r})\overrightarrow{\nabla_{\mathbf{r}}}\Big] + \ldots \Big\} . \tag{2.59}$$

It is commonly referred to as an interaction of *Skyrme type*, as T.H.R. Skyrme had suggested an effective interaction of that form as early as 1956 (1956), actually without using the background of the G-matrix. On the other hand, he had included a three-body term to which we will come below. This form may actually be used to derive from the average single particle potential $\hat{U}$ or the total potential energy from (2.35), into which antisymmetrized kets $|[\gamma\delta]\rangle$ enter.

To make these forces more realistic one has to account for the presence of the spin degrees of freedom, of course. We may recall from the discussion of the two-body densities of the Fermi gas the great impact the latter have on the symmetry of the space part of the two-body wave function. Notice that *the effective interaction is extremely short range*, being proportional to the delta function $\delta(\mathbf{r})$. Therefore, one gets finite values of the matrix elements (involving antisymmetrized wave functions) *only* when the *space part of this wave function is symmetric*. Moreover, the $t_\nu^s(\rho)$ may be operators in spin space, in principle. In practice, however, this feature is kept only for the very first term with $\nu = 0$. Likewise, a possible dependence on the density ρ is commonly discarded. Rather, a density-dependent term is added as it would come out from a simple three-body interaction. However, its form might as well be traced back to a ρ-dependent $t_0(\rho)$. We will comment on this point later. Finally, the coupling to the spin degree of freedom will lead to spin-orbit forces which must be taken into account in any applications, see below (for more details see (de Shalit and Feshbach, 1974)).

Due to the presence of the δ functions in (2.58) the two-body matrix elements can be brought to simpler form. The integral over $\mathbf{r}$ is readily performed such that the final result for *the total energy can be expressed in terms of an energy density* like

$$E_{\text{kin}}^{\text{BHF}} + E_{\text{pot}}^{\text{BHF}} \quad \Rightarrow \quad E_{\text{kin}} + E_{\text{pot}}^{\text{eff}} = \int d^3R\, \mathcal{H}(\mathbf{R}) \equiv \int d^3r\, \mathcal{H}(\mathbf{r}) \qquad (2.60)$$

where $E_{\text{pot}}^{\text{eff}}$ is obtained by inserting the approximated form (2.58) into (2.35) and $E_{\text{kin}}^{\text{BHF}} = E_{\text{kin}}$ is just the kinetic energy already accounted for in ordinary Hartree–Fock (see (20.20)). In writing this integral we at first have followed our previous notation according to which it is the center of mass coordinate of the two particle system which is left under the integral. As the integration variable is dummy one often replaces there the $\mathbf{R}$ by $\mathbf{r}$.

The general form of the effective interaction and thus of $\mathcal{H}(\mathbf{r})$ will be discussed later in Section 3.1.1 for the application of Hartree–Fock to finite nuclei. The important point is that $\mathcal{H}(\mathbf{r})$ can be expressed by the space density $\rho(\mathbf{r})$ (the diagonal element of the density operator) and terms which involve derivatives of at most second order with respect to the components of $\mathbf{r}$. In the end the latter two can be represented by a kinetic energy density $\tau(\mathbf{r})$ and a spin-orbit density $\mathbf{J}(\mathbf{r})$. In this chapter the main topic is nuclear matter in equilibrium such that we may concentrate first on the simpler case of a homogeneous system.

2.6 The nuclear equation of state (EOS)

The nuclear equation of state is an important means classifying the behavior of infinite nuclear systems at and away from the stationary point. As we shall see in Section 2.7, it appears in the transport equation for the one-body density, in application of the local density approximation. As these transport equations are used for the description of heavy-ion reactions at medium or higher energies per particles this prospect delivers a means to get experimental access to the EOS; in fact, away from equilibrium it is the only one which is at one's disposal. Furthermore, the nuclear EOS is badly needed for astrophysical applications. Commonly this requires an extension to final thermal excitations, where the EOS has the common meaning known from thermostatics or statistical mechanics: It represents the behavior of quantities like the energy (free energy) or the pressure as function of density and entropy (temperature). Although in realistic applications one needs to know the EOS for asymmetric matter we will mostly restrict ourselves to outlining the basic features for systems with $N = Z$, for which the neutron and proton densities equal just half the total density, $\rho_n = \rho_p = \rho/2$. At first we want to address calculations with Skyrme forces.

2.6.1 *An energy functional with three-body forces*

For a homogeneous system all terms which involve gradients of the density vanish. Indeed, evaluating the $\mathcal{H}$ introduced in (2.60) with the interaction (2.59) one gets the following simple form

$$\mathcal{H}^{(2)} = \frac{\hbar^2}{2m}\tau + \frac{3}{8}t_0\rho^2 + \frac{1}{16}(3t_1 + 5t_2)\rho\tau\,. \qquad (2.61)$$

The kinetic energy is written in terms of the density τ. If expressed by the single particle wave functions it would have the form (3.3). For a homogeneous system we may take over the result found in (1.24), $E_{\rm kin} = A\, 3e_{\rm F}/5$, which implies $\tau = 3\rho k_{\rm F}^2/5$. Likewise, the space density ρ relates to the Fermi momentum as found in (1.23), namely $\rho = 2k_{\rm F}^3/(3\pi^2)$ (for $g_{\rm s.i.} = 4$).

The upper index in $\mathcal{H}^{(2)}$ is meant to indicate that only the contributions from the two-body part of the effective interaction are included here, as they could be obtained from the G-matrix. As indicated above, one likes to add a contribution from an *effective three-body force*. The latter might be visualized as

$$V_{\rm eff}^3 = \sum_{i<j<k} V(i,j,k). \tag{2.62}$$

Choosing the $V(i,j,k)$ according to the rule

$$V(1,2,3) = t_3 \delta(\mathbf{r}_1 - \mathbf{r}_2)\delta(\mathbf{r}_1 - \mathbf{r}_3) \tag{2.63}$$

one gets the following contribution to the energy density (see e.g. (Ring and Schuck, 2000)) $\mathcal{H}^{(3)} = t_3 \rho^3/16$. Thus the total energy density may be written as $\mathcal{H} = \mathcal{H}^{(2)} + \mathcal{H}^{(3)}$. Comparing the $\mathcal{H}^{(3)}$ with the second term on the right-hand side of (2.61) one realizes that this form of the "three-body" contribution could also have been obtained from a $\mathcal{H}^{(2)}$ if one took seriously that the t_0 might depend on density. All that one would have to do is to add to the constant t_0 a term linear in ρ. This feature shows that for the effective interaction the notion "three body" has to be taken with care. In the version used in practical applications, and as given by (2.63), this concept is more or less equivalent to "density dependent". The truly important aspect is that this density dependence cannot be deduced from a G-matrix approach in *quantitative fashion*.

2.6.2 The EOS with Skyrme interactions

Let us begin looking at the case of zero thermal excitations. There the EOS is simply given by the functional $\mathsf{E} = \mathsf{E}[\rho] = E[\rho]/A$, which is easy to deduce for the model just discussed. We only need to calculate the E from the $\mathcal{H}$ which measures the energy per volume (remember that for homogeneous matter $\mathcal{H}(\mathbf{r}) = \mathcal{H}$). After putting together all terms which involve the kinetic energy density τ one may write

$$\mathsf{E}[\rho] = \frac{E}{A} \equiv \frac{\mathcal{H}}{\rho} = \frac{\hbar^2}{2m^*}\frac{\tau}{\rho} + \frac{3}{8}t_0\rho + \frac{1}{16}t_3\rho^2, \tag{2.64}$$

with an effective mass

$$\frac{\hbar^2}{2m^*} = \frac{\hbar^2}{2m} + \frac{1}{16}(3t_1 + 5t_2)\rho. \tag{2.65}$$

In this version four parameters appear. In their pioneering work D. Vautherin and D. Brink (1972) suggested the following values: $t_0 = -1169.9$ MeV fm^3; $t_1 = 585.6$ MeV fm^5 ; $t_2 = -27.1$ MeV fm^5; $t_3 = 9333.1$ MeV fm^6. In the meantime many other forms of such interactions have been suggested, both for nuclear matter (see e.g. J.M. Lattimer in (1981)), as well as for finite nuclei, to which we will come later. In present-day calculations often the last term on the right of (2.64), which in a sense represents a three-body force, is replaced by $t_3\rho^\sigma/16$ with the additional parameter σ and t_3 measured i units of MeV fm$^{3\sigma}$ (see e.g. (Gleissl et al., 1990) or (Chabanat et al., 1997)). As an example we cite the version SkM* which was invented by (Bartel et al., 1982) to allow for a more reliable calculation of fission barriers: $t_0 = -2645.00$; $t_1 = 410.00$; $t_2 = -135.00$; $t_3 = 15595.0$; $\sigma = 7/6$ (with dimensions as given above). Let us return to the form (2.64) for homogeneous matter. Using (2.65) only three parameters are left, m^*, and t_0 and t_3. They may be fixed by the following three physical quantities, the equilibrium density ρ_0, the energy E_0 per particle at the minimum and the stiffness $\partial^2 \mathsf{E}/\partial\rho^2$ evaluated there. Actually, the latter may be related to the *nuclear incompressibility* coefficient κ. This may be done by analogy with thermostatics, where the following relation to the pressure $\mathcal{P}$ is used:

$$\frac{1}{\kappa_{\text{therm}}} = \kappa_{\text{therm}}^{\text{in}} = -\mathcal{V}\left(\frac{\partial \mathcal{P}}{\partial \mathcal{V}}\right)_T = \rho\left(\frac{\partial \mathcal{P}}{\partial \rho}\right)_T. \qquad (2.66)$$

In the nuclear case it has become customary to use

$$\kappa = 9\left(\frac{\partial}{\partial \rho}\mathcal{P}\right) \qquad (2.67)$$

instead, which at times is called the coefficient of "compressibility", although the latter is the inverse of the former. The origin of the choice (2.67) is seen in the fact that $\rho \propto k_\mathrm{F}^3$, for which reason at equilibrium and at zero thermal excitation one gets the simple form

$$\kappa(\rho_0, T=0) = k_\mathrm{F}^2 \left.\frac{\partial^2 \mathsf{E}}{\partial k_\mathrm{F}^2}\right|_{\rho_0} \qquad \text{where} \qquad \left.\frac{\partial \mathsf{E}}{\partial k_\mathrm{F}}\right|_{\rho_0} = 0. \qquad (2.68)$$

In the past the value of $\kappa(\rho_0, T=0)$ has been much discussed in the literature. Estimates have ranged from about 150 to 300 MeV, with some preference for a value of about $\kappa(\rho_0, T=0) \simeq 210$ MeV (Chabanat et al., 1997). Depending on the size of $\kappa(\rho_0)$ the corresponding equations of state are referred to as being "soft" or "hard".

The definition (2.67) can also be used at finite thermal excitations. One only needs to specify whether the latter are parameterized by the thermal entropy $\mathcal{S}$ or by the temperature T. From the general thermodynamic relation (22.21) the pressure may be expressed as

$$\mathcal{P} = \left(\rho^2 \frac{\partial \mathsf{E}}{\partial \rho}\right)_\mathcal{S} = \left(\rho^2 \frac{\partial \mathsf{F}}{\partial \rho}\right)_T, \qquad (2.69)$$

Now the internal energy per particle $\mathsf{E} = \mathcal{E}/\mathcal{A}$ and the free energy per particle $\mathsf{F} = \mathcal{F}/\mathcal{A}$ are involved, both calculated for thermally averaged quantities, say in the sense of the grand canonical ensemble. The latter is discussed in Section 22.1.2.2 where the variation of E is given by (22.54); the variation of the free energy follows immediately.

At first we need to find the functional relation between density $\mathcal{A}/\mathcal{V}$ and temperature. To this end we have to evaluate the particle number $\mathcal{A}$ in terms of the Fermi momentum or the chemical potential, rather. This may be done in generalization of the calculation performed in (1.22) at zero temperature. Instead of the $dn(k)$ we may use the $dn_{e_{\rm kin}}(e_{\rm kin})$ of (1.18). In addition the thermal occupation numbers $n(e(\mathbf{k}), T)$ come into play to get

$$\rho = \frac{1}{\mathcal{V}} \int_0^\infty n(e(\mathbf{k}), T) \, dn_{e_{\rm kin}} = \frac{g_{\rm s.i.}}{(2\pi)^2} \left(\frac{2m^*}{\hbar^2}\right)^{3/2} \int_0^\infty \frac{\sqrt{e_{\rm kin}} \, de_{\rm kin}}{1 + \exp((e-\mu)/T)} \, . \tag{2.70}$$

Notice that, unlike the $dn_{e_{\rm kin}}$, the $n(e(\mathbf{k}), T)$ is determined by the total single particle energy $e = e_{\rm kin} + U$. As before in (2.64) one may define U as that part of the energy which is independent of the momentum and put the rest into the $e_{\rm kin} = \hbar^2 k^2/(2m^*)$ with the effective mass m^*. In practical calculations the interactions are assumed to be insensitive to variations of temperature, a conjecture which is confirmed in variational calculations like that of (Schmidt and Pandharipande, 1979). This implies that it is only the relation of the density itself and the kinetic energy density which depend on T explicitly.

To proceed further it is convenient (Küpper et al., 1974), (Lattimer, 1981) to introduce "Fermi-integrals" defined as

$$F_{r/2}(y) = \int_0^\infty \frac{x^{r/2}}{1 + \exp(x-y)} \, dx \, . \tag{2.71}$$

Then the previous result can be expressed through the relation

$$2\pi^2 \rho = \frac{g_{\rm s.i.}}{2} \left(\frac{2m^*T}{\hbar^2}\right)^{3/2} F_{1/2}\left(\frac{\mu-U}{T}\right) \, . \tag{2.72}$$

The calculation of the kinetic energy can be done analogously. In generalization of (1.24) for the zero temperature case, one has

$$E_{\rm kin} = \mathcal{A} \int_0^\infty n(e(\mathbf{k}), T) \, e_{\rm kin} \, dn_{e_{\rm kin}}(e_{\rm kin}) \, . \tag{2.73}$$

Introducing again the density τ through $\tau = 2m^* E_{\rm kin}/\hbar^2$, by analogy with the notation in (2.64), one gets

$$2\pi^2 \tau = \frac{g_{\rm s.i.}}{2} \left(\frac{2m^*T}{\hbar^2}\right)^{5/2} F_{3/2}\left(\frac{\mu-U}{T}\right) \, . \tag{2.74}$$

In this way the total energy per particle takes on the same form as given in (2.64), whereas the total energy per volume may be written as

$$\mathcal{H} = \frac{\hbar^2}{2m^*}\tau + \frac{3}{8}t_0\rho^2 + \frac{1}{16}t_3\rho^3. \tag{2.75}$$

To get the EOS in the form pressure $\mathcal{P}$ versus density at given temperature one may exploit the relation (2.69) after having calculated the free energy par particle from $\mathsf{F} = \mathsf{E} - T\mathsf{S}$, with S being the entropy per particle. The latter might be found from the general form for independent particles, see (2.52) or (22.137). Another way is to make use of the grand potential defined as $\Omega_{\text{gce}} = -T\mathcal{S} + \mathcal{E} - \mu\mathcal{A}$, which for fermions is known to be in the following relation to the kinetic energy: $\Omega_{\text{gce}} = -2\mathcal{E}_{\text{kin}}/3$, see (22.52) and (22.145). Thus one gets

$$T\frac{\mathcal{S}}{\mathcal{V}} = \frac{5}{3}\mathcal{H}_{\text{kin}} + \mathcal{H}_{\text{pot}} - \mu\rho \equiv \frac{\hbar^2}{2m^*}\frac{5\tau}{3} + (U - \mu)\rho. \tag{2.76}$$

For the Skyrme force SkM* the function $\mathcal{P}(\rho)$ is shown for different temperatures in Fig. 2.1. Evidently, an isolated system can exist only for zero pressure. These curves would suggest that this is no longer possible for temperatures above a value of slightly less than 12 MeV. In the so-called *spinodal* region where the slope of these curves is negative the system is unstable; it corresponds to a negative incompressibility. Below a critical temperature of $T_c = 14.6$ MeV a gas and a liquid phase may coexist. Please notice that for small temperatures, say up to 4 MeV, and around the equilibrium density the functional dependence of $\mathcal{P}(T, \rho)$ on ρ is quite inert against variations in T.[5]

Let us come back once more to the application of chiral perturbation theory (ChPT) which was touched on in Section 2.1.3. Recently it has been applied to deduce the equation of state at finite temperature (Fritsch et al., 2002), (Fritsch et al., 2005) as well as to obtain density functionals of Skyrme type (Kaiser et al., 2003). Still discarding the possibility of exciting nucleons to the Δ resonance, an EOS came out in (Fritsch et al., 2002) for which the critical temperature T_c for the liquid–gas transition was too large. Correcting for that deficiency in (Fritsch et al., 2005) and extending the calculation to include terms up to $\mathcal{F}_6$ in the expansion (2.9) one now finds a value of $T_c \simeq 15$, which agrees very well with the value given above and which is in accord with other microscopic approaches.

2.6.3 Applications in astrophysics

The EOS of homogeneous nucleon matter is an important ingredient for the understanding of stellar phenomena, like the stability of neutron stars or the rebound in supernova explosions of type II. For matter densities below $\rho_{\text{nuc}} \simeq 2.7 \times 10^{14}$ g/cm^3 (which corresponds to the $\rho_0 \simeq 0.172$ fm^{-3} of nuclear matter) and above the so-called neutron drip of $\rho_{\text{nuc}} \simeq 4.3 \times 10^{11}$ g/cm^3 finite nuclei may

---

[5]For the Fermi gas it can be shown that to lowest order the relative correction $\mathcal{P}(T, \rho)/\mathcal{P}(T = 0, \rho)$ to $\mathcal{P}(T = 0, \rho)$ is given by $\simeq 4(T/e_{\text{F}})^2$; see exercise 22.12. As we shall see later, for such thermal excitations finite nuclei behave much more sensibly. At zero temperature their static energy is dominated by shell effects, which get diminished with increasing temperature to disappear entirely above about 3 MeV.

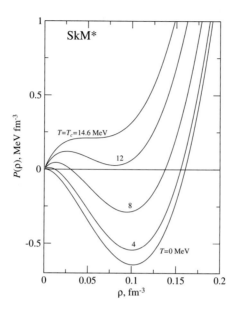

FIG. 2.1. The pressure as function of density for different temperatures, calculated for SkM* according to (Kolomietz et al., 2001); courtesy of A.I. Sanzhur.

coexist with neutrons and electrons. These nuclei are in statistical equilibrium with their surroundings and contribute essentially to the equation of state of the total system. The interested reader is referred to specialist literature like the book by S.L. Shapiro and S.A. Teukolsky (1983) or the recent review article by S.E. Woosley, A. Heger and T.A. Weaver (2002). The latter addresses all processes a massive stars undergoes as it proceeds through a gravitational collapses before it finally explodes. This includes the nuclear burning in the progenitor phases as well the final explosion itself. For the physics of supernova mechanisms we may also recommend the review by H.A. Bethe (1990) as well as his popular article with G.E. Brown in (1985) .

Let us return to the main topic of this section, the EOS of homogeneous matter. In astrophysics it is needed for temperatures ranging from $T \gtrsim 10$ MeV for the very early stages during the creation of neutron stars, down to zero for their final, fully equilibrated state. In practical applications Skyrme versions are often used, following the pioneering work of J.M. Lattimer (1981) and others, see (Lattimer et al., 1985), (Lattimer and Swesty, 1991), (Chabanat et al., 1997). Besides this approach results of more microscopic approaches like Brueckner theory or the variational method are also considered, for an overview see (Bombaci, 1999). In (Heiselberg and Pandharipande, 2000) for neutron star theories results of Brueckner theory are contrasted with variational calculations for Jastrow wave functions. Applications of this latter method to finite temperatures go back to

(Schmidt and Pandharipande, 1979) and (Friedman and Pandharipande, 1981), see also (Akmal et al., 1998). Finally, we should like to note that in the past two decades relativistic mean field theories have also been used to calculate the EOS, and more recently there have been speculations that neutron stars are made of or contain strange quark matter, for both subjects see (Morrison et al., 2004), (Heiselberg and Pandharipande, 2000), (Weber, 1999).

2.7 Transport phenomena in the Fermi liquid

In this section we will turn to time-dependent phenomena. Having convinced ourselves that the static energy may well be calculated on the basis of Hartree–Fock if only appropriate effective interactions are used, we want to take this approximation as the starting point. We are then going to argue for possible forms of equations of motion for the one-body density, which in the end will include terms responsible for irreversible features, as given by the effects of "real collisions" among the particles.

As described in Chapter 20.1 in the static case the Hartree–Fock approximation can be seen to be equivalent of finding a basis $|\lambda_\nu\rangle$ of single particle states which diagonalizes both the Hartree–Fock Hamiltonian to $\hat{h} = \sum_\nu e_\nu |\lambda_\nu\rangle\langle\lambda_\nu|$ as well as the one-body density operator to $\hat{\rho} = \sum_\nu |\lambda_\nu\rangle\langle\lambda_\nu|$. Evidently, both operators commute with each other $[\hat{h}, \hat{\rho}] = 0$. It is most natural to interpret this commutator in the sense of the von Neumann equation: For the static case the commutator of the Hamiltonian and the density operator vanishes. Consequently, one may expect a time-dependent situation which will be governed by the following equation of motion

$$\frac{d}{dt}\hat{\rho} = \frac{1}{i\hbar}[\hat{h}, \hat{\rho}]. \qquad (2.77)$$

This equation describes the dynamics of the many-body system if no correlations between the single particles are taken into account beyond those determined by the mean field (and the Pauli principle). In this case everything may be traced back to the one-body density operator $\hat{\rho}(t)$, the time dependence of which is found by solving (2.77). This task is far from trivial, as the $\hat{h}$ itself depends on the density. In this (important) sense (2.77) is much more difficult to solve than the *common von Neumann equation* which is *linear* in the density. Nevertheless, there exist numerical codes which allow one to obtain the $\hat{\rho}(t)$. In the 1970s it was fashionable to apply them even to such complicated events as collisions of heavy ions.

As the reader may have noticed, the arguments in favor of equation (2.77) are by no means restricted to the Hartree–Fock approximation. They would apply equally well to any other situation in which a Hamiltonian $\hat{h} = \hat{h}[\hat{\rho}]$ exists which describes the dynamics of excitations associated with motion of independent fermions. Actually, in Landau theory (Morel and Nozières, 1962), (Landau et al., 1981), (Baym and Pethick, 1991) the latter are represented by quasiparticles rather than physical particles. This theory can be applied to dense Fermi systems where Hartree–Fock may fail.

2.7.1 Semi-classical transport equations

Equation (2.77) has a severe deficiency: It does not account for "collisions" among the nucleons. The equation is time reversibly invariant and does not contain any stochastic features. To account for such features one *extends* (2.77) to the form

$$\frac{d}{dt}\hat{\rho}(t) - \frac{1}{i\hbar}[\hat{h},\,\hat{\rho}] = \hat{I}[\hat{\rho}(t)]\,. \tag{2.78}$$

The term on the right-hand side is meant to represent any effects of the interaction among the fermions beyond those contained in $\hat{h}$. As such effects will necessarily imply stochastic features, we may no longer anticipate the ρ to be an idempotent operator, which means to have $\rho^2 \neq \rho$. Below the $\hat{I}[\hat{\rho}]$ will be approximated in the simplest way possible to allow for incorporating some essential quantum features. This will involve semi-classical arguments. So let us first examine what the left-hand side of this equation looks like in this limit. The best way of obtaining the latter is by introducing the *Wigner transformation*, the properties of which are described in Section 19.3.

2.7.1.1 Exploiting the Wigner transformation

The Wigner transform of the density matrix in coordinate representation is defined by[6]

$$n(\mathbf{r},\mathbf{p},t) = \int d^3s \, \exp(-\frac{i}{\hbar}\mathbf{p}\mathbf{s})\, \rho(\mathbf{r}+\frac{\mathbf{s}}{2},\mathbf{r}-\frac{\mathbf{s}}{2},t)\,. \tag{2.79}$$

Please notice that $n(\mathbf{r},\mathbf{p},t)$ is a dimensionless quantity, in accord with the fact that the space density $\rho(\mathbf{r}_1.\mathbf{r}_2,t)$ needs to be of dimension one over volume. The $\langle \mathbf{r}|\hat{\rho}(t)|\mathbf{r}\rangle$ has the same meaning as the quantity we have used so far. However, in transport theory one usually normalizes the phase space density to unity, $\int d^3r \int d^3p\, n(\mathbf{r},\mathbf{p},t)/(2\pi\hbar)^3 = 1$. Then the integral over the space density becomes 1 instead of being identical to particle number A. We shall adopt this convention in what follows.

It may then be proven that to order zero in $\hbar$ the left-hand side of (2.78) simply reduces to a form known from the *Liouville equation*, namely

$$\frac{\partial}{\partial t}n(\mathbf{r},\mathbf{p},t) - \left\{h(\mathbf{r},\mathbf{p},t), n(\mathbf{r},\mathbf{p},t)\right\} = I[n(\mathbf{r},\mathbf{p},t)]\,. \tag{2.80}$$

Here the Hamiltonian, which may or may not depend on time explicitly, is given by the $h(\mathbf{r}_1,\mathbf{r}_2,t) \equiv \langle \mathbf{r}_1|\hat{h}(t)|\mathbf{r}_2\rangle$ through a relation similar to (2.79), namely

$$h(\mathbf{r},\mathbf{p},t) = \int d^3s \, \exp(-\frac{i}{\hbar}\mathbf{p}\mathbf{s})\, h(\mathbf{r}+\frac{\mathbf{s}}{2},\mathbf{r}-\frac{\mathbf{s}}{2},t)\,. \tag{2.81}$$

---

[6] Again, vectors of relative and center of mass motion are introduced, but in modified notation: In transport theory it is customary to use $\mathbf{s}$ and $\mathbf{r}$, respectively, and to write $\mathbf{r}_1 = \mathbf{r}+\mathbf{s}/2$ and $\mathbf{r}_2 = \mathbf{r}-\mathbf{s}/2$.

Notice, please, that this h is of dimension energy. Let us look at the potential part in $h = T + U$, for the case that the latter is diagonal $U(\mathbf{r}_1, \mathbf{r}_2) = U(\mathbf{r}_1, \mathbf{r}_1)\delta(\mathbf{r}_1 - \mathbf{r}_2)$. Then one gets $U(\mathbf{r}, \mathbf{p}) = U(\mathbf{r}, \mathbf{r}) \equiv U(\mathbf{r})$. To verify that for the kinetic energy one gets $T(\mathbf{r}, \mathbf{p}) = \mathbf{p}^2/2m$ is a bit more lengthy but straightforward, for details see Chapter 21. Here, it may suffice to appeal to the correspondence principle which requires that for $\hbar \to 0$ the commutator of the von Neumann equation has to turn into the Poisson bracket of the Liouville equation

$$\frac{1}{i\hbar}[\hat{A}, \hat{B}] \longrightarrow \{A, B\} \equiv \nabla_r A \nabla_p B - \nabla_p A \nabla_r B. \tag{2.82}$$

Although the $h(\mathbf{r}, \mathbf{p}, t)$ acts as a classical Hamilton function, the equation differs from the Liouville equation proper because this $h[n(\mathbf{r}, \mathbf{p}, t)]$ depends on the density. For $I[n(\mathbf{r}, \mathbf{p}, t)] = 0$ an equation of this type was first suggested by Vlasov, where the $h[n(\mathbf{r}, \mathbf{p}, t)]$ was calculated in mean field or Hartree approximation. However, the form (2.80) is more general. It applies as well to L. Landau's Fermi liquid theory, which goes beyond Hartree–Fock. This is true not only because there is a finite collision term $I[n(\mathbf{r}, \mathbf{p}, t)] \neq 0$, but also because the $h[n(\mathbf{r}, \mathbf{p}, t)]$ is defined in a way which differs essentially from Hartree–Fock. For such reasons an equation of the type (2.80) is sometimes called the *Landau–Vlasov equation*.

2.7.1.2 The effective mean field within Hartree–Fock

Applying the effective interaction of Skyrme type we should not worry about the exchange term of the one-body potential, but just take the direct term instead. In this sense the $h(\mathbf{r}, \mathbf{p}, t)$ may just be written as in the original Vlasov version. If spin degrees are not considered explicitly, the mean field is simply given by the folding integral over the two-body interaction $V(\mathbf{r} - \mathbf{r}')$ times the density $\rho(\mathbf{r}', \mathbf{r}')$ such that the single particle Hamiltonian reads

$$h(\mathbf{r}, \mathbf{p}, t) = \frac{\mathbf{p}^2}{2m} + U^D(\mathbf{r}) = \frac{\mathbf{p}^2}{2m} + \int d^3r' \, V_{\text{sky}}(\mathbf{r} - \mathbf{r}')\rho(\mathbf{r}', \mathbf{r}'). \tag{2.83}$$

In practical applications the interaction is often parameterized in the following simple, somewhat schematic, form $V_{\text{sky}}(\mathbf{r}) = a\delta(\mathbf{r}) + b\rho^\sigma \delta(\mathbf{r})$ with three parameters a, b and σ, which may be associated again with observables like E_0, ρ_0 and κ. Thus the $U^D(\mathbf{r})$ may be written as

$$U^D(\mathbf{r}) = a\rho(\mathbf{r}) + b\rho^\sigma(\mathbf{r}). \tag{2.84}$$

It may be worth reflecting a bit more on this ansatz. Recall the energy functional (2.60) obtained for a Skyrme interaction, of which the form used here is just another simple variant. The potential part of the $\mathcal{H}$ appearing there has to be given by

$$\mathcal{H}_{\text{pot}}(\mathbf{r}) = \frac{1}{2}a\rho^2(\mathbf{r}) + \frac{1}{\sigma + 1}b\rho^{\sigma + 1}(\mathbf{r}), \tag{2.85}$$

which may be verified along similar lines to those used in going from (2.60) to (2.61). The effective single particle potential must be obtained from the first variation of $\int d^3r \, \mathcal{H}(\mathbf{r})$ with respect to the density

$$U_{\text{pot}}^{\text{eff}} = \frac{\delta E_{\text{pot}}^{\text{eff}}}{\delta \rho} = \frac{\delta}{\delta \rho} \int d^3 r \, \mathcal{H}(\mathbf{r}) = \frac{\partial \mathcal{H}(\mathbf{r})}{\partial \rho}. \qquad (2.86)$$

Evidently, this leads to the form given in (2.84).

2.7.1.3 The collision term

The collision term is meant to portray how the density $n(\mathbf{r}, \mathbf{p}, t)$ at some given point $(\mathbf{r}, \mathbf{p})$ in phase space changes in time due to the collision of the particles, for which reason it is often written as

$$I[n(\mathbf{r}, \mathbf{p}, t)] \equiv \frac{\partial}{\partial t} n(\mathbf{r}, \mathbf{p}, t) \bigg|_{\text{coll}}. \qquad (2.87)$$

Within a semi-classical approximation it may be expressed as the balance between a "gain" and a "loss term" (see Section 21.2):

$$\frac{\partial}{\partial t} n(\mathbf{r}, \mathbf{p}, t) \bigg|_{\text{coll}} = -\int \frac{d^3 p_2}{(2\pi\hbar)^3} \int \frac{d^3 p_3}{(2\pi\hbar)^3} \int \frac{d^3 p_4}{(2\pi\hbar)^3} \, W(\mathbf{p}, \mathbf{p}_2; \mathbf{p}_3, \mathbf{p}_4) \times$$
$$\left[n(\mathbf{r}, \mathbf{p}, t) n(2) \left(1 - n(3)\right) \left(1 - n(4)\right) - \left(1 - n(\mathbf{r}, \mathbf{p}, t)\right) \left(1 - n(2)\right) n(3) n(4) \right], \qquad (21.10)$$

with $n(i) \equiv n(\mathbf{r}, \mathbf{p}_i, t)$ (all evaluated at $\mathbf{r}$). Here $W(\mathbf{p}, \mathbf{p}_2; \mathbf{p}_3, \mathbf{p}_4)$ is the transition rate for scattering from the initial momenta $\mathbf{p} \equiv \mathbf{p}_1, \mathbf{p}_2$ to the final $\mathbf{p}_3$ and $\mathbf{p}_4$. The first term stands for the depletion the density in a certain phase space cell suffers because of such a transition. The product $n(\mathbf{r}, \mathbf{p}, t) n(2)$ measures the occupation probability of this cell and $(1 - n(3))(1 - n(4))$ defines the probability that the final states are not occupied. For the gain term the reverse process has to be considered, with the same transition probability $W(\mathbf{p}_3, \mathbf{p}_4; \mathbf{p}, \mathbf{p}_2) = W(\mathbf{p}, \mathbf{p}_2; \mathbf{p}_3, \mathbf{p}_4)$, which is a consequence of the principle of microreversibility.

As discussed in Section 18.1.1 the transition probability for a scattering process is in principle determined by a T-matrix; from (18.12) one obtains

$$W(\mathbf{p}, \mathbf{p}_2; \mathbf{p}_3, \mathbf{p}_4) = \frac{2\pi}{\hbar} \left| \langle 12 | T | [34] \rangle \right|^2 \delta(e_1 + e_2 - e_3 - e_4). \qquad (2.88)$$

For the final state an antisymmetrized wave function $|[\mathbf{p}_3 \mathbf{p}_4]\rangle = (|\mathbf{p}_3 \mathbf{p}_4\rangle - |\mathbf{p}_4 \mathbf{p}_3\rangle)/\sqrt{2}$ is considered, simply because for identical particles the state $|\mathbf{p}_3 \mathbf{p}_4\rangle$ is equivalent to $|\mathbf{p}_4 \mathbf{p}_3\rangle$. The initial state, on the other hand, is fixed in the sense that particle "1" is the one whose dynamics we follow explicitly. The T-matrix which is involved here would have to be determined from the integral equation (2.32). There, the single particle Hamiltonians $\hat{h}_1$ and $\hat{h}_2$ appear which are associated with the energies e_i. The appearance of the T-matrix means that one is not bound to treat the interaction in low order perturbation. For the latter, the so-called Born approximation, only the first term on the right of (2.32) is to be considered, see Section 21.2.1. Since solutions of the full integral equation (2.32) are hard to find one may try a phenomenological approach by making use of measured differential cross sections. According to (18.15) the latter can be written as

$$\frac{d\sigma_{\mathbf{q}\to\mathbf{q}'}}{d\Omega_{\mathbf{q},\mathbf{q}'}} = \left(\frac{\mu}{q}\right)\frac{2\pi}{\hbar}|T_{\mathbf{q}\to\mathbf{q}'}|^2\,\Omega(E). \tag{2.89}$$

The $\mathbf{q}$ and $\mathbf{q}'$ are the relative momenta of the two particles, before and after the collision, and μ is the reduced mass. $\Omega_{\mathbf{q},\mathbf{q}'}$ measures the scattering angle and $\Omega(E)$ is the density of final states, essentially replacing the δ-function in (2.88).

The feature just described exhibits very clearly, on the other hand, one of the basic shortcomings if such a treatment is applied to the nuclear case. The T-matrix and the free cross section are reasonable estimates of the transition rate for a very dilute system. For instance, such a situation may be given for a classical gas for which the collision term becomes the one of the ordinary Boltzmann equation. In the nuclear case, however, where the scattering happens in the medium defined by the presence of all the other particles one would expect the G-matrix to appear here rather than the T-matrix. Often in practical applications this problem is discarded entirely. Sometimes one then tries to simulate medium effects by introducing modified cross sections. However, at this stage it should be made clear that such a procedure can be considered a very crude approximation at best. In the correct way one needs to exploit Green functions and self-energies and relate the latter to Brueckner theory. The concept of self-energies Σ in relation to the transport equation is elucidated in Section 21.3 within a simple model in relation to the relaxation to equilibrium. There one may see the intimate connection of the collision term with the imaginary part $\Sigma'' = -\Gamma/2$ of Σ. At this stage it is important to recall that the G-matrix itself may become *complex* self-energies already *in first order*, which in turn are in a sense related to the presence of a collision term. It is for such reasons that the problem cannot simply be solved by replacing in a form like (2.88) the T-matrix by the G-matrix. Indeed, in the Kadanoff–Baym theory the quantum mechanically correct transport equation is expressed entirely in terms of self-energies, without being forced to start with a semi-classical picture like the one explained above. Rather, the Kadanoff–Baym equation (1962) is derived from the general, fully quantal equation for the one-body Green function. Variants of it have been applied in nuclear physics for instance by P. Danielewicz (1984) and W. Bootermans and R. Malfliet (1990) in connection with or generalization of relativistic Brueckner theory.

Let us come back to the semi-classical versions of the transport equations, which in the literature are associated with the names L. Boltzmann, E.A. Uehling and G.E. Uhlenbeck (BUU) or to Landau and Vlasov, and add a few final remarks.

(1) It should be no surprise that both the form of the collision term as written here, as well as the transport equation (2.80) itself *do violate quantum mechanics* to a more or less large extent. Indeed, in this semi-classical version one assumes that in $n(\mathbf{r},\mathbf{p},t)$ both $\mathbf{r}$ as well as $\mathbf{p}$ can be specified together, at the same instant. This is not possible in a strict quantal sense: The Heisenberg uncertainty

relation would require a certain spread both in the momenta as well as in the coordinates. This feature is neither accounted for in (2.80)) nor in the perturbative version (21.16) of the collision term. Their applicability to finite nuclei has been questioned by V. Strutinsky and A.G. Magner (1976). According to them the very presence of the narrow nuclear surface implies too large an uncertainty in the momentum. Recall that for the form of the collision term given above it is assumed that *all* probability densities $n(i)$ are calculated at the *same* space point $\mathbf{r} = \mathbf{r}_1$. Moreover, quantum mechanics also *forbids specifying precisely both time and energy*, as assumed here working with well-defined energies at the given time t.

(2) Besides these apparent deficiencies, the form written in (21.16) may represent fairly well processes going on in collisions of fermions. But the quality of this approximation depends very much on the value of the system's space density. Take the time-energy uncertainty relation, for instance. There is no real need to regard the latter as a serious constraint if the next collision happens to occur only after a time lapse which is much longer than the duration of the previous collision. Whether or not all conditions of this type are met under realistic nuclear situations is questionable. Nevertheless, these assumptions are being made and collision terms of the kind given in (21.16) are frequently used to describe heavy-ion collisions, mainly because the associated transport equation is amenable to practical solutions.

(3) For situations close to thermal equilibrium the nonlinear integro-differential equation can be linearized by applying the so-called relaxation time approximation, described in Section 21.3.1. There one replaces the $I[n(\mathbf{r}, \mathbf{p}, t)]$ by the ratio $(n(\mathbf{r}, \mathbf{p}, t) - n_{\text{eq}})/\tau$ with n_{eq} being the equilibrium distribution, which in the simpler case represents the global one, but more realistically the local equilibrium. The relaxation time τ may be taken as a parameter, but it may also be related to the imaginary part of the self-energy $-\Gamma/2$. To lowest order in temperature T and the deviation $(e(p) - \mu)$ of the energies from the chemical potential μ one may make use of the formula

$$\Gamma(p) = \left[(e(p) - \mu)^2 + \pi^2 T^2\right]/\Gamma_0. \qquad (21.26)$$

Exercise

1. Estimate the contribution of the correlation energy $e_{\alpha\beta}$ to E_α. Discard any short-range correlations and calculate the matrix element $\langle\alpha\beta|\hat{V}_{ij}|\varphi_{\alpha\beta}\rangle \approx \langle\alpha\beta|\hat{V}_{12}|\alpha\beta\rangle$ with plane waves. Observing the existence of a hard core, one finds it to be proportional to $V_0 b^3/\mathcal{V}$, with V_0 and b measuring strength and range of the attractive part. Show that $\sum_{i<j} e_{\alpha_i \beta_j} \propto A^2/\mathcal{V}$.

3

INDEPENDENT PARTICLES AND QUASIPARTICLES IN FINITE NUCLEI

In this chapter we are going to pave the way for realistic descriptions of finite nuclei. Following the pioneering work of M. Goeppert-Mayer and J.H.D. Jensen (1955) all of them are based on the picture of independent particles moving in a mean field. We will use this notion no matter whether or not this single particle potential is determined self-consistently or simply introduced phenomenologically. At first we want to outline the most basic features of the self-consistent descriptions to refer to other textbooks like (de Shalit and Feshbach, 1974) or (Ring and Schuck, 2000) for more details.

3.1 Hartree–Fock with effective forces

3.1.1 H–F with the Skyrme interaction

The simplicity of the Skyrme-type interaction follows from its special, zero range character. This renders even the exchange term to a local one (in coordinate space). The version valid for a homogeneous system like nuclear matter has been given by (2.59), for the case of $N = Z$. When the density varies with $\mathbf{r}$ a term comes into play which is of the same order in the relative momenta or in ∇_s as those shown in (2.59) (In this chapter the relative coordinate is called $\mathbf{s}$ rather than $\mathbf{r}$ as in (2.59)). For a set of quantum numbers $\alpha_l \equiv \{\{n\}, m_s, m_\tau\}$ this term reads $+iw_0(\boldsymbol{\sigma}_1 + \boldsymbol{\sigma}_2) \cdot [\overleftarrow{\nabla}_s \times \delta(\mathbf{s})\overrightarrow{\nabla}_s]$. For the energy density

$$\mathcal{H}(\mathbf{r}) = \frac{\hbar^2}{2m}\tau + \frac{3}{8}t_0\rho(\mathbf{r})^2 + \frac{1}{16}t_3\rho(\mathbf{r})^3 + +\frac{1}{16}(3t_1 + 5t_2)\rho(\mathbf{r})\tau \\ + \frac{1}{64}(9t_1 - 5t_2)(\nabla\rho(\mathbf{r}))^2 - \frac{3}{4}w_0\rho(\mathbf{r})\nabla\mathbf{J} \quad (3.1)$$

it shows up in the last expression through the spin-orbit density

$$\mathbf{J}(\mathbf{r}) = i \sum_{nm_sm_tm'_s} \varphi^*_{nm_sm_t}(\mathbf{r}) \left[\langle m_s|\boldsymbol{\sigma}|m'_s\rangle \times \nabla_r\varphi_{nm'_sm_t}(\mathbf{r})\right] \quad (3.2)$$

(In principle, there could be a term like $(1/64)(t_1 - t_2)\mathbf{J}^2$, which usually is neglected). The other two densities are defined by

$$\rho(\mathbf{r}) = \sum_{nm_sm_t} |\varphi_{\alpha_l}(\mathbf{r})|^2 \qquad \tau(\mathbf{r}) = \sum_{nm_sm_t} |\nabla_r\varphi_{\alpha_l}(\mathbf{r})|^2 \quad (3.3)$$

For the sake of simplicity these equations are still written for a nucleus having an equal number of neutrons and protons. Like for nuclear matter, the energy functional $E_{\text{eff}}^{\text{HF}} = \int d^3r\, \mathcal{H}(\mathbf{r})$ contains the contribution $\mathcal{H}^{(3)} = t_3\rho^3/16$ from a three-body interaction (recall the discussion in Section 2.6.2).

The Hartree–Fock equations are determined from the variation $\delta E_{\text{eff}}^{\text{HF}}$. Writing it in the form

$$\delta E_{\text{eff}}^{\text{HF}} = \int d^3r \left[\frac{\hbar^2}{2m^*(\mathbf{r})} \delta\tau(\mathbf{r}) + U(\mathbf{r})\delta\rho(\mathbf{r}) + \mathbf{w}(\mathbf{r})\delta\mathbf{J}(\mathbf{r}) \right] \quad (3.4)$$

it is not difficult to convince oneself that one finds

$$U(\mathbf{r}) = \frac{3}{4}t_0\rho(\mathbf{r}) + \frac{3}{16}t_3\rho(\mathbf{r})^2 + + \frac{1}{16}(3t_1+5t_2)\tau + \frac{1}{32}(9t_1-5t_2)\nabla^2\rho(\mathbf{r}) - \frac{3}{4}w_0\rho(\mathbf{r})\nabla\mathbf{J} \quad (3.5)$$

for the potential,

$$\frac{\hbar^2}{2m^*(\mathbf{r})} = \frac{\hbar^2}{2m} + \frac{1}{16}(3t_1+5t_2)\rho(\mathbf{r}) \quad (3.6)$$

for the kinetic energy, as well as (after a partial integration) $\mathbf{w} = (3/4 w_0 \nabla \rho(\mathbf{r})$ for the spin-orbit term. Thus the Hartree–Fock equations finally read

$$\left(-\nabla \frac{\hbar^2}{2m^*(\mathbf{r})} \nabla + U(\mathbf{r}) + \mathbf{w}\left[\frac{\hbar}{i}\nabla \times \boldsymbol{\sigma} \right] \right) \varphi_\alpha(\mathbf{r}) = e_\alpha \varphi_\alpha(\mathbf{r}). \quad (3.7)$$

Like the common Schrödinger equation they are ordinary differential equations of order two. They have to be solved with the appropriate boundary conditions at $\mathbf{r} = 0$ and for $\mathbf{r} \to \infty$. Unlike the Schrödinger equation, however, the set (3.7) is *nonlinear* in the wave functions, simply because the mean field is a functional of the latter. Another difference is be found in the coordinate-dependent effective mass, which represents some effects of non-locality (in $\mathbf{r}$). For more details on techniques of solving these equations see (Ring and Schuck, 2000).

In case of a spherically symmetric density distributions the spin-orbit part in (3.7) takes on the common form

$$U_{ls}(r)\,\hat{\mathbf{l}}\cdot\hat{\mathbf{s}} = \frac{3}{2}w_0 \left(\frac{1}{r}\frac{\partial}{\partial r}\rho \right) \hat{\mathbf{l}}\cdot\hat{\mathbf{s}}. \quad (3.8)$$

This is obtained because of the identity $(\nabla\rho)[\hat{\mathbf{p}} \times \boldsymbol{\sigma}] = \boldsymbol{\sigma}[(\nabla\rho) \times \hat{\mathbf{p}}]$, which naturally leads to the angular momentum $\mathbf{l} = [\mathbf{r} \times \mathbf{p}]$. For a positive value of w_0 (it is of the order of 120 MeV fm^2) the spin-orbit becomes negative, opposite to the case in atomic physics.

3.1.2 Constrained Hartree–Fock

Often it may happen that the spherical configuration does *not* lead to the lowest energy when calculated within a mean-field approach. This phenomena is of very general nature, not restricted to Hartree–Fock for the strong interaction. For a rotationally invariant force the mean-field approximation then obviously *breaks this symmetry*. To see such a phenomenon happen an appropriate class of variational functions is required. For a thorough discussion of this point see Chapter 5 of (Ring and Schuck, 2000). If one needs the energy (or the wave functions etc.) away from the stationary points one must introduce constraining fields $\hat{F}$ and solve the Hartree–Fock equations under the constraint

$$q = \operatorname{tr} \hat{\rho}\hat{F} \equiv \langle \hat{F} \rangle. \tag{3.9}$$

This may be achieved by varying the modified energy functional $\langle \hat{H}' \rangle = \langle \hat{H} \rangle - \lambda \langle \hat{F} \rangle$ with the Lagrange multiplier λ being chosen to fulfill the constraint.

3.1.3 Other effective interactions

In the literature many other versions of effective interactions exist. We shall only mention two of them.

The Gogny interaction: To overcome problems encountered with the Skyrme interaction for pairing correlations (see later) D. Gogny has suggested another force of finite range (Gogny, 1975). Its dependence on $\mathbf{r}_1 - \mathbf{r}_2$ is expressed in terms of a sum over two Gaussians of different widths instead of the δ functions; more precisely this is done for those parts which involve the parameters t_0, t_1 and t_2. These Gaussians are multiplied by spin and isospin exchange operators of the form (26.40).

Separable forces: In second quantized form the general two-body interaction reads (see (26.51))

$$\hat{V} = \frac{1}{2} \sum_{\alpha\beta\gamma\delta} V_{\alpha\beta\gamma\delta} \, \hat{c}^\dagger_\alpha \hat{c}^\dagger_\beta \hat{c}_\delta \hat{c}_\gamma. \tag{3.10}$$

By making use of the anti-commutation relations and after interchanging indices this can be rewritten as

$$\hat{V} = \frac{1}{2}\left[\sum_{\alpha\beta}\left(\sum_\gamma V_{\alpha\gamma\beta\gamma}\right)\hat{c}^\dagger_\alpha \hat{c}_\beta - \sum_{\alpha\beta\gamma\delta} V_{\alpha\gamma\delta\beta}\,\hat{c}^\dagger_\alpha \hat{c}_\beta \hat{c}^\dagger_\gamma \hat{c}_\delta \right], \tag{3.11}$$

or by use of $\hat{n}_\mu = \hat{c}^\dagger_\alpha \hat{c}_\beta$ and $\hat{n}_\nu = \hat{c}^\dagger_\gamma \hat{c}_\delta$ in shorthand notation

$$\hat{V} = \frac{1}{2}\left[\sum_\mu \tilde{U}_\mu \hat{n}_\mu - \sum_{\mu\nu} \tilde{V}_{\mu\nu} \hat{n}_\mu \hat{n}_\nu\right]. \tag{3.12}$$

The first term is of one-body nature. The remaining two-body interaction $\hat{V}^{(2)}$ can be written in *separable form*

$$\hat{V}^{(2)} = \frac{1}{2} \sum_\mu k_\mu \hat{F}_\mu \hat{F}_\mu \tag{3.13}$$

after diagonalizing the matrix $\tilde{V}_{\mu\nu}$. Here, the summation extends over the square of the number of single particle states. It should be noted that rearrangements other than that in going from (3.10) to (3.13) are in use, depending on the application one has in mind, see e.g. (Negele and Orland, 1987). Instead of (3.11) one might write

$$\hat{V} = \frac{1}{2}\left[-\sum_{\alpha\beta}\left(\sum_{\gamma}V_{\alpha\gamma\gamma\beta}\right)\hat{c}_\alpha^\dagger\hat{c}_\beta + \sum_{\alpha\beta\gamma\delta}V_{\alpha\gamma\beta\delta}\,\hat{c}_\alpha^\dagger\hat{c}_\beta\hat{c}_\gamma^\dagger\hat{c}_\delta\right], \quad (3.14)$$

or simply

$$\hat{V} = \frac{1}{2}\sum_{\alpha\beta\gamma\delta}\hat{P}_{\alpha\beta}^\dagger V_{\alpha\beta\gamma\delta}\hat{P}_{\gamma\delta} \quad \text{with} \quad \hat{P}_{\alpha\beta}^\dagger = \hat{c}_\alpha^\dagger\hat{c}_\beta^\dagger, \quad \hat{P}_{\gamma\delta} = \hat{c}_\delta\hat{c}_\gamma. \quad (3.15)$$

Such separable forces may be of great help for practical calculations, for instance within the technique of functional integration, see the discussion in Section 24.4.1. This is true in particular if the very existence of this separable form is used for phenomenological parameterizations. Often one introduces forms where the $\hat{F}_\mu$ themselves are not Hermitian, in which case the product $\hat{F}_\mu\hat{F}_\mu$ has to be replaced by $\hat{F}_\mu^\dagger\hat{F}_\mu$, mind (3.15). Prominent examples of this type are the quadrupole–quadrupole interaction or the pairing interaction. For schematic studies often a simplified version is used, namely

$$\hat{H} = \hat{H}_0 + (k/2)\hat{F}\hat{F} \quad (3.16)$$

Here, $\hat{H}_0$ is meant to be some Hamiltonian in which some *unperturbed* mean field exists already and the $\hat{F}$ is some one-body Hamiltonian. The second term on the right then mimics a two-body interaction. As we shall see later forms of this type are extremely useful to study generic features of collective motion. Moreover, it may be clear already here that such a Hamiltonian is particularly useful when one needs to deal with a constraint of the type given in (3.9), with the $\hat{F}$ being the same operator in both cases.

3.2 Phenomenological single particle potentials

For practical applications it is often necessary to replace the self-consistent mean field $U(\mathbf{r})$ by a phenomenological one. For instance, this holds true to the present day in applications to deformed nuclei and to collective motion, which will be addressed below. In fact, after the discovery of the shell model such models were indispensable for many decades, before Hartree–Fock calculations became feasible in the 1970s. In the following we are going to complete the discussion started in Section 1.3.1.1.

3.2.1 The spherical case

In Section 1.3.2.2 the optical potential has been introduced with its real part being of Woods–Saxon type (1.56),

$$U_{\text{WS}}(r) = -U_0 \left(1 + \exp\left(\frac{r - R_u}{a_u}\right)\right)^{-1}, \qquad (3.17)$$

with the radius $R_u(A) = r_u A^{1/3}$ given in (1.59). The radius parameter was said to be about $r_u = 1.25$ fm and the diffuseness parameter a_u specifying the fall off of the density in the surface to be of order 0.65 fm. A solution of the Schrödinger equation for the single particle Hamiltonian $\hat{h}_0(\hat{\mathbf{r}}, \hat{\mathbf{p}}) = \hat{p}^2/2m_N + U(r)$ may then be expected to lead to single particle wave functions $\varphi_\nu(\mathbf{r})$ such that the density of the many-body ground state follows the density distribution of real nuclei. Indeed, for a two-body interaction $V(\mathbf{r} - \mathbf{r}')$ in Hartree approximation the mean field is given by

$$U(\mathbf{r}) = \int d^3 x'\, V(\mathbf{r} - \mathbf{r}')\rho(\mathbf{r}', \mathbf{r}'). \qquad (3.18)$$

For spherical symmetry the $\rho(\mathbf{r}', \mathbf{r}')$ would simply reduce to the density distribution $\rho(r)$ which may be expected to have the approximate form given in Section 1.3.3.2. For a truly short-range interaction $V(\mathbf{r} - \mathbf{r}')$ the $U(r)$ would indeed follow the shape of the density. In practice relation (3.18) is often used without going through the iterations required by Hartree or Hartree–Fock. Such models are of particular use when the density is deformed. Models of this type are the folded Yukawa model, which we shall address shortly. In this context one may also mention the Thomas–Fermi approach of W.D. Myers and W.J. Swiatecki who apply the Seyler–Blanchard interaction (1996).

Solutions of the Schrödinger equation with the Woods–Saxon potential (3.17) are possible only numerically. Simpler analytic models are the spherical square well or the oscillator. In the former case the potential is constant $U(r) = -U_0$ for $r \leq R$ and zero in the outer region. For the oscillator model one has

$$U_{\text{osc}}(r) = U_c + m_N \omega_0^2 r^2 / 2. \qquad (3.19)$$

The U_c is an immaterial constant, but it may be chosen such that for negative energies the $U_{\text{osc}}(r)$ and the $U_{\text{WS}}(r)$ are not too different. More important is the ω_0 for which often the following estimate is used (see exercise 1):

$$\hbar\omega_0 \approx \frac{5}{4}\left(\frac{3}{2}\right)^{1/3} \frac{\hbar^2}{m_N r_u^2} A^{-1/3} \approx 41 A^{-1/3} \text{ MeV}. \qquad (3.20)$$

In (Bohr and Mottelson, 1969) it is demonstrated (see also (Siemens and Jensen, 1987)) that in first approximation the wave functions for the oscillator and the Woods–Saxon potential look very similar. The influence of the shape of the two potentials is different for high or low values of the angular momentum. In any case this comparison demonstrates that the wave functions of the oscillator may very well serve as a decent basis for calculations in more realistic potentials. This is true in particular for deformed systems to which we will return below.

From the discussion in previous sections we know that the mean field has to be completed by adding to the $\hat{h}_0$ a spin-orbit interaction

$$\hat{h}_{ls} = U_{ls}(r)\,\hat{\mathbf{l}}\cdot\hat{\mathbf{s}} \quad \text{with} \quad U_{ls}(r) = u_{ls}\frac{1}{r}\frac{\partial}{\partial r}U(r) \qquad (3.21)$$

The actual single particle Hamiltonian thus becomes

$$\hat{h}(\hat{\mathbf{r}},\hat{\mathbf{p}}) = \frac{\hat{p}^2}{2m_N} + U(r) + \hat{h}_{ls}. \qquad (3.22)$$

The implication of $U_{ls}(r)$ can be seen as follows. One needs to add the particle's orbital angular momentum $\hat{\mathbf{l}}$ and its spin to get the total angular momentum $\hat{\mathbf{j}} = \hat{\mathbf{l}} + \hat{\mathbf{s}}$. The possible eigenvalues are $j_\pm = l \pm 1/2$ with eigenstates $|jmls\rangle$ (see Section 26.7), which satisfy the relation

$$2\hat{\mathbf{l}}\cdot\hat{\mathbf{s}}|jmls\rangle = \left(\hat{\mathbf{j}}^2 - \hat{\mathbf{l}}^2 - \hat{\mathbf{s}}^2\right)|jmls\rangle = (j(j+1) - l(l+1) - 3/4)\,|jmls\rangle. \qquad (3.23)$$

Since

$$j(j+1) - l(l+1) - 3/4 = \begin{cases} l & \text{for } j_+ = l + 1/2 \\ -l - 1 & \text{for } j_- = l - 1/2 \end{cases} \qquad (3.24)$$

the form factor $U_{ls}(r)$ of the spin-orbit interaction has to be *negative* to concur with the experimental fact that states with higher angular momentum j_+ are of lower energy. The splitting $\Delta_{ls} = e_+ - e_-$ of the two levels is proportional to $2l+1$ and of order $\Delta_{ls} \simeq -20A^{-2/3}(2l+1)$ (Bohr and Mottelson, 1975).

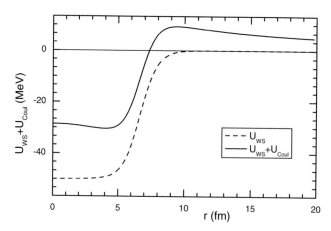

FIG. 3.1. Modification of the Woods–Saxon potential by the Coulomb potential of (3.25), calculated for ^{152}Dy with $U_0 = 50$ MeV, $r_u = 1.25$ fm, $a_u = 0.65$ fm.

Modifications from the Coulomb interaction: Assuming the nucleus to be a homogeneously charged sphere the Coulomb potential reads

$$U_{\text{Coul}}(r) = \frac{Ze^2}{2R_c}\left[3 - \left(\frac{r}{R_c}\right)^2\right] \qquad r \leq R_c$$
$$= \frac{Ze^2}{r} \qquad r \geq R_c, \qquad (3.25)$$

where often one takes $R_c \simeq R_u$. For protons this potential has to be added to the $U(r)$ of (3.22). As compared to the potential for neutrons it leads to two important modifications: (i) The overall potential becomes less deep in the interior by the amount $3Ze^2/2R_C$. This feature reflects directly the fact that in heavy nuclei there are more neutrons than protons. (ii) The potential acquires a Coulomb barrier, which for a medium heavy nuclei already is of the order of 10 MeV, as demonstrated in Fig. 3.1 for ^{152}Dy. This barrier hinders the emission of proton over that of neutrons.

3.2.2 The deformed single particle model

Knowing that nuclei may exhibit deformed shapes (recall the discussion in Section 1.3.3) it is almost inescapable to allow the mean field to become deformed as well. As we shall see later this feature not only has important consequences for static properties it also lays the physical basis for understanding the coupling between the dynamics of such shapes and single particle motion. In the latter sense this may be understood as a unification of the shell model with the liquid drop model described in Section 1.3.3.3. Indeed, in the literature this notion has been used to classify the pioneering work of A. Bohr and B.R. Mottelson which started in the early 1950s and which in 1975 was awarded the Nobel Prize.

Again, because of the short-range nature of the two-body interaction the shape of the potential should follow the one of the density. In practice one often goes the other way round and introduces shape degrees of freedom for the mean field. Evidently, self-consistency can be fulfilled only approximately. Nevertheless even on such a level it has great implications for the dynamics of the nucleus as a whole. In the following we want to describe a few of the most prominent examples, beginning with the simplest possibility, the deformed oscillator.

3.2.2.1 *The oscillator potential and the Nilsson model* The potential for the deformed oscillator reads:

$$U_{\text{osc}}(\mathbf{r}) = \frac{m_N}{2}\left(\varpi_1^2 x^2 + \varpi_2^2 y^2 + \varpi_3^2 z^2\right). \qquad (3.26)$$

Equipotential surfaces follow the formula $x^2/c_1^2 + y^2/c_2^2 + z^2/c_3^2 = const$ for an ellipse. Since nuclear matter is largely incompressible, the density distribution of the system should obey volume conservation. Self-consistency requires the same condition for the potential, which implies $c_i \varpi_i = R_0 \varpi_0$ for $i = 1, 3$, where R_0 and ϖ_0 are radius and frequency of the corresponding spherical configuration, respectively.

In the following we shall restrict ourselves to spheroidal deformations by putting $\varpi_1 = \varpi_2 = \varpi_\perp$ and $c_1 = c_2 = c_\perp$. The size of the deformation may then be measured by a single parameter, for instance by the ratio $c_\perp/c_3 = \varpi_3/\varpi_\perp$ or by a δ given by $c_3 - c_\perp = R_0 \delta \approx c_3 \delta$, where the last expression is valid to linear order in δ. Following (1.70) one may introduce an angle dependent radius parameter

$$R(\vartheta) = R_0 \left(1 + (2/3)\,\delta\, P_2(\cos\vartheta)\right) \tag{3.27}$$

by making use of the Legendre polynomial $P_2(x) = (3\cos^2 x - 1)/2$. Here, the angle ϑ is measured from the z-axis such that one has

$$\begin{aligned} c_3 &= R(\vartheta = 0) = R_0\left(1 + 2\delta/3\right) \\ c_\perp &= R(\pi/2) = R_0\left(1 - \delta/3\right), \end{aligned} \tag{3.28}$$

implying for the ratio

$$\frac{c_\perp}{c_3} = \frac{1 - \delta/3}{1 + 2\delta/3} \tag{3.29}$$

which for positive δ is the ratio of the shorter axis to the longer one. For a positive δ the shape is prolate, and oblate otherwise. For the frequencies the relations (3.28) imply

$$\begin{aligned} \varpi_3 &= \varpi_0 \left(1 + 2\delta/3\right)^{-1} \approx \varpi_0\left(1 - 2\delta/3\right) \\ \varpi_\perp &= \varpi_0 \left(1 - \delta/3\right)^{-1} \approx \varpi_0\left(1 + \delta/3\right), \end{aligned} \tag{3.30}$$

where the expressions on the right are again valid only to linear order. If these are plugged into the form (3.26) one gets for $U_{\rm osc}(\mathbf{r})$:

$$U_{\rm osc}(\mathbf{r}) = U_{\rm osc}(\vartheta) \approx \frac{m_N \varpi_0^2}{2} R_0^2 \left(1 - \frac{4}{3}\delta\, P_2(\cos\vartheta)\right). \tag{3.31}$$

As mentioned, for the oscillator potential the Schrödinger equation can be solved analytically. For the potential (3.26) the problem separates into one for three one-dimensional oscillators. Their wave functions are given in terms of Hermite polynomials. In this way one is able to construct basis states for practically any deformation. Of particular simplicity are the single particle energies for spheroidal shapes, which read

$$e(n_3, n_\perp) = \left(n_3 + 1/2\right)\hbar\varpi_3 + \left(n_\perp + 1\right)\hbar\varpi_\perp \tag{3.32}$$

with quantum numbers n_3 and $n_\perp = n_1 + n_2$ in the two independent directions. To leading order in δ this becomes

$$e(n_3, n_\perp) = \hbar\varpi_0\left(N + 3/2 - \delta\left(2n_3 - n_\perp\right)/3\right), \tag{3.33}$$

where $N = n_3 + n_\perp$ is the total oscillator quantum number. Actually, replacing δ by $\delta_{\rm osc} = 3(\varpi_\perp - \varpi_3)/(2\varpi_\perp + \varpi_3)$ the linear relation (3.33) can be seen to hold true for all values of $\delta_{\rm osc}$.

Nilsson model: Perhaps the most successful early deformed shell model is the one developed by S.G. Nilsson in (1955). It is based on the oscillator model but adds two spin-dependent terms to get

$$\hat{h}(\hat{\mathbf{r}}, \hat{\mathbf{p}}) = \hat{p}^2/2m_N + U_{\rm osc}(\mathbf{r}) + u_{ls}\hbar\omega_0 \hat{\mathbf{l}}\cdot\hat{\mathbf{s}} + u_{ll}\hbar\varpi_0\left(\hat{\mathbf{l}}^2 - \langle\hat{\mathbf{l}}^2\rangle_N\right). \qquad (3.34)$$

The very last term is considered to simulate effects seen for a potential of a more pronounced surface. For the latter states with larger l values have smaller energies than those in the pure oscillator potential $U_{\rm osc}$. A similar behavior can be achieved by an appropriately chosen negative value of u_{ll}. The subtraction of the constant $\langle\hat{\mathbf{l}}^2\rangle_N$, after being adapted to each oscillator shell with main quantum number N, warrants that the $\hat{\mathbf{l}}^2$ term does not modify the average energy difference between major shells. More details may be found in (Bohr and Mottelson, 1975).

3.2.2.2 Finite potential wells The oscillator may serve as a model case for possible procedures of constructing potentials of finite depth. Indeed, the result (3.31) may be viewed in a more general context. As we know from previous discussions, like that in Section 1.3.2.2, typically the single particle potential is of the form

$$U(\mathbf{r}) = U_u f\left(r - R(\vartheta, \phi; Q_\mu)\right). \qquad (3.35)$$

It is written for spherical polar coordinates which come in through the radius $R(\vartheta, \phi; Q_\mu)$. In detail this dependence is characterized by a set Q_μ of shape parameters. One such possibility is given by the

$$R(\vartheta, \phi) = R_0\left(1 + \sum_{\lambda\mu}\alpha_{\lambda\mu}Y^*_{\lambda\mu}(\vartheta, \phi)\right) \qquad (1.70)$$

used before in Section 1.3.3 to parameterize the deformation of the space density. Let us first look at the simpler case that the $\alpha_{\lambda\mu}$ are small. Then one may expand the $f(r-R)$ to obtain

$$U(\mathbf{r}) = U_u f\left(r - R_0\right) - U_u \left(R\frac{\partial f}{\partial R}\right)_{R=R_0}\sum_{\lambda\mu}\alpha_{\lambda\mu}Y^*_{\lambda\mu}(\vartheta, \phi). \qquad (3.36)$$

Putting for $U_u f(r)$ the quadratic dependence shown in (3.19) (with $U_c = 0$) one gets the form (3.31) after considering (3.27) as a variant of (1.70). The prime example of a potential of the form (3.35) but of finite depth is the Woods–Saxon potential (3.17). For given $R(\vartheta, \phi; Q_\mu)$ (3.35) may be used to generate the deformed potential independently of the size of the deformation. For small deformations one may make use of (3.36).

Cassini ovaloids: In order to describe fission one needs to have shapes which allow for the development of a neck which in the end may become thinner and thinner. One such possibility is offered by Cassini ovaloids pictured in Fig. 3.2.

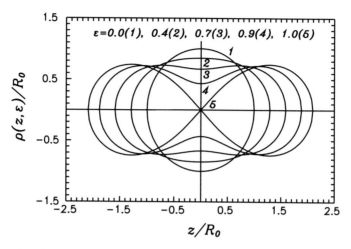

FIG. 3.2. Axially symmetric Cassini ovaloids specified by the dependence $\rho = \rho(z, \epsilon)$ of eqn(3.37) for cylindrical coordinates, for different values of the parameter ϵ, see also (Ivanyuk et al., 1997).

They are of rotationally symmetric shape about the z-axis. In cylindrical coordinates this shape can be expressed by specifying the radial distance $\rho(z, \epsilon)$ as follows (Pashkevich, 1971):

$$\rho(z,\epsilon) = R_0 \left[\sqrt{a^4 + 4\epsilon z^2/R_0^2} - z^2/R_0^2 - \epsilon \right]^{1/2}. \tag{3.37}$$

The constant a is defined by volume conservation and R_0 is the radius of the corresponding sphere, which is obtained if the deformation parameter ϵ is put $\epsilon = 0$. For small ϵ the ovaloids resemble those of spheroids with the ratio of the axes given by $c_\perp/c_3 = (1 - 2\epsilon/3)/(1 + \epsilon/3)$. At $\epsilon \approx 0.5$ a neck appears and at $\epsilon = 1.0$ the nucleus separates into two fragments. It may be mentioned in passing that in (Pashkevich, 1971) a generalization is suggested by allowing deviations from the pure surface specified by Legendre polynomials.

The single particle potential $U_{\text{C.O.}}(\mathbf{r})$ is constructed from the Woods–Saxon form (3.17). One has only to replace there the $r - R_u$ by a $l(\rho, z)$ which measures the shortest distance of the point $\mathbf{r} \equiv (\rho, z)$ to the surface. Finally the spin-orbit interaction is chosen to be proportional to the gradient of $U_{\text{C.O.}}(\mathbf{r})$. For any ϵ the solutions of the Schrödinger equation are obtained numerically by diagonalizing the Hamiltonian in the basis given by the deformed oscillator at the same value of ϵ. The dependence of a selection of single particle energies on ϵ is shown in Fig. 3.3 for the neutrons of ^{224}Th.

Folded Yukawa model of the Berkeley–Los Alamos group: This model allows for shapes which are more realistic for larger deformations and which for fissile

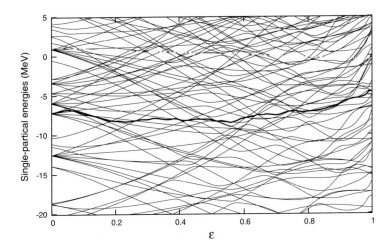

FIG. 3.3. Typical single particle energies as function of deformation, evaluated for Cassini ovaloids for the neutrons of ^{224}Th; the thick line represents the Fermi energy.

nuclei may even be used down to the scission region; more details of this parameterization are described in Section 5.3 with examples being shown in Fig. 5.4. The single particle potential is constructed from (3.18) with the two-body interaction being of Yukawa type. One gets

$$U_{\rm FY}({\bf r}) = -\frac{U_0}{4\pi a_{\rm pot}} \int_{\mathcal{V}} d^3 x' \, \frac{\exp\left(-|{\bf r}-{\bf r}'|/a_{\rm pot}\right)}{|{\bf r}-{\bf r}'|/a_{\rm pot}}, \qquad (3.38)$$

with the integration being extended over the (constant) volume $\mathcal{V}$ which corresponds to a given shape of the density. With this $U_{\rm FY}({\bf r})$ the following spin-orbit interaction is defined

$$\hat{h}_{ls} = -\frac{2\lambda}{\hbar}\left(\frac{\hbar}{2m_N c}\right)^2 \widehat{\bf s} \cdot \left[\widehat{\nabla} U_{\rm FY} \times \widehat{\bf p}\right]. \qquad (3.39)$$

The parameters involved in both potentials have been determined from adjustments of the resulting single particle levels to those found experimentally in the rare-earths and actinide regions. For the range of the interaction the value $a_{\rm pot} = 0.8$ fm was found. Both the depth U_0 as well as the spin-orbit strength λ are different for protons and neutrons and vary with particle number. Finally, for protons the Coulomb potential

$$U_{\rm C}({\bf r}) = e\rho_{\rm C} \int_{\mathcal{V}} d^3 x' \, \frac{1}{|{\bf r}-{\bf r}'|} \qquad (3.40)$$

is added, with $\rho_{\rm C}$ being the charge density $Ze/\mathcal{V}$. Details may be found in (Möller et al., 1995) with references to earlier work.

3.3 Excitations of the many-body system

3.3.1 The concept of particle-hole excitations

Suppose we are given a system with N particles. Their quantum states $|m; N\rangle$ may be classified by a set of quantum numbers m_i where i runs from 1 to N, such that we ,may write $|m; N\rangle = |m_1, m_2, m_3, \ldots, N\rangle$. For independent fermions these states may be given by Slater determinants (see Section 19.2), with the m_i representing single particle states. In the language of second quantization, with each such state $|m_i\rangle$ one associates a creation operator $\hat{c}_i^\dagger$ defined through $|m_i\rangle = \hat{c}_i^\dagger |\text{vac}\rangle$ where $|\text{vac}\rangle$ stands for the particle vacuum. For N independent particles the state $|m; N\rangle$ is thus given by

$$|m; N\rangle = \hat{c}_{m_N}^\dagger \cdots \hat{c}_{m_2}^\dagger \hat{c}_{m_1}^\dagger |\text{vac}\rangle. \tag{3.41}$$

States of different particle numbers are obtained by either adding or removing one particle. Hence, one may simply write $|m; N+1\rangle = \hat{c}_{m_k}^\dagger |m; N\rangle$ in the first case and $|m; N-1\rangle = \hat{c}_{m_k} |m; N\rangle$ in the second case. The $\hat{c}_{m_k}$ is the "annihilation" operator whose action on the vacuum must necessarily be zero, $\hat{c}_{m_k} |\text{vac}\rangle = 0$. Because we are dealing with fermions the creation and annihilation operators must anti-commute amongst themselves and they fulfill the relation (26.46) which here reads $\left[\hat{c}_{m_k}^\dagger, \hat{c}_{m_l} \right]_+ = \delta_{kl}$.

Of particular interest is the ground state $|0; N\rangle$ of the N-body system for which the m_i must correspond to the lowest possible single particle energies e_{m_i}. All these states are filled up to the last level associated with the Fermi energy e_F. Excited states $|m; N\rangle$ (of the system with the same number of particles) may then be created by first removing a particle from some state below the Fermi level and adding one in a state above the Fermi level. In a pictorial way one speaks of the *creation of a hole* h with $e_h \leq e_\text{F}$ and of a *particle* p with $e_p > e_\text{F}$; interpretations of more physical nature will be given in Section 4.3.4. Such 1p-1h-states may thus be defined as

$$|1p1h; N\rangle = \hat{c}_{m_p}^\dagger \hat{c}_{m_h} |0; N\rangle \quad \text{with} \quad \begin{cases} e_{m_p} > e_\text{F} \\ e_{m_h} \leq e_\text{F} \end{cases} \tag{3.42}$$

They are eigenstates of the Hamiltonian

$$\hat{H}_\text{ipm} = \sum_k e_k \hat{c}_k^\dagger \hat{c}_k. \tag{3.43}$$

of energy $E_0 + e_p - e_h$, with E_0 being the energy of the many-body ground state,

$$\hat{H}_\text{ipm} |0; N\rangle = E_0 |0; N\rangle \quad \text{with} \quad E_0 = \sum_{e_{m_k} \leq e_\text{F}} e_{m_k}. \tag{3.44}$$

This is easily proven from the basic relations of second quantization described in Section 26.8. This construction may be continued to obtain many-particle many-hole states and their associated energies.

3.3.2 Pair correlations

It is known that nucleons outside closed shells group together in pairs. This effect is similar to the appearance of superfluidity or superconductivity in macroscopic systems, which is explained in textbooks like (Fetter and Walecka, 1971) or (Bohr and Mottelson, 1975) and (de Shalit and Feshbach, 1974), where also due account is given on the historical development. In the following we shall treat such correlations on the basis of the so-called BCS-theory (according to J. Bardeen, L.N. Cooper and J.R. Schrieffer (Bardeen *et al.*, 1957)). This allows one to understand on a qualitative level, at least, basic features of the phase transition to the normal nuclear liquid. There is no room to discuss subtleties which originate from the finiteness of real nuclei and which may be seen on the basis of Hartree–Fock–Bogoliubov treatments (de Shalit and Feshbach, 1974), (Ring and Schuck, 2000).

The nature and the physical origin of pairing correlations can be understood assuming the general two-body interaction (3.10) to have a special form. Only matrix elements V_{klmn} of type $V_{k\bar{k}l\bar{l}}$ are considered where $|\bar{k}\rangle$ is a the state which is the time reversed one of $|k\rangle$, and likewise for $|\bar{l}\rangle$. This coupling is assumed to be attractive, but finite only within a certain range of the single particle energies in which it is taken to be constant. This amounts to say $V_{k\bar{k}l\bar{l}} \equiv (k\bar{k}|V|l\bar{l}) = -G$ and implies that this *pairing interaction* $\hat{V}^{\mathcal{P}}$ can be written as

$$\hat{V}^{\mathcal{P}} = -G\hat{P}^\dagger \hat{P} \qquad \text{with} \qquad \hat{P}^\dagger = {\sum_k}' \hat{c}_k^\dagger \hat{c}_{\bar{k}}^\dagger . \qquad (3.45)$$

The summation over single particle states k is meant not to include the time reversed ones. Moreover, it is restricted to the "pairing window" mentioned before, for which reason we put a prime as an upper index.

This two-body interaction is to be added to the Hamiltonian $\hat{H}_{\text{ipm}}$ for independent particles such that the total Hamiltonian may be written as

$$\hat{H}^{\mathcal{P}} = \hat{H}_{\text{ipm}} - G\hat{P}^\dagger \hat{P} \qquad \text{with} \qquad \hat{H}_{\text{ipm}} = {\sum_k}' e_k (\hat{c}_k^\dagger \hat{c}_k + \hat{c}_{\bar{k}}^\dagger \hat{c}_{\bar{k}}), \qquad (3.46)$$

and the e_k being the unperturbed single particle energies. Notice, that in $\hat{H}_{\text{ipm}}$ the summation does not underlie the same restriction as in the pairing interaction. The $\hat{H}^{\mathcal{P}}$ may be rewritten as

$$\hat{H}^{\mathcal{P}} = \hat{H}_{\text{ipm}} - \Delta(\hat{P}^\dagger + \hat{P}) + \Delta^2/G + \hat{H}_{\text{res}} \qquad (3.47)$$

by simply introducing the c-number parameter

$$\Delta \equiv G\langle \hat{P} \rangle = G\langle \hat{P}^\dagger \rangle \qquad (3.48)$$

and the "residual interaction"

$$\hat{H}_{\text{res}} = -G{\sum_{kj}}' \left(\hat{c}_k^\dagger \hat{c}_{\bar{k}}^\dagger - \langle \hat{c}_k^\dagger \hat{c}_{\bar{k}}^\dagger \rangle \right) \left(\hat{c}_{\bar{j}} \hat{c}_j - \langle \hat{c}_{\bar{j}} \hat{c}_j \rangle \right). \qquad (3.49)$$

Tentatively, it was assumed here that the averages of P and $P^\dagger$ are identical, which will prove to be correct shortly; otherwise one might work with a complex Δ. Discarding this $\hat{H}_\text{res}$ one reaches the level of the *mean-field approximation* with its Hamiltonian

$$\hat{H}^\mathcal{P}_\text{mf} = \hat{H}_\text{ipm} - \Delta(\hat{P}^\dagger + \hat{P}) + \Delta^2/G. \tag{3.50}$$

Here, the c-number term Δ^2/G has been included in anticipation of similar procedures in chapters 5 and 6. In this way the average total energy in mean-field approximation is represented by the expectation value of $\hat{H}^\mathcal{P}_\text{mf}$. Arguments why the residual interaction may be neglected may be found in the literature, see e.g. (Siemens and Jensen, 1987). So far the Δ is unknown. In the static considerations we aim at here it may be considered a variational parameter finally to be determined by minimizing the total energy. Later in Chapter 6 we will turn to time-dependent descriptions where the $\Delta(t)$ is treated as a collective variable for the pairing degree of freedom.

Because of the special structure of the operator $\hat{P}^\dagger$ for the pairing field the mean field Hamiltonian (3.50) is not of normal form with the creation operators being on the left of the annihilation operators. This may be achieved by a canonical transformation to new operators $\hat{\alpha}_k = u_k \hat{c}_k + v_k \hat{c}^\dagger_{\bar{k}}$ and $\hat{\alpha}_{\bar{k}} = u_k \hat{c}_{\bar{k}} - v_k \hat{c}^\dagger_k$, for which we have to require the normalization of the parameters u_k and v_k in the sense $u_k^2 + v_k^2 = 1$. Then the $\hat{\alpha}_k$ and $\hat{\alpha}^\dagger_k$, too, fulfill the common anti-commutation relations for fermion operators, see Section 26.8. Moreover, the inverse transformation is seen to be given by $\hat{c}_k = u_k \hat{\alpha}_k - v_k \hat{\alpha}^\dagger_{\bar{k}}$ and $\hat{c}_{\bar{k}} = u_k \hat{\alpha}_{\bar{k}} + v_k \hat{\alpha}^\dagger_k$. The coefficients of this *Bogoliubov transformation* may now be determined such that the transformed Hamiltonian is normally ordered with respect to the $\hat{\alpha}_k$ and $\hat{\alpha}^\dagger_k$. Of course, this property is not given for the transformed $\hat{H}_\text{ipm}$ itself and neither for the number operator

$$\hat{N} = {\sum_k}' (\hat{c}^\dagger_k \hat{c}_k + \hat{c}^\dagger_{\bar{k}} \hat{c}_{\bar{k}}). \tag{3.51}$$

In fact this number operator does not commute with the mean field Hamiltonian $\hat{H}^\mathcal{P}_\text{mf}$ anymore. Hence, in this approximation particle number can be fixed only on average $\langle \hat{N} \rangle = \mathcal{N}$, similar to the situation in the *grand canonical ensemble*. For this reason one prefers to work with the Hamiltonian

$$\hat{H}^\mathcal{P}_\text{GC} = \hat{H}^\mathcal{P}_\text{mf} - \mu \hat{N} \tag{3.52}$$

instead of the original $\hat{H}^\mathcal{P}_\text{mf}$. The Lagrange multiplier μ can be used to ensure particle number conservation on average.

After the transformation the $\hat{H}^\mathcal{P}_\text{GC}$ turns into

$$\hat{H}^\mathcal{P}_\text{GC} = U^\text{GC}_0 + \hat{H}^\text{GC}_{11} + \hat{H}^\text{GC}_{20}, \tag{3.53}$$

with the individual terms given by

$$U_0^{\text{GC}} = {\sum_k}' 2v_k^2(e_k - \mu) - 2\Delta {\sum_k}' u_k v_k + \Delta^2/G, \tag{3.54}$$

$$\hat{H}_{11}^{\text{GC}} = {\sum_k}' [(e_k - \mu)(u_k^2 - v_k^2) + 2\Delta u_k v_k](\hat{\alpha}_k^\dagger \hat{\alpha}_k + \hat{\alpha}_{\bar{k}}^\dagger \hat{\alpha}_{\bar{k}}), \tag{3.55}$$

$$\hat{H}_{20}^{\text{GC}} = {\sum_k}' [(e_k - \mu)2u_k v_k - \Delta(u_k^2 - v_k^2)](\hat{\alpha}_k^\dagger \hat{\alpha}_{\bar{k}}^\dagger + \hat{\alpha}_{\bar{k}} \hat{\alpha}_k). \tag{3.56}$$

It is observed that the first two of them are of the desired nature but not the third one given in (3.56). Of course, this $\hat{H}_{20}^{\text{GC}}$ can be made to vanish by simply putting the square brackets equal to zero,

$$(e_k - \mu) 2u_k v_k - \Delta \left(u_k^2 - v_k^2\right) = 0. \tag{3.57}$$

Considering normalization these equations may be solved uniquely to

$$u_k^2 = (1/2)\left(1 + (e_k - \mu)/E_k\right), \qquad v_k^2 = (1/2)\left(1 - (e_k - \mu)/E_k\right), \tag{3.58}$$

where

$$E_k = \sqrt{(e_k - \mu)^2 + \Delta^2}. \tag{3.59}$$

These u_k and v_k satisfy the following relations

$$u_k^2 - v_k^2 = (e_k - \mu)/E_k \qquad \text{and} \qquad u_k v_k = \Delta/2E_k, \tag{3.60}$$

from which it becomes evident that the $\hat{H}_{\text{GC}}^{\mathcal{P}}$ takes on the form for *independent quasiparticles* (of Bogoliubov type), namely

$$\hat{H}_{\text{GC}}^{\mathcal{P}} = U_0^{\text{GC}} + {\sum_k}' E_k \left(\hat{\alpha}_k^\dagger \hat{\alpha}_k + \hat{\alpha}_{\bar{k}}^\dagger \hat{\alpha}_{\bar{k}}\right). \tag{3.61}$$

The ground state of such a system $|\text{vac}_{\text{qp}}\rangle$ is the vacuum of the quasiparticles defined by

$$\hat{\alpha}_{\bar{k}}|\text{vac}_{\text{qp}}\rangle = \hat{\alpha}_k|\text{vac}_{\text{qp}}\rangle = 0. \tag{3.62}$$

In terms of the particle's vacuum $|\text{vac}\rangle$ with $\hat{c}_{\bar{k}}|\text{vac}\rangle = \hat{c}_k|\text{vac}\rangle = 0$, the $|\text{vac}_{\text{qp}}\rangle$ can be constructed as the so-called BCS-state

$$|\text{vac}_{\text{qp}}\rangle = \prod_k \left(u_k + v_k \hat{c}_k^\dagger \hat{c}_{\bar{k}}^\dagger\right)|\text{vac}\rangle. \tag{3.63}$$

Applying any of the $\hat{\alpha}_k^\dagger$ to this ground state one gets a state of excitation energy E_k. This state consists of a superposition of states with odd particle numbers and hence describes a nucleus with an odd number of nucleons. For even particle numbers the excited states are of type $\hat{\alpha}_{k_1}^\dagger \hat{\alpha}_{k_2}^\dagger |\text{vac}_{\text{qp}}\rangle$ with the excitation energy of $E_{k_1} + E_{k_2}$ being larger than the minimal value 2Δ determined by the *gap parameter* Δ.

As yet the value of Δ is still unknown. Before we are going to deduce an equation from which it may be calculated we should like to remark that there always exists the unpaired or *normal solution* for $\Delta = 0$:

$$u_k v_k = 0 \quad \text{and} \quad \begin{cases} u_k^2 = (1/2)(1 + (e_k - \mu)/E_k) \equiv \Theta(e_k - \mu) \\ v_k^2 = (1/2)(1 - (e_k - \mu)/E_k) \equiv \Theta(\mu - e_k) \end{cases}. \quad (3.64)$$

In this case the state $|\text{vac}_{\text{qp}}\rangle$ of (3.63) turns into the ground state of a *normal* system of fermions for which for given particle number N only the N lowest single particle states are occupied, with the chemical potential μ being identical to the Fermi energy.

For the following derivation of the "gap equation" we may take advantage of the fact that the discussion presented so far is also valid at finite temperature. We only have to interpret the expectation value as that of the grand canonical ensemble to be performed with the density operator

$$\hat{\rho}_{\mathcal{P}}(\Delta, T, \mu) = \frac{1}{Z_{\mathcal{P}}} \exp(-\hat{H}_{\text{GC}}^{\mathcal{P}}/T) \quad \text{with} \quad Z_{\mathcal{P}} = \text{tr} \exp(-\hat{H}_{\text{GC}}^{\mathcal{P}}/T). \quad (3.65)$$

With respect to the variable Δ this density should be considered as the one in a quasi-static picture: For any Δ it represents a kind of "local" or partial equilibrium for the particles or quasiparticles. The "true" or global equilibrium is found after minimizing either the free energy $\mathcal{F}$ or the internal energy $\mathcal{E}$. Here we may take advantage of eqn(22.22) which states that the first derivatives of both quantities with respect to a parameter like Δ are identical and equal to the average *partial* derivative of the Hamiltonian to the same Δ. In the following we prefer to work with the concept of fixing temperature rather than entropy. Hence, the relevant relation is

$$\left(\frac{\partial \mathcal{F}(\Delta, T, \mu)}{\partial \Delta}\right)_{T,\mu} = \left\langle \frac{\partial \hat{H}_{\text{GC}}^{\mathcal{P}}}{\partial \Delta} \right\rangle. \quad (3.66)$$

From (3.52) together with (3.50) the derivative of the Hamiltonian is seen to be

$$\frac{\partial \hat{H}_{\text{GC}}^{\mathcal{P}}(\Delta)}{\partial \Delta} = -(\hat{P} + \hat{P}^\dagger) + \frac{2\Delta}{G}. \quad (3.67)$$

The equilibrium value $\Delta_{\text{eq}} = \Delta_{\text{eq}}(T, \mu)$ is determined by the extremum of the free energy and thus to be found from the equation

$$\langle \hat{P} \rangle_{\text{eq}} \equiv \langle \hat{P}^\dagger \rangle_{\text{eq}} = \Delta_{\text{eq}}/G \quad (3.68)$$

where the averages have to be calculated for fixed T and μ from the density operator $\hat{\rho}_{\mathcal{P}}(\Delta, T, \mu)$ of (3.65).

To render this a useful equation we must first evaluate the expectation values of the pairing fields. This can be done for any Δ by making use of a few simple relations which the reader may prove (see exercise 2). One finds

$$\langle \hat{P}^\dagger \rangle_{\text{eq}} = \sideset{}{'}\sum_k \langle \hat{c}_k^\dagger \hat{c}_{\bar{k}}^\dagger \rangle_{\text{eq}} = \sideset{}{'}\sum_k (\Delta/2E_k)(1 - 2n_k^{\text{qp}}) = \langle \hat{P} \rangle_{\text{eq}} \qquad (3.69)$$

with

$$n_k^{\text{qp}} = \langle \hat{\alpha}_k^\dagger \hat{\alpha}_k \rangle_{\text{eq}} = (1 + \exp(E_k/T))^{-1} \qquad (3.70)$$

being the average occupation number of the quasiparticles in state k. Inserting this into (3.68) one finds the *gap equation*

$$\frac{2}{G} = \sideset{}{'}\sum_k \left. \frac{1 - 2n_k^{\text{qp}}}{E_k} \right|_{\Delta_{\text{eq}}} = \sideset{}{'}\sum_k \left. \frac{\tanh(E_k/(2T))}{E_k} \right|_{\Delta_{\text{eq}}} \qquad (3.71)$$

as an implicit equation for $\Delta_{\text{eq}} = \Delta_{\text{eq}}(T, \mu)$, for given T and μ. It is almost evident that there must be a critical temperature $T_c^{\mathcal{P}}$ beyond which no solution of (3.71) exists. Indeed, because of the restriction of the sum through the "pairing window" the right-hand side becomes smaller and smaller with increasing T. This feature may easily be seen in the degenerate model (see exercise 2). Another possibility is to transform (3.71) to the continuum limit with a single particle level density $g(e)$; in (Siemens and Jensen, 1987) and (Ring and Schuck, 2000) this is done for the case of zero temperature. In (Fetter and Walecka, 1971) such a model is used to show that the critical temperature is of the same order as the gap parameter at $T = 0$, more precisely one finds $T_c^{\mathcal{P}} \simeq 0.57 \Delta(T = 0)$.

Finally, we shall return to the problem of particle number conservation within the BCS formulation. As mentioned before, the $\hat{H}_{\text{mf}}^{\mathcal{P}}$ does not commute with $\hat{N}$. Therefore, projection techniques are required if one wants to get wave functions or states with proper particle number. For zero thermal excitations this problem can be solved by different techniques (Ring and Schuck, 2000), but they generally fail at $T \neq 0$; for an extensive discussion with references to earlier work see (Rossignoli et al., 1995).

Exercises

1. Convince yourself about the estimate (3.20). To this end calculate the mean square radius $\langle r^2 \rangle$ for the oscillator model and compare it with the result for a homogeneous density inside the radius R; it is sufficient to consider only terms to leading order.
2. (A) For any Δ (i) prove relation (3.70) for the average occupation number of the quasiparticles in state k. (ii) Show that the corresponding occupation number of the particles becomes

$$n_k = \langle \hat{c}_k^\dagger \hat{c}_k \rangle_{\text{eq}} = v_k^2 + (u_k^2 - v_k^2) n_k^{\text{qp}} = \frac{1}{2}\left(1 - \frac{e_k - \mu}{E_k} \tanh \frac{E_k}{2T}\right),$$

and (iii) prove

$$\langle \hat{c}_k^\dagger \hat{c}_{\bar{k}}^\dagger \rangle_{eq} = u_k v_k (1 - 2n_k^{qp}) = (\Delta/2E_k)(1 - 2n_k^{qp}) = \langle \hat{c}_{\bar{k}} \hat{c}_k \rangle_{eq}.$$

(B) Solve the gap equation (3.71) for the model of γ degenerate states $|k\rangle$ with $e_k = \mu$. Calculate the critical temperature T_c^P and compare it with the value of the gap parameter $\Delta(T=0)$ at zero temperature.

(C) Show that the grand potential grand potential is given by

$$\Omega = -T \ln Z_P = \sum_k{}' (e_k - \mu - E_k) - 2T \sum_k{}' \ln(1 + \exp(-E_k/T)) + \Delta^2/G.$$

(D) Calculate the particle number to

$$\mathcal{N} = -\frac{\partial \Omega}{\partial \mu} = 2 \sum_k{}' n_k = \sum_k{}' \left(1 - \frac{e_k - \mu}{E_k} \tanh \frac{E_k}{2T}\right).$$

Why is it not possible to apply formula (22.44) for the density operator (3.65) to get the relative fluctuation $\Delta \mathcal{N}/\mathcal{N}$ within this approximation?

4

FROM THE SHELL MODEL TO THE COMPOUND NUCLEUS

In this chapter we are going to study nuclear properties which are related to the presence of a two-body interaction $\hat{V}_{\text{res}}$ beyond the mean-field approximation. As we shall be discussing aspects of more generic nature the precise form of this $\hat{V}_{\text{res}}$ will not be very important. This is so in particular when we will be dealing with statistical aspects. Such concepts are unavoidable as soon as the level density has grown to values where it does not make sense to follow the excitations individually. As we may anticipate from the discussion in Section 10.1 this already happens at excitation energies $\mathcal{E}^* = E - E_0$ of a few MeV only.

4.1 Shell model with residual interactions

With $\hat{V} = \hat{V}_{\text{res}}$ the general many-body Hamiltonian may be written in the form

$$\hat{H} = \hat{H}_0 + \hat{V} \quad \text{with} \quad \hat{H}_0 = \hat{H}_{\text{ipm}} = \sum_k^A \hat{h}(\mathbf{r}_k, \mathbf{p}_k) \tag{4.1}$$

representing independent particle motion. The solution of the Schrödinger equation

$$\hat{H}|\Psi_i\rangle = E_i|\Psi_i\rangle. \tag{4.2}$$

may be traced back to the problem of diagonalizing the matrix $H_{nn'} \equiv \langle n|H|n'\rangle$ calculated with some model states $|n\rangle$. In practice the latter may be chosen as the solutions of the simpler problem $\hat{H}_0|n\rangle = E_n^0|n\rangle$. The matrix form of eqn(4.2) then may be written as

$$\sum_{n'} \langle n|(\hat{H} - E_i)|n'\rangle\langle n'|\Psi_i\rangle = 0 \tag{4.3}$$

with the eigenvalues E_i being determined by the secular equation

$$\det\left[\hat{H}_{nn'} - E_i \delta_{nn'}\right] = 0. \tag{4.4}$$

The feasibility of this problem depends very much on the dimension N of the model space. With present-day computers very large matrices can be diagonalized and meaningful results may be obtained for not too heavy nuclei. But as indicated above we will mostly be interested in statistical properties and thus not dwell any further on such numerical calculations. At first we want to get some information on the spacing of the levels E_i.

4.1.1 Nearest level spacing

Suppose the levels E_i to be given and ordered in a sequence like $E_1 \leq E_2 \leq \ldots$. Their distance shall be called $S_i = E_{i+1} - E_i$ with the average spacing being D. We ask for the probability density function $P(S)$, with $P(S)dS$ being the probability that any S_i lies between S and $S + dS$.

Let us consider first the case where the levels are *not correlated*. This means supposing that the probability of any of the E_i to be found in the interval $(E, E + dE)$ is *independent* of E_i and simply given by dE/D. To find the $P(S)$ we first look for the probability of having no level in the interval $(E, E+S)$ and one level in the interval $(E+S, E+S+dS)$. To this end the interval of length S will be partitioned into m equal parts, from E to $E+S/m$ to $E+2S/m$ and so forth. As the levels are independent, the probability $Q_{E,E+S}$ of having *no* level in $(E, E+S)$ is the product of the probabilities of having no level in any of these parts. For large m this means writing

$$Q_{E,E+S} = (1 - S(Dm))^m \quad \text{implying} \quad \lim_{m \to \infty} Q_{E,E+S} = \exp(-S/D); \quad (4.5)$$

To calculate the probability that there is no level in $(E, E+S)$ and one level in $(E+S, E+S+dS)$ we only need to multiply $Q_{E,E+S}$ by dS/D, which leads to the *Poisson* distribution

$$P(S)dS = \exp(-S/D)dS/D, \quad (4.6)$$

relevant for random, uncorrelated levels.

Such a situation is given for the many-body states of the independent particle model, for which the levels are many-particle many-hole states. After ordering them successively the chance that level i is of the same complexity as level $i+1$ is practically zero, in the sense that both levels would consist of the same number of holes and particles. Therefore, the neighboring levels will be statistically *in*dependent, and, hence, the conditions for the Poisson distribution are given. Notice please that this argument holds true no matter what the nature of the single particle motion is. It is known, however, that rule (4.6) does *not* apply to measured nuclear levels of the same spin (or total angular momentum) and parity, at least not above some finite but comparatively small excitation. Information about this question has been gained from the analysis of neutron s-wave resonances already in the early days of nuclear physics, recall the discussion in Section 1.3 including Fig. 1.2. In the 1950s E.P. Wigner suggested classifyng the level spacing by the distribution

$$P(S) = (\pi S/(2D^2)) \exp(-\pi S^2/(4D^2)) \quad (4.7)$$

now named after him. As in the early days experimental evidence was scarce, this suggestion is sometimes referred to as the "Wigner surmise". In the meantime this hypothesis has been verified to a very high degree. A closer look indicates that the form (4.7) and, hence, the distribution of *nuclear levels* is governed by stochastic features. Indeed, as we are going to demonstrate now, the form (4.7) may be derived by treating the matrices of the residual interaction, or those of the total Hamiltonian, as stochastic variables.

4.2 Random Matrix Model

This model was proposed by E.P. Wigner (1958) as a possible method to describe the spacings of nuclear levels. In the meantime this concept has very successfully been applied to many other physical problems. We will confine ourselves discussing the nuclear level problem in a more or less schematic way. Elaborate expositions of the method are given in the book by M.L. Mehta (Mehta, 2004) or in the review article by T. Brody *et al.* (1981). For a survey of many modern applications we recommend the review article by T. Guhr *et al.* (1998).

The RMM may be used to simulate solutions of the Schrödinger equation (4.2). According to Wigner it may not be necessary actually to diagonalize the matrix $\langle n|H|n'\rangle$. After all, one is not only confronted with a tremendous numerical problem, the Hamiltonian and the residual interaction, in particular, are not very well known either. It was for this reason, in particular, that Wigner suggested treating the $\langle n|H|n'\rangle$ as *stochastic variables*. No detailed information about these matrices would then be necessary to understand the spacing distribution. It might suffice to know how they behave under symmetry transformations. The form (4.7), for instance, comes out if the Hamiltonian matrix is invariant under *orthogonal transformations*. Indeed, the latter is guaranteed if the system is *time-reversal invariant and invariant under rotations*. In this case the matrix $H_{nn'}$ can be chosen to be *real symmetric*: $H_{nn'} = H_{n'n}$. If *time-reversal invariance is broken* the matrix $H_{nn'}$ will be complex Hermitian, irrespective of its behavior under rotations. The corresponding transformation group then is that of *unitary transformations*. Actually, there is one and only one more class of symmetry transformations, called the *symplectic group*. In the following we will look in some detail only at the first example, the "orthogonal group" which plays the most important role in nuclear physics. With respect to the others, as well as to the proofs of the statements made before concerning the reality or complexity of the matrix elements we like to refer to the literature (Haake, 1991), (Mehta, 2004). Fortunately, it turns out that the essential features can be understood on the basis of either 2×2 matrices in case of the orthogonal or unitary group, or a 4×4 matrix for the symplectic group. It may suffice to address in the following the orthogonal group, for which the matrix is symmetric implying $H_{12} = H_{21}$.

4.2.1 *Gaussian ensembles of real symmetric matrices*

Given the orthogonal transformation

$$O = \begin{pmatrix} \cos\Theta & \sin\Theta \\ -\sin\Theta & \cos\Theta \end{pmatrix} \qquad \text{where} \quad \widetilde{O} = O^{-1}, \qquad (4.8)$$

with $\widetilde{O}$ being the transposed matrix. Given further the probability density $P(H)$ for the distribution of the three independent elements H_{11}, H_{22}, H_{12}, which shall be normalized as $\int dH_{11} dH_{22} dH_{12} P(H) = 1$. Invariance with respect to this O implies that the $P(H)$ remains unchanged, which is to say $P(H) = P(H')$ for $H' = \tilde{O} H O$. The elements H_{ij} are assumed to be *statistically uncorrelated*, implying that $P(H)$ can be written in factorized form as $P(H) = P_{11}(H_{11}) P_{22}(H_{22}) P_{12}(H_{12})$. Besides the requirement of invariance this is the only other condition which one needs to impose! It may be interpreted as an assumption of maximum disorder. For an infinitesimal transformation with $\cos\Theta \approx 1$ and $\sin\Theta \approx \Theta$ the H'_{ij} become

$$H'_{11} = H_{11} - 2\Theta H_{12}, \quad H'_{22} = H_{22} + 2\Theta H_{12}, \quad H'_{12} = H_{12} + \Theta(H_{11} - H_{22}). \tag{4.9}$$

This may also be used to treat the invariance property to first order. For the factorized form one gets, after using the abbreviations $H_{ik} = H_\mu$ and $P_{ik} = P_\mu$,

$$P(H) = P(H') = P(H)\left(1 + \sum_\mu \frac{\partial \ln P_\mu}{\partial H_\mu} \Delta H_\mu\right). \tag{4.10}$$

Inserting (4.9) this reduces to

$$P(H) = P(H)\left(1 - \Theta\left[2H_{12}\frac{d\ln P_{11}}{dH_{11}} - 2H_{12}\frac{d\ln P_{22}}{dH_{22}} - (H_{11} - H_{22})\frac{d\ln P_{12}}{dH_{12}}\right]\right). \tag{4.11}$$

Evidently, the expression in square brackets must vanish. The resulting differential equation can be treated by separating variables, which implies

$$\frac{1}{H_{12}}\frac{d\ln P_{12}}{dH_{12}} = \frac{2}{H_{11} - H_{22}}\left(\frac{d\ln P_{11}}{dH_{11}} - \frac{d\ln P_{22}}{dH_{22}}\right) = -4A = \text{const}, \tag{4.12}$$

with the solution $\ln P_{12} = -2A H_{12}^2 + \ln C_{12}$. Likewise one has

$$\frac{d\ln P_{11}}{dH_{11}} + 2A H_{11} = \frac{d\ln P_{22}}{dH_{22}} + 2A H_{22} = -B = \text{const}$$

and, hence, $\ln P_{11} = -A H_{11}^2 - B H_{11} + \ln C_{11}$, with a similar result for $\ln P_{22}$. Putting things together one thus gets

$$P(H) = C \exp\left[-A(H_{11}^2 + H_{22}^2 + 2H_{12}^2) - B(H_{11} + H_{22})\right]. \tag{4.13}$$

Altogether, there are three integration constants, namely A, B, C. Actually, the B can be made to vanish by a clever choice of the zero of energy. Thus we may write without loss of generality

$$P(H) = C \exp(-A \operatorname{tr}\{H^2\}). \tag{4.14}$$

The constant A defines the unit of energy and C may be fixed by normalization.

Obviously, the distribution $P(H)$ is Gaussian. The set of such matrices which are invariant under *orthogonal* transformations is called the *Gaussian orthogonal ensemble*, in short GOE. The analogous set being invariant under *unitary transformations* is referred to as *Gaussian unitary ensemble GUE*. In this case the non-diagonal matrix elements are not real such that all four elements H_{11}, H_{22}, H_{12} and H_{21} of the 2×2 matrix are to be treated as independent. Actually, there is a third ensemble, the GSE which is invariant under the *symplectic group*, for which the smallest Hilbert space is four dimensional.

4.2.2 Eigenvalues, level spacings and eigenvectors

The eigenvalues are found by diagonalizing the matrix H_{ij} to become

$$E_\pm = \frac{1}{2}\left\{(H_{11} + H_{22}) \pm \left([H_{11} - H_{22}]^2 + 4H_{12}^2\right)^{1/2}\right\}. \quad (4.15)$$

To get their distribution we simply need to invoke the transformation (4.8) to relate the H_{ik} to the $E_\pm$ by

$$H = \tilde{O}\begin{pmatrix} E_+ & 0 \\ 0 & E_- \end{pmatrix} O, \quad (4.16)$$

which leads to the following set of equations:

$$\begin{aligned} H_{11} &= E_+ \cos^2\Theta + E_- \sin^2\Theta \\ H_{22} &= E_+ \sin^2\Theta + E_- \cos^2\Theta \\ H_{12} &= (E_+ - E_-)\cos\Theta\sin\Theta. \end{aligned} \quad (4.17)$$

These relations may be understood as a transformation from the three independent variables H_{11}, H_{22}, H_{12} to the $E_\pm$ and Θ. The associated Jacobian can be seen to attain the value

$$J = \det\frac{\partial(H_{11}, H_{22}, H_{12})}{\partial(E_+, E_-, \Theta)} = E_+ - E_-. \quad (4.18)$$

It is needed for calculating the distribution $P(E_+, E_+, \Theta)$ through the relation

$$dH_{11}dH_{22}dH_{12} = \left|\det\frac{\partial(H_{11}, H_{22}, H_{12})}{\partial(E_+, E_-, \Theta)}\right| dE_+ dE_- d\Theta \quad (4.19)$$

between the volume elements. Because of the invariance of the trace with respect to orthogonal transformations one has $\mathrm{tr}\{H^2\} = E_+^2 + E_-^2$ such that the $P(H)$ of (4.14) turns into

$$P(E_+, E_-, \Theta) = C\,|E_+ - E_-|\,e^{-A(E_+^2 + E_-^2)}. \quad (4.20)$$

It is seen to be independent of the angle Θ, which clearly is a consequence of the invariance with respect to rotations. Notice, please, that for our notation $|E_+ - E_-| = E_+ - E_-$. The form (4.20) may be generalized to the other groups, as well as to N dimensions. One still finds the Gaussian form, which is to be multiplied by different powers $(n-1)$ in the energy differences (see e.g. (Haake, 1991))

$$P(E) = \text{const} \prod_{\mu<\nu}^{1...N} |E_\mu - E_\nu|^{n-1} \exp\left(-A \sum_{\mu=1}^{N} E_\mu^2\right), \qquad (4.21)$$

where the so-called "co-dimension" n is given by $n = 2, 3$ or 5 for the GOE-, GUE- or GSE-ensembles, respectively.

Next we turn to the level spacing, but concentrate again to the GOE. In this case the distribution $P(S)$ can be obtained from (4.20) as

$$P(S) = C \int_0^\pi d\Theta \int_{-\infty}^\infty dE_+ \int_{-\infty}^\infty dE_- \delta(S - |E_+ - E_-|) |E_+ - E_-| e^{-A(E_+^2 + E_-^2)}. \qquad (4.22)$$

Both the C as well as the A may be fixed by requiring normalization according to $\int_0^\infty dS\, P(S) = 1$ and by identifying the first moment as the average level distance D

$$\langle \hat{S} \rangle = \int_0^\infty dS\, S P(S) = D. \qquad (4.23)$$

In this way one finds

$$P(S) = \frac{S\pi}{2D^2} \exp\left(-\frac{\pi}{4}\frac{S^2}{D^2}\right) \qquad \text{for the GOE} \qquad (4.24)$$

in agreement with (4.7). The distributions for the other two ensembles can be seen to become

$$P(s) = \begin{cases} \frac{s^2 32}{\pi^2} \exp\left(-s^2 4/\pi\right) & \text{GUE} \\ \frac{s^4 2^{18}}{3^6 \pi^3} \exp\left(-s^2 64/9\pi\right) & \text{GSE}, \end{cases} \qquad (4.25)$$

if written for a $P(s)$ with $s = S/D$ and $P(s)ds = P(S)dS$.

In the limit $N \to \infty$ the pre-factors of these distributions, which define the magnitude of the first non-vanishing derivatives at $s = 0$, differ only very little from the values given in (4.24) and (4.25): For the GOE (GUE) this factor changes from $\pi/2$ to $\pi^2/6$ ($32/\pi^2$ to $\pi^2/3$), which is larger by a factor of 1.047 (1.015). For the GSE the corresponding factor is smaller than one, namely 0.995 (see Table 2 of (Bohigas, 1991)). In the nuclear case, and for atomic and molecular systems alike, the limit $N \to \infty$ does not apply, for instance when one has to look at nucleons in some given shell, see (Benet et al., 2001), (Pluhař and Weidenmüller, 2002) and references therein to earlier work. In this case only a finite number of available levels is given, which according to J.B. French can be described by the so-called *embedded ensembles*. One should note that a behavior like the one just discussed for the nearest level spacing becomes generic *only* for $N \to \infty$. Besides the problem of the available finite number of states there is another one resulting from the fact that the interaction of the particles is only of two-body nature. This implies that there are far fewer independent two-body matrix elements than the number of random variables needed in the canonical Random Matrix Theory.

Finally, we shall study the distribution of the eigenvectors. Suppose that in some basis their amplitudes are a_1 and a_2, looking at the case of $N = 2$ again. They are given by an orthogonal transformation between the eigenvectors and this basis. Because of the rotational invariance all angles Θ between 0 and π are equally likely. Accounting further for the normalization, the distribution of the amplitudes must read

$$P(a_1, a_2) = \frac{1}{\pi} \delta\left(1 - a_1^2 - a_2^2\right). \tag{4.26}$$

Suppose one is interested in the amplitude a_1 associated with one particular basis state. Its probability distribution then is given by

$$P(a_1, a_2) = \int d a_2 \, P(a_1) = \frac{1}{\pi} \left(1 - a_1^2\right)^{-1/2}. \tag{4.27}$$

The final result follows from elementary properties of the δ-function.

Again we save ourselves from deriving the result for dimensions $N > 2$, which involves a discussion of the solid angle Ω_N in such spaces. For any one amplitude a_μ one finds (Bohr and Mottelson, 1969), (Feshbach, 1992)

$$P(a_\mu) = \frac{\Omega_{N-1}}{\Omega_N} \left(1 - a_\mu^2\right)^{(N-3)/2} \quad \text{with} \quad \Omega_N = \frac{2\pi^{N/2}}{\Gamma(N/2)}, \tag{4.28}$$

which involves the Γ-function. In the limit of $N \gg 1$ and for $a_\mu^2 \ll 1$ this turns into

$$P(a_\mu) = \left(\frac{N}{2\pi}\right)^{1/2} \exp\left(-\frac{N}{2} a_\mu^2\right). \tag{4.29}$$

4.2.3 Comments on the RMM

4.2.3.1 Comparison with experimental data
For the resonances seen in the scattering of slow nucleons on a target mentioned in Section 4.1.1 the "Wigner surmise" may be considered proven fact–provided one takes levels of identical quantum numbers for spin and parity at not too small excitations (Guhr et al., 1998). An extensive study has been performed by R.U. Haq, A. Pandey and O. Bohigas in 1982 (see (Bohigas et al., 1983) with references to earlier work) in which altogether 1726 nuclear spacings have been considered from data then available. In Fig. 4.1 the histogram of the distribution is shown together with the Poisson and the Wigner distribution.

In the meantime other tests have been performed with levels corresponding to nuclear states of different nature and in different (low) energy regimes. At very low energies, where the spectrum is influenced by collective effects, one may see other forms of the distribution which may be described as mixtures between the ones of Poisson and Wigner. This problem is still under examination; for a summary see e.g. (Guhr et al., 1998).

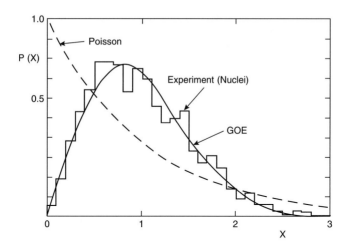

FIG. 4.1. The nearest neighbor spacing of nuclear levels according to R.U. Haq, A. Pandey and O. Bohigas (Bohigas et al., 1983).

4.2.3.2 *Chaotic behavior of complex systems* The RMM has been applied extensively in studies which address the nature of quantum chaos. This is largely due to the so-called "Bohigas conjecture" which relates spectra of time-reversal invariant systems showing the GOE distribution to chaotic behavior of the analogous classical system. Mostly these studies concentrate on systems having only very few degrees of freedom, unlike those given by heavier nuclei which define typical many-body systems. Prime examples consist of the so-called stadiums, which imply *free motion of one* particle with *specular reflection* at some given boundary. This boundary is determined by a two-dimensional surface whose shape is similar to a stadium. Let us mention two of them: (i) The "Bunimovich stadium" consists of two half-circles set apart by a rectangle (with two opposing sides being replaced by the half-circles), (ii) the "Sinai billiard" which consists of a square with a circle of smaller radius centered in the middle, and motion being permitted only in the space between the circle and the square. In both cases the classical motion shows chaotic behavior:[7] Trajectories which start with different but arbitrarily close initial conditions will fly apart to finally fill the total phase space allowed by the boundary conditions, except for certain "islands" perhaps (where regular orbits may exist, see below). This behavior is to be distinguished from a regular one which is dominated by closed orbits in phase space (resembling "Lissajous figures"). A circular or elliptic stadium would show this type of *integrable* motion. For such systems the Schrödinger equation (for a free particle inside an infinitely deep wall of the shape of the stadium) has been solved and the levels have been analyzed in their spacings. The result obtained by O.

---

[7]Actually, to avoid any symmetries the motion has eventually to be restricted to a subspace of the full billiard.

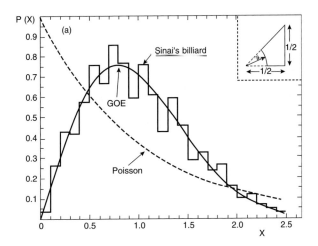

FIG. 4.2. The nearest neighbor spacing of levels for the Sinai billiard, according to O. Bohigas, M.-J. Giannoni and C. Schmit (Bohigas et al., 1984).

Bohigas, M.-J. Giannoni and C. Schmit (Bohigas et al., 1984) is shown in Fig. 4.2 for the Sinai billiard. We should like to stress that in these calculations the system's Hamiltonian is *fixed*. It is due to the inherent structure of the problem that the level spacings follow the Wigner law. To the best of my knowledge a formal proof of the Bohigas conjecture does not exist, not even for systems of low dimensionality; a discussion of this problem may be found in (Guhr et al., 1998). For *many-body systems* different features come into play, as there may be regularities even in cases that the underlying one-body motion is chaotic. As discussed previously, the level statistics of the nuclear many-body system follows the Poisson distribution if treated within the picture of independent particles. Whereas the latter may deliver an adequate description of nuclear dynamics at very small excitation energies it obviously fails already at the excitation brought into the nucleus by a captured neutron of practically zero initial kinetic energy. This observation is very important as it is a clear hint on the limitation of the independent particle model.

4.3 The spreading of states into more complicated configurations

We return to the general problem of getting information on the solution of the Schrödinger equation (4.2) of the full problem, $\hat{H}|\Psi_i\rangle = E_i|\Psi_i\rangle$ with $\hat{H} = \hat{H}_0 + \hat{V}$. Unlike the RMM, here all terms of the Hamiltonian are supposed to be known operators with their matrix elements being no stochastic variables. In this spirit we shall aim at finding detailed properties of the system, not only those of more inclusive nature like the density of levels or the distribution of their spacings. In particular, we will be interested in studying the influence of the interaction $\hat{V}$ on a particular state. This will be done within a schematic but realistic model widely used in the literature.

4.3.1 A schematic model

To pursue the diagonalization procedure outlined at the beginning of Section 4.1 we restrict ourselves to a situation where the set of unperturbed states $|n\rangle$ (with $n = 0, 1, 2...$) is split into one particular state $|a\rangle$ and the remainder, viz

$$\hat{H}_0|a\rangle = E_a^0|a\rangle; \qquad \hat{H}_0|m\rangle = E_m^0|m\rangle \qquad \text{with} \qquad m = 1\ldots N. \qquad (4.30)$$

The matrix elements $\langle n|V|n'\rangle$ of the interaction V will be assumed to be given by

$$\langle a|V|m\rangle = \langle m|V|a\rangle = V_{am}, \qquad \langle a|V|a\rangle = \langle m|V|m'\rangle = 0. \qquad (4.31)$$

Hence, the determinant (4.4) reduces to

$$\det\left[\hat{H}_{nn'} - E_i \delta_{nn'}\right] = \begin{vmatrix} E_a^0 - E_i & V_{an_1} & V_{an_2} & \cdots & V_{an_N} \\ V_{an_1} & E_{n_1}^0 - E_i & 0 & \cdots & 0 \\ V_{an_2} & 0 & E_{n_2}^0 - E_i & \cdots & 0 \\ \vdots & \vdots & \vdots & \ddots & \vdots \\ V_{an_N} & 0 & 0 & \cdots & E_{n_N}^0 - E_i \end{vmatrix}. \qquad (4.32)$$

It may be modified to triangle form with the first diagonal element $\mathcal{D}_{aa}$ being given by

$$-\mathcal{D}_{aa}(E_i) = \mathcal{D}(E_i) = E_i - E_a^0 - \sum_m \frac{V_{am}^2}{E_i - E_m^0} \qquad (4.33)$$

and the remaining ones unchanged, such that the determinant becomes

$$\det\left[\hat{H}_{nn'} - E_i \delta_{nn'}\right] = -\mathcal{D}(E_i) \prod_m \left(E_m^0 - E_i\right).$$

As we expect $E_m^0 \neq E_i$, the secular equation reduces to

$$\mathcal{D}(E_i) = 0 \qquad \text{or} \qquad E_i - E_a^0 = \sum_m \frac{V_{am}^2}{E_i - E_m^0}. \qquad (4.34)$$

Expanding the state $|\Psi_i\rangle$ into the basis states,

$$|\Psi_i\rangle = c_a(i)|a\rangle + \sum_m c_m(i)|m\rangle, \qquad (4.35)$$

the coefficients are easily seen to fulfill the relations

$$c_m(i) = -\frac{V_{am}}{E_m^0 - E_i} c_a(i) \qquad (4.36)$$

by noting that the orthogonality relations of the $|n\rangle$, together with (4.31) imply $\langle m|H|a\rangle = V_{am}$ and $\langle m|H|m'\rangle = \delta_{mm'} E_m^0$. From the normalization condition $\langle \Psi_i|\Psi_i\rangle = 1$ one finds

$$c_a(i) = \left(1 + \sum_m \frac{V_{am}^2}{(E_m^0 - E_i)^2}\right)^{-1/2} \equiv \left[\frac{d}{dE}\mathcal{D}(E)\right]_{E=E_i}^{-1/2}. \quad (4.37)$$

The expression on the very right simply follows from the definition of $\mathcal{D}(E)$ in (4.33). Because of $c_a(i) \equiv \langle a|\Psi_i\rangle = \langle \Psi_i|a\rangle^*$ the (real) coefficients $c_a(i)$ also determine the expansion of the special state $|a\rangle$ into the eigenstates $|\Psi_i\rangle$ of the Hamiltonian $\hat{H}$,

$$|a\rangle = \sum_i |\Psi_i\rangle\langle\Psi_i|a\rangle = \sum_i c_a(i)|\Psi_i\rangle. \quad (4.38)$$

Hence, the square of the amplitude $c_a(i)$ measures the *strength* S_i with which the state $|a\rangle$ distributes or *spreads* over the exact states $|\Psi_i\rangle$. If the $|a\rangle$ represents a single particle configuration the $S_i = (c_a(i))^2$ is also called the *spectroscopic factor*. We will return to this example below but first want to continue with some general observations (see also exercise 1).

4.3.2 Strength functions

Often one is not interested in the variation of the strength S_i with energy in greater detail. One may then introduce an averaged quantity $\overline{\varrho_a}$ defined as

$$\overline{\varrho_a}(E) = \frac{\Delta}{2\pi} \sum_{i=0}^{N} \frac{S_i}{(E - E_i)^2 + (\Delta/2)^2}. \quad (4.39)$$

This *strength function* measures the probability per unit of energy to find parts of the state $|a\rangle$ at energy E. In order for (4.39) to represent a genuine average, the smoothing interval Δ must be larger than the mean level spacing $D(E)$ of the states E_m^0 (although later on we may still wish to choose a smaller value). For functions $G(E)$ which are analytic in the upper half plane a Lorentzian smoothing implies evaluating them in the upper half plane, $\overline{G}(E) = G(E + i\Delta/2)$ (see exercise 2). Please observe that the strength function is normalized to unity,

$$\int_{-\infty}^{\infty} dE\,\overline{\varrho_a}(E) = \sum_i S_i = 1. \quad (4.40)$$

In a strict sense the continuum representation of the S_i is given by the function $\overline{S_a}(E) = \overline{\varrho_a}(E)D(E)$ rather than by the $\overline{\varrho_a}$, as only then one may define the transition $\sum_i \to \int dE/D(E)$ properly.

As an intermediate step for our derivation we may benefit from calculating the inverse of the $\mathcal{D}(E)$. As the residues of $1/\mathcal{D}(E + i\epsilon)$ are given by (4.37) one may exploit a pole expansion to obtain[8]

$$\frac{1}{\mathcal{D}(E + i\epsilon)} = \sum_{i=0}^{N} \frac{S_i}{E - E_i + i\epsilon}. \quad (4.41)$$

---

[8] According to (Whittaker and Watson, 1992) (see Chapter VII.7.4) there could at most be an additional constant term which is ruled out as $1/\mathcal{D}(E + i\epsilon)$ vanishes for $E \to \infty$.

To get the averaged quantity one may then replace the ϵ by the $\Delta/2$. A simple calculation shows that the $\overline{\varrho_a}$ of (4.39) may be written as

$$\overline{\varrho_a}(E) = \frac{i}{2\pi}\left([\overline{\mathcal{D}(E)}]^{-1} - [\overline{\mathcal{D}^*(E)}]^{-1}\right). \tag{4.42}$$

This follows because of the simple identity

$$L(\omega;\gamma) = \frac{1}{2\pi}\frac{\gamma}{\omega^2 + (\gamma/2)^2} = \frac{1}{2\pi i}\left[\frac{1}{\omega - i\gamma/2} - \frac{1}{\omega + i\gamma/2}\right] \tag{4.43}$$

valid for the Lorentz function, which is normalized to unity as $\int d\omega L(\omega;\gamma) = 1$. To get the $\overline{\mathcal{D}}(E + i\Delta)$ we first evaluate the term on the right of (4.33) with the sum. It may simply be written as

$$\sum_m \frac{V_{am}^2}{E - E_m^0 + i\Delta/2} = \Delta E_a(E) - i\Gamma_a/2, \tag{4.44}$$

with

$$\Delta E_a(E) \equiv \sum_m \frac{V_{am}^2(E - E_m^0)}{(E - E_m^0)^2 + (\Delta/2)^2} \tag{4.45}$$

and

$$\Gamma_a(E) \equiv \Delta \sum_m \frac{V_{am}^2}{(E - E_m^0)^2 + (\Delta/2)^2} \tag{4.46}$$

The physical meaning of this shorthand notation becomes clear from the final result for the averaged strength, which reads

$$\overline{\varrho_a}(E) = \frac{1}{2\pi}\frac{\Gamma_a(E) + \Delta}{(E - E_a^0 - \Delta E_a(E))^2 + (1/4)(\Gamma_a(E) + \Delta)^2}. \tag{4.47}$$

The $\Delta E_a(E)$ represents the shift the unperturbed energy E_a^0 acquires because of the coupling and the $\Gamma_a(E)$ determines the *spreading width*, i.e. the width in energy over which the strength is distributed. It is worth determining the limits of these quantities for vanishing Δ, namely

$$\lim_{\Delta \to 0} \Delta E_a(E) = \sum_m V_{am}^2 \frac{\mathcal{P}}{E - E_m^0} \tag{4.48}$$

$$\lim_{\Delta \to 0} \Gamma_a = 2\pi \sum_m V_{am}^2 \delta(E - E_m^0). \tag{4.49}$$

In (4.48) the $\mathcal{P}$ stands for *principal value*. In this sum over discrete states it is written simply to indicate that E must not be equal to one of the E_m^0. It may be mentioned that in Section 18.2.4 these results are derived as an example of an application of Feshbach's projection technique. This has been done for the expressions which do not involve any smoothing; how the latter may be performed is described in Section 18.3.

Replacing these sums by appropriate integrals one gets

$$\Delta E_a(E) = \mathcal{P} \int_{E_1^0}^{E_N^0} \frac{dE'}{2\pi} \frac{\Gamma_a(E')}{E - E'} \quad \text{with} \quad \Gamma_a(E) = 2\pi \frac{v^2(E)}{D(E)}. \qquad (4.50)$$

Here, the $v^2(E)$ is meant to represent an average of the squared coupling matrix elements V_{am}^2. The expression on the left of (4.50) is written in the form of a Kramers–Kronig relation (details of which can be found in Section 23.3). In exercise (1) a schematic model is presented which allows one to study in analytic fashion many of the essential features of this discussion from above.

The model set up in Section 4.3.1 is of sufficiently general nature that it may be applied to various problems. For instance, the state $|a\rangle$ might represent some excitation of some given nucleus described in terms of particle hole excitations. This is to say, we might wish to study a situation where the initial state is the one where several (or many) nucleons have been removed from levels below the Fermi surface and put into those above. As we will see in the chapters to come, of particular interest are states built of a "coherent" sum of excitations of 1p-1h type such that many nucleons participate "collectively" (with certain phase relations): For instance, this will be the case for the so-called giant resonances. Then the strength distribution parameterizes how such a state "dissolves" into the many microscopic eigenstates of the system, and the width Γ represents the damping of this (eventually mostly harmonic) mode. We will study such types of excitations in great detail later, where we will actually develop a different "language" allowing us to elaborate on dissipative effects in more explicit fashion.

4.3.2.1 *Pole structure of the strength function* For the following it is instructive to analyze first the pole structure of the strength function (4.47). Depending on the functional form of both the shift $\Delta E_a(E)$ and the $\Gamma_a(E)$, it may be quite complex. To get the poles of $\overline{\varrho_a}(E)$, notice first that this function can also be written as a sum of two terms, very much as was done in (4.43) for the Lorentz function proper. Each one of them has poles at the positions $E_i^\pm$ determined from the secular equation

$$E_i^\pm = E_a^0 + \Delta E_a(E_i^\pm) \pm i \left(\Gamma_a(E_i^\pm) + \Delta\right)/2, \qquad (4.51)$$

the solution of which may not be easy at all. To simplify matters, let us assume that in this equation the imaginary part may be discarded, say in a sense of weak damping. In this way both values of $E_i^\pm$ coincide into one E_i. Eventually, it might even suffice to apply perturbation theory in writing $E_i \approx E_a^0 + \Delta E_a(E_a^0)$. In the strength function itself, on the other hand, the information about the damping of the poles will be taken into account. In accord with the approximation just described we may calculate it at the (fixed) energy E_i, which means replacing in (4.47) the $\Gamma_a(E)$ by $\Gamma_a(E_i)$. In its immediate neighborhood the contribution of this pole may then be described by a strict Lorentzian, with its energy denominator being quadratic in energy, multiplied by the "strength factor"

$$Z_i = \left(1 - \frac{d}{dE}\Delta E_a(E)\bigg|_{E=E_i}\right)^{-1}, \tag{4.52}$$

namely

$$\overline{\varrho_a}^{(i)}(E) = \frac{Z_i}{2\pi} \frac{\Gamma_i}{(E-E_i)^2 + (\Gamma_i/2)^2}, \tag{4.53}$$

where

$$\Gamma_i = Z_i \left(\Gamma_a(E_i) + \Delta\right). \tag{4.54}$$

This result follows after calculating the residues of the poles at $E_i \pm i\Gamma_a(E_i)/2$. Notice that the forms found above for the $\Delta E_a(E)$ (as e.g. in (4.45)) imply the Z_i to be positive. By simply integrating the function (4.53) it is seen that the overall strength this particular pole contributes is equal to Z_i. If *all* poles are well separated the Z_i may be identified with the S_i introduced previously. The real virtue of the manipulations just described shows up when poles lie densely in the range (E_1^0, E_N^0). One example of such a situation is the influence of complicated configurations on the nature of single particle excitations, which we are going to discuss below.

4.3.3 *Time-dependent description*

Next we want to examine how the coupling exhibits itself in a time-dependent picture. Suppose that at some initial time $t=0$ we prepare our system to be in the state $|a\rangle$, which we know is not an eigenstate of the correct Hamiltonian. However, as the latter determines time evolution, the amplitude of finding the system still in this state after time $t>0$ is given by

$$A_a(t) = \left\langle a \left| \exp\left\{-\frac{i}{\hbar}Ht\right\} \right| a \right\rangle = \sum_i S_i \exp\left\{-\frac{i}{\hbar}E_i t\right\}. \tag{4.55}$$

Here, the sum extends over the exact states of the system.

For a system of discrete unperturbed levels also the correct spectrum with energies E_i will have this property. This implies that in such a case the function given in (4.55) will be of *quasi-periodic* nature, meaning that it shows *recurrence* after some time τ_{rec}. The average level spacing of the E_i may be expected to be similar to the D of the unperturbed spacing. Therefore, the τ_{rec} will be approximately given by $\tau_{\text{rec}} \simeq 2\pi\hbar/D$ Any observation of the system over a finite period τ_{obs} is related to a finite uncertainty in energy. The latter may be identified by the averaging interval Δ introduced above. Genuine averaging requires this to be large compared to the level spacing. For the timescales involved this implies the observation time much smaller than the recurrence time,

$$\tau_{\text{obs}} \ll \tau_{\text{rec}} \quad \Longleftrightarrow \quad h/\tau_{\text{obs}} = \Delta \gg D \simeq \hbar/\tau_{rec}. \tag{4.56}$$

Under these conditions the function $A_a(t)$ of (4.55) may be calculated as $A_a(t) = \int dE\, \overline{\varrho_a}(E) \exp(-iEt/\hbar)$. This integral can easily be evaluated if the $\overline{\varrho_a}(E)$ is of Lorentzian (or Breit–Wigner) form (4.43). For (4.53), for instance, one would get (see exercise 2)

$$A_a^{(i)}(t) = Z_i \exp\{-(\Gamma_i/2 - iE_i)t/\hbar\}, \tag{4.57}$$

after integrating from $-\infty$ to ∞. This latter manipulation is justified if $E_i \gg \Gamma_i$. As can be seen, these assumptions naturally lead to the typical decaying behavior, which in practice may be valid for an arbitrary length of time provided the spectrum is sufficiently dense. In the present discussion this decay shows up for a given state, but there can be little doubt that it is intimately related to the relaxation of statistical density distributions, a phenomenon which is discussed further in Section 21.3. This feature shows up whenever one encounters coupling to a dense system of final states. We will return to this problem once more below.

4.3.4 Spectral functions for single particle motion

We are now going to apply this formalism to examine the nature of single particle excitations in real nuclei. It must be expected that correlations beyond those of the independent particle model will become important. They will manifest themselves in the *spectral function* $\overline{S}(E) = \overline{\varrho}D(E)$, as the strength function for such excitations is often called. In the following the simple state $|a\rangle$ will represent a single particle configuration and the $|m\rangle$ are supposed to be of more complicated nature. Let us suppose a situation where one particle of angular momentum j, m is added to a nucleus in its ground state $|\Phi_{I_0=0}\rangle$; which for the sake of simplicity is assumed to have a vanishing total spin I_0. The state $|a\rangle$ may then be written as $|a\rangle = \hat{a}^\dagger_{njm}|\Phi_{I_0=0}\rangle$, with the $\hat{a}^\dagger_{njm}$ being the operator for "creating" a particle in the state with quantum numbers njm on top of the many-body state $|\Phi_{I_0=0}\rangle$. The total angular momentum of such a system will be given by $I = j$. Assuming the interaction to conserve angular momentum the states $|\Psi_i\rangle$ must have $J_{\text{tot}} = j$.

If the independent particle model were correct, the $|\Phi_{I_0=0}\rangle$ and, hence, the $|a\rangle$ as well would be simple Slater determinants. Since in such a case the interaction V were to be put equal to zero, the strength distribution $P_a(E)$ given in (4.47) would reduce to a sum of δ-functions $\delta(E - E_i)$. Moreover, no shift in energy would be present and the E_i would simply be given by their original values ($E_a = E_a^0$ and $E_m = E_m^0$). These δ-functions would result from the quasi-Lorentzian of (4.53) for $\Gamma_i \to 0$, and with $Z_i = 1$. (This situation corresponds to a case where $\Delta \ll D$–with D being the spacing in the independent particle model–such that any smoothing becomes meaningless.)

4.3.4.1 *Particles and holes* It turns out that the property just described is more or less given in the valence shells around the Fermi energy. This situation may be simulated by assuming the energy E_a^0 to be much smaller than the energy E_1^0 of the first complicated state. The energy of the "dressed" state $|\Psi_a\rangle$ may then be estimated to

$$E_a \approx E_a^0 + \int_{E_1^0}^\infty \frac{dE'}{2\pi} \frac{\Gamma_a(E')}{E_a^0 - E'}. \qquad (4.58)$$

Notice that in the second term the $E = E_a$ has been approximated by the unperturbed one. As this term is negative the E_a is smaller than the E_a^0. From (4.46) the width Γ_a can be seen to be small such that one may still associate a δ-function to this state, as a limit of the function $\overline{\varrho}_a$ in case the Γ_a is small compared to the spacing to the next level. Remember, however, that its strength Z_a is less than unity, as follows from (4.52). The remaining strength is distributed over the other states. An excitation of this type is called a *single particle excitation*. Its energy may be deduced from the following relation

$$e_p(a) = E_a^{(A+1)} - E_0^{(A)}, \qquad (4.59)$$

where $E_a^{(A+1)}$ is the energy in the state $|\Psi_a\rangle$ for the $A + 1$ particle system and $E_0^{(A)}$ the ground state energy of the A particle system. The negative of $e_p(a)$ is the separation energy $s_p = -e_p(a)$ from that single particle state. For ^{209}Pb, for instance, the least bound neutron has a separation energy $s_p \simeq 3.94$ MeV.

Suppose now a particle is removed from the same "parent nucleus", creating a "hole" in the latter. The analog of the model just discussed would be the one where the E_a lies well above the interval (E_N^0, E_1^0), now supposedly ordered with the E_m^0 to decrease with m. Then both (4.34) or (4.50) tell us that the E_a is larger than the corresponding unperturbed energy E_a^0. By analogy with (4.59) the energy of a hole is given by

$$e_h(a) = E_0^{(A)} - E_a^{(A-1)} = -s_h.$$

For the example of ^{208}Pb the separation energy of the least bound nucleon is 7.37 MeV $\equiv -e_h$. One might call this the Fermi energy of ^{208}Pb. Sometimes for finite nuclei one adopts another definition (see e.g. (Mahaux and Sartor, 1991)) namely $e_F = (e_F^+ + e_F^-)/2$, where e_F^+ and e_F^- are the energies of the first particle and the first hole state (counting from top), respectively. For this example of ^{208}Pb this gives $e_F = -(7.35 + 3.94)/2 = -5.65$ MeV.

4.3.4.2 *Quasiparticles and quasiholes* The situation described above for single particle excitations changes as a function of E_a^0. The closer the E_a^0 comes to the interval (E_1^0, E_N^0) the smaller the Z_a will be, which is to say the more the distribution will be spread out. If the level E_a^0 lies within the "band" (E_1^0, E_N^0) one can no longer speak of individual excitations. Rather, the excitation spreads smoothly over a whole range of energies, which may be determined by the Γ_a given in (4.54) together with (4.52). The excitation itself is called a *quasiparticle state*, or that of *quasihole* if a hole state gets dressed by the analogous mechanism. By their very construction, associated with these quasiparticle or -hole states are the finite *spreading widths* which according to (4.57) imply a finite life time of this type of excitation.

4.3.4.3 *Contributions from decays into the continuum* So far the widths introduced in Section 4.3.2 resulted from a coupling of our simple state to the more complicated ones representing compound states. At the end of this section it was remarked that basic results such as (4.48) for the shift and (4.49) for the widths are in accord with those from nuclear reaction theory, as derived in Section 18.2.4 together with Section 18.3. Above some minimal energy, the "threshold energy", the state $|a\rangle$ may also couple directly to the continuum representing the physical process of a possible decay of this state by particle emission. Evidently, this mechanism must lead to an *increase of the state's overall width*. How the model of Section 4.3.1 may be extended to incorporate this situation may be seen in (Mahaux and Sartor, 1991). For our purpose it may suffice to draw on the analogy with the *doorway state mechanism* discussed in Section 1.3.1.3 and in more detail in Section 18.3.2. Indeed, provided the direct coupling of the compound states to the continuum can be considered small, the state $|a\rangle$ may be considered a "doorway state" for the particular decay process at stake. In a Breit–Wigner type formula such as the (4.47), to the $\Gamma_a(E)$ in the denominator, now called $\Gamma_a^\downarrow(E)$, one has to add the width for particle emission, then called $\Gamma_a^\uparrow(E)$, which is to say one should replace $\Gamma_a(E)$ by $\Gamma_a^\downarrow(E) + \Gamma_a^\uparrow(E)$. Actually, if more than one "decay channel" is open the $\Gamma_a^\uparrow(E)$ must be the sum over the corresponding partial widths.

4.3.4.4 *Theoretical predictions versus experimental evidence* Experimental information about the spectroscopic function can be obtained in "pick up" or "stripping reactions". In the first case one nucleon is removed by the projectile from the target nucleus thus creating a hole in the latter. In a stripping reaction a nucleon is added to the target populating particle states, for instance in a (d,p) reaction where a deuteron hits a nucleus and transfers its neutron to the latter. There is no space to describe these reactions in detail. The general theoretical apparatus is discussed in Chapter 18 but for practical applications we have to refer to textbooks like (Bohr and Mottelson, 1969), (Feshbach, 1992) or (Hodgson, 1994); see also (Mahaux and Sartor, 1991). Such experiments allow one to get not only the spectroscopic factors but also information about the widths, at least for states not too far away from the Fermi surface. Here, we like to concentrate on the widths Γ_{nlj}, which in Fig. 4.3 are shown for quasiparticles. These Γ_{nlj} correspond to the form (4.54) (calculated for $\Delta = 0$) with the i standing for the quantum numbers nlj of single particle motion in an optical potential. The latter is obtained in a microscopic approach involving the G-matrix; recall the discussion in Section 2.3. It is seen that the dependence on energy is repre-

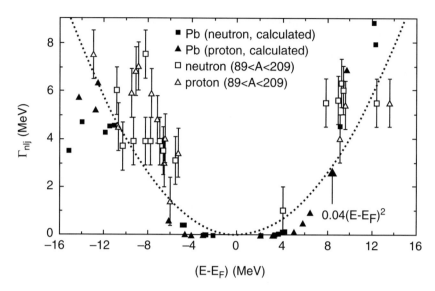

FIG. 4.3. Dependence of quasiparticle widths on energy. The open symbols represent empirical values. [Fig.7.20 of (Mahaux and Sartor, 1991)]

sented quite well by a quadratic form in the deviation from the Fermi energy. According to sections 2.7.1.3 and 21.3 such a dependence is suggested for infinite Fermi systems by the very form of the collision term in the transport equation for the one-body density. For finite systems deviations of this behavior are to be expected already by the presence of shell effects. From Fig. 4.3 it is seen that this quadratic behavior *overestimates* the correct values in the valence shells (up to about 8 MeV above and below the Fermi energy) but it *underestimates* them at larger deviations. This may be taken care of by parameterizing the functional dependence of the $\Gamma_a(E)$ on energy by more general functions. We will return to this question in Section 6.4.3 when setting up the microscopic damping mechanism for collective motion.

4.3.4.5 *Inferences on the validity of the independent particle model* The features just discussed indicate that the picture of independent particle motion is valid only in the valence shells below and above the Fermi level. Not only do such properties follow on theoretical grounds, they are confirmed by experimental findings. This fact casts much doubt on the use of Hartree–Fock calculations of the nucleus's total energy or of any other mean field approximation which discards couplings to more complicated states. On the other hand, it may be said that the Strutinsky method, to which we will come shortly, elegantly circumvents assumptions of this type.

Evidently the *correct* nuclear ground state $|\Psi_0\rangle$ *cannot be* an eigenstate of the Hamiltonian $\hat{H}_0$ representing independent particle motion, even if the latter had been determined through a Hartree–Fock procedure. The residual interaction must be expected to destroy the single particle motion as given by the picture of independent particles. Quite recently one has been able to even see this effect experimentally by electron scattering. Modern techniques allow one to measure the density distribution in the nuclear interior to high precision. Measuring differences in the densities of neighboring nuclei, such as ^{206}Pb and ^{205}Tl, give information about specific orbits; for the case mentioned it is the orbit for a proton in a 3s configuration. Such experiments, in combination with theoretical considerations, lead to the conclusion that the particles spend only about 2/3 of their time in states which correspond to independent particle motion, i.e. in states either found in Hartree–Fock calculations or in those of a shell model without residual interaction. For more details see the article by V.R. Pandharipande, I. Sick and P.K.A. de Witt Huberts in (1997).

Exercises

1. Assume that the levels are equidistantly spaced $E_m^0 = mD$ (for $m = 0, \pm1, \pm2, \ldots$) and that all non-vanishing matrix elements are equal to a given constant $V_{am} = v > 0$.

   (i) Show that the secular equation (4.34) can be brought to the form
   $$E_a^0 - E_i = -\frac{\pi v^2}{D} \cot \frac{\pi E_i}{D},$$

   (ii) and that the amplitude $c_a(i)$ turns into
   $$c_a(i) = \left[1 + \left(\frac{\pi v}{D}\right)^2 + \frac{(E_a^0 - E_i)^2}{v^2}\right]^{-1/2}.$$

   (iii) For $v > D$ deduce the *Breit–Wigner* form
   $$\overline{\varrho_a}(E) = \frac{1}{D} c_a^2(E_i \approx E) = \frac{1}{2\pi} \frac{\Gamma}{(E_a^0 - E)^2 + (\Gamma/2)^2},$$
   with the width $\Gamma = 2\pi v^2/D$.

2. (i) Lorentzian smoothing: Prove the validity of the relation
   $$\overline{f}(E) \equiv \frac{1}{2\pi} \int_{-\infty}^{\infty} f(z) \frac{\Gamma}{(z-E)^2 + (\Gamma/2)^2} = f(E + i\Delta/2) \qquad (4.60)$$
   for any function $f(z)$ which is analytic for $\operatorname{Im} z > 0$ and which for $|z| \to \infty$ drops to zero at least like $1/|z|$.

   (ii) Prove (4.57) by applying the residue theorem.

5

SHELL EFFECTS AND STRUTINSKY RENORMALIZATION

Nuclei are particularly stable when their numbers of neutrons or protons attain special values. This feature, originally recognized for spherical configurations, also holds true for deformed nuclei and is reflected in the total static energy. At a time when Hartree–Fock calculations were still neither numerically feasible nor could they be based on a promising "effective" interaction, V.M. Strutinsky suggested a method which allowed one to understand these facts microscopically (Strutinsky, 1967), see also (Brack et al., 1972). This procedure combines in an astute way essentials of the independent particle model with those of the liquid drop model, both being extended to finite deformations. On account of our discussions in the previous chapters the reader may perhaps be prepared to accept such a point of view, and we shall dwell further on it below.

In recent times the application of this method has become rather unfashionable. To a large extent this is due to the existence of effective interactions whose forms have been developed to such a degree of sophistication that they allow one to apply the Hartree–Fock method at a quantitative level. As we have seen in Section 2.5 these interactions may be traced back to Brueckner theory at the level of a local density approximation. Moreover, there is the approach of relativistic mean field theory (Serot and Walecka, 1986). This is based on the hypothesis that the form of Lagrangians, which are known to describe well the exchange of mesons between free nucleons, can be taken over to nuclear many-body problems merely by modifying the values of their parameters. Despite the obvious success of these approaches, we feel it worthwhile to discuss in greater detail the Strutinsky method, for the following reasons.

(1) In principle, the procedure is more general than the Hartree–Fock approximation. Whereas the latter presupposes independent particle motion for *all states*, for the *Strutinsky procedure such a hypothesis is necessary only close to the Fermi energy*. Indeed, recalling the discussion in Section 4.3.4 this assumption concurs very well with experimental facts.

(2) Already for this reason such a picture may serve better as a good starting point for a nuclear transport theory in which there is room for dissipation. In a strict sense the latter comes up only due to effects beyond the independent particle model.

(3) The Strutinsky method allows for a simple physical understanding of the key points of the *shell corrections*. This is true not only for their variation with deformation but also with respect to their disappearance with increasing thermal excitation.

(4) An understanding of these aspects opens the possibility for deeper insight

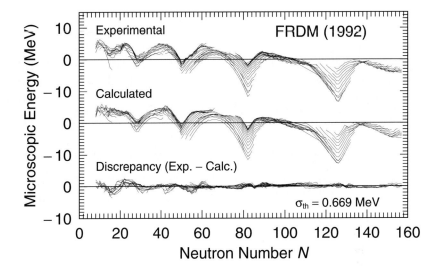

FIG. 5.1. The deviation of nuclear masses from their macroscopic values in the finite-range liquid droplet model, see the text [from the 1992 version of the model of the Berkeley–Los Alamos group (Möller et al., 1995)].

into the question of the macroscopic limit. Put differently, one may ask to what extent properties of an atomic nucleus can be attributed to a drop of nuclear matter?

(5) In fact, some of the most precise computations of nuclear masses have been performed on this combination of *macroscopic and microscopic* pictures. The level which has been achieved in applications of the models developed and applied by P. Möller, J.R. Nix , W. D. Myers and W. J. Swiatecki is demonstrated in Fig. 5.1. There, the deviation from the macroscopic limit of the total mass of nuclei is plotted, called "microscopic energy". The "macroscopic energy" is defined through a refined version of the Bethe–Weizsäcker formula (1.65) which leads to the "finite range droplet model" (FRDM), for more details see the discussion in Section 5.3. The figure presents the experimental values, together with the theoretical predictions and the differences between both. The lines correspond to identical proton numbers Z. It is seen that for heavier nuclei the error is less than 0.5 MeV.

(6) In Section 17.2 Strutinsky smoothing will be applied to response functions which in turn allow one to deduce dynamical properties and, hence, to shed light on the transition to the macroscopic limit of collective dynamics.

(7) The shell correction associated with motion in a single particle potential can also be understood in terms of "periodic orbit theory". Nowadays the latter is

often exploited to understand generic properties of complex, mesoscopic systems, last but not least when one wants to understand relationships between classical and quantum chaos. Recently it has been argued by O. Bohigas and P. P. Leboeuf (Bohigas and Leboeuf, 2002) that the remaining uncertainties in the mass-decrement seen in Fig. 5.1 may perhaps be attributed to the contribution from chaotic orbits.

(8) In the past one to two decades the shell correction method has been applied to metal clusters. This example, which we will address in Section 16.1, allows one to see the effects of so-called "super shells", which are predicted by the Strutinsky procedure but for which nuclei are simply too small to become manifest.

Before we elaborate on the most basic features it may be mentioned that many excellent review articles on the shell correction method exist (see the classic paper (Brack et al., 1972), as well as (Brack, 1992)) or textbooks like (Bohr and Mottelson, 1975), (Siemens and Jensen, 1987), (Brack and Bhaduri, 1997)).

5.1 Physical background

The basic issue of the whole Strutinsky method is the concept of separating smooth components from fluctuating ones. One of the clearest examples for demonstrating its usefulness has already been mentioned: the nuclear masses as a function of particle number. There is the very striking feature of a smooth trend plus (very) small fluctuations around it. Expressed in terms of energy one may write

$$E = \overline{E} + \delta E \equiv E_{\mathrm{LD}} + \delta E, \qquad (5.1)$$

with $\overline{E}$ representing the smooth part. Considered as function of particle number it is conceivable to associate it with the Bethe–Weizsäcker formula, commonly referred to as the "liquid drop" energy. As we recall from (1.64) and (1.65), the latter represents an expansion of the total nuclear energy in terms of powers of $A^{-1/3}$, and thus actually results from an expansion to the inverse radius of the system reflecting truly macroscopic properties. According to Strutinsky the fluctuating part δE is attributed to shell effects which in the end are determined by the single particle states in the neighborhood of the Fermi energy. For spherical nuclei such a concept had been developed independently by W.D. Myers and W.J. Swiatecki (Myers and Swiatecki, 1966).

In the previous chapter we learned about many indications that *independent particle motion cannot be the correct picture of nucleonic dynamics, nor for any other complex Fermi system.* For states away from the immediate neighborhood of the Fermi energy e_{F}, residual interactions become effective. One such trace is the dressing of the single particle states $|k\rangle$ through self-energies. Their imaginary part increases with the deviation Δe_k of these energies from e_{F}, at least for not too large values of Δe_k. Eventually, deeply bound states might be of such complicated nature that it would be very difficult, if at all possible, to describe them in practical terms. On the other hand, such properties ought already be hidden in the liquid drop model. This is why one may speak of a "renormalization" of the energy by the value given within the liquid drop model.

5.1.1 The independent particle picture, once more

The basic assumption of the *deformed* shell model developed in Section 3.2.2 is the hypothesis that one is able to guess the possible sequence of shapes Q the nucleus may undergo in the course of time evolution. This concept of the shape is used to define equipotential surfaces of a single particle potential for the one-body Hamiltonian $\hat{h}(\hat{\mathbf{x}}, \hat{\mathbf{p}}, Q)$. Taking such a $\hat{h}(\hat{\mathbf{x}}, \hat{\mathbf{p}}, Q)$ for granted and solving for the corresponding Schrödinger equation $\hat{h}(Q_0) \mid k, Q_0\rangle = e_k(Q_0) \mid k, Q_0\rangle$ one gets the single particle energies e_k, at any fixed Q_0. The simplest guess for the total energy E of the *many*-body system is the sum over all contributions from the individual states occupied by the A nucleons. In principle, this can be done for any many-body state. For the present purpose it is better just to look at the energy of the Q-dependent ground state, which is given by

$$E_{\mathrm{ipm}}(Q) = \sum_{e_k \leq e_{\mathrm{F}}} e_k(Q) \tag{5.2}$$

and which for given Q_0 represents an eigenvalue of the many-body Hamiltonian

$$\hat{H}_{\mathrm{ipm}}(\hat{x}_i, \hat{p}_i, Q_0) = \sum_{\alpha=1}^{A} \hat{h}(\hat{\mathbf{x}}_\alpha, \hat{\mathbf{p}}_\alpha, Q_0). \tag{5.3}$$

There are at least two reasons why the E_{ipm} of (5.2) cannot be considered a reasonable estimate of the total energy E. The first is intimately related to a basic feature of the Hartree–Fock approximation (discussed in detail in Section 20.1.2)): The sum over the single particle energies over counts the contributions from two-body interactions. The second reason has already been mentioned and is found in the fact that states far away from the Fermi energy e_{F} cannot be described adequately as those of independent particles. Both these points may be dealt with the help of the so-called *Strutinsky energy theorem* (Strutinsky, 1967), (Bethe, 1971b). In essence it implies that the ground state energy can be written as:

$$E = E_{\mathrm{ipm}} - \overline{E}_{\mathrm{pot}} \tag{5.4}$$

with the first term on the right being given by the shell model estimate. The quantity $\overline{E}_{\mathrm{pot}}$ accounts for contributions from two-body interactions. The bar on top of the $\overline{E}_{\mathrm{pot}}$ refers to the fact that in connection to formula (5.4) it is the *average* of E_{pot} which matters here. We shall soon see that in practical computations this quantity can be evaluated from models which numerically are easily accessible, and which in addition allow for a direct physical interpretation. A closer look at the energy theorem will give us hints of how the δE of (5.1) can be defined in terms of microscopic quantities and how it may actually be calculated.

5.1.2 The Strutinsky energy theorem

The essence of this theorem may be exhibited starting from the Hartree approximation applied to a two-body interaction $V(\mathbf{x},\mathbf{x}')$ which is independent of density. The total energy then reads

$$E_{\rm H} = \sum_{e_k^H \leq e_{\rm F}} e_k^H(Q) - \frac{1}{2}\int {\rm d}\mathbf{x}'\, {\rm d}\mathbf{x}\, V(\mathbf{x},\mathbf{x}')\rho(\mathbf{x}')\rho(\mathbf{x})\,, \qquad (5.5)$$

with the first term on the right being given by a formula like (5.2) but with the energies being replaced by those of the Hartree approximation, $e_k^{\rm H}$. The density $\rho(\mathbf{x})$ may be split into smooth and fluctuating parts, $\rho = \overline{\rho}+\delta\rho$. With the smooth part $\overline{\rho}$ one may define a "smooth" mean field, which in Hartree approximation would be $\overline{U}(x) = \int {\rm d}\mathbf{x}'\, V(\mathbf{x},\mathbf{x}')\overline{\rho}(\mathbf{x}') \equiv U_{\rm sm}(x)$. This $\overline{U}$ can be identified as the shell model potential $U_{\rm sm}$, which in turn allows one to calculate the total energy in the shell model by

$$E_{\rm sm} = E_{\rm ipm} - \frac{1}{2}\int {\rm d}\mathbf{x}\, U_{\rm sm}(\mathbf{x})\overline{\rho}(\mathbf{x}) \equiv E_{\rm ipm} - \overline{E}_{\rm pot}\,. \qquad (5.6)$$

By the last expression on the right we then have identified the $\overline{E}_{\rm pot}$ which appeared first in (5.4). The deviation of $E_{\rm H}$ from $E_{\rm ipm}$ may be evaluated from perturbation theory treating $\delta\rho$ to first order. If combined with the contributions of the potential one obtains the Strutinsky energy theorem in the form

$$E_{\rm H} = E_{\rm sm} + O(\delta\rho^2) \qquad (5.7)$$

which then proves the conjecture (5.4) to the extent that the ground state energy of the many-body system can be evaluated within Hartree approximation.

An extension of the proof to Hartree–Fock is straightforward (Strutinsky, 1967), (Brack et al., 1972) (see also (Siemens and Jensen, 1987) and (Bohr and Mottelson, 1975)). The case of density-dependent forces has been studied theoretically by H. Bethe (who also coined the name for this theorem) in (Bethe, 1971b), (Bethe, 1971a), G.G. Bunatian, V.M. Kolomietz and V.M. Strutinsky in (1972). Generalizations to pairing correlations and finite temperatures are discussed in (Brack and Quentin, 1981). Moreover, several numerical tests have been undertaken, largely also with density-dependent forces, by M. Brack and P. Quentin (1974), (1981) (see also (Brack, 1992)). As a final remark we would like to refer to (Bunatian et al., 1972) according to which in (5.7) the terms of second order amount to a contribution of the order of 0.5 MeV. It is claimed that such a contribution is to be expected from the interactions residual to Hartree–Fock. At times this is referred to as the "second order shell effect".

The formulas presented so far do not really elucidate why the Strutinsky method reduces so greatly the numerical efforts needed to calculate the total energy. Two more steps are necessary which to large extent take their justification from the energy theorem, again. Take the right-hand side of (5.6) and combine it with (5.7) to get, after neglecting the second order terms $O(\delta\rho^2)$,

$$E = E_{\rm H} - E_{\rm sm} = E_{\rm ipm} \quad \overline{E}_{\rm pot} = \overline{E}_{\rm ipm} - \overline{E}_{\rm pot} + E_{\rm ipm} - \overline{E}_{\rm ipm}, \quad (5.8)$$

where on the very right we have just added and subtracted $\overline{E}_{\rm ipm}$. The physical meaning of the first term evolving this way is evident: It just measures the *average* total energy and may thus be identified phenomenologically as the liquid drop energy, i.e.

$$\overline{E}_{\rm ipm} - \overline{E}_{\rm pot} = \overline{E} = E_{\rm LD}. \quad (5.9)$$

Obviously, this expression would allow for an easy calculation of the quantity $\overline{E}_{\rm pot}$, in case one were really interested in it. Fortunately, this is not necessary as in (5.1) one may directly use the liquid drop energy $E_{\rm LD}$. All that is needed then is the shell correction δE which is defined by the second term on the right of (5.8),

$$\delta E = E_{\rm ipm} - \overline{E}_{\rm ipm} = E - \overline{E}. \quad (5.10)$$

The expression in the middle may easily be computed from the independent particle model.

Both (5.9) and (5.10) require, of course, a decent formalism to actually perform the averages. This is partly a technical question, but it also contains much physical background. For the moment, let us just mention two aspects, first of all the relation to the liquid drop model. There is no doubt that the *association of $\overline{E}$ with the liquid drop energy $E_{\rm LD}$ is an additional assumption*. It is this step which leads the Strutinsky method beyond that of Hartree–Fock. In its very spirit the liquid drop model contains effects from configurations much higher than just 1p-1h. The same can be said, of course, about the $\overline{E}$ if we recall the very fact of the averaging procedure involving spreading in energies over more than one shell. The second aspect we like to mention is found in the possibility of relating average quantities to thermal properties of the system. This opens interesting physical interpretations, in particular in connection to the description of transport processes.

5.2 The Strutinsky procedure

5.2.1 Formal aspects of smoothing

For any function $G(x)$ of x its average $\overline{G}(x)$ [9] over an interval $\Delta x = \gamma_{\rm av}$ may be defined through the convolution integral

$$\overline{G}(x) = \int_{-\infty}^{\infty} G(x') f\left(\frac{x'-x}{\gamma_{\rm av}}\right) \frac{dx'}{\gamma_{\rm av}} \quad \text{where} \quad \int_{-\infty}^{\infty} \frac{dx'}{\gamma_{\rm av}} f\left(\frac{x'-x}{\gamma_{\rm av}}\right) = 1. \quad (5.11)$$

The smoothing function $f((x'-x)/\gamma_{\rm av})$ ought to have some bell-like shape of width $\gamma_{\rm av}$, with its maximum lying at $x' = x$. Eventually, one might use

---

[9] A short remark on notation: Unlike the convention often used we want to classify averaged quantities by $\overline{O}$ rather than by $\widetilde{O}$. Otherwise, one might get mixed up with our notation for response functions where $\sim$ always refers to the time-dependent ones.

Lorentzians or Gaussians, but they are less favorable with respect to stability of the method. To warrant the latter, Strutinsky (1967) chose a sum over Hermite polynomials $H_n(x)$ multiplied by a Gaussian,

$$f(x) = \frac{e^{-x^2}}{\sqrt{\pi}} \sum_{n=0,2,\ldots}^{2M} \alpha_n H_n(x) \quad \text{with} \quad \alpha_{n+2} = \frac{-\alpha_n}{2(n+2)}, \tag{5.12}$$

with $\alpha_0 = 1$. This ansatz represents the first $2M$ terms of an expansion of the δ-function into a series of Hermite polynomials. Smoothing with such a function restores polynomials in x of order k up to $k = 2M$,

$$P_k(x) = \int_{-\infty}^{\infty} dy \, f(x-y) \, P_k(y), \quad \text{if} \quad k \leq 2M. \tag{5.13}$$

In this sense, the $f(x)$ of (5.12) behaves as a δ function for any $G(x)$ which is "sufficiently" smooth in x. This property is essential to warrant stability, in the sense that a quantity $\overline{G}(x)$ smoothed once is recovered after applying the procedure again,

$$\overline{\overline{G}}(x) = \overline{G}(x). \tag{5.14}$$

One only needs to find an optimal regime of values for the averaging parameter γ_av. Within this regime, which may be different for different Ms, the result $\overline{G}(x)$ will not depend on the γ_av. Remember that this γ_av itself does not have any physical meaning, other than to warrant that (5.14) is fulfilled and that shell effects disappear, as will be discussed now.

5.2.2 Shell correction to level density and ground state energy

Let us begin looking at the single particle level density

$$g(e) = \sum_k \delta(e - e_k), \tag{5.15}$$

the average of which reads

$$\overline{g}(e) = \frac{1}{\gamma_\mathrm{av}} \int f\left(\frac{e' - e}{\gamma_\mathrm{av}}\right) g(e') de' = \frac{1}{\gamma_\mathrm{av}} \sum_k f\left(\frac{e_k - e}{\gamma_\mathrm{av}}\right). \tag{5.16}$$

The deviation of $g(e)$ from $\overline{g}(e)$ is called the fluctuating part $\delta g(e)$ of the level density,

$$\delta g(e) = g(e) - \overline{g}(e). \tag{5.17}$$

Note that its average vanishes provided condition (5.14) is fulfilled. The decomposition (5.17) allows one to deduce shell effects for any quantity Θ which is expressible by an integral (or sum) over single particle properties.

Let us look at the case of the total energy, for which the E_{ipm} and its average $\overline{E}_{\text{ipm}}$ are given by:

$$E_{\text{ipm}} = \sum_{e_k \leq e_F} e_k \equiv \int^{e_F} de\, g(e)\, e \quad \Longrightarrow \quad \overline{E}_{\text{ipm}} = \int^{\overline{\mu}} de\, \overline{g}(e)\, e\,. \qquad (5.18)$$

The quantity $\overline{\mu}$ introduced on the right is to be determined from particle number conservation. In combination with the smooth part of the level density this condition requires the actual particle number A to be given by the common relation

$$A = \int^{\overline{\mu}} de\, \overline{g}(e)\,. \qquad (5.19)$$

Although one is dealing with zero thermal excitations there is a close analogy with the chemical potential $\overline{\mu}$ of thermostatics; like there, here too, the $\overline{\mu}$ differs from the quantal Fermi energy e_F. The shell correction energy δE then is given by (5.10), which implies:

$$\delta E = \int^{\overline{\mu}} de\, \delta g(e)\, e\,. \qquad (5.20)$$

As mentioned earlier, the smooth part $\overline{E}$ is to be associated with the "liquid drop" energy, which in turn may be calculated from the Bethe–Weizsäcker formula. Thus the total energy is given through (5.1) as $E = \overline{E} + \delta E \equiv E_{\text{LD}} + \delta E$. The replacement of $\overline{E}_{\text{ipm}}$ by E_{LD} is justified only if the former does not contain shell effects anymore. This puts a stringent condition on the size of the averaging interval. Typically, the $g(e)$ shows oscillations on the scale of the average shell distance $\hbar\Omega_0^{\text{sh}}$. If shell effects are to be washed out completely the size of the averaging parameter γ_{av} has to be of that order. That this aim may actually be achieved is demonstrated in Fig. 5.2 for the shell correction energy δE for the nucleus ^{150}Er.

One observes that a decent plateau is found for not too small Ms and in the range of $\gamma_{\text{av}} \approx 10-15$ MeV which corresponds to values of the order of $1-2\hbar\Omega_0^{\text{sh}}$. The calculation was performed with the code of V.V. Pashkevich (1971).[10] The underlying single particle potential is of Woods–Saxon type, details of which are explained elsewhere in the text. For a potential of finite depth the shell correction necessarily gets contributions from single particle states at positive energies, simply because the Fermi level is away from zero energy less than the γ_{av}. In principle, levels with positive energies correspond to scattering states, but it would be difficult to calculate their density, in particular for deformed systems. Following (Strutinsky et al., 1980), in practical computations one therefore uses a different prescription. Rather than accounting for genuine scattering states, one diagonalizes the single particle Hamiltonian in an oscillator basis. This basis is restricted to states below some finite main quantum number $N_{\max}$. The latter

---

[10] I am grateful to F.A. Ivanyuk for his help and advice in preparing this section.

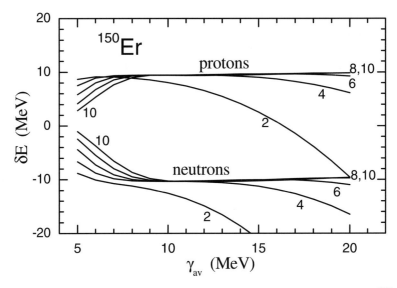

FIG. 5.2. The shell correction energy for the medium heavy nucleus ^{150}Er as function of the averaging interval γ_{av} and for different orders $2M = 2, 4 \ldots 10$ of the Hermite polynomial of (5.12).

has to be chosen such that the single particle density becomes sufficiently smooth at $e = 0$. Only then has one a chance to find a decent plateau. This is easily understood recalling that according to (5.16) for $\bar{g}(e)$ being stable also around the Fermi energy e_F one needs to have an appropriate $g(e)$ also at $e \approx e_F + \gamma_{\text{av}}$. This precaution is sufficient for the zero temperature case. At finite T one has to extend this procedure accordingly. Naturally, such problems do not appear if for the underlying single particle Hamiltonian a potential of infinite depth is chosen from the start. As a final remark on this issue, again following (Strutinsky et al., 1980), the oscillator frequency which defines this basis is taken as $\hbar\Omega_0^{\text{sh}} \simeq 55/A^{1/3}$ rather than the usual value $\simeq 41/A^{1/3}$ of the Nilsson model. For the case of ^{150}Er considered in the figure, this corresponds to a value of $\hbar\Omega_0^{\text{sh}} \simeq 10$ MeV.

It is important to keep in mind that the Strutinsky method was devised to evaluate the shell correction δE, or the $\delta\Theta$ in the general case. Quite generally, this $\delta\Theta$ is determined solely by levels in the neighborhood of the Fermi surface; this feature may be clear from the very construction of the method but we will return to this problem below. This important fact implies that only those properties of the independent particle model are exploited which are founded by experimental evidence. All the contributions from states far away from the Fermi energy are to be attributed to the smooth part $\overline{\Theta}$. As this quantity is insensitive to small-scale variations its value may be deduced from phenomenological pictures—provided they exist like the liquid drop model for the example of the total energy. For an elaborate discussion of these basic elements of the Strutinsky procedure we refer to the classic paper (Brack et al., 1972).

Often one is not interested in the finer details which some special shell model potential may deliver, say on the scale of an uncertainty Δe in the energy. For instance, for some given quantity $\Theta(e)$ it may suffice to understand its *gross shell* behavior, discarding any structure in the variations with energy on a scale smaller than Δe. This structure may be averaged out by applying (5.11) not with the $\gamma_{\rm av}$ but with a $\gamma \simeq \Delta e$ which is small compared to $\hbar\Omega_0^{\rm sh}$. We will return to this question below in connection to periodic-orbit theory.

5.2.3 Further averaging procedures

The smoothing procedure discussed in Section 5.2 is intimately related to the distribution of the single particle levels, for which reason it sometimes is referred to as "energy smoothing". This is not the only possibility. Depending on the specific problem one wants to look at, other procedures may be more adequate. Of particular interest are those associated with thermal averaging or those related to particle number smoothing, both of which have direct physical bearings.

Averaging over occupation number: Both the shell correction as well as the smooth part of any quantity may be traced back to the occupation numbers $n(e)$ of the single particle states by writing (Strutinsky, 1967), (Brack et al., 1972) $n = \bar{n} + \delta n$ and by relating the smooth occupation number to the smoothing function by way of

$$\bar{n}(x) = \int_x^\infty f(x')\,dx' \qquad \text{or} \qquad \frac{d\bar{n}}{dx} = -f(x)\,. \qquad (5.21)$$

Let us demonstrate this possibility for the case of the total energy at zero temperature, again. With the help of (5.16) and (5.18) its smooth part can be seen to become

$$\overline{E}_{\rm ipm} = \int_{-\infty}^{\bar{\mu}} de \sum_k \frac{1}{\gamma_{\rm av}} f\left(\frac{e_k - e}{\gamma_{\rm av}}\right) e \approx \sum_k e_k\,\bar{n}(e_k)\,. \qquad (5.22)$$

Indeed, after changing the integration variable $x = (e_k - e)/\gamma_{\rm av}$ one gets the expression

$$\overline{E}_{\rm ipm} = \sum_k e_k \int_{(e_k-\bar{\mu})/\gamma_{\rm av}}^\infty dx\,f(x) - \sum_k \gamma_{\rm av} \int_{(e_k-\bar{\mu})/\gamma_{\rm av}}^\infty dx\,xf(x) \qquad (5.23)$$

whose first term is easily recognized to deliver the desired expression given on the right of (5.22). Numerically, the second term turns out small for such $\gamma_{\rm av}$ for which the shell correction has its plateau, which is to say for which both the $\overline{E}_{\rm ipm}$ and the δE are insensitive to changes in the smoothing interval (see (Brack et al., 1972) and further references given therein). As recognized by M. Brack in his thesis (see (1972), (1992)), the second term of (5.23) can be written as $\gamma_{\rm av} d\overline{E}_{\rm ipm}/d\gamma_{\rm av}$, which for a perfect plateau vanishes by definition.

With the help of the separation of the occupation number we may understand which states contribute mainly to the shell correction δE. The latter is determined by the $\delta n(e_k) = n(e_k) - \overline{n}(e_k)$. The $n(e_k)$ is equal to unity up to the Fermi energy and zero beyond. The $\overline{n}(e_k)$, on the other hand, drops from unity to zero within a range of $\pm \gamma_{\mathrm{av}}$ around $\overline{\mu} \approx e_{\mathrm{F}}$. Hence it is this range which determines possible contributions to δE.

Averaging over particle number: Evidently, this method is most directly related to the macroscopic limit. In this sense it might be expected to deliver the most straightforward connection to the liquid drop model (of the static energy), as the latter consists in writing the total asymptotically as an expansion in powers of $A^{-1/3}$ (Bohr and Mottelson, 1975). However, as discussed in (Strutinsky and Ivanyuk, 1975), (Ivanyuk and Strutinsky, 1979), the equivalence to energy smoothing can be established only under certain conditions. Formally, particle number smoothing follows quite naturally from occupation number smoothing. The smooth occupation number, as given in (5.21), depends on energy in the same way as it does on the chemical potential, which by (5.19) is intimately connected to particle number.

Temperature smoothing: The form found in (5.22) for the average energy reminds one of thermostatics–provided the $\overline{n}$ can be interpreted like a Fermi function at some appropriate "temperature" T_s. The clue to establish such a relation is suggested by (5.21). It is known that the derivative of the Fermi function proper is of bell-like shape,

$$\frac{dn(e)}{de} = -\left(4T_s \cosh^2\left(\frac{e-\overline{\mu}}{2T_s}\right)\right)^{-1}. \tag{5.24}$$

Comparing its width with the γ_{av} of the smoothing function f one may actually deduce the following relation $T_s \approx \gamma_{\mathrm{av}}/4$. It is important to notice, however, that the temperature T_s introduced in this way has nothing to do with the parameter T, which in the canonical ensemble measures the thermal excitation. Indeed, the energy to which the smoothing is applied to is that of the ground state, rather than the internal energy at a physical temperature. This method goes back to (Bhaduri and Gupta, 1973), where the equivalence of temperature smoothing to the original version of Strutinsky was nicely demonstrated.

Application of Wigner–Kirkwood expansion: To complete the list of work where techniques from statistical mechanics are exploited to obtain the smooth part of the level density, and thus the shell correction, we should like to mention the Wigner–Kirkwood expansion. It was first applied to this problem by B.K. Jennings in his doctoral thesis (1976). This expansion delivers a series of the partition function $Z(\beta)$ into powers of $\hbar$, see the discussion in Section 24.2.2. The level density $g(e)$ may then be obtained from the inverse Laplace transform $g(e) = \mathcal{L}_e^{-1}[Z(\beta)]$, see e.g. Section 22.2.3. For a three-dimensional system and to second order in $\hbar$ the partition function is given by

$$Z_{\mathrm{WK}}(\beta) = \left(\frac{m}{2\pi\hbar^2\beta}\right)^{3/2} \int d^3x\, e^{-\beta V(\mathbf{x})} \left\{1 - \frac{\beta^2}{12}\frac{\hbar^2}{2m}\nabla^2 V + \mathcal{O}(\hbar^4)\right\}. \quad (24.75)$$

The first term just represents the $Z_{\mathrm{cl}}(\beta)$ of classical statistical mechanics. Applying basic rules of Laplace transforms, the corresponding level density becomes

$$\mathcal{L}^{-1}[Z_{\mathrm{cl}}(\beta)] \equiv g_{\mathrm{TF}}(e) = \frac{1}{4\pi^2}\left(\frac{2m}{\hbar^2}\right)^{3/2} \int d^3x\, \Theta\left(e - V(\mathbf{x})\right) \sqrt{(e - V(\mathbf{x}))}, \quad (5.25)$$

with the step function $\Theta(e - V(\mathbf{x}))$. This is nothing else but the $g(e)$ in Thomas–Fermi approximation. The contribution from the term of second order turns out to be

$$g^{(2)}_{\mathrm{WK}}(e) = -\frac{1}{96\pi^2}\left(\frac{2m}{\hbar^2}\right)^{1/2} \frac{\partial}{\partial e} \int d^3x\, \Theta(e - V(\mathbf{x})) \frac{\nabla^2 V(\mathbf{x})}{\sqrt{(e - V(\mathbf{x}))}}. \quad (5.26)$$

The sum of the two terms is commonly referred to as *extended Thomas–Fermi approximation* (ETF) (in second order),

$$g_{\mathrm{ETF}}(e) = g_{\mathrm{TF}}(e) + g^{(2)}_{\mathrm{WK}}(e). \quad (5.27)$$

It turns out that this level density is smooth both in energy as well as in particle number. In this sense the application of the Wigner–Kirkwood expansion may be considered an alternative to the Strutinsky method (Jennings, 1973), at least from a theoretical point of view. Practical applications are more cumbersome, however, and their realizations depend very much on the single particle potential one looks at. It may be mentioned that it is not only possible to obtain analytic expressions for the level density itself but for the total energy and particle number as well. They follow from basic properties of Laplace transforms, say as applied to (5.18) in case of the total energy. Eventually, an expansion of the partition function to fourth order in $\hbar$ is required to get stable results. For more details we have to refer to the literature, in particular to the book by M. Brack and R.K. Bhaduri (1997). One should not forget, however, that here also here these smooth parts are the ones obtained within an independent particle model.

It is interesting to note that the macroscopic limit of these quantities, as represented by their averaged values involve terms of at least order two in $\hbar$. Looking at (24.75) it is recognized that this feature is due to the finiteness of the system. The $\hbar$ goes along with variations in the single particle potential. For a nucleus it is predominantly the variation of the latter in the surface which matters.

5.3 The static energy of finite nuclei

In this section we will concentrate on examples from nuclear physics to discuss applications to mesoscopic systems in Part III. The main consequences of the

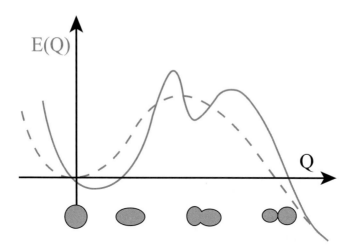

FIG. 5.3. A schematic picture of the implication of the shell correction on the deformation energy. The dashed curve shows the behavior of the liquid drop model, for the fully drawn line shell corrections are added.

shell correction to the static energy E are pictured schematically in Fig. 5.3 for a heavy nucleus. There, the variation of $E(Q)$ is plotted along a typical fission path, with the shapes associated with the collective variable Q being shown at the bottom. The dashed curve represents the liquid drop energy $E_{\text{LD}}(Q)$ and the fully drawn one the $E(Q)$ after shell corrections have been added. In the former case the energy exhibits one minimum at the spherical configuration and one barrier at elongated shapes. Typically, the shell correction not only shifts the value of the minimal energy but also its position; this feature is valid not only for fissile systems. As a matter of fact, it is only by this mechanism that one can understand why in its ground state most nuclei are deformed. The quantum effects hidden in the single particle level density $g(e)$ lead to enhanced binding whenever the $\delta g(e)$ is negative, which implies a negative contribution of δE to $E(Q)$. This effect is particularly large for the ground state energy at closed shells, which is to say when the nucleon number is "magic", but it may also exist at finite deformations. As a consequence the total energy may attain one or more other minima, which often happens in the region where the liquid energy has its maximum. On its way to fission such a system has to overcome (at least) two barriers rather than only one. Moreover, it might be trapped for some time in the second minimum, thus populating isomeric states. Indeed, states of this type have been measured experimentally and it was the first great success of the Strutinsky method to be able explaining such facts. Actually, in the meantime third minima have been established (Thirolf and Habs, 2002), (Csatlós et al., 2005).

Altogether, the Strutinsky method has proven capable in describing on quantitative levels (i) ground state energies, (ii) their associated equilibrium deformations, (iii) fission barriers as well as (iv) the existence and properties of fission isomers. In the following we like to report on two applications, namely the calculation of nuclear masses and of potential landscapes for large-scale collective motion.

Binding energies and deformations of finite nuclei: The most detailed and most precise calculations of nuclear ground-state masses and deformations with the Strutinsky procedure have been carried out by the Berkeley–Los Alamos collaboration. In the past few decades the models underlying their "macroscopic-microscopic method" have been continuously improved and refined. In the version published in 1995 (Möller *et al.*, 1995) an astonishingly high accuracy of better than 10^{-3} has been achieved (for heavy nuclei). By adjusting nine additional constants over those where confidence had been reached before the ground state masses of 1654 known nuclei ranging from ^{16}O to ^{263}106 could be calculated with an error less than 0.669 MeV, which for nuclei with neutron number $N \geq 65$ even was below 0.448 MeV; recall the presentation of these results in Fig. 5.1. For the macroscopic part of these calculations one uses a modified version of the droplet model mentioned below the Bethe–Weizsäcker formula (1.65). For instance, the droplet is allowed to have a diffuse surface, associating this diffuseness to the finite range of the two-body interaction (taken to be of Yukawa-form), for which reason the notion "finite range droplet model" (FRDM) has been coined. Other modifications involve an average pairing energy, as well as the so-called "congruence-" or "Wigner term" and one which is proportional to A^0. There is no room here to elaborate on the physical reasons why such terms are to be included; a thorough discussion may be found in this paper (Möller *et al.*, 1995) itself, in the literature cited therein as well as in (Myers and Swiatecki, 1996). The shell correction is calculated with a single particle potential obtained by folding the density distribution with a Yukawa interaction. This model has been described in Section 3.2.2 together with the deformation degrees of freedom. The tables and figures presented in (Möller *et al.*, 1995) for masses and deformations include nuclei from ^{16}O up to super-heavy elements. As a matter of fact, the latter can exist only because of the presence of shell corrections. More work in this field has been presented in other publications by this group together with different collaborators, see e.g. (Iwamoto *et al.*, 1996). An experimental review about the importance of shell effects in fusion and fission can be found in (Armbruster, 1999).

Potential landscapes The model just described for the calculations of the ground state masses has been applied extensively to explore the variation of the static energy with shapes. However, instead of the FRDM the "finite range liquid drop model" is used, which works better for shapes closer down and around the scission region. We confine ourselves to report on the most recent results, like those published in (Möller *et al.*, 2004*a*), (Möller *et al.*, 2004*b*). There, a five-dimensional deformation space is employed defined by the following degrees of freedom: elongation, neck diameter, mass asymmetry and deformations of both fragments.

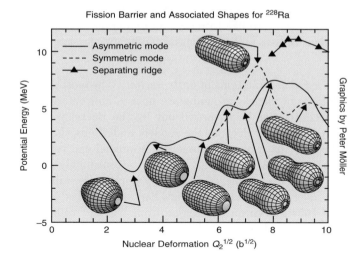

FIG. 5.4. The potential energy of ^{228}Ra presented in (Möller *et al.*, 2004*a*) as function of the elongation of the nuclear shape, see text.

Special care is payed to find all possible paths to the various scission modes including an appropriate search for the saddle points. This is achieved by applying the procedure of a "hypothetical water flow" (Mamdouh *et al.*, 1998) which is argued to lead to more reliable results than searching for saddle points by minimizing the total energy employing constraints, for a detailed discussion of this issue see also (Möller and Iwamoto, 2000). One of the most prominent features of the landscape is the existence of different valleys for symmetric and asymmetric fission modes and how their mutual relevance varies with particle number. As an example we show in Fig. 5.4 the case of ^{228}Ra. The two existing paths for symmetric and asymmetric fission are seen to have different barriers, the height for the asymmetric mode being lower by about 2 MeV. The separation of both paths only concerns the outer regions, with the corresponding deformations nicely visualized. Also shown is the ridge which exists between the two valleys beyond the outer barrier. For the case of lighter nuclei like the one shown in the figure this ridge is higher than the symmetric barrier such that the symmetric and asymmetric paths are well separated, which is no longer true for heavier nuclei like the U-isotopes. These findings for Ra are in accord with experimental results (see e.g. (Kudyaev *et al.*, 1976)). In these calculations progress has also been made to obtain energies which remain smooth at scission. This is to say that the energy $E(Q)$ calculated for one complex before scission turns into that calculated for two joining fragments after scission as smoothly as possible. This is achieved by allowing in the liquid drop energy the Wigner term and the A^0 term to be shape dependent.

5.4 An excursion into periodic-orbit theory (POT)

It may be intuitively evident that the appearance of shells in the quantum spectrum of finite systems is related to closed orbits in classical mechanics. One may think of the Bohr–Sommerfeld quantization of integrable systems. However, the issue is of more general nature and has also been well established in the meantime for non-integrable systems. There is no room to elaborate on such properties in any greater detail, but this topic is of too great importance to leave it untouched entirely. In the following we first want to discuss the semi-classical version of the single particle level density, which will then be applied to deduce information on the gross shell behavior. Finally we wish to address the question about the relevance of chaotic motion for the fluctuating part of the energy.

In the early 1970s it was shown by M.G. Gutzwiller (see (1990)) that in POT the fluctuating part $\delta g(e)$ of the level density (5.17) obtains the form

$$\delta g(e) = \sum_{p.p.o} \sum_{k=1}^{\infty} \mathcal{A}_{p.p.o}^{(k)}(e) \cos\left(\frac{k}{\hbar} W_{p.p.o}(e) - \sigma_{p.p.o}^{(k)} \frac{\pi}{2}\right). \tag{5.28}$$

The amplitudes $\mathcal{A}_{p.p.o}^{(n)}(e)$, the actions $W_{p.p.o}(e)$ as well as the phase indices $\sigma_{p.p.o}^{(n)}$ depend on details of the trajectories. The summation extends over all periodic orbits, the "primitive" ones, classified in the formula by "p.p.o." and those where the former are traversed k times. The essential steps needed for the derivation of (5.28) from the Green function in WKB approximation are discussed in Section 24.3. Equation (5.28) is commonly called the "Gutzwiller trace formula". It allows one to deduce valid information on the quantal level structure just from knowledge of the classical orbits. This is possible both for integrable systems as well as for such which show chaotic behavior on the classical level. The latter feature is particularly intriguing as in this way one may study quantization of chaotic dynamics. For details please see the textbooks mentioned as well as an ever increasing volume of literature.

Gross shell effects within POT: It has been mentioned before that important features of shell effects may still be seen after small and often unimportant details have been averaged out. This gross shell behavior may be seen if the average is performed over an interval $\Delta x = \gamma$ of a size considerably smaller than the γ_{av} of the shell correction method, where any shell structure is washed out totally. Suppose one applies Gaussian smoothing to the level density. Evidently, applying (5.11) and (5.16) to the present case one gets $g_\gamma(e) = (\gamma\sqrt{\pi})^{-1} \sum_k \exp(-[(e-e_k)/\gamma]^2)$. For periodic orbits the time it takes for *one* revolution (notice, that a given orbit may be traversed many times, so one revolution determines the "shortest" periodic orbit of a given class) can be calculated from $T_{P.O} = (dW(e)/de)_{P.O}$. Whenever the shortest orbits dominate the sum the relevant time is that of these primitive orbits which matters most. The latter is to be associated with the frequency Ω_0 which determines the shell spacing $\hbar\Omega_0^{\text{sh}}$, such that the condition on the γ amounts to say

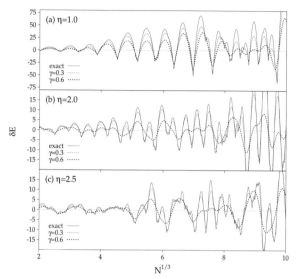

FIG. 5.5. The shell correction energy for a spheroidal cavity is shown as function of the cubic root of particle number and at different deformations η, taken from (Magner et al., 2002). Results from fully quantal calculations are compared with those obtained within POT; for more details see text.

$$\Delta e_k \leq \gamma \ll \hbar\Omega_0^{sh} \simeq 2\pi\hbar/T_{P.P.O.}. \qquad (5.29)$$

If this average is performed for the oscillating part of the level density $\delta g_\gamma(e)$ of POT one gets (see e.g. (Brack and Bhaduri, 1997))

$$\delta g_\gamma(e) = \sum_{P.O.} \exp\left(-\left(\gamma T_{P.O.}/2\hbar\right)^2\right) \delta g_{P.O.}(e). \qquad (5.30)$$

In principle, the sum over P.O. is meant to include orbits where the fundamental revolution is repeated arbitrarily many times, but longer ones will die out rapidly, in accord with condition (5.29). It should be said that one excludes the very short ones which would only give back the smooth level density $\bar{g}(e)$. In general, this amounts to neglecting the diametric trajectory; see (Strutinsky and Magner, 1987) (and (Strutinsky et al., 1977) for an extension to deformed systems). Whereas the sum in (5.30) is rapidly converging, such a feature is usually *not* guaranteed for the $\hbar$ expansions of the semi-classical approximation, as for the Gutzwiller trace formula. Convergence may also be obtained by applying Lorentzian smoothing, as done in the independent approach by R. Balian and C. Bloch (1974). As we know from Section 4.3.2 (see exercise 2) Lorentzian smoothing can be simulated by just including a finite imaginary part $i\gamma$ to the energy.

The gross shell behavior of the total energy is obtained after plugging the $\delta g_\gamma(e)$ of (5.30) into formula (5.20). In Fig. 5.5 results are shown from (Magner et al., 2002) for a spheroidal, infinitely deep square well potential. For two different values of γ the δE_γ obtained within POT is compared with the correct quantal calculation of δE.[11] This comparison is performed for three different deformations, measured by the ratio η of the two axes, and as function of the cubic root of particle number $A \equiv N$. The cases $\eta = 2$ and 2.5 correspond to the so-called "super- and hyperdeformed" configurations, respectively. Whereas for the spherical system ($\eta = 1$) (gross) shell structure survives even a smoothing with $\gamma = 0.6$ this is not the case for the strongly deformed systems. It may be mentioned that this effect is influenced by orbits exhibiting bifurcation.

Contributions to the total energy from chaotic orbits: Above it has been shown to which precision masses of real nuclei may be calculated. O. Bohigas and P. Leboeuf (2002) have raised the claim that the remaining discrepancy is not due to eventual deficiencies of the underlying single particle and liquid drop models. Rather, the authors try to attribute it to an inherent uncertainty in the picture of independent particle motion itself, which has its origin in the presence of chaotic features in the nucleon's dynamics. In the following we are going to sketch the main ideas and results without describing it in any great detail. The authors aim at calculating the fluctuating parts of the energy, which come from the splitting of the level density $\overline{g}(e) + \delta g(e)$ into smooth and oscillating parts. Unlike the common Strutinsky procedure, they understand the notion of a "fluctuating part" in the sense of a genuine statistical analysis: Having an averaging procedure at one's disposal one may calculate the mean square deviation $\overline{G^2} - \overline{G}^2$ of any quantity G of interest. Their averaging procedure is defined like but not identical to that of the original Strutinsky method. In the notation of (5.11) their averaging "function" is simply given by a constant of value unity inside the interval from $e - \gamma_{av}/2$ to $e + \gamma_{av}/2$ and zero outside. Let us leave untouched the question of the extent to which this kind of smoothing satisfies the stability criterion (5.14). However, the authors rely on the condition that the average over the oscillating quantity vanishes, which in the notation from before means to say $\overline{\delta G} = 0$. They apply this smoothing to evaluate the mean square deviation (or variance) $\overline{\delta E^2}$ of the ground state energy. The calculation of such fluctuations requires knowledge of the correlation function of the oscillating part $\delta g(e)$ of the level density as given by (5.28).

The contributions from regular and chaotic orbits, δE_{reg} and δE_{ch} to $\delta E = \delta E_{\text{reg}} + \delta E_{\text{ch}}$, are treated separately, arguing that the cross term $\overline{\delta E_{\text{reg}} \delta E_{\text{ch}}}$ is to vanish because of statistical reasons. An estimate of δE_{reg} leads to values of the same order as those known from the Strutinsky calculations. Interpreting the δE as the experimentally determined microscopic correction the difference

---

[11]The relation of the $\gamma \equiv \gamma_{\text{POT}}$ used in the figure to the γ for $g_\gamma(e)$ and that in (5.29) is given approximately by $\gamma = r \gamma_{\text{POT}} \hbar \Omega_0^{\text{sh}}$, with the factor r being of order 0.8 (and actually determined by the ratio of the length L of the most important orbit to the circumference $2\pi R$ of the sphere out of which the spheroid develops). The author is grateful to A.G. Magner for explaining details of (Magner et al., 2002).

$\delta E - \delta E_{\rm reg} = \delta E_{\rm ch}$ may be identified as the remaining discrepancy shown in Fig. 5.1. An estimate of the mean square deviation along the lines sketched before leads to $\overline{\delta E_{\rm ch}^2} = e_c^2/(8\pi^4)$. The e_c is the single particle energy associated with the shortest periodic orbit of length $4R$ with $R = 1.1 A^{1/3}$ fm. Expressed by the orbit's period $\tau_{\min}$ one has $e_c = h/\tau_{\min} = \pi e_{\rm F}/k_{\rm F} R$ which leads to values of $e_c = 77.5/A^{1/3}$ MeV. Compared to the 55 mentioned in Section 5.2.2 in connection to Fig. 5.2 the pre-factor of the $A^{-1/3}$ is a bit larger. It turns out that the square root of this $\overline{\delta E_{\rm ch}^2}$ is in astonishingly good agreement with the discrepancy between the experimental and theoretical microscopic deviations shown in the lower part of Fig. 5.1. This holds true not only with respect to the A-dependence but also in an absolute sense. Finally, it may be said that (Bohigas and Leboeuf, 2002) leave open the precise origin of the chaotic behavior. Although the estimates given are based on single particle motion, they claim the main final result to reflect motion in the many-body phase space. In such a case chaotic behavior may be induced by residual interactions, as discussed in Chapter 4. Fact is that the size of $\sqrt{\delta E_{\rm ch}}$ is pretty small. Incidentally, it is of similar order to that of the "second order shell effect" mentioned in Section 5.1.2.

5.5 The total energy at finite temperature

In this section we shall address the case of finite thermal excitations, as produced in nuclear reactions, predominantly in those involving heavy ions. Quite generally, quantum features tend to disappear with increasing thermal excitations. Such behavior can be expected to manifest itself in the disappearance of shell effects with increasing temperature. Indeed, this phenomenon was recognized quite early (Ramamurthy et al., 1970), (Bhaduri and Ross, 1971), (Ramamurthy and Kapoor, 1972), (Bhaduri and Gupta, 1973) to deliver another means for evaluating the smooth part formally, and much work has been done in this direction. In fact, this feature is the physical reason why the procedures behind temperature smoothing or the Wigner–Kirkwood expansion works at all, as explained in Section 5.2.3. In clear distinction to these formal treatments, in the discussion to come temperature is taken as a real physical quantity measuring the system's thermal excitation. For all problems associated with such a hypothesis we refer to the discussion in Chapter 22.

At $T = 0$ the shell correction method is based on the simple expression the total energy takes on in the independent particle model, say as given by (5.18). A natural generalization is found by applying this model to the grand canonical ensemble, for which the internal energy may be written as may be written as

$$\mathcal{E}_{\rm ipm}(T) = \int_0^\infty de\, n(e;T,\mu)\, g(e)\, e \quad \text{with} \quad n(e;T,\mu) = \frac{1}{e^{\beta(e-\mu)}+1} \quad (5.31)$$

defining the occupation probability for given temperature $T = 1/\beta$ and chemical potential μ. The latter is to be found from the expression

$$\mathcal{A} = \int_0^\infty de\, n(e;T,\mu)\, g(e) \quad (5.32)$$

for particle number. Notice that the zero of energy has been identified as the bottom of the single particle potential. Again, one may take advantage of the separation of the level density into smooth and fluctuating parts, $g(e) = \bar{g}(e) + \delta g(e)$, as introduced in (5.17). Writing the total internal energy as the sum of the energy of the ground state plus that of the excitation, $\mathcal{E} = E_0 + \mathcal{E}^*$, this splitting will influence both terms. In this spirit the internal energy may be written as

$$\mathcal{E} = \bar{\mathcal{E}} + \delta \mathcal{E}, \quad \text{with} \quad \bar{\mathcal{E}} = \bar{E}_0 + \bar{\mathcal{E}^*}, \quad \delta \mathcal{E} = \delta E_0 + \delta \mathcal{E}^*. \quad (5.33)$$

Note that we have written $\bar{\mathcal{E}^*}$ rather than $\bar{\mathcal{E}}^*$, adhering strictly to the convention of the bar to indicate smooth behavior of the associated quantity. In this spirit one may introduce a smooth part $\bar{\mu}$ of the chemical potential by requiring (5.32) to be fulfilled by the $n(e; T, \bar{\mu})$, at the same time replacing g by $\bar{g}$ but without changing the particle number $\mathcal{A}$. Doing the same replacements in (5.31) one gets the average total energy $\bar{\mathcal{E}}$. This procedure is a natural generalization of the one described in Section 5.2.2 for zero thermal excitations. In practical applications (see e.g. (Ivanyuk and Hofmann, 1999)) it turns out simpler to calculate $\bar{\mathcal{E}}$ approximately as $\bar{\mathcal{E}}_{\rm ipm} \approx \sum_k \bar{n}(e_k) e_k$ with the $\bar{n}(e_k)$ defined by

$$\bar{n}(e_k) = \frac{1}{\gamma_{\rm av}} \int_{-\infty}^{\infty} f\left(\frac{e' - e_k}{\gamma_{\rm av}}\right) n(e') de'. \quad (5.34)$$

Compared with the discussion in Section 5.2.3 this procedure is seen to be in close relation to the "occupation number smoothing", which earlier has been shown to be exact at $T = 0$, if only the averaging parameter is in the plateau region. At finite temperature the corrections are known to be of order $T^2/\gamma_{\rm av}^2$.

5.5.1 The smooth part of the energy at small excitations

The quantity $\bar{\mathcal{E}}$ is governed by $\bar{g}$, which, on the other hand, is sufficiently smooth to warrant application of the Sommerfeld expansion, as outlined in detail in Section 22.5.1. Within the independent particle model the average internal energy may be written in the form

$$\bar{\mathcal{E}}_{\rm ipm} = \bar{E}_0^{\rm ipm} + \bar{a}_{\rm ipm} T^2 \quad \text{with} \quad \bar{a}_{\rm ipm} = (\pi^2/6)\, \bar{g}(e_F). \quad (5.35)$$

Hence, the excitation energy of the smooth part is simply $\bar{\mathcal{E}^*}_{\rm ipm} = \bar{a}_{\rm ipm} T^2$. At this stage we have to remember again that this smooth part will involve states far away from the Fermi level, which is to say in regions where formula (5.31) cannot be trusted anymore. As for the situation at $T = 0$, one may try to get information from experimental evidence. For the $\bar{\mathcal{E}}$ this implies not only replacing $\bar{E}_0^{\rm ipm}$ by the liquid drop energy $E_{\rm LD}$ (at $T = 0$) but also the level density parameter $\bar{a}_{\rm ipm}$ by some phenomenological quantity $\bar{a}$,

$$\bar{\mathcal{E}}_{\rm ipm} \quad \Longrightarrow \quad \bar{\mathcal{E}} = E_{\rm LD}(T=0) + \bar{a} T^2 \equiv \bar{\mathcal{E}}_{\rm LD}(T). \quad (5.36)$$

Clearly, this $\bar{a}$ should reflect macroscopic behavior. One might try to use for $\bar{g}(e_F)$ the estimate of the normal Fermi gas given in (1.25), namely $g(e) =$

$(3A/2)\sqrt{e/e_{\rm F}^3}$. In this case one would obtain the value $a_{\rm F.G.} \approx A/15$, which unfortunately turns out to be too small. This may not be such a big surprise, however, as the estimate (1.25) corresponds to an infinite system of fermions, and thus not to the smooth level density of a real finite nucleus. Indeed, in (Tōke and Swiatecki, 1981) Tōke and Swiatecki have shown that the $a_{\rm F.G.}$ gets enlarged to near the realistic value simply by taking into account surface effects. In fact, the influence of space variations of the mean field on the single particle level density was seen already above in the Wigner–Kirkwood expansion. Tōke and Swiatecki apply the relation (1.23) between space density and Fermi momentum to the local density $\rho(r) = g_{\rm s.i.}(k_{\rm F}(r))^3/(6\pi^2)$ of a system of nucleons in a finite spherical potential $V(r)$. Relating the $k_{\rm F}(r)$ to the local Fermi energy by $(\hbar k_{\rm F}(r))^2 = 2m(e_{\rm F} - V(r))$ the $\rho(r)$ follows the functional form of the potential. Integrating $\rho(r)$ over the volume enclosed by the surface at which $e_{\rm F} - V(r)$ vanishes, one gets the total particle number $A = \int d^3x \rho(r)$ as functional of $e_{\rm F} - V(r)$. The energy density of states (at the Fermi surface) then is simply given by the derivative $g(e) = dA/de$ taken at $e_{\rm F}$. To distinguish contributions from the bulk from those of the surface region (as well as from curvature terms etc.) the authors evaluate the integral by applying a "leptodermous expansion".

Returning to the thermostatic aspects, it is evident that smooth parts of other quantities may be calculated from the first law of thermostatics, as given in (22.54). Assuming for the moment particle number, volume and collective variables to be fixed one gets

$$\overline{S} = 2\overline{a}T \quad \text{and} \quad \overline{\mathcal{F}} = E_{\rm LD} - \overline{a}T^2 \qquad (5.37)$$

for entropy and free energy at low temperatures where $\overline{a}_{\rm ipm}$ is independent of T. At times, in the literature the relation $\overline{S} = 2\overline{a}T$ is used even for a T-dependent $\overline{a}$; this question is addressed again in Section 10.1.1.

5.5.2 Contributions from the oscillating level density

In Section 22.5.2 this problem is treated for the general case that the oscillating part of the level density is expressed by the form

$$\delta g(e) = \sum_{n=1}^{\infty} A_n(e) \cos\left(W_n(e)/\hbar\right), \qquad (22.153)$$

which encompassed that of the Gutzwiller trace formula (5.28). From this level density one may evaluate the variation $\delta\Omega_{\rm gce}^{\rm ipm}$ of the grand potential which then allows one to deduce other thermal quantities of interest.

Here we like to specify to the schematic model described in (Bohr and Mottelson, 1975). One looks at a nucleus with $\mathcal{A}_{\rm cs}$ particles sitting in closed shells and $\mathcal{A} - \mathcal{A}_{\rm cs}$ particles in the valence shell. To simplify matters one assumes the smooth part of the level density to be a constant $\overline{g} = g_0$ and the fluctuating part to be given by

$$\delta g(e) = f \cos\bigl(2\pi(e - e_0)/(\hbar\Omega_0^{\rm sh})\bigr). \qquad (5.38)$$

Here, $\hbar\Omega_0^{\rm sh}$ defines the width of the open shell and e_0 its center. The chemical potential $\mu_{\rm cs}$ associated with the closed shells will then be given by $\mu_{\rm cs} = (e_0 - \hbar\Omega_0^{\rm sh})/2$ such that $\mathcal{A}_{\rm cs} = g_0 \mu_{\rm cs}$. In the same spirit we may introduce the smooth part $\overline{\mu} = \overline{\mu}(\mathcal{A})$ of the actual chemical potential $\mu(\mathcal{A})$ which must fulfill the condition $g_0 \overline{\mu} = g_0 \mu_{\rm cs} + (\mathcal{A} - \mathcal{A}_{\rm cs}) \equiv \mathcal{A}$. The relation of the particle number $\mathcal{A}$ to the chemical potential μ, on the other hand, is given by

$$\mathcal{A} = \int_0^\mu de\, g_0 + \delta\mathcal{A} = g_0 \mu + \delta\mathcal{A}. \tag{5.39}$$

This implies that the oscillating part $\delta\mu$ of the chemical potential is given by $\delta\mu = \mu - \overline{\mu} = -\delta\mathcal{A}/g_0$.

Having set up the model, we may proceed by applying formula (22.158) for the oscillating part of the grand potential. Comparing (5.38) with (22.153) we may identify $A(e) = f$ and $W(e) = 2\pi(e-e_0)/(\hbar\Omega_0^{\rm sh})$. With $\tau(T) = T 2\pi^2/(\hbar\Omega_0^{\rm sh})$ and since there is only one term formula (22.158) reduces to

$$\delta\Omega_{\rm gce}^{\rm ipm}(T,\mu) = \delta\Omega_0(\mu)\, R(T) = T\, \frac{f}{2}\, \hbar\Omega_0^{\rm sh}\, \frac{\cos\left(\frac{2\pi}{\hbar\Omega_0^{\rm sh}}\left(\frac{\alpha}{\beta} - e_0\right)\right)}{\sinh\tau(T)}. \tag{5.40}$$

The two factors $\delta\Omega_0(\mu)$ and $R(T)$ are found from (22.157) and (22.156) to be

$$\delta\Omega_0(\mu) = f\left(\frac{\hbar\Omega_0^{\rm sh}}{2\pi}\right)^2 \cos\left(\frac{2\pi}{\hbar\Omega_0^{\rm sh}}(\mu - e_0)\right) \qquad R(T) = \frac{\tau(T)}{\sinh(\tau(T))}, \tag{5.41}$$

respectively. The $R(T)$ is seen to be *independent* of μ. In (5.40) $\alpha = \mu\beta = \mu/T$ has been introduced to facilitate the calculation of the oscillating parts of particle number $\delta\mathcal{A}$ and internal energy $\delta\mathcal{E}$, as described in Chapter 22. In particular, one may make use of (22.135) and (22.136) replacing there $\ln Z_{\rm gce}^{\rm ipm}$ by $-\beta\Omega_{\rm gce}^{\rm ipm}$. Hence, the fluctuating part for particle number becomes

$$\delta\mathcal{A} = -\left(\frac{\partial\beta\delta\Omega_{\rm gce}^{\rm ipm}}{\partial\alpha}\right)_\beta = f\frac{\hbar\Omega_0^{\rm sh}}{2\pi}\frac{\tau(\beta)}{\sinh\tau(\beta)}\sin(W(\mu)) \tag{5.42}$$

and for the internal energy one gets

$$\delta\mathcal{E}(T,\mu) = \left(\frac{\partial\beta\delta\Omega_{\rm gce}^{\rm ipm}}{\partial\beta}\right)_\alpha \tag{5.43}$$

$$= \mu\frac{\hbar\Omega_0^{\rm sh}}{2\pi} f \frac{\tau}{\sinh\tau}\sin(W(\mu)) + \left(\frac{\hbar\Omega_0^{\rm sh}}{2\pi}\right)^2 f\frac{\tau^2\cosh\tau}{\sinh^2\tau}\cos(W(\mu)).$$

These relations are valid for given temperature and chemical potential. For practical applications one would like to express the latter by the number of particles, as given by (5.39), which is an implicit equation for μ. Without solving that explicitly we may proceed as follows. Firstly, one needs to include the contributions from the smooth level density. In this way the total energy becomes

$$\mathcal{E}_{\mathrm{ipm}}(T,\mu) = \int_0^\mu de\, e\, g_0 + \frac{\pi^2}{6} g_0 T^2 + \delta\mathcal{E}(T,\mu) = \overline{\mathcal{E}}(T,\mu) + \delta\mathcal{E}_{\mathrm{ipm}}(T,\mu). \quad (5.44)$$

Its smooth and oscillating parts are given by

$$\overline{\mathcal{E}}(T,\mu) = \overline{E}_{\mathrm{ipm}}(T=0) + \frac{\pi^2}{6} g_0 T^2 \quad (5.45)$$

$$\delta\mathcal{E}_{\mathrm{ipm}}(T,\mu) = g_0 \overline{\mu}\delta\mu + \frac{1}{2} g_0 (\delta\mu)^2 + \delta\mathcal{E}(T,\mu), \quad (5.46)$$

respectively, which is in accord with (5.35). Considering $\delta\mu = -\delta\mathcal{A}/g_0$, together with (5.43) for $\delta\mathcal{E}(T,\mu)$ the shell correction $\delta\mathcal{E}_{\mathrm{ipm}}(T,\mu)$ becomes

$$\delta\mathcal{E}_{\mathrm{ipm}}(T,\mu) = \left(\frac{\hbar\Omega_0^{\mathrm{sh}}}{2\pi}\right)^2 f \left(\frac{\tau^2 \cosh\tau}{\sinh^2\tau} \cos(W(\mu)) - \frac{f}{2g_0} \frac{\tau^2}{\sinh^2\tau} \sin^2(W(\mu))\right). \quad (5.47)$$

Let us turn to entropy and free energy. The former may be obtained either from the general relation $\Omega_{\mathrm{gce}}^{\mathrm{ipm}} = -T\mathcal{S} + \mathcal{E} - \mu\mathcal{A}$ of thermostatics (see Section 22.1.2.2) or by applying directly (22.159) to get

$$\delta S_{\mathrm{ipm}} = \hbar\Omega_0^{\mathrm{sh}} \frac{f}{2} \cos(W(\mu)) \frac{\tau \coth\tau - 1}{\sinh\tau} \quad (5.48)$$

for the shell correction of the entropy, with the macroscopic part given by $\overline{S}_{\mathrm{ipm}} = (\pi^2/3)g_0 T$. Similarly, for the free energy $\delta\mathcal{F} = \delta\mathcal{E} - T\delta S$ one has

$$\delta\mathcal{F}_{\mathrm{ipm}} = \left(\frac{\hbar\Omega_0^{\mathrm{sh}}}{2\pi}\right)^2 f \left(\cos(W(\mu)) \frac{\tau}{\sinh\tau} - \frac{f}{2g_0} \sin^2(W(\mu)) \frac{\tau^2}{\sinh^2\tau}\right). \quad (5.49)$$

In (Bohr and Mottelson, 1975) all formulas are written in a form where $W(\mu)$ is replaced by $2\pi(y - 1/2)$ with the parameter y being defined through $\hbar\Omega_0^{\mathrm{sh}} y = \mu - \mu_{\mathrm{cs}}$. It may be expressed by particle number through the implicit equation

$$\frac{\mathcal{A} - \mathcal{A}_{\mathrm{cs}}}{g_0 \hbar\Omega_0^{\mathrm{sh}}} = y - \frac{f}{2\pi g_0} \frac{\tau}{\sinh\tau} \sin(2\pi y), \quad (5.50)$$

which follows from $g_0\overline{\mu} \equiv \mathcal{A}$ and $\delta\mu = -\delta\mathcal{A}/g_0$, together with (5.42).

The results for all three quantities $\delta\mathcal{E}_{\mathrm{ipm}}$, δS_{ipm} and $\delta\mathcal{F}_{\mathrm{ipm}}$ exhibit very nicely not only the disappearance of shell effects with increasing τ in general, but also that this happens exponentially fast. In Fig. 5.6 a numerical calculation is shown in which the approximation given by (5.49) is compared with a full calculation along the lines described at the beginning of this section; for details see (Ivanyuk and Hofmann, 1999). It is seen that shell effects disappear at temperatures around 1.5 MeV, and also that this schematic model reproduces the essential properties quite well. A comparison of this model to computations within a realistic single particle model has already been done in 1973 by A. S. Jensen and J. Damgaard (1973). For easy practical applications one may still simplify formulas (5.47) and (5.49), in particular. In (5.49) it is seen that the second term in the brackets decreases with τ much faster than the first one. Considering only this first term one may rewrite this formula as

$$\delta\mathcal{F}_{\text{ipm}}(Q,T) = \delta\mathcal{F}_{\text{ipm}}(Q,T=0)\,\frac{\tau}{\sinh\tau}\,. \tag{5.51}$$

The $\delta\mathcal{F}(Q,T=0)$ is nothing else but the shell correction $\delta\mathcal{E}(T=0) = \delta E$ at zero thermal excitation. Knowing this one may easily get a fair estimate for the T-dependence of the shell correction to the free energy.

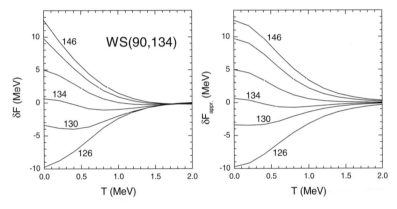

FIG. 5.6. The shell correction of the free energy of (5.49) for different numbers of particles in the valence shell above $\mathcal{A}_{\text{cs}} = 126$. The calculations on the left have been performed for the spectra of a Woods–Saxon potential (WS) adjusted to the nucleus Th^{224} with 90 protons and 134 neutrons.

The disappearance of shell effects with increasing thermal excitation has important consequences for the potential landscape over which collective motion proceeds. In Section 5.3 it has been seen that at $T = 0$ the static energy exhibits detailed structure. In one-dimension this manifests itself in the possible existence of more than one minimum and barrier. In multidimensional spaces different valleys may exist which lead to different fission modes.

Exercises

1. Show that for an oscillator in three dimensions (of frequency ϖ but without spin-orbit potential) to second order in $\hbar$ the level density becomes $g_{\text{ETF}}(e) = (e^2 - (\hbar\varpi)^2/4)/(2\,(\hbar\varpi)^3)$. One may start from the fully quantal expression $Z(\beta) = (8\sinh^3(\beta\hbar\varpi/2))^{-1}$ for the partition function.

6
AVERAGE COLLECTIVE MOTION OF SMALL AMPLITUDE

In the previous chapter we convinced ourselves that the static energy of the system may be calculated on the basis of the Hamiltonian of independent particle motion, even if the latter is approximated by that of the phenomenological shell model. The ultimate reason for this feature holding true is the Strutinsky energy theorem of Section 5.1.2. To render the latter a powerful tool for practical applications one only has to replace the smooth part of the energy by that of the liquid drop model. In this way contributions from states far away from the Fermi level are parameterized phenomenologically. Now we are going one step further in assuming that these features still apply to time-dependent processes. Indeed, we will be concerned with *slow collective motion* for which the additional kinetic energy put into the system is very small, in particular in comparison to the Fermi energy e_F if calculated per particle. At first we will only be looking at zero thermal excitations, in which case this additional energy is simply the kinetic energy for collective motion. Later we are going to generalize to finite temperatures T, which again will be assumed small in the sense $T \ll e_F \approx \mu$. Thus it should be quite safe to assume that the *total energy* $E_{\rm tot}$ of the system can be calculated as the average $E_{\rm tot} = \langle \hat{H} \rangle$ of a Hamiltonian which may be constructed like

$$\hat{H} = \hat{H}(\hat{x}_i, \hat{p}_i, Q(t)) = \hat{H}_{\rm ipm}(\hat{x}_i, \hat{p}_i, Q) - \overline{E}_{\rm pot} = \sum_{\alpha=1}^{A} \hat{h}(\hat{\mathbf{x}}_\alpha, \hat{\mathbf{p}}_\alpha, Q) - \overline{E}_{\rm pot} \quad (6.1)$$

by exploiting (5.3) and (5.4). The "ground state" of the "nucleonic degrees of freedom" is the solution of the set

$$\hat{H}(\hat{x}_i, \hat{p}_i, Q)|n; Q\rangle = E_n(Q)|n; Q\rangle \quad (6.2)$$

which corresponds to the lowest energy $E_0(Q)$ called $E(Q)$ henceforth, in accord with (5.4). We will soon see that the derivative of $\hat{H}(\hat{x}_i, \hat{p}_i, Q(t))$ with respect to Q will play a dominant role: It acts as a kind of a "generator of collective motion" and will be abbreviated by

$$\hat{F}(\hat{x}_i, \hat{p}_i, Q) = \frac{\partial \hat{H}(\hat{x}_i, \hat{p}_i, Q)}{\partial Q} = \frac{\partial \hat{H}_{\rm ipm}(\hat{x}_i, \hat{p}_i, Q)}{\partial Q} - \frac{\partial \overline{E}_{\rm pot}}{\partial Q}. \quad (6.3)$$

Please observe that in the expression on the right-hand side only the first part is essential. The additional c-number might at best contribute to diagonal matrix elements of the $\hat{F}(\hat{x}_i, \hat{p}_i, Q)$, which will not play any role for the quantities which determine dynamical properties of the system. As the $\hat{H}_{\text{ipm}}(\hat{x}_i, \hat{p}_i, Q)$ is of *one-body nature* so is the $\hat{F}(\hat{x}_i, \hat{p}_i, Q)$. Hamiltonians of this type have frequently been used in theories of nuclear collective motion. We would just point to (Bohr and Mottelson, 1975) and (Siemens and Jensen, 1987), but many other textbooks may be referred to as well. Later on we may wish to generalize this concept by allowing the $\hat{H}(\hat{x}_i, \hat{p}_i, Q)$ to contain a two-body interaction which will be assumed insensitive to variations of Q. It is to represent the effects of collisional damping, which in the end will be responsible for the microscopic damping mechanism.

To simplify the discussion we will restrict ourselves to the example of one collective variable Q. The generalization to the multidimensional case will be sketched later in Section 6.2.2. Furthermore, in this chapter we will only be looking at the linear case; treatments of collective motion of large scale will be deferred to Chapter 7. Although in practical applications the Q is usually identified as the deformation parameter of the shell model potential the formal developments described below can be considered to be more generic. One may often just interpret the $Q(t)$ as representing a parameter specifying the time dependence of the nuclear density. The self-consistency problem behind this hypothesis will be touched upon explicitly later for separable interactions, both on the level of average motion in Section 6.2.3 as well as for quantal fluctuations in Section 7.5. In this context we should also like to refer to Section 24.4 where this problem is studied with functional integrals.

6.1 Equation of motion from energy conservation

For an *isolated* system like a nucleus the total energy E_{tot} must be a conserved quantity. This fact may be used as one possible way to obtain the equation of motion for $Q(t)$. Indeed, differentiating with respect to t we get from Ehrenfest's theorem:

$$\frac{d}{dt} E_{\text{tot}} = \frac{d}{dt} \langle \hat{H} \rangle_t = \left\langle \frac{\partial \hat{H}}{\partial Q} \right\rangle_t \dot{Q} = 0 \quad \Longrightarrow \quad \left\langle \frac{\partial \hat{H}}{\partial Q} \right\rangle_t = 0. \qquad (6.4)$$

The last equation follows for any dynamical situation for which the collective velocity is finite.

6.1.1 Induced forces for harmonic motion

As it stands, eqn(6.4) is not yet of great practical use. This situation changes if we stick to motion of small amplitude. In this case we may use an expansion of the Hamiltonian to *second order* and write

$$\hat{H}(\hat{x}_i, \hat{p}_i, Q) \approx \hat{H}(\hat{x}_i, \hat{p}_i, Q_0) + (Q - Q_0) \left. \frac{\partial \hat{H}}{\partial Q} \right|_{Q_0} + \frac{1}{2}(Q - Q_0)^2 \left\langle \left. \frac{\partial^2 \hat{H}}{\partial Q^2} \right|_{Q_0} \right\rangle_{Q_0}^{\text{gs}}. \qquad (6.5)$$

Remember, please, that the Q_0 need not necessarily correspond to the minimum of the static energy; this point will be clarified later. For linear motion it is necessary to keep the operator form for the term of first order but not for that of second order. There, it suffices to take the expectation value as evaluated at $Q = Q_0$. The field operator $\partial \hat{H}/\partial Q$ needed in (6.4) is then given by

$$\frac{\partial \hat{H}}{\partial Q} = \hat{F}(\hat{x}_i, \hat{p}_i, Q_0) + (Q - Q_0) \left\langle \frac{\partial^2 \hat{H}}{\partial Q^2} \bigg|_{Q_0} \right\rangle^{gs}_{Q_0}. \tag{6.6}$$

If plugged into the condition on the right of (6.4) one gets the equation

$$\left\langle \hat{F}(\hat{x}_i, \hat{p}_i, Q_0) \right\rangle_t = -(Q - Q_0) \left\langle \frac{\partial^2 \hat{H}}{\partial Q^2} \bigg|_{Q_0} \right\rangle^{gs}_{Q_0}, \tag{6.7}$$

which is of much simpler structure. Indeed, the term on the right-hand side is already of first order in $(Q - Q_0)$. The left-hand side is more subtle. Here we need to calculate the expectation value of $\hat{F}(\hat{x}_i, \hat{p}_i, Q_0)$ at time t, again to first order in $(Q - Q_0)$. In zero order there will the term

$$\langle \hat{F} \rangle^{gs}_{Q_0} = \frac{\partial E(Q)}{\partial Q} \bigg|_{Q_0} \tag{6.8}$$

which represents a constant static force. Likewise, for the second factor on the right of (6.7) one has

$$\left\langle \frac{\partial^2 \hat{H}}{\partial Q^2} \bigg|_{Q_0} \right\rangle^{gs}_{Q_0} = \frac{\partial^2 E(Q)}{\partial Q^2} \bigg|_{Q_0} + \chi(0). \tag{6.9}$$

Both formulas are derived in Section 22.1.1 for finite thermal excitations, see (22.20) and (22.26), respectively; one only has to observe that for the ground state both the free energy and the internal energy become identical to $E(Q) = \text{tr}\, \hat{\rho}_{\text{gs}}(Q_0)\hat{H}$, where $\hat{\rho}_{\text{gs}}(Q_0) = |0; Q_0\rangle\langle 0; Q_0|$. The $\chi(0)$ is given by the expression (22.25) with $X = T = S = 0$ and is nothing else but the static response. With the abbreviation $\hbar\Omega_{n0} = E_n - E_0$, for $T = 0$ it may be written as as

$$\chi(0) = \frac{1}{\hbar} \sum_n \langle 0|\hat{F}|n\rangle \langle n|\hat{F}|0\rangle \left[\mathcal{P} \frac{2}{\Omega_{n0}} \right], \tag{6.10}$$

where the symbol $\mathcal{P}$ stands for "principal value" and is meant to prevent summation over terms with $\Omega_{n0} = 0$. In writing these formulas the shorthand notation $\hat{F} \equiv \hat{F}(\hat{x}_i, \hat{p}_i, Q_0)$ has been introduced. To avoid confusion we will continue to write $\partial \hat{H}/\partial Q$ or $\hat{F}(\hat{x}_i, \hat{p}_i, Q)$ whenever this derivative has to be evaluated at a $Q \neq Q_0$. Finally, we note that the relations (6.8) and (6.9) are of general nature. They may be derived by calculating from the Hamiltonian (6.5) the static energy $E(Q) = E(Q_0) + (Q - Q_0)(\partial E(Q)/\partial Q)_{Q_0} + (1/2)(Q - Q_0)^2(\partial^2 E(Q)/\partial^2 Q)_{Q_0}$ within second order perturbation theory.

To evaluate the time-dependent average $\langle \hat{F} \rangle_t$ we may take advantage of the benefits of linear response theory described in Chapter 23. This is possible because the relevant part of the perturbation factorizes into the product $(Q(t) - Q_0)\hat{F}$. The $(Q(t) - Q_0)$ may be considered as a time-dependent "external" field, in the sense of being "external" to the nucleonic system for whose operator $\hat{F}$ the expectation value is to be calculated. The corresponding unperturbed density operator is the $\hat{\rho}_{\rm qs}(Q_0, T_0)$ which represents the system at Q_0 and is specified by the Hamiltonian $\hat{H}(\hat{x}_i, \hat{p}_i, Q_0)$. In the terminology of Section 23.3 it is the FF response function which matters here. In the interaction picture with $\hat{F}^I(t) = \exp(i\hat{H}(Q_0)t)\hat{F}\exp(-i\hat{H}(Q_0)t)$ its causal part can be defined as

$$\tilde{\chi}(t-s) = \Theta(t-s)\frac{i}{\hbar}{\rm tr}\left(\hat{\rho}_{\rm qs}(Q_0,T_0)\left[\hat{F}^I(t), \hat{F}^I(s)\right]\right), \qquad (6.11)$$

such that the induced force becomes

$$\langle \hat{F} \rangle_t = \langle \hat{F} \rangle_{Q_0}^{\rm gs} - \int_{-\infty}^{+\infty} \tilde{\chi}(t-s)(Q(s)-Q_0)\,{\rm d}s\,. \qquad (6.12)$$

These formulas follow after applying eqs. (23.1), (23.44) to 23.46). As in Section 23.3 when going from (23.43) to (23.46), the lower limit of the integral in (6.12) has been set equal to $-\infty$.

6.1.2 Equation of motion

Above we succeeded in expressing $\langle \hat{F} \rangle_t$ in terms of the collective variable. From (6.12) together with (6.7) and (6.8) we thus obtain:

$$\left.\frac{\partial E(Q)}{\partial Q}\right|_{Q_0} - \int_{-\infty}^{\infty}\tilde{\chi}(t-s)(Q(s)-Q_0)\,{\rm d}s + (Q-Q_0)\left\langle\left.\frac{\partial^2 \hat{H}}{\partial Q^2}\right|_{Q_0}\right\rangle_{Q_0}^{\rm gs} = 0\,. \quad (6.13)$$

The harmonic nature of this equation is more readily seen for the variable

$$q(t) = Q(t) - Q_m \quad {\rm with} \quad \left.\frac{\partial E(Q)}{\partial Q}\right|_{Q_0} + \left.\frac{\partial^2 E(Q)}{\partial Q^2}\right|_{Q_0}(Q_m - Q_0) = 0, \quad (6.14)$$

where Q_m defines the position of the center of the local oscillator. As Q_m and Q_0 do not depend on time and because of $\int_{-\infty}^{\infty}\tilde{\chi}(t-s)\,{\rm d}s = \chi(0)$ one has

$$\int_{-\infty}^{\infty}\tilde{\chi}(t-s)(Q(s)-Q_0)\,{\rm d}s = \int_{-\infty}^{\infty}\tilde{\chi}(t-s)q(s)\,{\rm d}s + (Q_m - Q_0)\chi(0)\,. \quad (6.15)$$

To proceed further it is convenient to introduce the quantity

$$-k^{-1} = \left\langle\left.\frac{\partial^2 \hat{H}}{\partial Q^2}\right|_{Q_0}\right\rangle_{Q_0}^{\rm gs} = \left.\frac{\partial^2 E(Q)}{\partial Q^2}\right|_{Q_0} + \chi(0) \equiv C(0) + \chi(0)\,, \qquad (6.16)$$

which actually will serve as a kind of coupling constant later on. Here, $C(0)$ is the local stiffness of the total static energy. A short calculation shows that the equation for $q(t)$ reduces to

$$k^{-1} q(t) + \int_{-\infty}^{\infty} \widetilde{\chi}(t-s) q(s) \, \mathrm{d}s = 0 \,. \tag{6.17}$$

A Fourier transformation leads to the *secular equation* for the possible local frequencies of the harmonic motion:

$$k^{-1} + \chi(\omega) = 0 \,, \tag{6.18}$$

which because of (6.16) may be rewritten as

$$\chi(\omega) - \chi(0) = C(0) \,. \tag{6.19}$$

The response function for the operator $\hat{F}$ needed here is given by

$$\chi(\omega) = -\frac{1}{\hbar} \sum_n |F_{no}|^2 \left[\frac{1}{\omega - \Omega_{n0} + i\epsilon} - \frac{1}{\omega + \Omega_{n0} + i\epsilon} \right] = \chi'(\omega) + i\chi''(\omega) \,. \tag{6.20}$$

The quantities introduced on the very right are the reactive and dissipative parts of $\chi(\omega)$, which (for real frequencies) can be identified as its real and imaginary parts, respectively. This form follows directly from those presented in Section 23.3.4 for the multidimensional case, see e.g. eqn(23.70). The states $|n\rangle$ are those of eqn(6.2) with excitation energies $\hbar\Omega_{n0}$, all calculated at $Q = Q_0$. To shorten the presentation, the Q_0 is not being written explicitly.

Before we discuss solutions of the secular equation (6.19) in more detail a first orientation may be gained from a *degenerate model*. Suppose the operator $\hat{F}$ only allows for transitions from the ground state $|0\rangle$ to a set of N degenerate states $|n\rangle$ with excitation energy $\hbar\Omega$. Then the solutions of (6.19) can be seen to be given by

$$\left(\frac{\omega^\pm}{\Omega}\right)^2 = \frac{C(0)}{\chi(0) + C(0)} \,. \tag{6.21}$$

As this ratio differs from unity for any non-vanishing $\chi(0)$ the imaginary part of the response vanishes identically: $\chi''(\omega^\pm) = 0$ (see exercise 1). This is the reason *why* in this model the collective motion comes out undamped: There is no overlap of its frequency (or energy) with the Ω of the "nucleonic" degrees of freedom; hence there can be no *real* transfer of the energy; recall formula (23.62). The excitations of the nucleonic degrees of freedom play some role only in a "virtual sense". Often the static response is much larger than the stiffness of the static energy, $\chi(0) \gg C(0)$, in which case one gets $\omega^\pm \ll \Omega$. Realize, please, that this model may even be applied to the case of a negative stiffness, if only $\chi(0) > |C(0)|$, rendering the $\omega^\pm$ purely imaginary as it should be for motion across a barrier. We shall return to this important case below.

Before we address damping of collective motion it is worthwhile to first continue with the undamped case. This is done best in the single particle picture in which the relevant excitations are of one-particle one-hole nature. At zero temperature this has been the classic example of collective motion in nuclei and may be found in all textbooks, e.g. (Bohr and Mottelson, 1975), (Siemens and Jensen, 1987) or (Ring and Schuck, 2000).

6.1.3 One-particle one-hole excitations

For the pure independent particle model the response function $\chi(\omega)$ takes on the form

$$\chi(\omega) = \sum_{jk} |F_{kj}|^2 \, \frac{n(e_k) - n(e_j)}{\hbar\omega - e_{kj} + i\epsilon}, \tag{6.22}$$

proven in Section 23.3.4 for the grand canonical ensemble. At $T = 0$ the Fermi occupation numbers reduce to $n(e_l) = 1$ when $e_l \leq e_F$ and $n(e_l) = 0$ for $e_l > e_F$. Again, the shorthand notation $|F_{kj}|^2 = \langle j|F|k\rangle\langle k|F|j\rangle$ is used for the squared matrix elements of the single particle operator $\hat{F} = \partial \hat{h}(\hat{\mathbf{x}}, \hat{\mathbf{p}}, Q)/\partial Q|_{Q_0}$, and $e_{kj} = e_k - e_j$ for the energy differences. The states $|k\rangle$ and the single particle energies e_k are determined by the one-body Schrödinger equation, $\hat{h}(Q_0) \mid k; Q_0\rangle = e_k(Q_0) \mid k; Q_0\rangle$. Notice that for the $\hat{F}$ the c-number term $\partial \overline{E}_{\rm pot}/\partial Q$ has been omitted as it does not contribute to the commutator in the response function (6.11). For the reactive and dissipative parts of $\chi(\omega) = \chi'(\omega) + i\chi''(\omega)$ one gets from (6.22)

$$\chi'(\omega) = \sum_{jk} n(e_j) |F_{kj}|^2 \, \frac{2e_{kj}}{e_{kj}^2 - (\hbar\omega)^2} \tag{6.23}$$

$$\chi''(\omega) = -\pi \sum_{jk} |F_{kj}|^2 \left(n(e_k) - n(e_j)\right) \delta(\hbar\omega - e_{kj}). \tag{6.24}$$

From these forms it becomes apparent that the response functions are determined by the 1p-1h excitations introduced in Section 3.3.1.

Let us turn to collective excitations, the frequencies of which are determined by the secular equation (6.18) or, more conveniently, by (6.19). For stable modes with $C(0) > 0$, after separating real and imaginary parts and using (6.23) and (6.24), eqn(6.19) is seen to be equivalent to

$$\chi''(\omega_\nu) = 0 \quad \Longleftrightarrow \quad \begin{cases} \chi'(\omega_\nu) - \chi(0) = C(0) \\ \omega_\nu^* = \omega_\nu \quad \neq e_{kj}. \end{cases} \tag{6.25}$$

The dissipative part vanishes identically just because the real part of (6.19) leads to real solutions which are different from the frequencies of the (original) nucleonic modes. The solutions of (6.25) are commonly described graphically. For this method our way of expressing the coupling constant through the static response and the stiffness becomes particularly useful. This can be seen from Fig. 6.1 in which a situation with three nucleonic modes is sketched. The solution

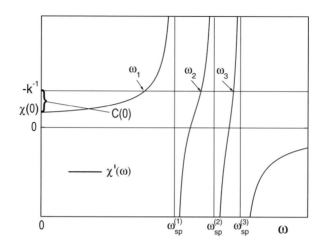

FIG. 6.1. The reactive part of the response function $\chi'(\omega)$ for a schematic model of three undamped intrinsic modes at $\omega = \omega_{sp}^{(i)}$ (from (Kiderlen et al., 1992)); the physical meaning of the coupling constant can be seen on the left vertical axis; the collective modes ω_ν are found from $\chi'(\omega_\nu) = -1/k$.

ω_1 defines the frequency of the truly collective mode, the nature of which will become more apparent below. The plot clearly exhibits the sensitivity of ω_1 on the stiffness of the static energy.

For unstable modes the $C(0)$ is negative. In this case *one real* solution gets lost but it shows up as a *purely imaginary* one. To examine this property let us write $\omega = i\omega_b$ with $\omega_b = \omega_b^*$ such that one gets from (6.22) and (6.23)

$$\chi(i\omega_b) \equiv \chi'(i\omega_b) = \sum_{jk} |F_{kj}|^2 \, (n(e_j) - n(e_k)) \, e_{kj} \left(e_{kj}^2 + (\hbar\omega_b)^2 \right)^{-1}.$$

For $\omega_b > 0$ this $\chi(i\omega_b)$ is a *monotonically decreasing* function of ω_b starting from $\chi(i\omega_b = i0) = \chi(0)$. Indeed, its derivative with respect to ω_b can be written as

$$\partial \chi(i\omega_b)/\partial \omega_b = -2\omega_b \hbar^2 \sum_{jk} |F_{kj}|^2 \, (n(e_j) - n(e_k)) \, e_{kj} \left(e_{kj}^2 + (\hbar\omega_b)^2 \right)^{-2}.$$

For positive ω_b this expression is negative, simply because $(n(e_j) - n(e_k)) \, e_{kj} \geq 0$. In this way it is readily verified that (6.19) has *just one* solution $\omega_b^1 > 0$.

6.2 The collective response function

In the previous section the technique of linear response theory served as an ideal means of formulating a perturbation approach. An appropriate response function was introduced to parameterize the influence of the collective degrees of freedom on the nucleonic system, calculated to linear order in $(Q - Q_0)$. However, traditionally, linear response theory is used to describe the excitations of the system induced by *truly external* sources. We are now going to define such a scheme also for the present case. To this end a coupling to a time-dependent external field $f_{\text{ext}}(t)$ is introduced by adding $f_{\text{ext}}(t)\hat{F}$ to our Hamiltonian $\hat{H}(Q)$ to define a

$$\hat{H}_{\text{sc}} = \hat{H}(\hat{x}_i, \hat{p}_i, Q) + f_{\text{ext}}(t)\hat{F}. \tag{6.26}$$

The $\hat{H}(Q)$ will be taken in the approximate form (6.5) obtained by expanding to second order in $(Q - Q_0)$. The coupling is chosen to have the same form as that between the two "subsystems" of collective and nucleonic degrees of freedom, with only the $(Q(t) - Q_0)$ replaced by the "external field" f_{ext}. The important feature is that the operator part is the same in both cases: it represents the quantity or property of our system whose average $\langle \hat{F} \rangle_t$ we want to study. Although this "external field" may not always represent a measurable quantity the form (6.26) delivers a formal basis for identifying possible excitation modes.

For any finite $df_{\text{ext}}(t)/dt$ the energy associated with the Hamiltonian $\hat{H}$ will no longer stay constant. Above, this energy was denoted by E_{tot} to indicate that it represents the *total* energy of our nucleus consisting both of the collective part plus an eventual "heat" associated with nuclear excitations. The rate of change of E_{tot} can be obtained from Ehrenfest's equation, again, which in the present case reads

$$\frac{d}{dt}E_{\text{tot}} \equiv \frac{d}{dt}\langle \hat{H} \rangle_t = \frac{i}{\hbar}\langle [\hat{F}, \hat{H}] \rangle_t f_{\text{ext}} + \left\langle \frac{\partial \hat{H}}{\partial Q} \right\rangle_t \dot{Q}, \tag{6.27}$$

simply because time evolution now is determined by $\hat{H}_{\text{sc}}$. Tentatively, the last term on the right of (6.27) has been included. However, in the previous discussion for $f_{\text{ext}} = 0$ we were able to deduce a complete scenario for collective motion by putting this term equal to zero. In fact there is no reason why we should give up the condition $\langle \partial \hat{H} / \partial Q \rangle_t = 0$ which would imply (6.27) to reduce to

$$\frac{d}{dt}E_{\text{tot}} = -f_{\text{ext}}\frac{d}{dt}\langle \hat{F} \rangle_t. \tag{6.28}$$

Firstly, in this way (6.28) becomes a natural generalization of (6.4), and we may take over many of the equations derived above. More important is, however, the physical consequence we gain from such a condition: The system then behaves as if *effectively* the $\hat{H}(Q(t))$ were time *independent*. In this way the collective coordinate $Q(t)$ is rendered an *internal* variable, simply representing some particular excitation of the nucleus, but which per se has nothing to do with an *external* quantity.[12] Moreover, it is conceivable that collective excitations could be described by a time-*independent* Hamiltonian; actually we will elaborate on this feature in Section 6.2.3.

---

[12] Unfortunately, it must be said that such a misinterpretation of the nature of shape degrees of freedom is found too often in the literature. A discussion of collective motion in the same spirit as used here may be found in (Siemens and Jensen, 1987).

After these preliminary but important remarks we may now proceed to discuss the *collective response function*. It will be defined through the relation

$$\delta\langle\hat{F}\rangle_\omega = -\chi_{\text{coll}}(\omega) f_{\text{ext}}(\omega), \tag{6.29}$$

where for convenience we have switched to frequency representation. The difference between the $\delta\langle\hat{F}\rangle_\omega$ and the $\langle\hat{F}\rangle_\omega$ should at best have a term proportional to $\delta(\omega)$, representing some static force. To simplify matters let us assume here that the Q_0 corresponds to the minimum of the potential where this conservative force (6.8) vanishes,

$$Q_0 = Q_m \quad \Longleftrightarrow \quad \langle\hat{F}\rangle_{Q_0}^{\text{gs}} = \left.\frac{\partial E(Q)}{\partial Q}\right|_{Q_0} = 0. \tag{6.30}$$

The construction of $\chi_{\text{coll}}(\omega)$ may be performed following the procedure of Clausius and Mossotti in deriving the polarizability function for electric media, which (at zero thermal excitations) has previously been applied to the nuclear case (Bohr and Mottelson, 1975). Realize, please, that the external perturbation $f_{\text{ext}}(t)\hat{F}$ ought to be added to the "induced" one given by $q(t)\hat{F}$, both acting on the system defined by $\hat{H}(Q_0)$. Hence, in generalization of (6.12) the $\langle\hat{F}\rangle_t$ can be calculated as

$$\langle\hat{F}\rangle_t = -\int_{-\infty}^{+\infty} \tilde{\chi}(t-s)\left(f_{\text{ext}}(s) + Q(s) - Q_0\right) ds. \tag{6.31}$$

Next we may take advantage of the fact that (as a consequence of (6.4)) eqn(6.7) is still valid, which together with (6.16) implies

$$k\langle\hat{F}\rangle_t = Q - Q_0. \tag{6.32}$$

From (6.32) and (6.31), together with the definition (6.29) of the collective response function, one easily gets the final and well-known result

$$\chi_{\text{coll}}(\omega) = \frac{\chi(\omega)}{1 + k\chi(\omega)}. \tag{6.33}$$

The notion "collective" response function may now become clear. The $\chi_{\text{coll}}(\omega)$ has its poles at the solutions of the secular equation (6.18), which define the collective excitations. For undamped motion we might instead have used the more conventional notation "RPA" (Random Phase Approximation), as often done in the literature, see (Siemens and Jensen, 1987), for instance. Actually, the form (6.33) allows one to understand better the distinct physical meaning of the collective and intrinsic (or nucleonic) response functions. Both differ whenever k is finite but the $\chi_{\text{coll}}(\omega)$ will become identical to $\chi(\omega)$ for $k \to 0$. As seen from (6.16), the limit $1/k \to \infty$ is reached in case that, as a function of Q, the static energy becomes extremely stiff. Then no collective motion will be possible any more, such that an external force may only excite the purely "nucleonic modes" manifested in $\chi(\omega)$. This feature is most directly visible at the result (6.21) for the degenerate model. Finally we wish to add a comment on (6.32). This form reminds one of a *self-consistency* condition. As will be discussed below, it results from treating an effective two-body interaction $(k/2)\hat{F}\hat{F}$ in mean-field approximation. However, such a notion is justified on more general grounds. It is this relation (6.32) which expresses most clearly that the $Q(t)$ ought to be considered an *internal* variable–measuring some properties of the nucleus and clearly to be distinguished from truly external fields like the f_{ext}.

6.2.1 Collective response and sum rules for stable systems

Important information is contained in the pole structure of the collective response function (6.33), which we are now going to examine. To be specific let us concentrate on stable systems for which the solutions ω_ν of the secular equation (6.25) are real; generalizations to unstable modes may be found in (Hofmann, 1997), (Hofmann and Kiderlen, 1998). As can be seen from (6.23) the poles come in pairs: For each ω_ν the $-\omega_\nu$ is a solution as well. The residues of these poles are easily seen to be

$$[\text{Res}\{\chi_{\text{coll}}(\omega)\}]^{-1}_{\omega_\nu} = -k^2 \frac{\partial \chi'}{\partial \omega}\bigg|_{\omega_\nu}. \qquad (6.34)$$

Since χ' is an even function of ω the residue at $-\omega_\nu$ is the negative of that at ω_ν. With the abbreviation $R_\nu = \text{Res}\{\chi_{\text{coll}}(\omega_\nu)\} = -\text{Res}\{\chi_{\text{coll}}(-\omega_\nu)\}$ and assuming only poles of first order, one may write down a pole expansion in a form analogous to (4.41),

$$\chi_{\text{coll}}(\omega) = \sum_\nu R_\nu \left[\frac{1}{\hbar\omega - \hbar\omega_\nu + i\epsilon} - \frac{1}{\hbar\omega + \hbar\omega_\nu + i\epsilon} \right]. \qquad (6.35)$$

In the sum (6.35) each term looks like the response function of an undamped oscillator (see Section 23.1), if we only express the residue of mode ν through its inertia M_ν^F as

$$R_\nu = \frac{\hbar}{M_\nu^F 2\omega_\nu} \quad \text{with} \quad M_\nu^F = \frac{\hbar}{2\omega_\nu} k^2 \frac{\partial \chi}{\partial \omega}\bigg|_{\omega_\nu}. \qquad (6.36)$$

The value of the residues depends on the slope of the function $\chi'(\omega)$, the larger the latter the smaller will be the former. For the inertia it is just the opposite. Typically the closer the ω_ν lie to their corresponding unperturbed excitation energies e_{kj}, the larger will be the slope. This feature can be read off from Fig. 6.1. On the other hand, for a small $C(0)$ the lowest pair of modes $\pm\omega_1$ becomes to lie in a region where the slope is very small. In the ideal case all strength may be concentrated in this mode, which then is as collective as may be: Since no other excitations are relevant all particles in a sense contribute to this motion. This feature can be followed more closely by exploiting "sum rules", like that associated with the energy weighted sum. The significance of the latter may be seen by referring to the formula for the energy transfer studied in Section 23.3.3. Written for a one-dimensional case with $E_{\rm sc} = \langle \hat{H}_{\rm sc} \rangle$ eqn(23.62) reads

$$\overline{E_{\rm sc}} = \int_{-\infty}^{\infty} \frac{d\omega}{2\pi} \, |f_{\rm ext}(\omega)|^2 \, \omega \, \chi''_{\rm coll}(\Omega) \, . \tag{6.37}$$

Suppose the external field excites all frequencies with equal probability, meaning that $f_{\rm ext}(\omega)$ is *in*dependent of ω. For convenience we may choose it equal to unity to get "energy weighted sum"

$$\overline{E_{\rm sc}} = \int_{-\infty}^{\infty} \frac{d\omega}{2\pi} \, \omega \, \chi''_{\rm coll}(\Omega) \equiv S_F^{(1)} \, . \tag{6.38}$$

We are now going to prove that it has the same value no matter whether we use the nucleonic $\chi(\omega)$ or the collective response $\chi_{\rm coll}(\omega)$, provided they are related by (6.33).

As shown in Section 23.3.2, for any response function $\chi_{\mu\nu}(\omega)$ the negative value of the energy weighted sum is identical to the limit $\lim_{\omega \to \infty} \omega^2 \chi_{\mu\nu}(\omega)/2$, provided some physically reasonable conditions are fulfilled. Obviously, the form (6.33) of the collective response function implies the following relation between two limits, $\lim_{\omega \to \infty} \omega^2 (\chi_{\rm coll}(\omega) - \chi(\omega)) = 0$, from which it follows that

$$S_F^{(1)} = \int_{-\infty}^{+\infty} \frac{d\Omega}{2\pi} \chi''_{\rm coll}(\Omega) \, \Omega = \int_{-\infty}^{+\infty} \frac{d\Omega}{2\pi} \chi''(\Omega) \, \Omega = \frac{i}{2\hbar} \langle [\hat{\dot{F}}, \hat{F}] \rangle \, , \tag{6.39}$$

with $\hat{\dot{F}} = (i/\hbar)[\hat{H}(Q_0), \hat{F}]$. The last expression on the right may be proved from the definition of the response function together with elementary properties of Fourier transforms, for details see Section 23.3.2. In this way the energy weighted sum is expressed by a double commutator of one-body operators and may thus be calculated without greater difficulties. By analogy with the sum rule of the oscillator (see eqn(23.14)), let us write the value of $2S_F^{(1)}$ as an inverse inertia,

$$2 \, S_F^{(1)} = i \, \langle [\hat{\dot{F}}, \hat{F}] \rangle / \hbar \equiv 1/m_0^F \, . \tag{6.40}$$

There is an interesting relation between the m_0^F introduced here and the M_ν^F of the pole expansion (6.35): As is easily inferred from (6.39), each individual term of (6.35) contributes a value of exactly $1/M_\nu^F$ to the overall strength. Since all masses must be positive it follows that $M_\nu^F \geq m_0^F$. Evidently, the value $M_1 = M(\omega_1)$ can be approximated by m_0^F the better the more the strength of the single mode "1" exhausts the sum.[13] It may be noted that these features

---

[13]It can be shown (Bohr and Mottelson, 1975) that the m_0^F is related to the inertia of the liquid drop model, more precisely to the irrotational flow value, which for multipole vibrations of polarity l about the sphere of radius R is given by $M_{\rm irr} = (3/4\pi) A m_N R^2 / l$, for A nucleons of mass m_N. The strict identity $m_0^F = M_{\rm irr}$ requires certain assumptions about the quantal evaluation of the moments in the nuclear radii.

remain essentially valid even for damped motion which will be discussed below. Then the pole expansion (6.35) simply has to be replaced by (6.113). Finally we wish to add that the $1/m_0^F$ of (6.40) can be expressed as the expectation value of the operator $(i/\hbar)[\hat{\dot{F}}, \hat{F}] = \sum_{\alpha=1}^{A}(\partial \hat{F}/\partial \mathbf{x}_\alpha)^2$ in case there is no momentum dependence in $\hat{H}(Q_0)$ other than that of the kinetic energy. For any $\hat{F}$ which varies linearly in the $\mathbf{x}_\alpha$ this operator reduces to a mere c-number; this feature is the basis for the famous "f-sum rule" (at times also-called "oscillator sum rule" or "Thomas–Reiche–Kuhn sum rule").

Next let us look at the sum which is weighted by Ω^{-1}. Its value can be expressed by the static response (recalling eqns(23.47) and (23.48))

$$2\, S_F^{(-1)} = \int_{-\infty}^{+\infty} \frac{d\Omega}{\pi} \frac{\chi''_{\text{coll}}(\Omega)}{\Omega} = \chi''_{\text{coll}}(\Omega = 0) = \frac{\chi(0)}{1+k\chi(0)}\,. \tag{6.41}$$

Using here the expansion (6.35) each term contributes an amount C_ν^{-1}, the stiffness of the oscillator associated with $\pm\omega_\nu$ (see (23.14)) which for stable systems must be positive. This leads to the inequality $C_1^F \geq (1+k\chi(0))/\chi(0)$. After multiplication by k^{-2} and observing relation (6.16) one gets

$$\frac{1}{k^2} C_1^F \geq -\frac{C(0)}{k\chi(0)} \geq C(0)\,. \tag{6.42}$$

6.2.2 Generalization to several dimensions

By analogy with the one-dimensional case (6.29) the collective response tensor is to be defined as

$$\delta\langle \hat{F}_\mu \rangle_\omega = -\sum_{\nu=1}^{n} \chi^{\text{coll}}_{\mu\nu}(\omega) f^{\text{ext}}_\nu(\omega)\,. \tag{6.43}$$

This equation is written for an example of n degrees of freedom. The f^{ext}_ν represent generalized "external fields" with the coupling to the system being given by $\delta \hat{H} = \sum_{\nu=1}^{n} \hat{F}_\nu\, f^{\text{ext}}_\nu(t)$. The construction of the collective response tensor can be performed by complete analogy with the one-dimensional case, see e.g. (Siemens and Jensen, 1987). As the essential input one needs the intrinsic response tensor $\chi_{\mu\nu}$ and the tensor for the coupling constants. The former is obtained by applying the methods discussed in Section 23.3, to the coupling $\sum_{\nu=1}^{n} \hat{F}_\nu(\hat{x}_i, \hat{p}_i; Q_1, Q_2, \ldots)\, q_\nu(t)$ with the unperturbed Hamiltonian being given by $\hat{H}(\hat{x}_i, \hat{p}_i; Q_1^0, Q_2^0, \ldots)$. The $\chi_{\mu\nu}(t-s)$ are then defined by (23.46) with the A_ν replaced by F_ν and the a_ν^{ext} replaced by q_ν. As for the coupling constants it turns out more convenient to work with the inverse of the k used so far. This means writing

$$-\left(\frac{1}{k}\right)_{\mu\nu} \equiv -\kappa_{\mu\nu} = \left. \frac{\partial^2 E(Q_\mu)}{\partial Q_\mu^0 \partial Q_\nu^0} \right|_{Q_\mu^0} + \chi_{\mu\nu}(\omega=0)\,, \tag{6.44}$$

in generalization of the expressions derived in Section 6.1.2, see (Samhammer et al., 1989). The collective response tensor turns out to be given by

$$\underline{\chi}^{\text{coll}}(\omega) = \underline{\kappa} \left(\underline{\kappa} + \underline{\chi}(\omega)\right)^{-1} \underline{\chi}(\omega) \qquad (6.45)$$

by analogy with (6.33). Here the convention is used to mark tensors by underlining the corresponding symbol. The frequencies of the possible collective modes must satisfy the following equations:

$$\det\left((\underline{\chi}^{\text{coll}}(\omega))^{-1}\right) = 0 \quad \Longrightarrow \quad \det\left(\underline{\kappa} + \underline{\chi}(\omega)\right) = 0\,. \qquad (6.46)$$

The first follows directly from (6.43), after inverting there the $\chi_{\mu\nu}^{\text{coll}}$ and putting the f_μ^{ext} equal to zero. Notice that we are seeking modes for self-sustained motion which will exist even for vanishing external fields. The second, more convenient version of the secular equation follows from (6.45). When dealing with these response tensors, it is important to know whether they obey certain symmetry relations. We may suppose all fields $\hat{F}_\mu$ to have the same symmetry with respect to time reversal. According to (23.53) this implies that for each component of the tensor $\underline{\chi}(\omega)$ the reactive and dissipative parts are even and odd functions of frequencies, respectively, and the tensor itself is easily seen to be symmetric $\chi_{\mu\nu}(\omega) = \chi_{\nu\mu}(\omega)$.

Multipole vibrations: Prime examples for nuclear collective modes (Bohr and Mottelson, 1975), (Siemens and Jensen, 1987) are found by multipole expansions of the mean field described in Section 3.2.2.2. In this case the shape degrees of freedom are given by the α_{lm} which according to (1.70) determine the deviation of the radius $R(\vartheta, \phi)$ from that of the corresponding sphere of the same volume. For potentials of the type $U(r - R(\vartheta, \phi))$ and according to eqn(3.36), the perturbing field counted per particle becomes

$$F_{lm} = -R_0 \frac{\partial U}{\partial r} Y_{lm}(\vartheta, \phi)\,. \qquad (6.47)$$

For the coupling constants needed for the secular equation one often follows the suggestion of A. Bohr and B.R. Mottelson who argue for the form (1975)

$$\kappa_l \equiv \frac{1}{k_l} = -\frac{R_0}{2\pi^2} \int d^3r \, \frac{1}{r^2} \frac{\partial}{\partial r} \left(\frac{\partial U}{\partial r}\right) \rho_0\,, \qquad (6.48)$$

where ρ_0 represents the spherical density distribution of the ground state. As shown in (Siemens and Jensen, 1987), for $T = 0$ this expression can be related to the forms given above in (6.16) or (6.44). The advantage of our version is the possibility of its generalization to vibrations away from the potential minimum as well as for finite thermal excitations.

In the Nilsson model, or for oscillator potentials, the F_{lm} are often taken as $F_{lm} = r^l Y_{lm}$. Then the following structure for the unperturbed response function

$\chi''(\omega)$ appears. The transitions are mainly determined by the change ΔN of the major quantum number N (for the oscillator). Whereas for the quadrupole with $l \equiv \lambda = 2$ (see exercise 2) only excitations occur which are associated with $\Delta N = 0$ and $\Delta N = 2$, for $l \equiv \lambda = 3$ one has $\Delta N = 1, 3$. Such a behavior reflects simple selection rules. If there were no spin-orbit interactions the states would be degenerate. With a finite spin-orbit coupling transitions within a group of given ΔN are also possible. It may be said that according to the rule $\hbar\Omega_0 = \hbar\omega_0 = 41 \text{ MeV}/(A)^{1/3}$ the excitation energies may reach are quite large values. For this reason the application of the pure independent particle model may become doubtful even at zero thermal excitations. Indeed, experimentally the so-called "giant resonances", which are collective modes of excitation more than about 10 MeV, exhibit considerable widths. It is only for the low frequency excitations, which for the quadrupole are determined by $\Delta N = 0$, that the collective mode may be treated without damping.

6.2.3 Mean field approximation for an effective two-body interaction

So far our discussion has been based entirely on the claim to have at one's disposal a reasonable guess for the single particle field $\hat{F}(\hat{x}_i, \hat{p}_i, Q)$–as defined by the derivative of the basic Hamiltonian $\hat{H}(\hat{x}_i, \hat{p}_i, Q)$. No reference has been made to an explicit two-body interaction. It may be instructive to establish such a connection within a schematic, albeit widely used model before we move on to describe dissipative motion. Indeed, such an $\hat{H}(\hat{x}_i, \hat{p}_i, Q)$ may be interpreted as a mean-field approximation to a separable two-body interaction appearing in a Hamiltonian like (remember the discussion in Section 3.1.3)

$$\hat{H}^{(2)} = \hat{H}_0 + (k/2)\hat{F}\hat{F}, \qquad (6.49)$$

if only $\hat{H}_0 = \hat{H}(Q_0)$ and $\hat{F}$ are taken to be the one-body operators introduced in (6.5) and (6.6). Applying the Hartree approximation to (6.49), discarding the exchange terms, one gets the following expression (see exercise 3):

$$\hat{H}^{(2)}_{\text{mf}} = \hat{H}_0 + k\langle\hat{F}\rangle\hat{F}. \qquad (6.50)$$

For the second term we may again introduce a collective variable by writing

$$k\langle\hat{F}\rangle = (Q - Q_0), \qquad (6.51)$$

thus recovering the first two terms of our linearized version of the Hamiltonian $H(Q)$ as given in (6.5). Where is the third one? It is easily recognized as the term one needs to regain the feature which we had exploited before, namely that the total energy of the system can be expressed by the average of a one-body Hamiltonian. Indeed, from the Hartree procedure one knows that in mean-field approximation the total energy of the system is given by

$$E_{\text{tot}} = \langle\hat{H}^{(2)}_{\text{mf}}\rangle - (k/2)\langle\hat{F}\rangle\langle\hat{F}\rangle. \qquad (6.52)$$

The last term on the right-hand side is easily seen to be identical to the last term of (6.5) by only accounting for the definition (6.51) as well as for the identity (6.16) for the coupling constant.

Please bear in mind that the physics contained in the form (6.49) goes beyond average dynamics. In fact, this Hamiltonian may be taken as the basis for deriving a Hamiltonian for the system of *all* relevant degrees of freedom, which must then have a part representing collective motion in *quantized* form. One way to achieve this goal is to apply the Bohm–Pines procedure, which will be addressed in Section 7.5.1. Another possibility is given by the Hubbard–Stratonovich transformation within the functional integral approach explained in Section 24.4.

So far the nature of the expectation values in (6.50) to (6.52) has not been specified. As a matter of fact, the discussion applies both to the time-dependent case as well as to the static situation. In the first case the "choice" (6.51) is demanded to make this equation identical to (6.32). In the second case we would just have to apply static perturbation theory to get the total static energy to second order in $Q - Q_0$. Such a calculation should *not be mixed up* with the calculation of a *constrained* energy $E_{tot}^{const}(Q)$, that energy which is calculated from the wave functions of a Hartree approximation obeying a subsidiary condition like (6.51). Notice, please, that there is no need to consider (6.51) if one wants to get (6.52) from (6.49) together with (6.50). Moreover, as we know from above, our Q is *not an external* variable, so it would not make sense trying to fix it from outside. Rather it is the dynamical situation which decides which values of Q will be reached in the course of motion.

To understand this point of view more clearly, it may be worthwhile for the reader actually to carry through the constraining procedure to get the energy $E_{tot}^{const}(Q)$. The latter has to be defined as in (6.52)

$$E_{tot}^{const}(Q) = \langle \hat{H}_{mf}^{(2)} \rangle^{const} - (k/2)\langle \hat{F} \rangle^{const} \langle \hat{F} \rangle^{const}, \qquad (6.53)$$

but with the averages to be evaluated from the *constrained* mean field. The latter is obtained by adding to (6.50) a static *external* field, say $f_{ext}\hat{F}$. It has the same meaning as before in eqn(6.26), exept that now the f_{ext} is time *in*dependent and that it can be chosen such as to guarantee (6.51) to be fulfilled for static averages. All the formulas derived in the previous subsection apply, except that now the response functions have to be taken at zero frequency. A straightforward calculation leads to the following answer:

$$\Delta E_{tot}^{const}(Q) \equiv E_{tot}^{const}(Q) - E(Q_0) = \frac{\left(\langle \hat{F} \rangle^{const}\right)^2}{2\chi_{coll}(\omega = 0)} = \frac{(Q - Q_0)^2}{2k^2\chi_{coll}(\omega = 0)}. \qquad (6.54)$$

It is the inverse static collective response which determines the stiffness of the constrained energy. By exploiting the general form (6.33), together with the relation (6.16), which lead to $k^2\chi_{coll}(\omega = 0) = -k\chi(0)/C(0)$, the variation of the constrained energy becomes

$$\Delta E_{tot}^{const}(Q) = \frac{1}{2}C_{const}(Q - Q_0)^2 \quad \text{with} \quad C_{const} \equiv \frac{C(0)}{-k\chi(0)} \geq C(0). \qquad (6.55)$$

It is seen that for bound motion the effective stiffness of the constrained energy is *larger than* $C(0)$. The inequality (6.42) implies that the effective stiffness of a typical collective *mode* cannot be smaller than the C_{const}. Here, we anticipate that according to (6.117) the C_1^F/k^2 is the stiffness of the lowest mode ω_1 if described in terms of the collective variable itself.

6.2.4 Isovector modes

It goes without saying that the examples of collective dynamics treated so far were those for which neutrons and protons *move in phase*. This happens to be the relevant type for large-scale motion with an eventual behavior typical of transport phenomena in which we are primarily interested. For small amplitude motion there is, however, another class of modes where neutrons and protons vibrate against each other. In this case one speaks of *isovector modes* in distinction from the *isoscalar* ones discussed above. This notion is related to the value of isospin associated with these modes. Indeed, the driving force results from the symmetry (or charge excess) terms in the total energy, which typically are proportional to the square of the z-component of total isospin $T_z = (Z - N)/2$. Such terms appear both in eqn(1.28) for the kinetic energy as well as in the Bethe–Weizsäcker formula (1.65). Following A. Bohr and B.R. Mottelson (1969, 1975) one may deduce a corresponding mean field δU for single particle motion as $\delta U = (1/2)\hat{t}_z U_{\text{sym}}(N - Z)/A$. Restricting oneself to the volume term (1.66), the constant U_{sym} is seen to be positive and of order 100 MeV. One may then proceed and extend the δU to multipole deformations. In this way one obtains coupling constants $\kappa_l = 1/k_l$ analogous with the ones encountered above for multipole vibrations of isoscalar type, but with the important difference that they now are positive. In Bohr and Mottelson (1975) the simple formula $\kappa_l = \pi U_{\text{sym}}/(A\langle r^{2l} \rangle)$ is suggested from which the dependence on the size of the system can be read off. For the dipole this leads to frequencies of the order of $\hbar\varpi_D \approx 80 A^{-1/3}$ MeV, which for medium and heavy nuclei agrees rather well with experimental findings. Deviations from that rule are due to shell effects.

Looking back at the secular equation (6.18) and its graphical realization in Fig. 6.1 one recognizes that the frequencies of such modes will lie *above* those for the intrinsic excitations. It must be said, however, that so far an interpretation of the coupling constant in terms of a static energy is still lacking, as it is given by (6.16) for the examples of isoscalar modes. This is unfortunate, as it is the latter form which leads to the nice generalization to heated systems described in Section 6.5.1. The problem at stake can easily be seen from the results of the previous subsection. For a $k > 0$ the stiffness of the total energy (6.52) is negative. Only the stiffness of the constraint energy (6.55) is positive.

6.3 Rotations as degenerate vibrations

There is no room to discuss rotations in greater detail. However, treating rotations on a similar footing as vibrations one may understand the most important features. To simplify matters we will stick to zero thermal excitations here and refer to Chapter 6.7 for the interesting features which show up at finite temperatures.

Suppose the shell model potential U is deformed but has rotational symmetry about the x-axis, say. Suppose further that this symmetry axis is rotated by an angle α in the x, y plane. This rotation will be described by the operator $\hat{R} = \exp(-i\hat{j}_z\alpha/\hbar)$ such that the single particle Hamiltonian transforms according to

$$\hat{h}' = e^{-i\hat{j}_z\alpha/\hbar}\hat{h}e^{+i\hat{j}_z\alpha/\hbar} \simeq i\alpha[\hat{h},\hat{j}_z]/\hbar, \qquad (6.56)$$

where the last expression holds true for an infinitesimally small α. On the other hand, the linear dependence of $\hat{h}'$ on α may be obtained by appealing to a geometric interpretation. One may express the coordinates of the rotated frame in terms of the lab frame by rotating with the negative angle $-\alpha$. This means writing

$$\hat{h}' \equiv \hat{h}(\mathcal{R}^{-1}\mathbf{r};\mathbf{p}) = \hat{h}(\mathbf{r};\mathbf{p}) - \alpha\partial\hat{h}/\partial\alpha. \qquad (6.57)$$

This last expression becomes even more clear if the vector $\mathbf{r}$ is expressed in polar coordinates, which implies to have $\hat{h}(\mathcal{R}^{-1}\mathbf{r};\mathbf{p}) = \hat{h}(r,\vartheta,\phi-\alpha;\mathbf{p})$.

Obviously, such expansion may be identified with the one we had before in terms of the collective variable. This means evaluating the energy of the rotation simply as the kinetic energy put into the $Q = \alpha$. Evidently, in this case there is no restoring force, as the rotation is supposed to be a free one. Hence, there will be no collective potential and the total collective energy will simply be given by $\Delta E \equiv E_{\rm rot} = \mathcal{J}\dot{\alpha}^2/2$. Moreover, the "effective frequency" of the associated "vibration" is zero, for which reason the inertia will be the one in the "zero frequency limit". (This "frequency" is not to be mixed up with the rotational frequency $\omega_{\rm rot} = \dot{\alpha}$)). Anticipating results from Section 6.5.2 for the collective "inertia" one gets from (6.94):

$$\mathcal{J} = 2\hbar^2 \sum_{n\neq 0} \frac{|\langle n|\partial H/\partial\alpha|0\rangle|^2}{(E_n-E_0)^3} = 2 \sum_{n\neq 0} \frac{|\langle n|\hat{J}_z|0\rangle|^2}{(E_n-E_0)}. \qquad (6.58)$$

On the very right the total angular momentum $\hat{J}_z = \sum_{k=1}^{A}\hat{j}_z^k$ was used, together with the relation

$$i\hbar\langle 0|\partial\hat{H}/\partial\alpha|n\rangle = \langle 0|[\hat{H},\hat{J}_z]|n\rangle = (E_0-E_n)\langle 0|\hat{J}_z|n\rangle$$

and $\hat{H}$ is the many-body Hamiltonian. The rotational moment of inertia (6.58) was first derived by D.R. Inglis. Often the picture behind this formula is referred to as the cranking model, simply because a "cranking" of the mean field is involved. In practical evaluations the value of $\mathcal{J}$ often comes out close to that of a rigid body; for details we may refer to the very enlightening discussion in §4-3 of (Bohr and Mottelson, 1975). Measured values however are smaller by something like a factor of two or even more. Theoretically, this may be understood by taking into account *pairing correlations*. Experimental values of $\mathcal{J}$ can be obtained from the rotational energy through the common relation $2\mathcal{J} E_{\rm rot} = \hbar^2 J(J+1)$, for more details see Bohr and Mottelson (1975) or other textbooks like (Siemens and Jensen, 1987), (Ring and Schuck, 2000). The formulas given above allow one to understand why the nucleus cannot rotate collectively about an axis of symmetry. In this case the ground state would be an eigenstate of $\hat{J}_z$ and the $\mathcal{J}$ would be identical to zero.

6.4 Microscopic origin of macroscopic damping

In different contexts damping and relaxation phenomena have already been encountered in previous sections. In Section 4.3 the time-dependent amplitude for the transition of a quantal state into itself was studied as well as the conditions under which this dependence shows exponential decay. Related to this problem is the decrease of flux seen for the optical model in Section 1.3.2. We are now going to address damping of average collective motion, the latter being represented by the collective variable $Q(t)$. There, a new feature comes into play as this damping may eventually be understood as dissipation in the sense of non-equilibrium statistical mechanics associated with the production of entropy. It is this property which we are aiming at. The friction forces used in descriptions of heavy-ion collisions or of nuclear fission are interpreted that way. Indeed, the presence of genuine friction implies the existence of fluctuating forces which show up in transport equations of Fokker–Planck type; this topic will be addressed in the next chapter.

The existence of dissipative collective motion is intimately connected to its irreversible behavior. A necessary condition for this to happen is that $Q(t)$ does not show recurrence at large times. But this property may not be sufficient. Eventually, it may be obtained by dealing with a continuous set of internal excitations. One prime example of this type is Landau damping in infinite systems. However, as shown in Vol. X of Landau and Lifshitz (and Pitajewski) on "Physical Kinetics" (Landau *et al.*, 1981) this kind of damping actually conserves entropy. Often finite systems are treated by approximating a discrete set of states by a quasi-continuous one. In this way results are obtained for which an interpretation in the sense of dissipation is very tempting but may not be justified in a true sense. Examples of this type are the DC-conductivity of metal clusters and "wall friction" for slow nuclear collective motion, which is sometimes interpreted as the analog of the former. These cases will be discussed in Chapter 16 and Chapter 17, where it will be shown in which sense a quasi-continuous limit imitates features not warranted by a genuine quantal treatment of the finite system, if for the latter no couplings to configurations more complicated than those of independent particle motion are considered.

In the following we want to demonstrate how the latter may lead to a genuine, finite friction force for nuclear collective motion. We begin with a simple, pedagogical model to establish essential features. In this example couplings to more complicated states are simulated by the presence of finite widths of the

146   *Average collective motion of small amplitude*

unperturbed or undressed states. These widths may be obtained by just performing energy smearing of response functions at zero temperature. In doing so we will leave aside questions of more fundamental nature, which are addressed in (Mackey, 1991), (Dorfman, 1999), for instance, or in (Cohen, 2000). As a matter of fact, later in this section we will be able to demonstrate that heavy nuclei are sufficiently complex to warrant collective dissipation, in particular at finite thermal excitations, if only basic features of nucleonic dynamics are taken into account which had been discussed in previous sections.

6.4.1 *Irreversibility through energy smearing*

Let us suppose the nucleonic excitations are described by the response function (6.20). The associated dissipative parts, related to each other by Fourier transformation, are given by

$$\chi''(\omega) = \frac{\pi}{\hbar} \sum_n \left| \hat{F}_{n0} \right|^2 \left[\delta(\omega - \Omega_{n0}) - \delta(\omega + \Omega_{n0}) \right], \quad (6.59)$$

$$\widetilde{\chi}''(t) = -\frac{i}{\hbar} \sum_n \left| \hat{F}_{n0} \right|^2 \sin(\Omega_{n0} t). \quad (6.60)$$

In writing these expressions a discrete level spacing has tacitly been assumed. Let us for the moment dwell further on this picture and let us introduce the average level spacing $D = \Delta \hbar \Omega_{n0} / \Delta N$, with ΔN being the number of levels in the interval $\Delta \hbar \Omega_{n0}$. The $\widetilde{\chi}''(t)$ is a quasi periodic function with a more or less

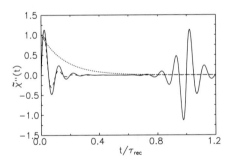
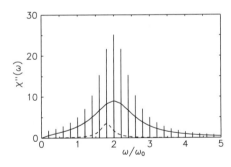

FIG. 6.2. Response functions for a schematic model (see text and (Hofmann, 1997)): time dependence on the left and Fourier transform on the right. The fully drawn lines correspond to (6.60) and (6.59) for a discrete spectrum; the dashed line on the left is obtained after averaging the frequency spectrum with the g_{av} of (6.63), with $\tilde{g}_{\mathrm{av}}(t)$ and $g_{\mathrm{av}}(\omega)$ represented by the dotted lines.

pronounced recurrence behavior. The recurrence time τ_{rec} is related to the inverse level spacing by $\tau_{\mathrm{rec}} \sim 2\pi\hbar/D$. To make the following discussion as transparent as possible let us *suppose to be given the schematic case of equally spaced, nondegenerate levels*, for which one has $\hbar\Omega_{n0} = nD$. For this model (6.60) is nothing else but a Fourier sine expansion with $\tau_{\rm rec}$ being the fundamental period such that the relation $\tau_{\rm rec} = 2\pi\hbar/D$ becomes *exact*.

The behavior of the function $\widetilde{\chi}''(t)$ inside the interval $0 \leq t \leq \tau_{\rm rec}$ depends on the distribution of the squared matrix elements $\left|\hat{F}_{n0}\right|^2$ as function of n or energy. A beautiful idealization is the Lorentzian

$$\left|\hat{F}_{n0}\right|^2 = \frac{D}{2\pi}\overline{F^2}\,\frac{\Gamma_{\rm F}}{\hbar^2(\Omega_{n0} - \Omega_{\rm F})^2 + (\Gamma_{\rm F}/2)^2} \equiv \mathcal{F}^2(\Omega)\Big|_{\Omega_{n0}}, \qquad (6.61)$$

with $\overline{F^2}$ being an averaged squared matrix element. The $\Gamma_{\rm F}$ measures the width of this distribution and $\Omega_{\rm F}$ its centroid. If both parameters are larger than the average spacing one gets for $\widetilde{\chi}''(t)$ the typical behavior shown on the left panel of Fig. (6.2) by the fully drawn line. It shows both relaxation behavior at small times $t \ll \tau_{\rm rec}$, as well as the recurrence at larger times $t \gtrsim \tau_{\rm rec}$. Clearly, this nice relaxation behavior of $\widetilde{\chi}''(t)$ for $t \ll \tau_{\rm rec}$ is a consequence of the particular choice of the parameters. But whatever the distribution of the $\left|\hat{F}_{n0}\right|^2$, there is no doubt that for a *discrete* level spacing the system will never show relaxation behavior in *true sense*: For large times the $\widetilde{\chi}''(t)$ will never really vanish. (A result of a computation within a realistic shell model is presented in Fig. 6.4 below.) The reason for such a behavior can easily be understood physically. Recall please that the $\chi''(\omega)$ stands for possible transfers of energy to the internal degrees of freedom. The form (6.59) represents a case where these excitations go into the specific modes $|n\rangle$, and they stay there because of lack of coupling to additional degrees of freedom. All that happens is that the probability of exciting the individual modes changes with t such that eventually the oscillations from those cancel each other for some *finite* interval of time, but never for ever.

This consideration indicates under which circumstances the situation will be different. There are several possibilities.

(i) For a *given* level spacing we may *introduce irreversible* features, here of the nucleonic dynamics, by *abandoning the exact description* of the excitations. Formally this may be done by associating with each level some uncertainty, in other words, by performing an appropriate energy smearing. Actually this procedure can be carried out–and has been done so in the literature–just for the shell model part of $\hat{H}(\hat{x}_i, \hat{p}_i, Q)$, called $\hat{H}_{\rm ipm}(\hat{x}_i, \hat{p}_i, Q)$ in (6.1). As discussed there, the difference $-\overline{E}_{\rm pot} = \hat{H}(\hat{x}_i, \hat{p}_i, Q) - \hat{H}_{\rm ipm}(\hat{x}_i, \hat{p}_i, Q)$ is a c-number which is immaterial here.

(ii) We may take seriously the fact of having to deal with a residual interaction $\hat{V}_{\rm res}$ which couples the simple states $|n\rangle$ to more complicated ones and, in a sense, makes the excitation "flow" into other modes. Formally, this interaction might be accounted for by replacing our Hamiltonian

$$\hat{H}(\hat{x}_i, \hat{p}_i, Q) \quad \Longrightarrow \quad \hat{H}_{\rm ipm}(\hat{x}_i, \hat{p}_i, Q) + \hat{V}_{\rm res} \qquad (6.62)$$

and by trying to evaluate the influence of $\hat{V}_{\rm res}$ on the response functions, eventually in the way described in Section 4.3. Generally speaking the $\hat{V}_{\rm res}$ couples

1p-1h states to more complicated states. The latter may lie much more densely such that a distinction of individual excitations may become infeasible. Actually, the compound states already form a true continuum at the quite small total excitations of $10-15$ MeV. In fact this situation is given only because at such excitations the nucleus cannot be considered a closed system anymore, as there will always be open channels which couple the system to genuine continuum states.

(iii) Eventually, in cases where the motion of independent particles is chaotic this property in itself may lead to irreversibility. In fact, in (Cohen, 2000) and (Cohen, 2003) it has been argued that non-linear effects in the "collective" velocity $\dot{Q}(t)$ may to some extent be handled by introducing a width Γ the magnitude of which depends on the velocity.

Let us go back to our simple and schematic considerations. Suppose the dissipative response function $\chi''(\omega)$ of (6.59) is smoothed by the Lorentzian

$$g_{\rm av}(\omega - \Omega) = \frac{\Omega_{\rm av}}{(\omega - \Omega)^2 + (\Omega_{\rm av}/2)^2} \qquad (6.63)$$

of width $\Omega_{\rm av} = \Gamma_{\rm av}/\hbar$ according to the rule $\overline{f}(\omega) = \int_{-\infty}^{\infty} (d\Omega/2\pi) f(\Omega) g_{\rm av}(\omega - \Omega)$. One gets the form

$$\overline{\chi}''(\omega) = \frac{1}{2\hbar} \sum_n |\hat{F}_{n0}|^2 \left[g_{\rm av}(\omega - \Omega_{n0}) - g_{\rm av}(\omega + \Omega_{n0}) \right], \qquad (6.64)$$

in which the δ functions simply have been replaced by "smooth" ones. Evidently, the smoothing interval must be larger than the average spacing D, but it should be small compared to the scale Γ_χ for the typical variation of the response function $\chi''(\omega)$ with ω, which means writing

$$D \ll \Gamma_{\rm av} \ll \Gamma_\chi. \qquad (6.65)$$

In the model presented in (6.61) one would have $\Gamma_\chi = \Gamma_{\rm F}$, and all that remains of the functional form of $\overline{\chi}''(\omega)$ is the structure given in (6.61). In (6.65) the left condition implies that the spectrum of the levels $|n\rangle$ can effectively be considered continuous. Thus in a formula like (6.64) we may replace the sum over the discrete states by an integral over the energies $\hbar\Omega \equiv \hbar\Omega_{n0}$. We need only care about the correct normalization. The latter is obtained by the following consideration. From the $\Delta N(\Omega_0)$ states $|n\rangle$ in the interval $\hbar\Omega_0 \leq \hbar\Omega_{n0} \leq \hbar\Omega_0 + \Delta\hbar\Omega_0$ one may evaluate an average matrix element by just summing the $|\hat{F}_{n0}|^2$ over all these states and dividing by $\Delta N(\Omega_0)$. The same result should be obtained if we integrate the function $\mathcal{F}^2(\Omega)$ over the corresponding interval in Ω. Considering the density of states D this implies the following identification

$$\sum_{\Omega_0 \leq \Omega_{n0} \leq \Omega_0 + \Delta\Omega_0} |\hat{F}_{n0}|^2 \implies \int_{\Omega_0 \leq \Omega \leq \Omega_0 + \Delta\Omega_0} \frac{d\hbar\Omega}{D} \mathcal{F}^2(\Omega) \qquad (6.66)$$

which is meaningful under the following condition. The interval $\Delta\Omega_0$ must be sufficiently large in order to contain sufficiently many states, and it must be small enough that the matrix elements may be considered constant in this interval. Because of (6.65) such an interval may be found.

Going back to the response function, by way of (6.66) the form (6.64) turns into

$$\overline{\chi}''(\omega) = \frac{1}{2\hbar} \int \frac{d\hbar\Omega}{D} \mathcal{F}^2(\Omega) \left[g_{\mathrm{av}}(\omega - \Omega) - g_{\mathrm{av}}(\omega + \Omega) \right] \tag{6.67}$$

For the model presented above the $\mathcal{F}^2(\Omega)$ is given by (6.61) and the integral (6.67) turns into a folding integral over two Lorentzians. The latter can be performed analytically, the result being a Lorentzian whose width is just the sum of those of the two factors, $\Gamma = \Gamma_F + \Gamma_{\mathrm{av}}$. That is to say we get:

$$\overline{\chi}''(\omega) = \frac{\overline{F^2}}{2} \left[\frac{\Gamma}{\hbar^2(\omega - \Omega_F)^2 + (\Gamma/2)^2} - \frac{\Gamma}{\hbar^2(\omega + \Omega_F)^2 + (\Gamma/2)^2} \right]. \tag{6.68}$$

Because of (6.65) the width is essentially that of the distribution of the matrix elements in (6.61), $\Gamma \simeq \Gamma_F$. A graphical demonstration can be seen in the right panel of Fig. 6.2, where the original strength distribution (6.59) is shown, together with the smoothed one (6.68) as well as the smoothing function $g_{\mathrm{av}}(\omega)$ of (6.63) (dashed line).

Let us have a first look at consequences for the physics of collective motion. The two different forms (6.59) and (6.68) may lead to a completely different behavior of this motion. In the former case the secular equation (6.18) will be satisfied by real frequencies ω_ν fulfilling eqn(6.25) and implying collective motion to be *undamped*. This is no longer possible as soon as the dissipative response function is a *continuous function* of frequency, in which case the ω_ν become complex. The existence of a finite imaginary part reflects irreversible behavior of the collective degrees of freedom. This is not possible without having irreversible behavior of the nucleonic degrees of freedom in the first place. To understand this feature it may be worthwhile to re-examine the time dependence of the response functions. According to the folding theorem the Fourier transform of $\overline{\chi}''(\omega)$ is just given by

$$\widetilde{\overline{\chi}}''(t) = \widetilde{\chi}''(t)\, \tilde{g}_{\mathrm{av}}(t) \tag{6.69}$$

with $\tilde{g}_{\mathrm{av}}(t)$ being the Fourier transform of $g_{\mathrm{av}}(\omega)$, which for the case of our model is just an exponential function with decay time $2\hbar/\Gamma_{\mathrm{av}}$. Because of the condition (6.65) the factor $\tilde{g}_{\mathrm{av}}(t)$ will change only very little the behavior of the response function for small times, but it cuts out the oscillations at large t. Evidently, $\widetilde{\overline{\chi}}''(t)$ shows "relaxation" behavior, in the sense of having

$$\lim_{|t| \to \infty} \exp\left(\frac{\Gamma |t|}{2\hbar}\right) \widetilde{\overline{\chi}}''(t) = 0. \tag{6.70}$$

Notice that the "relaxation" time is determined by $\Gamma = \Gamma_\chi$ *and not by* Γ_{av}.

6.4.2 Relaxation in a Random Matrix Model

One possibility of studying the influence of the residual interaction introduced in (6.62) is by applying the concept of Random Matrix Theory. In Section 23.4.4 it is demonstrated how the RM-model described in (Weidenmüller, 1980) may be adjusted to linear response. For this purpose one needs to work with finite statistical excitations, as otherwise the RMM is not applicable. In this case the response functions involve transition matrix elements $\langle m|F|n\rangle$ between excited states. Their ensemble average $\overline{\langle m|F|n\rangle}$ is assumed to vanish and for the average of their squares one takes the following Gaussian form

$$\overline{\langle m|F|n\rangle \langle n|F|m\rangle} = \overline{F_{\rm R}^2} \sqrt{D(E_n) D(E_m)} \; \exp\left(-\frac{(E_n - E_m)^2}{2\Gamma_{\rm R}^2}\right) \equiv \mathcal{F}_{\rm R}^2(E_n, E_m) \tag{6.71}$$

instead of the Lorentzian (6.61); bear in mind that the dimension of $\overline{F_{\rm R}^2}$ differs from that of the $\overline{F^2}$. The width $\Gamma_{\rm R}$ of the Gaussian measures the range in energy over which the operator $\hat{F}$ couples. The dissipative part of the response function $\chi''(\omega)$ is shown in (23.110). A Fourier transformation back to time leads to

$$\widetilde{\chi}''(t) = -(i/\hbar)\overline{F^2} \, \exp\left(-t^2/2\tau_{\rm R}^2\right) \sin \Omega_{\rm R} t, \tag{6.72}$$

with the relevant timescales being $\tau_{\rm R} = \hbar/\Gamma_{\rm R}$ and $\Omega_{\rm R} = \Gamma_{\rm R}^2/(2T\hbar)$. For the sake of consistency with the notation used previously in this section the overall pre-factor of the time-dependent functions has again been written in terms of an $\overline{F^2}$ being given by $\overline{F^2} = \overline{F_{\rm R}^2} \sqrt{2\pi} \, \Gamma_{\rm R} \exp\left(\Gamma_{\rm R}^2/8\,T^2\right)$. It is seen that the $\widetilde{\chi}''(t)$ falls off like a Gaussian with a relaxation time $\tau_{\rm R}$. As the $\Gamma_{\rm R}$ is supposed to take on values in the range of $5-9$ MeV the $\tau_{\rm R}$ becomes of the order of 10^{-22} sec.

6.4.3 The effects of "collisions" on nucleonic motion

Let us now address another method of accounting for the residual interaction $\hat{V}_{\rm res}$ of (6.62). As discussed in Section 4.3, such a $\hat{V}_{\rm res}$ leads to a spreading of single particle excitations over more complicated states. A simple way of considering this effect in response functions is by replacing in expressions like (6.24) for the $\chi''(\omega)$ the single particle strength $\varrho(\omega) = \delta(\hbar\omega - e_k)$ of the bare independent particle model by one with a finite width. As suggested by (4.47) we may use the form[14]

$$\overline{\varrho_k}(\omega) = \frac{1}{2\pi} \frac{\Gamma_k(\omega)}{\left(\hbar\omega - e_k - \Sigma'_k(\omega)\right)^2 + \left(\Gamma_k(\omega)/2\right)^2}, \tag{6.73}$$

with $\Sigma'_k(\omega)$ being given by the dispersion relation

$$\Sigma'_k(\omega) = \mathcal{P} \int_{-\infty}^{\infty} \frac{d\Omega}{2\pi} \frac{\Gamma_k(\Omega)}{\omega - \Omega}. \tag{6.74}$$

It represents the real part of the self-energy $\Sigma_k(\omega+i\epsilon) = \Sigma'_k(\omega) - \Gamma_k(\omega)/2$ for the one-body Green function $\mathcal{G}_k(\omega + i\epsilon) = \left(\hbar\omega - e_k - \Sigma'_k(\omega) + \frac{i}{2}\Gamma_k(\omega)\right)^{-1}$. Rather

---

[14] We refrain from indicating explicitly the Δ which in (4.47) has been written to remind one of the fact that a smooth function $\Gamma_k = \Gamma_k(\omega)$ requires energy averages.

than using the original form (6.24) of $\chi''(\omega)$ for these manipulations one may first rewrite this expression as

$$\chi''(\omega) = \pi\hbar \sum_{jk} |F_{kj}|^2 \int_{-\infty}^{\infty} d\Omega \left(n(\hbar\Omega - \frac{\hbar\omega}{2}) - n(\hbar\Omega + \frac{\hbar\omega}{2}) \right) \overline{\varrho_k}(\Omega + \frac{\omega}{2}) \overline{\varrho_j}(\Omega - \frac{\omega}{2})$$
(6.75)

and insert here the $\overline{\varrho_k}(\omega)$ of (6.73). This expression can then be used to obtain the full response function $\chi(\omega)$ from the general relation (23.47) or the reactive part $\chi'(\omega)$ from the Kramers–Kronig relation (23.49).

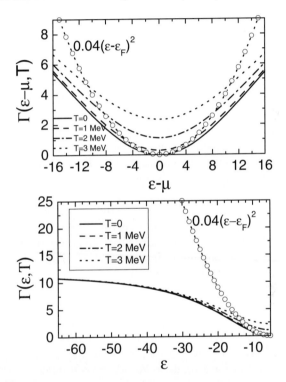

FIG. 6.3. The width Γ of (6.76) as function of energies (NB the different scales on bottom and top) for the parameters $\Gamma_0 = 33$ MeV^{-1} and $c = 20$ MeV. The dashed curves with circles represent the form given in (Mahaux and Sartor, 1991).

The form (6.75) is valid at finite temperature, for which reason also the self-energies should be calculated accordingly. A hint of how that can be done comes from the evaluation of the width on the basis of the collision term of the Landau–Vlasov or Boltzmann–Uehling–Uhlenbeck equations. For infinite systems this was discussed in Section 2.7, with more details to be found in Section 21.3. The width of a particle at energy $e(p)$ is given by (21.26). For the width needed in (6.73), on the other hand, the value of the energy should not be fixed but ought to be taken as a variable quantity, in accord with the form given in (4.46). In generalization of (21.26) one might thus write $\Gamma_k(\omega) = [(\hbar\omega - \mu)^2 + \pi^2 T^2]/\Gamma_0$. In this way the folding integral in (6.75) gets contributions from "off shell". One must be aware, however, that this simple form is bound to become incorrect at larger excitations, be it in temperature or in the $(\hbar\omega - \mu)$. Therefore, in (Siemens et al., 1984) the following modification has been suggested and used thereafter in numerical applications:

$$\Gamma(\omega, T) = \frac{1}{\Gamma_0} \frac{(\hbar\omega - \mu)^2 + \pi^2 T^2}{1 + (1/c^2)\left[(\hbar\omega - \mu)^2 + \pi^2 T^2\right]}. \qquad (6.76)$$

The two parameters appearing here are usually chosen to be $c \simeq 20$ MeV and $\Gamma_0 \simeq 33$ MeV, by applying arguments (Siemens et al., 1984), (Hofmann, 1997) which involve both information on the density dependence of the width as well as on the nucleon's effective mass.

In Section 4.3 the quadratic variation with $e - e_F$ was also seen to reflect very nicely the average behavior in finite nuclei at $T = 0$. In Fig. 6.3 the form (6.76) is shown for different temperatures. The $T = 0$ curve is compared with that of (Mahaux and Sartor, 1991) shown in Fig. 4.3. The right panel of Fig. 6.3 exhibits the behavior at larger negative energies where the increase of the width must diminish on general grounds; the interested reader may compare this figure with Figs. 7.19 and 7.21 of (Mahaux and Sartor, 1991). As a matter of fact, the $\Gamma(\omega, T)$ perhaps ought to decrease more than given by (6.76). Unfortunately, experimental information on this property is not available. Evidently, also the numbers of the parameters are subject to uncertainties, not to mention the very fact that in practical applications any consideration of the influence of shell effects on these widths is out of reach.

The presence of single particle widths can be expected to influence greatly the relaxation behavior of excitations of the nuclear many-body system represented by the response function. In Fig. 6.4 we present a calculation of $\tilde{\chi}''(t)$ performed in (Ivanyuk et al., 1997) for a realistic Woods–Saxon model, with the collective coordinate chosen to represent motion along a fission path for ^{224}Th. The figure displays the $\tilde{\chi}''(t)$ at the minimum and at the saddle of the fissioning system. In both cases two curves are shown. The dashed one corresponds to the pure independent particle model. In this case no relaxation in a true sense is observed. After the first big oscillation there appear many of smaller amplitude which can be seen for a very long time. Actually, there is no reason why they should die out at all and the fact that in the calculation they seem to get weaker may eventually be due to "numerical viscosity". On the other hand, the fully drawn curve decays to zero in a true sense. It is obtained by taking into account the damping mechanism just explained with the finite single particle width $\Gamma(\omega)$ of (6.76). Concerning the relaxation at small t several interesting features can be

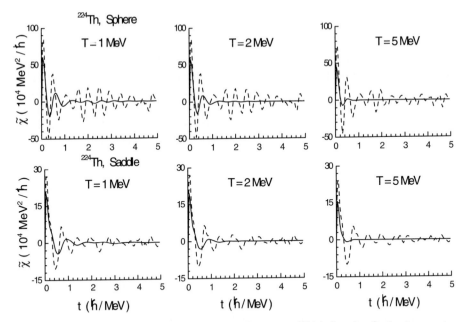

FIG. 6.4. The time-dependent response function $\tilde{\chi}''(t)$ for the fissioning system of ^{224}Th at two characteristic deformations. The dashed curve corresponds to a model of independent particles; for the fully drawn curve collisional damping is taken into account, with the parameters of $\Gamma(\omega)$ as specified in Fig. 6.3, see also (Ivanyuk et al., 1997).

noted. (i) The "relaxation time" τ_{intr} is very similar to that which one may wish to associate with the decay time seen also for the case of the pure shell model (of independent particles). (ii) Its value is of the order of 0.2 to 0.5$\hbar$/ MeV. iii) The smaller values are reached at higher temperatures and correspond roughly to those found previously for the RMM. Commonly, the collective timescale τ_{coll} is several times larger, though not by orders of magnitude.

Finally we should like to comment on the physical nature of the approximations behind formula (6.75) as well as on some of its deficiencies. By way of construction it is assumed that particles propagate independently of each other, even under the presence of the residual interaction $\hat{V}_{\text{res}}$. The latter dresses the particles' state but in such a way that the self-energy only depends on the energy of the state but not on its quantum numbers. In this way degeneracies are not lifted, as one would expect from the presence of an interaction $\hat{V}_{\text{res}}$. This fact has consequences for the behavior of response and correlation functions at very small frequencies, as will be exhibited in Section 6.6.3.1, whereas it may not be very relevant at larger frequencies. It is important, however, to recognize that it is the *nucleonic response* $\chi(\omega)$ which is calculated in this way. This function is *not* supposed to be influenced by *collective* motion, which is to say, none of its poles are to be associated with collective modes. The latter are represented by *collective* response functions of the type shown in (6.33). In the following we will focus on studying their properties as far as they are related to transport behavior. It is worth mentioning, though, that in the past approximations similar to those exploited here have been applied also to other problems, like the description of giant resonances (at $T = 0$) or of deeply lying hole states (Dover et al., 1972), (Klevansky and Lemmer, 1983). Often it has been argued that the assumption of independent propagation of particles and holes would overestimate the total width (see e.g. (Bertsch et al., 1983) and (Wambach, 1988)) claiming that the latter were given by the sum of the widths of the particle and hole individually. However, such a feature is given only if the integral in (6.75) is computed by an "on shell" approximation. Without such a restriction the folding integral leads to a considerable reduction of the total width. This has been demonstrated in (Alkofer et al., 1988) for an analytically solvable model.

6.5 Damped collective motion at thermal excitations

In this section we will be looking for the consequences of microscopic damping on macroscopic motion. From the previous discussion it may be clear that finite collective damping may require finite excitations of the nucleonic degrees of freedom. We shall assume them to be those of thermal equilibrium. This implies replacing the density operator for the ground state by a statistical one,

$$\hat{\rho}_{\text{gs}}(Q_0) \implies \hat{\rho}_{\text{qs}}[H(Q_0), X]. \tag{6.77}$$

The index "qs" refers to the quasi-static picture assumed to hold true here. The X is the parameter by which we may specify the local equilibrium by either entropy $\mathcal{S}$ or temperature T. The concept "local" is to be understood in the sense of the variable Q. Relevant thermostatic relations to be applied below can be found in Section 22.1.1, the quasi-static picture is described in greater detail in sections 23.6 and 23.7. It requires a clear separation of the relevant microscopic (intrinsic) and macroscopic (collective) timescales, $\tau_{\text{micro}} \ll \Delta t \ll \tau_{\text{macro}}$, in particular when applied to large-scale motion. In accord with this assumption the dynamics in X may also be considered to be local in time. In other words, with respect to this variable non-Markovian features will be discarded. Otherwise, it should be treated on the same footing as the collective variable Q, meaning that an *adequate equation of motion for the $Q(t)$ can only be found if its variation is studied consistently with that in X*. Locally, the collective motion will again be described by making use of linear response theory, with the response function $\chi(\omega)$ evaluated as described in the previous section.

6.5.1 The equation of motion at finite thermal excitations

The derivation of the equation of motion may largely follow the scheme already used at $T = 0$. We may still make use of the conservation of the total energy $E_{\text{tot}}(t) = \langle \hat{H}(\hat{x}_i, \hat{p}_i, Q(t)) \rangle_t$. Hence, eqn(6.4) is still valid and after performing the expansion of $\hat{H}(\hat{x}_i, \hat{p}_i, Q(t))$ to second order we get the equation

$$\langle \hat{F}(\hat{x}_i, \hat{p}_i, Q_0)\rangle_t = -(Q(t) - Q_0)\left\langle \left.\frac{\partial^2 \hat{H}}{\partial Q^2}\right|_{Q_0}\right\rangle^{qs}_{Q_0, S_0} \equiv (Q(t) - Q_0)\,\kappa, \quad (6.78)$$

by analogy with (6.7). The right-hand side is to be interpreted as a quasi-static force. Those which involve dynamic properties are hidden in the $\langle \hat{F}(\hat{x}_i, \hat{p}_i, Q_0)\rangle_t$. This implies that the expectation value on the right ought to be calculated with the quasi-static density operator $\hat{\rho}_{\rm qs}[H(Q_0), S_0]$. Since the nucleus is an isolated system entropy can change *only* because of *dynamical* effects. Hence, the static forces are to be evaluated at *constant entropy* rather than at constant temperature. Therefore the expectation value on the right of (6.78) must be expressed through the stiffness of the internal energy together with the adiabatic susceptibility

$$-\kappa \equiv \left\langle \left.\frac{\partial^2 \hat{H}}{\partial Q^2}\right|_{Q_0}\right\rangle^{qs}_{Q_0, S_0} = \left(\frac{\partial^2 \mathcal{E}}{\partial Q^2}\right)_{Q_0, S_0} + \chi^{\rm ad} \quad (6.79)$$

(remember (22.26)). The $\langle \hat{F}\rangle_t$ on the left will involve dynamic and static forces, all of them are to be evaluated up to order one in $Q(t) - Q_0$. The zero order term, the locally constant force (6.8) will now be given by

$$\langle \hat{F}\rangle^{qs}_{Q_0, X_0} = \left.\frac{\partial \mathcal{E}(Q, S_0)}{\partial Q}\right|_{Q_0} = \left.\frac{\partial \mathcal{F}(Q, T_0)}{\partial Q}\right|_{Q_0}. \quad (6.80)$$

Separating the first order corrections into a part $\delta_Q \langle \hat{F}\rangle^{qs}_{S_0}$ of static nature and a time-dependent part $\delta_Q \langle \hat{F}\rangle_t$ one may write

$$\langle \hat{F}\rangle_t = \left.\frac{\partial \mathcal{E}(Q, S_0)}{\partial Q}\right|_{Q_0} + \delta_Q \langle \hat{F}\rangle^{qs}_{S_0} + \delta_Q \langle \hat{F}\rangle_t, \quad (6.81)$$

Following the definition of $\chi^{\rm ad}$ (see Section 23.6) the $\delta_Q \langle \hat{F}\rangle^{qs}_{S_0}$ is given by

$$\delta_Q \langle \hat{F}\rangle^{qs}_{S_0} = -\chi^{\rm ad}(Q - Q_0). \quad (6.82)$$

For the $\delta_Q \langle \hat{F}\rangle_t$, on the other hand, and with $\Delta Q(s) = Q(s) - Q_0$ we may write

$$\delta_Q \langle \hat{F}\rangle_t = \chi(\omega = 0)\Delta Q(t) - \int_{-\infty}^{+\infty} \tilde{\chi}(t - s)\Delta Q(s)\,{\rm d}s. \quad (6.83)$$

The first term ensures that the static forces are indeed represented correctly by those at constant entropy. Indeed, whenever the $\Delta Q(s)$ can be replaced by $\Delta Q(t)$ the $\delta_Q \langle \hat{F}\rangle_t$ vanishes identically. In passing it may be worth noticing that the introduction of the dynamic response function does not necessarily require the use of a canonical ensemble, see Section 23.3.1. The final expression of eqn(6.81) can be written in the condensed form

$$\Delta \langle \hat{F} \rangle_t \equiv \langle \hat{F} \rangle_t - \langle \hat{F} \rangle_{Q_0, S_0}^{qs} = -\left(\chi^{ad} - \chi(0)\right) \Delta Q(t) - \int_{-\infty}^{+\infty} \tilde{\chi}(t-s) \Delta Q(s) \, ds \,. \tag{6.84}$$

Putting things together (6.78) turns into

$$\left.\frac{\partial \mathcal{E}(Q, S_0)}{\partial Q}\right|_{Q_0} - \int_{-\infty}^{\infty} \tilde{\chi}(t-s) \Delta Q(s) \, ds + \Delta Q(t) \left(\left.\frac{\partial^2 \mathcal{E}(Q, S_0)}{\partial Q^2}\right|_{Q_0} + \chi(0)\right) = 0 \,. \tag{6.85}$$

The alert reader might wish to compare the form (6.84) for $\Delta \langle \hat{F} \rangle_t$ with those discussed in Section 23.7.1 in connection with the so-called *relaxation functions*. The latter represent the behavior of averages like the $\Delta \langle \hat{F} \rangle_t$ in future time once they are brought to a finite initial value in static processes, at either constant entropy or constant temperature. These static elongations are specified by *static parameters* ξ_ν. Depending on the properties of the system the elongation $\Delta \langle \hat{F} \rangle_t$ then relaxes back to zero or not. These properties are specified by the differences of the corresponding susceptibilities, $\chi^{ad} - \chi(0)$ in the adiabatic case and $\chi^T - \chi(0)$ in the isothermal one. For situations where these differences vanish the systems are called *ergodic* (as mentioned elsewhere, we want to reserve this notation for the adiabatic case). The situation we are facing in our problem is manifestly different: The $Q(t)$ is *not a free* parameter (like the ξ_ν are in the other case)! Rather, by relation (6.78) it is intimately connected to the $\langle \hat{F} \rangle_t$ itself. This feature is related to the fact that we are neither dealing with an initial value problem nor (at this place) with an external field problem.

Before we discuss the Markov approximation to eqn(6.85) let us rewrite it again in the form (6.17),

$$-k^{-1} q(t) - \int_{-\infty}^{\infty} \tilde{\chi}(t-s) q(s) \, ds = 0 \,, \tag{6.86}$$

which is valid for the variable $q(t) = Q(t) - Q_m$ with the Q_m determined from

$$\left.\frac{\partial \mathcal{E}(Q, S_0)}{\partial Q}\right|_{Q_0} + \left.\frac{\partial^2 \mathcal{E}(Q, S_0)}{\partial Q^2}\right|_{Q_0} (Q_m - Q_0) = 0 \,. \tag{6.87}$$

Here, the effective coupling constant k turns out to be

$$-\frac{1}{k} = \left\langle \left.\frac{\partial^2 \hat{H}}{\partial Q^2}\right|_{Q_0} \right\rangle_{Q_0, S_0}^{qs} + (\chi(0) - \chi^{ad})$$

$$= \left.\frac{\partial^2 \mathcal{E}(Q, S_0)}{\partial Q^2}\right|_{Q_0} + \chi(0) \equiv C(0) + \chi(0) \,. \tag{6.88}$$

The possible local frequencies of harmonic motion are again given by the secular equation (6.18), $k^{-1} + \chi(\omega) = 0$, obtained by Fourier-transforming (6.86). Please observe the subtle difference from the (6.16) found before for zero temperature. Here, the $-1/k$ is not given by the expectation value of $\partial^2 \hat{H}/\partial Q^2$ called $-\kappa$ in (6.78) and (6.79). This $-\kappa$ is identical to the $-1/k$ only for ergodic systems, for which, as just mentioned, $\chi(0) = \chi^{\text{ad}}$; see also the discussion in Section 23.6. The origin of this difference lies in the construction of (6.84). Had we discarded there the correction $-\left(\chi^{\text{ad}} - \chi(0)\right) \Delta Q(t)$ we would have ended up with (6.86) but with the coupling constant given by $-k^{-1} = -\kappa$ with the κ of (6.79).

In practical calculations one may want to use temperature rather than entropy. As discussed in Chapter 22 this is also possible for such a small, isolated system as a finite nucleus. The worst that could happen is that the value of T may be subject to some inherent uncertainty. Ignoring this problem here, one may simply make use of general relations of thermostatics. Indeed, as seen from (6.80) the locally constant static force may either be evaluated from the internal energy $\mathcal{E}$ or from the free energy $\mathcal{F}$. For the relation (6.79), on the other hand, the adiabatic susceptibility χ^{ad} should by replaced by the isothermal one χ^{T}, see (22.26). Thus for the coupling constant one has $-k^{-1} = -k_T^{-1} + \chi^{\text{T}} - \chi^{\text{ad}}$ where

$$-k_T^{-1} = \left(\partial^2 \mathcal{F}(Q, T_0)/\partial Q^2\right)_{Q_0} + \chi(0) \equiv C_{\mathcal{F}}(0) + \chi(0). \qquad (6.89)$$

Actually, this version of the coupling constant comes out if the derivation from before is repeated assuming temperature to stay constant rather than entropy. The two constants k and k_T are strictly identical for $\chi^{\text{T}} - \chi^{\text{ad}} = 0$. In Section 22.1.1.2 formula (22.32) is derived by which this difference is expressed in terms of the derivatives $\partial^2 \mathcal{F}/\partial T^2$ and $\partial^2 \mathcal{F}/\partial T \partial Q$ of the free energy. Typically, in the nuclear case the mixed $\partial^2 \mathcal{F}/\partial T \partial Q$ is much smaller than the stiffness of the free energy $C_{\mathcal{F}}(0)$. At $T = 0$ all susceptibilities are identical anyhow. For doubly magic nuclei, when calculated at the "ground state" deformation, $\partial^2 \mathcal{F}/\partial T \partial Q$ vanishes identically at all temperatures, simply because these nuclei remain spherical no matter how big the intrinsic excitation is. Numerical studies have also shown that this difference is not very large in most other cases. They were obtained in (Kiderlen et al., 1992) for the deformed oscillator without spin-orbit force. It was seen that for the liquid drop model $\chi^{\text{T}} - \chi^{\text{ad}}$ is completely negligible compared with the local stiffness of the free energy. With shell effects included the situation is different at intermediate temperatures but becomes similar to the liquid drop model already at $T \simeq 1.5 - 2$ MeV. Below in Section 6.6.3.2 we will discuss properties of the difference $\chi^{\text{ad}} - \chi(0)$ of the adiabatic susceptibility and the static response. Because of the inequalities $\chi^{\text{T}} \geq \chi^{\text{ad}} \geq \chi(0)$ information on the value of $\chi^{\text{ad}} - \chi(0)$ can be obtained from $\chi^{\text{T}} - \chi(0)$. Some properties of this difference which may be considered generic for independent particle motion will be discussed in Section 6.6.2 with the help of the same model just mentioned. Whereas it is feasible to evaluate numerically formula (23.161) for $\chi^{\text{T}} - \chi(0)$ this is not the case for $\chi^{\text{ad}} - \chi(0)$.

6.5.2 The strict Markov limit

Let us return to equation (6.85) and rewrite it as

$$-\int_{-\infty}^{\infty} \tilde{\chi}(s)(Q(t-s) - Q_0) \, ds + (Q(t) - Q_0)\chi(0) + \frac{\partial \mathcal{E}(Q, \mathcal{S}_0)}{\partial Q} = 0, \quad (6.90)$$

where the static force is correct to order one in $(Q - Q_0)$. Notice the change of the integration variables from s to $t - s$. This is useful for slow collective motion because in this case the integral can be evaluated expanding the $Q(t-s)$ to second order in s. In this way one gets the common differential form of the equation of motion,

$$M(0)\ddot{Q} + \gamma(0)\dot{Q} + \frac{\partial \mathcal{E}(Q, \mathcal{S}_0)}{\partial Q} = 0, \quad (6.91)$$

Here, the following coefficients for friction and inertia have been introduced

$$\gamma(0) = \int_{-\infty}^{\infty} ds\, \tilde{\chi}(s)\, s = -i \left(\frac{\partial \chi}{\partial \omega}\right)_{\omega=0} = \left(\frac{\partial \chi''}{\partial \omega}\right)_{\omega=0} = \frac{1}{2T}\psi''(0), \quad (6.92)$$

$$M(0) = -\frac{1}{2}\int_{-\infty}^{\infty} ds\, \tilde{\chi}(s)\, s^2 = \frac{1}{2}\left(\frac{\partial^2 \chi}{\partial \omega^2}\right)_{\omega=0} = \frac{1}{2}\left(\frac{\partial^2 \chi'}{\partial^2 \omega}\right)_{\omega=0}, \quad (6.93)$$

which will later be referred to as those of the *zero frequency limit*. For the expression on the very right of (6.92) we have made use of the fluctuation dissipation theorem (23.91), with $\psi''(\omega)$ representing the correlation function. This form will be of interest for the transport theory to be developed in the next chapter. It is easily verified that eqn(6.91) is in accord with the secular equation $k^{-1} + \chi(\omega) = 0$ evaluated in second order in ω (see exercise 4).

It may be worth adding a few comments on the inertia. In Section 6.2.1 formula (6.36) was derived for undamped motion. It is easy to see that (6.93) is nothing else but (6.36) in the limit $\omega_\nu = \omega_1 \to 0$ (if only transformed to the q-mode by way of (6.117)). In terms of eigenstates of the unperturbed Hamiltonian one gets:

$$M(0) = 2\hbar^2 \sum_{E_n \neq E_m} \rho(E_m) \frac{|F_{nm}|^2}{(E_n - E_m)^3} = 2\hbar^2 \sum_{e_k \neq e_j} n(e_j) \frac{|F_{kj}|^2}{(e_{kj})^3}. \quad (6.94)$$

Both expressions are valid only in the single particle picture, with the first written for many-body states and the second for single particle states. The restriction of the summations is still a relic of the fact that the reactive part of the response function is to be calculated as a principal value. In the literature these expressions are associated with the "cranking model", which originally has been invented for rotations (see Section 6.3). Commonly they are applied at zero temperature (which in the first version of (6.94) means that in the sum over m only the ground state survives). Below we shall argue that the presence of damping will drastically modify the value of the inertia. Actually, if used bluntly for damped intrinsic motions formula (6.94) may lead to negative values. This is most easily seen by using for $\chi'(\omega)$ the schematic form (6.68). Below in (6.112) a form will be given where this deficiency is largely cured.

Let us now address the problem mentioned earlier that the equation of motion for the $Q(t)$ is to be combined with one for the thermal quantity X used to parameterize the excitation of the internal degrees of freedom. Above we opted to choose for X the entropy $\mathcal{S}$ and ended up with the fact that the conservative force is determined by the internal energy $\mathcal{E}(Q,\mathcal{S})$. Please notice that this expected and correct feature was obtained only by subtracting in (6.83) the term $\chi(\omega = 0)\,(Q(t) - Q_0)$ from the common linear response formula for $\delta_Q \langle \hat{F} \rangle_t$. The differential equation (6.91) was obtained microscopically by requiring the system's *total energy* E_{tot} to be conserved. In a "macroscopic" interpretation we may identify this total energy as $E_{\text{tot}} = E_{\text{kin}} + \mathcal{E}(Q,\mathcal{S})$ where now the entropy $\mathcal{S}$ itself is subject to variation. The condition $dE_{\text{tot}} = 0$ leads to (Hofmann, 1976)

$$-dE_{\text{kin}} = d\mathcal{E}(Q,\mathcal{S}) = T\,d\mathcal{S} + \left(\partial\mathcal{E}/\partial Q\right)_{\mathcal{S}} dQ \quad \text{with} \quad T = \left(\partial\mathcal{E}/\partial\mathcal{S}\right)_Q. \tag{6.95}$$

Restricting ourselves to the harmonic case and after dividing by dt we indeed get

$$-\frac{d}{dt}E_{\text{coll}} \equiv -\frac{d}{dt}\left(\frac{M(0)}{2}\dot{Q}^2 + \frac{C(0)}{2}\Delta Q^2\right) = T\frac{d}{dt}\mathcal{S}, \tag{6.96}$$

with the conservative force being given by the derivative of the internal energy at *constant entropy*:

$$-C(0)\Delta Q = -\left(\frac{\partial \mathcal{E}}{\partial Q}\bigg|_{Q_0}\right)_{\mathcal{S}}. \tag{6.97}$$

As a consequence of the equation of motion (6.91) for the variable $Q(t)$ the variation of entropy in a time-dependent process then is given by

$$T\frac{d}{dt}\mathcal{S} = \gamma(0)\dot{Q}^2. \tag{6.98}$$

In accord with the second law the rate of the entropy change is quadratic in the velocity: As $\gamma(0)$ cannot be negative the entropy must either stay constant or increase in time. This problem is discussed in Section 23.7 for general quasi-static processes, where the macroscopic variables may involve coordinates and currents alike. In the linear case the corresponding equations of motion are of first order in the derivatives. The set corresponding to (6.91) would involve Q and the kinetic momentum $P_{\text{kin}} = M(0)\dot{Q}$. It is not difficult to see how such an approach in the end again leads to (6.98); see e.g. (Landau and Lifshitz, 1980), (Brenig, 1989) where this calculation is done explicitly.

From (6.98) and the knowledge of $Q(t)$ and $T(t)$ the entropy can be calculated as function of time. It is more convenient, of course, to remove the entropy entirely by expressing it as function of the two variables Q and T. This leads to the differential $d\mathcal{S} = (\partial\mathcal{S}/\partial Q)_T dQ + (\partial\mathcal{S}/\partial T)_Q dT$. Within the quasi-static picture the first law may be applied, from which it follows that the entropy can be expressed by the free energy $\mathcal{S}(Q,T) = -(\partial \mathcal{F}/\partial T)_Q$. In this way it can be seen that $(\partial \mathcal{S}/\partial T)_Q$ is identical to $(\partial \mathcal{P}_Q/\partial T)_Q$, where $\mathcal{P}_Q$ is the analog of the ordinary pressure $\mathcal{P}$ of a medium inside the volume $\mathcal{V}$, with the latter here being substituted by Q; for more details see Section 22.1.1. In the same spirit $T(\partial \mathcal{S}/\partial T)_Q$ is nothing else but the specific heat of the nucleonic degrees of freedom at constant Q. In passing we should like to mention the case in which the entropy production is either identical to zero or sufficiently small that it may be discarded. Then the variation of temperature can be expressed by formula (22.34) derived in Section 22.1.1.2.

6.5.3 The collective response for quasi-static processes

In Section 6.2 the collective response function has been introduced and shown afterwards to take on the form

$$\chi_{\text{coll}}(\omega) = \frac{\chi(\omega)}{1 + k\chi(\omega)}. \tag{6.99}$$

In (Kiderlen et al., 1992) this derivation was extended to finite thermal excitation for an ergodic system, which is to say for $\chi(0) = \chi^{\text{ad}}$, see also (Hofmann, 1997). Indeed, for such a system eqn(6.78) together with (6.79) have the same form as the corresponding ones at $T = 0$. In general we must expect the coupling constant k to be given by (6.88) and thus to differ from the $1/\kappa$ of (6.79). The most direct possibility of deriving a collective response function like the $\chi_{\text{coll}}(\omega)$ is by modifying the derivation given in Section 6.5.1 for eqn(6.86). As we are aiming at a relation between $\langle \hat{F} \rangle_\omega$ and $f_{\text{ext}}(\omega)$ we have to introduce a coupling to the external field through the $f_{\text{ext}}(t)\hat{F}$ introduced in (6.26). This additional coupling term will not modify the internal forces of the system. Hence, in the expression (6.81) for $\langle \hat{F} \rangle_t$ it should appear only in the $\delta_Q \langle \hat{F} \rangle_t$ and in (6.83) only the integrand has to be modified by the replacement:

$$-\int_{-\infty}^{+\infty} \widetilde{\chi}(t-s)\,\Delta Q(s)\,ds \quad \Longrightarrow \quad -\int_{-\infty}^{+\infty} \widetilde{\chi}(t-s)\bigl(\Delta Q(s) + f_{\text{ext}}(s)\bigr)\,ds\,. \tag{6.100}$$

This implies that (6.85) changes into an equation where on the right-hand side the 0 is replaced by $\int_{-\infty}^{+\infty} \widetilde{\chi}(t-s) f_{\text{ext}}(s)\,ds$. The left-hand side may still be treated as before such that we end up with (6.86) but where on the right again the term $\int_{-\infty}^{+\infty} \widetilde{\chi}(t-s) f_{\text{ext}}(s)\,ds$ appears. Fourier-transforming this equation one gets the equation

$$\bigl(-k^{-1} - \chi(\omega)\bigr)\,q(\omega) = \chi(\omega) f_{\text{ext}}(\omega)\,, \tag{6.101}$$

which is valid for all frequencies. If we want to use it for obtaining the desired relation between $\langle \hat{F} \rangle_\omega$ and $f_{\text{ext}}(\omega)$ we need to invoke (6.78) in which the Q_0 appears. Considering the definition of $\Delta Q(t)$ and of $q(t)$ we may write

$$\langle \hat{F} \rangle_\omega = (Q_m - Q_0)\,\kappa\,\delta(\omega) + q(\omega)\kappa\,. \tag{6.102}$$

This implies the result

$$\begin{aligned}\langle \hat{F}\rangle_\omega &= (Q_m - Q_0)\,\kappa\,\delta(\omega) - \{k\kappa\,\chi_{\text{coll}}(\omega)\}\,f_{\text{ext}}(\omega)\\ &= \langle \hat{F}\rangle^{\text{qs}}_{Q_m,S_0} - \{k\kappa\,\chi_{\text{coll}}(\omega)\}\,f_{\text{ext}}(\omega)\,.\end{aligned} \quad (6.103)$$

The second line follows from equations (6.79) for κ and (6.87) for Q_m and because of the obvious quasi-static relation for the $\langle \hat{F}\rangle^{\text{qs}}_{Q,S_0}$ at constant entropy

$$\langle \hat{F}\rangle^{\text{qs}}_{Q_m,S_0} - \langle \hat{F}\rangle^{\text{qs}}_{Q_0,S_0} = -\chi^{\text{ad}}(Q_m - Q_0)\,. \quad (6.104)$$

The final form obtained in (6.103) can be interpreted as the usual definition of a response function. In this spirit the $\{k\kappa\chi_{\text{coll}}(\omega)\}$ is nothing else but the collective response for non-ergodic motion, which for $\chi(0) = \chi^{\text{ad}}$ turns into the old one. The only difference appears in the overall strength. This strength is important only for a finite coupling to an external field. Let us look at the change of energy once more. Using (6.78) in (6.28) it can be expressed as

$$\frac{d}{dt}E_{\text{tot}}(t) = -f_{\text{ext}}(t)\frac{d}{dt}\langle \hat{F}\rangle_t = -\kappa f_{\text{ext}}(t)\dot{q}(t)\,. \quad (6.105)$$

Here, the time-dependent external field $f_{\text{ext}}(t)$ may be removed by employing relation (6.101). In this way one gets

$$\frac{d}{dt}E_{\text{tot}}(t) = \mathcal{FT}\left\{\frac{1+k\chi(\omega)}{k^2\chi(\omega)}q(\omega)\right\}_t\,\dot{q}(t)\,\kappa k = \mathcal{FT}\left\{\frac{q(\omega)}{\chi_{qq}(\omega)}\right\}_t\,\dot{q}(t)\,\kappa k\,, \quad (6.106)$$

where $\mathcal{FT}\{g(\omega)\}_t = \tilde{g}(t)$ is the Fourier transform of $g(\omega)$. Here, the same notation was used as in previous publications (see (Hofmann, 1997) and others) putting $\chi_{qq}(\omega) = k^2\chi_{\text{coll}}(\omega)$. Remember that (6.106) is valid both with or without a finite f_{ext}. In the latter case this expression has to vanish, of course, such that the factor κk becomes irrelevant. Let us finally look at its value. From (6.79) and (6.88) one gets

$$\kappa\,k = 1 + \frac{\chi^{\text{ad}} - \chi(0)}{C(0) + \chi(0)} \geq 1 \quad (6.107)$$

It cannot be smaller than unity because the adiabatic susceptibility is larger than the static response. For a simple estimate one may replace χ^{ad} by χ^{T}, and at not too small temperatures one may neglect the $C(0)$ compared to $\chi(0)$.

6.5.3.1 *Equations of motion from the collective response* The equations for self-sustained motion can be obtained by imposing energy conservation to the relation (6.106) which leads to

$$(\chi_{qq}(\omega))^{-1}\,q(\omega) = 0\,. \quad (6.108)$$

This equation allows for approximations which go beyond those behind eqn(6.91). One possibility consists in expanding the inverse collective response function $\chi_{qq}(\omega)$ to second order, which is to say

$$(\chi_{qq}(\omega))^{-1} \approx -\omega^2 M - \omega i\gamma + C. \tag{6.109}$$

Evidently, when inserted into (6.108) the Fourier transform of this equation has again the form of that for a damped oscillator. The transport coefficients are now given by

$$C \equiv \frac{1}{\chi_{qq}(\omega)}\bigg|_{\omega=0} = -\left(\frac{1}{k\chi(0)}\right) C(0) \simeq C(0) \tag{6.110}$$

$$\gamma \equiv i\frac{\partial(\chi_{qq}(\omega))^{-1}}{\partial\omega}\bigg|_{\omega=0} = \left(\frac{1}{k\chi(0)}\right)^2 \gamma(0) \simeq \gamma(0) \tag{6.111}$$

$$M \equiv -\frac{\partial^2(\chi_{qq}(\omega))^{-1}}{2\,\partial\omega^2}\bigg|_{\omega=0} = \left(\frac{1}{k\chi(0)}\right)^2 \left(M(0) + \frac{\gamma^2(0)}{\chi(0)}\right) \simeq \left(M(0) + \frac{\gamma^2(0)}{\chi(0)}\right). \tag{6.112}$$

Here, $C(0) = \partial^2 \mathcal{E}(Q, \mathcal{S}_0)/\partial Q^2|_{Q_0}$ is the static stiffness introduced in (6.88) and $\gamma(0)$ and $M(0)$ were given in (6.92) and (6.93). From the definition (6.88) of $-1/k$ it follows that $-1/(k\chi(0)) = 1 + C(0)/\chi(0)$. The approximation performed on the very right of (6.110) to (6.112) is valid under the condition $|C(0)| \ll \chi(0)$. Numerical evaluations show this to be largely fulfilled, at least at higher temperatures, see Section 6.6.3. However, at smaller thermal excitation this may not be given (Kiderlen et al., 1992).

A further way to reduce the Fourier transform of (6.108) to differential form is possible if the collective response function $\chi_{qq}(\omega)$ can be reduced to a single pair of modes. In Section 6.2.1 for the function $\chi_{\text{coll}}(\omega)$ an expansion into its pole terms was discussed. A similar form can of course be written for the $\chi_{qq}(\omega)$, but now we are interested in a generalization to damped motion. To this end let us recall first the response for a single damped oscillator given in (23.6), for instance. Assuming the same form for each mode we may write

$$\chi_{qq}(\omega) = -\sum_\nu \frac{1}{2M_\nu \mathcal{E}_\nu}\left(\frac{1}{\omega - \omega_\nu^+} - \frac{1}{\omega - \omega_\nu^-}\right), \tag{6.113}$$

where the frequencies of the poles are supposed to be given by:

$$\omega_\nu^\pm = \pm\mathcal{E}_\nu - i\Gamma_\nu/2 = \varpi_\nu \left(\pm\sqrt{\text{sign}C_\nu - \eta_\nu^2} - i\eta_\nu\right) \tag{6.114}$$

In this form transport coefficients for the individual modes are hidden. They can be deduced whenever in some region of frequency the strength distribution $\chi''_{\text{coll}}(\omega)$ has a pronounced peak. Suppose we take the lowest one. Through

$$\omega_1^\pm = \pm\sqrt{\frac{C_1}{M_1} - \left(\frac{\gamma_1}{2M_1}\right)^2} - i\frac{\gamma_1}{2M_1}, \tag{6.115}$$

together with the height of the peak the coefficients of (local) stiffness, inertia and friction, $C_1 = C(\omega_1)$, $M_1 = M(\omega_1)$, $\gamma_1 = \gamma(\omega_1)$ may be determined uniquely. In

(Samhammer et al., 1989) this has been demonstrated even for a two-dimensional system (involving two operators $\hat{F}_\mu$). In the following we shall often assume that the $\chi_{qq}(\omega)$ of (6.113) may be replaced by the oscillator response

$$\chi_{\text{osc}}(\omega) = -\frac{1}{M_1(\omega_1^+ - \omega_1^-)}\left(\frac{1}{\omega - \omega_1^+} - \frac{1}{\omega - \omega_1^-}\right). \tag{6.116}$$

Because of the relation $\chi_{qq}(\omega) = k^2 \chi_{\text{coll}}(\omega)$ to the $\chi_{\text{coll}}(\omega)$ of (6.99) one may introduce analogous transport coefficients for the $\langle \hat{F} \rangle_\omega$-mode, namely

$$C^F = k^2 C(\omega_1), \quad \gamma^F = k^2 \gamma(\omega_1), \quad M^F = k^2 M(\omega_1). \tag{6.117}$$

Let us return to the problem of getting a differential equation for the $q(t)$. After inserting the $\chi_{\text{osc}}(\omega)$ of (6.116) into (6.108) one ends up with

$$M(\omega_1)\ddot{q}(t) + \gamma(\omega_1)\dot{q}(t) + C(\omega_1)q(t) = -q_{\text{ext}}(t). \tag{6.118}$$

The transport coefficients introduced in this way depend on frequency, more precisely on that frequency which we have chosen among the solutions of the dispersion relation $k^{-1} + \chi(\omega) = 0$ which defines the position of the poles of $\chi_{qq}(\omega)$. The very fact that the same ω_1 also solves the secular equation (6.118) of the damped oscillators simply reflects the self-consistency of the procedure. The reason we chose the lowest of all possible ones is dictated by the desire that the differential equation can be understood as representing local properties of slow collective motion. The formal steps necessary to achieve this goal will be discussed below in chapters 7 and 11. Also in such a case one must be able to treat the intrinsic excitations in a consistent way. For this reason let us look again at the variation of entropy. After reducing the full response function to that of the oscillator one gets:

$$\frac{d}{dt}E_{\text{tot}} = \kappa k \left\{\frac{d}{dt}\left(\frac{M(\omega_1)}{2}\dot{q}^2 + \frac{C(\omega_1)}{2}q^2\right) + \gamma(\omega_1)\dot{q}^2\right\}. \tag{6.119}$$

This relation is valid both with or without an external force. Let us look first at the case of $f_{\text{ext}} \neq 0 \neq q_{\text{ext}}$. By proper choice of the external field we may force the system to perform strict harmonic vibrations in $q(t)$, for instance with the frequency $\varpi = \sqrt{C/M}$. Integrating in time over periods of this vibration, the contribution from the first term on the left of (6.119) vanishes identically. The contribution from the second term, on the other hand, remains finite and shows the typical feature of *irreversible behavior*. For a *vanishing external field the total energy stays constant* such that (6.119) can be written as

$$-\frac{d}{dt}E_{\text{coll}} \equiv -\frac{d}{dt}\left(\frac{M(\omega_1)}{2}\dot{q}^2 + \frac{C(\omega_1)}{2}q^2\right) = \gamma(\omega_1)\dot{q}^2, \tag{6.120}$$

which states that *collective energy is lost irreversibly*. As in the case discussed in Section 6.5.2 the rate at which that happens is determined by the friction coefficient times the squared velocity. One may still interpret this result as the production of heat per unit of time in the spirit of eqn(6.98). This is possible as long as the frequencies $w_1^\pm$ still imply collective motion to be slow in comparison to that of the nucleonic degrees of freedom. Actually, the fact that the transport coefficients depend on these $w_1^\pm$ does not necessarily invalidate this assumption.

6.5.4 An analytically solvable model

At this stage it may be worthwhile to study some of our essential approximations at a simple model. For this we may take the form (6.68) for the dissipative intrinsic response function. As is easily verified, the corresponding total response reads (setting $\hbar = 1$ temporarily and replacing the Ω_F of (6.68)) by Ω)

$$\chi(\omega) = -\overline{F^2} \left[\frac{1}{\omega - \Omega + i\Gamma/2} - \frac{1}{\omega + \Omega + i\Gamma/2} \right]$$
$$= \frac{-2\Omega \overline{F^2}}{\omega^2 + i\Gamma\omega - (\Omega^2 + (\Gamma/2)^2)} = \frac{-1/M_{\rm int}}{\omega^2 + i\Gamma_{\rm int}\omega - \varpi_{\rm int}^2} \,. \qquad (6.121)$$

In the second line $\chi(\omega)$ is written in the form of an oscillator response, with the transport coefficients for intrinsic motion $M_{\rm int} = 1/(2\Omega \overline{F^2})$ for the inertia, $\Gamma_{\rm int} = \Gamma$ for the damping width and $\varpi_{\rm int}^2 = \Omega^2 + (\Gamma/2)^2$ as the square of an effective internal frequency with a corresponding stiffness $C_{\rm int} = M_{\rm int}\varpi_{\rm int}^2$. The model behind (6.121) may be interpreted as a generalization of the degenerate one introduced at the end of Section 6.1.2. Now the one-body operator $\hat{F}$ couples to a full "band" of states still centered at Ω but spread over a width Γ. The fact of having damped collective motion is reflected in the collective response function, for which we like to choose the one for the Q-mode given by $\chi_{qq}(\omega) = k^2 \chi_{\rm coll}(\omega)$ with the $\chi_{\rm coll}(\omega)$ of (6.99) involving the appropriate transport coefficients, namely

$$\chi_{qq}(\omega) = \frac{k^2 \chi(\omega)}{1 + k\chi(\omega)} = \left(\frac{1}{k^2 \chi(\omega)} + 1/k \right)^{-1} = \frac{-1}{M\omega^2 + i\gamma\omega - \varpi^2} \,, \qquad (6.122)$$

with stiffness, friction and inertia being expressed as

$$C = \frac{C_{\rm int}}{k} \left[\frac{1}{k} + \frac{1}{C_{\rm int}} \right] = \left(\frac{1}{k\chi(0)} \right) C(0)$$

$$\gamma = \frac{M_{\rm int}}{k^2} \Gamma_{\rm int} = \left(\frac{1}{k\chi(0)} \right)^2 \gamma(0) \qquad (6.123)$$

$$M = \frac{M_{\rm int}}{k^2} = \left(\frac{1}{k\chi(0)} \right)^2 \frac{1}{C_{\rm int}\varpi_{\rm int}^2} > 0 \,.$$

The static response reads

$$\chi(0) = \frac{2\Omega \overline{F^2}}{\Omega^2 + (\Gamma/2)^2} = \frac{1}{C_{\rm int}} \qquad (6.124)$$

and for the zero frequency limit one has from (6.92) and (6.93)

$$\gamma(0) = -i\left(\frac{\partial \chi}{\partial \omega}\right)_{\omega=0} = \frac{2\Omega \overline{F^2}}{(\Omega^2 + (\Gamma/2)^2)^2}\Gamma = \frac{\Gamma_{\rm int}}{C_{\rm int}\varpi_{\rm int}^2} \qquad (6.125)$$

$$\begin{aligned} M(0) &= \left(\frac{1}{2}\frac{\partial^2 \chi}{\partial \omega^2}\right)_{\omega=0} = \frac{1}{C_{\rm int}\varpi_{\rm int}^2} - \frac{\gamma^2(0)}{\chi(0)} \\ &= \frac{1}{C_{\rm int}\varpi_{\rm int}^2}\left(1 - \frac{\Gamma^2}{\Omega^2 + (\Gamma/2)^2}\right). \end{aligned} \qquad (6.126)$$

Let us interpret these results. The possible intrinsic excitations as manifested by one pair of damped intrinsic modes call for just one pair of damped collective modes $\omega_1^\pm$. Therefore, the transport coefficients introduced in (6.122) and (6.123) represent the full collective response. There is no need for a reduction like the one employed in going from the non-Markovian equation (6.108) to the Markovian form (6.109). Hence, the coefficients (6.123) and the corresponding differential equation of the damped oscillator incorporate non-Markovian properties. Whereas for this model these features represent the true situation exactly, for the more realistic case the degree of approximation depends on the strength distribution of all modes. In fact in the next section we are going to demonstrate for the case of quadrupole vibrations that with increasing temperature their strength more and more concentrates in one prominent mode at small frequencies.

In this context it may be worthwhile to mention in passing that the response functions of the RMM of Section 6.4.2 essentially represent just one intrinsic mode, although the precise functional form of the dissipative part $\chi''(\omega)$ given in (23.110) is not of Lorentzian type. However, to get the one corresponding to (6.121) one would only have to replace the Gaussian decay behavior for the $\widetilde{\chi}''(t)$ of (6.72) by an exponential one.

Let us examine next the quality of the zero-frequency approximation to the transport coefficients. From the discussion above it is evident that within our model the forms found in (6.110) to (6.112) are exact; these equations are identical to those in (6.123). The situation is different for the coefficients introduced in (6.5.2) for slow collective motion. As seen from (6.123), stiffness and friction may well be represented by $C(0)$ and $\gamma(0)$ if only the product $|k|\chi(0)$ is sufficiently close to unity, which is the case for $|C(0)| \ll \chi(0)$. For the inertia, on the other hand, a further condition is required:

$$\left(k\chi(0)\right)^2 M \simeq M(0) \quad \text{only for} \quad \frac{\Gamma^2}{\Omega^2 + (\Gamma/2)^2} \ll 1. \qquad (6.127)$$

Its meaning is easily understood. For too large a width the distribution in frequency spreads too far away from zero and $\chi(\omega)$ can no longer be represented by an expansion to second order around $\omega = 0$.

166   *Average collective motion of small amplitude*

6.6 Temperature dependence of nuclear transport

As demonstrated in sections 6.2 and 6.5 there is an intimate relation between the transport properties of collective motion and the strength distribution of its various modes. This feature becomes of particular importance at finite temperature. As we shall elaborate in the following, this is one of the main features in which the various models on nuclear transport differ in their predictions.

6.6.1 *The collective strength distribution at finite T*

According to (6.103) the collective response function, which defines the variation of $\langle \hat{F} \rangle_\omega$ with an external field $f_{\text{ext}}(\omega)$, is given by $-\{k\kappa\chi_{\text{coll}}(\omega)\}$, with the $\chi_{\text{coll}}(\omega)$ of (6.99). For non-ergodic systems the product $k\kappa$ differs from unity but it is always larger than one. Its value may vary with temperature but it has no influence on the way how the individual modes are distributed over frequencies. This property is entirely determined by the dissipative part of $\chi_{\text{coll}}(\omega)$. A typical example are (isoscalar) quadrupole vibrations of ^{208}Pb around the potential minimum. This case has been examined in (Hofmann et al., 1992) where the response functions have been evaluated with the numerical code developed earlier by S. Yamaji, for more details see e.g. (Yamaji et al., 1988). On the left of Fig. 6.5 the result for $\chi''_{\text{coll}}(\omega)$ is presented. At the small temperature of $T = 0.5$ MeV one observes essentially two modes.[15] They are still visible at $T = 2$ MeV but dissolve themselves into a broad one being peaked at very small frequencies. The dashed lines shown in the figure represent Lorentzian fits to the two main peaks, and thus define the oscillator response functions introduced in (6.113). Notice, please, that the transition to essentially one peak occurs in the range of temperatures where shell effects disappear. In fact, above this point the strength distribution resembles the one of a strongly damped oscillator like shown in the lower left part of Fig. 23.1.

As described above, knowledge of the three parameters of the Lorentzians allows one to deduce the transport coefficients which specify the oscillator modes. Let us concentrate here on the inertia. It has been shown in Section 6.2.1 that the inertia of some mode will be that of the energy weighted sum (6.40) if this mode exhausts the sum. Exactly this is what happens here at larger T, as demonstrated on the right of Fig. 6.5, where the inertia $M(\omega_1) \equiv M(\Omega_0)$ of the low frequency mode is shown as function of T. It starts off from values close to that of the cranking model M_{cr} given in (6.94). Above $T \simeq 1$ MeV the $M(\omega_1)$ decreases smoothly but strongly down to the much smaller value given by the sum rule, which for this model happens to be identical to the inertia M_{irr} of the liquid drop model for irrotational flow, see the discussion below eqn(6.40). The dashed curve shown on the right is the inertia in the zero frequency limit (6.93) (scaled by

---

[15] The reader who expects the high-lying mode at somewhat larger energies may be reminded that the model just described is in some sense only of schematic nature: No attempt was made to renormalize the constants of the underlying model–which is necessary as soon as one takes serious the imaginary parts of the self-energies. Remember also that the mechanism for producing the shift to smaller frequencies is effective already at small T.

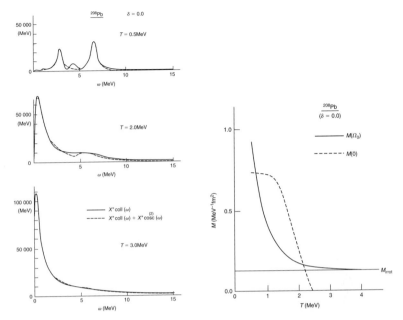

FIG. 6.5. The left part shows the collective strength distribution for quadrupole excitations of ^{208}Pb about the spherical shape (corresponding to the deformation parameter $\delta = 0.0$) at different temperatures. On the right the inertia of the low frequency mode ω_1 (called Ω_0 in the figure) is presented together with the inertia in the zero-frequency limit, see text and (Hofmann et al., 1992).

$k^2(T=0)$ such that it may be compared with the $M(\omega_1)$ which belong to the F-mode). It is seen that it fails almost completely, the reason being the large width of the low mode (recall the discussion in Section 6.5.4). As expected from the presentation given above, the transition of $M(\omega_1)$ to $M_{\rm irr}$ happens to take place exactly at that temperature where shell effects in the static energy disappear. This is quite a remarkable result as it demonstrates that features typical for the Strutinsky renormalization may also hold true for dynamic quantities: When shell effects disappear the nucleus shows the behavior associated with macroscopic dynamics, and not only for static properties. Similar observations of the behavior of the inertia have also been made in other calculations. It must be said though that the transition to the sum rule value does not always come out as clearly as seen in Fig. 6.5. This feature is related to the fact that the mean field used in (Hofmann et al., 1992) is constructed from deformed oscillators, for which case it can be proven analytically (Bohr and Mottelson, 1975) that the sum rule value is identical to that of irrotational flow. For other models like those based on the Woods–Saxon potential the situation is less clean, see e.g. (Ivanyuk et al., 1997).

Let us examine next where this intriguing temperature dependence comes

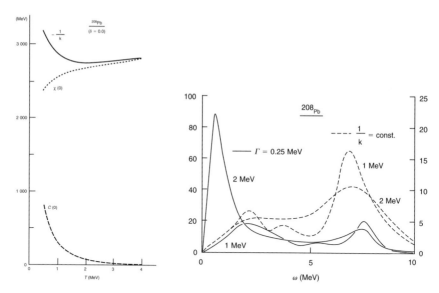

FIG. 6.6. Left panel: The stiffness, the static response and the coupling constant as function of temperature (Yamaji et al., 1994). Right panel: The strength distribution for constant width and constant coupling constant at two different temperatures $T = 1$ and 2 MeV; for details see text.

from. This can be done with the aid of Fig. 6.6. On the left the inverse coupling constant $-1/k = C(0) + \chi(0)$ is shown as function of temperature, together with the behavior of its two ingredients, the static response $\chi(0)$ and the stiffness $C(0)$ of the free energy, the latter being calculated with the approximation (5.51). (Remember that the potential minimum of ^{208}Pb stays at the spherical configuration independent of temperature. Consequently, the stiffness of the free and the internal energies coincide, see Section 22.1.1.2.) To understand the essential mechanism for the shift to lower frequencies it is sufficient to look at the dramatic decrease of the stiffness by several orders of magnitude. Indeed, as the secular equation for collective modes can be written as $\chi(\omega) - \chi(0) = C(0)$ it is obvious that the decrease of $C(0)$ with T will make the collective modes very soft. This shift to lower frequencies is much less influenced by the T-dependence of the self-energies involved. This fact is exhibited on the right of Fig. 6.6. There, two different calculations are portrayed. (i) For the dashed curves the coupling constant was chosen to be independent of T, identical to that of the Copenhagen model for vibrations at zero thermal excitation; for the present case a formula like (6.48) leads to $-1/k = (4/9)A\langle m_0 \varpi_0^2 r^2 \rangle$ (recall Section 3.2.2.1). All other quantities were evaluated as for the curves in Fig. 6.5. In particular the particles and holes were dressed with temperature-dependent self-energies with the imaginary parts given by (6.76). (ii) For the fully drawn lines, on the other hand, the widths of the single particle states were chosen to be constant, independent both of frequency and temperature. Nevertheless the shift to lower frequencies is present, due to the temperature-dependent coupling constant as determined from the relation $-1/k = C(0) + \chi(0)$. The temperature dependence of the stiffness is, of course, intimately connected to shell effects. They are large at low excitations but disappear exponentially with increasing thermal excitations. These features are not restricted to the case of quadrupole vibrations. In principle, they also apply to the local modes encountered for large-scale motion like fission within the "locally harmonic approximation" and hence have great impact on the T-dependence of the transport coefficients, details of which will be described below. Of course, there may be modes for which this shift is not as strong as in the case of quadrupole excitations. In (Yamaji et al., 1994) monopole vibrations of Pb have been studied, for which the free energy was assumed to be given by $\mathcal{F}(Q,T) = (A/18)\kappa(T)(1-\rho(Q)/\rho_0)^2$ with the T-dependence solely coming through the incompressibility $\kappa(T)$. Using the ansatz $\kappa(T) = 216.7(1-0.006T^2)$, as motivated by the extended Thomas–Fermi calculation of (Guet et al., 1988), the T-dependence of $\mathcal{F}(Q,T)$ turns out to be quite weak and the frequency shift was seen be quite small.

6.6.1.1 *The macroscopic limit through Strutinsky smoothing* The subtle temperature dependence of the collective strength distribution found above is intimately related to the form of $\chi_{\text{coll}}(\omega)$ given in (6.99). As we have seen, the origin of this effect is mainly due to the large change the stiffness of the static energy experiences when shell effects disappear. This feature can also be simulated by applying smoothing procedures to the static energy of the ground state in the independent particle model. It may therefore be expected that curves like those shown in Fig. 6.5 can be reproduced through Strutinsky smoothing, at least qualitatively. An increasing temperature may then be simulated by using smoothing functions for which the width is treated as a parameter γ. Its value should vary from small numbers to the γ_{av} for which shell effects are washed out completely. This is more or less straightforward as far as the static energy is concerned. However, it would not be very meaningful if the response functions were left untouched. In the following we will discuss these problems reporting on studies performed in (Hofmann and Ivanyuk, 1993).

Let us begin with the stiffness. Noting that in the single particle model the static force is given by $\partial E/\partial Q = \sum_k n_k \partial e_k/\partial Q$ the stiffness we aim at can be written in the form

$$C_\gamma = C_{\text{LD}} + \frac{\partial}{\partial Q} \sum_k (n_k^\gamma - \overline{n}_k) \frac{\partial e_k}{\partial Q}. \qquad (6.128)$$

Here, we apply the smoothing over occupation number described in Section 5.2.3. The $\overline{n}_k \equiv \overline{n}(e_k)$ is the smooth one leading to the smooth energy $\overline{E}_{\text{ipm}}$ which may be identified as that of the liquid drop model. Having this in mind, the C_γ is just given by the terms in the sum which are proportional to n_k^γ. These smoothed occupation numbers are to be calculated like the $\overline{n}$ in (5.21) with the only difference of replacing the smoothing width $\gamma_{\rm av}$ by $\gamma \leq \gamma_{\rm av}$. When calculated as a function of γ the C_γ starts from the large values given by the shell correction to the much smaller values typical of the liquid drop model.

Next we turn to the response functions. A convenient strategy is first to smooth the dissipative part $\chi''(\omega)$ and evaluate the $\chi(\omega)$ through the Kramers–Kronig relation. As the starting point for $\chi''(\omega)$ one may choose eqn(6.24). In principle one may apply the same technique as just described for the stiffness, namely replacing the $n(e_k)$ in eqn(6.24) by the n_k^γ. In this way, however, one does not account for the fact that in the realistic case the single particle widths increase with temperature. Indeed, without such an effect both the internal as well as the collective strength distribution gets additional structure which is unimportant for the main property we are after. Therefore, the δ-functions of (6.24) were replaced by Lorentzians with constant width Γ for the ph-excitations. The value of this Γ ought to be related to the smoothing width γ used for the stiffness. Realizing that $\delta(\hbar\omega - e_k + e_j)$ can be written as the folding integral $\delta(\hbar\omega - e_{kj}) = \int de\,\delta(e - \hbar\omega - e_j)\delta(e - e_k)$ one may replace the two single particle δ-functions by smooth ones denoted by $\overline{\delta_\gamma}(x)$ in the sequel. They can be viewed as representatives of the single particle density. To be consistent with the smoothing of the static energy one ought to use for $\overline{\delta_\gamma}(x)$ the smoothing functions $f(x)$ defined in (5.12) for the original Strutinsky procedure. On the other hand, the folding integral can be calculated much more easily if at this stage the $\overline{\delta_\gamma}(x)$ is simply replaced by a Lorentzian. As the folding of two such functions results in a new one with twice the width one gets the response function for "density smoothing" (as it was called in (Hofmann and Ivanyuk, 1993))

$$\chi_\gamma''(\omega) = -\pi \sum_{jk} |F_{kj}|^2 \left(n(e_k) - n(e_j)\right) \overline{\delta_{2\gamma}}(\hbar\omega - e_{kj}). \tag{6.129}$$

Fig. 6.7 shows the final result for the collective strength distribution. It is seen that with increasing smoothing width γ the strength shifts to smaller energies. For the value of $\gamma = \hbar\Omega_0$, for which the shell effects are washed out, only one broad peak at small frequencies survives–in very close analogy with the situation shown in Fig. 6.5. It is tempting to refer to this result as the *macroscopic limit for dynamical quantities*, in generalization of the Strutinsky method if applied to the level density and the static energy. There is (at least) one essential difference, however. The Strutinsky procedure applied to the calculation of nuclear masses can be interpreted in such a way that the smooth part represents an average over masses for different nuclei. This is definitely not so for the message behind the behavior of the response function seen in Fig. 6.7. The peak which survives large smoothing represents a collective mode of low frequency which is *strongly damped*. It may thus *not be associated* with an average of properties of low frequency vibrations of actual nuclei, which are practically *undamped*. In this sense one may conclude that the "macroscopic limit" for *dynamic quantities* behaves very differently: It is reached at higher temperatures only, namely at those temperatures where shell effects have disappeared. In this regime collective motion

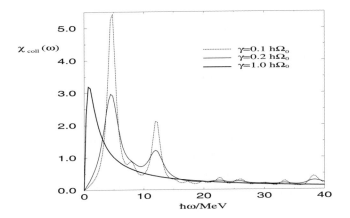

FIG. 6.7. The collective strength distribution for averaged stiffness and intrinsic response function, see text and (Hofmann and Ivanyuk, 1993).

is damped which for slow modes is not the case at $T=0$. In this sense properties attained for larger T do not reflect average quantities at zero excitation–as is the case for static quantities. We will return to this issue in Section 6.6.3.4.

6.6.2 Diabatic models

In the previous sections one of the basic assumptions was that the static energy is defined in the quasi-static picture: At any Q the potential energy was the one which for given entropy (or temperature eventually) corresponds to the minimal internal excitation. In "diabatic models" the situation is different: There the static density operator $\hat{\rho}_{\rm di}(Q)$ is defined in the common way but where the probabilities for the population of a given internal state $|m;Q\rangle$ are *fixed numbers*. In this sense one may write

$$\hat{\rho}_{\rm di}(Q) = \sum_m \rho_m^{\rm di} |m;Q\rangle\langle m;Q| \qquad \text{with} \qquad \rho_m^{\rm di} = \rho_m^{\rm di}(Q_0, T_0) \qquad (6.130)$$

Still the summation extends over all many-body states of the Hamiltonian $\hat{H}(Q)$. The $\rho_m^{\rm di}$ vary neither with the collective variable nor with the excitation energy. In the following we will look at the case of vibrations about Q_0 and where the $\rho_m^{\rm di}$ may be parameterized by a fixed temperature T_0.

The important point is that unlike the quasi-static case the $\hat{\rho}_{\rm di}(Q)$ depends on Q *only through the states* $|m;Q\rangle$. In this sense the situation resembles that of vibrations at zero temperature, where the system always stays in the Q-dependent ground state $|m=0;Q\rangle$. For this reason the derivation of the equation of motion parallels completely the one at zero temperature. We only have to replace the ground state $|m=0;Q\rangle$ everywhere by the statistical mixture represented by the density operator of (6.130). The static energy is therefore to be calculated as $E_{\rm di}(Q) = \sum_m \rho_m^{\rm di}(Q_0; T_0) E_m(Q)$. For harmonic motion the coupling constant is given by a formula like (6.16) but where the ground state energy $E_0(Q)$ (called $E(Q)$ in (6.16)) is replaced by $E_{\rm di}(Q)$. In this way one gets

$$-k_{\rm di}^{-1} = \left\langle \left.\frac{\partial^2 \hat{H}}{\partial Q^2}\right|_{Q_0}\right\rangle^{\rm qs}_{Q_0} = \left.\frac{\partial^2 E_{\rm di}(Q)}{\partial Q^2}\right|_{Q_0} + \chi(0) \equiv C_{\rm di}(0) + \chi(0). \qquad (6.131)$$

In writing this equation the following feature was taken into account. All quantities which appear in terms of first order in the $Q - Q_0$ may be calculated with the unperturbed density operator $\hat{\rho}_{\rm di}(Q_0)$. Through the choice of the occupation probabilities in (6.130) these quantities are then analogous to those used above for the quasi-static case. This holds true for the expectation value of $\partial^2 \hat{H}/\partial Q^2$ as well as for the static response $\chi(0)$ in (6.131) as well as the dynamic response $\chi(\omega)$ in the secular equation, be it in the form $1 + k_{\rm di}\chi(\omega) = 0$ or in $\chi(\omega) - \chi(0) = C_{\rm di}(0)$. This feature facilitates a direct comparison between the two cases. The difference of the two (inverse) coupling constants, for instance, can be expressed by the difference of susceptibilities. Indeed, from (6.131) and (6.88) one gets

$$-k_{\rm di}^{-1} + k^{-1} = C_{\rm di}(0) - C(0) = \chi^{\rm ad} - \chi(0) \lesssim \chi^{\rm T} - \chi(0). \qquad (6.132)$$

In (Kiderlen et al., 1992) such quantities have been evaluated within the same purely independent particle model mentioned already above. In Fig. 6.8 the static energy is shown on the left: The fully drawn curve corresponds to the $E_{\rm ipm}$ of (5.18) considering only the configurations of lowest possible energy. The cusps seen at level crossings are unphysical of course. This is one of the deficiencies of the independent particle model, which can be dealt with by considering residual interactions. In a sense, an effect of this type is provided by Strutinsky smoothing (see the discussion in Section 5). Applied to the present problem it leads to the curve shown in Fig. 6.8 by the dashed line. Let us concentrate on the region of the minimum at $Q = 1$ which corresponds to a spherical shape; the calculation was done for the example of closed shells. For diabatic motion the occupation numbers would be fixed, which implies that no cusps appear. The potential energy would just continue to rise and for small amplitude vibrations the stiffness $C_{\rm di}(0)$ would be given by the large value calculated at $Q = 1$. Evidently, the situation for the smooth curve is drastically different: The stiffness is much smaller such that vibrations would reach much farther out, even if they are treated in linearized form. Actually, a Strutinsky smoothed curve somehow represents a heated nucleus of about $T = 2 - 3$ MeV. The property of a diabatic energy $E_{\rm di}(Q)$ increasing ever with Q tells one that the pure diabatic model is unable to describe nuclear fission.

In the curves in the right part of Fig. 6.8 the temperature dependence is studied in detail. On the bottom the static response is compared with the stiffness $C_{\mathcal{F}}(0)$ of the free energy. It is seen that above $T \simeq 0.5$ MeV the $C_{\mathcal{F}}(0)$

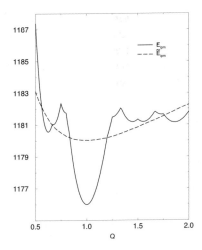 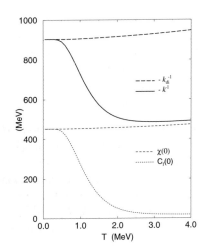

FIG. 6.8. Left panel: Static energy for a model of independent particles in the deformed oscillator with no spin-orbit force for the schematic case of 4×56 nucleons; fully drawn line: sum over single particle energies for the Q-dependent ground state; dashed line: the same energy after Strutinsky smoothing. Right panel: The T-dependence of the coupling constants for quasi-static and diabatic motion (the difference of which also presents an estimate for $\chi^T - \chi(0)$), as well as that of the static response and of the stiffness of the free energy; all for a spherical configuration. [From (Kiderlen et al., 1992)]

quickly deviates from $\chi(0)$ to become much smaller when the shell effects have disappeared. It may be mentioned in passing that it is only for this oscillator model that both quantities are identical at small thermal excitations. In more realistic cases $\chi(0)$ is larger than the stiffness even at $T=0$. This strong decrease of $C_\mathcal{F}(0)$ with increasing T implies a similar behavior of the inverse coupling constant $-1/k$. Please observe that according to (6.132) the difference between the curves in the upper part represents the difference in the stiffnesses and that in the susceptibilities, namely $\chi^T - \chi(0)$. Remember that, quite generally, for a non-degenerate ground state χ^T approaches $\chi(0)$ when T turns to zero.

Next we may draw the reader's attention to the fact of the very weak temperature dependence of the diabatic coupling constant. From the discussion in the previous subsection it should be evident that the strong shift of the strength distribution with increasing temperature will be absent. The dissipative part of the collective response function for diabatic motion resembles the one represented by the dashed curves of Fig. 6.6 rather than those of the fully drawn ones, or even the more realistic ones of Fig. 6.5.

Finally, we would like to briefly add a comment on RPA at finite temperature. Indeed, in the commonly used versions there are great similarities to the diabatic model as treated here. There, too, temperature simply enters as a fixed parameter to define the single particle's occupation number of the unperturbed density operator. Moreover, in RPA nucleonic dynamics is restricted to the 1p-1h level, for which the difference of the susceptibilities to the static response is largest. It can be expected that residual interactions will reduce this value simply because it will lift the many degeneracies present in the pure independent particle model. We will return to this question in Section 6.6.3.1.

6.6.2.1 *The DDD model of Nörenberg et al.* In the 1980s W. Nörenberg and collaborators have developed a model which the authors labelled as "dissipative diabatic dynamics" (DDD), see e.g. (1982), (1989). To some extent it incorporates both diabatic as well as quasi-static (adiabatic) features. To facilitate comparison with the treatment from above we follow (Hofmann et al., 1996), (Hofmann, 1997) and apply DDD to small amplitude vibrations about the potential minimum at Q_0 and describe them by nucleonic and collective response function. Essentially, all one needs to do is to linearize the basic equations of motion of DDD, for instance those presented in Nörenberg's review article (1989). The e.o.m. of DDD are restricted to average dynamics but include non-Markovian features. The latter are embodied in a force of the type

$$\int_{t_0}^{t} ds\, K(t,s)\dot{q}(s) = C_D \int_{t_0}^{t} ds\, \exp\left(-\frac{(t-s)}{\tau_{\text{intr}}}\right) \dot{q}(s). \tag{6.133}$$

It replaces the Markovian friction force $\propto \dot{q}(t)$ into which the right-hand side of (6.133) turns for slow motion. This will be clarified below where the nature of the constant C_D will also be discussed. As before, $q = Q - Q_0$ specifies the amplitude of the vibration. The τ_{intr} represents the intrinsic relaxation time, which in the linearized version is taken to be independent of $q(t)$. Close to thermal equilibrium the τ_{intr} can be estimated to be (Bertsch, 1978)

$$\frac{\tau_{\text{intr}}}{\hbar} \simeq \frac{2\times 10^{-22}\,\text{sec}}{T^2 \times 10^{-1}\hbar} \simeq \frac{3}{T^2\,\text{MeV}} \simeq \frac{30}{\pi^2 T^2\,\text{MeV}} \qquad (T\text{ in MeV}). \tag{6.134}$$

Actually, this value of τ_{intr} is identical to $\hbar/\Gamma(\hbar\omega = \mu, T)$ with the $\Gamma(\hbar\omega, T)$ given by (6.76) for $c = \infty$. In (Nörenberg, 1989) the τ_{intr} is introduced as the relaxation time for the occupation numbers $n_k(t)$ for single particle states $|k\rangle$. Its time dependence is governed by an equation of the same type as the (21.27) for the Wigner function of the one-body density. Here, the equilibrium $n_k^{eq}(q,\mu,T)$ is represented by the Fermi occupation number. In addition to chemical potential μ and temperature T, it may in principle depend on q, which implies a coupling to the e.o.m. for the $q(t)$.

In the linearized version close to equilibrium this coupling may be discarded such that the effective forces only depend on a constant temperature. These forces consist of the kinetic one plus that of (6.133) together with the derivative $-C(0)q$ of the static energy in adiabatic approximation; see eqns(2.30–32) of (Nörenberg, 1989). Since we want to work with response functions we need to add a term $-q_{\text{ext}}(t)$ on the right-hand side of the equation of motion (say to eqn(2.30) of (Nörenberg, 1989)), in complete analogy with eqn(23.3). This additional term represents an external force, with the coupling to the system given by $\delta H = q_{\text{ext}}(t)\, q$. For applying Fourier transforms in ordinary fashion it is more convenient to put the initial time t_0 of (6.133) equal to $-\infty$, which is possible for a sufficiently small τ_{intr}. Defining the response function for collective motion in the usual way one easily derives the following expression:

$$-\delta q(\omega)/\delta q_{\text{ext}}(\omega) \equiv \chi^D_{\text{coll}}(\omega) = -\left(m_0\omega^2 - C(0) - C_D \chi^D_{\dot q}(\omega)\right)^{-1}. \qquad (6.135)$$

Here, m_0 is the collective inertia and C_D represents the difference between the stiffnesses C_{di} and $C(0)$ of the diabatic (see (6.131)) and the adiabatic potentials:

$$C_D = C_{\text{di}} - C(0) \quad \simeq \quad \chi^{\text{T}} - \chi(0). \qquad (6.136)$$

On the very right use has been made of the statement (6.132) from above that this difference of the stiffnesses can be identified as the one of susceptibilities. As we shall see below this interpretation sheds some light on the nature of the whole approximation. In (6.135) there also appears the response function

$$\chi^D_{\dot q}(\omega) = -i\omega \chi^D(\omega) \equiv -i\omega \int_{-\infty}^{\infty} d(t-s)\, \Theta(t-s)\, \exp\bigl[(i\omega - 1/\tau_{\text{intr}})(t-s)\bigr]$$
$$= -i\omega \bigl(i/(\omega + i/\tau_{\text{intr}})\bigr) = \omega\tau_{\text{intr}}/(\omega\tau_{\text{intr}} + i). \qquad (6.137)$$

Here, the Θ-function has been introduced to take care of the upper integration limit in (6.133); it renders $\chi^D(\omega)$ and $\chi^D_{\dot q}(\omega)$ to be causal functions. They take up the part which has been called the "nucleonic" response in previous sections. It is obvious that the present functions have much less structure than those used in Section 6.5 which base on (6.75) for damped motion. In fact, this is even true for the simplified, schematic model of Section 6.5.4. The $\chi^D(\omega)$ given in (6.137) represents a single, over-damped intrinsic mode specified through one parameter. In the $\chi(\omega)$ of (6.121), on the other hand, there is the additional quantity Ω which stands for possible excitations to intrinsic states of energy $\hbar\Omega$.

Before we analyze (6.135) let us mention that the derivation of the e.o.m. in the DDD, and thus that of (6.135), is based on an application of the scaling model. For this reason it is no surprise that the effective inertia, called m_0 here, becomes identical to that of irrotational flow–independent of the intrinsic excitation (Hofmann and Sollacher, 1988), (Hofmann, 1997). This feature may be explained looking at the Hamiltonian $\hat{\mathcal{H}}_{\text{a}}$ to be discussed below in Section 7.5.1.1. With the simplification (7.54) for the mass operator taken into account, the $\hat{\mathcal{H}}_{\text{a}}$ is given by (7.50). If the renormalization of this unperturbed inertia m_0 to the final M (as it appears in eqn(7.57) as well as in subsequent equations) were discarded, one would end up with a response function of the type (6.135). This would be found in a simple Born–Oppenheimer approach, in which the equation for collective motion is deduced by just averaging the $\hat{\mathcal{H}}_a$ over the (unperturbed) nucleonic degrees of freedom. Because of $\langle\hat{F}\rangle=0$ in such an averaged Hamiltonian the coupling term would drop out which is proportional to the canonical momentum $\hat{\Pi}$.

Let us turn now to the possible modes described by the response function (6.135). Two limiting cases are readily singled out for which the $\chi_{\hat{q}}^D(\omega)$ reduces to simple forms: High and low frequency modes distinguished by the conditions $\omega\tau_{\rm intr} \gg 1$ and $\omega\tau_{\rm intr} \ll 1$, respectively. In the first case $\chi_{\hat{q}}^D(\omega)$ simply turns into the 1 such that $\chi_{\rm coll}^D(\omega)$ takes on the structure of an *undamped* oscillator with renormalized stiffness $C(0)+C_D = C_{\rm di}$. For $\omega\tau_{\rm intr} \ll 1$, on the other hand, $\chi_{\hat{q}}^D(\omega)$ becomes equal to $-i$ and the resulting $\chi_{\rm coll}^D(\omega)$ represents motion of a damped oscillator with the quasi-static stiffness $C(0)$ and the friction coefficient $\gamma_D = \tau_{\rm intr} C_D$, which will be analyzed further in Section 6.6.3.1. From (6.134) it is seen that this limit is given only for temperatures which are definitely larger than $T \gtrsim 1$ MeV. For the relevant scale for the frequency one ought to take the $\varpi_D = \sqrt{C(0)/m_0}$ of the oscillator. Hence, the condition $\omega\tau_{\rm intr} \ll 1$ can be written as $3\hbar\varpi_D \ll T^2$, but the $\hbar\varpi_D$ is typically of the order of a few MeV (remember that one is dealing with a comparatively small inertia m_0).

A system of such properties behaves like an elasto-plastic material, and the founders of DDD claim the nucleus to behave like that. It must be said, however, that the structure of the response functions (6.135) is closely linked to the applicability of the independent particle model, for practically all temperatures. Indeed, traces of the residual interaction $\hat{V}_{\rm res}$ only show up in the finite value of the relaxation time $\tau_{\rm intr}$. But the latter only parameterizes relaxation to the equilibrium distribution described for a system of independent particles–and *not* to that for *genuine compound states*. One of the essential parameters of the model, the C_D which specifies the strength of the force (6.133) is proportional to $\chi^T - \chi(0)$. This relation given in (6.136) is valid within the independent particle model for which the difference $\chi^T - \chi(0)$ is particularly large, see the discussion in Section 6.6.3.2. It can be expected to be much smaller for ergodic systems, which is to say if it were evaluated for states of more complicated nature. Such effects are not accounted for in the model, but they may have large impact on the value of the friction coefficient, as will be shown below. Of course, there may be processes for which there is not enough time for the system to build up such states of more complex nature, analogous in a sense to the typical pre-equilibrium reactions discussed in Chapter 9. Indeed, the DDD model has mainly been applied to describe the entrance phase of heavy-ion collisions (Nörenberg, 1989) where such a situation may be given.

Finally, it may be mentioned that the strength distribution for the response $\chi_{\rm coll}^D(\omega)$ also varies with temperature, not unlike the situation discussed in Section 6.6.1. This has been demonstrated in (Hofmann et al., 1996) in a numerical calculation, which showed the existence of a shift to lower frequencies. However, the transition appears at temperatures not smaller than 4 MeV, which is to say at values which are definitely larger than the ones found in Section 6.6.1. But there are other differences. It is seen that at small T in the DDD model practically no low frequency mode exists. This may be a consequence of the scaling assumption, for which the inertia is very small. Actually, for zero temperature the difference $C_{\rm di} - C(0)$ vanishes identically (if evaluated for a situation typical for a vibration about a spherical configuration). It is only above $T \simeq 1$ MeV that $\chi^{\rm T} - \chi(0) \gtrsim C_{\rm di} - C(0)$ starts to grow; these features may already be seen in the schematic model presented in Fig. 6.8. All these properties also have important consequences for the friction coefficients which we are going to address in the next section.

6.6.3 T-dependence of transport coefficients

6.6.3.1 Contributions to friction from very small frequencies
As indicated in (6.92) the friction coefficient $\gamma(0)$ in the zero frequency limit can be expressed by the correlation function $\psi''(\omega)$ evaluated at $\omega = 0$. Basic properties of the latter are described in Section 23.4. There it is shown in (23.84) that at $\omega = 0$ the $\psi''(\omega)$ gets contributions from states of the same energy. If evaluated from the (exact) spectral representation associated with the Hamiltonian which specifies the quantal states this contribution shows up in a δ-function singularity. As derived in Section 23.6.3 this term is proportional to $T(\chi^{\rm T} - \chi(0)) \equiv \psi^0$. The whole function $\psi''(\omega)$ may then be written in the form

$$\psi''(\omega) = T(\chi^{\rm T} - \chi(0))\, 2\pi\, \delta(\omega)\; +\; {}_R\psi''(\omega) \equiv {}_0\psi''(\omega) + {}_R\psi''(\omega)\,, \qquad (6.138)$$

where ${}_R\psi''(\omega)$ is the regular part, which gets contributions only from states of different energies. In application to a dissipative system the first term is to be treated with some caution. It is evident that *dissipation* can come only from transitions between states of *different energies*. (Interested readers may acquaint themselves with this feature by looking at the description of irreversible processes within the Kubo formalism presented in Section 23.7.) On the other hand, to obtain a finite friction force one needs to go beyond Hamiltonian dynamics. For instance, one may apply the energy smearing described in Section 6.4.1. Within the single particle model this means nothing else but introducing a constant width for each level. In a sense this is equivalent to a relaxation time approximation in the equation of motion for the single particle density. The method introduced in Section 6.4.3 is more general as the dressing of particles and holes is done with frequency and temperature-dependent self-energies. In any case such an introduction of widths for the single particle levels is meant as an approximation to simulate the existence of more complicated states. If this were done correctly, it would be the energy difference of these states which matters. Later in Section 6.6.3.2 we will discuss some severe problems behind such simplified procedures of handling residual interactions when treating dynamics at small excitations. At first we shall continue using such prescriptions and discuss applications published in the literature.

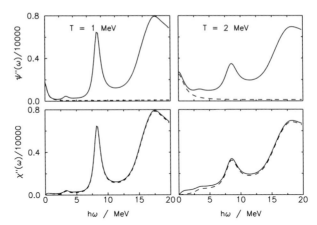

FIG. 6.9. The correlation function $\psi''(\omega)$ and the response function $\chi''(\omega)$ for two temperatures (Hofmann et al., 1996). The contribution from $_0\psi''(\omega)$ to $\psi''(\omega)$ is shown explicitly. The dashed curve in the lower parts represents $\chi''(\omega)$ without contributions from the diagonal elements.

In (Hofmann et al., 1996) the correlation function $\psi''(\omega)$ has been evaluated in the same way as described in Section 6.4.3 for the response function. For the underlying single particle model an infinitely high square well potential was chosen. Details of the application of such a microscopic model to transport problems are described in Section 16.2.3; they are not of greater importance here. It should only be mentioned that for the evaluation of response and correlation functions for a square well potential requires the introduction of cutoffs in the energy, which typically was chosen to be 30 MeV above the Fermi level. In Fig. 6.9 correlation and response functions $\psi''(\omega)$ and $\chi''(\omega)$ are shown as functions of ω for two different temperatures. It is seen that because of collisional damping the δ-function peak at $\omega = 0$ has changed into a smooth one, which is to say[16]

$$_0\psi''(\omega) = \psi^0\, 2\pi\, \delta(\omega) \implies {}_0\psi''(\omega) = \psi^0 \frac{\hbar \Gamma_T}{\hbar^2 \omega^2 + \Gamma_T^2/4}. \qquad (6.139)$$

From these calculations both the strength ψ^0 as well as the width Γ_T of the peak can be deduced. In this way it was seen that the $\chi^{\mathrm{T}} - \chi(0) = \psi^0/T$ obtained from the calculation with collisional damping can very well be reproduced by the $\chi^{\mathrm{T}} - \chi(0)$ of the independent particle model. For the latter one finds

---

[16] In (Hofmann et al., 1996) and (Hofmann, 1997) this contribution has been called the "heat pole". This notion was chosen because of analogies with density-density correlation functions of infinite systems, like Fermi liquids, for instance, where this peak is also referred to as Landau-Placzek peak.

$$\psi_{\text{ipm}}^0 = \sum_{jk,\ e_j=e_k} |F_{kj}|^2\, n(e_k)(1-n(e_k)) = T\sum_k \left|\frac{\partial n(e_k)}{\partial e_k}\right| \left(\frac{\partial(e_k-\mu)}{\partial Q}\right)^2. \tag{6.140}$$

The behavior of $\chi^T - \chi(0)$ as a function of T is similar to that seen already in the right part of Fig. 6.8: It starts from zero at $T=0$ and above $T \simeq 0.5$ MeV rises quickly to a practically constant value for $T \gtrsim 2$ MeV. The origin of $\psi^0 \approx \psi_{\text{ipm}}^0$ lies in the deficiency in the way in which in these calculations sp-states are dressed. As explained at the end of Section 6.4.3, this procedure does not lift the degeneracies of the states, and therefore, according to Section 23.6.3, cannot be expected to diminish the difference $\chi^{\text{ad}} - \chi(0)$. As explained at the end of Section 6.5.1, usually in the nuclear case $\chi^T - \chi^{\text{ad}}$ is small such that $\chi^{\text{ad}} - \chi(0)$ contributes most to $\chi^T - \chi(0)$. Up to about $T \simeq 5$ MeV the Γ_T found from the numerically evaluated function $_0\psi''(\omega)$ can very well be represented by twice the value $\Gamma(\hbar\omega = \mu, T)$ the single particle width has at the chemical potential. From (6.76) it is seen to be given by

$$\Gamma(\hbar\omega=\mu,T) = \frac{1}{\Gamma_0}\frac{c^2\,\pi^2 T^2}{c^2+\pi^2 T^2} = \frac{1}{\Gamma_0}\frac{\pi^2 T^2}{1+\pi^2 T^2/c^2}. \tag{6.141}$$

For a $c \simeq 20$ MeV above about $T=1$ MeV the Γ_T rises almost linearly with T.

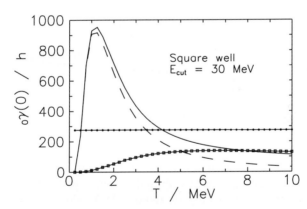

FIG. 6.10. Contribution from diagonal matrix elements to friction in the independent particle model (Hofmann et al., 1996). The T of $_0\gamma(0)$ is shown by the fully drawn and the dashed curves. For the first, the c of (6.141) is chosen to be 20 MeV and the second corresponds to $1/c = 0$. The horizontal line with stars represents the value of the wall formula (at $T=0$). The line with squares shows the contribution from non-diagonal matrix elements.

Let us return to the friction coefficient γ and let us assume the zero frequency limit $\gamma \approx \gamma(0)$ to be applicable. Roughly speaking this requires the collective frequency ω_1 to be small compared to the typical intrinsic one. This condition may be looked at for the schematic but illustrative case that the correlation function $\psi''(\omega)$ is dominated by the part $_0\psi''(\omega)$ given in (6.139). A simple calculation shows the condition to be similar to (6.21) found for the undamped case (see exercise 5). From $\gamma(0) = \psi''(0)/2T$ one identifies the contribution from $_0\psi''$ as (remember that $\Gamma_T = 2\Gamma(\mu,T)$)

$$_0\gamma(0) = \frac{\hbar}{\Gamma(\mu,T)}(\chi^T - \chi(0)) = \frac{\hbar}{\Gamma(\mu,T)}\sum_k \left|\frac{\partial n(e_k)}{\partial e_k}\right|\left(\frac{\partial(e_k-\mu)}{\partial Q}\right)^2. \quad (6.142)$$

Since for large T the difference $\chi^T - \chi(0)$ approaches a constant value, there the variation of $_0\gamma(0)$ is entirely determined by the choice of c. For any finite value of c the width $\Gamma(\mu,T)$ turns into c^2/Γ_0 and $_0\gamma(0)$ becomes independent of T. For $1/c = 0$, on the other hand, $\Gamma(\mu,T)$ becomes proportional to T^2 such that $_0\gamma(0)$ tends to zero like T^{-2}, a behavior which usually is associated with "hydrodynamic dissipation" or "two-body viscosity". For small T the pre-factor $1/\Gamma(\mu,T)$ of $_0\gamma(0)$ behaves like T^{-2} in both cases. However, this divergence is cancelled by the factor $\chi^T - \chi(0)$, which is expected to approach zero exponentially. Consequently $_0\gamma(0)$ starts from zero and increases quickly to some maximum value at $T \approx 1$ MeV. Fig. 6.10 shows the results of the numerical calculations for the square well. Also presented is the contribution $\gamma(0)$ from the regular part $_R\psi''(\omega)$ of the correlation function in (6.138). This corresponds to the first derivative of the response function which in Fig. 6.9 is represented by the dashed curve.

With the aid of Fig. 6.10 it is possible to discuss generic features of results of theories for nuclear dissipation suggested in the literature. Indeed, more characteristic than the absolute value of the friction coefficient itself is its variation with thermal excitation. The wall formula, for instance, only shows a very weak dependence on T. In particular, it predicts a finite and large value even for zero temperature, as given by (17.2) and shown in the figure by the straight line. The physics behind the wall formula is described in detail in Chapter 17. There it will be shown that it comes out as a certain macroscopic limit of the linear response formula, but where the contributions from diagonal matrix elements require special care. Hydrodynamic behavior is obtained for slow motion in the DDD model of Nörenberg et al. (Ayik and Nörenberg, 1982) described in Section 6.6.2.1. For temperatures above about 1 MeV, slow motion in this model is governed by a friction force with its coefficient being $\gamma_D = \tau_{\text{intr}} C_D$, which is identical to the form (6.142). This follows because the τ_{intr} is to be taken from (6.134), which renders it equal to $\hbar/\Gamma(\mu,T)$, and the C_D from (6.136). In (Ivanyuk, 1989) a similar result was derived on the basis of a von Neumann type equation for the one-particle density operator. The streaming term was defined through the shape dependent mean field Hamiltonian and the dissipation mechanism was traced to a collision term defined in relaxation time approximation (with a constant relaxation time $\tau = \hbar/\Gamma$). Finally, we may mention the model of (Bush et al., 1992) in which collective dynamics itself is assumed to be governed by two-body collisions, rather than by the mean field itself. It is not astonishing that the friction coefficient of this model also goes like T^{-2}.

6.6.3.2 Problems with relaxation time approximations

Let us return to formula (6.142) and examine the origin of the strong overshoot even over the wall formula, which itself is known to represent a strong nuclear friction force, see below. This feature can be traced back to the value the factor ψ^0 obtains in the model of independent particles. As discussed earlier this ψ^0_{ipm} is dominated by the largely overestimated difference of the adiabatic susceptibility from the static response, $\chi^{\text{ad}} - \chi(0)$. In the independent particle model this difference turns out to be particularly large. This is due to the many degeneracies which occur in this model. As described in Section 23.6.3, the fewer degeneracies a system has, the smaller the value of $\chi^{\text{ad}} - \chi(0)$. Simply dressing single particle states by a self-energy with the finite width $\Gamma(\omega, T)$ of (6.76) does not lift the degeneracies of the independent particle model. Of course, neither is this achieved if this energy-dependent width is replaced by a mere constant, as is the case in the various versions of the relaxation time approximation to the dynamics of the one-body density. Such approximations are not capable of accounting for effects of level repulsion due to residual interactions. In more general contexts, in this way one cannot treat correctly the behavior of two-body (or ph-) Green functions at small excitations. One may look at the problem from a different point of view. In a formula like (6.142) the residual interaction $\hat{V}_{\text{res}}$ only influences the first factor $\hbar/\Gamma(\mu, T)$ but not the $(\chi^{\text{T}} - \chi(0))$, which is still expressed by its value in the independent particle model. The width $\Gamma(\mu, T)$, however, turns to zero like the averaged squared matrix element of $\hat{V}_{\text{res}}$. Considered as a functional of $\hat{V}_{\text{res}}$ this leads to a very strange divergence for $_0\gamma(0)$. Remember that the expression on the right of (6.142) represents $(\chi^{\text{T}} - \chi(0))$ to order zero in $\hat{V}_{\text{res}}$. To solve that problem one will have to evaluate $(\chi^{\text{T}} - \chi(0))$ in non-perturbative fashion.

Indications that such a procedure may lead to significant changes can be obtained by the following argument. We know from Section 4.2.3.1 that the nearest neighbor spacing of the genuine many-body states of real nuclei is of Wigner type–if considered within a certain group (or block) of states with the same symmetries. These latter symmetries are those of the nucleus as a whole and thus correspond to a much smaller manifold than the largely arbitrary ones of single particle models used in practical computations. These statements are confirmed by the findings discussed at the end of Section 23.4.4 where response and correlation functions are evaluated for a Random Matrix Model. In this case no sign of any substantial peak of $\psi''(\omega)$ at $\omega = 0$ is seen, which in other words says that ψ^0 is practically zero.

6.6.3.3 Analytic formulas from a schematic model

There are numerous papers where numerical computations of the response functions and the transport coefficients introduced and described in Section 6.5 have been presented, see e.g. (Yamaji et al., 1988), (Samhammer et al., 1989), (Hofmann et al., 1996), (Ivanyuk et al., 1997), (Yamaji et al., 1997), (Ivanyuk and Hofmann, 1999), (Hofmann and Ivanyuk, 1999). In most of them for friction only the contributions from states of different single particle energies were considered. In this way this very dubious part $_0\gamma(0)$ of (6.142) is discarded. Evidently this problem plays an important role at such deformations where the energies of two single particle states become identical or nearly identical, which is to say at level crossings or avoided crossings. This subject will be discussed in greater detail in Section 11.2 where examples for the Q-dependence of transport coefficients will also be shown. Here, we shall concentrate on the temperature dependence of coefficients evaluated at the potential minimum and at the barrier. Indeed, in comparisons with experimental results on fission Kramers' rate formula is often taken for which only information on the dynamics at these extremal points is needed. In fact, rather than the transport coefficients for inertia M, stiffness C and friction γ themselves only special combinations are of interest, namely

$$\tau_{\text{coll}} = \frac{\gamma}{|C|} = \frac{\hbar}{\Gamma_{\text{coll}}}, \quad \tau_{\text{kin}} = \frac{M}{\gamma} = \frac{\hbar}{\Gamma_{\text{kin}}} \quad \text{and} \quad \varpi^2 = \frac{|C|}{M}. \quad (6.143)$$

The τ_{coll} sets the scale for (local) relaxation of collective motion in a given potential of (local) stiffness C. The τ_{kin}, on the other hand, measures the relaxation of the kinetic energy to the equilibrium value of the Maxwell distribution. Typically, for slow collective motion we expect this time to be smaller than the former. The limit of over-damped motion applies for $\tau_{\text{kin}} \ll \tau_{\text{coll}}$. In terms of the dimensionless quantity η defined by

$$2\eta = \sqrt{\tau_{\text{coll}}/\tau_{\text{kin}}} = \Gamma_{\text{kin}}/(\hbar\varpi) = \varpi\tau_{\text{coll}} \quad (6.144)$$

the over-damped limit is reached for $\eta \gg 1$. For a positive stiffness ($C > 0$) and under-damped motion, the ϖ would be the frequency of the vibration and the Γ_{kin} its width. It should be noted that in the literature a different notation is sometimes used where γ stands for η, and often the $\Gamma_{\text{kin}}/\hbar$ is referred to as β.

In (Hofmann et al., 2001) analytic formulas have been suggested, which were suggested by microscopic, numerical calculations for the example of ^{224}Th. The Q was chosen to represent the distance between the centers of mass of the left and right parts of the whole nucleus, normalized to twice the radius of the spherical shape of the same volume. The static energy was approximated by the free energy $\mathcal{F}(Q,T)$. The temperature dependence of the shell correction $\delta\mathcal{F}(Q,T)$ was assumed to follow the simplified law (5.51). The $\delta\mathcal{F}(Q,T=0)$ was evaluated through a Strutinsky procedure but with averages performed such that only gross shell features survived. In this way all necessary properties of the potential energy were obtained. At its maximum and minimum friction and inertia were calculated in (the modified) zero frequency approximation of Section 6.5.3.1. To obtain analytical forms of their functional dependence on T the schematic model of Section 6.5.4 was employed, where the ph-width was approximated by twice the estimate (6.141) for the sp-width, $2\Gamma(\hbar\omega = \mu, T)$. Comparing these analytical forms with actual numerical calculations the following formulas have been verified:
(1) For a first orientation the $\hbar\varpi$ can be approximated as $\hbar\varpi_a \simeq \hbar\varpi_b \simeq 1$ MeV,

with deviations being within a factor of 2 or less. This appears to be the case even when pairing is included at smaller T (Ivanyuk and Hofmann, 1999). This estimate is in accord with general expectations as well as with previous numerical calculations (Yamaji et al., 1997), (Ivanyuk et al., 1997). It is worth noticing that this feature implies that, if considered as function of T, the product $\tau_{\text{coll}}\tau_{\text{kin}}$ remains constant.

(2) For temperatures up to $T \simeq 2-3$ MeV the ratio of friction to inertia γ/M may be approximated as (with T measured in MeV):

$$\frac{\gamma}{M}\hbar = \Gamma_{\text{kin}} \approx 2\Gamma(\mu, T) = \frac{2}{\Gamma_0} \frac{\pi^2 T^2}{1+\pi^2 T^2/c^2} \approx \frac{0.6T^2}{1+T^2/40} \text{ MeV}. \tag{6.145}$$

For larger T motion gets over-damped and the value of M becomes irrelevant.

(3) For the ratio of friction to stiffness the following formula was suggested:

$$\tau_{\text{coll}} = \frac{\gamma}{|C|} \approx \frac{2}{\hbar\varpi^2 \Gamma_0} \frac{\pi^2 T^2}{1+\pi^2 T^2/c^2_{\text{macro}}}$$
$$\approx \frac{0.6T^2}{1+\pi^2 T^2/c^2_{\text{macro}}} \frac{\hbar}{\text{MeV}} \quad (T, c \text{ macro in MeV}). \tag{6.146}$$

Here, the c of the previous formulas has been changed to a parameter c macro which may be fixed in such a way that for large T the τ_{coll} reaches a macroscopic limit. For the stiffness evidently the latter would be given by the liquid drop model $C_{\text{LDM}}(T)$. Previous computations (Hofmann et al., 1996), (Yamaji et al., 1997) for the friction coefficient have shown that it reaches a plateau, very similar to the line marked with squares in Fig. 6.10. In first approximation this high temperature limit may be represented by about half the value of wall friction, although the factor $1/2$ is only to be considered a rough rule of thumb.

(4) In (Ivanyuk and Hofmann, 1999) modifications have been studied which are due to pairing correlations. For small temperatures the latter prohibit "scattering" of quasiparticles into states which differ in energy less than the gap. As a consequence, the effective width of these states is small (zero at $T=0$) below the energy of the gap, which in turn implies slow collective modes to be only weakly damped, and strictly undamped at $T=0$. With increasing temperature this hindrance mechanism becomes less and less effective such that friction will show a strong increase at small temperatures. The modifications for the τ_{kin} and τ_{coll} from above can be summarized in a factor $f_{\text{pair}} = \left(1+\exp(-a(T-T_0))\right)^{-1}$ by which they have to be multiplied to get $\tau_{\text{kin}} = f_{\text{pair}}\tau_{\text{kin}}(\Delta=0)$ and τ_{coll} similarly. The parameters appearing here are: $a = 10$ MeV^{-1} and $T_0 = 0.55$ MeV at the barrier, and $a = 12$ MeV^{-1}, $T_0 = 0.48$ MeV at the minimum.

6.6.3.4 Comparison with the macroscopic limit

The relation of the macroscopic limit to the Strutinsky procedure has already been discussed in Section 6.6.1.1. For the static energy it is known (see Section 5.5) that this macroscopic limit is closely related to the high temperature limit. In this context the latter is simply to be understood as that limit where shell effects have disappeared. Indeed, both for the inertia as well as for friction a behavior of this type has been observed above. The inertia was seen to reach the sum rule limit. As this is in some relation to the inertia of irrotational flow this property in a sense confirms the claim from before. For friction the situation is very different. In the quasi-static model presented in the previous sections a value close to the wall formula was obtained for large temperatures. It must be said though that the precise behavior of friction as function of T depends on the parameters which specify the single particle widths. Truth is that in most applications of transport models to nuclear fission, and in some to heavy-ion collisions as well, one often simply assumes these limits to be given: Irrotational flow value for the inertia (obtained through the so-called Werner–Wheeler method) and the wall formula for friction. In doing so any temperature dependence in inertia and friction is discarded, only that of the free energy is sometimes taken into account. The microscopic results presented above, on the other hand, show a very sensitive dependence on temperature. In Fig. 6.11 a comparison of microscopic and macroscopic coefficients is presented. Moreover, it contains a comparison between the analytic formulas of Section 6.6.3.3 and full microscopic computations. Notice, please, that in the microscopic case collective motion changes from under-damped to over-damped at a temperature $T \approx 2$ MeV, whereas in the macroscopic limit collective motion is over-damped ($\eta_b > 1$) even at very small temperatures. Remember that with increasing η_b the inertia M becomes less and less important, to drop out entirely for the truly over-damped case. For this reason the estimates given above should be understood to represent those quantities which involve M only for smaller temperatures, up to $T \approx 2-3$ MeV at best.

Although no definite answer can at present be given it seems that at temperatures above $T \simeq 1-2$ MeV slow collective motion is strongly damped or even over-damped–in contrast to the case of zero thermal excitation for which low frequency modes are undamped. In fact, in (Back et al., 1999) experimental evidence from fission has been presented that η is very small at low excitations to rise more or less strongly with T depending on the system. In (Diószegi et al., 2000) a linear and quadratic variation with T has been suggested. These findings are in qualitative agreement with the theoretical predictions presented above, recall e.g. the combination of (6.144) with (6.146). One must say, however, that in such analyses of experimental data questionable assumptions about fission timescales are made, a problem to which we will return to in Section 12.4. This puzzle will have to be clarified before a more quantitative understanding of the precise temperature dependence of nuclear transport can be achieved. But already the present knowledge allows for a statement about the nature of the macroscopic limit. Unlike the case of static energies, for dynamic quantities the high temperature limit does not represent an average behavior of quantities observed in finite nuclei at $T = 0$.

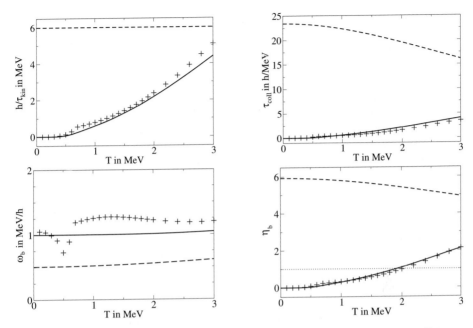

FIG. 6.11. Temperature dependence of relevant ratios of transport coefficients (Rummel and Hofmann, 2003). Critical damping $\eta_b = 1$ is marked by the dotted line in the bottom right panel. Crosses refer to the microscopic computations of (Hofmann *et al.*, 2001) and (Ivanyuk and Hofmann, 1999). The solid line represents the fits (6.145) and (6.146) to these data. Macroscopic transport coefficients are given by the dashed line (Ivanyuk *et al.*, 1997).

6.7 Rotations at finite thermal excitations

In Section 6.3 it was indicated that in the pure independent particle model the value of the rotational moment of inertia $\mathcal{J}$ is close to the $\mathcal{J}_{\rm rig}$ of the rigid body. Effects of pairing reduce this value to about its half. This behavior is largely confirmed in microscopic computations which involve advanced formal techniques like angular momentum and particle number projection (Ring and Schuck, 2000). Generalizations to finite thermal excitations seem to confirm these features. Above the critical temperature for pairing correlations $\mathcal{J}$ increases to close the $\mathcal{J}_{\rm rig}$. Amongst others such studies are important in connection with the nuclear level densities to which we will return to in Section 10.1.2. Here we would like to restrict ourselves discussing some points of interest in the light of nuclear transport.

In previous sections in this chapter it was pointed out that with increasing temperature nuclear collective motion will become dissipative, essentially because of the presence of interactions residual to the mean field. This implies the intriguing question of the relevance of such effects on rotational motion. Indeed, as described in Section 1.3.3.3, based on the hypothesis of the validity of the compound model for nuclear reactions N. Bohr and F. Kalckar (1937) argued that nuclear dynamics may resemble that of a liquid drop. In this case, and assuming irrotational flow, the rotational moment of inertia $\mathcal{J}_{\mathrm{irr}}$ would be smaller than $\mathcal{J}_{\mathrm{rig}}$ by about a factor of 10. As described in Chapter 18 by now it is known that the compound or statistical model is valid at best in some restricted sense, in particular at lower energies. However, it is also clear that with increasing energy, collisions among nucleons become more and more relevant. Viewed in the light of the Boltzmann equation for the one-body density one may argue that with increasing temperatures the concept of local equilibrium may become more and more applicable. In this sense one could expect that in such a regime of excitations Bohr and Kalckar's conjecture is also valid for rotational motion.

At this stage it is appropriate to recall the findings described in Section 6.6.1. There it was found that the vibrational inertia turns into the liquid drop value (taken here as synonymous with the sum rule value) already at the relatively small temperatures between 1 and 2 MeV. For this feature the effects of two-body collisions are somewhat less important. Rather, it is the disappearance of shell effects and the ensuing large decrease in the stiffness for this vibration. One may thus raise the question whether or not a similar behavior would be given for rotations as well. So far this intriguing problem has not been studied. One could think of the well-known example of the five-dimensional model for quadrupole vibrations introduced at the end of Section 1.3.3.2. The vibrations discussed in Section 6.6.1 can be understood to represent those in the β degree of freedom. As shown in App.6B-2 of (Bohr and Mottelson, 1975) the (constant) inertia for this β-mode also determines the rotational moments of inertia. There are some other trivial geometrical factors which depend on β and the other parameter γ, which parameterizes triaxial deformations. Actually, this inertial coefficient would be the one common to all the $\alpha_{2\mu}$ of (1.77) and the geometrical factors only result from the transformation to the body fixed frame.

So far we have been speaking about rotational motion of collective nature. For this the moment of inertia is strictly zero for rotations about any symmetry axis. This feature is given both for formula (6.58) of the cranking model as well as for the liquid drop model with irrotational flow. However, there may be a situation where the angular momenta $\mathbf{j}_k$ of the individual nucleons align along a symmetry axis such that the $\mathbf{j}_k$ add up to a total angular momentum $\hat{I}_{\mathrm{int}}$ about this axis; here the index "int" indicates that we are dealing with the property of the *intrinsic degrees of freedom*. Under certain circumstances one may then find a term in the systems' total energy which goes like $\hat{I}^2_{\mathrm{int}}/(2\mathcal{J})$, but where the $\mathcal{J}$ is found to be given by the rigid body moment of inertia $\mathcal{J}_{\mathrm{rig}}$ (see §4-3 of (Bohr and Mottelson, 1975)). Such terms appear in level density formulas such as (10.9).

This discussion hints at the problem of distinguishing between intrinsic and collective degrees of freedom, and, hence, at that for a proper definition of collective variables for rotational motion. With such a separation given the total angular momentum (about a given axis) would appear in the form $\hat{I}_{\rm tot} = \hat{I}_{\rm int} + \hat{I}_{\rm coll}$. A famous example in which this problem can be studied is the rotor plus particle model, which is commonly associated with the names of A. Bohr and B.R. Mottelson (1958), but to which since the 1950s many others have contributed (see the citations in this reference or in (Bohr and Mottelson, 1975) as well as in the other textbooks mentioned above). It can be said that in this model the inherent self-consistency problem is not fully accounted for. Later in Section 7.5 we are going to discuss general questions behind this task and present the Bohm–Pines procedure by which in principle the self-consistency problem can be tackled. It must be said, however, that rotations have never been dealt with in this way. This would require a multidimensional treatment whereas practical applications have so far been restricted to vibrations in one degree of freedom.

From the discussion above it may be conjectured that the rotational moment(s) of inertia $\mathcal{J}$ might have a sensitive temperature dependence. This leads to an additional self-consistency problem for the determination of the value of T. The variation of the rotational energy $E_{\rm rot}(Q,T;I) = \hbar^2 I(I+1)/(2\mathcal{J}(Q,T))$ with T influences the energy balance $E = \mathcal{E}(Q,T;I) + E_{\rm rot}(Q,T;I)$ in a sensible way. It is the energy $\mathcal{E}(Q,T;I)$ stored in the degrees of freedom other than rotation from which the temperature is to be calculated, which in turn influences the moment of inertia $\mathcal{J}(Q,T)$, the latter being a property of this "intrinsic" system. The problem at stake can be formulated as a proper generalization of the usual treatment at zero thermal excitations to one which is in compete accord with the common procedure of thermostatics. Within the mean-field approach of the cranking model rotational excitations are generated by adding to the Hamiltonian $\hat{H}(Q)$ terms like $-\omega_{\rm rot}\hat{I}$. Writing the Hamiltonian as $\hat{H}(Q)$ it is implied that it corresponds to a deformed system thus violating rotational invariance. In principle this deformation must be found self-consistently, but we may ignore this problem for the moment, assuming a sufficiently large stiffness for this deformation, for instance. The $\omega_{\rm rot}$ is a Lagrange multiplier which is to be fixed by requiring the expectation value $\langle \hat{I} \rangle$ to have a given value. At finite thermal excitation this problem is that of finding the appropriate density operator for a generalized ensemble under certain constraints. It is based on the maximum entropy principle and is discussed in greater detail in Section 22.1.2. The point is that for this construction the averages are to be evaluated with the generalized density operator, which is to say the one at finite temperature. In this way also the inertial coefficient $\mathcal{J}(Q,T)$) will be the one at the temperature T, which in turn is determined from the average internal energy $\mathcal{E}(Q,T;I)$. It goes without saying that one may of course wish to include particle number as another constraint. The total energy E may be plotted versus the angular momentum I. The line which in such a diagram specifies the minimal energy at given I has been called "yrast line" (it represents the "dizziest" situation the nucleus may be in, having maximal angular momentum at given total energy E). In the literature this picture is mostly applied to small excitations. Then the rotational moment of inertia may be replaced by the one which corresponds to the deformed ground state. Indeed, for very small $\mathcal{E}(Q,T;I)$ it would not make sense to apply canonical or grand canonical ensembles.

In descriptions of fission and heavy-ion collisions by transport equations rotations are usually assumed to be rigid. Use of the $\mathcal{J}_{\rm irr}$ instead of $\mathcal{J}_{\rm rig}$ might very well influence the numerical results. This is so in particular for small deformations. As shown in App.6A-5 of (Bohr and Mottelson, 1975) one has $\mathcal{J}_1^{\rm irr}/\mathcal{J}_1^{\rm rig} = \left((R_2^2 - R_3^2)/(R_2^2 + R_3^2)\right)^2$ for rotations about the 1-axis. The rigid body value is given by $\mathcal{J}_1^{\rm rig} = Am(R_2^2 + R_3^2)/5$ with Am being the total mass of the nucleus. For small deformations with $R_2 = R_3 + \delta R_2$ one gets $\mathcal{J}_1^{\rm irr} = \mathcal{J}_1^{\rm rig}(\delta R_2/R_2)^2$. If R_2 is much larger than R_3 both inertias approach each other like $\mathcal{J}_1^{\rm irr} = \mathcal{J}_1^{\rm rig}(1 - 4(R_3/R_2)^4)$.

Next we should like to take up the problem addressed in (Hofmann and Pomorski, 1991) (Ivanyuk and Yamaji, 2001), namely the influence of rotations on the transport coefficients for motion in the collective variable Q. In the first paper the friction coefficients for β and γ vibrations were evaluated and seen to have a sensitive dependence on the rotational frequency $\omega_{\rm rot}$. In the second reference only the elongation mode was considered for axially symmetric deformations. The strong dependence on $\omega_{\rm rot}$ seen before could be attributed to spurious contributions from the violation of rotational invariance. In (Ivanyuk and Yamaji, 2001) the latter was taken care of within a linear response approach. Then both friction and inertia turned out to be only weakly dependent on frequency for temperatures above the critical value for the disappearance of shell effects. Below that value a more moderate increase with $\omega_{\rm rot}$ was observed.

Finally, we ought to address the possibility that rotational motion behaves dissipatively. Think again of the analogy of rotations and vibrations, say for the five-dimensional oscillator mentioned above. As described earlier in this chapter, at not too small temperatures vibrational energy will dissipate into heat and so may collective rotations. Of course, the latter case is more sensitive as the total angular momentum $\hat{I}_{\rm tot} = \hat{I}_{\rm int} + \hat{I}_{\rm coll}$ is conserved. Again, for the sake of simplicity let us consider the model for which only rotations about one given axis are possible. For dissipative collective motion, in the course of time $\hat{I}_{\rm coll}$ will become smaller such that $\hat{I}_{\rm int}$ will increase until it has reached the maximal value for which $\hat{I}_{\rm tot} = \hat{I}_{\rm int}$. Along with such a change in the nature of the rotation will go the one for the associated moment of inertia, implying again a sensitive variation of the rotational energy. Remember that the effective inertias connected to $\hat{I}_{\rm int}$ and $\hat{I}_{\rm coll}$ are very different. An in a sense analogous mechanism has been used in descriptions of heavy-ion collisions, see e.g. (Berlanger et al., 1978b; Deubler et al., 1980; Schröder and Huizenga, 1984; Fröbrich and Lipperheide, 1996). There two fragments approaching each other may begin to rotate, but this rotational motion may undergo damping; such processes are the origin of the so-called "tangential friction" for radial motion.

The phenomenon just described is to be distinguished from the concept of *rotational damping* introduced in the mid 1980s. In the discussion above rotations were meant to be described in semi-classical fashion with an explicit appearance of a friction force for dissipation of angular momentum. "Rotational damping", on the other hand, refers to the spreading of quantal rotational levels into compound states in the spirit discussed in Section 4.3. It is applied to the decay from states $|\mu(I)\rangle$ of one rotational band to the next lower one $|\mu(I-2)\rangle$. This transition is governed by matrix elements of an operator $\mathcal{M}(E2)$ of quadrupole type. Experimentally such transitions are seen at very high spins and thus lie in a regime where couplings $V_{\mu'\mu}$ to more complicated compound states cannot be neglected. Still the overall excitation above the yrast line is so small that one is essentially dealing with the zero temperature case. The theoretical description involving the interplay of the two interactions $\mathcal{M}(E2)$ and $V_{\mu'\mu}$ has been worked out by B. Lauritzen, T. Døssing and R.A. Broglia in (1986); for further reading see also (Bortignon et al., 1998).

Exercises

1. Degenerate model: For $\Omega_{n0} = \Omega$ for all n one may introduce the average transition probability $F^2 = \sum_n |F_{no}|^2 / N$. Show that the reactive and the dissipative parts of the response functions are given by $\chi'(\omega) = (1/\hbar)NF^2 2\Omega/(\Omega^2 - \omega^2)$ and $\chi''_f(\omega) = (\pi/\hbar)NF^2(\delta(\omega-\Omega) - \delta(\omega+\Omega))$, respectively, and prove relation (6.21) for the solutions of $\chi'(\omega) - \chi(0) = C(0)$. Why is the latter equation identical to (6.19)? Discuss the examples of positive and negative stiffness $C(0)$.

2. Study quadrupole vibrations around the sphere in the pure oscillator model of Section 3.2.2.1: (i) Take the potential (3.26) and calculate the static stiffness $C(0)$ for spheroidal deformations for a coordinate defined as $q = \varpi_\perp/\varpi_3$. (ii) Show that the dissipative part of the response function is given by $\chi''(t) = -iC(0)\varpi_0 \sin(2\varpi_0 t)$; interpret the result. (iii) Calculate real and dissipative parts of the response function in frequency representation. (iv) Find the collective frequencies from the secular equation (6.19).

3. Derive (6.50) and (6.52) from (6.49) in the Hartree approximation discussed in Section 20.1 and 20.1.2.

4. Take the secular equation $k^{-1} + \chi(\omega) = 0$ and discuss the conditions under which its solutions are in accord with (6.91).

5. (i) Apply the fluctuation dissipation theorem (23.96) to the $_0\psi''(\omega)$ of (6.139) and show that for $\hbar\omega \ll T$ the resulting response function reads:
$_0\chi(\omega) = (\chi^T -_0 \chi(0))i\Gamma(\mu, T)/(\omega + i\Gamma(\mu, T))$
(ii) Show that a secular equation of the form $-C(0) -_0 \chi(0) +_0 \chi(\omega) = 0$ leads to the collective frequency $\omega_1 = -i\Gamma(\mu, T)C(0)/(C(0) +_0 \chi(0))$ and compare this with the frequency which in this case specifies internal dynamics.

7

TRANSPORT THEORY OF NUCLEAR COLLECTIVE MOTION

In this chapter we are going to develop a transport theory for large-scale motion which is applicable to fission or heavy-ion collisions. This discussion will be restricted to situations where the motion in the collective variables is sufficiently slow. In such cases it may be meaningful to follow the dynamics in terms of the shape degrees of freedom Q_μ introduced before. The time evolution of this subset of variables can then be expected to show typical features of transport processes. One basic condition must be the separation of the timescales between intrinsic (or nucleonic) and collective motion:

$$\tau_{\text{intr}} \leq \delta t \ll \tau_{coll} . \tag{7.1}$$

Often, but not always, this property simply reflects the presence of a large effective inertia for the collective modes. In addition, their velocity is reduced by the large value of friction which seems to be given at not too small temperatures.

Condition (7.1) may be interpreted in the following way: To obtain the effective equation of motion it may suffice to *study the time propagation over a macroscopically small interval δt*. Likewise, it suggests concentrating on Markovian types of motion. This means that the time evolution of the probability density in collective phase space might also be described by the Chapman–Kolmogorov equation, which is an integral equation. On the classical level, a transformation to the associated Fokker–Planck equation is then possible on the basis of appropriately chosen short-time propagators. Details of such a derivation are described in Section 25.2.3. In the quantum case one may wish to get equations of motion for the Wigner functions introduced in Section 19.3. Also in this case propagators are a useful tool which may allow for an inclusion of quantum effects on the level of a locally harmonic approximation.

To set up the transport equation for the short-time propagator one may proceed in two steps: First one has to find an adequate description of average motion and then the one for the dynamics of fluctuations. For the former we may take over the results from the previous chapter, through which the "drift" terms of the transport equation are specified. The fluctuations, on the other hand, are governed by stochastic forces. They will be determined by equilibrium fluctuations, which are to be defined by an appropriate application of the fluctuation dissipation theorem. However, this concept of "equilibrium" will not always be taken in its literal, physical meaning. Indeed, for motion across barriers the system may be far away from an equilibrium in genuine sense.

For the sake of simplicity we will again restrict ourselves to one collective variable Q. Moreover, our discussion will be concentrated on the Fokker–Planck approach leaving aside formulations in terms of Langevin equations. The relation of the latter to the former is discussed in Chapter 25. At first we shall be guided largely by heuristic arguments, which at the end of this chapter will be put on a sounder microscopic basis.

7.1 The locally harmonic approximation

In the previous chapter average motion was described within a linear response approach. The latter is applied to describe the local interplay between nucleonic and collective motion which takes place around a certain value Q_0 of the collective variable. In that region the coupling between the two sets of degrees of freedom may be treated to second order in the deviation $Q - Q_0$. In this way a harmonic approximation for collective dynamics is introduced whereas the intrinsic degrees of freedom are treated to all orders. This scheme works provided condition (7.1) is fulfilled. It allows one to include correlations of RPA-type, although for genuine transport processes one must account for dissipation. This scheme will now be extended to treat the dynamics of the full phase space distribution.[17]

Let $d(Q_f, P_f, t_f)$ be the Wigner function which represents the collective density at time t_f. It may be obtained from that at some initial time t_i by multiplying the initial distribution with a propagator $K(Q_f, P_f, t_f; Q_i, P_i, t_i)$ and by integrating over all phase space:

$$d(Q_f, P_f, t_f) = \int dQ_i dP_i \ K(Q_f, P_f, t_f; Q_i, P_i, t_i) \ d(Q_i, P_i, t_i). \qquad (7.2)$$

This process can be repeated, say with $N - 1$ intermediate steps. With the shorthand notation $K(Q_k, P_k, t_k; Q_{k-1}, P_{k-1}, t_{k-1}) = K(k; k-1)$ the propagation from time t_0 to time t_N will then be given by

$$K(N;0) = \int \left(\prod_{k=1}^{N-1} dQ_k dP_k \right) K(N; N-1) K(N-1; N-2) \cdots K(2;1) K(1;0).$$
$$(7.3)$$

For obvious reasons the individual propagators ought to satisfy the initial condition

$$\lim_{t_k \to t_{k-1}} K(Q_k, P_k, t_k; Q_{k-1}, P_{k-1}, t_{k-1}) = \delta(Q_k - Q_{k-1}) \delta(P_k - P_{k-1}) \qquad (7.4)$$

and they have to be normalized to unity with respect to the final variables,

$$\int dQ_k dP_k \ K(Q_k, P_k, t_k; Q_{k-1}, P_{k-1}, t_{k-1}) = 1. \qquad (7.5)$$

This latter property warrants conservation of the norm of the density. A remark is in order with respect to the initial condition (7.4). In Section 19.3.3 it is shown

---

[17] One should notice that the general collective coordinate is not of dimension length. This implies that the phase space volume does not have the dimension of an action. Therefore, we will omit factors of $(2\pi\hbar)^{-1}$ in the integral measure, in distinction from the procedure in Section 19.3.3.

that for the propagation of pure states this condition is in accord with quantum mechanics. Some care is required, however, for density distributions which correspond to statistical mixtures within the Markov approximation. Eventually, one may have to give up the strict condition (7.4) and replace the δ-functions by smooth ones of some finite widths; we will return to this problem below.

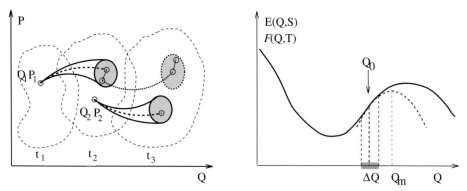

FIG. 7.1. Left panel: Propagation of a global density distribution (dashed lines) in small time steps. The individual propagators are determined by an average trajectory plus the fluctuations around it. For the step from t_2 to t_3 a Langevin description is indicated: propagation along the average trajectory and selection of one end point from all possible values inside the shaded area. Right panel: Linearization of the static energy (fully drawn line), which might be the internal or the free energy. The dashed line represents the harmonic approximation around Q_0 valid within a ΔQ. [From (Hofmann, 1997).]

The method suggested by (7.2) to (7.5) was presented first in (Hofmann, 1980). It may be pictured as shown in Fig. 7.1. The left part contains the essential elements of the procedure and it demonstrates in illustrative manner the main underlying hypothesis: The individual propagators restrict to small areas and are thus defined by local properties of the complex system. The figure on the right explains in a pictorial way the linearization procedure for the example of the static energy over which collective motion proceeds. As discussed in Section 6.5, from a more fundamental point of view it should be the internal energy $\mathcal{E}(Q, S)$ at given entropy, but in practical application one may want to express the latter by the free energy $\mathcal{F}(Q, T)$ at fixed temperature; in the following we will simply speak of a potential $V(Q)$. Because of the localization in phase space the propagators may be assumed to be of Gaussian form, which means writing

$$K_G(Q, P, t; Q_0, P_0, t_0) = (2\pi)^{-1} \Delta^{-1/2} \exp\left\{-\frac{1}{2\Delta} \times \right. \quad (7.6)$$
$$\left. \left[(P - P^c(t))^2 \Sigma_{qq}(t) - 2(P - P^c(t))(Q - Q^c(t))\Sigma_{qp}(t) + (Q - Q^c(t))^2 \Sigma_{pp}(t)\right]\right\}.$$

Here, $Q^c(t)$ and $P^c(t)$ represent that classical trajectory which at t_0 starts at Q_0, P_0. The $\Delta \equiv \det \Sigma_{\mu\nu}(t)$ represents the determinant of the fluctuation matrix

The locally harmonic approximation 193

$\Sigma_{\mu\nu}(t)$ whose elements for $A_\mu = Q$ or P are defined as

$$\Sigma_{\mu\nu}(t) = \int dQ dP \left(A_\mu - A_\mu^c(t)\right)\left(A_\nu - A_\nu^c(t)\right) K_G(Q, P, t; Q_0, P_0, t_0). \quad (7.7)$$

If they start off from zero, obeying $\Sigma_{\mu\nu}(t \to t_0) = 0$, the propagator $K_G(t; t_0)$ is easily seen to fulfill a condition like (7.4). The $Q^c(t)$, $P^c(t)$, on the one hand, and the fluctuations $\Sigma_{\mu\nu}(t)$ on the other hand are simply given by the first and second moments in Q and P, see Chapter 25.

At this stage a few comments may be in order.
(i) Without any doubt, for genuine transport problems the restriction to the Markov limit and the use of propagators imply limitations to the time evolution one wants to describe. Although the method has similarities to that of path integrals, there is the essential difference that δt cannot be made *arbitrarily small*. As a consequence of condition (7.1) quantum effects in collective transport can at best be described on a semi-classical level. We will return to this problem later in Section 12.5 where the decay of metastable systems will be treated.
(ii) On the other hand, the propagator method has been shown to be numerically feasible, even for such complex situations as the motion across a fission barrier (Scheuter and Hofmann, 1983). For multidimensional cases Monte-Carlo treatments of Langevin equations seem to be more appropriate (Lamm and Schulten, 1983), (Fröbrich and Xu, 1988; Abe et al., 1996).
(iii) Within the Langevin approach one follows individual trajectories in time rather than the full density distribution. The essential idea is pictured in Fig. 7.1: Starting at some initial point one first follows average motion for a short time interval δt to then "throw the dice" to get one end point, which serves as the initial point for the next step. The statistical aspect is brought in by the presence of *fluctuating forces*. To find the final distribution reached after a large time lapse one needs to repeat such calculations for many initial starting points. Formal aspects of the relation of Langevin and Fokker–Planck equations can be found in Chapter 25.

Let us proceed now in constructing the propagators. The dynamics of both the local trajectory and that of the fluctuations around it will be governed by the potential $V(Q) \approx V(Q_0) + (Q - Q_0)V'(Q_0) + (Q - Q_0)^2 V''(Q_0)/2$ which for $q = Q - Q_m$ is equivalent to $V(Q_m) + C(Q_0)q^2/2$, the potential of the oscillator which osculates the global potential. As in Section 6.1.2 the center Q_m of this local oscillator is defined such that terms linear in q drop out, which implies

$$\left.\frac{\partial V(Q)}{\partial Q}\right|_{Q_0} + \left.\frac{\partial^2 V(Q)}{\partial Q^2}\right|_{Q_0}(Q_m - Q_0) = 0 \quad (7.8)$$

$$V(Q_m) = V(Q_0) - (Q_0 - Q_m)^2 V''(Q_0)/2.$$

Notice, please, that the stiffness $C(Q_0) = V''(Q_0)$ may be positive or negative. Since we are looking for the propagation in phase space the local equation of average motion will be written in the form (with $M = M(Q_0)$ and $\gamma = \gamma(Q_0)$)

$$P^c(t) = M\frac{dq^c}{dt} \qquad \frac{dP^c(t)}{dt} = -Cq^c(t) - \gamma\frac{dq^c}{dt}. \qquad (7.9)$$

With the combinations $\Gamma = \gamma/M$, $\varpi^2 = |C|/M$ and $2\eta = \gamma/\sqrt{M|C|}$ of the constant transport coefficients the solutions (23.5) of the secular equation $-\omega^2 M - i\omega\gamma + C = 0$ may be written as $\omega^\pm = \pm\mathcal{E} - i\Gamma/2 = \varpi(\pm\sqrt{\text{sign}C - \eta^2} - i\eta)$. They determine fully the response function $\chi_{\text{osc}}(\omega) = (-\omega^2 M - i\omega\gamma + C)^{-1}$ of (23.6) for the locally defined damped oscillator.

The transport coefficients vary with the Q_0 around which the dynamics has been linearized. Let us try, therefore, to get a *global equation of average motion* by invoking the arguments behind the linearization procedure, but now in reverse order. First we should introduce the $\Delta Q = Q - Q_0 = q - (Q_0 - Q_m)$, which trivially implies $dQ^c(t)/dt = d\Delta Q^c(t)/dt = dq^c(t)/dt$. Because of the definition of Q_m in (7.8) one gets

$$\frac{dP^c(t)}{dt} = -\left.\frac{\partial V(Q)}{\partial Q}\right|_{Q_0} - \left.\frac{\partial^2 V(Q)}{\partial Q^2}\right|_{Q_0}(\Delta Q)^c(t) - \gamma(Q_0)\frac{dQ^c}{dt}. \qquad (7.10)$$

The first two terms can be combined to give the conservative force evaluated at $Q^c(t)$, which is to say $\partial V(Q^c(t))/\partial Q^c(t)$. Recalling that (7.10) is *correct to first order in* $\Delta Q^c(t)$ *and* $dQ^c(t)/dt$ one may replace $\gamma(Q_0)dQ^c/dt$ by $\gamma(Q^c(t))dQ^c/dt$. The same manipulation may be done in the relation of $P_c(t)$ to the velocity, which also means replacing there the Q_0 in $M(Q_0)$ by $Q^c(t)$. Hence, the equations for *global average motion* take on the form

$$P^c(t) = M(Q^c(t))\frac{dQ^c}{dt} \qquad \frac{dP^c(t)}{dt} = -\frac{\partial V(Q^c(t))}{\partial Q^c(t)} - \gamma(Q^c(t))\frac{dQ^c}{dt}. \qquad (7.11)$$

Unfortunately, for variable inertia the second equation is not completely correct, as the term $(1/2)(P/M)^2(\partial M(Q)/\partial Q)$ is missing. For zero friction this term is part of Hamilton's equations. The reason why it gets lost in the LHA is very clear: It is of order two in the velocity and hence discarded in any harmonic approximation. We will return to this problem in Section 11.1 where we will try to get global transport equations for the density $d(Q,P,t)$ itself.

7.2 Equilibrium fluctuations of the local oscillator

Knowing the response function equilibrium fluctuations can be obtained through the fluctuation dissipation theorem (FDT), as shown by (23.97) and (23.98). If written in terms of the full response function for q-motion constructed in Section 6.5 the equilibrium fluctuations are given by

$$\Sigma_{qq}^{\text{eq}} = \int \frac{d\omega}{2\pi} \hbar \coth\left(\frac{\hbar\omega}{2T}\right) \chi''_{qq}(\omega), \qquad (7.12)$$

$$\Sigma_{qp}^{\text{eq}} = iM \int \frac{d\omega}{2\pi} \hbar \coth\left(\frac{\hbar\omega}{2T}\right) \omega\chi''_{qq}(\omega) = 0, \qquad (7.13)$$

$$\Sigma_{pp}^{\rm eq} = M^2 \int \frac{d\omega}{2\pi} \hbar\coth\left(\frac{\hbar\omega}{2T}\right) \omega^2 \chi''_{qq}(\omega). \qquad (7.14)$$

For the fluctuations which involve the momentum P use has been made of the fact that $P = M\dot{q}$. As demonstrated in Section 23.3.2 this implies that the response functions $\chi''_{pp}(\omega)$ and $\chi''_{qp}(\omega)$ can be be expressed as $\chi''_{pp}(\omega) = (M\omega)^2 \chi''_{qq}(\omega)$ and $\chi''_{qp}(\omega) = iM\omega \chi''_{qq}(\omega)$. The fact that the cross fluctuation vanishes derives from the usual symmetry argument: $\omega \chi''_{qq}(\omega)$ is an even function in ω. These expression have to be compatible with the differential equations of average motion. The latter have been obtained by replacing the true response function $\chi_{qq}(\omega)$ by the (23.6) of the damped oscillator. It is only in this way that the transport coefficients for average motion M, γ and C show up. For such reasons it would not be consistent to evaluate the diffusion coefficients from (7.12) to (7.14) using the *full strength function* $\chi''_{qq}(\omega)$. Rather one should replace in (7.12) and (7.14) this $\chi''_{qq}(\omega)$ by the strength function of the oscillator response (23.6) to get

$$\begin{aligned}\Sigma_{qq}^{\rm eq} &\simeq \Sigma_{qq}^{\rm osc} = \int \frac{d\omega}{2\pi} \hbar\coth\left(\frac{\hbar\omega}{2T}\right) \chi''_{\rm osc}(\omega) \\ \Sigma_{pp}^{\rm eq} &\simeq \Sigma_{pp}^{\rm osc} = M^2 \int \frac{d\omega}{2\pi} \hbar\coth\left(\frac{\hbar\omega}{2T}\right) \omega^2 \chi''_{\rm osc}(\omega)\end{aligned} \qquad (7.15)$$

leaving (7.13) *untouched*, of course, as it results from a fundamental symmetry. Details of a possible evaluation of these integrals are described in Section 23.5.3. Unfortunately, the integral for $\Sigma_{pp}^{\rm osc}$ diverges and requires regularization. In (Hofmann et al., 1989) this has been done by introducing a cutoff. Another possibility is offered by the Drude form of the oscillator response (23.127). Numerical calculations show both methods to be equivalent; likewise no change is observed if a Gaussian cutoff is used. In Fig. 7.2 the result of a calculation with Drude regularization is shown (Hofmann, 1997), (Hofmann and Kiderlen, 1998). One observes that with increasing damping strength the distribution is squeezed in the coordinate and stretched in momentum. Evidently, this feature is a consequence of the Heisenberg uncertainty relations. To get further inside it may be worth to look at a few simple limiting cases.

Limit of zero damping: Let us begin with the limit of zero friction defined in the following way: Take the oscillator response (23.6) and let the imaginary part Γ of $\omega_1^{\pm}$ go to zero. Then its dissipative part takes on the form (23.12), namely $\lim_{\gamma\to 0} \chi''_{\rm osc}(\omega) = (\pi/(2M\varpi))(\delta(\omega - \varpi) - \delta(\omega + \varpi))$. Using this expression for (7.15) one gets

$$\frac{\Sigma_{pp}^{\rm eq}}{2M} = \frac{T^*}{2} = \frac{C}{2}\Sigma_{qq}^{\rm eq} \qquad \text{with} \qquad T^* = \frac{\hbar\varpi}{2}\coth\left(\frac{\hbar\varpi}{2T}\right). \qquad (7.16)$$

This result is to be expected from the quantal version of the equipartition theorem, simply because the equilibrium fluctuations of the oscillator represent the average values of the kinetic and potential energies. In Section 19.3.5 these results are derived from the Wigner transform of the canonical density operator.

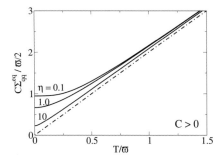 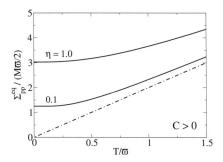

FIG. 7.2. Equilibrium fluctuations in the coordinate (left panel) and in the momentum (right panel) as functions of temperature normalized to the frequency ϖ of the (Drude) oscillator (with cutoff frequency $\varpi_D = 10\varpi$), for different damping rates η.

Limit of high temperature: In such a situation we expect the *classical equipartition theorem* $\sum_{pp}^{eq}/M = T = C\sum_{qq}^{eq}$ to apply. Indeed, this result follows if in (7.12) and (7.14) the integrals are calculated after replacing $\hbar\coth(\hbar\omega/(2T))$ by $2T/\omega$. An interesting and important question is the applicability of this limit. In the case of weak damping the condition is simply $T \gg \hbar\varpi$, as may easily be verified from (7.16). With increasing damping, however, the width of the oscillator strength increases as well and, hence, the minimal temperature will have to be higher, too: In general, T will have to be larger than the largest frequency involved in the response function. In the present case this is the Drude frequency ϖ_D which unfortunately is of the order of 10 MeV. A similar observation was seen in the transport model of A. Bulgac, D. Kusnezov and G. Do Dang (1996, 2001, 2001), which is based on a non-perturbative application of RMM. There the relevant energy scale is the "band width" within which the essential coupling strength concentrates and which is of the same order of magnitude. For practical applications to nuclear physics this sounds like a condition which in practice may never be fulfilled. Fortunately, Fig. 7.2 indicates that the situation need not be taken as stringently as it sounds. It is true that for large damping the curves approach the classical limit very slowly. However, even at smaller T the deviations are not really large. Moreover, in the case of the coordinate they even become smaller with increasing damping. As we shall see later, this feature is essential for the treatment of dynamics, simply because for over-damped motion the momentum distribution in a sense becomes irrelevant.

7.3 Fluctuations of the local propagators

To find the Gaussian form (7.6) of the local propagator we need to know the equations of motion for the fluctuations $\sum_{\mu\nu}(t)$ around the classical trajectory which is determined by the equations of motion (7.9). The $\sum_{\mu\nu}(t)$ associated with the propagator starts from small values, if not from zero in the ideal case, then at least from values small compared to those of equilibrium. Hence, within the short time interval δt they can be expected to remain small. Thus, their dynamics will essentially be determined by the drift related to the classical trajectory. The $\Sigma_{\mu\nu}(t)$ may thus be expected to fulfill the following set of equations:

$$\frac{d}{dt}\Sigma_{qq}(t) - \frac{2}{M}\Sigma_{qp}(t) = 2D_{qq} \tag{7.17}$$

$$\frac{d}{dt}\Sigma_{qp}(t) - \frac{1}{M}\Sigma_{pp}(t) + C\Sigma_{qq}(t) + \frac{\gamma}{M}\Sigma_{qp}(t) = 2D_{qp} \tag{7.18}$$

$$\frac{d}{dt}\Sigma_{pp}(t) + 2C\Sigma_{qp}(t) + 2\frac{\gamma}{M}\Sigma_{pp}(t) = 2D_{pp}. \tag{7.19}$$

Here, diffusion coefficients, which originate from the presence of fluctuating forces, have been introduced as inhomogeneity terms. To understand the structure of the left-hand sides two comments are in order. Firstly, if there is no damping the analogous equations follow from the Liouville equation, which according to the discussion in sections 19.3 and 21.1 defines the leading term in an expansion of the Wigner transform of the von Neumann equation into powers of $\hbar$ (see exercise 1). Secondly, one may realize that the structure of the homogeneous set of equations just reflects the Newton equation (7.9) for average motion. One way to see this is to replace the $\Sigma_{\mu\nu}(t)$ by just the second moments defined as given in (7.7) but without the $A_\mu^c(t)$ and calculated for the sharp propagator $K_G(q, P, t; q_0, P_0, t_0) = \delta(q - q^c(t; t_0))\,\delta(P - P^c(t; t_0))$. The reader who is not satisfied by these arguments may be referred to Section 7.5.2, where a microscopic proof will be discussed.

The factor of two on the right-hand sides of (7.17) to (7.19) is in accord with the general concept described in Section 25.2.3. To first order in δt the set (7.17)-(7.19) implies the correct dependence on δt,

$$\Sigma_{\mu\nu}(\delta t) \simeq 2D_{\mu\nu}\delta t \qquad \text{for} \qquad \Sigma_{\mu\nu}(\delta t = 0) = 0. \tag{7.20}$$

For the $\Delta = \det \Sigma_{\mu\nu}$ of (7.6) this means

$$\Delta(t) = \Sigma_{qq}(t)\Sigma_{pp}(t) - \Sigma_{qp}(t)\Sigma_{pq}(t) \simeq 4\delta t^2 \left(D_{qq}D_{pp} - D_{qp}^2\right) \equiv 4\delta t^2 \det D_{\mu\nu}. \tag{7.21}$$

Hence, the Gaussian propagator (7.6) is also well-defined at short times if the determinant $\det D_{\mu\nu}$ of the diffusion tensor turns out positive semi-definite.

The set (7.17)–(7.19) has stationary solutions which for a bound system correspond to the fluctuations in equilibrium:

$$\begin{aligned}
D_{qq} &= \Sigma_{qp}^{eq} \\
D_{qp} &= \frac{C}{2}\Sigma_{qq}^{eq} - \frac{1}{2M}\Sigma_{pp}^{eq} + \frac{\gamma}{2M}\Sigma_{qp}^{eq} \\
D_{pp} &= \frac{\gamma}{M}\Sigma_{pp}^{eq} + C\Sigma_{qp}^{eq}.
\end{aligned} \tag{7.22}$$

These equations can be viewed in two different ways:
(a) For given diffusion coefficients $D_{\mu\nu}$ they allow one to calculate the equilibrium fluctuations.

(b) If the latter are given, for instance through the fluctuation dissipation theorem, the diffusion coefficients are determined uniquely.

Unfortunately, both ways do not always lead to the same result. Indeed, in (Hofmann and Jensen, 1984), (Hofmann et al., 1989) expressions for the $D_{\mu\nu}$ have been found within the linear response approach which violate the FDT; we will return to this question below in Section 7.5.2. This choice has never been pursued any further, although it is conceivable that it might give reasonable results for motion farther away from local equilibrium; such a situation may be given for the initial stage of heavy collisions, see Section 13.4. Rather, in practicable applications to nuclear fission, choice (b) was preferred. In the sense relevant here, fission may be considered a slow process. Indeed, as we shall see later, this choice leads to results which are in agreement with those obtained in practically analytic fashion for the Caldeira–Leggett model.

In the following we will assume the $\sum_{\mu\nu}^{\text{eq}}$ to be those given in (7.15) through the FDT. There, the $\sum_{qp}^{\text{eq}}$ was seen to vanish such that one gets

$$D_{pp}^{\text{FDT}} = \frac{\gamma}{M}\sum_{pp}^{\text{eq}}, \qquad D_{qq}^{\text{FDT}} = 0 \quad \text{and} \quad D_{qp}^{\text{FDT}} = \frac{C}{2}\sum_{qq}^{\text{eq}} - \frac{1}{2M}\sum_{pp}^{\text{eq}}. \tag{7.23}$$

The fact of $D_{qq}^{\text{FDT}} = 0$ implies a problem with the propagator at short times. Indeed, for a non-vanishing D_{qp}^{FDT} the determinant introduced in (7.21) becomes negative:

$$\Delta = -4\delta t^2 \left(D_{qp}^{\text{FDT}}\right)^2 < 0. \tag{7.24}$$

This does not necessarily imply that this choice (7.23) of the diffusion coefficients is useless. It only means that the Gaussian propagator (7.6) is not well defined at very short times if one requires it to start with zero fluctuations (see exercise 2). Instead, one may replace the strict δ-function in (7.4) by a smooth one of some finite width. Remember that for the harmonic oscillator both the set of equations (7.17) to (7.19) as well as the Gaussian form can also be used for finite initial values. There are various reasons for the appearance of this problem. One should not forget that at very short times genuine quantal motion must be treated as non-Markovian. Already condition (7.1) tells one that for a transport process of the type we have in mind time zero must not be understood in a real, mathematical sense. Below in Section 7.4.2 we will return to this problem and demonstrate the essential features with the aid of numerical solutions for the fluctuations and for the determinant $\Delta(t)$. It may be mentioned that problems with a non-positive definite diffusion matrix are known from the equations used in quantum optics, see e.g. (Yuen and Tombesi, 1986), (Vogel and Risken, 1988) as well as Supplement 11 of (Risken, 1989). There the associated Fokker–Planck equation is referred to as "Pseudo-FPE" and the distribution a pseudo- or quasi-distribution.

Over-damped motion: When the friction force becomes so strong that the oscillator just "creeps" along the potential surface, no trace of the inertia is left. Then the equations for the first and second moment become

$$\gamma \frac{dq^c(t)}{dt} + Cq^c(t) = 0, \tag{7.25}$$

$$\frac{d}{dt}\Sigma_{qq}(t) + 2\frac{C}{\gamma}\Sigma_{qq}(t) = 2D_{qq}^{\text{ovd}} \tag{7.26}$$

respectively. To see this for the fluctuations it is useful to replace momentum by velocity through $P = M\dot{q}$ also in the $\Sigma_{qp}(t)$ and $\Sigma_{pp}(t)$. Suppose this is done in the homogeneous parts of the set (7.17) to (7.19). In the resulting equations one may then safely put the inertia equal to zero. This leads to the homogeneous part of (7.26). The diffusion term may then be added with the same justification as discussed before for the general case. Evidently, it must be in the same relation to the equilibrium fluctuation, namely

$$D_{qq}^{\text{ovd}} = \frac{C}{\gamma}\Sigma_{qq}^{\text{eq}} \quad \longrightarrow \quad \frac{T}{\gamma} \quad \text{high } T. \tag{7.27}$$

In principle, the Σ_{qq}^{eq} can be calculated as described before. However, as just ex-

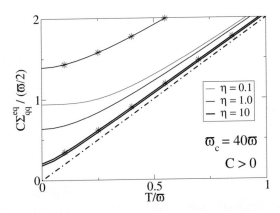

FIG. 7.3. The equilibrium fluctuations of the coordinate as function of temperature for different damping strengths. The lines without crosses correspond to those of Fig. 7.2, the other two are calculated for the response function of over-damped motion for $\eta = 1$ and 10, see text.

plained, for over-damped motion inertial effects should not be expected anymore. Indeed, the dissipative part of the response function which corresponds to the equation of average motion (7.25) can be seen to be equal to $i\gamma\omega/((\gamma\omega)^2 + C^2)$, see (23.17). Unfortunately, for this function the integral in (7.12) cannot be regularized anymore by the Drude method. The result of calculations with a sharp cutoff are shown in Fig. 7.3 by the lines with crosses and compared with those from before in which the full oscillator response function is used. For strongly over-damped motion it is seen that both computations agree. The figure demonstrates once more that and how the $\sum_{qq}^{\mathrm{eq}}$ and thus the D_{qq}^{ovd}, too, turns to the high temperature limit, for which the corresponding Fokker–Planck equation is of Smoluchowski type. One recognizes that with increasing damping the high T limit is reached at lower and lower temperatures. This implies that for the over-damped case quantum effects are not distinguishable from the classical result already at quite small temperatures. We will return to this question below.

Finally, we like to draw the attention to a very intriguing feature which may turn out to be important for practical applications. As long as the diffusion coefficient D_{qq}^{ovd} is positive semi-definite the problem with the short-time propagator behind (7.24) *does not show up for over-damped motion*. For positive stiffness this condition is fulfilled at all temperatures. For instabilities, on the other hand, it is fulfilled only above a critical temperature; this problem will be discussed in detail in Section 7.3.1.1.

7.3.1 Quantal diffusion coefficients from the FDT

The values of the diffusion coefficients D_{pp}^{FDT} and D_{qp}^{FDT} given in (7.23) may be read off from Fig. 7.2. One may deduce, for instance, how they approach the high temperature limit

$$D_{pp} = \gamma T \qquad D_{qp} = D_{qq} = 0 \,, \tag{7.28}$$

which represents Kramers' case, and which is often referred to as the *Einstein relation*. For weak damping, on the other hand, one gets from (7.16)

$$D_{pp}^{\mathrm{W.D.}} = \gamma T^*(\varpi) \equiv \gamma \frac{\hbar\varpi}{2} \coth\left(\frac{\hbar\varpi}{2T}\right) \quad \text{and} \quad D_{qp}^{\mathrm{W.D.}} = D_{qq}^{\mathrm{W.D.}} = 0 \,. \tag{7.29}$$

Here, the ϖ is to be calculated in zero order in γ. In a perturbative approach like that of Section 25.6.2 it is to be identified with the frequency ϖ_0 of the unperturbed oscillator.

7.3.1.1 Unstable modes
So far it has been tentatively assumed that the local modes are stable. Evidently, large-scale motion may pass over regions where this is not the case, a situation which is exhibited in Fig. 7.1. Then the FDT loses its original meaning in delivering equilibrium fluctuations. However, as we shall soon see, relations like those given in (7.23) may still be used to find appropriate diffusion coefficients. All we need to do is to exploit an analytic continuation to negative stiffness $C = -|C| < 0$, which for ϖ implies the replacement

$$\varpi = \sqrt{|C|/M} \quad \Longrightarrow \quad i\varpi_{\mathrm{b}} = \sqrt{-|C|/M} \,. \tag{7.30}$$

Since in this case the frequency is purely imaginary, it is convenient to introduce for the solutions of the secular equation $-\omega^2 M - i\omega\gamma + C = 0$ the real quantity $z = -i\omega$. Then the frequencies $\omega^\pm$ of (23.5) turn into

$$z^{\pm} = \pm \mathcal{E}_{\mathrm{b}} - \Gamma/2 = \varpi_{\mathrm{b}}\left(\pm\sqrt{1+\eta_{\mathrm{b}}^{2}} - \eta_{\mathrm{b}}\right), \quad (7.31)$$

where $\varpi_{\mathrm{b}}^{2} = |C|/M$ and $2\eta_{\mathrm{b}} = \gamma/\sqrt{M|C|}$. In fact, all expressions derived in Section 7.2 are analytical functions of the stiffness parameter C. Hence, we may take over all formulas derived there by simply applying an analytical continuation defined through the transformation (7.30). At this stage it may be worthwhile to recall once more from where the information about the value of C is obtained; more generally how we get the three transport coefficients for average motion. The basic element is the response function $\chi_{\mathrm{coll}}(\omega)$, as given by eqn(6.33) for instance, which parameterizes local collective motion in a self-consistent fashion. The transport coefficients are then introduced by approximating the $\chi_{\mathrm{coll}}(\omega)$, in some chosen, supposedly relevant regime of frequencies, by the response function $\chi_{\mathrm{osc}}(\omega)$ of a damped oscillator. In practice this is done by comparing the two associated strength distributions. This "fit" allows one to distinguish *uniquely* the cases of positive and negative stiffnesses. Please recall the discussion in Section 6.1.3 where an analytical treatment was discussed.

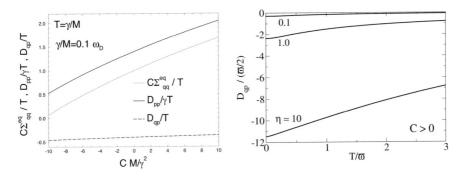

FIG. 7.4. Left panel: Diffusion coefficients (diagonal: thick fully drawn line, off diagonal: dashed line) and analytic continuation of the equilibrium fluctuations in q (thin full line) as functions of the stiffness C, for fixed values of M and γ (Hofmann and Kiderlen, 1998). Right panel: Non-diagonal diffusion coefficient D_{qp}^{FDT} as function of temperature for different values of η.

The implications of this transformation for the equilibrium fluctuations themselves are described in Section 23.5.2. Before we take over this information for the general case let us look at the simpler example of weak damping. Applying (7.30) to (7.29) one gets $D_{qp}^{\mathrm{W.D.}} = D_{qq}^{\mathrm{W.D.}} = 0$ and

$$D_{pp}^{\mathrm{W.D.}} = \gamma T^{*}(i\varpi_{\mathrm{b}}) \equiv \gamma \frac{\hbar \varpi_{\mathrm{b}}}{2} \cot\left(\frac{\hbar \varpi_{\mathrm{b}}}{2T}\right). \quad (7.32)$$

As the D_{pp}^{FDT} ought to be positive, it becomes obvious that the application of the analytic continuation to the diffusion coefficients will break down at a critical temperature T_c. For the zero friction limit this puts the following condition on temperature:

$$T_c(\gamma = 0) \equiv \hbar \varpi_b / \pi < T. \qquad (7.33)$$

For the general case one simply has to replace the $\sum_{pp}^{eq}$ and $\sum_{qq}^{eq}$ in (7.23) by their analytic continuations $\sum_{pp}^{eq,b}$ and $\sum_{qq}^{eq,b}$, respectively. In Fig. 7.4 we present the results of a computation performed in (Hofmann and Kiderlen, 1998). Three typical quantities $C\sum_{qq}^{eq}$, D_{pp}^{FDT}/T and D_{qp}^{FDT}, all normalized to temperature, are shown as functions of the stiffness parameter C measured in appropriate units. The result clearly supports our conjecture about the smoothness or analyticity across the point $C = 0$. From this figure, too, the limitations of the analytical continuation may be seen. Quantities which need to be positive, like $C\sum_{qq}^{eq,b}$ and D_{pp}^{FDT} may become zero or even negative. Remember that $C\sum_{qq}^{eq,b} > 0$ implies that the "fluctuation" $\sum_{qq}^{eq,b}$ itself is negative. As we shall recognize soon this property does not cause any problem for distributions which reflect motion across the barrier. Also shown in Fig. 7.4 is the non-diagonal diffusion coefficient D_{qp}, both as function of C and T; it is negative at all temperatures.

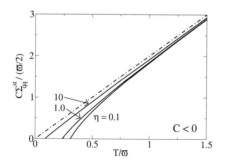
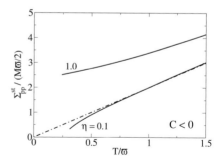

FIG. 7.5. Analytic continuations of the (diagonal) equilibrium fluctuations to the case of negative stiffness, as functions of temperature for different values of the damping rate η: Left panel for the coordinate, right panel for the momentum above the critical temperature T_c for $\sum_{qq}^{eq,b}$.

The property of $C\sum_{qq}^{eq,b}$ and D_{pp}^{FDT} becoming zero at the critical temperature can be seen more clearly in Fig. 7.5 where we plot the "equilibrium fluctuations" of coordinate and momentum, $\sum_{qq}^{eq,b}$ and $\sum_{pp}^{eq,b}$ as functions of temperature divided by the bare frequency, T/ϖ. The two quantities are normalized such that in the high temperature limit they tend to the same value, namely $2T/\varpi$. For the weak damping of $\eta = 0.1$ the $C\sum_{qq}^{eq,b}$ becomes negative slightly below the value of T_c given in (7.33). Three important features are observed: (a) The T_c gets *smaller* with increasing γ, such that $T_c(\gamma \neq 0) \leq T_c(\gamma = 0)$. Notice, please, that the limiting value of T is quite small, at least for practical nuclear applications, where the typical absolute value of $\hbar\varpi_b$ is of order 1 MeV. (b) The value of T_c is larger than the critical (or "cross over") temperature one finds for imaginary time propagation both within the Caldeira–Leggett model and within the PSPA to nuclear dynamics; this problem will be discussed in greater detail in Section 12.5. (c) The T_c actually is determined by the $C\sum_{qq}^{\text{eq,b}}$. The momentum fluctuations turn to zero at smaller temperatures. One should recall, however, that for larger values of η the motion becomes over-damped. In this case only the diffusion coefficient D_{qq}^{ovd} matters, which according to (7.26) is given by $D_{qq}^{\text{ovd}} = C\sum_{qq}^{\text{eq,b}}/\gamma$.

7.4 Fokker–Planck equations for the damped harmonic oscillator

For the diffusion coefficients (7.23) the transport equation is expected to read

$$\frac{\partial}{\partial t} d(Q,P,t) = \Big(-\frac{\partial}{\partial Q}\frac{P}{m_0} + \frac{\partial}{\partial P}CQ + \frac{\partial}{\partial P}\frac{P}{m_0}\gamma + 2D_{qp}\frac{\partial^2}{\partial Q \partial P} + D_{pp}\frac{\partial^2}{\partial P \partial P}\Big) d(Q,P,t). \quad (7.34)$$

This follows more or less from the equations of average motion (7.9), together with those for the second moments (7.17) to (7.19). Indeed, if one were to start from (7.34) it is easy to see that the set of equations just mentioned is recovered after applying the general technique described in Section 25.1. For Gaussian functions this association then is unique, implying that it holds true for the Gaussian propagator (7.6)–provided we ignore for the moment the problem at very short times. At first we shall be looking at stationary solutions for which such a problem does not appear.

7.4.1 Stationary solutions for oscillators

7.4.1.1 The phase space distribution of equilibrium
By the very construction of the diffusion coefficients $D_{\mu\nu}^{\text{FDT}}$ it is clear that (7.34) correctly describes the Gaussian equilibrium distribution. Indeed, (7.9) implies that the first moments vanish and the $D_{\mu\nu}^{\text{FDT}}$ of (7.23) are chosen such that (7.17) to (7.19) are valid in the stationary case. In generalization of (19.85) to damped motion one thus gets the following function, which is normalized to $\int dq dP \, d_{\text{eq}}(q,P;\beta) = 1$,

$$d_{\text{eq}}(q,P) = \frac{(2\pi)^{-1}}{\sqrt{\sum_{qq}^{\text{eq}}\sum_{pp}^{\text{eq}}}} \exp\Big(-\frac{P^2}{2\sum_{pp}^{\text{eq}}} - \frac{q^2}{2\sum_{qq}^{\text{eq}}}\Big) \equiv \frac{1}{Z} k_{\text{eq}}(q,P;\beta) \quad (7.35)$$

The $k_{\text{eq}}(q,P;\beta)$ introduced on the very right will turn out to be useful for unstable modes. In Section 19.3.5 the case of undamped motion is studied, for which the $d_{\text{eq}}(q,P)$ can be constructed as the Wigner transform of the quantal density operator, with the Z being the common partition function. The damped case is more delicate. Indeed, as we know from the previous discussion, for finite damping the equipartition theorem may not be fulfilled; this happens whenever the diffusion coefficient $D_{qp}^{\rm FDT}$ of (7.23) is finite. Then the form (7.35) does not correspond to the usual equilibrium distribution as defined by the Hamiltonian of the oscillator. From the previous discussion we know that this latter feature is warranted only for weak damping and for large temperatures.

7.4.1.2 The inverted oscillator

Formally, one possible stationary solution of (7.34) would be given by transforming (7.35) to the unstable case with the $\sum_{qq}^{\rm eq}$ and $\sum_{pp}^{\rm eq}$ replaced by $\sum_{qq}^{\rm eq,b}$ and $\sum_{pp}^{\rm eq,b}$. Likewise, this can be done for the $k_{\rm eq}$. Then the $\mathcal{Z}_{\rm b}$ would appear as the analytical continuation of the Z from positive to negative stiffness. To understand the basic elements of this game let us for the moment think of *undamped* motion. All we have to do then is to use the transformation (7.30) in (19.79). The effective temperature $T^*(i\varpi_{\rm b})$ is given by (7.32) and the partition function $Z^{-1} = 2\sinh(\hbar\varpi/2T)$ turns into

$$(\mathcal{Z}_{\rm b})^{-1} = 2i\sin(\hbar\varpi_{\rm b}/2T). \qquad (7.36)$$

Of course, the definition of $\mathcal{Z}_{\rm b}$ (and likewise the normalization of the density) exhibits a problem. The relation $Z = \int dq dP\, k_{\rm eq}(q,P;\beta)$ *cannot* simply be taken over. For the inverted oscillator the q-integral would highly diverge if it were to be performed along the real q-axis! To make the integral well-behaved an analytical continuation to imaginary q is required.[18] For the case of undamped motion this may simply be performed such that the result (7.36) is obtained. Actually, as described in Section 12.5, the appearance of an imaginary part in the partition function is a general property of unstable or metastable systems. The special feature for the inverted oscillator is simply that here no real part exists at all.

Distributions like the (7.35) correspond to zero flux, which only makes sense for a bound system. For the inverted oscillator, on the other hand, one is interested in a *steady state* solution with a *finite* current density j. It is easy to check by direct differentiation that the form (Ingold, 1988)

$$k_{\rm b}(q,P) = \zeta_{\rm b} \exp\left(-\frac{P^2}{2\sum_{pp}^{\rm eq,b}} - \frac{q^2}{2\sum_{qq}^{\rm eq,b}}\right) \int_{-\infty}^{P-Aq} \frac{du}{\sqrt{2\pi\sigma}} \exp\left[-\frac{u^2}{2\sigma}\right]$$

$$\text{with} \qquad \sigma = -\sum_{qq}^{\rm eq,b} A^2 - \sum_{pp}^{\rm eq,b} \qquad \text{and} \qquad A = \frac{1}{z_{\rm b}^+ M_{\rm b}} \frac{\sum_{pp}^{\rm eq,b}}{(-\sum_{qq}^{\rm eq,b})} \qquad (7.37)$$

is a stationary solution of (7.34). It is convenient to choose the normalization factor $\zeta_{\rm b}$ as

---

[18] Analytical continuations of this type are well known for evaluations of functional integrals. This trick goes back to (Langer, 1967), has been exploited in (Callan and Coleman, 1977) (see also (Coleman, 1977)) for instantons and finally to transport problems in various papers. The one which comes closest to our case is found in (Affleck, 1981) (see also (Ingold, 1988)). Whereas in (Affleck, 1981) a problem only occurs in the upper limit of the integral over coordinate, here one has to "regularize" at both ends.

$$\zeta_{\rm b} = \mathcal{Z}_{\rm b}/\left(\textstyle\sum_{qq}^{\rm eq,b}\textstyle\sum_{pp}^{\rm eq,b}\right)^{-1/2} = |\mathcal{Z}_{\rm b}|/\left(-\textstyle\sum_{qq}^{\rm eq,b}\textstyle\sum_{pp}^{\rm eq,b}\right)^{-1/2}, \qquad (7.38)$$

with which the current may be written as

$$j_{\rm b} = \int_{-\infty}^{\infty} dP \, \frac{P}{M_{\rm b}} \, k_{\rm b}(q,P) = \frac{1}{2\pi} z_{\rm b}^{+} |\mathcal{Z}_{\rm b}| \,. \qquad (7.39)$$

This formula may for instance be derived by applying to the integral over P the transformation $P = s + Aq$ to the new variable s. The remaining integrals are then easily expressed by the two moments $J_0 = J$ and J_1 given by (26.6). The flux is seen to be *independent* of q–as it has to be according to the continuity equation (see Chapter 25), which for the stationary case reduces to $\partial j/\partial q = 0$.

We still need to address the question of convergence of the integral in (7.37). To make it well-behaved one must require the σ to be positive, which implies

$$\sigma \equiv \frac{A}{z_{\rm b}^{+}} \left(\frac{\sum_{pp}^{\rm eq,b}}{M_{\rm b}} - M_{\rm b}(z_{\rm b}^{+})^2 (-\textstyle\sum_{qq}^{\rm eq,b}) \right) > 0 \quad \Longleftrightarrow \quad C_{\rm b} \textstyle\sum_{qq}^{\rm eq,b} > 0. \qquad (7.40)$$

The formal proof of the equivalence of both conditions is lengthy, but with some patience the correctness of the statement can be read off from the numerical results presented in Section 7.3.1. Please notice that (7.40) is fulfilled for temperatures above the T_c introduced earlier. As there the A is a positive quantity, in (7.37) the factor with the integral tends to zero for $q \to +\infty$ and to $+1$ for $q \to -\infty$. In other words for large negative values of q, the distribution function becomes proportional to that $k_{\rm b}(q,P;\beta)$ which is the analytic continuation of the one introduced in eqn(7.35) for the ordinary oscillator.

Let us add a few remarks on the historical development. In the classical case a stationary solution for the inverted oscillator was first derived by Kramers in 1940 (1940). Indeed, the form (7.37) is easily seen to be a generalization of Kramers' result to quantum physics. His form is recovered in the high temperature limit if one only uses the analytic continuation

$$\frac{C_{\rm b}}{2} \textstyle\sum_{qq}^{\rm eq,b} \equiv -\frac{|C_{\rm b}|}{2} \textstyle\sum_{qq}^{\rm eq,b} = \frac{T}{2} \qquad (7.41)$$

of the classical equipartition theorem. The A and the σ introduced in (7.37) become

$$\sigma = M_{\rm b} T \left(\left[\sqrt{1+\eta_{\rm b}^2} + \eta_{\rm b} \right]^2 - 1 \right) \quad \text{and} \quad A = |C_{\rm b}|/z_{\rm b}^{+} \qquad (7.42)$$

and are easily seen to be positive. Similar results are obtained for the limit of zero friction. One has to replace T by $T^*(i\varpi_{\rm b})$ and make sure that T^* stays positive, i.e. that condition (7.33) is fulfilled. The form (7.37) was first constructed by G.-L. Ingold (1988) within the Caldeira–Leggett model; an extended version

of this theory may be found in (Ankerhold et al., 1995). The connection with
the transport equation (7.34) was first noted in (Hofmann and Ingold, 1991).
The possibility of establishing contact with the analytic results found within the
Caldeira–Leggett model is quite important. It shows the relevance of our choice
of the diffusion equation through the fluctuation dissipation theorem as well as
its analytic continuation to negative stiffness. As shall be demonstrated later in
Section 12.5.1, Kramers' picture of the decay rate can in this way be extended
to include quantum effects within the locally harmonic approximation.

7.4.2 Dynamics of fluctuations for stable modes

The set of equations (7.17) to (7.19) (with constant transport coefficients) allows
for analytic solutions; for $D_{qq} = 0$ they can be found in Section 25.5.2. Some
typical examples are presented in Fig. 7.6 for zero initial widths corresponding
to the propagator $K_G(Q, P, t; Q_0, P_0, t = 0)$ of (7.6), and for microscopically calculated transport coefficients. The ones of average motion are those described in
Section 6.6.3. For the diffusion coefficients one only needs ϖ and $\eta = \Gamma_{\text{kin}}/(2\hbar\varpi)$,
which were fixed according to the prescription given in Section 6.6.3.3. The fig-
ure shows results for three different choices of the diffusion coefficients: (i) the
fully quantal case of (7.23), (ii) those of (7.29) for weak damping, and (iii) the
Einstein relation (7.28).

It is seen that for the first case the $\sum_{qq}(t)$ becomes negative at very short
times. This is in accord with the analytic result found in formula (7.76) of exercise
2. The figure shows that after a time lapse of the order of $1/\varpi$ the $\sum_{qq}(t)$ turns
positive. Notice please that this time is small on the scale of the period of one
oscillation which is not smaller than $2\pi/\varpi$. Such a problem does not appear
for the $\sum_{pp}(t)$. It is worth noticing that for the two temperatures chosen the
motion is under-damped with an effective damping rate η being 0.08 and 0.3,
respectively. For larger T the problem becomes less severe as then the motion
becomes more and more damped becoming over-damped at $T \simeq 2$ MeV. As may
also be seen from Fig. 7.6, no problem appears for the diffusion coefficients in
the weak damping or high temperature limit. This may easily be understood
on the basis of (7.76) (see exercise 2), simply because in these two cases the
equipartition theorem is fulfilled, in the sense that the average potential energy
equals that of the kinetic energy.

Next we turn to the question whether or not for a physical distribution
$d^W(q, P, t)$ the fluctuations satisfy the Heisenberg uncertainty relation. This requires that the $\Delta(t)$ of (7.21) must always be equal or larger than $\hbar^2/4$. In Fig.
7.7 the determinant $\Delta(t)$ is shown as a function of time, for an initial distribution
which corresponds to a "minimal packet" with $\Delta(t_0) = \hbar^2/4$. It is seen that for
a small initial fluctuation in the coordinate $\Delta(t)$ decreases below the minimal
value, indicating violation of the uncertainty relation. From the analytic result
(7.77) to be found in exercise 2 it is seen that such a behavior is to be expected
for finite $\sum_{qq}(t_0)$ whenever $\sum_{pp}(t_0) > \sum_{pp}^{\text{eq}}$.

These observations allow for some general conclusions. (a) For the quantal

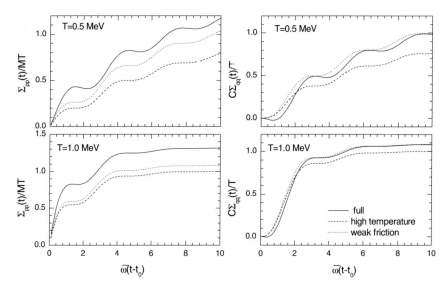

FIG. 7.6. Average kinetic and potential energies as function of time, and for two different temperatures, with the equilibrium values being normalized to the high T limit. The diffusion coefficients are chosen according to the FDT as given in (7.24) for the following cases: (i) full quantum equilibrium (solid curve) with Drude regularization for $\varpi_D = 10\varpi$, (ii) weak damping (dotted curve) and (iii) Kramers' case (dashed curve).

diffusion coefficients of the FDT with $D_{qp}^{\rm FDT} \neq 0$ the $\sum_{qq}(t)$ first decreases in time before it starts to rise. This feature rules out actually starting from a vanishing initial width $\sum_{qq}(t_0) = 0$. This problem disappears, however, with increasing damping. Indeed, for over-damped motion the solution of (7.26) behaves correctly. (b) For special initial conditions the $\Delta(t)$ may decrease linearly in time. This problem occurs no matter whether the D_{qp} is zero or not. Indeed, this feature is also seen for weak damping and, hence, for the equations used in quantum optics, see also the discussion in Section 25.6.2. To circumvent such problems one has to start from sufficiently broad initial distributions, see also the comment in the appendix of (Vogel and Risken, 1988).

7.4.3 The time-dependent solutions for unstable modes and their physical interpretation

Let us look first at the average value of the coordinate, which according to (25.130) and for $q^c(t_0 = 0) = q_0$ is given by:

$$q^c(t) = \frac{e^{-\Gamma 2 t}}{2\mathcal{E}_{\rm b}} \left[\left(\frac{P_0}{M} - q_0 z^-\right) \exp(\mathcal{E}_{\rm b} t) - \left(\frac{P_0}{M} - q_0 z^+\right) \exp(-\mathcal{E}_{\rm b} t) \right] \quad (7.43)$$

For large times with $t \gg \tau_{\rm inv} \equiv (|\,z^-\,|)^{-1}$ only the first term survives such that

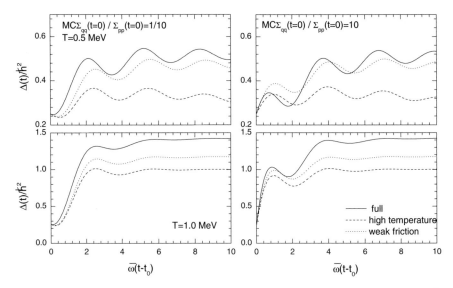

FIG. 7.7. The time evolution of the uncertainty $\Delta(t)$ of (7.21) for $\Delta(t_0) = \hbar^2/4$ but with $\sum_{qp}(t_0) = \sum_{pq}(t_0) = 0$, for two different relative values of the initial fluctuations in q and p.

$$q^c(t) \implies \frac{1}{2\mathcal{E}_b}\left(\frac{P_0}{M} - q_0 z^-\right)\exp(z^+ t) \equiv \xi_\infty^q \exp(z^+ t). \quad (7.44)$$

As q measures the distance between Q and Q_m, the center of the oscillator, for the singular case of $\xi_\infty^q = 0$ the trajectory hits the top of the barrier, otherwise $q^c(t)$ tends to $\pm\infty$. The trajectory which starts on the left side of the barrier ($q_0 < 0$) with a positive P_0, will always stay on the same side for $\xi_\infty^q < 0$, but will cross the barrier for $\xi_\infty^q > 0$. The time dependence of the second moments is given in (25.135). We are interested only in the values the second moment in q takes on at large times. Whenever the fluctuations start out from zero, as it is the case for the propagators, one finds:

$$\lim_{t\to\infty}\sum_{qq}(t) = \frac{-e^{(-\Gamma+2\mathcal{E}_b)t}}{4\mathcal{E}_b^2}\left[(\mathcal{E}_b + \frac{1}{2}\Gamma)^2\sum_{qq}^{eq,b} + \frac{\sum_{pp}^{eq,b}}{M^2}\right] \equiv \xi_\infty^{qq}\exp(2z^+ t) \quad (7.45)$$

Apparently it grows with t exactly as $(q^c(t))^2$, the square of the mean value of the coordinate itself. The coefficient ξ_∞^{qq} can be rewritten as

$$4M\mathcal{E}_b^2\,\xi_\infty^{qq} = -M(z^-)^2\sum_{qq}^{eq,b} - \frac{\sum_{pp}^{eq,b}}{M} = \frac{(z^-)^2}{-C}\left[M2D_{qp} + \frac{D_{pp}}{(-z^-)}\right] > 0. \quad (7.46)$$

It must be positive since we certainly want the *physical* fluctuation to be a positive quantity, which imposes a condition on the diffusion coefficients. Apparently, it is fulfilled for Kramers' case with $D_{pp} = \gamma T$ and $D_{qp} = 0$. For low temperatures (7.46) implies certain restrictions which the "equilibrium" fluctuations may take on, and thus a condition on temperature below which our approximations will break down (see (Hofmann and Kiderlen, 1998), (Hofmann, 1997)). For the weak damping limit the diffusion coefficients are given by (7.32) such that (7.46) will be fulfilled as long as the stays positive, which for T itself implies the lower limit specified in (7.33). As we know from the discussion above, finite values of friction actually *diminish* this critical temperature.

Next we address the question of finding the system on one side of the barrier. The probability of having it to the right is determined by the formula:

$$\Pi(t) = \int_0^\infty dq \int_{-\infty}^\infty dP \, d(q,P,t) = \int_0^\infty dq \, n(q,t). \qquad (7.47)$$

As the density $n(Q,t)$ takes on a Gaussian form (see Chapter 25) the remaining q-integral leads to an error function. The latter approaches a constant value as time grows to infinity:

$$\Pi(t) = \frac{1}{2}\left(1 + \mathrm{erf}\left[\frac{q^c(t)}{\sqrt{2\Sigma_{qq}(t)}}\right]\right) \overset{t \to \infty}{\Longrightarrow} \frac{1}{2}\left(1 + \mathrm{erf}\left[\frac{\xi_\infty^q}{\sqrt{2\xi_\infty^{qq}}}\right]\right). \qquad (7.48)$$

The interpretation of this result is as follows: From (7.44) it is recognized that for $\xi_\infty^q = 0$ the trajectory, i.e. the time-dependent mean value of q, hits the top of the barrier (and stays there). For such a condition, and a finite ξ_∞^{qq}, the argument of the error function becomes zero, rendering Π equal to $\frac{1}{2}$. Apparently, Π is larger than $\frac{1}{2}$ for $\xi_\infty^q > 0$ and smaller for $\xi_\infty^q < 0$. In the first case the center of the distribution $f(q,P,t)$ moves to the right, whereas in the second it moves to the left. As should be the case for a probability, the asymptotic value $\Pi = \Pi(t \to \infty)$ varies between zero and one depending on the ratio $\xi_\infty^q / \sqrt{2\xi_\infty^{qq}}$. It reaches 1 for very large positive values of this quantity and it vanishes for very large negative values. In the first case the center of the distribution has moved to the right whereas in the first one it moves to the left. How fast Π varies between these two limiting values depends on the size of the fluctuation, and hence on the diffusion coefficients D_{pp}, D_{qp}. In the case where the ξ_∞^{qq} approaches zero, the $\Pi(\xi_\infty^q)$ becomes a step function of zero width. This situation corresponds to the deterministic time evolution of Newtonian mechanics. Conversely, the larger the ξ_∞^{qq} the smoother this step-like function will be. The picture developed here was used in (Hofmann and Siemens, 1975) and (Rummel and Hofmann, 2003) to evaluate fusion cross sections for heavy-ion collisions (within the high-temperature limit for the diffusion coefficients). This problem will be addressed in more detail in Section 13.3.

7.5 Quantum features of collective transport from the microscopic point of view

At last we turn to the question of deriving the transport equations on a genuinely microscopic basis. Traditionally, one starts from a Hamiltonian of the full system of all degrees of freedom. We shall adhere to this concept too, although we must be aware that the final equation of collective motion may reflect the presence of strong friction forces, eventually in the limit of strong damping. In this case a Hamiltonian description may be bound to fail. Indeed, for a Hamiltonian of the form $\hat{\mathcal{H}} = \hat{H}_{\rm B} + \hat{H}_{\rm SB} + \hat{H}_{\rm S}$ it may become impossible to write the average $\text{tr}_{\rm B}\hat{\mathcal{H}}$ over the bath variables in the conventional form of an effective Hamiltonian $\hat{H}_{\rm S}^{\rm eff}$ for the system variables. This must be expected for a strong coupling $\hat{H}_{\rm SB}$, in particular if this coupling implies irreversible behavior of the system's motion. For the nuclear case this problem becomes even more delicate as the collective degrees of freedom are *not independent* of those of the "heat bath" delivered by the intrinsic or "nucleonic" degrees of freedom. Then the very definition of $\hat{\mathcal{H}}$ or its construction is a problem in itself. The solution suggested below is based heavily on the application of the locally harmonic approximation. Before we come to that some basic problems related to the form of $\hat{\mathcal{H}}$ shall be exhibited.

7.5.1 Quantized Hamiltonians for collective motion

In nuclear physics the Hamiltonian for all (effective) degrees of freedom would be of the form

$$\hat{\mathcal{H}} = \hat{H}_{\rm nucl}(\mathbf{x}_i, \mathbf{p}_i) + \hat{H}_{\rm coup}(\mathbf{x}_i; Q) + \hat{H}_{\rm coll}(Q_\mu, \Pi_\mu)\,, \tag{7.49}$$

with unperturbed parts for nucleonic and collective motion, $\hat{H}_{\rm nucl}$ and $\hat{H}_{\rm coll}$ respectively, and the coupling $\hat{H}_{\rm coup}$. Very often one just sticks to this naive ansatz *assuming* that (i) the $(\mathbf{x}_i, \mathbf{p}_i)$ represent coordinates and momenta of *all* nucleons, (ii) the collective degrees of freedom with coordinates Q_μ and their conjugate momenta Π_μ are *independent* of the former, and (iii) the $\hat{H}_{\rm coup}(\mathbf{x}_i; Q)$ *depends only on the coordinates*. The hypothesis of both sets of variables not to be correlated implies the Q_μ to be *additional and thus superfluous variables*, reflecting lack of any information about self-consistency between nucleonic and collective motion. Indeed, if one is to account for this important property the collective degrees of freedom necessarily become functionals of the nucleonic ones. This cannot be obtained appropriately by simply introducing the Q_μ alone, say by a point transformation from the set of the $\mathbf{x}_i$ without involving the momenta. It has been known since the 1950s that in this way one gets an unperturbed kinetic energy with an inertia being that of irrotational flow (Tolhoek, 1954), (Süssmann, 1954), (Tomonaga, 1955), which at low temperatures is about one order of magnitude too small. Evidently, because the inertia has important implications on timescales it is very questionable to derive conditions of adiabaticity on this ansatz (Nemes and Weidenmüller, 1981*a*), (Nemes and Weidenmüller, 1981*b*).

Despite such shortcomings, an ansatz as just described has been the basis for several attempts to derive nuclear transport equations, as for instance in the first two papers on the application of linear response theory (Hofmann and Siemens, 1976), (Hofmann and Siemens, 1977), in applications of the RMM (see (Weidenmüller, 1980) for an early report and (Bulgac et al., 1996), (Bulgac et al.,

2001) for improved versions) as well as in application of the Caldeira–Leggett model (Caldeira and Leggett, 1983). In the latter one makes the additional simplifying assumption of the $\hat{H}_{\rm nucl}(x_i, p_i)$ to be quadratic in the x_i and p_i and the $\hat{H}_{\rm coup}(x_i; Q)$ to be linear in the x_i. A Hamiltonian of such a structure allows for straightforward applications of functional integrals, as the x_i degrees of freedom may be integrated out analytically, see Section 24.1.3. It is true that in this way one was able to clarify many important features of quantum transport. On the other hand, it is almost evident that such an ansatz *should not be used in nuclear physics*. For instance, amongst several other deficiencies, it is not possible to obtain a meaningful effective collective inertia. An unperturbed inertia m_0 has to be put into $\hat{H}_{\rm coll}$, which then might get renormalized to an $M_{\rm eff}$ due to the coupling $H_{\rm coup}$. It may be mentioned in passing that such a renormalization of the inertia is usually not considered in applications of the RMM. But it is this $M_{\rm eff}$ which ought to be compared with that of the cranking model or those in more fancy self-consistent theories. In more practical terms, in this way it is impossible to find an $M_{\rm eff}$ which shows the features discussed in Section 6.6.1. A similar problem occurs for the conservative forces, of course, for which reasonable models exist on the level of mean field as described in Chapter 5.

This critique mainly concerns applications to the dynamics of compact nuclear complexes as encountered in fission or the later stages of a heavy-ion collision. In the early stages of the latter the problem is less severe. As long as the two fragments have only a small overlap the Q_μ may very well represent the vector of relative motion with the inertia being justly approximated by the reduced mass. The x_i and p_i could then represent coordinates and momenta of the two practically separated fragments (Balantekin and Takigawa, 1998). Another application of a non-self-consistent Hamiltonian like (7.49) is conceivable if the nuclear part only refers to nucleons outside an inert core with the latter defining the collective part. In essence, this is the spirit behind the Bohr Hamiltonian often used in the past, see (Bohr and Mottelson, 1975).

We are now going to attack the problem of finding a decent Hamiltonian by exploiting the Bohm–Pines method, restricting ourselves again to one collective degree of freedom. In the next section then the effective equation of motion for the collective degrees of freedom is derived within a projection method based on the Nakajima–Zwanzig procedure.

7.5.1.1 *The self-consistent version of the Bohm–Pines method* In Chapter 6 the dynamics of average collective motion was treated on a microscopic basis in which the collective variable Q was introduced as a c-number parameter specifying the deformation of the mean field Hamiltonian $\hat{H}(\hat{x}_i, \hat{p}_i, Q)$. To describe the dynamics of *fluctuating collective variables*, one first needs to introduce collective degrees of freedom as genuine dynamical variables with an associated canonical momentum Π. The method we are going to sketch now tries again to benefit from the locally harmonic approximation. In other words, the $\hat{\mathcal{H}}$ we aim at is supposed to describe fluctuating collective dynamics in a restricted area of phase space. The key point is to make use of the fact that the two-body Hamiltonian of the type

$$\hat{H}^{(2)}(\hat{x}_i,\hat{p}_i) = \hat{H}(\hat{x}_i,\hat{p}_i;Q_0) + (k/2)\hat{F}\hat{F}, \qquad (6.49)$$

is in close relation to average local motion on the level of RPA-correlations. As described in Section 6.2.3, this connection is established if only the $\hat{F} = \hat{F}(\hat{x}_i, Q_0)$ is chosen to be the field operator defined in (6.3) as the derivative of the $\hat{H}(\hat{x}_i, \hat{p}_i, Q(t))$ with respect to the collective coordinate. In order to cope with the requirements put up in Section 6.5.1 the coupling constant $k = k(T, Q_0)$ itself must be allowed to vary both with the collective coordinate as well as with the thermal excitation. If this latter aspect is discarded the Hamiltonian of (6.49) is typical of separable two-body interaction frequently used in nuclear physics to describe generic features of undamped collective motion, see e.g. (Bohr and Mottelson, 1975). In our case damping may be introduced along the lines described in Section 6.4. In principle, interactions like that of (6.49) allow one to exploit the technique of functional integrals but only if the "coupling constant" k is constant in a true sense (see Section 24.4).

The construction of a Hamiltonian $\hat{\mathcal{H}}$ of the form (7.49) was suggested in (Hofmann, 1982), (Siemens, 1982) and later worked out in greater detail in (Hofmann and Sollacher, 1988), see also Section 3.2 of (Hofmann, 1997). One exploits the procedure which was introduced by D. Bohm and D. Pines in a series of papers to treat collective motion of the electron gas (see (1953) and earlier papers cited therein). There, too, a separable force appears but of repulsive nature, whereas the coupling constant k of (6.49) is negative, meant to represent nuclear dynamics of isoscalar nature.

The essence of the method is to introduce operators for a collective coordinate $\hat{Q}$ as well as for its conjugate momentum $\hat{\Pi}$. In this way the number of degrees of freedom gets enlarged. But it can be shown that the Hamiltonian

$$\hat{\mathcal{H}} = \frac{\hat{\Pi}^2}{2}\widehat{\frac{1}{m_0}} + \left(\frac{\tilde{\beta}}{k^2} + \frac{1}{2k}\right)\hat{q}^2 + k\hat{\Pi}\widehat{\hat{F}} - \frac{\tilde{\beta}}{k}\hat{q}\hat{F} + \hat{H}(\hat{x}_i,\hat{p}_i;Q_0) \qquad (7.50)$$

is equivalent to the original one given in (6.49) if only the following subsidiary condition is fulfilled:

$$k\hat{F} - (\hat{Q} - Q_0) \equiv k\hat{F} - \hat{q} = 0. \qquad (7.51)$$

This is to say that after formally introducing a larger Hilbert space physics will be described only by the states in the subspace in which (7.51) is fulfilled exactly. Evidently, one will be able to achieve this goal only approximately. By mere inspection one realizes that the time-dependent average of this condition is nothing else but the condition (6.51) which for the time-dependent mean field description warrants self-consistency.

To achieve the equivalence of (6.49) and (7.50) in a strict sense the inverse inertia $\widehat{1/m_0}$ of the (unperturbed) collective part

$$\hat{H}_{\text{coll}} = \frac{\hat{\Pi}^2}{2} \widehat{\frac{1}{m_0}} + \left(\frac{\tilde{\beta}}{k^2} + \frac{1}{2k}\right) \hat{q}^2 \qquad (7.52)$$

of $\hat{\mathcal{H}}$ ought to be chosen as the following operator in x-space:

$$\widehat{\frac{1}{m_0}} = ik^2\,[\widehat{\hat{F}}, \hat{F}] \quad \text{where} \quad \widehat{\hat{F}} = i\,[\hat{H}_{\text{nucl}}, \hat{F}]. \qquad (7.53)$$

Restricting to harmonic motion the operator $\widehat{1/m_0}$ may be replaced by its unperturbed average, which for the Hamiltonian implies a reduction to the approximate form $\hat{\mathcal{H}}_a$ defined by the replacements

$$\widehat{\frac{1}{m_0}} \longrightarrow \frac{1}{m_0} \equiv \left\langle \widehat{\frac{1}{m_0}} \right\rangle_0 \qquad \hat{\mathcal{H}} \longrightarrow \hat{\mathcal{H}}_a. \qquad (7.54)$$

The c-number m_0 serves as unperturbed inertia for the q-mode. According to (6.40) its value is related to the sum rule limit of the energy weighted sum (remember the relations (6.117)). However, one must not forget that this replacement is crucial for the question of the subsidiary condition (7.51). One may easily check that the original Hamiltonian $\hat{\mathcal{H}}$ of (7.50) commutes with this condition,

$$\left[\hat{\mathcal{H}},\, k\hat{F} - \hat{q}\right] = 0. \qquad (7.55)$$

This implies a special symmetry of $\hat{\mathcal{H}}$ for which there will be an exact zero-mode.

It is interesting to compare the structure of the Bohm–Pines Hamiltonian (7.50) with that of the $\hat{\mathcal{H}}$ of (7.49). It is observed that the unperturbed intrinsic or nucleonic part $\hat{H}_{\text{nucl}}$ is simply given by the $\hat{H}(\hat{x}_i, \hat{p}_i; Q_0)$, as was the case when discussing average motion in the last chapter. The coupling term

$$\hat{H}_{\text{coup}} = k\,\widehat{\hat{F}}\,\widehat{\Pi} - \frac{\tilde{\beta}}{k}\,\hat{q}\hat{F}, \qquad (7.56)$$

on the other hand, now consists of two terms. The first and new one involves the canonical momentum. The second one almost looks like before, with the exception of the appearance of the constant $\tilde{\beta}$. As this $\tilde{\beta}$ does not interfere with the subsidiary condition its value may later be chosen at one's convenience.

We should like to draw the reader's attention to the nature of the collective coordinate. Its introduction was guided by dynamical aspects, namely to reduce the impact of the nucleon's mutual interaction. Indeed, in going from the Hamiltonian (6.49) to that of (7.50) we succeeded in getting rid of the two-body interaction simply by defining the collective coordinate through the subsidiary condition (7.51). Evidently, if the $\hat{F}$ depends on the particles' momenta, the relation (7.51) will be part of a *canonical* transformation rather than of a simple coordinate transformation (see also (Bohr and Mottelson, 1975)).

Let us return once more to the method of functional integrals which is described in Section 24.4. There, the collective degree of freedom, together with the associated mean field is introduced via a Hubbard–Stratonovich transformation. Very like the Bohm–Pines method, in this way the two-body interaction is removed. In both cases this happens at the expense of introducing a (locally) quadratic potential energy for the collective degree of freedom. One essential difference of the two methods is, however, that the Hubbard–Stratonovich transformation does not allow for a proper introduction of a collective inertia for the q-mode. This has serious implications when applied to large-scale motion, like for fission, see the discussion in Section 12.5.2.

Undamped motion It is instructive to briefly discuss the undamped case. We know from (6.25) that this situation is given whenever *the nucleonic spectrum is discrete* (and must be treated as such), which implies the $\chi''(\omega)$ to vanish at the collective frequencies ω_ν. In such a case the dynamics of the system can be treated by effectively "diagonalizing" the Hamiltonian (7.50), or rather, transforming to canonical form those parts which involve collective motion. This has been demonstrated in (Hofmann and Sollacher, 1988) in greater detail following to some extent the original methods of (Bohm and Pines, 1953). The new form of the Hamiltonian is introduced by means of a unitary (canonical) transformation: $\hat{\mathcal{H}}_{\rm new} = \exp(-i\hat{S})\hat{\mathcal{H}}\exp(i\hat{S})$. The latter is constructed explicitly by exploiting special perturbative methods. In this way a "renormalized" Hamiltonian is obtained. When averaged over the nucleonic degrees of freedom an *effective* collective Hamiltonian $\hat{H}_{\rm coll}^{\rm eff}$ survives. To harmonic order it has the structure of the one for the common oscillator

$$\langle \hat{\mathcal{H}}_{\rm new}\rangle_0 = {\rm tr}\,\hat{\rho}_{\rm qs}\hat{\mathcal{H}}_{\rm new} \equiv E_0^c + \hat{H}_{\rm coll}^{\rm eff} = E_0^c + \frac{1}{2M}\hat{\Pi}^2 + \frac{M\omega^2}{2}\hat{q}^2\,. \qquad (7.57)$$

The c-number E_0^c contains a correlation energy. Its origin is found in the presence of the coupling to the collective degrees of freedom and may be interpreted as a correction to ${\rm tr}\,\hat{\rho}_{\rm qs}\hat{H}(\hat{x}_i,\hat{p}_i;Q_0)$. More interesting is the collective part which is expressed in terms of the unknown frequency ω and the renormalized inertia M. For both quantities the following two equations are found

$$1 + k\chi'(\omega) = 0 \qquad (7.58)$$
$$\tilde{\beta} + k = k^2 M \omega^2\,. \qquad (7.59)$$

The first one is the secular equation we know already from Section 6.1.3. The second one relates $\tilde{\beta}$ to M. Since we know how to construct the inertia from the collective response in this way the unknown parameter $\tilde{\beta}$ is fixed.

Linear response for damped motion: Introducing the coupling $\delta\hat{\mathcal{H}} = q_{\rm ext}(t)\hat{q} + p_{\rm ext}(t)\hat{\Pi}$ to some external, time-dependent "fields" $q_{\rm ext}(t)$ and $p_{\rm ext}(t)$ one may define response functions by writing the frequency-dependent averages of coordinate and momentum, $q_c(t) = \langle\hat{Q}\rangle_t$ and $\Pi_c(t) = \langle\hat{\Pi}\rangle_t$ in the form

$$q_c(\omega) = -\chi_{qq}(\omega) q_{\text{ext}}(\omega) - \chi_{q\pi}(\omega) p_{\text{ext}}(\omega) \tag{7.60}$$
$$\Pi_c(\omega) = -\chi_{\pi q}(\omega) q_{\text{ext}}(\omega) - \chi_{\pi\pi}(\omega) p_{\text{ext}}(\omega). \tag{7.61}$$

These expressions may be calculated from Ehrenfest's equations. For the approximate Hamiltonian $\hat{\mathcal{H}}_a$ they read:

$$\frac{dq_c}{dt} = \frac{\Pi_c(t)}{m_0} + k\,\langle \hat{F}\rangle_t + p_{\text{ext}}(t)\,, \qquad \frac{d\Pi_c}{dt} = -\frac{(2\widetilde{\beta}+k)}{k^2} q_c(t) + \frac{\widetilde{\beta}}{k}\langle \hat{F}\rangle_t - q_{\text{ext}}(t)\,. \tag{7.62}$$

To close these equations one needs to evaluate the averages $\langle \hat{F}\rangle_t$ and $\langle \hat{\dot F}\rangle_t$ as functionals of $q_c(s)$ and $\Pi_c(s)$. For harmonic motion it suffices to apply linear response theory considering the coupling (7.56) as a perturbation on the nucleonic system defined by $\hat{H}(Q_0)$, with the $q_c(t)$ and $\Pi_c(t)$ being unknown functions of time. This leads to

$$\frac{dq_c(t)}{dt} = \frac{\Pi_c(t)}{m_0} + \int_{-\infty}^{+\infty} ds \left(\widetilde{\beta}\chi_{\dot F F}(s) q_c(t-s) - k^2 \chi_{\dot F \dot F} \Pi_c(t-s)\right) + p_{\text{ext}}(t)$$

$$\frac{d\Pi_c(t)}{dt} = -\frac{2\widetilde{\beta}+k}{k^2} q_c(t) + \int_{-\infty}^{+\infty} ds \left(\frac{\widetilde{\beta}^2}{k^2}\chi_{FF}\, q_c(t-s) - \widetilde{\beta}\chi_{F\dot F}\, \Pi_c(t-s)\right) - q_{\text{ext}}(t),$$

if for the moment and for the sake of simplicity we assume the unperturbed average $\langle \hat{F}\rangle_0$ to vanish. Notice, please, that in applying linear response theory to evaluate the $\langle \hat{F}\rangle_t$ and $\langle \hat{\dot F}\rangle_t$ the actual time dependence of the $q_c(t)$ and $\Pi_c(t)$ has been inserted, and not that one which would be obtained if the collective parts, too, were treated perturbatively. After some straightforward calculation one gets:

$$\chi_{qq}(\omega) = \frac{k^2 \chi(\omega)}{1+k\chi(\omega)} \qquad \chi_{\pi\pi} = \frac{-\bigl(2\widetilde{\beta} + k + \widetilde{\beta}^2 \chi(\omega)\bigr)}{k^2 \omega^2 \bigl(1+k\chi(\omega)\bigr)} \tag{7.63}$$

and

$$\chi_{q\pi} = \frac{-i\bigl(1-\widetilde{\beta}\chi(\omega)\bigr)}{\omega\bigl(1+k\chi(\omega)\bigr)} = -\chi_{\pi q} \tag{7.64}$$

Here use was made of relations derived in Section 23.3.2 which for the present case lead to $\chi_{\dot F F} = -i\omega\chi(\omega) = -\chi_{F\dot F}(\omega)$ and $\chi_{\dot F \dot F}(\omega) = \omega^2 \chi(\omega) + 1/(k^2 m_0)$. It is seen that the q-q-response is defined by the same function $\chi_{qq}(\omega)$ encountered already in Chapter 6 for the description of average motion where no momentum was present. The functions involving Π have poles at the same frequencies as the $\chi_{qq}(\omega)$ determined by the secular equation (6.18) for damped motion, $1 + k\chi(\omega) = 0$. In addition there are poles at $\omega = 0$. Evidently, these are related to the zero frequency modes we expected from the earlier discussion because of the symmetry (7.55) of the Hamiltonian (7.50). In this sense they are unphysical and should not be considered as real modes, for instance when the fluctuation dissipation theorem is applied to evaluate fluctuations in momentum. How such a subtraction procedure has to be carried out in practice is described in (Hofmann and Sollacher, 1988). Using an unperturbed inertia $\widetilde{m}_0$ instead of the m_0 of (7.54) the zero mode takes on a finite ω_0. Its value may be controlled by the smallness parameter $\epsilon = 1/\widetilde{m}_0 - 1/m_0$.

Self-consistency condition: The symmetry given by (7.55) implies that the operators $k\hat{F}(t)$ and $\hat{q}(t)$ are identical at all times provided the time evolution is determined by the original Hamiltonian $\hat{\mathcal{H}}$ of (7.50). However, the actual calculations are done for the approximate Hamiltonian $\hat{\mathcal{H}}_a$, in particular so within the linear response approach to the dynamics of average values. In this case one may expect the subsidiary condition (7.51) to be fulfilled on average,

$$\frac{d}{dt}\left(k\langle\hat{F}\rangle_t - \langle\hat{q}\rangle_t\right) = 0, \qquad (7.65)$$

a conjecture which was proven in (Hofmann and Sollacher, 1988). Because of the intimate relation between average dynamics and response functions it may not be a big surprise to find $\chi_{qq}(\omega) \equiv k\chi_{Fq}^{\text{coll}} \equiv k\chi_{qF}^{\text{coll}} \equiv k^2\chi_{FF}^{\text{coll}}$. Here, functions with F are understood to represent those of collective motion, not the $\chi_{FF} \equiv \chi$ of nucleonic dynamics. As an immediate consequence one has

$$\chi_{q-kF,\,q-kF}^{\prime\prime\,\text{coll}} \equiv 0 \quad \text{implying} \quad \left\langle\left(\hat{q} - k\hat{F}\right)^2\right\rangle_{\text{eq}} = 0 \qquad (7.66)$$

by the fluctuation dissipation theorem. We may thus conclude that the approximation used ensures that the subsidiary condition be fulfilled on a time-dependent average as well as for the fluctuations in equilibrium. This also has important implications for the dynamics of fluctuations, at least if the diffusion coefficients are chosen according to (7.23); we will return to this question below.

Shifted oscillator: In (7.51) the q is defined as the difference of the actual Q to the c-number Q_0 around which linearization is to be performed. In the calculations discussed thereafter it was tentatively assumed that the unperturbed value $\langle\hat{F}\rangle_0$ would vanish. Otherwise, in the equations for $d\Pi_c(t)/dt$ one would have to add the term $\widetilde{\beta}\langle\hat{F}\rangle_0$ on the right-hand side. However, this may easily be dealt with by redefining the q as $Q - Q_m$ with the Q_m defined by the relation

$$k\langle\hat{F}\rangle_0 + \left(2\beta + k - \beta^2\chi(0)\right)(Q_m - Q_0) = 0. \qquad (7.67)$$

This relation may be seen to become identical to the (6.14) which involves the static energy $E(q)$ as function of the c-number q if only this static energy is calculated to second order from the $\hat{\mathcal{H}}_a$ of (7.54) by (i) treating q and Π as c-numbers and (ii) discarding the momentum-dependent terms.

One pole approximation: In Section 6.5 transport coefficients were introduced by reducing the response for q-motion, like the $\chi_{qq}(\omega)$ of (7.63) to the $\chi_{\text{osc}}(\omega)$ of (6.116) for the damped oscillator. Actually, this latter function is determined by the *pair* of poles located at $\omega_1^{\pm}$ but the phrase "one pole approximation" was used in (Hofmann and Sollacher, 1988). There, consistency with the treatment of average motion along the lines described in Chapter 6 was shown. In this procedure the following two relations proved useful: Firstly, in generalization of (7.59), $\widetilde{\beta} + k = -k^2 M \omega_1^+ \omega_1^-$ and, secondly, a relation between the kinetic and canonical momentum,

$$M \frac{dq(t)}{dt} = \Pi(t) - \gamma q(t). \qquad (7.68)$$

7.5.2 A non-perturbative Nakajima–Zwanzig approach

In the following the essential steps will be sketched by which a transport equation may be derived which allows for an inclusion of those quantum features of collective motion which have been described above. In full length the derivation can be found in (Hofmann et al., 1989). It is based on the Bohm–Pines Hamiltonian (7.50) in the approximation specified by (7.54), and, very much like this Hamiltonian itself, it is specifically adapted to the locally harmonic approximation. One takes advantage of the fact that the coupling $\hat{H}_{\text{coup}}$ factorizes in intrinsic and collective parts; actually it is a sum of two such terms. Furthermore, it is sufficient to find the transport equation which describes propagation within a limited time interval only. As in this case the collective propagator (for Wigner functions) will be localized in phase space, one may assume that the initial distribution factorizes into collective and intrinsic parts.

The formal means of the derivation is the Nakajima–Zwanzig projection technique, which is presented in Section 25.6.1, together with the perturbation approach in which the coupling is treated to second order. As described in Section 25.6.2, in this case one ends up with a quantal transport equation which finds application in quantum optics, for instance. However, this equation does not meet the conditions required for nuclear applications. Amongst others, it cannot account for the RPA-type correlations which in Chapter 6 turned out to be important for the treatment of average motion within the LHA.

The final goal is to obtain a transport equation for the reduced density $\hat{d}(t) = \text{tr}_{\text{nucl}} \hat{W}(t)$ of the subsystem of the collective degrees of freedom from the von Neumann equation (25.140) $i\hbar \, d\hat{W}(t)/dt = [\hat{\mathcal{H}}, \hat{W}(t)] \equiv -i\hat{L}\,\hat{W}(t)$ for the density $\hat{W}(t)$ of the full system, given a Hamiltonian of the type (7.49). The Liouville operator $\hat{L}$ introduced on the very right may simply be understood as a shorthand notation for the corresponding commutator. A formal solution to this problem of finding $\hat{d}(t)$ is described in Section 25.6.1. One ends up with the integro-differential equation (25.145) which here reads

$$\frac{d}{dt}\hat{d}(t) = -i\hat{L}_{\text{coll}}^{(1)}\hat{d}(t) + \int_{t_0}^{t} ds\, \hat{\mathcal{K}}(t-s)\,\hat{d}(s) + \hat{I}(t). \qquad (7.69)$$

The $\hat{L}_{\text{coll}}^{(1)} = \hat{L}_{\text{coll}} + \text{tr}_{\text{nucl}} \hat{L}_{\text{coup}} \hat{\rho}_{\text{qs}}$ represents the Liouvillian which is correct to first order in the coupling, with $\hat{\rho}_{\text{qs}}$ being the density operator for the unperturbed nuclear system. The kernel $\hat{\mathcal{K}}$ is at least of order two. In almost all applications one sticks to that order, in which case the $\hat{\mathcal{K}}$ is given by

$$\hat{\mathcal{K}}(t-s) = \operatorname{tr}_{\text{nucl}} \hat{L}_{\text{coup}} \exp\bigl(-i(\hat{L}_{\text{coll}} + \hat{L}_{\text{nucl}})(t-s)\bigr) \hat{L}_{\text{coup}} \,\hat{\rho}_{\text{qs}}\,. \qquad (7.70)$$

The inhomogeneity $\hat{I}(t)$, finally, vanishes if one assumes the system initially to factorize like $\hat{W}(t=t_0) = \hat{\rho}_{\text{qs}} \hat{d}(t=t_0)$.

As mentioned before, the coupling $\hat{H}_{\text{coup}}$ appears in factorized form, where the *collective parts are linear in the operators for coordinate and momentum*, unlike the schematic model discussed in Section 25.6.2 where only the coordinate $\hat{q}$ appears. When calculating the kernel $\hat{\mathcal{K}}$ in perturbation theory one ends up with an equation which is totally analogous to (25.153) or, after calculating the time evolution (backward) in $-s$ of $\hat{q}$ and Π, to (25.156). For the realistic nuclear case perturbation theory does not suffice. It is necessary to invent a method which overcomes this shortcoming but which nevertheless is simple enough to be carried out in practical applications. One possibility has been suggested and worked out in (Hofmann et al., 1989) and discussed further in (Hofmann, 1997). There, one is guided by the structure by which average motion is treated within the LHA. As described in Section 6.2, the basic element is the collective response function (6.33) of the form $\chi_{\text{coll}}(\omega) = \chi(\omega)/(1 + k\chi(\omega))$, which is recovered also when average motion is derived from the Bohm–Pines Hamiltonian. The typical feature is that the nucleonic or intrinsic response $\chi(\omega)$ is the one that follows after applying the usual linear response argument to evaluate average values for the nucleonic operator $\hat{F}(\hat{x}_i, \hat{p}_i, Q_0)$ at given and fixed Q_0. The collective response $\chi_{\text{coll}}(\omega)$, on the other hand, is obtained by summing the geometric series up to infinity.

A natural translation of this feature to the Nakajima–Zwanzig method is to introduce an *effective Liouvillian* $\hat{\mathcal{L}}_{\text{coll}}$ for collective evolution by substituting

$$\exp\bigl(-i(\hat{L}_{\text{coll}} + \hat{L}_{\text{nucl}})s\bigr) \quad \Longrightarrow \quad \exp\bigl(-i(\hat{\mathcal{L}}_{\text{coll}} + \hat{L}_{\text{nucl}})s\bigr)\,. \qquad (7.71)$$

The $\hat{\mathcal{L}}_{\text{coll}}$ is supposed to account for the non-perturbative features. Its action on the collective density is meant to describe time evolution fully, such that one may write

$$\exp\bigl(-i\hat{\mathcal{L}}_{\text{coll}}s\bigr)\hat{d}(t-s) = \hat{d}(t) \qquad (7.72)$$

instead of the only approximate relation $\exp\bigl(-i\hat{L}_{\text{coll}}s\bigr)\hat{d}(t-s) \approx \hat{d}(t)$ of (25.152) in perturbation theory. For undamped motion one might go ahead and even evaluate the time evolution of the operators like $\hat{q}$. In this case the $\hat{\mathcal{L}}_{\text{coll}}$ may simply be taken from common RPA expressed in terms of operators, which in the harmonic case have the same time evolution as their averages. For damped motion this is no longer true because then there will be fluctuating forces which are not present in average dynamics.

Unfortunately, exactly this feature makes the resulting Nakajima–Zwanzig equation intractable. As to be expected from its origin, the problem appears first in the equations for the second moments. There quantities of the following type appear

$$\widetilde{\chi}''_{AB}(-s;t) = (1/2)\mathrm{tr}\left[\hat{A}(-s), \hat{B}\right]\hat{d}(t)$$
$$\widetilde{\psi}''_{AB}(-s;t) = (1/2)\mathrm{tr}\left[\hat{A}(-s), \hat{B}\right]_+\hat{d}(t) \quad (7.73)$$

with the time dependence being defined as $\hat{A}(-s) = \exp(-i\hat{L}_{\mathrm{coll}}s)\hat{A}$ and where A, B are either q or Π. The functions introduced on the left have a structure similar to collective response and correlation functions—with the *essential and important* difference that they are defined for the actual, time-dependent collective density operator $\hat{d}(t)$ and not for the $\hat{d}_{\mathrm{eq}}$ of collective equilibrium. To calculate functions of this type it is not even sufficient to know the solution $\hat{d}(t)$ of the transport equation. One also needs the $\hat{A}(-s)$ themselves. Unfortunately, these functions cannot be traced back to moments, either to the second moments or to any other.

The situation is of course different for a system already in equilibrium. In this case the $\widetilde{\chi}''_{AB}(-s;t)$ and $\widetilde{\psi}''_{AB}(-s;t)$ simply become the usual collective response and correlation functions $\widetilde{\chi}''_{AB}(-s)$ and $\widetilde{\psi}''_{AB}(-s)$, respectively. Their Fourier transforms can be expected to satisfy the fluctuation dissipation theorem in the form (23.91). Indeed, it can be shown that this theorem is in accord with the non-perturbative approach explained above, more specifically with the full, non-Markovian equations of motion for the second moments in their stationary limit. One only needs to assume this theorem to hold true for the nucleonic degrees of freedom. It may be mentioned that in these equations nucleonic response functions always appear in connection with the collective correlation functions and vice versa.

Let us return now to the problem of getting an appropriate approximation to the true time dependence of the fluctuations. In (Hofmann et al., 1989), (Hofmann, 1997) the following strategy has been exploited:

(1) The integro-differential equations for the first moments, which are closed in themselves, have been reduced to differential form by pursuing the same steps as applied already in (Hofmann and Sollacher, 1988) for Ehrenfest's equations for the Bohm–Pines Hamiltonian. This implies exploiting the "one pole approximation" by which the collective response function is replaced by that of the oscillator. Recall, please, that the same approximation was used in Section 7.2 to evaluate the equilibrium fluctuations.

(2) Next it is assumed that the same type of equations can be taken over to the operators which appear in the functions $\widetilde{\chi}''_{AB}(-s;t)$ and $\widetilde{\psi}''_{AB}(-s;t)$ introduced in (7.73). This means that these functions are calculated ignoring any effects of the fluctuating force. In this way the terms representing the "drift" take on exactly the forms of the left-hand sides of the set (7.17) to (7.19). They are given by terms only which involve nucleonic response functions combined with the $\widetilde{\psi}''_{AB}(-s;t)$ specified by collective anti-commutators.

(3) The diffusion coefficients, on the other hand, are determined by the other combinations, namely those of commutators of collective operators with intrinsic correlation functions. In principle, one might proceed by evaluating these terms like done for the drift terms. Indeed, in this way they turn out to become *constants, not proportional* to the fluctuations themselves–as should be the case for diffusion coefficients. Interestingly enough they turn out to be non-linear functions of the frequencies $\omega^{\pm}$ for (the in general damped) local motion given in (23.5) (Hofmann and Jensen, 1984), (Hofmann et al., 1989). It can be seen, however, that this choice violates the FDT, for details see (Hofmann et al., 1989), (Hofmann, 1997). This may be understood from the very construction which for the $\hat{A}(-s)$ discards any information about the fluctuating force. It has therefore been argued that in this "one pole approximation" the diffusion coefficients are to be determined from (7.23), which is to say from the FDT known to be in accord with the non-perturbative procedure. For possible consequences for the equations of motion the reader is reminded of the discussion below eqns(7.22).

Let us finally turn to the equation of motion for the density operator $\hat{d}(t)$. To the set of equations (7.17) to (7.19), there should exist an equation for the operator $\hat{d}$, which is in one to one correspondence with the equation for the Wigner transform $d(Q, P, t)$. Indeed, for constant transport coefficients and with the diffusion coefficients chosen according to (7.23), the equation for $\hat{d}$ reads

$$\frac{d}{dt}\hat{d}(t) = -\frac{i}{\hbar}\left[\frac{1}{2M}\hat{P}^2 + \frac{C}{2}\hat{q}^2, \hat{d}(t)\right] - \frac{i}{\hbar}\gamma\left[\hat{q}, \frac{1}{2}\left[\frac{\hat{P}}{M}, \hat{d}(t)\right]_+\right] + \frac{1}{2\hbar^2}\left[\hat{q}, \left[\left(-D_{pp}^{\mathrm{FDT}}\hat{q} + D_{qp}^{\mathrm{FDT}}\frac{\hat{P}}{M}\right), \hat{d}(t)\right]\right] \quad (7.74)$$

The conjecture that (7.74) corresponds to (7.34) can be made apparent by simply exploiting basic properties of the Wigner transform discussed in Section 19.3. There, it is demonstrated that the following correspondence rules hold true (see eqns(19.75) and (19.76)):

$$-\frac{i}{\hbar}[\hat{A}, \hat{B}] \iff \{A, B\} = \frac{\partial A}{\partial Q}\frac{\partial B}{\partial P} - \frac{\partial A}{\partial P}\frac{\partial B}{\partial Q}$$
$$\frac{1}{2}[\hat{A}, \hat{B}]_+ \iff AB. \quad (7.75)$$

With their help the supposition from above is easily seen to be correct. Thus for the harmonic oscillator eqn(7.74) may be considered a quantal version of Kramers' equation.

Exercises

1. (i) Derive the set of eqns(7.17) to (7.19) for the pure quantal harmonic oscillator (without dissipative and stochastic forces). (ii) Is this set still

closed if the potential is of higher order? (iii) Take a $V(q)$ with $d^3V/dq^3 \neq 0$ and derive the set of equations for the third moments; one may take advantage of eqn(21.8). (iv) For short times $\tau = t - t_0$, to which order in τ do they progress when they start from zero values. Interpret the result in the light of the Kramers–Moyal expansion of Section 25.2.3.

2. (a) From the set of equations (7.17) to (7.19) for the second moments show for the diffusion coefficients of (7.23) (i) that

$$\frac{M}{2}\frac{d^2\Sigma_{qq}(t)}{dt^2} = C\left(\Sigma_{qq}^{eq} - \Sigma_{qq}(t)\right) - \frac{1}{M}\left(\Sigma_{pp}^{eq} - \Sigma_{pp}(t)\right) - \frac{\gamma}{M}\Sigma_{qp}(t), \tag{7.76}$$

discuss the implication for the fluctuations of the propagator at the initial time t_0 where $\Sigma_{\mu\nu}(t_0) = 0$ (mind the results shown in Fig. 7.4). (ii) Prove that

$$\left.\frac{d\Delta(t)}{dt}\right|_{t_0} = \frac{2\gamma}{M}\Sigma_{qq}(t_0)\left(\Sigma_{pp}^{eq} - \Sigma_{pp}(t_0)\right) \quad \text{for} \quad \Sigma_{qp}(t_0) = 0. \tag{7.77}$$

(b) For positive stiffness C, the $\Delta(t)$ turns to $\Delta^{eq} \equiv \Sigma_{qq}^{eq}\Sigma_{pp}^{eq}$ for $t \to \infty$. Convince yourself that for weak damping $\Delta^{eq} \geq \hbar^2/4$. What happens in the high temperature limit?

PART II

COMPLEX NUCLEAR SYSTEMS

8

THE STATISTICAL MODEL FOR THE DECAY OF EXCITED NUCLEI

Since the seminal work of N. Bohr (1936) the statistical model has played an immensely important role for the understanding of nuclear reactions. Originally devoted to describing the sharp resonances seen in elastic scattering of neutrons at nuclei at low energy, it has found much wider application in the course of time. Physically, the essential point is seen in the *Bohr hypothesis* that the decay of the compound nucleus is independent of how it was created before, except for the information contained in conserved quantities. This independence hypothesis implies that at an intermediate stage the system reaches a statistical equilibrium, which naturally is to be characterized as a microcanonical ensemble. Details of this model and under which conditions it evolves quite naturally within a microscopic formulation of reaction theory are described in Section 18.4. At the end of this section limits of applicability are also mentioned. Below in Chapter 9 we will address pre-compound reactions which are initiated by light particles at higher energies. In such experiments emission of light particles occurs in stages before the composite system has reached its equilibrium.

A situation where the Bohr hypothesis may be more appropriate are collisions between heavy ions at lower energy. If measured per particle this energy may be of the order of a few MeV, and thus definitely less than the typical corresponding energy for pre-equilibrium reactions. The Bohr hypothesis is more likely to be applicable to portraying the decay of such systems. This is of great practical importance as it is very difficult to describe both the initial, the intermediate as well as the final stages of the reaction by the same transport model. Mainly this is so because of the very different timescales involved.

8.1 Decay of the compound nucleus by particle emission

In this section we are going to work out rate formulas largely by drawing on analogies to procedures from statistical mechanics. It will be seen that to some extent the concept of temperature may be used without actually leaving the realm of microcanonical ensembles.

8.1.1 *Transition rates*

At first we need to derive an expression for the rate of the decay of a compound nucleus $C \to b + B$ into fragments B and b. Such a rate formula can be obtained in terms of the formation cross section $\sigma_\beta^{(C)}$ for the inverse reaction. In these processes the energy balance can be written as

… The statistical model for the decay of excited nuclei

$$E_C = E_\beta + \epsilon_\beta \quad \text{with} \quad E_\beta = E_b^\beta + E_B^\beta. \tag{8.1}$$

Here, β is meant to represent all quantum numbers which specify this particular "channel". The energies of the *separated* fragments may be excited to total energies E_b^β and E_B^β, and the energy of relative motion is represented by ϵ_β. To simplify matters let us first refrain from specifying the channel β any closer, which essentially means ignoring angular momentum couplings. This will facilitate the physical interpretation of the rate formula, in particular in the terminology of statistical mechanics.

Transition probabilities can be expressed by T-matrix elements (see Section 18.1.1). The ones appropriate for the formation process can be written as

$$P_{\beta C} = \frac{2\pi}{\hbar} |T_{\beta C}|^2 \, \Omega_C(E_C). \tag{8.2}$$

Here, $\Omega_C(E_C)$ is the density of states of the compound nucleus. In terms of the formation cross section $\sigma_\beta^{(C)}$ this $P_{\beta C}$ can be written in the form

$$P_{\beta C} = \frac{v_\beta}{\mathcal{V}} \sigma_\beta^{(C)}. \tag{8.3}$$

One imagines the system to be enclosed in a volume $\mathcal{V}$ in which both fragments approach each other with relative velocity v_β. Equating both expressions, the following formula is obtained

$$\frac{2\pi}{\hbar} |T_{\beta C}|^2 = \frac{v_\beta}{\mathcal{V}} \frac{\sigma_\beta^{(C)}}{\Omega_C(E_C)} \tag{8.4}$$

for the squared transition matrix.

For the decay rate one may also employ the transition probability for the reverse process. With Ω_β being the density of the final channels it reads

$$P_{C\beta} = \frac{2\pi}{\hbar} |T_{C\beta}|^2 \, \Omega_\beta = \frac{v_\beta \sigma_\beta^{(C)}}{\mathcal{V}} \frac{\Omega_\beta}{\Omega_C(E_C)}. \tag{8.5}$$

To get the expression on the very right the concept of microreversibility was used, which is discussed in Section 18.1.1. For the present case it says that the T-matrices appearing in (8.2) and (8.5) are identical, $T_{C\beta} = T_{\beta C}$. Equations (8.3) and (8.4) then allow one to express $T_{C\beta}$ through the formation cross section. Next we need to evaluate the density of final states Ω_β. For this we may take over the discussion presented in Section 22.2.1 for composite systems treated in a microcanonical ensemble. Notice that we are looking at a situation where the decay products are sufficiently far apart for neglecting their mutual interaction to be justified. Moreover, as we may assume the energy interval Δ to be sufficiently small the $\Omega(\mathcal{E})$ can be written in factorized form, $\Omega(\mathcal{E}) = \Omega_1(\mathcal{E}_1)\,\Omega_2(\mathcal{E}_2)\,\Delta$ when $\mathcal{E} = \mathcal{E}_1 + \mathcal{E}_2$. Applying this formula twice the Ω_β becomes

$$\Omega_\beta(E_C) = \Omega_B(E_B^\beta)\Omega_b(E_b^\beta)\,dE_b^\beta\,\Omega_{\mathrm{kin}}^{\mathcal{V}}(\epsilon_\beta)\,d\epsilon_\beta\,. \tag{8.6}$$

Using (8.6) one obtains the double differential cross section

$$d^2 P_{C\beta} = \frac{v_\beta \sigma_\beta^{(C)}}{\mathcal{V}}\,\frac{\Omega_B(E_B^\beta)\Omega_b(E_b^\beta)\,dE_b^\beta\,\Omega_{\mathrm{kin}}^{\mathcal{V}}(\epsilon_\beta)\,d\epsilon_\beta}{\Omega_C(E_C)}\,. \tag{8.7}$$

The ratio of the level densities allows one to introduce probability densities for the subsystems by making use of general formulas of thermostatics. Indeed, eqn(22.73), $w_1(\mathcal{E}_1;\mathcal{E}) = (\Omega_1(\mathcal{E}_1)\Omega_2(\mathcal{E} - \mathcal{E}_1))/\Omega(\mathcal{E})$, determines this probability density for subsystem "1" to have energy $\mathcal{E}_1$ when the total energy of the combined system is given by $\mathcal{E} = \mathcal{E}_1 + \mathcal{E}_2$. This interpretation may be exploited to introduce the concept of temperature without any further approximations; details are presented in Section 10.3. One may wonder at what point these analogies to concepts of statistical mechanics come into play. First of all, the very notion of the "compound nucleus" refers to a fully equilibrated system. Another "source" is found at the stage when in the original formulas for the transition rates the level densities are introduced. In this way *all processes* which happen within given energy intervals $d\epsilon_\beta$ and dE_b^β are *treated with equal probability*. The T-matrices are thus assumed to represent average values, implying that the formation cross section, too, is an averaged one.

To render the rate formulas applicable to actual problems one needs to know the absorption or formation cross section $\sigma_\beta^{(C)}$ for the particular channel β. Apart from the exceptional situation where this channel represents two nuclei approaching each other sitting in their ground states the $\sigma_\beta^{(C)}$ is not accessible experimentally. For most cases of practical interest, where at least the heavier fragment is excited, one has to rely on theoretical estimates. The simplest (and perhaps crudest) possibility is given by evaluating $\sigma_\beta^{(C)}$ within a geometrical picture behind a "black box" approximation. In this case it might become feasible even to take into account deformations. As a possible improvement, if available, one may take the optical model, in accord with the spirit of the Hauser–Feshbach theory, see also (Gadioli and Hodgson, 1992). For rotating, deformed nuclei a direct evaluation of the transition rate within a semi-classical approach was given in (Dietrich et al., 1995). For spherical nuclei, later in (Pomorski et al., 2000) the result was shown to be equivalent to Weisskopf's formula.

Still present in (8.7) is the volume $\mathcal{V}$, which one should like to get rid of. It was introduced in (8.3) looking at the relative motion of the two fragments. However, if combined with the level density $\Omega_{\mathrm{kin}}^{\mathcal{V}}$ for these degrees of freedom one may make use of the fact that the ratio $\Omega_{\mathrm{kin}}^{\mathcal{V}}/\mathcal{V}$ can be identified with the $\Omega_{\mathrm{kin}}(E)$ of (18.16). Indeed, (18.16) corresponds to a normalization of plane waves in free space, say according to (18.3). For a particle in a box the wave function gets the additional factor $1/\sqrt{\mathcal{V}}$, see (1.10). Finally, since we are interested in transition rates for angle integrated processes we may introduce another factor of 4π to get

$$\frac{\Omega_{\text{kin}}^{\mathcal{V}}(\epsilon_\beta)}{\mathcal{V}} = 4\pi \frac{\mu_\beta p_\beta}{(2\pi\hbar)^3} = \frac{\mu_\beta}{\pi^2 \hbar^3} \epsilon_\beta \,. \tag{8.8}$$

Let us assume the $d\epsilon_\beta$ to be the only independent variation, obeying the constraint (8.1), of course. Then (8.7) can be rewritten as

$$dP_{\{E,J\} \to \{I_b, I_B, \epsilon_\beta\}} = \sigma_{\{I_b, I_B, \epsilon_\beta\}}^{(C)} \frac{\Omega_B(E_B^\beta, I_B) \Omega_b(E_b^\beta, I_b)}{\Omega_C(E,J)} \frac{\mu_\beta}{\pi^2 \hbar^3} \epsilon_\beta d\epsilon_\beta \,. \tag{8.9}$$

It is worth relating this expression to the average width known from statistical reaction theory, namely $\overline{\Gamma_\beta} = (\overline{\sigma_\beta^{(C)}}/\Omega_C)(\mu_\beta \epsilon_\beta/\pi^2 \hbar^2)$ (see (18.148)). To get (8.9) one simply needs to multiply $\overline{\Gamma_\beta}/\hbar$ by the number of states ΔG_{Bb} in the system Bb, namely $\Delta G_{Bb} = \Omega_B(E_B^\beta) \Omega_b(E_b^\beta) dE_b^\beta d\epsilon_\beta$. Notice that in the derivation of formulas (18.148) the density of states of relative motion has been taken care of already. This comparison shows that the transition rates discussed here are nothing else but the average widths Γ_β the compound nucleus gets through channel β. Indeed, the branching ratio $G_\beta^{(C)} = \overline{\Gamma_\beta}/\overline{\Gamma}$ of (18.150) to all other open channels might be evaluated from (8.9), see e.g. (Ericson, 1960) or Section IV.9 of (Feshbach, 1992).

For the evaluation of the energies it is useful to introduce the excitation energy ΔE_X^γ of the fragments by writing $E_X^\gamma = E_X^0 + \Delta E_X^\gamma$ with E_X^0 being the ground state of fragment X. Indeed, the level densities will only depend on the ΔE_X^γ. Applying such a relation also to the compound system (8.1) can be rewritten as

$$\Delta E_b^\beta + \Delta E_B^\beta + \epsilon_\beta = \Delta E_C - B_{\{b,B\} \leftrightarrow \{C\}} \,, \tag{8.10}$$

with

$$B_{\{b,B\} \leftrightarrow \{C\}} = E_b^0 + E_B^0 - E_C^0 \tag{8.11}$$

being the binding energy of the compound nucleus in the formation from the fragments b and B (for the relation to the Q-value see Section 18.2.1). We may note in passing that this relation may be considered the analog of the formula (1.64) for the total energy or mass $M(N,Z)$ of a nucleus when considered to be formed or split into N neutrons and Z protons. Here, $B(N,Z)$ is the binding energy of the a nucleus in the Bethe–Weizsäcker formula (1.65).

8.1.2 Evaporation rates for light particles

Above threshold and for not too large energies, the most important decay modes are those of the evaporation of neutrons, protons and α particles, besides fission which will be addressed below. These light particles have in common that they are emitted without having any internal excitation, such that the ΔE_b is to be put equal to zero. Since we are mostly interested in larger excitations we will not look at individual channels β anymore and may thus replace the ΔE_B^β simply by the thermal excitation energy $\mathcal{E}_B^*$. The energy balance may then be written as

$$\mathcal{E}_B^* + \epsilon_\beta = \mathcal{E}_C^* - B_{\{b,B\}\leftrightarrow\{C\}} \qquad (8.12)$$

with the binding energy being defined in (8.11). For zero excitation of fragment b its level density Ω_b just reduces to the spin degeneracy factor $\Omega_b = 2I_b + 1$. To get the total evaporation rate we need to sum over all possible spins I_B of fragment B and integrate over all possible energies

$$P_{\{E,J\}\to\{I_b,I_B,\epsilon_\beta\}} = \frac{\mu_\beta}{\pi^2\hbar^3}\frac{2I_b+1}{\Omega_C(E,J)} \times$$
$$\sum_{I_B}\int_0^{\mathcal{E}^*-B_{\{b,B\}\leftrightarrow\{C\}}} d\epsilon_\beta\, \epsilon_\beta\, \sigma^{(C)}_{\{I_b,I_B,\epsilon_\beta\}}\, \Omega_B(\mathcal{E}_C^* - B_{\{b,B\}\leftrightarrow\{C\}} - \epsilon_\beta, I_B). \qquad (8.13)$$

From the very fact that the formation cross section appears in this formula it may immediately be concluded that the decay into charged particles is strongly prohibited as compared to the evaporation of neutrons. When the former try to penetrate into a nucleus they encounter the Coulomb barrier mentioned in Section 3.2 which reduces the fusion probability.

8.2 Fission

8.2.1 The Bohr–Wheeler formula

From a merely formal point of view the results derived in Section 8.1.1 might also be applied to the breakup of the compound nucleus in two large pieces. However, for this situation the formation or fusion cross section which enters (8.9) is of no use. It is impossible to realize a reaction which might mimic the reverse of a fission process. This follows already from the fact that fusion and fission proceed along different paths in a multidimensional landscape and are associated with different timescales. But it is not difficult to modify the arguments from above to find the following formula (see exercise 1)

$$P_{\{E,J\}\to f} \equiv R_{\mathrm{BW}} = \frac{1}{2\pi\hbar\Omega_C(E,J)}\int_0^{E-B_f} d\epsilon_f\, \Omega^{\mathrm{sad}}_{C-1}(E-B_f-\epsilon_f, J) \qquad (8.14)$$

for the fission rate which was published by N. Bohr and J.A. Wheeler (1939) shortly after the discovery of fission. To understand better the meaning of the individual quantities it is instructive to follow the original derivation which involves the transition state picture.

Assume one is given a microcanonical ensemble of fissioning nuclei having energies between E and $E+dE$ and angular momentum J (which is conserved through the process). With $\Omega_C(E,J)$ being the level density at the ground state deformation there are $\Omega_C(E,J)dE$ nuclei altogether. The number of nuclei which decay by fission per unit of time is then given by $\Omega_C(E,J)R_{\mathrm{BW}}dE$, with an obvious definition of the rate $R_{\rm BW}$. Treating collective motion within a classical picture, this number must be equal to the number of nuclei $\dot{N}_f$ which pass the saddle point per unit of time. Per unit of momentum the $d\dot{N}_f$ will be given by

$$d\dot{N}_f = \frac{\dot{q}_f dp_f}{2\pi\hbar}\Omega^{\rm sad}_{C-1}(E - B_f - \epsilon_f, J) = \frac{d\epsilon_f}{2\pi\hbar}\Omega^{\rm sad}_{C-1}(E - B_f - \epsilon_f, J). \quad (8.15)$$

All quantities appearing here refer to the position of the saddle point. The ϵ_f is the (classical) kinetic energy in the direction of fission (with $d\epsilon_f = \dot{q}_f dp_f$) and $\Omega^{\rm sad}_{C-1}$ represents the (quantal) density of levels of that system which consists of all degrees of freedom of C but the fissioning one. This residual system has an excitation energy

$$E^{\rm sad}_{C-1} = E - B_f - \epsilon_f, \quad (8.16)$$

with B_f being the barrier height. The latter measures the difference of the static energy of the compound nucleus between its value at the barrier and that at the potential minimum.

Now the important question arises how this static energy should be evaluated. One might be inclined (Charity, 1996) to choose it equal to that for the intrinsic ground states (defined for each deformation). In such a case the B_f would be influenced by the full shell effects seen at $T = 0$. However, one must not forget the basic assumption behind the transition state method, namely that of *complete equilibrium for all degrees of freedom* inside the barrier. This conjecture *implies that the static energy itself*, and hence the B_f, does depend on the $E^{\rm sad}_{C-1}$. As a consequence, with increasing excitation the shell effects will disappear and the barrier height approaches that of the liquid drop model. Evidently, taken literally, for $B_f = B_f(E^{\rm sad}_{C-1})$ eqn(8.16) becomes an implicit equation for the $E^{\rm sad}_{C-1}$, a feature which in practical applications is commonly ignored. Finally we should like to note that the very construction of (8.14) requires that only two prominent stationary points exist, one representing the ground state minimum and the other one the maximum of the static energy.

The Bohr–Wheeler formula in the version of (8.14) only employs the microcanonical ensemble. Of course, further assumptions are possible to simplify the general formula. Let us go through a few steps of progressive simplification to end up with a version where a temperature appears explicitly. This will facilitate comparisons with Kramers' formula in Chapter 12. Let us assume the excitation energy to be sufficiently large that shell effects may be neglected. Then the density of levels at excitation energy $\mathcal{E}^* = E - B_f - \epsilon_f$ may be approximated by a formula of the type $\Omega(\mathcal{E}^*) = F(\mathcal{E}^*)\exp(2\sqrt{a\mathcal{E}^*})$ with an energy independent parameter a. The pre-factor $F(\mathcal{E}^*)$ is a smooth function of $\mathcal{E}^*$, details of which depend on the system but are unimportant for the following; the same holds true for an eventual "back shift" of the energy or the dependence on angular momentum, for details see Chapter 10. With such a functional form used for the $\Omega^{\rm sad}_{C-1}$ the integral in (8.14) can be handled analytically. The integrand $\Omega^{\rm sad}_{C-1}(E - B_f - \epsilon_f)$ will be largest at $\epsilon_f = 0$ and its variation with ϵ_f will be

Fission

dominated by the exponential. Thus the $F(E - B_f - \epsilon_f)$ can safely be replaced by $F(E - B_f)$ and pulled out of the integral which can then be calculated to

$$\int_0^{E-B_f} d\epsilon_f \, \Omega_{C-1}^{\text{sad}}(E - B_f - \epsilon_f)$$
$$\approx F_{\text{sad}}(E - B_f) \int_0^{E-B_f} d\epsilon_f \, \exp\left(2\sqrt{a_{\text{sad}}(E - B_f - \epsilon_f)}\right) \qquad (8.17)$$
$$\approx \frac{F_{\text{sad}}(E - B_f)}{2 a_{\text{sad}}} \left(2\sqrt{a_{\text{sad}}(E - B_f)} - 1\right) \exp\left(2\sqrt{a_{\text{sad}}(E - B_f)}\right).$$

Here, a term has been neglected which is of order unity as compared with the exponential. Likewise, the 1 inside the bracket may usually be neglected as well.

Next we need to look at the $\Omega_C(E)$ in the denominator of (8.14). As compared with the $\Omega_{C-1}^{\text{sad}}$ this level density corresponds to a system (the "mother" nucleus) which has one degree of freedom more. The latter ought to be attributed to that collective motion by which the decaying system moves towards the saddle point. *Assuming* this motion to be *weakly coupled* to the intrinsic degrees of freedom the $\Omega_C(E)$ will be given by (22.74),

$$\Omega_C(E) = \int_0^E de_q \, \Omega_{C-1}^{\min}(E - e_q) \Omega_q^{\min}(e_q) \approx \frac{1}{\hbar\varpi_{\min}} \int_0^E de_q \, \Omega_{C-1}^{\min}(E - e_q). \qquad (8.18)$$

Here, the following assumptions have been made: (i) The level density $\Omega_C(E)$ corresponds to dynamics in the neighborhood of the potential minimum (with the ground state energy $E_{\min}^0$ being put equal to zero). (ii) There collective motion may be described by an oscillator for which quantum effects may be neglected such that (iii) according to (24.95) its level density is given by $\Omega_q^{\min}(e_q) \approx 1/\hbar\varpi_{\min}$. The remaining integral in (8.18) can then be calculated following (8.17). One obtains the same final expression with the only differences that the level density parameter $a_{\min}$ is that for the deformation around the minimum and that the energy E appears rather than $E - B_f$.

To get the decay rate we need to build the ratio of the two expressions. For a weak dependence of the level density parameter on deformation one may put $a_{\min} = a_{\text{sad}} = a$. If, furthermore, $E \gg B_f$ one may replace $E - B_f$ by E in the pre-factors such that they cancel, besides the important one, namely $\varpi_{\min}/2\pi$. In the exponentials, however, which carry the strongest energy dependence, one must not neglect the difference $2\sqrt{a(E - B_f)} - 2\sqrt{aE}$. To leading order it turns into $-B_f\sqrt{a/E} \equiv -B_f/T$, with the temperature being defined through $E = aT^2$. The final formula thus becomes

$$R_{\text{BW}} \approx \frac{\varpi_{\min}}{2\pi} e^{-B_f/T}, \qquad (8.19)$$

which is governed by the Arrhenius factor and where the pre-factor represents the "assault frequency" for overcoming the barrier. In this version the rate depends on angular momentum insofar as the rotation influences the static energy which defines the potential landscape.

Let us reflect once more how the result (8.19) was obtained. One of the essential steps was the evaluation of the level density $\Omega_C(E)$ through (8.18). The pre-factors in (8.17) only dropped out because in (8.18) a similar integral appeared. Suppose the $\Omega_{C-1}^{\text{sad}}$ had been identified *incorrectly* with Ω_C, making the same approximations with respect to the level density parameters and the energy dependencies otherwise. Then in the rate formula only the pre-factor

$$\sqrt{a(E-B_f)}/(2\pi\hbar a) \approx \sqrt{aE}/(2\pi\hbar a) = T/(2\pi\hbar) \qquad (8.20)$$

would have survived. The resulting decay rate differs from the more appropriate result (8.19) by the factor $T/\hbar\varpi_{\min}$. According to (Thoennessen, 1996) in some statistical codes the incorrect formula is used.

8.2.2 Stability conditions in the macroscopic limit

An essential element of the Bohr–Wheeler formula is the static energy of the fissioning nucleus. How the potential landscape is to be calculated microscopically has been described in Chapter 5. Often (8.14) is used at larger excitations where shell effects are not important anymore. Based on such a hypothesis let us examine how the landscape and thus the fission barrier is influenced by angular momentum J. Then a new instability might occur which results from a strong centrifugal force at larger J. This effects is similar to the one studied in Section 1.3.3.3 where the fissility parameter x was introduced in (1.76). Whenever its value is larger than unity spherical shapes are already unstable against quadrupole deformations. For a rotating nucleus the energy $E(Q_\nu)$, with the Q_ν being some general shape variables, has to be modified by adding the rotational energy E^{rot}. For a rotation about the axis 1 this is given by $E^{\text{rot}} = \hbar^2 J^2/(2\mathcal{J}_1(Q_\nu))$ (see e.g. Section 6.3). The $\mathcal{J}_1(Q_\nu)$ is the rotational moment of inertia. Within a semi-classical approximation one might replace the J^2 by $J(J+1)$. Problems associated with the calculation and the properties of the moment of inertia are discussed in Section 6.7. Often it is assumed to be given by the value $\mathcal{J} = \mathcal{J}^{\text{rig}}$ for rigid rotation.

Neglecting the symmetry term the static energy can be written as

$$E_{\text{LD}}(Q_\mu; J) = E_{\text{LD}}^{\text{surf}}(Q_\mu) + E_{\text{LD}}^{\text{Coul}}(Q_\mu) + E_{\text{rig}}^{\text{rot}}(Q_\mu; J). \qquad (8.21)$$

For zero angular momentum the deformation energy for quadrupole deformations turns out to be given by the formula

$$\frac{E_{\text{LD}}(\beta,\gamma) - E_{\text{LD}}(\beta=0)}{a_{\text{surf}} A^{2/3}} = \frac{1}{2\pi}\left((1-x)\beta^2 - \frac{2}{21}\left(\frac{5}{4\pi}\right)^{1/2}(1+2x)\beta^3\cos 3\gamma\right), \qquad (8.22)$$

which goes back to (Bohr and Wheeler, 1939) (see App.6A-2 of (Bohr and Mottelson, 1975) for more details of this model and (Cohen and Swiatecki, 1963) for more realistic treatments). Notice that the energy is taken in units of the spherical surface energy. This formula is valid to third order in the deformation parameter β (see Section 1.3.3.2) and is derived for x close to unity. As expected, the minimum is seen to lie at $\beta = 0$. The shape of the saddle point turns out prolate ($\gamma = 0$) and lies at

$$\beta_{\text{sad}} \equiv \beta_f = \frac{7}{3} \left(\frac{4\pi}{5}\right)^{1/2} (1-x), \tag{8.23}$$

with a barrier height of

$$B_f = \frac{98}{135} (1-x)^3 \, a_{\text{surf}} A^{2/3}. \tag{8.24}$$

The structure of (8.22) is such that the stiffness $C(\beta_f)$ at the barrier is just minus that at the minimum, namely

$$-C(\beta_f) = C(\beta = 0) = \frac{1}{\pi} a_{\text{surf}} A^{2/3}((1-x). \tag{8.25}$$

To measure the influence of rotations it is convenient to introduce the parameter

$$y = \frac{J^2 \hbar^2}{2 \mathcal{J}_1(Q_\nu)} \left(a_{\text{surf}} A^{2/3}\right)^{-1}, \tag{8.26}$$

which is of order $2.1 A^{-7/3} J^2$. The potential energy now reads (to lowest order in $(1-x)$)

$$\frac{E_{\text{LD}}(\beta, \gamma; y) - E_{\text{LD}}(\beta=0; y)}{a_{\text{surf}} A^{2/3}} = \tag{8.27}$$

$$\frac{1}{2\pi}\left((1-x)\beta^2 - \frac{2}{21}\left(\frac{5}{4\pi}\right)^{1/2}(1+2x)\beta^3 \cos 3\gamma\right) + y\left(\frac{5}{4\pi}\right)^{1/2} \beta \cos\left(\gamma + \frac{2\pi}{3}\right).$$

Remember that for rigid rotation there is a finite rotational energy of $E_{\text{LD}}(\beta = 0) = E_{\text{rig}}^{\text{rot}} = y \, a_{\text{surf}} A^{2/3} > 0$. As an effect of the centrifugal force, the stationary points of the potential change. The minimum is now attained for an oblate spheroid ($\gamma = \pi/3$) (similar to the shape of the earth). The barrier attains a γ which is linear in y. With increasing angular momentum the shape of the minimum becomes unstable with respect to triaxial deformations. But for $x > 0.8$ the system first becomes unstable with respect to fission. That a finite rotation reduces the barrier height can be calculated perturbatively in y. One simply adds to (8.24) the contribution from the last term on the right of (8.27) when calculated for $\gamma = 0$ and at the deformation of the barrier for $y = 0$, as given by (8.23). In this way one gets

$$B_f(y > 0) = B_f(y = 0)\left(1 - \frac{45}{28} \frac{y}{(1-x)^2}\right). \tag{8.28}$$

This formula would suggest the critical y to be given by $y_{\text{crit}} \approx (4/9)(7/5)(1-x^2)$. Actually the system becomes unstable for $y \geq (7/5)(1-x)^2 = y_{\text{crit}}$ where minimum and barrier coalesce at the deformation defined by $\gamma = \pi/3$ and $\beta = 7/6(4\pi/5)^{1/2}(1-x)$.

The mass asymmetry degree of freedom: The splitting up of a compound nucleus of mass A into two fragments does not necessarily happen in symmetric fashion, meaning that the masses of the two fragments may be very different from $A/2$. A first crude estimate can be made from the liquid drop model and the Bethe–Weizsäcker formula. Let us rewrite (8.11) as

$$E^0(A, Z) = E^0(A_1, Z_1) + E^0(A_1, Z_1) - B_{\{A_1, A_2\} \leftrightarrow \{A, Z\}} . \qquad (8.29)$$

It determines the ground state energy of the combined system with $A = A_1 + A_2$ and $Z = Z_1 + Z_2$. According to this formula, fission of the compound nucleus is possible into such combinations of A_i, Z_i for which the binding energy $B_{\{A_1, A_2\} \leftrightarrow \{A, Z\}}$ is positive. One may calculate this binding energy as function of the asymmetry parameters $\mathcal{A}_{\text{asym}} = A_1/A$ and $\mathcal{Z}_{\text{asym}} = Z_1/Z$. To simplify matters one may fix Z by either having $Z = N$ or take it along the valley of β-stability. Then $B_{\{A_1, A_2\} \leftrightarrow \{A, Z\}}$ simply becomes a function of $\mathcal{A}_{\text{asym}}$, with the fissility parameter x of (1.76) being the relevant parameter. Plots of that type may for instance be found in (Wilets, 1964) (for $N = Z$); see also (Hasse and Myers, 1988)). Usually, the asymmetry parameter is then visualized by two touching spheres and the plots are done for $0 \leq \mathcal{A}_{\text{asym}} \leq 1$ although the left–right symmetry $\mathcal{A}_{\text{asym}} = 1 - \mathcal{A}_{\text{asym}}$ has no real physical meaning. It is seen that $B_{\{A_1, A_2\} \leftrightarrow \{A, Z\}}$ may become negative for x larger than about 0.7. For x larger than about 0.4 the curve develops a minimum at $\mathcal{A}_{\text{asym}} = 0.5$ implying that the splitting favors a symmetric configuration.

Often the $\mathcal{A}_{\text{asym}}$ is used as a dynamical variable to describe the *mass exchange modes* for processes of fusion or fission. In this case one should add the Coulomb energy for two touching spheres $V_C = Z_1 Z_2 e^2 / R_C$ with R_C being the distance between the centers of the charges. In (Fröbrich and Lipperheide, 1996) a value of $R_C = 0.5 + 1.36(A^{1/3} + A^{1/3})$ is taken and the charge of the fragments is chosen according to $Z = (A - 0.006 A^{5/3})/2$. Then the symmetric configuration is favored for $A \geq 120$.

Exercises

1. Derive the Bohr–Wheeler formula starting from eqn(8.9). One may assume that the "reaction" takes place along a one-dimensional path q_f in the potential landscape and that all events allowed by energy conservation occur with probability one.

9

PRE-EQUILIBRIUM REACTIONS

Processes which involve the compound nucleus (in strict sense) proceed through a stage of complete equilibrium described in terms of the microcanonical ensemble, see Chapter 8 and Chapter 18. Since the 1960s there has been growing evidence that reactions of light particles with medium to heavy nuclei show features which may be associated neither with such an "evaporation" from an equilibrated system nor with direct reactions. These two types of reactions correspond to very different timescales. Measured inelastic spectra for (p, p') reactions, for instance, exhibit that protons are emitted with energies intermediate between the ones which result from prompt or delayed processes. Likewise, the cross sections for the emission of several nucleons, say in (n, xn) reactions, could not be explained by the evaporation model either. There is no room to go into any details here, for which we refer to the monograph by E. Gadioli and P.E. Hodgson (1992) or to other review articles cited there, like the one by M. Blann (1975). Truth is that reactions of this type exist which show an angular distribution which is symmetric about 90, as is the case for compound reactions. This fact suggests that the random phase hypothesis may be applied here as well and that interference effects may largely be discarded.

In pioneering papers J.J. Griffin (1966), (1967) suggested describing such reactions as a series of subsequent transitions between different classes of sets of configurations. As one basic assumption, the principle of *equiprobability* both for the population of individual configurations within one class and with respect to the transition from one class to the next was applied. In one way or another, assumptions of such type also underlie the more formal theories developed after Griffin's papers had appeared and which will be addressed below. It turns out that the handling of decay modes or the calculation of transition rates are not very different from those of the statistical model described in Section 18.4. As we shall see, basic relations used there can simply be taken over. The essential difference can be seen in the configurations between which transitions take place or from which decay into the continuum is seen. These configurations are those through which the system passes before some initial excitation has "relaxed" to equilibrium. There exists a hierarchy of such configurations from simple to more and more complex structure. In this sense pre-compound processes are intimately related to equilibration and may thus give hints on relaxation times. Superimposed upon these "internal" processes are the decays to the outside. The relevant timescales are commonly expressed in terms of two widths, the $\Gamma^\downarrow$ and $\Gamma^\uparrow$, respectively. Whereas the former is an (average) measure for the coupling of simpler configurations to the more complex ones, the latter just describes the (average) width for the decay into the continuum. One basic element in theoretical approaches is the evaluation of such quantities. Experimentally, the relaxation process is accessible only indirectly. The emitted particles may originate either from the pre-equilibrium stages or from genuine compound states.

Unfortunately, the relaxation times one may get out of such experiments cannot be taken over without great reservations to those one needs to know for transport models for slow collective degrees of freedom. As explained in previous chapters, for pragmatic reasons one needs to assume fast intrinsic relaxation at given shape. The relaxation of the latter is described by the appropriate transport equation: there exist two typical relaxation times, the τ for the nucleonic degrees of freedom and those for collective motion, named τ_{kin} and τ_{coll} in Section 6.6.3. It must be said that practically all information known from the typical pre-equilibrium models only delivers hints at τ but not at collective motion. Moreover, as these models do not take into account shape degrees explicitly no shape dependence of the $\Gamma^{\downarrow}$ is known.

In the spirit of these comments one may say that any description of nuclear dynamics by transport theory falls in the realm of pre-equilibrium processes. For fissile nuclei a complete equilibrium of the total system may then exist at best around the global minimum of the potential landscape, and if so only in a restricted sense as a model assumption. Indeed, this picture delivers the background both of the Bohr–Wheeler formula and of the application of Kramers' model in the original sense, as discussed in Section 8.2 and Chapter 12. It is also for this more or less obvious reason that it should be worthwhile to elaborate on the experience one has gained on pre-equilibrium processes induced by reactions with light particles.

9.1 An illustrative, realistic prototype

The essence of the physics of pre-compound reactions can be understood within a relatively simple but intuitively clear model, which we take over from previous presentations but adapt to our purpose. It will allow us to understand basic physical pictures and concepts, including those behind the more fundamental and microscopic theories, in particular those of D. Agassi, H.A. Weidenmüller and G. Mantzouranis (AWM) (1975) and H. Feshbach, A.K. Kerman and S.E.Koonin (FKK) (1980). Partly this model has already been introduced in (Agassi et al., 1975) and developed further by K.W. McVoy and X.T. Tsang (1983) (see also (McVoy, 1985)) to explain similarities and differences of (AWM), (FKK) and the theory of W.A. Friedman et al. (1981).

Assuming that a statistical approach is adequate let us introduce the probability $\Pi_m(t)$ that states of class m are occupied at time t. Having r classes altogether, a master equation may be set up following basic elements of such an equation as described in Section 25.2.4. Denoting by P_{nm} the probability per unit of time for the transition from $n \to m$ one may write:

$$\dot{\Pi}_m(t) = -\Pi_m(t)\Gamma_m/\hbar + \sum_{n\neq m}^{r} \Pi_n(t)P_{nm}\,. \qquad (9.1)$$

The $\Gamma_m/\hbar$ is meant to represent the total rate for depletion of class m, including both transitions to other classes as well as into all possible open channels γ. Hence, one may write $\Gamma_m = \Gamma_m^{\uparrow} + \Gamma_m^{\downarrow}$, with $\Gamma_m^{\uparrow} = \sum_\gamma \Gamma_m^{\gamma}$ and $\Gamma_m^{\downarrow} = \hbar \sum_{n\neq m}^{r} P_{mn}$. (In the microscopic formulation of (Feshbach et al., 1980) these relations only hold true for weak coupling to the open channels.) The $\Gamma_m^{\uparrow}$ is the total escape width from class m into all open channels and the $\Gamma_m^{\downarrow}$ summarizes the widths for transitions from a specific class m into all other classes n. As all these widths correspond to the whole set of individual states (within a class) it is understood that they represent average quantities. To simplify the notation we leave out the bar over these symbols, unlike Chapter 18.4 where bars are used to distinguish average quantities from bare ones.

Before continuing further it is worth studying the situation that for some given m the escape widths can be neglected, which is to say $\Gamma_m^{\uparrow} \ll \Gamma_m^{\downarrow}$. Evidently, it is only in such a case that the nucleus may establish equilibrium at all. Indeed, under this condition the master equation (9.1) reduces to one which for

$$\Pi_m^{(\mathrm{eq})} \left(\sum_n^r P_{mn} \right) = \sum_n^r \Pi_n^{(\mathrm{eq})} P_{nm} \qquad (9.2)$$

has a stationary solution. This condition is indeed fulfilled if the transitions obey the relation of detailed balance

$$\Pi_m^{(\mathrm{eq})} P_{mn} = \Pi_n^{(\mathrm{eq})} P_{nm}, \qquad (9.3)$$

written here in the form (18.155). After introducing the relaxation matrix

$$\Gamma_{mn} = \delta_{nm}\Gamma_m - \hbar P_{nm}\,, \qquad (9.4)$$

the master equation (9.1) can be written in compact form as

$$\dot{\Pi}(t) = -\Pi(t)\Gamma/\hbar\,, \qquad (9.5)$$

with Π being understood as the row vector $\Pi \equiv (\Pi_1, \Pi_2 \ldots \Pi_r)$. The solution of (9.5) may be written as

$$\Pi(t) = \Pi(t=0)\exp\left(-\Gamma t/\hbar\right)\,. \qquad (9.6)$$

To establish contact with the statistical reaction theory it is suggested that we turn to a time-independent description. This may be performed by introducing appropriate time averages. A relevant timescale in each class is delivered by the

Poincaré recurrence time $\tau_{\text{rec}}^{(n)} = 2\pi\hbar/D_n$, with D_n being the mean spacing of levels in class n. Taking $\tau_{\text{rec}}^{(n)}$ as the relevant averaging interval we may introduce

$$\overline{\Pi_n} = \left(\tau_{\text{rec}}^{(n)}\right)^{-1} \int_0^{\tau_{\text{rec}}^{(n)}} dt\, \Pi_n(t) \approx \int_0^\infty dt\, \Pi_n(t)/\tau_{\text{rec}}^{(n)}. \tag{9.7}$$

The approximation on the right is valid if in each class the recurrence time is much larger than the largest decay time associated by (9.6) to the matrix $\hbar/\Gamma$. The integral over the $\Pi(t)$ of (9.6) is readily performed to give

$$\int_0^\infty dt\, \Pi(t) = \Pi(t=0)\, \hbar\, \Gamma^{-1}. \tag{9.8}$$

The vector $\overline{\Pi}$ is obtained by multiplying this result from the right by the diagonal matrix $D/(2\pi\hbar)$ leading to

$$\overline{\Pi} = \Pi(t=0)\, \Gamma^{-1} D/(2\pi).$$

To apply this result for a description of a nuclear reaction we need a decent interpretation of the probability $\Pi_m(t=0)$ of the initial population of class m states out of the entrance channel α. This property is given by nothing else but the transmission coefficient $\mathcal{T}_m^\alpha$, which leads one to identify the former by the latter, $\Pi_m(t=0) = \mathcal{T}_m^\alpha$. Likewise, since every configuration is supposed to couple to the continuum, a transmission coefficient $\mathcal{T}_n^\beta$ to channel β may be associated with any class n state. The fluctuation cross section for a reaction from channel α to channel β may thus be written as

$$\sigma_{\alpha\beta}^{(\text{fl})} = \frac{\pi}{k_\alpha^2} \sum_n^r \overline{\Pi}_n \mathcal{T}_n^\beta = \frac{\pi}{k_\alpha^2} \sum_{mn}^r \mathcal{T}_m^\alpha \left[\Gamma^{-1}\frac{D}{2\pi}\right]_{mn} \mathcal{T}_n^\beta. \tag{9.9}$$

The pre-factor π/k_α^2 is anticipated because of the close relation to the statistical reaction theory developed in Section 18.4. Indeed we shall soon be able to connect to the Hauser–Feshbach version described in Section 18.4.3. In the microscopic formulations like those of (Agassi et al., 1975) or (Feshbach et al., 1980) this factor comes up because the wave functions are "normalized to energy" (see Section 18.2.6).

At this stage it may be worthwhile to briefly present a central result of the time-dependent formulation of (McVoy and Tsang, 1983). With our notation their time-dependent cross section may be written in the form (McVoy, 1985)

$$\sigma_{\alpha\beta}^{(\text{fl})}(t) = \frac{\pi}{k_\alpha^2} \sum_{mn} \mathcal{T}_m^\alpha \left[\left(1 - e^{-\Gamma t/\hbar}\right)\Gamma^{-1}\frac{D}{2\pi}\right]_{mn} \mathcal{T}_n^\beta. \tag{9.10}$$

For sufficiently large times it turns into the stationary result (9.9). Notice that in (9.10) only *time-independent decay constants* appear, but the additional factor accounts for the fact that the population of intermediate classes first grows in time. One must say, however, that the decay times are very short such that it is difficult to exploit this explicit time dependence experimentally(McVoy and Tsang, 1983).

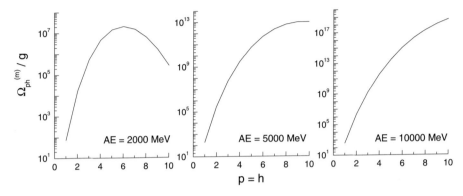

FIG. 9.1. The Ericson estimate (9.11) of the particle-hole level density as function of $p = h$ for different combinations of particle number and excitation energy.

Formula (9.9) is still too complicated to allow for easy and direct comparisons with cross sections of processes which pass through states of equilibrium. This may be achieved by applying two simplifying assumptions, which are also made in the FKK-theory:

(i) *Chaining hypothesis:* Transitions from $n \to m$ take place only for $n = m \pm 1$, which is to say the complexity of states changes only by one unit. This assumption is natural in a particle-hole picture when transitions are caused by two-body interactions.

(ii) *The never come back approximation:* Transitions from $n \to m$ may be discarded whenever $n > m$. Indeed, the transition probability P_{nm} is proportional to the level density $\Omega^{(m)}$ in class m, see (18.13). But this level density grows strongly with the complexity of the wave functions, implying $\Omega^{(n)} \gg \Omega^{(m)}$ for $n > m$ and hence $P_{nm} \simeq 0$. As a matter of fact, this statement is correct only for not too large n. Indeed, the assumption $P_{nm} \simeq 0$ for $n > m$ must break down close to the statistical equilibrium where according to (9.3) the transition from $n+1$ to n becomes equally likely. It turns out that this situation happens before the level density $\Omega_{\text{ph}}^{(n)}(E)$ starts to decline (as function of n).

The $\Omega_{\text{ph}}^{(n)}(E)$ may easily be calculated in a model where the class m is defined by the number $N_m = p+h$ of particle and hole states, assuming a constant single particle level density g. According to T. Ericson (1960) one gets (see exercise 1)

$$\Omega_{\text{ph}}^{(m)}(E) = \frac{g(gE)^{N_m-1}}{p!h!(N_m-1)!}. \qquad (9.11)$$

The dependence of this function on $N_m = p + h$ is demonstrated in Fig. 9.1 for the case when $p = h$, with the single particle level density being that of the Fermi gas, $g = eA/(2e_F)$ for $e_F = 40$ MeV. A critical discussion of (9.11) may be found in (Blann, 1975) and (Gadioli and Hodgson, 1992)), where for equidistant levels a more correct formula is presented. It is seen that (9.11) adequately describes the rising part of the curve but overestimates the density around and beyond its maximum. To large extent this is due to the fact that (9.11) does not treat correctly constraints imposed by the Pauli principle.

Let us continue applying this model. The two assumptions made imply that the $\Gamma_n^\downarrow$ is simply given by $\Gamma_n^\downarrow = \hbar P_{n,n+1}$, such that the matrix Γ of (9.4) becomes

$$\Gamma = \begin{pmatrix} \Gamma_1 & -\Gamma_1^\downarrow & 0 & 0 & 0 & . & . & 0 \\ 0 & \Gamma_2 & -\Gamma_2^\downarrow & 0 & 0 & . & . & 0 \\ 0 & 0 & \Gamma_3 & -\Gamma_3^\downarrow & 0 & . & . & 0 \\ . & . & . & . & . & . & . & . \\ . & . & . & . & . & \Gamma_{r-1} & \Gamma_{r-1}^\downarrow & . \\ . & . & . & . & . & . & \Gamma_r \end{pmatrix} \qquad (9.12)$$

Let us assume further that the entrance channel couples only to the first stage, namely to class $m = 1$. For the cross section (9.9) therefore only the matrix elements $\left[\Gamma^{-1}D\right]_{1n}$ are needed, which for the diagonal D is simply given by $\Gamma_{1n}^{-1}D_n$. From (9.12) one gets

$$\Gamma_{1n}^{-1} = \frac{1}{\Gamma_n} \prod_{k=1}^{n-1} \frac{\Gamma_k^\downarrow}{\Gamma_k}. \qquad (9.13)$$

Applying the well-known relation $\mathcal{T}_\alpha = 2\pi \overline{\Gamma_\alpha}/D$ (see (18.157)) the transmission coefficients needed in (9.9) can be written as

$$\mathcal{T}_1^\alpha = \frac{2\pi \Gamma_1^\alpha}{D_1} \quad \text{and} \quad \mathcal{T}_n^\beta = \frac{2\pi \Gamma_n^\beta}{D_n}. \qquad (9.14)$$

with associated widths Γ_1^α and Γ_n^β. The cross section thus becomes

$$\sigma_{\alpha\beta}^{(\text{fl})} = \frac{\pi}{k_\alpha^2} \frac{2\pi \Gamma_1^\alpha}{D_1} \sum_{n=1}^{r} \left(\prod_{k=1}^{n-1} \frac{\Gamma_k^\downarrow}{\Gamma_k}\right) \frac{\Gamma_n^\beta}{\Gamma_n}. \qquad (9.15)$$

So far the final stage r has not yet been specified, but we may identify it with the compound state associated with equilibrium. Then the only contribution to the diagonal term Γ_r of the master equation (9.1) will be the

$$\Gamma_r = \Gamma_r^\uparrow = \sum_\gamma \Gamma_r^\gamma. \qquad (9.16)$$

This is in accord with relation (9.2), as close to equilibrium detailed balance will hold true implying that the $\Gamma_m^\downarrow$ no longer affect the time development. The cross section of (9.15) may thus be split up as

$$\sigma_{\alpha\beta}^{(\text{fl})} = \sigma_{\alpha\beta}^{(\text{PE})} + \sigma_{\alpha\beta}^{(\text{EQ})}, \qquad \text{where the} \qquad \sigma_{\alpha\beta}^{(\text{EQ})} = \sigma_{\alpha\beta}^{(\text{r})} \qquad (9.17)$$

belongs to the equilibrated system of the compound nucleus although it is *not identical* to the compound cross section in strict sense. From (9.15) it is seen to be given by

$$\sigma_{\alpha\beta}^{(\text{r})} = \frac{\pi}{k_\alpha^2} \frac{2\pi \Gamma_1^\alpha}{D_1} \left(\prod_{k=1}^{r-1} \frac{\Gamma_k^\downarrow}{\Gamma_k} \right) \frac{\Gamma_r^\beta}{\Gamma_r}. \qquad (9.18)$$

To proceed further let us introduce two transmission factors, a $\mathcal{T}_\gamma = 2\pi\Gamma_r^\gamma/D_r$ for the coupling of the final stage r to the open channels γ and another one for the entrance channel to this class r,

$$\mathcal{T}_\alpha^{(\text{r})} = \frac{2\pi\Gamma_1^\alpha}{D_1} \prod_{k=1}^{r-1} \frac{\Gamma_k^\downarrow}{\Gamma_k}. \qquad (9.19)$$

After identifying the $\mathcal{T}_n^\beta$ of (9.14) as $\mathcal{T}_\beta$, the $\sigma_{\alpha\beta}^{(\text{r})}$ of (9.18) takes on the form of the Hauser–Feshbach theory,

$$\sigma_{\alpha\beta}^{(\text{r})} = \frac{\pi}{k_\alpha^2} \frac{\mathcal{T}_\alpha^{(\text{r})} \mathcal{T}_\beta}{\sum_\gamma \mathcal{T}_\gamma}. \qquad (9.20)$$

However, its value differs from that of the Hauser–Feshbach expression (18.159) for compound reactions by a so-called *depletion factor* $\prod_{k=1}^{r-1} \Gamma_k^\downarrow/\Gamma_k$,

$$\sigma_{\alpha\beta}^{(\text{r})} = \sigma_{\alpha\beta}^{(\text{HF})} \prod_{k=1}^{r-1} \frac{\Gamma_k^\downarrow}{\Gamma_k}, \qquad (9.21)$$

by which the $\mathcal{T}_\alpha^{(\text{r})}$ differs from the $\mathcal{T}_\alpha \equiv \mathcal{T}_1^\alpha$ in (18.159). This factor accounts for the loss of probability in the first $r - 1$ stages due to escape into the continuum. Obviously, the result of the cross section for the genuine compound reaction is obtained for $r = 1$, which is to say if the equilibrium is already given in the first stage. Only then is Bohr's independence hypothesis fulfilled, which is described in Section 18.4.2. In the general case this hypothesis is violated not only because of the presence of this depletion factor being different from unity. It also depends on the initial stage and hence on the nature of the entrance channel.

This essential difference can also be seen by rewriting (9.18) in the form

$$\sigma_{\alpha\beta}^{(\text{r})} = \sigma_\alpha^{(\text{r})} \frac{\Gamma_r^\beta}{\Gamma_r} \qquad \text{with} \qquad \sigma_\alpha^{(\text{r})} = \frac{\pi}{k_\alpha^2} \frac{2\pi\Gamma_1^\alpha}{D_1} \left(\prod_{k=1}^{r-1} \frac{\Gamma_k^\downarrow}{\Gamma_k} \right). \qquad (9.22)$$

which is analogous to (18.149). Whereas the "formation cross section" $\sigma_\alpha^{(\text{r})}$ depends on the initial stage, this is not true for genuine compound reactions where according to (18.147) the corresponding $\overline{\sigma_\alpha^{(C)}}$ can be expressed in terms of the level density $1/D_C$ of the compound nucleus and the average decay width $\overline{\Gamma_\alpha}$. On the other hand, if stage r represents an equilibrated system its decay is still given by the branching ratio Γ_r^β/Γ_r which according to (9.16) has the same meaning as in (18.150) for the genuine compound reaction.

9.2 A sketch of existing theories

In this section we shall make connections with some of the models and microscopic theories which have been developed in the past. We shall concentrate on those aspects which are of greater interest with respect to the main topics of our book, without aiming at completeness or any greater rigor.

In the Feshbach–Kerman–Koonin theory (FKK) the form of the cross section (9.15) is identical to that derived microscopically in (Feshbach et al., 1980) for the so-called *multi-step compound reactions*;[19] its derivation can also be found in Section VII.5 of (Feshbach, 1992). The formal means used there is the projection technique discussed in Section 18.2.3 for compound reactions. The essential difference is that the Q-space is split up into r subspaces with projection operators $Q = \sum_n Q_n$. The Q_n are ordered in the hierarchy of the complexity of the associated wave functions, beginning at the "simple level" Q_1. The entrance channel is assumed only to feed this Q_1 class, as suggested above for the derivation of (9.15). On the other hand, at each stage m emission into possible exit channels γ is allowed and formally taken care of by a coupling via P space. The Γ_m^γ introduced above is given by

$$\Gamma_m^\gamma = 2\pi \overline{\left| \langle \chi_\gamma^{(-)} | \hat{P}\hat{H}\hat{Q}_m | \psi_{mq} \rangle \right|^2}, \qquad (9.23)$$

where averaging is understood over the states ψ_{mq} in Q_m space. This form may be interpreted as a generalization of that for the escape width found in (18.122) for the doorway mechanism. Indeed, this analogy is not accidental, as in a sense every one of the Q_m spaces acts as a doorway to the next hierarchy. The chaining hypothesis by which the interaction is assumed to change the complexity of the states by at most one unit is considered a generalization of the doorway hypothesis. The *spreading widths* are found to be

$$\Gamma_m^\downarrow = \frac{2\pi}{D_{m+1}} \overline{\left| \langle \widetilde{\psi}_{(m+1)p} | \hat{Q}_{(m+1)} \hat{H} \hat{Q}_m | \psi_{mq} \rangle \right|^2}, \qquad (9.24)$$

with averages over states in the $(m+1)$-st and m-th subspaces.

In the theory of D. Agassi, H.A. Weidenmüller and G. Mantzouranis (AWM) the Random Matrix Model is used to specify the two-body interaction $\hat{V}$ which couples different classes of states. As they avoid the chaining hypothesis the

---

[19] We concentrate on discussing this case and leave out the multi-step direct reactions which involve "fast" processes without any relation to equilibrium.

general form of the cross section is given by (9.9) with (9.15) being a special case. In the matrix Γ the non-diagonal terms do not vanish but decrease in magnitude with their distance from the diagonal. The reason is the same as that which above led us to the (more stringent) never come back approximation. The doorway states appearing in each stage are introduced as eigenstates of the model Hamiltonian $\hat{H}_0$, which together with $\hat{V}$ makes up the total Hamiltonian $\hat{H} = \hat{H}_0 + \hat{V}$. In this context we may mention the theory of W.A. Friedman, M.S. Hussein, K.W. McVoy and P.A. Mello (1981). There formulas are derived which are identical to those of AWM with the difference that the relevant matrix elements are those of the full Hamiltonian $\hat{H}$ (for a comparison see also (McVoy, 1985)). In (1980) W.A. Friedman had pointed out that these approaches are essentially based on the assumption of having Markovian processes, which for a time-independent picture is to be understood in the sense "that at each stage the next step is governed by probability-conserving branching ratios independent of the previous history".

As indicated earlier there exist phenomenological models which have been in great use in practical applications. The transition rates are calculated within semi-classical pictures which in one way or another apply the concept of detailed balance or *equiprobability*. Unlike the statistical model outlined in Chapter 8 one studies the decay from "exciton-like" configurations rather than those of the compound nucleus. Following Griffin (1966) an exciton is defined by the number $m = p + h$ of particles and holes. Here we shall just sketch the basic ideas of Griffin's original work and refer to the literature mentioned in the introduction. In this exciton model the probability density for having a single particle p in a state of energy ϵ_p when the composite nucleus of energy E is in a $m = p + h$ state may be written as

$$w_{\rm p}(p, h, E, \epsilon_p) = \frac{\Omega^{(m-1)}_{\rm p\text{-}1\,h}(E - \epsilon_p) g(\epsilon_p)}{\Omega^{(m)}_{\rm ph}(E)}, \qquad (9.25)$$

which is nothing else but an application of formulas like (10.20) to this special situation. The $g(\epsilon_p)$ is the single particle level density and $\Omega^{(m)}_{\rm ph}(E)$ may be calculated from (9.11). Evidently, this formula reflects the *principle of equiprobability* for states of a given class m. Likewise, the decay rate may be calculated by analogy with (8.7):

$$\Gamma^{\uparrow}_m = w_{\rm p}(p, h, E, \epsilon_p) \frac{\sigma_{\rm inv} v \Omega^{\mathcal{V}}_{\rm kin}}{g(\epsilon_p) \mathcal{V}}. \qquad (9.26)$$

Here, v is the velocity of the emitted particle, $\Omega^{\mathcal{V}}_{\rm kin}$ the level density of this particle if confined to a certain volume $\mathcal{V}$, where from (8.8) one has $\Omega^{\mathcal{V}}_{\rm kin}/\mathcal{V} = \mu^2 v/(2\pi^2\hbar^3)$. The $\sigma_{\rm inv}$ is the cross section for the inverse process. How the latter can be estimated, eventually by applying Fermi's golden rule can be found in the literature, see e.g. (Gadioli and Hodgson, 1992).

9.2.1 Comments

Finally we should like to comment on the fact that starting from a time-dependent description we ended up with formulas for the cross sections which are identical to those obtained in time-*in*dependent reaction theory. In our presentation this feature came up after introducing a time average in (9.7). The final expression for the cross section as given in (9.15) sums up contributions from all the stages which the system undergoes, and which in any semi-classical picture would be successive in time. Nevertheless, only the "stationary" widths appear without any reference to transient times which certainly exist for any class n which is first populated before it may decay. This feature is all the more remarkable as such transient times have been introduced when describing nuclear fission by way of Fokker–Planck equations, see Section 12.4.

It can be said that reactions of light particles with medium heavy nuclei have been analyzed successfully on the basis of the theories described above, in particular with FKK. There is no room for more details which actually may be found in (Gadioli and Hodgson, 1992), for instance, with reference to the original literature, see also (Feshbach, 1992). It should be sufficient to summarize the most important features, say for reactions with neutrons or protons of about $10-20$ MeV. One has been able to measure energy spectra of the emitted neutrons and protons, not only for single ones but also for pairs. Typically one sees a high energy component superimposed onto the evaporation spectrum. In a theoretical analysis one may then study the contributions from the different stages. As to be expected, particles with higher energies are those which are emitted first. The finally stage r is usually reached after a few steps only. Essentially, similar features are also seen in reactions induced by composite particles or heavy ions. The higher the energy per particle the higher the multiplicities of promptly emitted particles. One theoretical means of describing such reactions is given by the Boltzmann–Uehling–Uhlenbeck equation. At energies lower than about 10 MeV/A the evaporation mechanism is the dominant one. From this observation one may conclude that in this energy regime the assumption of a "local" equilibrium may be quite realistic. As described elsewhere the notion "local" here is to be defined with respect to the shape degrees of freedom. It is only the nucleonic degrees of freedom which equilibrate, the dynamics of the nucleus as a whole is one of "pre-equilibrium" and is to be described by transport theory.

Exercises

1. For constant level density g the probability that the energy of a particle lies between x and $x+dx$ simply is gdx, and likewise for a hole. Hence, the probability that the energy E is shared by $N_m = p+h$ particles and holes is given by:

$$\Omega_{\rm ph}^{(m)}(E) = \frac{1}{p!h!} \int_0^\infty dx_1 \cdots dx_p \int_0^\infty dy_1 \cdots dy_h \, g^{N_m} \, \delta\left(E - \sum_i^p x_i - \sum_i^h y_i\right).$$

From common integral representations of the δ-function (see Section (26.3)) show that this expression takes on the form:

$$\Omega_{\text{ph}}^{(m)}(E) = \frac{1}{2\pi} \frac{g^{N_m}}{p!h!} \lim_{\epsilon \to 0} \int_{-\infty}^{\infty} dk \, \frac{e^{ikE}}{(i(k-i\epsilon))^{N_m}}.$$

Prove (9.11) by applying the residue theorem.

10

LEVEL DENSITIES AND NUCLEAR THERMOMETRY

Knowledge of the level density of the nucleus as a many-body system is of greatest importance for the understanding of any phenomenon which involves statistical aspects. Theoretical approaches date back as early as the 1930s, but this problem has aroused the interest of both theoreticians and experimentalists up to the present time. Early reviews were given in (Bethe, 1937), (Ericson, 1960), (Gilbert and Cameron, 1965), (Bloch, 1968), (Huizenga and Moretto, 1972) and a comprehensive exposition can be found in chapter 2 of (Gadioli and Hodgson, 1992), which gives due account of the literature up to the beginning of the 1990s. In the following we will restrict ourselves elaborating on those aspects which relate more or less directly to the main topic of this book, the description of nuclear dynamics at finite thermal excitations. Moreover, we shall briefly present some of the modern theoretical methods.

10.1 Darwin–Fowler approach for theoretical models

Practically all theoretical treatments are based on the application of the Darwin–Fowler method. The latter makes use of the intimate connection between the level density and the partition function of generalized canonical ensembles. Basic features of this aspect are presented in Section 22.2, with a critical discussion of the use of the concept of temperature in nuclear physics in Section 22.3. In the following we want to apply the formula

$$\Omega_{\rm DF}(\mathcal{A}_0, \ldots \mathcal{A}_n) = \frac{\exp(\mathcal{S}(\lambda_0, \cdots \lambda_n))}{\sqrt{(2\pi)^n |\mathcal{D}(\lambda_0, \cdots \lambda_n)|}}$$

$$\mathcal{D} = \det\left\{\frac{\partial^2}{\partial \lambda_i \partial \lambda_j} \ln Z\right\} = (-)^n \det\left\{\frac{\partial \langle \hat{A}_i \rangle}{\partial \lambda_j}\right],$$

(10.1)

by which the level density $\Omega_{\rm DF}$ is expressed as a function of the set $\mathcal{A}_0, \mathcal{A}_1 \ldots \mathcal{A}_n$ of averages of the n observables $\hat{A}_i$. The quantity $\mathcal{S}$ is the thermal entropy which corresponds to the partition function $Z(\lambda_0, \cdots \lambda_n) = \text{tr} \exp(-\sum_i \lambda_i \hat{A}_i)$. We take over the notation of Section 22.1.2 in which the index $i = 0$ corresponds to energy, with $\lambda_0 = \beta$ and $\hat{A}_0 = \hat{H}$, and where for $i \geq 1$ it is convenient to extract the factor β in writing $\lambda_i = \beta a_i$. To find the functional dependence of the level density on the $\mathcal{A}_i$ one needs to express their conjugate variables λ_i in terms of the former. The equations for this inversion are delivered by the set (22.111) which determine the saddle point. In fact, these relations are identical to the set of equations (22.38) which fix the parameters λ_i of the general canonical ensemble in terms of the average thermal quantities $\mathcal{A}_i = \langle \hat{A}_i \rangle$. For this reason one is confronted with the same fundamental problem–which was recognized already by H.A. Bethe in his celebrated article in (1937)–namely that an application of canonical ensembles to small systems may lead to non-negligible errors, in particular at low excitations. This issue is outlined in sections 22.2.3 and 22.3 for the case of temperature and energy, with the size of the uncertainties estimated in Section 22.2.2. On the other hand, besides this question, (10.1) is completely general, in the sense that so far no further approximation is made on the evaluation of the thermal quantities themselves.

10.1.1 Level densities and Strutinsky renormalization

Commonly formula (10.1) is evaluated in the independent particle model. Very often one restricts oneself to the Fermi gas model (Ericson, 1960), (Bohr and Mottelson, 1969), (Siemens and Jensen, 1987), (Feshbach, 1992), where any shell structure is discarded. More involved approaches rely on numerical computations of the thermal quantities within particular single particle models for real nuclei, like that of the Woods–Saxon potential used in (Døssing and Jensen, 1974). In such procedures a basic quantity is the single particle level density $g(e)$, which may be split into its smooth and fluctuating parts, $g(e) = \overline{g}(e) + \delta g(e)$. Such an ansatz is in accord with both the Strutinsky method and with periodic orbit theory. In this way a clear separation becomes possible between the macroscopic limit related to $\overline{g}(e)$ and the contribution from shell effects represented by $\delta g(e)$. Unfortunately, the independent particle model alone is not very reliable for getting the smooth part $\overline{g}(e)$. Such questions have already been discussed in Section 5.3 in connection with the calculations of nuclear masses within the Strutinsky method. Of course, here one needs to consider its extension to finite thermal excitations described in Section 5.5.1.

Results for the macroscopic limit: Let us restrict ourselves to the usual grand canonical ensemble where besides energy $\mathcal{E}$ only particle number $\mathcal{A}$ is considered, which in (10.1) takes the part of A_1, with $\lambda_1 = -\beta\mu$. Assuming the $\overline{g}(e)$ to be given all thermal quantities necessary for (10.1) can be found in Section 22.5.1. To get the determinant one may use (22.148) for energy and (22.150) for particle number. In lowest order derivatives of $\overline{g}(e)$ may be ignored. In terms of the excitation energy $\overline{\mathcal{E}}^* = \overline{\mathcal{E}} - \overline{E}_0 = \overline{a}T^2$ introduced in Section 5.5 the determinant is found to be $\mathcal{D} = (12/\pi^2)(\overline{\mathcal{E}}^*)^2$. As the entropy is given $\overline{S}(T) = (\pi^2/3)\overline{g}(e_\text{F})T \equiv 2\sqrt{\overline{a}\overline{\mathcal{E}}^*}$ the level density becomes:

$$\Omega(\mathcal{A}, \overline{\mathcal{E}^*}) = \frac{1}{\overline{\mathcal{E}^*}\sqrt{48}} \exp\left(2\sqrt{\overline{a}\overline{\mathcal{E}^*}}\right). \qquad (10.2)$$

Within such an approximation it is *only* the single particle level density *at the chemical potential* (Fermi energy) which appears. For this reason it is no surprise that the result (10.2) is identical to the formula obtained by H.A. Bethe (1937) for a constant level density $\overline{g}(e) = g_0$.

Let us return to independent particle model. Often it is desirable to consider neutrons and protons separately. The total single particle level density then is given by the sum $\bar{g} = \bar{g}_n + \bar{g}_p$. In this case the determinant becomes $\mathcal{D} = (\pi^2/3)\bar{g}_p\bar{g}_n\bar{g}(\overline{\mathcal{E}^*}/a)^{5/2}$ but the entropy remains unchanged with still $a = (\pi^2/6)\bar{g}$ (see e.g. (Bohr and Mottelson, 1969), (Gadioli and Hodgson, 1992)). In the literature the resulting level density is often written in the simplified version one gets for $\bar{g}_p = \bar{g}_n = 2\bar{g}$, namely

$$\Omega(\mathcal{A}, \overline{\mathcal{E}^*}) = \frac{\sqrt{\pi}}{12} \frac{\exp\left(2\sqrt{a\overline{\mathcal{E}^*}}\right)}{\bar{a}^{1/4}(\overline{\mathcal{E}^*})^{5/4}} \qquad \text{for} \quad \mathcal{Z} = \mathcal{N} = 2\mathcal{A}. \qquad (10.3)$$

As it turns out, good fits to experimental evidences can be obtained if in this expression the following modification is applied

$$\overline{\mathcal{E}^*} \longrightarrow \overline{\mathcal{E}^*} - \Delta_{\mathcal{E}}. \qquad (10.4)$$

In the literature the expression obtained in this way is commonly referred to as the *back-shifted Bethe formula*. The $\Delta_{\mathcal{E}}$ is used as a fit parameter, which may turn out either positive or negative with the absolute value being of order of MeV, for more details see below.

Formally, expressions like (10.3) or (10.2) diverge when the excitation energy turns to zero. Comparisons with exact calculations performed in schematic single particle models (Ericson, 1960), (Ghosh and Jennings, 2001) show (10.2) to deviate from the correct one only at rather low excitation energies. However, it is clear from the very construction that in such regimes serious problems are to be expected. First of all, there is this question of fundamental nature mentioned before which concerns the applicability of canonical ensembles for small systems. Secondly, one should bear in mind that the Fermi gas model can be expected to work only at excitation energies where shell effects are washed out. To account for the latter one must consider the fluctuating part of the single particle level density.

Before we address this problem a few remarks are in order about the use of formulas like (10.2) or (10.3) in practical applications. Usually, one starts from such expressions and treats quantities like the *level density parameter* $\bar{a}$ as a free parameter which is to be fitted to empirical knowledge. To some extent this procedure is in accord with the spirit of the Strutinsky method–in the sense that the independent particle model ought to be trusted only in the neighborhood of the Fermi energy and that contributions from states which lie farther apart are to be taken from experiments. For the present problem the situation is manifestly different from that of the total static energy at zero thermal excitation. There one may rely on the liquid drop model to define the macroscopic limit. For the level density one needs valid information about the entropy and its *variation with energy* (and particle number). Unfortunately, direct information from experimental results is limited to small excitations. For this reason it is not possible to check the functional form $\overline{S} = \overline{S}(\overline{\mathcal{E}^*})$. On the other hand, there are strong hints both from experiment as well as from theory that the proportionality to $(\overline{\mathcal{E}^*})^{1/2}$ is not valid at larger excitations (even discarding shell corrections). In practice this feature is often accounted for by allowing the level density parameters $\overline{a}(\overline{\mathcal{E}^*})$ to vary with energy. One may then start from $\overline{S}(\overline{\mathcal{E}^*}) = 2(\overline{a}(\overline{\mathcal{E}^*})\,\overline{\mathcal{E}^*})^{1/2}$ (as actually done by Ignatuk's ansatz to account for shell corrections) and deduce from here other thermal quantities from the *differential form* of the first law, like e.g. the variation of the energy $\overline{\mathcal{E}^*}(T)$ with temperature. Evidently, one can no longer expect the form $\overline{\mathcal{E}^*} = \overline{a}T^2$ to hold true. Conversely, if a temperature-dependent parameter $\overline{a}(T)$ is introduced from knowledge of the excitation energy through the relation $\overline{\mathcal{E}^*}(T) = \overline{a}(T)T^2$ the entropy will not be given by $\overline{S} = 2\overline{a}(T)T$.

Inclusion of shell effects: The evaluation of $\Omega(\mathcal{A}, \mathcal{E}^*)$ from (10.1) can be traced back to that of the grand potential Ω_{gce}. Section 22.5.2 describes how one may get its fluctuating part $\delta\Omega_{\text{gce}}^{\text{ipm}}(T,\mu)$ from $\delta g(e)$. For the latter the general form (22.153) is used which includes that of the Gutzwiller trace formula. The result for $\delta\Omega_{\text{gce}}^{\text{ipm}}(T,\mu)$ is given in eqn(22.158). In (10.1) the entropy appears, whose fluctuating part $\delta\mathcal{S}_{\text{gce}}^{\text{ipm}}(T,\mu)$ is found in eqn(22.159). In this way one can quite easily get the level density Ω_{DF} as function of T and μ. To obtain the Ω_{DF} as function of energy and particle number requires further efforts. This inversion may be completed numerically, for instance. In (Leboeuf and Monastra, 2002) an analytic procedure has been developed by employing an expansion similar to that which led to (22.158); the resulting formula has recently been applied in (Leboeuf et al., 2005).

An approach of this type had already been suggested by S.K. Kataria, V.S. Ramamurthy and S.S. Kapoorin in (1978). There, an expansion like (22.153) for $\delta g(e)$ was assumed to represent a Fourier series. It was found that to very good approximation only one term in the sum is sufficient, in which case the calculation of the thermal quantities goes along the lines described in Section 5.5.2. Within this approach only a few parameters had to be fitted to experimental evidence to obtain a semi-empirical level density formula which has been successfully applied to a large range of nuclei from Ca to Pu. Other important effects, like pairing correction to the ground state energies, deformations as well as rotations were included as well.

A. Ignatyuk and collaborators (Ignatyuk et al., 1975) have developed another method by which shell effects can be accounted for. It is easier to handle but leads to results similar to those of Kataria et al. (Gadioli and Hodgson, 1992). Rather than first evaluating thermal quantities as function of temperature (and chemical potential) they are parameterized directly in terms of energy (and particle number). For entropy one starts with the ansatz

$$\mathcal{S}(\mathcal{E}^*) = 2\sqrt{a(\mathcal{E}^*)\,\mathcal{E}^*} \qquad (10.5)$$

and uses the following phenomenological ansatz for the level density parameter

$$a(\mathcal{E}^*) = \bar{a}\left(1 - \delta B \frac{1 - \exp(-\gamma_{\mathrm{IG}}\mathcal{E}^*)}{\mathcal{E}^*}\right). \tag{10.6}$$

It accounts for the shell correction and approaches the macroscopic limit $a(\mathcal{E}^*) \to \bar{a}$ for $\delta B/\mathcal{E}^* \ll 1$. The δB is the shell correction to the binding energy $B(N,Z) = Nm_n + Zm_p - M(N,Z)$, which is determined by subtracting from the experimental "mass defect" $B(N,Z)$ of (1.64) the mass $M_{\mathrm{LD}}(N,Z)$ (times c^2) of the liquid drop model. This δB corresponds to the negative shell correction energy $-\delta E$ at zero thermal excitation. The decrease of shell effects with increasing excitation energy is governed by the parameter γ_{IG}. Its value is chosen to be of the order of $\gamma_{\mathrm{IG}} \simeq 0.05 - 0.06$ MeV^{-1}. From (10.6) the following relation between δB and the asymptotic value of $\mathcal{S}(\mathcal{E}^*)$ may be deduced. At large $\mathcal{E}^*$ one has $S^2/4 \to \bar{a}(\mathcal{E}^* - \delta B)$. This implies that δB is equal to that value of $\mathcal{E}^*$ where the extrapolation of the asymptotic form of $S^2(\mathcal{E}^*)/4$ crosses the $\mathcal{E}^*$-axis. The correct behavior at small $\mathcal{E}^*$ is given by

$$a(\mathcal{E}^*) = \bar{a}\left(1 + \gamma_{\mathrm{IG}}\delta E\right), \tag{10.7}$$

if written in terms of the shell correction energy. It may be said that on the qualitative level, at least, the $a(\mathcal{E}^*)$ has the right behavior with the single particle level density $g(e_{\mathrm{F}})$ at the Fermi surface: $a(\mathcal{E}^*)$ gets smaller for negative values of δE and larger in the opposite case, recall the discussion in Section 5.3.

This method is widely used in experimental analyses. One should note, however, that although the form (10.5) reminds one of the simple Fermi gas model the relation between excitation energy and temperature is *not* given by $\mathcal{E}^* = aT^2$. This may simply be seen by applying (for constant particle number) the basic relation $1/T = dS/d\mathcal{E}^*$ which leads to

$$\frac{1}{T} = \frac{\partial \mathcal{S}}{\partial \mathcal{E}^*} = \frac{2}{\mathcal{S}}\left(a(\mathcal{E}^*) + \frac{\partial}{\partial \mathcal{E}^*}a(\mathcal{E}^*)\right). \tag{10.8}$$

In the region where shell effects are important the second term cannot be neglected implying a definite deviation from the conventional Fermi gas result.

Application to collective motion: In connection with large-scale collective motion one does not need the level density of the nucleus as a whole, but that for the system of intrinsic or nucleonic degrees of freedom. A prime example is the Bohr–Wheeler formula for the fission decay rate given in (8.14). There, the density of those levels which the fissioning system encounters appears on top of the fission barrier. This density may be considerably different from that of the entire mother nucleus. A minor problem here is that one is dealing with a system whose degrees of freedom are reduced by the few collective ones. Much more important is the fact that the single particle level density has a marked dependence on deformation. This feature will modify not only the internal or free energy, an effect which is important for the overall energy balance, but the functional dependence of the partition function as well, and, hence, the dependence of the level density on temperature or energy. Evidently, such features are more important at lower temperatures. From a fundamental point of view they can be accounted for by the procedures described at the beginning of this section on the basis of an underlying single particle model. In the approach of J. Toke and W.J. Swiatecki (1981) mentioned in Section 5.5.1 a deformation-dependent level density parameter was derived within the (leptodermous) expansion in powers of $A^{-1/3}$. In (Hasse and Myers, 1988) this has been generalized to deformations at the saddle point.

10.1.2 Dependence on angular momentum

Let us look first at the Fermi gas model for which the modifications of the grand canonical ensemble (for $\mathcal{A}$ nucleons) are described in Section 22.5.3. The application to the basic formula (10.1) is more or less straightforward. Let M be the component of the total angular momentum along a symmetry axis. In the grand canonical ensemble its average can be written as the sum over the magnetic quantum numbers m_k of the individual nucleons weighted by the appropriate occupation numbers, $\mathcal{M}_{\rm ipm} = \sum_k n(e_k, m_k) m_k$. With the help of formulas like (22.160)-(22.162) the result (10.2) is easily generalized to

$$\Omega(\mathcal{A}, \overline{\mathcal{E}^*}, \overline{\mathcal{M}}) = \frac{1}{24} \left(\frac{6}{\pi}\right)^{1/2} \left(\frac{\overline{a}^3}{\mathcal{J}^{\rm eff}}\right)^{1/2} \frac{\exp\left(2\sqrt{\overline{a}\overline{\mathcal{E}}^*_M}\right)}{\left(\overline{a}\overline{\mathcal{E}}^*_M\right)^{5/4}}. \qquad (10.9)$$

where $a = (\pi^2/6)\overline{g}(\mu)$ and

$$\overline{\mathcal{E}}^*_M = \overline{\mathcal{E}}^* - \frac{\overline{\mathcal{M}}^2}{2\mathcal{J}^{\rm eff}}, \qquad (10.10)$$

$$\mathcal{J}^{\rm eff} = \overline{g}(\mu)\langle m^2 \rangle \equiv \int_{-\infty}^{\infty} \overline{g}(e, m)\, m^2\, dm. \qquad (10.11)$$

The $\overline{g}(e, m) = \overline{g}(e, -m)$ is the smooth component of the density of single particle states $g(e, m) = \sum_k \delta(e - e_k) \delta_{m, m_k}$ where besides energy the magnetic component m_k is also specified. The $\mathcal{J}^{\rm eff}$ may be interpreted as an effective rotational moment of inertia. It turns out to be given by the rigid body value $\mathcal{J}_{\rm rig}$, see §4-3 of (Bohr and Mottelson, 1975). The $\overline{\mathcal{E}}^*_M$ would then be the excitation energy on top of that energy stored in the particle's motion around the quantization axis. As we may recall from Section 6.7 this motion should not be mixed up, however, with a collective rotation.

For not too large angular momentum, formula (10.9) can be simplified by expanding the entropy in the exponent like

$$\mathcal{S}(\overline{\mathcal{E}^*}, \overline{\mathcal{M}}) = 2\sqrt{\overline{a}\overline{\mathcal{E}}^*_M} = \mathcal{S}(\overline{\mathcal{E}^*}, \overline{\mathcal{M}} = 0) - \frac{\overline{\mathcal{M}}^2}{2\sigma^2}, \quad \text{with} \quad \sigma^2 = \mathcal{J}^{\rm eff}\left(\frac{\overline{\mathcal{E}^*}}{\overline{a}}\right)^{1/2} \qquad (10.12)$$

being the *spin cutoff parameter*. Discarding further the weak $\overline{\mathcal{M}}$-dependence in the denominator one obtains

$$\Omega(\mathcal{A},\overline{\mathcal{E}^*},\overline{\mathcal{M}}) = \frac{1}{\sqrt{2\pi}\sigma}\Omega(\mathcal{A},\overline{\mathcal{E}^*})\exp\left(-\frac{\overline{\mathcal{M}}^2}{2\sigma^2}\right), \qquad (10.13)$$

where $\Omega(\mathcal{A},\overline{\mathcal{E}^*})$ is taken from (10.2), which itself is recovered after integrating over $\overline{\mathcal{M}}$, viz $\Omega(\mathcal{A},\overline{\mathcal{E}^*}) = \int_{-\infty}^{\infty} d\overline{\mathcal{M}}\, \Omega(\mathcal{A},\overline{\mathcal{E}^*},\overline{\mathcal{M}})$.

Rather than in the dependence of the level density on $\overline{\mathcal{M}}$ one is interested more in that on the total angular momentum $\overline{J}$. To a given $\overline{\mathcal{M}}$ all states with $\overline{J} \geq \overline{\mathcal{M}}$ contribute to $\Omega(\mathcal{A},\overline{\mathcal{E}^*},\overline{\mathcal{M}})$. The $\Omega(\mathcal{A},\overline{\mathcal{E}^*},\overline{\mathcal{M}})$ may thus be obtained from the difference

$$\Omega(\mathcal{A},\overline{\mathcal{E}^*},\overline{J}) = \left(\Omega(\mathcal{A},\overline{\mathcal{E}^*},\overline{\mathcal{M}}=\overline{J}) - \Omega(\mathcal{A},\overline{\mathcal{E}^*},\overline{\mathcal{M}}=\overline{J}+1)\right)$$
$$\approx -\left(\frac{\partial \Omega(\mathcal{A},\overline{\mathcal{E}^*},\overline{\mathcal{M}})}{\partial \overline{\mathcal{M}}}\right)_{\overline{\mathcal{M}}=\overline{J}+1/2}, \qquad (10.14)$$

applying a common semi-classical approximation for the angular momenta. In this way one obtains from (10.13) (Bloch, 1968)

$$\Omega(\mathcal{A},\overline{\mathcal{E}^*},\overline{J}) = \frac{2\overline{J}+1}{2\sigma\overline{a}^3\sqrt{2\pi}}\Omega(\mathcal{A},\overline{\mathcal{E}^*})\exp\left(-\frac{(\overline{J}+1/2)^2}{2\sigma^2}\right). \qquad (10.15)$$

Notice, please, that the pre-factor $2\overline{J}+1$ results from differentiation and should at this place not be interpreted as the usual degeneracy factor. The latter must be considered in addition if we want to sum or integrate over all $\overline{J}$ to get back to the $\Omega(\mathcal{A},\overline{\mathcal{E}^*})$ through

$$\sum_{\overline{J}=0}^{\infty}(2\overline{J}+1)\Omega(\mathcal{A},\overline{\mathcal{E}^*},\overline{J}) \approx \int_{-1/2}^{\infty} d\overline{J}(2\overline{J}+1)\Omega(\mathcal{A},\overline{\mathcal{E}^*},\overline{J}) = \Omega(\mathcal{A},\overline{\mathcal{E}^*}). \qquad (10.16)$$

10.1.3 Microscopic models with residual interactions

In the past two decades it has become possible to work out microscopic results for the partition function and thus for the level density beyond the mean field level. The residual interactions taken into account are commonly of separable nature, like that given in (3.13). Typically one restricts oneself to quadrupole–quadrupole (QQ) and pairing (PP) forces. In the following we shall briefly discuss two different methods.

The functional integral approach: The method of functional integrals is described in Chapter 24, with its application to many-body systems given in Section 24.4. This is done for a two-body interaction of the type $k\hat{F}\hat{F}/2$ specified in (3.16), which allows one to understand the essential features of this method. The discussion in Section 24.4 concentrates on the calculation of the partition function, but the application to the Darwin–Fowler approximation to the level density is more or less straightforward. Within both the "Static Path Approximation" (SPA) and the "Perturbed-SPA" (PSPA) discussed in Section 24.4 such calculations have been presented in (Puddu et al., 1991) (with PSPA being called RPA-SPA). These approximations have been tested within the *Lipkin model*, which allows for an exact solution. Later, in works by A. Ansari and his group interactions of more realistic nature have been employed. In (Agrawal and Ansari, 1998) calculations have been done for a quadrupole–quadrupole force (QQ) (for a typical medium heavy nucleus ^{110}Sn and with excitation energies $\mathcal{E}^*$ up to 45 MeV), within both the SPA and the PSPA, which in (Agrawal et al., 1999) were extended to include a typical pairing interaction of BCS-type (for ^{152}Sm and ^{160}Yb for $\mathcal{E}^* \leq 50$ MeV). Comparing their results with formulas like (10.13) an essential energy dependence was found for the parameters involved. Let us concentrate on the level density parameter $\bar{a}$, which in fact was obtained not by a direct comparison of the level densities. Rather, two versions were presented, one by employing the equation $\overline{\mathcal{E}^*}(T) = \bar{a}(T)T^2$ and the other one through the relation $\overline{S} = 2\bar{a}(T)T$ to entropy. Both versions exhibit some structure for $T \lesssim 0.5$ MeV, in particular when pairing correlations are considered. Above this the $K = A/\bar{a}$ increases monotonically to values of about $10-12$ at $T = 2$ MeV.

In (Canosa et al., 1994) spin-dependent level densities have been evaluated for ^{164}Er employing spin projections for two models, one corresponding to a deformed mean field and a second one treating a quadrupole–quadrupole interaction within the SPA. In both cases, no big influence of spin on the level density parameter a was observed (in calculations for J-values up to $40\hbar$). Within the mean-field approach the a was also found to be quite insensitive to the excitation energy (for $\mathcal{E}^*$ up to 100 MeV measured on top of the rotational energy); for K a value of $10-12$ was found. The K of the SPA, on the other hand, came out markedly smaller, about 6 for $\mathcal{E}^* \simeq 10$ MeV and about 10 for $\mathcal{E}^* \simeq 100$ MeV.

Typically, when compared to the SPA, all calculations show the PSPA to be superior, in particular at lower energies. This is expected as the essential difference between the PSPA to the SPA is that the former includes quantal fluctuations. For systems for which the free energy exhibits a barrier as function of the collective variable the PSPA breaks down at a critical temperature T_0 (see Section 24.4.3). In (Rummel and Ankerhold, 2002) and (Rummel and Hofmann, 2005b) two different methods have been developed to overcome this problem, with both being tested with the Lipkin model. Since in all these approaches the level density is calculated within the Darwin–Fowler approximation all of them suffer from the insufficiencies of this method at small energies, at which even the very definition of temperature becomes questionable. This is directly visible whenever the specific heat C is calculated. Typically, when plotted as a function of temperature, C is seen to have a peak (or a shoulder in some cases) around the temperature at which pairing correlations dissolve. It is most questionable, however, that such a structure is physically meaningful. As demonstrated in Section 22.3.2, for an isolated system the temperature can be defined only up to an uncertainty of $\Delta T/T = C^{1/2}$. It turns out that this uncertainty is of the same order as the width of the bump seen in the calculations (see e.g. (Rossignoli et al., 1998), (Liu and Alhassid, 2001)). Of course, such a problem does not exist for nuclei in a decent heat bath, as may be found in the interior of stars, for instance. We will return to this question below after having discussed empirical information about level densities.

The shell model Monte Carlo (SMMC) method: In the shell model Monte Carlo approach one assumes that a mean field Hamiltonian is given and adds separable interactions similar to the ones mentioned above for applications of SPA or PSPA to real nuclei. The essential difference is that here this Hamiltonian (or the residual interaction) is diagonalized numerically by exploiting Monte Carlo methods. For details we may refer to the review by S.E. Koonin, D.J. Dean and K. Langanke (1997). In (Nakada and Alhassid, 1997), (Alhassid et al., 2003) (with further references given therein) this method has been applied to calculating thermal properties and level densities. Because calculations with the SMMC require a finite and not too large configuration space, thermal properties can only be calculated when temperatures are limited to values below about 2 MeV. In (Alhassid et al., 2003) a method is described overcoming this problem, where in some way continuum states of the underlying single particle model were considered to model an enlarged configuration space. This approach is demonstrated for the two iron nuclei ^{56}Fe and ^{57}Fe and it is claimed to be adequate for temperatures $T \sim 1-4$ MeV. In this range the results were seen to follow well the back-shifted Bethe formula (10.3) with (10.4). A fit to the microscopically determined densities led to the following values of the two parameters involved: $\bar{a} \simeq 5.9$ MeV and $\Delta_\varepsilon \simeq 1.37$ MeV for ^{56}Fe and 6.05 MeV and -1.3 MeV (for $\bar{a} = A/K$ these values of $\bar{a}$ correspond to $K \simeq 10.5$). In (Alhassid et al., 2005) the model was extended to obtain the parameter σ in formula (10.15) for the spin distribution of levels. This requires a microscopic computation of the rotational moment of inertia at finite temperature which is to replace the $\mathcal{J}^{\text{eff}}$ of (10.11).

10.2 Empirical level densities

It would be well beyond the scope of the book to discuss in greater detail possible procedures to deduce valid information about level densities from experiments; an extensive overview may for instance be found in (Gadioli and Hodgson, 1992). We shall concentrate on two approaches published more recently which allow for a more or less direct deduction from known spectra. Naturally, such procedures are restricted to low excitations. Possible extensions to larger energies will be touched on in the next section.

The work of T. von Egidy et al.: Systematic studies have been performed by T. von Egidy and collaborators since the mid 1980s (1986, 1988) and about which

new results were published in (2005) for nuclei ranging from ^{18}F to ^{251}Cf. These studies are based on knowledge of the complete level schemes at low energies and the s-wave neutron resonance spacings at the neutron binding energies. Three different parameterizations have been used for the spin-independent factor of the level density:
(1) That of the back-shifted Fermi gas model (BSFG) given in (10.3).
(2) The same formula but with Ignatyuk's ansatz (10.6) for the level density parameter to account for shell effects explicitly.
(3) The so-called "constant temperature" model (CT) (Gilbert and Cameron, 1965) for which one has

$$\Omega(E) = \frac{1}{T} \exp\left(\frac{E - \Delta_\varepsilon}{T}\right) \quad \text{where} \quad \frac{1}{T} = \frac{\partial \ln \Omega(E)}{\partial E} \triangleq \text{const.} \quad (10.17)$$

The definition of the parameter T follows the introduction of temperature in a microcanonical ensemble. The basic assumption here is that $\ln \Omega(E)$ varies *linearly* with energy E. In fact, formula (10.17) is applied only at very small energies and has to be joined smoothly to other representations of Ω at higher energies. For this reason a construction of the canonical ensemble with this T has to be done with some caution.

In all three cases the spin dependence is parameterized by the formula which is obtained if the first line of (10.14) is used together with (10.13) (for $\overline{\mathcal{E}^*} = E$). For the spin cutoff factor σ a special ansatz is taken over from the literature which expresses σ in terms of particle number A, the level density parameter a and the back shift Δ_ε of the energy. Finally, for the γ_{IG} of (10.6) a value of $\gamma_{\text{IG}} = 0.06$ MeV^{-1} is assumed. Hence, in all three cases only two free quantities remained to be found, either the $\overline{a}$ (in our notation) or the T and the Δ_ε. As all of them must be expected to vary from nucleus to nucleus these quantities were not taken as fit parameters themselves. Rather, they were expressed in terms of particle number, the shell correction to the binding energy and a pairing correction by simple, physically motivated forms. In this way, seven parameters for each model (six in case of Ignatyuk's ansatz) were introduced which were then fixed by least square fits to all nuclei under study.
The results may be summarized as follows.
(a) Level density parameter: (i) For the Ignatyuk ansatz a practically linear dependence on particle number was found with only a small quadratic correction,

$$\overline{a} = A/K \quad \text{with} \quad K \simeq 7.87 + 18 \times 10^{-5} A. \quad (10.18)$$

(ii) These two terms were also obtained for the BSFG model. However, since here shell corrections are not accounted for explicitly a third term was introduced, which was assumed to be proportional to the shell correction energy. Although the coefficient in front of the latter was found to only be of the order of 5×10^{-3} this term modifies the $\overline{a}$ or K considerably.
(iii) The parameter T of the CT model turns out to be strongly correlated to the

$\bar{a}$ of the BSFG model. In (Bucurescu and von Egidy, 2005) a result was found which may be summarized by the simple rule $\mathcal{T} \approx 5\bar{a}^{-3/4}$.

(b) Back shift energy: The $\Delta_{\mathcal{E}}$ shows a very similar behavior for all three models. For even-even nuclei above $A \simeq 40$ its value lies between 1 MeV and practically zero (for the very heavy nuclei). For odd-odd nuclei $\Delta_{\mathcal{E}}$ is negative with the absolute value lying between about 2 MeV (for lighter nuclei) and about 1 MeV. The values for even-odd nuclei lie somewhat in between, with strong correlations to (deuteron) pairing.

Results of the Oslo group: Their method consists in measuring γ spectra of nuclei which are excited in neutron pick-up reactions with ^{3}He as projectile. It is assumed that the γ emission is delayed sufficiently long that the initially populated states may relax to true compound states. If this condition is fulfilled the distribution of the emitted photons reflects the level density of the genuine many-body states (Henden et al., 1995). It is claimed that in this way all compound states are found and that their density $\Omega(E)$ can be obtained up to $E = 6$ MeV, for the medium heavy nuclei ^{162}Dy, ^{166}Er and ^{172}Yb considered in (Melby et al., 1999). The $\Omega(E)$ obtained in this way is not completely smooth but shows fluctuations. They exhibit themselves more clearly in the logarithmic derivative which by $(\partial \ln \Omega(E)/\partial E) = 1/T(E)$ can be used to introduce the inverse temperature of a microcanonical ensemble. Between $E = 1$ and 5 MeV the $T(E)$ displayed in Fig. 2 show about three oscillations with a "period" of about $\Delta E \simeq 1$ MeV, seen best for ^{172}Yb and less clearly for the other two nuclei. The differences between the maximal and minimal values of T are smaller than $0.3 - 0.4$ MeV (with 0.4 MeV being the largest value for the ^{172}Yb case between $E = 1$ and 2 MeV).

This structure appears to be different from the results of von Egidy et al. reported above. It might of course come from an incomplete compilation of states. In any case, as far as the level density itself is concerned the result is very interesting and deserves further studies. Before one may accept the derived $T(E)$ as a real measure of the nuclear temperature, however, one must consider its inherent uncertainty, already mentioned above in connection with theoretical calculations. From the heat capacities calculated as $C(E) = (\partial E/\partial T)$ and shown in Fig. 4 of (Melby et al., 1999) the $\Delta T/T = C^{1/2}$ lies between $1/3$ at $E = 1$ MeV and $1/5$ at 5 MeV.

In a subsequent paper (Schiller et al., 2001), such structure anticipated for the heat capacity is used to demonstrate the quenching of pairing correlations for three Yb and Dy nuclei. Rather than from the definition used before (which is now referred to as the "microcanonical heat capacity"), the $C(T)$ is calculated within the canonical ensemble as the derivative of the average energy, $C(T) = (\partial \mathcal{E}/\partial T)$; the canonical partition function is defined in the usual way by inserting into a formula like (22.69) (for small enough temperatures) the measured level density. Between $T = 0$ and $T \simeq 0.7$ MeV the $C(T)$ shows an "S-shaped" behavior with the peak or shoulder lying around $T = 0.5$ MeV and having a width of about 0.2 MeV. In that region the value of C lies between 10 and 20 implying a relative uncertainty of about 0.3 or 0.2. For this reason it appears questionable whether the functional form of $C(T)$ can be taken as a clear signal for the pairing phase transition. A similar objection holds true for the results obtained more recently in (Kaneko and Hasegawa, 2005) for ^{184}W.

10.3 Nuclear thermometry

Let us now address the problem of how temperature may be measured experimentally. It should suffice to concentrate on more fundamental questions, to refer to the literature for further details, in particular for experimental techniques etc. An extensive and elaborate presentation of the methods used, together with their problems and shortcomings can be found in the review article by D.J. Morrissey, W. Beneson and W.A. Friedman (1994).

Let us begin by issuing a few general warnings. One ought to bear in mind that the definition of a nuclear temperature can only be done for intermediate stages of the evolution of the nuclear complex, and hence is restricted to a *certain length of time*. The reasons are as follows: (i) The very use of the temperature concept makes sense only at not too small excitations. Typically, they have to be larger than the threshold for particle emission. But each emitted particle takes away energy (and changes particle number as well as spin etc.) and thus modifies the value of T–provided the system reaches another equilibrium sufficiently fast. (ii) Whenever collective motion (of large scale) is involved, temperature refers to the intrinsic degrees of freedom and requires the applicability of the quasi-static picture. But even under such conditions the intrinsic excitation may vary both with shape (the collective degrees of freedom) and with time (because of dissipation). A "measured" temperature will not always reflect the excitation energy of the *whole* nucleus shared amongst all degrees of freedom. Often *only a subsystem* may be excited and be *in equilibrium with itself*. Here, the notion "subsystem" can be understood in several ways. For instance, it may refer to a *subsystem of nucleons*, like the more loosely bound ones in the valence shell. Even if *all nucleons participate* it may be such that only *simple configurations are involved*, say those of 1p-1h nature and not those of the compound nucleus. After all it takes some time before the system may relax to the "exact" many-p many-h states. Evidently, the problems behind these features are intimately related to the field of pre-equilibrium reactions discussed in Chapter 9. From the theoretical point of view it may be said that the value of T does not only depend on the choice of the (model) Hamiltonian but on other quantities $\hat{A}_i$ as well whose averages are fixed by subsidiary conditions. In this case one has to apply the generalized ensemble defined through (22.36). Then T is not simply determined by the average energy, but from a set of equations like (22.38) for *all* Lagrange multipliers λ_i.

Temperature from measured level densities Ω_{exp}: As mentioned earlier, experimental information on the level density can be obtained only up to energies

E of a few MeV. Let us assume the $\Omega_{\rm exp}(E)$ to be given within this range. One may then define first the microcanonical ensemble as described in Section 22.2. As a function of the variable ϵ its density distribution $\rho(\epsilon, E)$ would be given by the inverse of the number of states $\Delta G(E) \approx \Omega_{\rm exp}(E)\Delta$ found in the interval $E - \Delta \leq \epsilon \leq E$. For the definition of temperature in thermostatics one would commonly use one of the two possibilities, which are known to be equivalent in the thermostatic limit (and only then): Assuming the E to represent the *average energy* $\mathcal{E}$ of the associated *canonical* ensemble one may try to evaluate the partition function and deduce temperature from the relation $\mathcal{E} = -\partial \ln Z(\beta)/\partial \beta$. However, for the nuclear case this way is prohibited simply because $\Omega_{\rm exp}$ is not known for all energies which contribute to the integral for the partition function. Therefore, it is more appropriate to make use of the relation $1/T = \partial \ln \Omega_{\rm exp}(E)/\partial E$. This choice is based on the assumption that E defines the most likely energy of the canonical distribution which may be constructed from this T. A detailed discussion of the problems inherent to an application of these procedures to small systems is given in Section 22.3.1.

In nuclear physics a formula of the type $1/T = \partial \ln \Omega(E)/\partial E$ is often used in practice, but with a slightly but essentially different interpretation. For the energy dependence of the level density one uses (see e.g. (Siemens and Jensen, 1987), (Feshbach, 1992), (Morrissey et al., 1994)) the form given by the $\Omega_{\rm DF}(E)$ of the Darwin–Fowler method, say in equations (10.2) or (10.3). There a prefactor of the exponential appears which varies with energy and thus influences the value of temperature. In a sense this contradicts the very construction of the Darwin–Fowler result and its relation to thermostatics, as outlined in Section 22.2.3. Thermostatics only deals with *average* properties, but they are contained in the exponent of (10.1)–and only there! In the general form (10.1) it is the entropy expressed as function of the intensive variables λ_i. As demonstrated for the simple case of considering only the $\lambda_0 = \beta$, one may introduce the extensive variables as well, which in (10.2) is the excitation energy. The *pre-factor* of the Darwin–Fowler formulas, on the other hand, is determined by the *fluctuations* about the mean values. Take again the simple case that only energy and temperature are involved. The final result (22.104) simply reflects the Gaussian integral (22.103) over (quadratic) deviations from the mean value of β.

Of course, one may take the opposite point of view and take the function $\Omega_{\rm DF}(E)$ to represent the genuine $\Omega_{\rm exp}(E)$ after the (few) parameters having been *fixed by experimental evidences*. Interpreting this $\Omega_{\rm DF}(E)$ as being the level density of microcanonical ensembles one might proceed as described above using the formula $1/T_{\rm DF} = \partial \ln \Omega_{\rm DF}(E)/\partial E$. Nevertheless–in particular when used at larger energies where the individual levels cannot be counted anymore–it is evident that the $\Omega_{\rm DF}$ can at best be considered (an eventually crude) approximation to the correct level density.

Temperature from measured evaporation rates: The methods commonly used to determine nuclear temperatures experimentally are based (i) on the application of the Weisskopf formula to the emission of particles in a binary decay of a compound nucleus (derived in Section 8.1), and (ii) *on a logarithmic expansion of the level density of the daughter nucleus* (Blatt and Weisskopf, 1952). In the following we are going to present the essential features, referring once more to (Morrissey et al., 1994) for more details. Later we are going to present an argument why the approximation behind (ii) is not really necessary.

For a compound system of total energy E the evaporation rate can be read off from eqn(8.5). With Ω_C being the level density of the compound nucleus (the mother nucleus) and Ω_β that of the final states one has:

$$P_{C\beta}(E) = \frac{v_\beta \sigma_\beta^{(C)}}{\mathcal{V}} \frac{\Omega_\beta(E)}{\Omega_C(E)}. \tag{10.19}$$

The Ω_β depends on the observation of the final products. In Section 8.1.1 different possibilities were considered, with the aim of obtaining double differential transition rates. For the present purpose of introducing temperature it is appropriate first to refer to the procedure commonly used in the literature. The decay products define the binary system consisting of the daughter nucleus B and of the emitted fragment b, where both may carry energies E_B and E_b. Let us add to the latter the kinetic energy of relative motion E_{kin} to get $E_{\tilde{b}} = E_b + E_{\text{kin}}$, thus specifying the "subsystem" $\tilde{b}$, besides that of B. Let us assume the two fragments to be sufficiently far apart such that the interaction between both may be discarded. Then the total energy E simply splits up like $E = E_B + E_{\tilde{b}}$.[20]

To obtain the density of final states we may take over the results obtained in Section 22.2.1 for the microcanonical ensemble. Indeed, the energy bin dE_b may be considered to be sufficiently small such that one has $\Omega_\beta(E) \simeq \Omega_{\tilde{b}}(E_b) \times \Omega_B(E - E_b) \, dE_b$. This implies that the transition rate becomes proportional to the common ratio of level densities (mind the discussion in Section 8.1.1),

$$P_{C\beta}(E) \propto \frac{\Omega_\beta(E)}{\Omega_C(E)} \propto \frac{\Omega_{\tilde{b}}(E_{\tilde{b}}) \, \Omega_B(E - E_{\tilde{b}})}{\Omega_C(E)} \equiv w_{\tilde{b}}(E_{\tilde{b}}; E), \tag{10.20}$$

which is nothing else but the probability $w_{\tilde{b}}(E_{\tilde{b}}; E)$ to find the "system" $\tilde{b}$ at energy $E_{\tilde{b}}$. As explained in Section 22.2.1, this ratio of level densities allows one to introduce the concept of temperature. If the daughter nucleus may be considered sufficiently large, one may apply eqn(22.77) with $1/T = \partial \ln \Omega_B(\mathcal{E}_B)/\partial \mathcal{E}_B$ to get

$$\Omega_B(E - E_{\tilde{b}}) \simeq \Omega_B(\mathcal{E}_B) e^{\mathcal{E}_{\tilde{b}}/T} e^{-E_{\tilde{b}}/T}. \tag{10.21}$$

Here, $\mathcal{E}_B$ and $\mathcal{E}_{\tilde{b}}$ are meant to represent the average energies of the subsystems B and $\tilde{b}$, respectively, with the underlying hypothesis that the total energy E may adequately be replaced by $\mathcal{E}_B + \mathcal{E}_{\tilde{b}}$. Following (22.78) and (22.79) one may

---

[20]To simplify the notation we leave out the channel index β here.

then introduce for $\widetilde{b}$ the canonical ensemble with $E_{\widetilde{b}}$ being the variable for the (fluctuating) energy of system $\widetilde{b}$.

Two comments are in order here: (i) The introduction of temperature hinges on the logarithmic expansion of the level density $\Omega_B(E - E_{\widetilde{b}})$, (ii) the value of temperature deduced in this way reflects the thermal excitation of the daughter nucleus. In fact, these restrictions can be eased by making use of an observation which actually goes back to L. Szilard and was later used by B.B. Mandelbrot and L. Tisza to justify the expected form of the canonical ensemble for isolated systems; for details see Section 22.2.1. In (10.20) the probability $w_{\widetilde{b}}(E_{\widetilde{b}}; E)$ for $\widetilde{b}$ to have energy $E_{\widetilde{b}}$ is formulated for microcanonical ensembles. According to eqn(22.76), the same $w_{\widetilde{b}}(E_{\widetilde{b}}; E)$ may as well be evaluated for appropriately defined canonical ensembles. For the present case this relations reads:

$$w_{\widetilde{b}}(E_{\widetilde{b}}; E) = \frac{w_{\widetilde{b}}(E_{\widetilde{b}}; T) \, w_B(E - E_{\widetilde{b}}; T)}{w_C(E; T)}. \tag{10.22}$$

Once more, the $w_{\widetilde{b}}(E_{\widetilde{b}}; E)$ is the *probability in the microcanonical* ensemble. In contrast, the $w_X(X; T)$ on the right-hand side are probabilities calculated for three *canonical* ensembles having the *same temperature* T. As this is a correct equation this step *justifies the use of canonical ensembles without invoking any further approximation*. In fact, for this relation to hold true no heat bath is required. Another question is the determination of the value of T. This problem is subject to the uncertainties inherent in small systems which have been mentioned before. Of course, here one may take the freedom the fix it through the largest system, namely that of the mother nucleus.

The definition of subsystem $\widetilde{b}$ allows for a further specification. As it includes both the excitation of fragment b and the kinetic energy of relative motion one may employ similar arguments as before to write $\Omega_{\widetilde{b}}(E_{\widetilde{b}}) \propto \Omega_b(E_b) \times \Omega^V_{\text{kin}}(E_{\text{kin}})$ and, hence,

$$\frac{\Omega_\beta(E)}{\Omega_C(E)} \propto \frac{\Omega_b(E_b)\Omega^V_{\text{kin}}(E_{\text{kin}})\,\Omega_B(E_B)}{\Omega_C(E)} \propto \frac{\Omega_{Bb}(E_B + E_b)\Omega^V_{\text{kin}}(E_{\text{kin}})}{\Omega_C(E)}$$
$$= w_{\text{kin}}(E_{\text{kin}}; E) \equiv \frac{w_{\text{kin}}(E_{\text{kin}}; T) \, w_{B+b}(E_{B+b}; T)}{w_C(E; T)}. \tag{10.23}$$

Here, for the degrees of freedom residual to those of relative motion, the level density $\Omega_{Bb}(E_{B+b}) \propto \Omega_b(E_b) \times \Omega_B(E_B)$ for the combined system of the two fragments at energy $E_B + E_b$ was introduced. To obtain the second line the same arguments are employed which above led us from (10.20) to (10.22). In the very last expression the energy distribution $w_{\text{kin}}(E_{\text{kin}}; T)$ appears, which represents that of the system of relative motion at a given temperature. For the evaporation of light particles, say of neutrons, not to get in conflict with the Coulomb repulsion with the remaining fragment, it practically means having a gas of neutrons outside the (mother or daughter) nucleus at *thermal equilibrium represented by a canonical ensemble at the same temperature as the nucleus itself.* Such a picture is often used in the literature, actually without caring much about proper justification. Clearly, in a genuine sense such a gas can hardly be created in any experimental setup in a lab. To be in thermal equilibrium one would need to make sure that the evaporated neutrons do not disappear at infinity, for instance by installing reflecting walls. The message of formula (10.23) is that, nevertheless, (i) such a picture is valid, and (ii) it is largely in accord with the basic assumptions underlying a microcanonical description of the decaying process.

As a consequence, if the energy distribution of the emitted particles can be measured one is able to deduce the associated temperature, which by way of (10.23) is identical to the temperature of the excited compound nucleus. To get this energy distribution it may actually suffice to study the particles' spectrum, which is proportional to $\exp(-E_{\rm kin})/T$. In fact, such a measurement not only involves the emission products of *one* nucleus but from many nuclei of those of the whole target. In this sense it does not underlie the insufficiencies of a *small system*. For this reason the uncertainties may eventually be kept small.

The formulas presented above do not only involve the distribution of the kinetic energy but also the probabilities for populating different excited states (as reflected in the values of $E_{\rm B}$ and $E_{\rm b}$). This implies that information about temperature may also be obtained by trying to identify from which excited states particles are emitted and how many. For details we refer to the review by D.J. Morrissey *et al.* (1994) and the references to original literature given therein.

11

LARGE-SCALE COLLECTIVE MOTION AT FINITE THERMAL EXCITATIONS

In this chapter we want to discuss some general aspects of damped collective motion of a global nature. It will largely be based on the LHA described in chapters 6 and 7. As discussed at the beginning of Chapter 7, large-scale motion may be described in practice by applying the propagator method or the associated Langevin approach. Actually, all applications to fission or heavy-ion collisions are done on that basis, more so with the Langevin equation than with the Fokker–Planck equation. This simply has numerical reasons, essentially because repeated integrations over phase space volume (as in (7.2)) turn out to be very time consuming, see e.g. (Scheuter and Hofmann, 1983), (Scheuter et al., 1984). The characteristic feature of all these approaches is that one proceeds in small steps. This has very important consequences. In this way it is easiest to account for evaporation of light particles or γs as well as changes of the temperature. Moreover, one should not forget that the local Fokker–Planck equation developed in Chapter 7 even contains certain quantum effects. We shall begin with the question to what extent the LHA allows one to set up a global Fokker–Planck equation at constant temperature. Later in Section 11.2.2 we will return to the general case.

11.1 Global transport equations

11.1.1 Fokker–Planck equations

A common procedure for deriving transport equations for global motion is the Kramers–Moyal expansion described in Section 25.2.3. All one needs to know is the short-time propagator of the type $K(Q_2, P_2, t_1 + \tau; Q_1, P_1, t_1)$ for which moments in $Q_2 - Q_1$ and $P_2 - P_1$ can be defined (with respect to the distribution in the end variables Q_2 and P_2). This feature allows one to exploit the Gaussian propagator $K_G(Q, P, t; Q_0, P_0, t_0)$ of (7.6) derived in sections 7.1 and 7.3 within the locally harmonic approximation. It is true that this propagator is specified by second moments which are defined around the trajectory, which itself represents the dynamics of the first moments. However, it is formally simple, albeit somewhat tedious, to rewrite it in the form necessary for the Kramers–Moyal expansion. To this end one needs to solve the set of equations (7.9) and (7.17) to (7.19) to first order in τ. As the result one obtains the following Fokker–Planck equation (Bundschuh, 2004)

$$\frac{\partial}{\partial t} f(Q,P,t) = \Bigl[-\frac{\partial}{\partial Q}\frac{P}{M(Q)} + \frac{\partial}{\partial P}\frac{\partial V(Q)}{\partial Q} + \frac{\partial}{\partial P}\frac{\gamma(Q)}{M(Q)}P \qquad (11.1)$$
$$+ \frac{\partial^2}{\partial Q^2} D_{qq}(Q) + 2\frac{\partial^2}{\partial Q \partial P} D_{qp}(Q) + \frac{\partial^2}{\partial P^2} D_{pp}(Q) \Bigr] f(Q,P,t).$$

The important feature is that all coordinate-dependent transport coefficients, as well as the conservative force, appear to the right of the differential operators. This structure will be elucidated below for the simpler example of over-damped motion. This equation is of the general form (25.57); the factor 2 in front of the cross term comes about because of the symmetry of differential operators of second order, remember (25.58). In (11.1) the coefficients were assumed to be independent of momentum. This is in line with the microscopic justification of the transport coefficients and goes back to the fact that the mean field is assumed to show non-linear behavior only with respect to the collective coordinate Q.

Inspecting eqn(11.1) more closely one may realize that for undamped motion and no diffusion terms it does not necessarily have the structure of the Liouville equation. If a variable inertia the latter must not vary too strongly with Q. Then the following replacement

$$\Bigl[-\frac{\partial}{\partial Q}\frac{P}{M(Q)} + \frac{\partial}{\partial P}\frac{\partial V(Q)}{\partial Q} \Bigr] f(Q,P,t) \quad \Longrightarrow \quad \{\mathcal{H}, f(Q,P,t)\} \qquad (11.2)$$

is allowed, with Poisson brackets for the density and the effective Hamiltonian

$$\mathcal{H}(Q,P) = \frac{P^2}{2M(Q)} + V(Q) \quad \text{with} \quad V(Q) = \begin{cases} \mathcal{E}(Q,S) \\ \mathcal{F}(Q,T). \end{cases} \qquad (11.3)$$

This manipulation involves both terms with $\partial f(Q,P,t)/\partial Q$ as factor as well as such with $\partial f(Q,P,t)/\partial P$. Thus one should have

$$\bigl|(\partial \ln M(Q)/\partial Q) f(Q,P,t)\bigr| \ll \bigl|\partial f(Q,P,t)/\partial Q\bigr|$$

as well as

$$\bigl|[\partial(P^2/2M(Q)/\partial Q]\partial f(Q,P,t)/\partial P\bigr| \ll$$
$$\bigl|[(\partial/\partial P)(-K(Q) + P\gamma(Q)/M(Q))f(Q,P,t)]\bigr|$$

A similar problem has already been encountered in (7.11) for average motion. It is due to the fact that in any harmonic approximation terms of order two in momentum or velocity are ignored.

Notice that (11.1) is written with a finite D_{qq}. In fact, the common application of the Kramers–Moyal expansion requires the short-time propagator $K(Q_2, P_2, t_1 + \tau; Q_1, P_1, t_1)$ to be a decent transition probability. This is warranted only if the determinant $\Delta(t)$ of the fluctuations is positive, which according to (7.21) is so to lowest order in τ provided the determinant of the diffusion coefficients is positive definite. As outlined in Section 7.3 this is not the case if the diffusion coefficients are chosen according to the fluctuation dissipation theorem, where the D_{qq} vanishes. However, as demonstrated there, this problem exists only for macroscopically small times and the fluctuations are described well at larger times. It is probably safe to argue, therefore, that for a vanishing D_{qq} the form of the global transport equation should have the same structure as given by (11.1). Evidently, as may be deduced from (7.24), the problem disappears in the limit of a vanishing cross diffusion term D_{qp}, which is generally the case in the classical, high temperature limit and for weak damping even for quantal systems. Indeed, for the $D_{\mu\nu}$ of (7.28) one obtains the form

$$\frac{\partial}{\partial t} f(Q,P,t) = \{\mathcal{H}, f(Q,P,t)\} + \frac{\partial}{\partial P}\left(\frac{P}{M(Q)}\gamma(Q) + \gamma(Q)T\frac{\partial}{\partial P}\right) f(Q,P,t), \tag{11.4}$$

which may be considered a generalization of Kramers' equation to variable transport coefficients. It ensures that the continuity equation is valid in conventional form (25.17), $\partial n(Q,t)/\partial t + \partial j(Q,t)/\partial Q = 0$, with the same expressions for spatial and current densities as in (25.18), $n(Q,t) = \int dP\, f(Q,P,t)$ and $j(Q,t) = \int dP (P/M(Q)) f(Q,P,t)$. A possible stationary solution of (11.4) is that of thermal equilibrium defined by the Hamiltonian $\mathcal{H}$ of (11.3), namely

$$f_{\rm eq}(Q,P) = \frac{1}{Z}\exp(-\beta\mathcal{H}) \quad \text{with} \quad Z = \int dQ\, dP \exp(-\beta\mathcal{H}). \tag{11.5}$$

For applications of transport equations outside nuclear physics the inertia is usually treated as being *independent* of the variables. In the nuclear case, however, the M is known to show a marked dependence on the coordinates Q_μ (in the multidimensional case). For only one variable Q it is possible to transform equations like (11.4) to a new system of coordinate and momentum for which $M(Q)$ turns into a constant m. This can be achieved by the transformation

$$(Q,P) \quad \leftrightarrow \quad (X, P_X) \quad \text{with} \quad \begin{cases} X = \int^Q \sqrt{\frac{M(Q')}{m_X}}\, dQ' \\ P_X = \sqrt{\frac{m_X}{M(Q)}}\, P \end{cases} \tag{11.6}$$

with respect to which the Hamiltonian $\mathcal{H}$ and the density $f(Q,P,t)$ are scalars and which leaves the Poisson brackets invariant (see exercise 1). Since the effective damping rate $\gamma(Q)/M(Q)$ must also be a scalar the Fokker–Planck equation (11.4) is form invariant.

Finally, we should like to comment on the extent to which a transport equation like (11.1) can be expected to contain quantum effects. That such are present is evident from the discussion of the harmonic cases in Section 7.4. Truth is, however, that for undamped motion and vanishing diffusion coefficients no quantum effects of a more global nature are present. This follows from the simple fact that the resulting equation is of Liouville form, without any explicit appearance of $\hbar$.

The corrections which arise from inharmonic forces in the von Neumann equation[21] are discussed in Section 21.1. Quantum effects for damped motion can be seen to exist already in steady state solutions, in particular in those of global equilibrium. It is difficult to demonstrate this feature analytically for the equations which involve momentum. For over-damped motion, on the other hand, such a proof is possible and will be presented below. As can be inferred from (11.9) the functional form of the density $n_{\rm eq}(Q)$ depends strongly on how the transport coefficients vary with Q. Later in Section 12.5.1 we are going to show how these quantum effects associated with the LHA lead to quantum corrections of Kramers' decay rate for metastable systems.

11.1.2 Over-damped motion

It is worthwhile to consider this case separately. First of all because nuclear collective motion may be strongly damped, for instance at higher temperatures. Secondly, as it involves only the coordinate we may use this simpler case to study a few important formal issues. Following the same line of arguments behind the Kramers–Moyal expansion one finds the following equation of global motion

$$\frac{\partial}{\partial t} n(Q,t) = \frac{\partial}{\partial Q}\left[\frac{1}{\gamma(Q)}\frac{\partial V(Q)}{\partial Q} + \frac{\partial}{\partial Q} D^{\rm ovd}_{qq}(Q)\right] n(Q,t). \qquad (11.7)$$

Here the basic ingredients are eqn(7.25) for average motion and (7.26) for the fluctuation. In this case a problem similar to that encountered in the previous section does not show up. Both the $\sum_{qq}(t)$ and the $D^{\rm ovd}_{qq}(Q)$ are positive definite; for negative local stiffness this is true as long as the temperature is above its critical value T_c discussed in Section 7.3.1. As before, all coordinate-dependent functions stand to the right of the differential operators. It is not difficult to convince oneself of the correctness of this statement (see exercise 2). Before we turn to the classical limit let us briefly touch upon the quantum aspects contained in (11.7).

As may be inferred from Fig. 7.5 together with (7.27), for large damping not only does the critical T_c become small, the deviation from the classical limit is also quite small. In this sense, for strongly damped motion, quantum effects are practically indistinguishable from the classical result already at low temperatures. This concurs with findings in (Ankerhold et al., 2001) albeit there the partial differential operator defining the current has an additional term involving the third order derivative of the potential. As these authors start from the Caldeira–Leggett model their friction is necessarily independent of the coordinate. Considering this fact, the modification to be performed in (11.7) consists in the following replacement

---

[21]In a cumulant expansion such corrections would show up as inhomogeneity terms in the equations for those of third order (Scheuter, 1982)–in contrast to the statistical diffusion coefficients which are present in the equations for the variances already.

$$\frac{\partial V(Q)}{\partial Q} \implies \frac{\partial V(Q)}{\partial Q} + \frac{\lambda}{2}\frac{\partial^3 V(Q)}{\partial Q^3} \quad \text{with} \quad \lambda = \Sigma_{qq}^{\text{eq}} - \frac{T}{C}$$

with the Σ_{qq}^{eq} and the stiffness C being defined locally; remember (7.27). As can be seen from Fig. 7.3, for large damping the λ introduced here as the difference of the quantal equilibrium fluctuation from its high temperature limit becomes small. One must say, however, that the λ in (Ankerhold et al., 2001) is calculated through a different formula. Whereas our Σ_{qq}^{eq} is the one for over-damped motion in which no trace of the inertia is found anymore, in (Ankerhold et al., 2001) one starts from the full response function of the oscillator and looks for a different limit in which the inertia is still present. In this way the consistency with the FDT in relation to the response of the over-damped oscillator is lost.

At this stage it may be instructive to look at the equilibrium solution $n_{\text{eq}}(Q)$ of (11.7). Writing this equation in the form of the continuity equation one easily verifies that the current can be expressed as

$$j(Q,t) = -\left[\frac{1}{\gamma(Q)}\frac{\partial V(Q)}{\partial Q} + \frac{\partial}{\partial Q}D_{qq}^{\text{ovd}}(Q)\right]n(Q,t). \tag{11.8}$$

For global equilibrium the current vanishes, which for $n_{\text{eq}}(Q)$ implies a differential equation of first order. Its solution is of the form

$$n_{\text{eq}}(Q) = \frac{\mathcal{N}}{D_{qq}^{\text{ovd}}(Q)}\exp\left(-\int^Q \frac{1}{D_{qq}^{\text{ovd}}(Q)\gamma(Q)}\frac{\partial V(Q)}{\partial Q}dQ\right), \tag{11.9}$$

with $\mathcal{N}$ being a normalization constant. In the high temperature limit one has $D_{qq}^{\text{ovd}}(Q)\gamma(Q) = T$. Assuming T to be constant one thus gets the classical version

$$n_{\text{eq}}(Q) = \mathcal{N}\frac{\gamma(Q)}{T}\exp\left(-\frac{V(Q)}{T}\right), \tag{11.10}$$

with the corresponding transport equation

$$\frac{\partial}{\partial t}n(Q,t) = \frac{\partial}{\partial Q}\left(\frac{1}{\gamma(Q)}\frac{\partial V(Q)}{\partial Q} + \frac{\partial}{\partial Q}\frac{T}{\gamma(Q)}\right)n(Q,t). \tag{11.11}$$

It is seen that a variable diffusion term in the transport equation (11.7) leads to a rather complex structure of the equilibrium distribution. At lower temperatures where $D_{qq}^{\text{ovd}}(Q)\gamma(Q) \neq T$ this structure is strongly influenced by quantum features. In addition there are the shell effects in the potential $V(Q)$ itself (which for constant T would have to be identified with the free energy of the internal system). Notice, please, that even in the classical, high temperature limit the $n_{\text{eq}}(Q)$ depends on the friction coefficient, provided this varies with Q.[22] This

---

[22]Situations where friction appears in the equilibrium distribution are known in the literature and discussed in connection with general equations of the form (25.89); see e.g. the textbook by J. Honerkamp (1994) or the article by R. Graham in (Moss and McClintock, 1989).

feature is in clear distinction from the distribution (11.5) which corresponds to Kramers' equation. Trivially, an equation of the form (11.4) makes sense only if a Hamiltonian can be defined at all. This requires weak coupling of the subsystem of collective variables to the heat bath. A strong damping force indicates that this condition may not be fulfilled. It may be worthwhile at this stage to recall the situation encountered in Section 7.2 for a quantal system. There, the equilibrium fluctuation of an oscillator is influenced even by a constant friction force.

Equations of Smoluchowski type may be deduced from that of Kramers by an expansion with respect to inverse friction. For constant inertia the case of variable friction $\gamma(Q)$ is treated in Section 10.4.4 of (Risken, 1989). In leading order one obtains

$$\frac{\partial}{\partial t}n(Q,t) = \frac{\partial}{\partial Q}\left(\frac{1}{\gamma(Q)}\frac{\partial V(Q)}{\partial Q} + \frac{T}{\gamma(Q)}\frac{\partial}{\partial Q}\right)n(Q,t), \qquad (11.12)$$

with correction terms being of order $(1/\gamma)^3$. This latter statement allows one to observe the essential difference between (11.12) and (11.11), which becomes apparent from the relation

$$\frac{1}{\gamma(Q)}\frac{\partial}{\partial Q} - \frac{\partial}{\partial Q}\frac{1}{\gamma(Q)} = \frac{1}{\gamma(Q)}\frac{\partial \ln \gamma(Q)}{\partial Q}. \qquad (11.13)$$

Equation (11.12) turns into (11.11) if the potential is changed to $V + T\ln\gamma(Q)$. As this additional term would show up in the equation of motion for $\langle Q \rangle_t$ it represents a "noise induced force". The latter is negligible if $|\partial V(Q)/\partial Q| \gg |T\partial \ln\gamma(Q)/\partial Q|$. In any case, eqn(11.12) would have the common equilibrium distribution, as given by (11.10) but without the factor $\gamma(Q)$.

For variable inertia $M(Q)$ the inverse friction expansion leads to an equation similar to (11.12) but with the following modification in the potential (for constant temperature)

$$V(Q) \quad \longrightarrow \quad V(Q) + T\ln\left(M(Q)\right)^{-1/2}. \qquad (11.14)$$

This can be inferred by first transforming eqn(11.4) into one with constant inertia by employing (11.6). For the transport equation in (X, P_X) (see exercise 11.1) the equation for the reduced density $\tilde{n}(X,t)$ is of the form (11.12). This latter equation can be transformed back to the original coordinate system observing $\sqrt{M(Q)}\tilde{n}(X,t) = \sqrt{m_X}n(Q,t)$.

11.1.3 Langevin equations

Starting from a Fokker–Planck description one may develop the corresponding Langevin equations by exploiting the Kramers–Moyal expansion. General aspects of such a relation are described in Section 25.2.3.1. For Kramers' equation in the version of (11.4) it is easily proven (see exercise 3) that the equivalent Langevin equations are

$$\dot{Q} = \frac{P}{M(Q)}$$
$$\dot{P} = -\gamma(Q)\frac{P}{M(Q)} - \frac{\partial}{\partial Q}\left(\frac{P^2}{2M(Q)} + V(Q)\right) + \sqrt{\gamma(Q)}L(t).$$
(11.15)

The $L(t)$ represents the stochastic force, which here has the properties of *white noise*: The ensemble average vanishes, $\langle L(t)\rangle = 0$, and the correlation function $\langle L(t)L(t+s)\rangle$ is sharply peaked at $s = 0$. As described in detail in Chapter 25, the value of $\langle L(t)L(t+s)\rangle$ is determined by the equilibrium fluctuations which in the high temperature limit implies $\langle L(t)L(t+s)\rangle = 2T\delta(s)$. In general the stochastic forces are vectors L_ν multiplied by a coordinate-dependent tensor $c_{\mu\nu}(Q,P)$. Kramers' case is special insofar as the only element $D_{PP}(Q)$ of this $c_{\mu\nu}$ depends on Q but not on P, which then implies the comparatively simple structure of eqns(11.15). This feature is no longer given for eqn(11.1) such that the associated Langevin equations would be more complex. Rather than examining this case further we shall concentrate on over-damped motion, where the same problem appears for any $\gamma = \gamma(Q)$.

In mathematical terms a Langevin equation belongs to the class of *stochastic differential equations*, which for *multiplicative noise* are ill defined. The notion multiplicative noise is used whenever the stochastic force varies with the coordinates. Questions of a more fundamental nature arising for such situations are discussed in Section 25.3.1 for the example of a one-dimensional system with equation (25.85), namely $\dot{Q} = a(Q) + c(Q)L(t)$. Different possibilities exist in handling such an equation. This has important consequences for the relation between the Fokker–Planck (or Smoluchowski) equations and the Langevin equations, see Section 25.2.3.1. Suppose we are given the Fokker–Planck equation of type (11.7) from which we want to find the corresponding Langevin equation; after all eqn(11.7) has some microscopic justification and includes the high temperature limit in the form of (11.11). Applying Itô's procedure one would find the following form of a Langevin equation:

$$\frac{dQ}{dt} = -\frac{1}{\gamma(Q)}\frac{\partial V(Q)}{\partial Q} + \sqrt{\frac{D_{qq}^{\text{ovd}}(Q)}{T}}\, L(t).$$
(11.16)

For Stratonovich's choice, on the other hand, one would get the form

$$\frac{dQ}{dt} = -\frac{1}{\gamma(Q)}\frac{\partial V(Q)}{\partial Q} - \frac{1}{2}\frac{\partial D_{qq}^{\text{ovd}}(Q)}{\partial Q} + \sqrt{\frac{D_{qq}^{\text{ovd}}(Q)}{T}}\, L(t).$$
(11.17)

Here, in addition to the force $-(1/\gamma(Q))(\partial V(Q)/\partial Q)$ another one appears whenever the diffusion coefficient depends on the coordinate, $-(1/2)(\partial D_{qq}^{\text{ovd}}(Q)/\partial Q)$.

Let us turn the argument around, assuming that a Langevin equation of the form (11.16) is given but with Stratonovich's interpretation of the stochastic force. In this case the corresponding Fokker–Planck equation would read

$$\frac{\partial}{\partial t}n(Q,t) = \frac{\partial}{\partial Q}\left(\frac{1}{\gamma(Q)}\frac{\partial V(Q)}{\partial Q} + \sqrt{D_{qq}^{\text{ovd}}(Q)}\frac{\partial}{\partial Q}\sqrt{D_{qq}^{\text{ovd}}(Q)}\right)n(Q,t). \quad (11.18)$$

Compared to equation (11.7) it contains the noise induced drift term

$$-\frac{1}{2}\frac{\partial D_{qq}^{\text{ovd}}(Q)}{\partial Q} \longrightarrow T\frac{\partial \ln\sqrt{\gamma(Q)}}{\partial Q} \quad \text{for} \quad D_{qq}^{\text{ovd}} = \frac{T}{\gamma(Q)}.$$

Hence, one may say that in the high temperature limit (11.18) is obtained from (11.11) if there the potential is modified to $V(Q) + T\ln\sqrt{\gamma(Q)}$. The corresponding equilibrium distribution $n_{\text{eq}}(Q)$ would be given by the form (11.10) but with $\gamma(Q)$ replaced by $\sqrt{\gamma(Q)}$.

From the physical point of view the Stratonovich method is to be preferred over that of Itô. Indeed, for physically realistic correlation functions the correlation time $\tau_{\text{micro}} > 0$ will be finite (instead of $\tau_{\text{micro}} = 0$ as for the delta function $\delta(s)$). Then the problem behind the stochastic differential equation is no longer present. As demonstrated in Section 25.3.1, its solution becomes identical to that obtained by the Stratonovich method. Having this in mind the important question is how the basic Langevin equation has to be established. Above, our arguments were based on the Fokker–Planck equation to start with, as the latter was obtained on a microscopic basis within the LHA.

The situation may be different in a phenomenological approach. Also this would have to be based on information about (i) the deterministic equation for the drift and the (ii) the fluctuating force. The first question to answer is whether or not there is a noise induced drift. An equation of Newtonian type, for instance, would not contain such a term. Secondly, for the nuclear system the stochastic force on the collective degrees of freedom is due to the (not precisely specified) dynamics of the nucleons. The problem at stake is reflected nicely in the warning given by van Kampen in Section IX.5 of his book (2001): *For internal noise (as would be given for the nuclear case) one cannot just postulate a nonlinear Langevin equation or a Fokker–Planck equation and then hope to determine its coefficients from macroscopic data.* Here, the notion "macroscopic" refers to a situation where (solely) the deterministic equation is fixed by comparison with experimental evidences (as sometimes done in nuclear physics) and where the fluctuating forces are put in thereafter.

11.1.4 *Probability distribution for collective variables*

For a time-dependent process the probability distribution $W(Q,t) \equiv n(Q,t)$ is determined by the (normalized) solution of the appropriate transport equation. Whereas for over-damped motion the $n(Q,t)$ is obtained directly, one first needs to integrate over the momentum in case the Fokker–Planck equation is the one for the density in phase space.

For many applications one only needs the distributions $W(Q)$ in equilibrium, in particular the one at given temperature. Often it is argued (see e.g. (Alhassid et al., 1987), (Alhassid, 1991)) that this $W(Q)$ is determined by the static free energy $\mathcal{F}_{\text{stat}}(Q,T)$ in combination with the corresponding partition function $Z_{\text{stat}}(T)$. Introducing some trivial normalization constant $\mathcal{N}$ to ensure $\int dQ\, W(Q) = 1$ one has

$$W(Q) = (\mathcal{N}/Z_{\text{stat}}(T))\exp\left(-\mathcal{F}_{\text{stat}}(Q,T)/T\right). \quad (11.19)$$

Evidently, this form contradicts our findings from above, where it was seen that the equilibrium may be influenced by transport coefficients, which is to say by quantities which reflect *dynamic* properties of the collective subsystem. For *strongly damped motion* the equilibrium distribution may involve information on *friction*, as given in (11.10) for instance. In the following we want to restrict the discussion to more conventional systems where the equilibrium for the collective degrees of freedom can be defined by a collective Hamiltonian of the type given in (11.3), $\mathcal{H} = P^2/2M(Q)) + \mathcal{F}_{\text{stat}}(Q,T)$. (It is assumed that the internal degrees of freedom can be treated within the *quasi-static* picture where the effective potential is specified by the free energy $\mathcal{F}_{\text{stat}}(Q,T)$.) The equilibrium distribution in phase space is then given by the $f_{\text{eq}}(Q,P)$ of (11.5), which for the $W(Q)$ implies

$$W(Q) \equiv n_{\text{eq}}(Q) = \int_{-\infty}^{\infty} dP\, f_{\text{eq}}(Q,P) = \frac{\sqrt{2\pi T M(Q)}}{Z(T)} \exp\left(-\frac{\mathcal{F}_{\text{stat}}(Q,T)}{T}\right), \quad (11.20)$$

where $Z(T) = \int dQ dP \exp(-\mathcal{H}/T)$ is the partition function for the *nucleus as a whole*, consisting of nucleonic *and* collective degrees of freedom. Notice, please, that for variable inertia the $\sqrt{M(Q)}\, dQ$ defines the proper volume element which is invariant under the coordinate transformation defined in (11.6). Evidently for *constant* $M(Q) = M$ one would find (11.19) with $\mathcal{N} = (2\pi T M)^{1/2}$ and $Z_{\text{stat}}(T) \equiv Z(T)$. For variable inertia $M(Q)$, on the other hand, the partition function has to be written as

$$Z(T) = \int dQ\, \sqrt{2\pi T M(Q)}\exp\left(-\mathcal{F}_{\text{stat}}(Q,T)/T\right). \quad (11.21)$$

Of course, the probability distribution $W(Q)$ of (11.20) can be brought into the form (11.19). One only has to replace $\mathcal{F}_{\text{stat}}(Q,T)$ by an *effective free energy* defined as $\mathcal{F}_{\text{eff}}(Q) = \mathcal{F}_{\text{stat}}(Q,T) + T\ln\left(M(Q)\right)^{1/2}$, recall (11.14).

11.2 Transport coefficients for large-scale motion

The Fokker–Planck equation (11.1) was obtained on the basis of the Kramers–Moyal expansion. The latter allows one to deduce an equation for global motion from information obtained through the local linearization procedure by which transport equations were derived in chapters 6 and 7. It is clear that such a scheme can work only if this implicit dependence on Q_0 is sufficiently weak. One needs to distinguish, however, the macroscopic point of view relevant for the Kramers–Moyal expansion itself from the microscopic formulation of linear response theory. Let us address this latter issue first. The conditions for linearization can be expressed in terms of response and correlation functions, say in their explicit dependence on frequency. Clearly only that range of ω is important which is relevant for collective motion. From a more pragmatic point of view it is easier to look at the transport coefficients themselves, which in fact represent essential properties of response functions. Let us symbolize them by a $\Xi = \Xi(Q)$. If this function is to be smooth on the scale δQ one may write

$$\left| \frac{1}{\Xi(Q)} \frac{\partial \Xi(Q)}{\partial Q} \delta Q \right| \ll 1. \tag{11.22}$$

For average motion the relevant δQ can be estimated as $\delta Q \simeq \dot{Q}\delta t$ where δt is that time interval over which we have to follow the dynamics microscopically. According to (7.1) this interval should fulfill the inequalities $\tau_{\text{intr}} \leq \delta t \ll \tau_{\text{coll}}$. In our case the Q is fluctuating with an associated width ΔQ^2, the precise definition of which depends on the problem one is treating. Evidently, the presence of such fluctuations implies a certain smoothing in Q. At first we shall ignore this phenomenon and treat Q as a c-number parameter which is "external" to the nucleonic degrees of freedom whose properties are characterized by response functions or by the transport coefficients of average motion, in particular. Since they depend on the thermal excitation of the internal system also the latter must not vary too much, which in practical applications is parameterized by temperature. The variation of T which results from collective motion cannot be stronger than that for the transport coefficients themselves. For these reasons we will restrict ourselves to discussing (11.22) applied to average motion.

Numerical calculations of transport coefficients show that condition (11.22) may only be badly fulfilled. Serious problems may occur in the neighborhood of crossings or avoided crossings. In the single particle model such events happen very frequently, as may for instance be seen from the example shown in Fig. 3.3. At genuine crossings the energies of two levels become exactly equal to each other whereas for quasi-crossings they only come close before they repel each other. Following (Landau *et al.*, 1977) this phenomenon can easily be understood for a model where all levels are discarded besides these special two. Suppose we know the solutions of the Schrödinger equation $\hat{\mathcal{H}}_0 \varphi_{1,2} = E^0_{1,2} \varphi_{1,2}$ for a general Hamiltonian $\hat{\mathcal{H}}_0 = \hat{\mathcal{H}}(\widetilde{Q}_0)$ taken at some point $\widetilde{Q}_0$ and where $\Delta E^0 = E^0_2 - E^0_1 > 0$. For the present purpose we may think of $\hat{\mathcal{H}}_0$ as identical with the $\hat{h}_0$ for a single particle, but the phenomenon we are talking about is of a general nature. To get the energies and eigenstates at some $Q = \widetilde{Q}_0 + \widetilde{q}$ we expand the $\hat{\mathcal{H}}(Q)$ to first order in $\widetilde{q}$ to get $\hat{\mathcal{H}}(Q) = \hat{\mathcal{H}}_0 + \widetilde{q}\hat{F}$. Fortunately, the two level problem can be solved exactly; one may recall the solutions found in Section 4.2.2. For the difference of the energies one gets $\Delta E(\widetilde{q}) = e_2(\widetilde{q}) - e_1(\widetilde{q}) = [(\Delta E^D(\widetilde{q}))^2 + 4F_{12}^2 \widetilde{q}^2]^{1/2}$. Here, the $E_i^D(\widetilde{q}) = E_i^0 + \langle \varphi_i | \hat{F} | \varphi_i \rangle \widetilde{q} \equiv E_i^0 + F_{ii}\widetilde{q}$ are the energies as they would come out in first order of the coupling, and the F_{12} is the matrix element of $\hat{F}$ between the two unperturbed states. Let us assume the slope F_{22} of $E_2^D(\widetilde{q})$ to be negative and the F_{11} to be positive. A *genuine crossing becomes possible only* for $F_{12} = 0$ and lies at $\widetilde{q}_c = -\Delta E^0/(F_{22} - F_{11}) > 0$. In the *opposite case of a finite coupling* $F_{12} \neq 0$ the levels cannot cross and the energy difference remains finite at and beyond the $\widetilde{q}_c$, where it is seen to be $\Delta E(\widetilde{q}_c) = 2|F_{12}\widetilde{q}_c|$. One may wish to refer to the $E_i^D(\widetilde{q})$ as the "diabatic" energies and call the perturbed levels the "adiabatic ones". The difference $\Delta E(\widetilde{q})$ of the actual energies can be written in the form $\Delta E^2(\widetilde{q}) = \Delta E^2(\widetilde{q}_{a.c.}) + (\Delta F^2 + 4F_{12}^2)(\widetilde{q} - \widetilde{q}_{a.c.})^2$. It is minimal at the position $\widetilde{q}_{a.c.}$ of the *avoided crossing* defined through the relation $\Delta E^D(\widetilde{q}_{a.c.})\Delta F + 4F_{12}^2\widetilde{q}_{a.c.} = 0$

Let us examine now how these features influence the transport coefficients, looking first at the static energy and turning to inertia and friction afterwards. In the first case only the single particle energies themselves are of interest. Without considering any residual interactions the static energy $E(Q)$ is discontinuous at level crossings such that neither the force nor the stiffness is well defined there. For the pure independent particle model this problem is demonstrated in Fig. 6.8 for the schematic case of the deformed oscillator. Even when $E(Q)$ is calculated with the Strutinsky method such discontinuities remain. Of course they drop out of the smooth part but they would definitely show up in the shell correction δE if no further measures were taken. This would imply that in $E(Q)$ additional structure would be seen on top of the one which is usually referred to as gross shell behavior. For the latter one would expect the $E(Q)$ to show just one or two local minima besides the one which at $T = 0$ corresponds to the ground state deformation. Evidently when temperature rises above the critical value beyond which shell effects become unimportant the free energy becomes smooth. To get smooth curves for lower temperatures as well one needs to perform additional smoothing. One possibility is to proceed as in Section 6.6.1.1 and use smoothed occupation number. One might also replace the $\delta g(e)$ of (5.17) by a $\delta g_\gamma(e) = g_\gamma(e) - \overline{g}(e)$ where the smoothing parameter γ is again chosen such that only gross shell structure survives. An alternative method is to apply the common Strutinsky codes and simply smooth the computed function $E(Q)$ in Q itself.

Let us turn now to the coefficients of inertia and friction, which are governed by the squared matrix elements of $\partial \hat{h}/\partial Q$ between two states and by the differences in their energies. Remember that here it is not the unperturbed values that are of interest but those calculated in the neighborhood of avoided crossings. One may convince oneself that the matrix elements are proportional to the F_{12}. Hence, no problems occur in the neighborhood of crossings where $F_{12} = 0$. Nor do greater problems appear for quasi-crossings as long as they show up sufficiently far away from the Fermi level. This may be seen with the help of the

$$\chi''(\omega) = \pi\hbar \sum_{jk} |F_{kj}|^2 \int_{-\infty}^{\infty} d\Omega \left(n(\hbar\Omega - \frac{\hbar\omega}{2}) - n(\hbar\Omega + \frac{\hbar\omega}{2}) \right) \overline{\varrho_k}(\Omega + \frac{\omega}{2})\overline{\varrho_j}(\Omega - \frac{\omega}{2})$$

(6.75)

for the dissipative part of the response function. Take one given pair k and j of levels. Their strengths $\overline{\varrho_k}$ and $\overline{\varrho_j}$ concentrate around the energies e_k and e_j. For small frequencies ω the difference of the occupation numbers takes on sizable values only if the e_k and e_j lie on different sides of the Fermi level, in which case these two levels may produce an especially large contribution to $\chi''(\omega)$. Evidently, this mechanism is the more effective the smaller the temperature. Examples from numerical computations can be found in (Ivanyuk et al., 1997). In this publication an example is shown where at some finite deformation and at $T = 1$ MeV for $\hbar\omega \lesssim 0.2$ MeV the $\chi''(\omega)$ increases strongly with ω. In the zero frequency limit this large slope leads to an extremely large friction coefficient of the same type as discussed in Section 6.6.3.1. Of course, here too the same arguments apply which in Section 6.6.3.2 were presented against such contributions to friction for vibrations about a potential minimum. In fact, for this example 99% of this large value of $_0\gamma(0)$ come from only *two* avoided crossings. This is a further, well-grounded hint against the application of the relaxation time approximation. Indeed, for just two excitation modes of the intrinsic system at such small frequencies the dressing of the single particle states described in Section 6.4.3 implies not much more than attributing to these excitation a finite and constant width. In (Ivanyuk and Hofmann, 1999) the same features have been studied with pairing correlations included, but essentially the same properties were found.

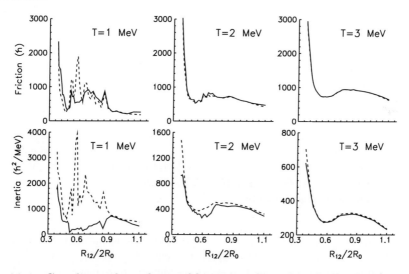

FIG. 11.1. Coordinate dependence of friction and inertia calculated in (Ivanyuk et al., 1997) for Cassini ovaloids and for ^{224}Th as a typical heavy nucleus: Dashed lines for zero frequency limits, fully drawn lines from fit to collective response.

In Fig. 11.1 the coordinate dependence of friction and inertia are shown for the model of Cassini ovaloids. As collective variable the distance R_{12} between the left and right centers of mass of a complex nucleus is used normalized to the diameter of the sphere of the same volume: $Q = R_{12}/(2R_0)$. The strong decrease with Q seen at smaller R_{12} may be related to the fact that this choice of Q may not be optimal at small distances. To get some feeling for the scale it may be mentioned that the sphere corresponds to $Q = 0.375$. There the liquid drop energy has its minimum and for ^{224}Th its maximum lies around 0.7. In the numerical evaluation contributions from all pairs of levels which lie closer in energy than their widths $\Gamma(\mu, T)$ calculated at the chemical potential were ignored, in accord with the suggestions made in Section 6.6.3.1. Whereas the dashed curves show the results of the zero frequency limit, the fully drawn lines correspond to the fit of the oscillator response (6.116) to the $\chi_{qq}(\omega)$ for collective motion, for details see Section 6.5.3.1. For temperatures above $T = 2$ MeV both versions practically coincide. At smaller values of T the zero frequency coefficients still show fluctuations but they are much larger than those of the fully drawn lines. The main reason for this feature may be seen in the fact that the fit involves properties of the response functions over a larger range of frequencies not just those around $\omega = 0$. In this way the somewhat irregular behavior around $\omega = 0$ is to some extent smoothed out.

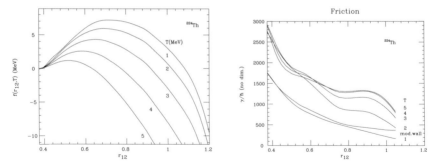

FIG. 11.2. Dependence of the free energy (left) and of friction (right) on the collective coordinate $Q = r_{12} = R_{12}/(2R_0)$ calculated for the two-center shell model (Yamaji et al., 1997).

Thinking of condition (11.22) it is clear that these fluctuations seen at smaller temperatures are unpleasant. Moreover, it is clear that it will be impossible to relate such fine structure in the transport coefficients to experimental evidences. Recall, please, that theoretical predictions must be made not solely on the basis of average motion. Rather, they must involve solutions of Fokker–Planck or Langevin equations. For nuclear fission, for instance, the relevant distribution in phase space will have a width in Q much larger than the typical small-scale variations seen in Fig. 11.1 for the transport coefficients. For this reason it is perhaps well justified to apply some (additional) smoothing in Q. Of course, such a procedure raises questions of more general or even fundamental nature.

Before we discuss such problems, let us demonstrate the usefulness of such a procedure. In Fig. 11.2 results are presented which were obtained in (Yamaji et al., 1997) with a two-center shell model. The free energy shown on the left was obtained through a Strutinsky calculation of the energy $E(Q)$ at zero temperature discarding pairing correlations. The resulting small scale variations were smoothed out by averaging over small intervals in Q. To get the free energy for five different temperatures the simplified expression (5.51) was employed for the shell correction. Then the response functions were computed at the following five points: $r_{12} = 0.375, 0.539, 0.725, 0.937, 1.048$. There, the transport coefficients were carefully deduced and afterwards their values were joined smoothly. The resulting friction coefficient is shown on the right of Fig. 11.2. The curve labelled "mod.wall" corresponds to the so-called "modified wall formula" $\gamma_{\mathrm{m.w.}} = 0.27\,\gamma_{\mathrm{w.f.}}$. In good agreement with the suggestions made in Section 6.6.3.3 it is seen that (i) the friction coefficient increases with T such that (ii) it saturates above $T \simeq 4 - 5$ MeV, and that (iii) $\gamma_{\mathrm{w.f.}}$ is still about 1.5 to 2 times larger than this high temperature limit. More important for the present context is the fact the coordinate dependence is very weak such that no contradiction to condition (11.22) is to be expected. In fact the same can be said about the inertia. In this context it may be of interest to note that also derivations of the wall formula for friction involve averaging procedures; details are presented in Section 17.1.1.

11.2.1 The LHA at level crossings and avoided crossings

The discussion above has shown that the *linearization on the microscopic level* may not always be justified. It is important to realize, however, that these problems are largely connected with the special features of the deformed shell model employed. There, residual interactions are commonly either ignored completely or treated only in a very restrictive way. Crossings of levels lead to cusps in the total static energy, which can easily be smoothed out, but do no harm for dynamical transport coefficients like friction and inertia. An example of this type is given by the pure oscillator model of Section 3.2.2.1. Avoided crossings show up as soon as spin-orbit interactions are taken into account, as in the Nilsson model, for instance. It appears that the density of avoided crossings is higher for single particle models which are based on the Woods–Saxon potential, like the one of (Ivanyuk et al., 1997) used for the results shown in Fig. 11.1. For practical reasons these models are commonly restricted to only a few shape degrees of freedom, in most cases to just one. On the other hand, the true fission path will lie in a complex multidimensional landscape for which the probability for two levels of the same symmetry to come close is very small. Recall that avoided crossing can occur only if the matrix element of $\partial \hat{h}(Q)/\partial Q$ between these two states is different from zero. This requires special conditions on their quantum numbers and/or symmetries.

As soon as residual interactions $\hat{V}_{\mathrm{res}}$ are taken into account the situation may change considerably–provided one allows this $\hat{V}_{\mathrm{res}}$ to influence both levels simultaneously rather than only to dress them individually. One important effect is that the two levels repel each other, which necessarily leads to a smaller absolute value of the curvature of the two levels. This is a general phenomenon and has been examined in various papers dealing with quantum chaos, such as (Zakrzewski and Delande, 1993), (von Oppen, 1994), (Persson and Aaberg, 1995) or in (Zelevinsky et al., 1996) for applications to the nuclear shell model. In fact, it is easily verified that at the avoided crossing the curvature (or stiffness) of $\Delta E^2(\tilde{q})$ is given by $K_{a.c.} = (\Delta F^2 + 4F_{12}^2)/\Delta E(\tilde{q}_{a.c.})$. In the references just mentioned Random Matrix Theory is applied to get information on the probability distribution of the levels. To understand the basic features behind such an approach we may take advantage of the discussion in Section 4.2. The results obtained there for a fixed Hamiltonian may be translated to the problem that our $\hat{\mathcal{H}}(Q) = \hat{\mathcal{H}}_0 + \tilde{q}\hat{F}$ depends in parametric way on the shape variable Q. For this purpose let us assume that for any Q the $\hat{\mathcal{H}}(Q)$ is a random Hamiltonian with an associated Gaussian ensemble, the character of which does not change for all Q in the region of interest. Then at some fixed Q the nearest level spacing is given by the distribution functions $P(S(Q))$ written in (4.24) for the GOE, and in (4.25) for the GUE and GSE. For the two level problem the value of $S(Q)$ is identical to the $\Delta E(Q)$ found above. Whereas the formulas found for $P(S)$ in the two level case represent quite well the true situation for large matrices, this is no longer true for the distribution $P_c(K)$ of the curvature (von Oppen, 1994). There is no place here to go into more details. Let us just mention the main result, namely that the distribution for a renormalized, dimensionless curvature k follows the law $P_c(k) \propto (1+k^2)^{-(2+\beta)/2}$, with $\beta = 1, 2, 4$ for Gaussian orthogonal, unitary and simplectic ensembles, respectively (Zelevinsky et al., 1996). It is seen that the most probable curvatures are small and that in the limit of large curvatures one has $P_c(k) \propto k^{-(2+\beta)}$. Because $S(Q) = \Delta E(Q) \propto 1/K(Q)$ this rule is intimately connected with the nearest neighbor distribution $P(S)$. For small S the latter is proportional to S^β, including in leading order the case of a regular spectrum where $\beta = 0$. At the avoided crossing one has $P_c(K_{a.c.} \gg 1) \propto P(S_{a.c.} \to 0) \propto K_{a.c.}^{-(\beta+2)}$. Looking only at the nearest level spacing or at the distribution of curvatures, evidently a chaotic spectrum helps to reduce the problems with avoided crossings. Unfortunately, these considerations are incomplete insofar as they do not tell much about combinations of the energy differences and the matrix elements for the coupling between the two states. Only such combinations contain information about the smoothness of transport coefficients.

The problems we have been discussing here may be viewed in a more general context. It is more or less evident that they arise because we have been *linearizing* the dynamics on the *strictly microscopic and quantal level*. This reminds one of the critique raised by van Kampen on linear response theory in general, see e.g. the discussion in sections 6.2 and 6.4 of the book by J.R. Dorfman (1999). The problem can be phrased by the claim that *macroscopic linearity is different from microscopic linearity*. Indeed, on the macroscopic level the dynamics can be expected to be sufficiently smooth to warrant the applicability of linearization. In their reply in Section 4.7.2 of (1991a) R. Kubo and his coauthors claim that "the interchange of stochastization and linearization ... is legitimate ... although difficult to prove". Let us consider our formulation in the light of such statements. Indeed, it would be preferable to have a microscopic Hamiltonian $\hat{H}(Q)$ at one's disposal for which the resulting quantities like energies, matrix elements and response functions do not suffer from such deficiencies as mentioned above. As mentioned above, the microscopic objects $\Xi(Q)$ carry more information than actually needed, it is definitely much beyond the level required for comparison with experimental evidences. At low internal excitations, a simple average over statistical ensembles of thermostatics does not deliver sufficient smoothing.

It may be said that in nuclear physics problems or phenomena of this type are known in other fields, not only in the context of collective motion. In theories for reactions of nuclei with light particles, for instance, it has long been recognized that fine structure may well be averaged out if it cannot be seen in experiments of only poor resolutions. In this particular case energy averages are appropriate. As described in Section 18.3, one is then able to derive Hamiltonians for which the solutions of the corresponding Schrödinger equation lead to sufficiently smooth cross sections. By analogy with this procedure the spreading of states into more complex configurations can also be treated, see the discussion in Section 4.3.2. Indeed, formula (4.47) has great similarities with those of the nucleonic response functions. In addition to the natural width $\Gamma_a(E)$ of the state $|a\rangle$ in (4.47) there explicitly appears a smoothing width Δ. In fact this kind of smoothing is totally analogous to the procedure applied in Section 6.6.1.1 for response functions of the pure independent particle model, where no natural widths appear at all. But since at low temperatures the presence of such $\Gamma(\omega, T)$ does not lead to sufficient smoothing one might well introduce an additional frequency smoothing, which in practice has not yet been done.

Finally, we need to return once more to the fact that collective motion is a stochastic process implying that *the variable Q is a random object*. In this sense the situation is very different from the examples mentioned above in connection with both (i) the critique of general linear response theory and (ii) the examples of RMT. Whereas in these examples this extra (or external) parameter is a c-number, in our case it is only the *distribution in Q and its velocity (or associated momentum)* which is physically significant. This very feature *implies additional stochasticity* beyond that given by the stochastic properties of the internal or nucleonic degrees of freedom. Hence, smoothing beyond the one determined by properties of the internal degrees of freedom is justified. It is worth noticing that in the Caldeira–Leggett model, which is very popular outside nuclear physics, problems of the type we are discussing here do not occur at all. This is because in that model the intrinsic Hamiltonian does not vary with the collective variable, for which reason no level crossings can appear.

11.2.2 Thermal aspects of global motion

In this section we should like to address the concept of "adiabatic motion", which has historically played an important role in nuclear physics, albeit mostly for the dynamics at zero thermal excitation. Commonly it is associated with "slow" motion in one way or other, but in thermostatics the concept acquires a different meaning. For our purpose it is of interest to discuss this problem in some more detail. We may not dwell much on the formulation in terms of the level density as an *adiabatic invariant*. In this version the theorem of *adiabatic invariance* says that for a change in an external variable Q which happens arbitrarily slowly, the number of levels $G(E)$ (say as defined in (22.62)) below a given energy stays constant: $G(E + \Delta E; Q + \Delta Q) = G(E; Q)$ for $dQ/dt \to 0$. Here, ΔE and ΔQ are in the following relation $\Delta E = \langle \partial \hat{H}/\partial Q \rangle \Delta Q$. For systems without level crossing a simple proof of this relation is found in (Kubo et al., 1991b). Related to this issue is the *adiabatic theorem of quantum mechanics*. It states that for very slow changes of the "external" field, i.e. for $dQ/dt \to 0$ or more precisely if the condition (see e.g. Vol.II of (Messiah, 1965))

$$\left| \langle m'|\partial \hat{H}/\partial t|m \rangle \right| \ll (E_m - E_{m'})^2/\hbar \tag{11.23}$$

were fulfilled for all states $|n\rangle$ and at any time, the occupation numbers ρ_m specifying the density operator $\hat{\rho}(t) = \sum_m |m(t)\rangle \rho_m \langle m(t)|$ *remain unchanged* when time progresses from t to $t + dt$, only the wave functions change from $|m(t)\rangle$ to $|m(t+dt)\rangle$. For the variation of the total energy $\mathcal{E}_{\rm QM}(t) = \sum_m \rho_m E_m(t)$ this implies $d\mathcal{E}_{\rm QM} = \sum_m \rho_m dE_m$; remember that $d\langle m|m\rangle = 0$ because of the normalization of the states.

This relation reminds one of thermostatics where in (22.17) the variation in energy is given by $d\mathcal{E} = \sum_m \rho_m dE_m + \sum_m d\rho_m E_m$. From the fact that in $d\mathcal{E}_{\rm QM}$ no term is present proportional to $d\rho_m$ one may be tempted to infer immediately the constancy of entropy. One should be aware, however, of the subtle difference of the adiabatic theorem of quantum mechanics from that of statistical mechanics (Kubo et al., 1991b), (Balian, 1991). In (1991) R. Balian argues that for a system, for which the concepts of thermostatics ought to be valid, a condition like (11.23) can never be applied: Too many levels simply lie too close for the characteristic time to become of meaningful length. Rather, for a quasi-static process in which the system passes through states of maximal disorder (see Section 22.1.2 where the equilibrium density is found from the maximum entropy principle), and for which the density operator is to be described by the canonical ensemble, one needs to have *changes in the temperature*. This problem is discussed in Section 22.1.1.2 for the case of one shape variable with a Hamiltonian $\hat{H}(Q)$ and in Section 23.6.2 for a general system coupled to several external fields. Here, it may suffice to quote the result stated in (22.34) which says that for $dS = 0$ the change in temperature is given by $dT = (\partial^2 \mathcal{F}/\partial Q \partial T)^{-1}(\chi^{\rm T} - \chi^{\rm ad}) dQ$; for the more general case see (23.156). It is interesting to realize that the dT/dQ is related to the difference of the isothermal and adiabatic susceptibilities, but that the static response $\chi(0)$ is not involved. Hence these formulas apply no matter whether or not the system behaves ergodically, in the sense defined at the end of Section 23.6.2.

Of course, this relation between dT an dQ is restricted to situations where the increase of entropy due to irreversible motion can be neglected and as long as no particles are emitted. Then the expression given for dT/dQ might be integrated to obtain a function $T = T(Q)$. The latter could be inserted into the global Fokker–Planck equation of the type (11.1). For all terms that involve derivatives with respect to Q there will be additional forces on top of those already present for constant temperature. It may be worth noticing that for Kramers' equation no change would be seen for the diffusion term. As discussed at the end of Section 6.5.2, for dissipative motion the temperature must be expected not only to vary with Q but with time t as well. In addition, any emission of particles or γ-rays from the nucleus will cool it and thus reduce its temperature. In this case a global equation loses its meaning in any literal sense. Rather, one has to apply transport equations for the *time steps in between successive emissions* of particles or γ quanta. In modern applications this is usually done by combining Langevin codes with codes which evaluate the emission of particles along the lines described in Chapter 8. The Langevin part is done either in locally linear or locally harmonic approximation, see e.g. (Pomorski et al., 1996), (Abe et al., 2003). In summary we may say that all three types of processes may well be described by the propagator method of the LHA.

Exercises

1. Given a Hamiltonian of standard form (11.3) with a variable inertia $M(Q)$:
   (a) Show that the transformation (11.6) (i) leads to the Hamiltonian $\mathcal{H}(X, P_X) = P_X^2/2m_X + V(X)$, (ii) leaves invariant both the Poisson bracket of $\mathcal{H}$ with $f(Q, P, t)$, (iii) as well as the volume element in phase space, $dQ dP = dX dP_X$ (remember the Jacobian).
   (b) Give reasons for $\gamma(Q)/M(Q) = \gamma_X/m_X$ and derive from (11.4) the equation for the variables (X, P_X).

2. Write eqn(11.7) in the form of (25.89) where $A(Q) = -(1/\gamma)(\partial V/\partial Q)$ and $D_{qq}^{\text{ovd}}(Q) = D(Q)$ and derive equations of motion for $\Delta Q(\tau) = \langle Q \rangle_\tau - Q_1$ and $\Sigma_{QQ}(\tau) = \langle (Q - Q_1)^2 \rangle_\tau$ after expanding $A(Q)$ and $D(Q)$ around the initial point Q_1 to second order in $(Q - Q_1)$. Solve these equations for the initial condition $n(Q, \tau = 0) = \delta(Q - Q_0)$ to leading order in τ and compare it with the solutions of (7.25) and (7.26). Use this result to convince yourself that the Kramers–Moyal expansion described in Section 25.2.3 leads back to (11.7).

3. Show the equivalence of (11.15) to (11.4) by exploiting (25.76) and (25.77).

12

DYNAMICS OF FISSION AT FINITE TEMPERATURE

In this chapter we want to address Kramers' picture of nuclear fission and discuss its consequences. His seminal paper (Kramers, 1940) was written as a response to the one by N. Bohr and J.A. Wheeler (1939), which as we have seen in Chapter 8 treated fission by means of a mere statistical approach discarding any dynamics. To overcome this drawback Kramers developed and applied his famous transport equation of Fokker–Planck type. Kramers, too, made the basic assumption that the system starts out of a (quasi-)equilibrium as suggested by N. Bohr's compound picture of nuclear reactions in that the system's decay is independent of its production. However, he argued that on the way out dynamical effects may influence the decay rate. Unlike the Bohr–Wheeler approach, Kramers assumed that the equilibrium may be described by the canonical distribution. This may, indeed, cause problems, but perhaps only at smaller thermal excitations, see Section 22.3. Moreover, it was assumed that temperature does not change along the fission path. Here, we will at first adhere to this classical model and refer to sections 11.2 and 12.5 where applications of transport equations to more general situations are discussed. In this spirit, we will concentrate on the one-dimensional case with a potential $V(Q)$, with the aim of getting the rate for the dynamics out of a minimum at Q_a over the barrier at Q_b.

12.1 Transitions between potential wells

Suppose we are given a Hamiltonian of the type $\mathcal{H}(Q,P) = P^2/(2M(Q)) + V(Q)$, with a potential $V(Q)$ in which a barrier at Q_b separates two regions, each specified by a minimum, one at $Q_a \equiv Q_m^l$ on the left of the barrier and one at Q_m^r to the right. For $V(Q_m^l) \gg V(Q_m^r)$ we may expect the rate for the transition from left to right to be similar to the case we are primarily interested in, namely where the potential is unbound to the right and drops (smoothly) to zero for $Q \gg Q_b$. For a bound system where $V(Q) \to \infty$ for $|Q| \to \infty$ there will be a genuine equilibrium. For the canonical ensemble it can be represented by the distribution given in (11.5), namely $f_{\rm eq}(Q,P) = (1/Z)\exp(-\beta\mathcal{H}(Q,P))$, which is normalized to unity by $\int_{-\infty}^{\infty} dP\, dQ\, f_{\rm eq}(Q,P) = 1$. From the second law of thermostatics we expect this distribution to be reached at $t \to \infty$—even if initially it is concentrated on the left of the barrier. The probability of finding the system to the right is given by

$$\mathcal{P}_> = \int_{-\infty}^{\infty} dP \int_{Q_b}^{\infty} dQ\, f_{\rm eq}) = \frac{\sqrt{2\pi T}}{Z} \int_{Q_b}^{\infty} \sqrt{M(Q)}\, dQ\, e^{-V(Q)/T}. \qquad (12.1)$$

Under the assumption made above, the equilibrium distribution will concentrate around the lower potential minimum. The rate at which that happens is a problem of *non-equilibrium* statistical mechanics. At first we want to tackle this problem for the over-damped limit.

12.1.1 Transition rate for over-damped motion

Later we want to establish connection with Kramers' rate formula. For this reason we will start from that version of a Smoluchowski equation which in Section 11.1.2 was seen to follow from Kramers' equation in leading order in inverse friction (assuming the inertia to be constant):

$$\frac{\partial}{\partial t} n(Q,t) = \frac{\partial}{\partial Q} \left(\frac{1}{\gamma(Q)} \frac{\partial V(Q)}{\partial Q} + \frac{T}{\gamma(Q)} \frac{\partial}{\partial Q} \right) n(Q,t). \qquad (11.12)$$

Following Kramers we will apply the following scenario: (i) The barrier height is much larger than temperature, $E_{\rm b} = V(Q_{\rm b}) - V(Q_{\rm a}) \gg T$; (ii) initially the number of particles found to the very right of the barrier is negligibly small. The conditions just imposed imply the "decay rate" to be small. In fact, for a sufficiently high barrier we may assume the system to first "relax" to a *local* equilibrium concentrated around $Q_{\rm a}$. We shall not care how this quasi-equilibrium is reached but simply take it for granted. As mentioned earlier such a hypothesis is in close analogy with Bohr's independence hypothesis for statistical reactions. This local equilibrium may be specified by the harmonic oscillator which at $Q_{\rm a}$ fits to the true $V(Q)$. In this sense we may write $V(Q) \approx V(Q_{\rm a}) + (C_{\rm a}/2)(Q - Q_{\rm a})^2$. The decay rate R will be determined by the flux across the barrier at $Q_{\rm b}$. It may be defined by the number of ensemble particles $-dN_{\rm a}/dt$ which leave the region around $Q_{\rm a}$ per unit of time, *relative* to the number of particles $N_{\rm a}$ sitting in the well. The $dN_{\rm a}/dt$ may be calculated from the current by employing the continuity equation, which means to write

$$-\frac{dN_{\rm a}(t)}{dt} \equiv -\int_{-\infty}^{Q_m^r} dQ\, \frac{\partial n(Q,t)}{\partial t} = \int_{-\infty}^{Q_m^r} dQ\, \frac{\partial j(Q,t)}{\partial Q} = j(Q,t)\Big|_{-\infty}^{Q_m^r} = j(Q_m^r, t).$$

The rate is thus given by

$$R \equiv -\frac{1}{N_{\rm a}} \frac{dN_{\rm a}}{dt} = \frac{j(Q_m^r, t)}{N_{\rm a}}. \qquad (12.2)$$

Under the hypothesis stated before the current $j(Q_m^r, t)$ will be a *quasi-stationary* one, $j(Q_m^r, t) = j_{\rm st}$. As it belongs to the transport equation (11.12) it can be expressed as

$$j_{\rm st} = -\frac{T}{\gamma(Q)} \exp\left(-\frac{V(Q)}{T}\right) \frac{\partial}{\partial Q} n_{\rm st}(Q) \exp\left(\frac{V(Q)}{T}\right).$$

Let us multiply this equation by $\gamma(Q) \exp(V(Q)/T)$ and integrate the resulting expression between some Q_1 and Q_2 to get:

$$j_{\rm st} \int_{Q_1}^{Q_2} \gamma(Q) \exp\left(\frac{V(Q)}{T}\right) dQ = -\left[T n_{\rm st}(Q) \exp\left(\frac{V(Q)}{T}\right)\right]_{Q_1}^{Q_2}. \quad (12.3)$$

To proceed further let us choose the Q_2 to lie way beyond the barrier on the right-hand side where (at finite times) the density $n_{\rm st}(Q_2)$ may be considered negligibly small. In this case the term proportional to $n_{\rm st}(Q_2) \exp(V(Q_2)/T)$ on the right-hand side may be neglected. For a Q_1 in the well around $Q_{\rm a}$, on the other hand, the $n_{\rm st}(Q)$ may be replaced by the equilibrium distribution. For eqn(11.12) the latter is given by $n_{\rm eq}(Q) = \mathcal{N}_{eq} \exp(-V(Q)/T)$ such that one has:

$$n_{\rm st}(Q) \exp(V(Q)/T) = \mathcal{N}_{st} \quad \text{for} \quad Q \simeq Q_{\rm a}. \quad (12.4)$$

The constant $\mathcal{N}_{st}$ defines an overall normalization factor for the stationary solution $n_{\rm st}(Q)$.

We still need to evaluate the integral on the left-hand side of (12.3). For a $Q_1 = Q_{\rm a}$ and $Q_2 = Q_m^r \gg Q_{\rm b}$ it will be dominated by the region around the barrier top, where $V(Q)$ is highest. We may thus apply the saddle point approximation, which means replacing in the exponent the potential by that of an "inverted" oscillator $V(Q) \approx V(Q_{\rm b}) + (C_{\rm b}/2)(Q - Q_{\rm b})^2$ with the negative stiffness $C_{\rm b} = d^2V/dQ^2|_{\rm b} = -|C_{\rm b}|$. The pre-factor $\sqrt{\gamma(Q)}$ may be considered a smoothly varying function such that its value at the barrier may be taken out of the integral. In this way the current becomes

$$j_{\rm st} = \frac{T}{\gamma_{\rm b}} \frac{\mathcal{N}_{st} \exp(-V(Q_{\rm b})/T)}{\int_{-\infty}^{+\infty} dQ \exp(-|C_{\rm b}|(Q - Q_{\rm b})^2/2T)} = \frac{T}{\gamma_{\rm b}} \sqrt{\frac{|C_{\rm b}|}{2\pi T}} \mathcal{N}_{st} \exp(-V(Q_{\rm b})/T). \quad (12.5)$$

The constant $N_{\rm a}$ needed in the rate formula (12.2) may be estimated by employing (12.4) to get $N_{\rm a} = \int_{Q_{\rm a}-\Delta}^{Q_{\rm a}+\Delta} n_{\rm st}(Q)\, dQ$. Here, the integration was assumed to extend over a finite interval of length 2Δ. The latter has to be sufficiently large such that its range encompasses the vast majority of the ensemble points sitting in the first well. The saddle point approximation then leads to

$$N_{\rm a} = \mathcal{N}_{st} \sqrt{2\pi T/C_{\rm a}} \exp(-V(Q_{\rm a})/T). \quad (12.6)$$

With $E_{\rm b} = V(Q_{\rm b}) - V(Q_{\rm a})$ the final formula for the decay rate thus becomes

$$R_{\rm ovd}^{\rm K} = \sqrt{C_{\rm a}|C_{\rm b}|} \exp(-E_{\rm b}/T)/(2\pi\gamma_{\rm b}). \quad (12.7)$$

This result is intimately related to the transport equation (11.12) we started with. In Section 11.1 two other equations for over-damped motion were discussed. They only differ in the order the variable friction coefficient $\gamma(Q)$ is to be put with respect to the differential operator $\partial/\partial Q$. Consequently, the only difference between these rate formulas is seen in the appearance of the friction coefficient. Had we started with (11.11) the $\gamma_{\rm b}$ in (12.7) would have to be replaced by $\gamma_{\rm a}$.

In the high temperature limit of (11.18), on the other hand, $1/\gamma_b$ would have to be changed to $1/\sqrt{\gamma_a \gamma_b}$ (see exercise 1).

Finally, we turn to the case that in Kramers' equation (11.4) the coordinate dependence of the inertia may not be neglected. We know from Section 11.1.2 that then eqn(11.12) has to be modified in such a way that the potential $V(Q)$ is replaced by $V(Q) + T \ln(M(Q))^{-1/2}$. This means that we may follow all steps in deriving (12.7) only observing this modification. Because of (12.4) the implication of the right-hand side of (12.3) on (12.5) remains unchanged. On the left-hand side, however, the $\gamma(Q)$ has to be replaced by $\gamma(Q)/\sqrt{M(Q)}$. Evaluating the integral still in saddle point approximation means replacing in (12.5) the γ_b by $\gamma_b/\sqrt{M_b}$. A similar factor is obtained for the number of particles sitting in the well. The N_a of (12.6) gets the additional factor $\sqrt{M_a}$. Altogether this means that the $R_{\text{ovd}}^{\text{K}}$ of (12.7) turns into

$$R_{\text{ovd}}^{\text{K,M}} = R_{\text{ovd}}^{\text{K}} \sqrt{M_b/M_a} \,. \tag{12.8}$$

We are now going to discuss Kramers' famous rate formula, for which the same limit will be seen to hold true for large damping.

12.2 The rate formulas of Kramers and Langer

In Kramers' paper (1940) the transport coefficients for friction and inertia were assumed constant. Under such circumstances all forms obtained for over-damped motion become identical. They should represent the leading order term in an expansion in $1/\eta$ of (23.5), the dimensionless coefficient which for the local oscillator indicates the strength of damping. For the so-called "high viscosity limit" limit Kramers obtained the formula

$$R_{\text{K}} = (\varpi_a/2\pi) \exp(-E_b/T) \left(\sqrt{1 + \eta_b^2} - \eta_b \right). \tag{12.9}$$

The notion "high viscosity limit" simply means having a friction coefficient big enough to ensure that the flux over the (sufficiently high) barrier can be associated with a diffusive process in phase space which is in a steady state. The ϖ_a may be interpreted as the "assault frequency" at the potential minimum at Q_a and η_b is the damping rate which governs the constant flux across the barrier at Q_b. In fact formula (12.9) is also valid for variable transport coefficients, although in the past this issue has caused some confusion, to which we will come below. In Section 11.1.1 the generalized equation was written in the form

$$\frac{\partial}{\partial t} f(Q,P,t) = \{\mathcal{H}, f(Q,P,t)\} + \frac{\partial}{\partial P} \left(\frac{P}{M(Q)} \gamma(Q) + \gamma(Q) T \frac{\partial}{\partial P} \right) f(Q,P,t) \,. \tag{11.4}$$

It was argued that this form is invariant with respect to the transformation (11.6) to variables X and P_X for which the inertia turns into the constant m_X and where friction becomes $\gamma_X(X) = \gamma(Q)(m_X/M(Q))$. For this transformation the density $f(Q,P,t)$, the volume element $dQ\,dP$, as well as the Hamiltonian $\mathcal{H}$ are scalars. Since the current can be written as

$$j(Q,t) = \int dP\,(P/M(Q))f(Q,P,t) \equiv \int d(P^2/2M(Q))\,f(Q,P,t) \quad (12.10)$$

it is seen to be form invariant, too. For this reason one may apply the basic formula for the decay rate,

$$R_{\mathrm{K}} = j_{\mathrm{st}}(Q_{\mathrm{b}})/N_{\mathrm{a}}, \quad (12.11)$$

in the system (X, P_X) and transform back to the original one. Please notice that in the final result (12.9) only ratios of transport coefficients appear.

To evaluate the stationary current $j_{\mathrm{st}}(Q_{\mathrm{b}})$ Kramers approximated the barrier by an inverted oscillator and solved his equation for this model. In (7.37) this solution was given for the extension to quantal diffusion coefficients. For the present case one may just take the high temperature limit (with the parameters σ and A entering (7.37) given in (7.42)); the inclusion of quantum effects will be addressed in Section 12.5. Likewise, the number N_{a} of particles in the well may also be calculated by approximating the Hamiltonian around the potential minimum by that of a harmonic oscillator. (As for the situation of over-damped motion, this makes sense if the two extremal points of the potential are sufficiently well separated.) To actually apply formula (12.11) it is necessary to know how the density around the barrier connects to the one at the minimum. In the classical case this problem is treated best by using the ansatz

$$f_{\mathrm{st}}(Q,P) = \xi(Q,P)\exp(-\mathcal{H}(Q,P)/T) \quad (12.12)$$

for a stationary, *globally valid* solution of eqn(11.4). The differential equation obtained for $\xi(Q,P)$ may then be solved with appropriate boundary conditions. In this way one gets an $f_{\mathrm{st}}(Q,P)$ which around Q_{a} and Q_{b} is of the structure just described. For details we may refer to the original work by Kramers (1940) or to reviews such as (Hänggi et al., 1990). Incidentally, one may as well start from the two possible solutions for the linearized forms of (11.4) and find the intermediate point Q_{m} (with $Q_{\mathrm{a}} < Q_{\mathrm{m}} < Q_{\mathrm{b}}$) where the two densities can be joined smoothly (van Kampen, 2001). All these treatments deal with constant inertia. The case of variable inertia has been studied in (Hofmann et al., 2001) with the approximation that at Q_{m} not the full solution $f_{\mathrm{st}}(Q,P)$ was joined but only the momentum integrated $n_{\mathrm{st}}(Q)$. In this way a modified rate formula was found which will be given below in (12.18) together with further comments.

In (1969) J.S. Langer presented a derivation for a multidimensional situation. In its most basic features it is similar to that of Kramers. He, too, applies saddle point approximations for both the dynamics around the potential minimum and the flux across the barrier. His derivation is based on a form of the equation where the conservative forces derive from an energy functional without explicit reference to the inertia being constant; the frictional and fluctuating forces, on the other hand, are assumed independent of the coordinates. With the parameter z^+ introduced in (7.31) Langer's formula may be written as

$$R_{\rm L} = z^+\,{\rm Im}\,\mathcal{F}/(\pi T) \equiv \varpi_{\rm b}\left(\sqrt{1+\eta_{\rm b}^2}-\eta_{\rm b}\right){\rm Im}\,\mathcal{F}/(\pi T)\,. \qquad (12.13)$$

The inverse of z^+ sets the scale for the behavior of $\langle(Q-Q_{\rm b})\rangle_t$ at large times, which goes as $\langle(Q-Q_{\rm b})\rangle_{t=0}\exp(z^+(t-t_{\rm b}))$ and describes the motion of the average trajectory away from the barrier top along the steepest descent, see Section 25.5.2. The most peculiar feature of (12.13) is that it involves the *imaginary part* of the free energy $\mathcal{F}=-T\ln Z$. Evidently, for an unstable system this free energy cannot have the same interpretation as that of canonical ensembles for equilibrium. With $V(Q)\equiv\mathcal{F}_{\rm stat}(Q)$ the partition function for the one-dimensional system would be given by

$$Z(T) = \int dQ\,\sqrt{2\pi T\,M(Q)}\,\exp(-\mathcal{F}_{\rm stat}(Q)/T)\,. \qquad (11.21)$$

Here, it was taken into account that for constant temperature the potential energy is given by the free energy $\mathcal{F}_{\rm stat}(Q,T)$ of the quasi-static picture for the internal degrees of freedom. Clearly, for the form of the $\mathcal{F}_{\rm stat}(Q)$ we have in mind the integral appearing there is ill defined: Beyond the barrier the free energy decreases with increasing Q. As suggested in (Langer, 1967) the integration contour $\mathcal{C}$ should be defined in such a way that beyond the top of the barrier it runs parallel to the purely imaginary Q-axis. In a saddle point approximation we may then replace the barrier in that region by an inverted oscillator, for which the stiffness becomes negative: $C_{\rm b}^{\rm stat}=\partial^2\mathcal{F}_{\rm stat}/\partial Q^2\,|\,Q_{\rm b}<0$. This leads to the following contribution to the $Z(\beta)$ of (11.21):

$$\mathcal{Z}_{>Q_{\rm b}}(\beta) = \sqrt{2\pi T}\int_{\mathcal{C}>Q_{\rm b}}\sqrt{M(Q)}\,\exp\left(-\beta\mathcal{F}_{\rm stat}(\beta,Q)\right)$$

$$\simeq \sqrt{2\pi T M_{\rm b}}\,\exp\left(-\beta\mathcal{F}_{\rm stat}(\beta,Q_{\rm b})\right)\int_0^{i\infty}d\xi\,e^{\beta|C_{\rm b}^{\rm stat}|\xi^2/2}$$

$$= \frac{i}{2}\sqrt{\frac{M_{\rm b}}{|C_{\rm b}^{\rm stat}|}}\,\exp\left(-\beta\mathcal{F}_{\rm stat}(\beta,Q_{\rm b})\right) \equiv \frac{i}{2\varpi_{\rm b}}\exp\left(-\beta\mathcal{F}_{\rm stat}(\beta,Q_{\rm b})\right)\,.$$

In the last step it was assumed that the local frequency $\varpi_{\rm b}$ relates to the stiffness of the free energy $C^{\rm stat}$ in the common form $C_{\rm b}^{\rm stat}=M_{\rm b}\varpi_{\rm b}^2$–although from the microscopic point of view this may only be given in a certain limit.

As expected the $\mathcal{Z}_{>Q_{\rm b}}(\beta)$ is purely imaginary. Actually, we may expect a real contribution from the region on the left of the stationary point at $Q_{\rm b}$. Its value, too, would be proportional to $\exp(-\beta\mathcal{F}_{\rm stat}(\beta,Q_{\rm b}))$ and thus exponentially small compared to the real contribution $Z_{Q_{\rm a}}(\beta)$ from the stationary point $Q_{\rm a}$ at the minimum (recall that we again want to assume $E_{\rm b}\equiv\mathcal{F}_{\rm stat}(\beta,Q_{\rm b})-\mathcal{F}_{\rm stat}(\beta,Q_{\rm a})\gg T$). The $Z_{Q_{\rm a}}(\beta)$ is readily evaluated to become

$$Z_{Q_a}(\beta) = (\varpi_a)^{-1} \exp(-\beta \mathcal{F}_{\text{stat}}(\beta, Q_a)) \ .$$

Adding both contributions the total $Z(\beta)$ may thus be written as

$$Z(\beta) \simeq Z_{Q_a}(\beta) + \mathcal{Z}_{>Q_b}(\beta) = Z_{Q_a}(\beta)\Big(1 + \mathcal{Z}_{>Q_b}(\beta)\Big/ Z_{Q_a}(\beta)\Big) .$$

Since the ratio appearing on the very right is much smaller than unity and because Z_{Q_a} is purely real from $\text{Im}\{\mathcal{F}(\beta)\} = -T \,\text{Im}\{\ln Z(\beta)\}$ the imaginary part of the free energy becomes

$$\text{Im}\{\mathcal{F}(\beta)\} \simeq -T\, |\mathcal{Z}_{>Q_b}(\beta)| \,/\, Z_{Q_a}(\beta) = -T(\varpi_a/\varpi_b) \exp\big(-E_b/T\big). \quad (12.14)$$

After inserting this result into (12.13) one gets Kramers' original formula (12.9). Most importantly, this version comes out *even for variable inertia*. However, had we chosen to use *only* the free energy $\mathcal{F}_{\text{stat}}(\beta, Q)$ of the quasi-static equilibrium *without the factor* $(M(Q))^{1/2}$ in (11.21) we would have ended up with the wrong formula (12.18) shown below but which had already been mentioned above.

As mentioned earlier Kramers was able to derive an analytic result for the rate also in the case of "low viscosity" by solving eqn(11.4) to leading order in γ. The result can be expressed as

$$R_K^{\text{l.v.}} = (\gamma/M)_b \,(E_b/T) \exp(-E_b/T) \ . \quad (12.15)$$

The correct rate, valid within classical physics but for all damping rates $\eta_b = (\gamma/(2M\varpi))_b$, first *increases* with η_b as given by (12.15) to reach a maximum and *decreases* afterward turning quickly into the form given by (12.9). Amongst other proofs (see (Hänggi et al., 1990)), numerical verifications of this feature have been found in (Scheuter and Hofmann, 1983) by solving Kramers' equation within the propagator method discussed in Section 7.1. The condition for the "high viscosity limit" to be applicable may be written as (see Kramers' article or (Hänggi et al., 1990))

$$\eta_b \gtrsim T/(2E_b) \ . \quad (12.16)$$

For nuclear physics, typically this value is quite small such that the notion "high viscosity limit" should by no means be mixed up with over-damped motion.

The relation to the transition state result: In transition state theory one assumes (i) to be given a system which is totally equilibrated inside the barrier and (ii) for which at the barrier current only flows outward ignoring any backflow. For the canonical distribution at constant temperature the current can be written as

$$j_b^{\text{trans}} = \int_0^\infty dP \, \frac{P}{M_b} f_{\text{eq}}(Q = Q_b, P) \propto \int_{-\infty}^\infty dP \, \frac{P}{M_b} \exp\left(-\frac{P^2}{2M_b T}\right) \Theta(P).$$

This expression is identical to the $j_b^K(\eta = 0)$ one gets for Kramers' stationary solution for zero damping. To see this one needs to take the high temperature limit of (7.37) and consider the following representation of the Theta function: $\Theta(P) = \lim_{\sigma \to 0} \int_{-\infty}^{P} du \exp(-u^2/2\sigma)/\sqrt{2\pi\sigma}$. Indeed, eqn(7.42) implies that for $\eta \to 0$ the parameter σ turns to zero as well. As a result, it is seen that the rate of the transition state method is identical to that of Kramers in the "high viscosity limit" but where the influence of damping on the rate itself may be neglected: $R_{\text{trans}} = R_K(\eta = 0)$. The fact that a finite damping is involved at all may not be too surprising. Otherwise there would be no reason for the system to equilibrate at all. Of course, for a general formulation of the transition state theory one does not need to restrict to the canonical ensemble. In fact, the Bohr–Wheeler formula (8.14) presented in Section 8.2.1 is based on a description for a microcanonical ensemble. To the extent that the latter may be approximated by a canonical one, with one and the same temperature at the minimum and at the barrier, (8.14) reduces to (8.19) which is to say $R_{BW} \to R_K(\eta = 0)$ (with $\varpi_{\min} \equiv \varpi_a$ and $B_f \equiv E_b$).

In all these forms presented so far collective motion is treated on the classical level only, but quantum effects might lead to measurable modifications. If no fluctuating forces were present at all, one might try to account for them by multiplying the $\Omega_{C-1}^{\text{sad}}$ in the integrand of (8.14) with the quantum probability that the system of energy ϵ_f may overcome the barrier. For this the Hill–Wheeler formula is sometimes used, see e.g. (Gadioli and Hodgson, 1992) or (Fröbrich and Lipperheide, 1996). This modification does not appear very realistic, as any influence of damping and hence of quantal *transport properties* are discarded.

The limit of over-damped motion: Expressed in terms of the effective, dimensionless damping rate η the over-damped limit is reached for $\eta \gg 1$. As $(1+\eta_b^2)^{1/2} - \eta_b = ((1+\eta_b^2)^{1/2} + \eta_b)^{-1}$ in leading order in $1/\eta_b$ formula (12.9) is easily seen to become

$$R_K \longrightarrow \frac{1}{2\pi} \frac{\varpi_a}{2\eta_b} \exp(-E_b/T) = \frac{\sqrt{C_a|C_b|}}{2\pi\gamma_b} \sqrt{\frac{M_b}{M_a}} \exp(-E_b/T) = R_{\text{ovd}}^{K,M},$$
(12.17)

and thus reduces to (12.8) rather than to (12.7). Here, a somewhat unexpected feature is seen in that traces of the inertia appear although one is dealing with over-damped motion. For this reason in (Hofmann et al., 2001) the following modification of Kramers' rate formula was suggested (and, as described above, to some extent verified):

$$R_K \Rightarrow \frac{\varpi_b}{2\pi} \sqrt{\frac{C_a}{|C_b|}} \left(\sqrt{1+\eta_b^2} - \eta_b \right) \exp(-E_b/T)$$
$$= \frac{\varpi_a}{2\pi} \sqrt{\frac{M_a}{M_b}} \left(\sqrt{1+\eta_b^2} - \eta_b \right) \exp(-E_b/T).$$
(12.18)

Then for $\eta_b \gg 1$ the factor with the inertias would drop out, indeed. A result of this type had been obtained before by V.M. Strutinsky in (Strutinsky, 1973) (see the discussion in (Hofmann et al., 2001)), as well as in (Rummel, 2000), (Rummel and Hofmann, 2001) within an extension of the Perturbed Static Path Approximation (PSPA), to which we will return below. In the spirit of the discussion in Section 12.1.1 and the arguments given above for the derivation of (12.9) the conjecture behind (12.18) cannot be maintained. As one may recall from Section 11.1.2, an inverse friction expansion of Kramers' equation in the form of (11.4) necessarily leads to an equation for the reduced density $n(Q,t)$ in which for variable inertia an additional term appears, see eqn(11.14).

12.3 Escape time for strongly damped motion

In the past few decades timescales have played an important role in the analysis of fission experiments. Before we address this issue further we shall introduce the concept of the *mean first passage time*. It delivers another possibility for deriving rate formulas (Hänggi et al., 1990), (Reimann et al., 1999) and helps to interpret them in view of timescales for the decaying process. The concept as such is explained in Section 25.4 for the one dimensional Fokker–Planck equation for which analytic solutions are possible. Along these lines we will restrict ourselves to over-damped motion and derive some useful formulas.

Different to the derivation in Section 12.1.1 let us start from the equation

$$\frac{\partial}{\partial t} n(Q,t) = \frac{\partial}{\partial Q}\left(\frac{1}{\gamma(Q)}\frac{\partial V(Q)}{\partial Q} + \frac{\partial}{\partial Q}\frac{T}{\gamma(Q)}\right) n(Q,t), \qquad (11.11)$$

for which the MFPT is given by the expression

$$\tau_{\mathrm{mfpt}}(Q_{\mathrm{in}}, Q_{\mathrm{ex}}) = \int_{Q_{\mathrm{in}}}^{Q_{\mathrm{ex}}} du\ \exp\left[\frac{V(u)}{T}\right] \int_{-\infty}^{u} dv\ \frac{\gamma(v)}{T}\ \exp\left[-\frac{V(v)}{T}\right]. \qquad (12.19)$$

It represents the average time a "particle" needs to reach the point Q_{ex} for the first time after it was released at $Q_{\mathrm{in}} < Q_{\mathrm{ex}}$ initially. The Q_{ex} is treated as an "exit point" in a literal sense. After it had been crossed the particle is removed from the whole ensemble whose dynamics is described by the density distribution $n(Q,t)$. Formula (12.19) is valid for a situation that no particle can escape to the left of Q_{in}, which in our case will be ensured by assuming the potential to rise to infinity for $Q \to -\infty$. Equation (12.19) is easily verified from (25.119) together with (25.120) after observing that for eqn(11.11) the two quantities A and D are given by $A(Q) = -(1/\gamma(Q))(\partial V(Q)/\partial Q)$ and $D(Q) = 2T/\gamma(Q)$.

Before we may apply (12.19) to a typical model a few remarks are in order.

(i) In principle, we may take any value for the Q_{ex}, but the most interesting examples will be those where it is larger than the Q_{a} of the minimum of the potential $V(Q)$ inside the (first) barrier. In the asymptotic region, when the exit point Q_{ex} is put sufficiently far away to the right of the (last) barrier, the $\tau_{\mathrm{mfpt}}(Q_0, Q_{\mathrm{ex}})$ can be identified with the mean escape time τ_{esc}, the inverse of which determines the fission rate.

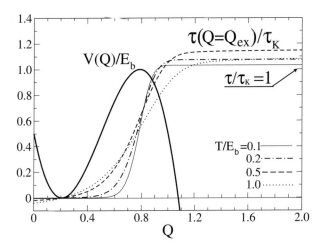

FIG. 12.1. The mean first passage time $\tau(Q = Q_{\rm ex}) \equiv \tau_{\rm mfpt}(Q_a, Q_{\rm ex})$ of (12.19) (normalized to the $\tau_{\rm K} \equiv \tau_{\rm esc}$ of (12.22)) for different temperatures measured in units of $E_{\rm b}$. The computations were done for a cubic potential and constant friction, see Fig.1 of (Hofmann and Magner, 2003).

(ii) Secondly, we should like to point out that so far no assumption had to be made neither on the form of the potential $V(Q)$ nor on the ratio V/T. The $V(Q)$ might have several humps as implied by shell corrections. For this it is important to recall that the "high temperature limit" for the diffusion coefficient of overdamped motion may already be valid at quite low thermal excitations where shell effects might still be present, see Section 11.1.2.

(iii) It is worth noticing that the result (12.19) also covers the other two versions of the Smoluchowski equation discussed in Section 11.1. The one for eqn(11.12) is obtained by replacing $V(Q)$ by $V(Q) + T \ln \gamma(Q)$, whereas for (11.18) (in the high temperature limit) one has to choose $V(Q) + T \ln \sqrt{\gamma(Q)}$.

Let us suppose the $V(Q)$ to be of the same form of the potential which underlies Kramers' model case for the decay rate, having one pronounced minimum and one barrier. For sufficiently small temperatures (in comparison with the barrier height) the integrals will be dominated by their contributions from the extrema. If the $Q_{\rm ex}$ lies sufficiently far to the right of the barrier at $Q_{\rm b}$ the integral in (12.19) over du will be dominated by the first factor. The latter is largest around the top of the barrier such that in a steepest descent approach (see Section 26.2) the potential in the exponent may be approximated by $V(Q) \approx V(Q_{\rm b}) - |C_{\rm b}|/2 \, (Q - Q_{\rm b})^2$. Also the inner integral will be dominated by the exponential, and may thus be expected to increase with u. Close to the barrier and beyond it will vary very little with u such that it may be approximated by its value at the stationary point of the u-integral. This is to say we may write

$$\tau_{\text{mfpt}}(Q_{\text{a}}, Q_{\text{ex}}) = \exp(V_{\text{b}}/T) \left\{ \int_{Q_{\text{a}}}^{Q_{\text{ex}}} du \, \exp\left[|C_{\text{b}}|(u - Q_{\text{b}})^2/2T\right] \right\} \times \int_{-\infty}^{Q_{\text{b}}} dv \, (\gamma(v)/T) \, \exp\left[-(V(v)/T)\right]. \quad (12.20)$$

Applying the steepest descent method also to the second integral one obtains

$$\int_{-\infty}^{Q_{\text{b}}} dv \, \frac{\gamma(v)}{T} \exp\left[-\frac{V(v)}{T}\right] \approx \frac{\gamma(Q_{\text{a}})}{T} e^{-V_{\text{a}}/T} \int_{-\infty}^{\infty} dv \, \exp\left(-\frac{C_{\text{a}}(v - Q_{\text{a}})^2}{2T}\right), \quad (12.21)$$

after writing the potential in the neighborhood of the minimum at Q_{a} as $V(Q) = V(Q_{\text{a}}) + C_{\text{a}}(Q - Q_{\text{a}})^2/2$ and replacing the upper limit of integration by $+\infty$. This latter step is allowed if, for given temperature, the minimum and the barrier are sufficiently well separated. In this way the remaining integral in (12.20) can also be handled such that finally one gets

$$\tau_{\text{esc}} \equiv \tau_{\text{mfpt}}(Q_{\text{a}}, Q_{\text{ex}} \gg Q_{\text{b}}) = R_{\text{ovd}}^{-1} \approx \frac{2\pi\gamma_{\text{a}}}{\sqrt{C_{\text{a}}|C_{\text{b}}|}} \exp\left(\frac{E_{\text{b}}}{T}\right). \quad (12.22)$$

For the other two equations for over-damped motion, (11.12) and (11.18), one would get the same result but with γ_{a} replaced by γ_{b} or $\sqrt{\gamma_{\text{a}}\gamma_{\text{b}}}$, respectively; remember the comments made above in item (iii) (see exercise 2). In all cases the τ_{esc} is the inverse of the corresponding decay rate. For equation (11.12) the τ_{esc} is identical to the decay time τ_{K} as the inverse of Kramers' rate $R_{\text{ovd}}^{\text{K}}$ of (12.7) for over-damped motion. In the following we will stick to the case of constant friction such that these differences do not matter. In fact, for the validity of the saddle point approximation to (12.19) the exponential factors involving the potential are more decisive.

We are now going to examine properties of the MFPT for a few typical examples. In Fig. 12.1 the results of calculations (Hofmann and Magner, 2003) of $\tau_{\text{mfpt}}(Q_{\text{a}} \to Q_{\text{ex}})/\tau_{\text{esc}}$ are shown for the schematic case of a cubic potential. The $\tau_{\text{mfpt}}(Q_{\text{a}} \to Q_{\text{ex}})$ was obtained by evaluating the integrals of (12.19) numerically; the τ_{K} is identical to the corresponding result τ_{esc} of (12.22). The potential used may be specified by its first derivative to be given by the form $V'(Q) \propto (Q - Q_{\text{a}})(Q - Q_{\text{b}})$, with extrema at $Q_{\text{a}} \simeq 0.2$ and $Q_{\text{b}} \simeq 0.8$ and a barrier height E_{b} of 8 MeV. It is observed that in the asymptotic region the ratio $\tau_{\text{mfpt}}/\tau_{\text{K}}$ becomes close to unity, indeed, if only the parameter temperature over barrier height becomes small enough. However, even for this case of exactly two well pronounced extrema, deviations from unity are clearly visible at higher temperatures, which is a definite indication of the failure of the saddle point approximation used for (12.22). This situation becomes more dramatic as soon as the potential shows additional structure. In Fig. 12.2 this is demonstrated for a schematic potential of fifth order with its derivative being $V'(Q) \propto (Q - Q_{\text{a}})(Q - Q_{\text{b}})(Q - Q_{\text{c}})(Q - Q_{\text{d}})$ but the values of E_{b}, Q_{a} and Q_{b} unchanged. The remaining two parameters Q_{c} and Q_{d} are chosen to be identical to one another, $Q_{\text{c}} = Q_{\text{d}} = Q_{\text{s}}$, with

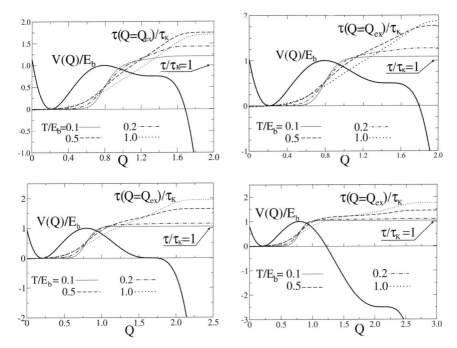

FIG. 12.2. Same as in Fig. 12.1 but for potentials having shoulders at some Q_s of heights V_s/E_b relative to the barrier: $V_s/E_b = 0.75$ top left, 0.5 top right, 0 bottom left, −2.5 bottom right, see Fig.2 of (Hofmann and Magner, 2003).

their values fixed such that the height V_s of the then existing shoulder takes on the values specified in the figure captions. In (Hofmann and Magner, 2003) a potential having second minima and maxima is also considered for which the effect is even larger.

12.4 A critical discussion of timescales

For excited nuclei fission is not the only decay mode. With increasing excitation the chance of decays through emission of photons or light particles, and of neutrons in particular, is rather high. Commonly, all these processes are being described with the statistical model, the essential features of which are presented in Section 18.4. The basic ingredients are the (partial) widths for the individual decay modes, say Γ_n or Γ_γ for the emission of neutrons and photons, and Γ_f for fission; formulas for these quantities may be found in Section 8.1.2 (neutrons), Chapter 14 (γ from dipole transitions) and Section 8.2.1 for the Bohr–Wheeler formula with $\Gamma_f = \Gamma_{BW}$. For more details of precisely how this model is employed to analyze such processes we have to refer to the literature, see e.g. (Vandenbosch and Huizenga, 1973), (Gadioli and Hodgson, 1992), (Fröbrich and Lipperheide,

1996), as well as to the specific papers mentioned below in which various statistical codes are used. For our purpose a more qualitative argument within a semi-classical picture may suffice.

Suppose we look at the competition between neutron evaporation and fission. The ratio $\Gamma_n/\Gamma_f = \tau_f/\tau_n$ may be taken as an indication of the number of neutrons emitted per fission event, the neutron "multiplicity". For heavy nuclei and at larger excitations typically this ratio is larger than one. Hence, one may say that during the process of fission several neutrons may be emitted. One may measure these numbers experimentally and compare the results with theoretical predictions. In this way it has long been known that the *multiplicity is underestimated* if the fission width Γ_f is identified with the Bohr–Wheeler result Γ_{BW} (in the conventional form where the static energy is identified as the liquid drop energy). Recalling Kramers' rate formula (12.9) (in the "high viscosity limit" limit) one notices that the decay rate decreases with increasing dissipation strength η_b. Under the assumption that $\Gamma_K(\eta_b = 0)$ may be replaced by the Γ_{BW} of the Bohr–Wheeler formula this gives rise to the interesting possibility to actually get more or less direct experimental access to the magnitude of nuclear dissipation. Indeed, the ratio Γ_f/Γ_{BW} between the measured fission rate Γ_f to that of the statistical model Γ_{BW} then is simply given by the "Kramers factor" $\Gamma_K(\eta_b)/\Gamma_K(\eta_b = 0) = (1 + (\eta_b)^2)^{1/2} - \eta_b$. Strangely enough, in this connection one has always been aiming at friction alone,[23] or for "reduced friction" as it once was termed. By that notion one meant the ratio of friction over inertia, which often were called β. It represents nothing else but the damping width $\Gamma_{\rm kin} = \gamma/M$ of the local oscillator, see (6.143) or (23.5). Indeed, by following these lines one may at best get information of how *collective motion behaves in the neighborhood of the barrier*. Here, the emphasis is on both "collective motion" as well as "at the barrier". The second point is evident from the very fact that any derivation of Kramers' rate formula involves saddle point approximations (like that of the Bohr–Wheeler formula). The first issue is no less important. The character of collective motion does not only depend on friction but on inertia and the potential energy alike. In fact, the evaluation of the effective inertia at finite thermal excitations and, hence, for dissipative behavior, is as great a theoretical challenge as that for friction.

Let us return to the discrepancies between the measured and the predicted multiplicities of neutrons and photons. Truth is that even the replacement of the Bohr–Wheeler formula by that of Kramers (with respect to its dependence on the dissipation strength) did not help to remove this problem. To remedy this situation special features of the time evolution of a fissioning system were considered. Before turning in more detail to the problems involved in such an analysis, we should like to recall one important feature of Kramers' picture. Although his equation is one of non-equilibrium statistical mechanics, devoted first of all to

---

[23] Often one refers to it as "nuclear viscosity", although this notion gives the wrong impression that nuclear dissipation is synonymous with the (two-body) viscosity of a nuclear fluid.

12.4.1 Transient- and saddle-scission times

Modern applications of Kramers' transport equation in nuclear physics began in the 1970s with descriptions of heavy-ion collisions. For this problem it is very natural to follow the reaction in time with the initial configuration being defined by the two approaching fragments; details will be explained in Section 13.2. Later, in attempts to describe fission processes this time-dependent picture was taken over. One only had to modify the initial condition to one which represents the state of the compound nucleus. This makes sense, indeed, if Bohr's independence hypothesis is valid in that the decay of the system is independent of its production. Strictly speaking, this requires an intermediate state with all degrees of freedom in equilibrium. Taken literally, the latter should be given also with respect to the collective degrees of freedom, including the metastable one along which fission progresses. Indeed, this was one of the assumptions behind Kramers' derivation of the rate formula.

Of course, these hypotheses may justly be questioned. After all, collective modes of this type are perhaps the slowest ones of all. For this reason, one may wish to choose an initial density distribution which deviates from that of equilibrium around the potential minimum.[24] We must assume, however, that it is *immaterial to know how (i) the system has reached this region* of deformation space and *(ii) during which time span this has happened*. Otherwise, *any relation to Kramers' picture would definitely not be meaningful at all*. Naturally, the initial configuration has an important impact on the way the density evolves over short time spans. In any case it will take some finite time before the current on top of the barrier reaches the stationary value from which according to (12.11) the rate formula follows. These features are demonstrated in Fig. 12.3 where the current across the barrier is presented for different initial conditions. The calculations have been performed in (Hofmann and Ivanyuk, 2003) by Langevin simulations exploiting a locally harmonic approximation which is equivalent to solutions of a transport equation for the density distribution. In our case it would be one of Smoluchowski type for over-damped motion; amongst other reasons this case is chosen because the effects we aim at increase with damping. The potential used consists of two oscillators, an upright one and one put upside down, both joined to each other with a smooth first derivative. In all cases the asymptotic value of the current is seen to follow the law $\Gamma_K \exp(-\Gamma_K t/\hbar)$, shown by the fully drawn straight line.

(i) For the dashed and dotted curves the system starts out of equilibrium, with the latter being defined for the oscillator centered at Q_a. The dashed curve

---

[24]The corresponding kinetic momentum $P = M\dot{Q}$ may safely be assumed to equilibrate much faster than the coordinate Q.

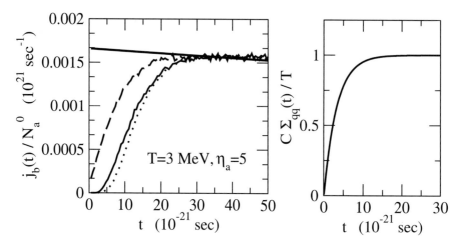

FIG. 12.3. The current across the barrier relative to the initial population at the minimum. It is calculated for different initial conditions, see text, and for the following parameters: $T = 3$ MeV, $E_b = 8$ MeV, $\hbar\varpi_a = \hbar\varpi_a = 1$ MeV and $\eta_a = \eta_b = 5$, see Fig.1 of (Hofmann and Ivanyuk, 2003).

corresponds to the current at the barrier $j_b(t) = j(Q_b, t)$ and the dotted one to that in the scission region $j_{sc}(t) = j(Q_{sc}, t)$.

(ii) For the fully drawn line the system starts at Q_a sharp. The obvious delay by about $6.6 \cdot 10^{-21}$ sec is essentially due to the relaxation of the initial distribution to the quasi-equilibrium in the well. This feature is demonstrated on the right by the t-dependence of the width in Q (exhibited in terms of fluctuations of the potential energy $C\Sigma_{qq}/2 = C(\langle Q^2 \rangle - \langle Q \rangle^2)/2$). For over-damped motion this relaxation time is given by the $\tau_{coll} = \gamma/C = 2\eta/\varpi$ of (6.143) or (6.144).

From this model calculation the physical meaning of two delay times which have been studied in the literature becomes apparent. (1) The "transient time" τ_{trans} it takes before the current at the barrier top reaches its stationary value. (2) The "saddle to scission time" τ_{ssc} during which the system moves from the saddle point to the scission region. Analytic formulas have been derived in (Grangé et al., 1983), (Bhatt et al., 1986) for τ_{trans} and in (Hofmann and Nix, 1983), (Nix et al., 1984) for τ_{ssc} (with numerical estimates based on microscopic input for the transport coefficients in (Hofmann and Kiderlen, 1997)). Incidentally, for over-damped motion both timescales increase linearly with the dissipation strength. In these original papers it was suggested that during these time intervals evaporation processes may take place in addition to those accounted for by Kramers rate formula. This has often been justified, arguing that the overall fission time τ_f can be obtained by just summing up all three contributions, $\tau_f = \tau_{trans} + \tau_K + \tau_{ssc}$, with $\tau_K = \hbar/\Gamma_K$, see e.g. (Hinde et al., 1989). Below we will return to this problem to demonstrate with the help of the "mean first passage time" that such a hypothesis is *not justified*. Truth is, however, that this model is in wide use for analyzing experimental data, and it was in this way that the results on the temperature dependence of nuclear dissipation were obtained which were mentioned at the end of Section 6.6.3.4. Actually it is quite impossible to give due account of all the papers in which this picture has been applied, for which reason we confine ourselves to mentioning a few reviews and original work (Hilscher and Rossner, 1992), (Lestone, 1993), (Paul and Thoennessen, 1994), (Thoennessen, 1996), in which references to earlier publications may be found. However, it must be said that there also exist many applications of the traditional statistical model, with very good results and decent arguments in favor of this approach; amongst them are: (i) (Forster et al., 1987) where experimental outcomes are explained which had been obtained by the channeling method, (ii) the systematic studies by L.G. Moretto and his group (1995b), (1995a) and (iii) new observations of fission properties obtained in reactions where the mother nucleus is excited by 2.5 GeV protons (Tishenko et al., 2005).

Let us return to the physical meaning of τ_{trans} and τ_{ssc}. From their very definition it is clear that both timescales are intimately related to the adopted picture in which the density is followed in time beginning with a certain initial distribution around the potential minimum. Let us look first at the transient time. Fig. 12.3 clearly demonstrates remarkable uncertainties in this concept. The τ_{trans} depends strongly on the initial conditions and there is considerable arbitrariness in choosing time zero. If the calculation were repeated at some later time $t_0 > \tau_{\text{trans}}$, very similar features would be seen! Indeed, the whole effect comes about only because initially there are favorable parts for which it is easiest to reach the barrier. This is demonstrated in Fig. 12.4 where those points of the initial equilibrium are sampled which cross the saddle after some time τ_s. For large values of τ_s *greater parts* of the initial distribution have already "fissioned"; this is shown on the right panel. For the much shorter time $\tau_s \simeq \tau_{\text{trans}}$ shown on the left, only a small fraction of points have succeeded in doing this, namely those which started close to the barrier. The vast majority of particles is still waiting to complete the same motion but at later times! This aspect is important, not only for an understanding of the essentials of the concept of the MFPT, but also with respect to the evaporation of neutrons. Indeed, even for $\tau_K \gtrsim t \gg \tau_{\text{trans}}$ there is ample time for them to be emitted from *inside* the barrier.

Similar properties hold true for the saddle to scission time τ_{ssc}. No matter whether it is calculated from Newton's equation or on the basis of Kramers' stationary solution as in (Hofmann and Nix, 1983), (Nix et al., 1984), it always defines the time lapse within which the system reaches the scission point having been released at the barrier at an arbitrarily chosen time zero. No information enters such considerations about how long the flux across the barrier will be maintained.

By their very definition, neither the τ_{trans} nor the τ_{ssc} carry any message about how long it takes before the system has decayed. In the mean free passage time τ_{mfpt}, on the other hand, such knowledge *is indeed contained*. Amongst others, this is a consequence of the special boundary conditions used for the

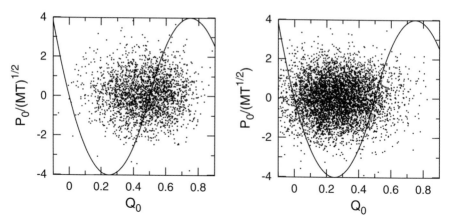

FIG. 12.4. Samples of initial points which overcome the saddle in time τ_s: Left part: $\tau_s \simeq \tau_{\text{trans}}$; right part $\tau_s \simeq \tau_{\text{K}}$. [Fig.2 of (Hofmann and Ivanyuk, 2003)]

solution of the transport equation to obtain formula (12.19). These conditions are different from those employed to get τ_{trans} or the τ_{ssc} from the same transport equation. To understand the importance of the information about the timescale of the source the following analogy may be useful. Think of a stream of water and consider the situation that at some point a piece of wood is dropped on its surface. The τ_{ssc} can then be used to find the average time it takes before that piece has moved a certain distance. But this τ_{ssc} does not contain any information about how long water is still supported by its source–and neither does the τ_{trans}. Back to the case of fission, exactly this timescale is essential to figure out for how long pre- or post-saddle emission of particles is possible.

12.4.2 Implications from the concept of the MFPT

The MFPT is defined in (25.104) as the first moment in time of a genuine probability distribution. The latter accounts properly for the statistics implied by (i) the underlying fluctuating forces, (ii) details of the potential, as well as (iii) appropriate boundary conditions for the solutions of the transport equation; as mentioned these conditions and, hence, the solutions are manifestly different from those used to deduce τ_{trans} or τ_{ssc}. In this probability distribution processes of relaxation in Q are accounted for which happen around the minimum and, for exit points chosen sufficiently far outside the barrier, the sliding down from saddle to scission. In the left part of Fig. 12.5 the $\tau_{\text{mfpt}}(Q_{\text{in}} \to Q_{\text{ex}})$ is shown for different temperatures and as a function of the exit point Q_{ex}, but normalized to its asymptotic value. As can be seen from (12.19), for constant friction this coefficient drops out of such a ratio. The calculation was done for the cubic potential already discussed in the previous section.

The following important features can be seen: If the exit point equals the saddle point the corresponding MFPT is exactly *half the total decay time*–in clear

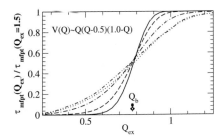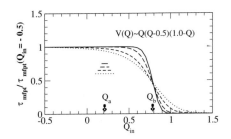

FIG. 12.5. MFPT $\tau_{\text{mfpt}}(Q_{\text{in}} \to Q_{\text{ex}} = 1.5)$ for a cubic potential normalized to its asymptotic values as function of the exit point Q_{ex} (left part, Fig.3 of (Hofmann and Ivanyuk, 2003)) and the initial point Q_{in} (right part). The solid, dashed, dotted-dashed and dotted curves correspond to $T/E_b = 0.1, 0.2, 0.5, 1.0$, see text. The dashed double dotted curve in the left part represents a calculation within the Langevin approach for $T/E_b = 1$.

distinction from the τ_{trans}. Likewise, on average the decaying system spends the same time in the region from saddle to scission. Furthermore, another difference is observed: Whereas τ_{trans} is quite sensitive to the initial starting point (see (Hofmann and Ivanyuk, 2003)) this is much less so for the τ_{mfpt}. This is demonstrated in the right part of Fig. 12.5 where $\tau_{\text{mfpt}}(Q_{\text{in}} \to Q_{\text{ex}})$ is shown as function of the initial point Q_{in}. The length of the horizontal lines represents the thermal fluctuations $\Delta Q = \sqrt{T/C_a}$ of the coordinate around the potential minimum at Q_a, calculated for temperatures which correspond to those used for the τ_{mfpt}. It is seen that the influence of these fluctuations on τ_{mfpt} is not larger than 10% even for temperatures where Kramers' rate formula definitely becomes very questionable.

So far we have been looking only at the relative times the system stays in the regions before or after the saddle. Let us turn now to the *escape time* τ_{esc} introduced in Section 12.3 as the asymptotic value of $\tau_{\text{mfpt}}(Q_{\text{in}} \to Q_{\text{ex}})$ reached when the exit point lies well beyond the saddle point. Actually, as seen from Figs.12.1 and 12.2 this τ_{esc} may considerably overshoot the τ_K given on the right of (12.22). (Recall that for the present purpose one should set $\gamma_a = \gamma_b = \gamma$ such that $\tau_{\text{esc}} \equiv \tau_K$ whenever the conditions for (12.22) on the saddle point approximation are given.) It is interesting to see this τ_{esc} to differ from τ_K even for the simple potential with just one minimum and one barrier. Naturally, the overshoot becomes larger for more complex structures, as we have seen in Fig. 12.2. The τ_{esc} may be identified as the fission decay time τ_f with the associated fission width $\Gamma_f = \hbar/\tau_f \equiv \hbar/\tau_{\text{esc}}$. Hence, larger values of τ_{esc} imply more time for emission processes to happen. Suppose we look at neutrons. Whenever their average width Γ_n may be used to calculate their multiplicity from Γ_n/Γ_f, the enhancement of this number over that given by Γ_n/Γ_K is determined by the ratio τ_{esc}/τ_K, viz $\Gamma_n/\Gamma_f = (\Gamma_n/\Gamma_K)(\tau_{\text{esc}}/\tau_K)$. To get some feeling for absolute values of this extra available time one may estimate the pre-factor $\gamma/\sqrt{C_a|C_b|} \simeq \gamma/C \equiv$

$\overline{\tau}_{\rm coll}$ of (12.22) using the microscopic transport coefficients from Section 6.6.3. They imply (see e.g. Fig. 6.11 or (Hofmann et al., 2001)) that for temperatures between 1 and 4 MeV this $\tau_{\rm coll}$ shows an almost linear dependence in T such that one may write $\overline{\tau}_{\rm coll} \simeq -(3/4 + (5/4)T(\hbar/{\rm MeV})$. Applying this to formula (12.22) and choosing $T/E_b \simeq 0.5$ the $\tau_{\rm K}(T)$ becomes about $200\,\hbar/{\rm MeV}$ large. For a $\tau_{\rm esc}/\tau_{\rm K}$ of about 1.5 the *additional time increment* $\Delta\tau_{\rm f} = \tau_{\rm esc} - \tau_{\rm K}$ takes on the sizable value of roughly $100\,\hbar/{\rm MeV}$, and, hence, is at least as large as a typical transient time.

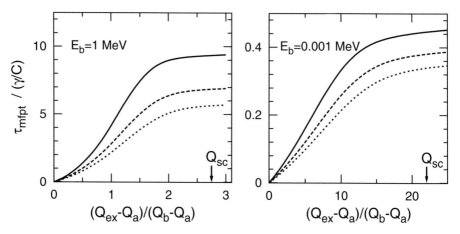

FIG. 12.6. The MFPT for the case of small and vanishing barriers for $T = 2, 3, 4$ MeV from top to bottom. The $Q_{\rm sc}$ corresponds to a scission point 20 MeV below the barrier. [Fig.4 of (Hofmann and Ivanyuk, 2003)]

Finally, let us look at cases which definitely do not allow for the application of the saddle point approximation, and for which thus Kramers' formula does not make any sense at all. There is no reason, however, why formula (12.19) for the MFPT should not be applicable. In Fig. 12.6 the results of calculations of this $\tau_{\rm mfpt}$ are shown (normalized to the local relaxation time $\tau_{\rm coll} = \gamma/C$ at the saddle) for the small barrier of $E_b = 1$ MeV as well as for a practically vanishing barrier (for constant friction, again). It is seen that even in the latter case the $\tau_{\rm mfpt}(Q_a \to Q_{\rm ex})$ reaches a plateau for sufficiently large $Q_{\rm ex}$, which may then be identified as an escape time $\tau_{\rm esc}$ again. Also this $\tau_{\rm esc}$ may be long enough for neutrons to be evaporated before scission, which may be seen as follows. The neutron width typically is of the order of $1 - 3$ MeV (for medium heavy nuclei (Pomorski et al., 1996)). For the γ/C we may again use the estimate just given. Taking for $\tau_{\rm esc}$ the values at the plateau from the right part of Fig. 12.6 one gets a width $\Gamma_{\rm esc} = \hbar/\tau_{\rm esc}$ of the order of $0.9 - 2.3$ MeV, which is comparable to the neutron width. From the left part of Fig. 12.6 it is seen that a small increase of the barrier by 1 MeV enlarges the $\tau_{\rm esc}$ drastically.

12.5 Inclusion of quantum effects

Possible extensions of Fokker–Planck equations to quantum physics have been described in Chapter 7. They are based on the locally harmonic approximation and manifest themselves in modified diffusion coefficients. As discussed in Section 7.3.1 in barrier regions these quantal diffusion coefficients can be defined only above a critical temperature T_c. In terms of the local frequency ϖ_b the T_c was seen to be smaller than $\hbar\varpi_b/\pi$ and to decrease with increasing friction. We are now going to describe how this modification may be employed to extend Kramers' formula for the decay rate to $R = f_{q.c.} R_K$ with a quantum correction factor $f_{q.c.}$.

Many problems of fundamental nature have been obtained with functional integral approaches for the Caldeira–Leggett model (Ingold, 1988), (Ankerhold et al., 1995). It was introduced in Section 7.5 and its applications to functional integrals are briefly described in Section 24.1.3. The great benefit of the model is seen in the fact that the "bath variables" x_i can be integrated out to obtain an effective action $S[q(t)]$ for the "system variable" which parameterizes the dynamics of the metastable degree of freedom. In the original, pioneering work by A.O. Caldeira and A.J. Leggett (1983) the authors concentrated on the decay of the ground state. Later this work was generalized to finite temperatures in (Larkin and Ochinnikov, 1984) and (Grabert and Weiss, 1984), (Grabert et al., 1984a); excellent reviews on this subject may be found in (Grabert et al., 1988), (Hänggi et al., 1990), (Weiss, 1993). If the path integral method is used for real time propagation (as done in the previously mentioned papers (Ingold, 1988), (Ankerhold et al., 1995)) one runs into the problem with the critical temperature T_c. This difficulty may be overcome if one deals with "imaginary time propagation" for which the $S[q(t)]$ turns into the Euclidean action $S_E[q(\tau)]$ described in Section 24.2.

In all these cases the remaining path integrals for the system variable cannot be evaluated exactly. Rather one has to rely on expansions of the (effective) action around the "classical trajectories" defined by the equation $\delta S_E[q(\tau)] = 0$. The quality of the approximation depends on how far this expansion of $S_E[q(\tau)]$ can be performed. In the usual Gaussian approximation the critical temperature reduces to the smaller value of the so-called *crossover temperature* $T_0 < T_c$. At T_0 quantum effects become dominant such that one may no longer speak of corrections to Kramers' rate formula. Formally this is seen by the fact that above T_0 the stationary points corresponding to $\delta S_E[q(\tau)] = 0$ just consist of the two points of the minimum and the maximum of the potential. The expansion of $S_E[q(\tau)]$ to second order (Gaussian approximation) just includes harmonic "vibrations" around the stationary points. This feature exhibits the close relation to the derivation of the formula $R = f_{q.c.} R_K$ within our LHA. At smaller temperatures the "stationary points", or "stationary functionals", rather, consist of genuine trajectories which connect the region of the potential minimum with that outside potential. Evidently, in such a case the solution of lowest order, namely the "classical" trajectory itself already carries important information about quantum effects of global motion.

As outlined before in Section 7.5.1, the Caldeira–Leggett model is of only limited use for applications in nuclear physics. The LHA in its quantal version delivers one possible way to incorporate essentials of nuclear structure. Below we will at first sketch details of the derivation of $R = \text{f}_{\text{q.c.}} R_\text{K}$ within the LHA. Thereafter we will address a functional integral approach which allows one to overcome the deficiencies one encounters for the real time propagation with which the LHA is irrevocably associated. It must be said, however, that, partly because of the much greater complexity of these nuclear models they do not allow for such detailed and formally far-reaching studies which in the past have been possible for the Caldeira–Leggett model.

12.5.1 Quantum decay rates within the LHA

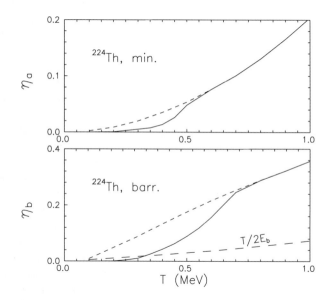

FIG. 12.7. The damping factor η at the potential minimum (top) and at the fission barrier (bottom) of ^{224}Th; for fully drawn (dashed) curves (no) pairing correlations are considered. Remember that Kramers' high viscosity limit is given for $\eta_\text{b} \gtrsim T/2E_\text{b}$ (12.16). [Fig.1 of (Hofmann and Ivanyuk, 1999)]

The derivation of formula $R = \text{f}_{\text{q.c.}} R_\text{K}$ largely follows that of its classical counterpart (12.9). Again we will assume we are given a system with a potential $V(q)$ with just one minimum and one barrier. As before, the precise shape of this $V(q)$ is not very important but one may construct a model case using two shifted oscillators, a normal one and one put upside down.[25] Refer-

---

[25]To have a potential with a continuous first derivative one may choose $V(Q) = -E_\text{b} + C_\text{a}(Q - Q_\text{a})^2/2$ for $Q \leq Q_\text{s}$ and $V(Q) = -|C_\text{b}|(Q - Q_\text{b})^2/2$ for $Q \geq Q_\text{s}$, with the E_b being

ring once more to a stationary situation the value of the rate may be deduced from the $R_K = j_{st}(Q_b)/N_a$ of (12.11). In principle, both the current $j_{st}(Q_b)$ as well as the number of particles in the well N_a ought to be calculated from a globally valid density distribution. Obviously for this an overall normalization factor drops out, for which reason we may apply a non-normalized Wigner function $k^W(Q, P)$, as a generalization of the type discussed in Section 7.4.1 for harmonic oscillators. By analogy with (12.12) one might wish to try the form $k^W_{st}(Q, P) = \xi(Q, P) k^W_{eq}(Q, P)$ where $k^W_{eq}(Q, P)$ represents global equilibrium.[26] However, as explained in Section 11.1.1 in the quantum case such a $k^W_{eq}(Q, P)$ will not be of the simple structure $\exp(-\mathcal{H}(Q, P)/T)$. For a general equation like (11.1) its construction is probably almost as difficult as the solution for the $k^W_{st}(Q, P)$ itself. This is different for the Caldeira–Leggett model where global solutions have been found in (Ingold, 1988) and (Ankerhold et al., 1995). Of course, under the usual assumptions for Kramers' rate formula the final result will be dominated by the linearized versions given in (7.35) and (7.37). These forms have been chosen in such a way that according to (Ingold, 1988) they allow for a direct application of (12.11). With the current from eqn(7.39) and the $k^W_{eq}(Q, P)$ around Q_a given by (7.35) one obtains (Hofmann et al., 1993)

$$R = \frac{|\mathcal{Z}_b|}{\mathcal{Z}_a} \frac{\varpi_b}{2\pi} \left(\sqrt{1+\eta_b^2} - \eta_b\right) e^{-E_b/T} = \frac{|\mathcal{Z}_b|}{\mathcal{Z}_a} \frac{\varpi_b}{\varpi_a} R_K \equiv f_{q.c.} R_K. \quad (12.23)$$

The partition functions appearing here are such that they belong to oscillators with their centers put at zero energy. The difference E_b in the energy between the barrier and the minimum is put into the Arrhenius factor. The last equation on the right defines the *quantum correction factor* $f_{q.c.}$ to the R_K of (12.9).

Before we discuss another derivation of this formula it may be of interest to address numerical evaluations presented in (Hofmann and Ivanyuk, 1999). They are based on the simplified expression (12.31) for the $f_{q.c.}$ which is discussed below. For the transport coefficients appearing there microscopically computed values have been considered with pairing correlations included (see Section 6.6.3.3); the Γ of (12.31) was defined as the average between the values at the minimum and at the barrier. In Fig. 12.7 for the case of ^{224}Th the damping parameter $\eta = \gamma/(2\sqrt{M|C|})$ is shown as a function of temperature both at the potential minimum and at the barrier. In the lower part the quantity $T/2E_b$ is exhibited which according to (12.16) must be smaller than η_b to ensure applicability of the picture employed for the derivation of the rate, which in Kramers' classic version is the "high viscosity limit". It is seen that this condition is definitely violated for temperatures below about 0.5 MeV where pairing correlations imply the motion to be practically undamped (see item (4) in Section 6.6.3.3. As

given by $E_b = |C_b|(Q_s - Q_b)^2(|C_b| + C_a)/(2C_a)$.

[26] Evidently, for a potential which is unbound beyond the barrier this $k^W_{eq}(Q, P)$ can be defined only up to a certain value of Q. Otherwise one must modify the potential at very large values of Q.

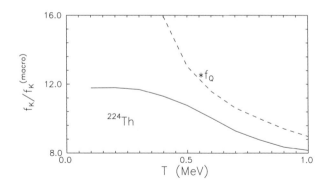

FIG. 12.8. The ratio of Kramers correction factor f_K with microscopic input to its macroscopic counterpart as a function of temperature. The dashed line represents the quantity $f_{q.c.} f_K / f_K^{macro}$ with $f_{q.c.}$ being the quantum correction factor (for $|\varpi_b| = \varpi_a$). [Fig.2 of (Hofmann and Ivanyuk, 1999)]

we may recall from Section 7.3.1 the crossover temperature T_c lies in the same regime. As a conclusion one may say that for such low temperatures the LHA is not capable of accounting for those quantum effects in the decay rate which are to be attributed to collective motion.

We should remember, however, that within the LHA *nucleonic motion* is treated *fully quantal*. The relative importance of these two, quite different effects varies with temperature. This is demonstrated in Fig. 12.8, which shows the ratio of two estimates of Kramers' correction factor $f_K = (1+\eta_b^2)^{1/2} - \eta_b$ calculated for two different models. Whereas for the numerator the η_b from Fig. 12.7 is used, the denominator is chosen to correspond to the macroscopic limit described in Section 6.6.1.1. As we may recall, in the latter any quantum effects of nucleonic dynamics are ignored. It is seen that quantum features in single particle motion enhance the decay rate by about one order of magnitude. Also shown in Fig. 12.8 is the influence of the correction factor $f_{q.c.}$ which accounts for quantum effects in collective motion. It leads to an additional increase in the rate. This increase is seen to be comparatively small; above $T \simeq 0.5$ MeV, where this factor can be trusted, it amounts to about 30%. Later on we will be able to demonstrate this feature to hold true also for other examples of collective motion like fusion of two heavy ions.

12.5.2 Rate formulas for motion treated self-consistently

In previous sections a close relation of the decay rate to the partition function was encountered. On the classical level such a feature was seen in Langer's version (12.13) of the rate formula. With a $T_0 = \hbar\varpi_b((1+\eta_b^2)^{1/2} - \eta_b)/(2\pi)$ it may be rewritten in the form

$$R(T \geq T_0) = -\frac{2}{\hbar} \frac{T_0}{T} \operatorname{Im} \mathcal{F}(\beta) \equiv -\frac{\varpi_b}{\pi T}\left(\sqrt{1+\eta_b^2} - \eta_b\right) \operatorname{Im} \mathcal{F}(\beta). \quad (12.24)$$

Its range of validity and the implication of the appearance of the T_0 will be discussed below. Partition functions also appeared in the quantum correction factor $f_{q.c.}$ in (12.23), which involves the ratio of two partition functions for linearized dynamics calculated at the barrier and the potential minimum.

An elegant way of treating collective motion fully self-consistently is given by the functional integral method described in Section 24.4. Whenever the two-body interaction of the many-body system can be modelled in separable form this procedure even allows for practical applications. As explained in chapters 6 and 7 the LHA is closely related to a Hamiltonian of the structure given in (6.49) where the two-body interaction is of the form $(k/2)\hat{F}\hat{F}$. In the following we will restrict ourselves to this example. It will allow us to give another proof of formula (12.23) by applying the so-called "Perturbed Static Path Approximation" (PSPA) of Section 24.4.3. In this approximation one treats fluctuations around stationary functionals to second order. Actually, to obtain the decay rate of a metastable system only *stationary points* of the potential will be considered. This effective potential is defined by the free energy of the system with the collective coordinate treated in the static limit. To facilitate comparisons with the results of the previous sections, we will assume the free energy to be of the same form as in Kramers' case, namely that there are only two such stationary points, one defining the minimum and one the barrier.

The following presentation is largely based on work done with C. Rummel, see (Rummel, 2000), (Rummel and Hofmann, 2001), (Rummel, 2004), (Rummel and Hofmann, 2005a). Whereas (Rummel, 2000), (Rummel and Hofmann, 2001) only deals with the application of the PSPA, in (Rummel, 2004), (Rummel and Hofmann, 2005a) methods have been developed to overcome deficiencies of the PSPA. Here, we will restrict ourselves to displaying the main features of the procedure for the example of the PSPA, and will only briefly mention results of the more advanced procedures at the end.

For the PSPA the partition function reads (see (24.108))

$$Z^{\text{PSPA}}(\beta) = \sqrt{\frac{\beta}{2\pi |k|}} \int d\bar{q} \, \exp\left(-\beta \mathcal{F}^{\text{SPA}}(\beta, \bar{q})\right) \mathcal{C}^{\text{PSPA}}(\beta, \bar{q}) \,. \qquad (12.25)$$

The $\mathcal{F}^{\text{SPA}}(\beta, \bar{q})$ is the free energy in SPA approximation. Its definition is given in (24.107) from where it can be seen that it is identical to the free energy calculated in mean-field approximation used throughout this book. The $\mathcal{C}^{\text{PSPA}}(\beta, \bar{q})$ specifies the quantum correction to the partition function in the classical limit to which the $Z^{\text{PSPA}}(\beta)$ of (12.25) reduces for $\mathcal{C}^{\text{PSPA}}(\beta, \bar{q}) = 1$. In the following we want to restrict ourselves to the model of having simply one local collective mode together with the corresponding intrinsic one. This model was introduced in Section 6.5.3.1 and is applied to the PSPA at the end of Section 24.4.3. From formulas (24.119) and (24.126) one gets

$$\mathcal{C}^{\text{PSPA}}(\beta, \bar{q}) = \prod_{r>0} \left(\lambda_r(\beta, \bar{q})\right)^{-1} = \prod_{r>0} \frac{\nu_r^2 + \nu_r \Gamma + \Omega^2}{\nu_r^2 + \nu_r \Gamma + \varpi^2} \,, \qquad (12.26)$$

with the frequency $\varpi(\beta,\bar{q})$ and the width $\Gamma(\beta,\bar{q})$ of the collective mode. For the intrinsic mode only the effective stiffness changes to $M\Omega^2 = M\varpi^2 - k$. It is worth stressing that in (12.26) the infinite product is convergent. This is ensured by the feature (Whittaker and Watson, 1992) that the factors of the terms linear in ν_r in the numerator and the denominator are identical.

According to (12.24) the decay rate is proportional to the imaginary part $\mathrm{Im}\{Z^{\mathrm{PSPA}}(\beta)\}$ of the free energy. For the present problem we may proceed as described in Section 12.2 below eqn(12.13), where Langer's suggestion as to how the integral in (12.25) is to be handled for metastable systems is presented. Translated to the PSPA the $\mathrm{Im}\{\mathcal{F}^{\mathrm{PSPA}}(\beta)\} = -T\,\mathrm{Im}\{\ln Z^{\mathrm{PSPA}}(\beta)\}$ is estimated to $-T\,|\,\mathcal{Z}^{\mathrm{PSPA}}_{>q_{\mathrm{b}}}(\beta)\,|\,/\,Z^{\mathrm{PSPA}}_{q_{\mathrm{a}}}(\beta)$, where the local partition functions are given by

$$\mathcal{Z}^{\mathrm{PSPA}}_{>q_{\mathrm{b}}}(\beta) = \sqrt{\frac{\beta}{2\pi|k|}} \exp\bigl(-\beta\mathcal{F}^{\mathrm{SPA}}(\beta,\bar{q}_{\mathrm{b}})\bigr)\, \mathcal{C}^{\mathrm{PSPA}}(\beta,\bar{q}_{\mathrm{b}}) \int_0^{i\infty} d\xi\; e^{\beta|C^{\mathrm{SPA}}_{\mathrm{b}}|\xi^2/2}$$

$$= \frac{i}{2}\sqrt{\frac{1}{|k|\,|C^{\mathrm{SPA}}_{\mathrm{b}}|}} \exp\bigl(-\beta\mathcal{F}^{\mathrm{SPA}}(\beta,\bar{q}_{\mathrm{b}})\bigr)\, \mathcal{C}^{\mathrm{PSPA}}(\beta,\bar{q}_{\mathrm{b}}), \qquad (12.27)$$

$$Z^{\mathrm{PSPA}}_{q_{\mathrm{a}}}(\beta) = \frac{1}{\sqrt{|k|\,C^{\mathrm{SPA}}_{\mathrm{a}}(\beta)}}\, \exp\bigl(-\beta\mathcal{F}^{\mathrm{SPA}}_{\mathrm{a}}(\beta)\bigr)\, \mathcal{C}^{\mathrm{PSPA}}(\beta,\bar{q}_{\mathrm{a}}). \qquad (12.28)$$

In these expressions the quantum corrections are contained in the $\mathcal{C}^{\mathrm{PSPA}}$. Let us first look at the classical limit where these factors have to be put equal to one. As this limit is identical to the "Static Path Approximation" we may denote the corresponding rate by $R^{\mathrm{SPA}}_{\mathrm{class}}$. It is obtained by using (12.27) and (12.28) for the basic formula (12.24) which then becomes (Rummel, 2000), (Rummel and Hofmann, 2001)

$$R^{\mathrm{SPA}}_{\mathrm{class}} = \bigl(\varpi_b/2\pi\bigr)\left(\sqrt{1+\eta_b^2}-\eta_b\right)\sqrt{C^{\mathrm{SPA}}_{\mathrm{a}}/\,|C^{\mathrm{SPA}}_{\mathrm{b}}|}\; \exp(-E^{\mathrm{SPA}}_{\mathrm{b}}/T). \qquad (12.29)$$

Here, the barrier height is defined as $E^{\mathrm{SPA}}_{\mathrm{b}} = \mathcal{F}^{\mathrm{SPA}}(\beta,\bar{q}_{\mathrm{b}})-\mathcal{F}^{\mathrm{SPA}}(\beta,\bar{q}_{\mathrm{a}})$. Formula (12.29) is very similar to that of Kramers of (12.9). Under the following conditions both become identical: (i) The stiffnesses of the static free energy C^{SPA} can be expressed by the frequencies of the local modes through the relation $C^{\mathrm{SPA}} = M\varpi^2$. (ii) The inertia at the potential minimum M_{a} is equal to the M_{b} at the barrier. The first condition may be fulfilled for slow motion as a consequence of the secular equation, see Section 6.5.2. If the corresponding inertia varies with the collective coordinate one would get the form (12.18), which is to say Kramers' result multiplied by the factor $\sqrt{M_a/M_b}$. Comments on this result have already been given below (12.18). Recall please that this additional factor does not show up if one starts with Langer's formula (12.13)–provided for the partition function the formula is chosen which is appropriate for variable inertia. To reach the same level of sophistication for the present problem one will have to modify the way the collective degree of freedom is introduced in the Hubbard–Stratonovich transformation. Unlike the Bohm–Pines method sketched in Section 7.5.1 the version used in the literature so far does not allow for a decent introduction of a collective inertia, see Section 24.4.1 and Section 24.4.2.

Next we may study the inclusion of quantum effects. By analogy with (12.23) we may write the total rate as $R^{\mathrm{PSPA}} = \mathrm{f}_{\mathrm{q.c.}} R^{\mathrm{SPA}}_{\mathrm{class}}$. According to the previous discussion and based on (12.27) and (12.28) the quantum correction factor $\mathrm{f}_{\mathrm{q.c.}}$ here is given by

$$\mathrm{f}_{\mathrm{q.c.}}(\beta) = \frac{\mathcal{C}^{\mathrm{PSPA}}(\beta, \overline{q}_{\mathrm{b}})}{\mathcal{C}^{\mathrm{PSPA}}(\beta, \overline{q}_{\mathrm{a}})} = \prod_{r>0} \left[\frac{\nu_r^2 + \nu_r \Gamma_{\mathrm{b}} + \Omega_{\mathrm{b}}^2}{\nu_r^2 + \nu_r \Gamma_{\mathrm{a}} + \Omega_{\mathrm{a}}^2} \times \frac{\nu_r^2 + \nu_r \Gamma_{\mathrm{a}} + \varpi_{\mathrm{a}}^2}{\nu_r^2 + \nu_r \Gamma_{\mathrm{b}} - \varpi_{\mathrm{b}}^2} \right] . \quad (12.30)$$

In this expression it was accounted for that even at the barrier, where the collective stiffness becomes negative, the corresponding intrinsic modes are positive. For simplified models where the intrinsic modes do not vary with the collective variable the $\mathrm{f}_{\mathrm{q.c.}}(\beta)$ reduces to the form of the Caldeira–Leggett model,

$$\mathrm{f}_{\mathrm{q.c.}}(\beta) = \prod_{r>0} \frac{\nu_r^2 + \nu_r \Gamma + \varpi_{\mathrm{a}}^2}{\nu_r^2 + \nu_r \Gamma - \varpi_{\mathrm{b}}^2} \quad \text{for} \quad \Gamma_{\mathrm{a}} = \Gamma_{\mathrm{b}} = \Gamma , \quad \Omega_{\mathrm{a}} = \Omega_{\mathrm{b}} . \quad (12.31)$$

More precisely, one should say that this expression is obtained for constant (Ohmic) damping. Within the Caldeira–Leggett model one may include non-Markovian effects where the damping coefficient Γ itself varies with ν_r. Please notice that (above T_0) the $\mathrm{f}_{\mathrm{q.c.}}(\beta)$ of (12.30) is well defined. This follows from the fact that it is the ratio of two quantities of the type given in (12.26) for which the infinite products involved converge. Such a property would not be given if in (12.31) different Γs were used for the barrier and for the minimum, as suggested in (Fröbrich and Tillack, 1992).

Let us turn now to properties of the $\mathrm{f}_{\mathrm{q.c.}}$ of (12.30). First of all, it is evident that this quantity diverges to $+\infty$ when the temperature approaches the critical value T_0 from above. Indeed, according to (24.127) at the barrier this temperature is given by $T_0 = \hbar \varpi_{\mathrm{b}} ((1+\eta_{\mathrm{b}}^2)^{1/2} - \eta_{\mathrm{b}})/(2\pi)$. It follows that for $T = T_0$ and $r = 1$ the factor which contains ϖ_{b} vanishes, $\nu_1^2 + \nu_1 \Gamma_{\mathrm{b}} - \varpi_{\mathrm{b}}^2 = 0$. All other factors are positive and different from zero for $T > T_0$. As we may recall, this T_0 is the critical temperature of the PSPA. Since T_0 is smaller than the critical temperature T_c encountered for the real time approach of the LHA the PSPA may in this sense be interpreted as an improvement. For the limiting case of zero damping the T_0 can be seen to be of just half the value of the T_c (see exercise 3). One should not forget, however, that the functional integral approach can only be used for processes where the temperature stays constant. This condition is hardly fulfilled in realistic fission events and is not required to hold true for the LHA.

The behavior for increasing temperature remains to be looked at. From (12.30) and (12.31) it is more or less obvious that $\mathrm{f}_{\mathrm{q.c.}}(\beta)$ turns to 1 as soon as T is larger than the relevant timescales involved, as specified by the various frequencies and widths. As a matter of fact, the widths Γ and the frequencies Ω for internal motion are typically of the same order when calculated at Q_{a} and Q_{b} each. Thus the most relevant factor for this question is the one involving ϖ_{a}

and $\varpi_{\rm b}$. From the numerical estimate given in the previous section and which was based on (12.31) it is clear that $f_{\rm q.c.}(\beta)$ turns to unity for quite small values of T, see again Fig. 12.8. Remember that according to Section 6.6.3 the value of collective frequencies is about 1 MeV/$\hbar$ large. The Γs can be estimated from (6.145) which implies that Γ/ν_r is small compared to one even for the lowest rs.

So far the whole discussion has been restricted to temperatures larger than the T_0 below which the PSPA breaks down. As a fortunate coincidence in that regime one may employ Langer's formula (12.24). For lower temperatures it is commonly understood to take

$$R(T < T_0) = -2 \, {\rm Im} \, \mathcal{F}(T) \, /\hbar \qquad (12.32)$$

instead, see e.g. (Hänggi et al., 1990), (Weiss, 1993). This formula goes back to I. Affleck who in (Affleck, 1981) has shown that it gives the same result as a Boltzmann average over the decay rate $R = -2 \, {\rm Im} \, E/\hbar$ known to be valid for pure quantum states. Notice that formulas (12.24) and (12.32) coincide at $\beta = \beta_0$. Based on (12.32) it has been a great challenge to evaluate quantum corrections also for lower temperatures. Indeed, as already mentioned in this section within the Caldeira–Leggett model is has been possible (see (Larkin and Ochinnikov, 1984) and (Grabert and Weiss, 1984), (Grabert et al., 1984a)) to solve this problem. In the region around T_0 one may expand the effective action around the stationary trajectories to orders higher than two. In fact, at lower and lower temperatures quantum effects become more and more important which reflect genuine quantum tunneling. For this reason the notion *crossover temperature* was coined because T_0 marks the transition from "thermal hopping" to quantum barrier penetration.

For the nuclear problem two improvements over the PSPA approach have been suggested. In (Rummel and Ankerhold, 2002), (Rummel, 2004) a cumulant expansion to fourth order was performed (see Section 24.4.2) following earlier work within the Caldeira–Leggett model (Grabert et al., 1988), (Hänggi et al., 1990), (Weiss, 1993). In this way one has been able to move down the critical temperature from T_0 to half this value. In (Rummel, 2004), (Rummel and Hofmann, 2005b), (Rummel and Hofmann, 2005a) a method was developed which allows the Feynman–Kleinert variational procedure (Feynman, 1988), (Feynman and Kleinert, 1986), (Kleinert, 1990) to be applied. In this method one tries to work with local oscillators the frequency of which is treated as a variational parameter. As the most important feature, these local frequencies turn out to be such that for the Gaussian integrals practically no divergence problems occur, even in the neighborhood of barriers. In (Rummel, 2004), (Rummel and Hofmann, 2005b), (Rummel and Hofmann, 2005a) the main problem was to generalize the original work to the many-body problem. So far only solutions for schematic models have been worked out which showed promising results.

12.5.3 Quantum effects in collective transport, a true challenge

Before closing this chapter it may be worthwhile to contemplate on the theoretical problems one is facing for a truly quantal transport theory for slow collective motion in nuclei. Let us begin by comparing advantages and disadvantages of the methods described before, starting with the LHA. The latter is applicable only above the critical temperature T_c, but where collective quantum effects may still be measurable although they are unobserved so far. The important benefits of the LHA are that it may account for various features which are essential for nuclear dynamics: (i) the implication of shell effects on the transport coefficients, (ii) the fact of a varying temperature, (iii) the evaporation of light particles (recall the discussion in Section 11.2.2). These latter two facts are probably out of range of any application of functional integrals. It is true that this method allows one to treat collective motion in global fashion but only at the expense of working with a constant temperature, and no simultaneous treatment of evaporation processes appears to be possible. This method may be more adequate in handling the self-consistency problem, but still there is the unresolved question of an appropriate introduction of the collective inertia. Moreover, the evaluation of decay rates for systems where the internal static energy is of more complex structure seems to be very difficult and has not been tackled so far; recall that at lower temperatures shell effects will lead to several humps. Evidently, none of these issues can be treated by the Caldeira–Leggett model.

These observations show that there are great challenges for future work. Indeed, the nuclear transport problem has many facets which are not found in other fields. Besides the more or less obvious problems already mentioned one ought to bear in mind also the following. It would be a major improvement if one were able to treat the transition from undamped to damped motion. The latter may occur practically along one trajectory, say in imaginary time. Indeed, there are situations where the system tunnels back and forth between regions of high level density to those of small level density. This happens in heavy nuclei for transitions between the first, second or even third well, which have been observed experimentally (Thirolf and Habs, 2002). As a major difficulty, perhaps, one should mention that such processes happen at very small internal excitation. This means that one may well be in a regime where the concept of temperature is not applicable at all.

Exercises

1. Calculate the rate formulas starting from the transport equations (11.11) and (11.18) in the high T-limit.
2. Calculate $\tau_{\rm esc}$ in saddle point approximation for the transport equations (11.12) and (11.18) (in the high T-limit).
3. Calculate the critical temperature T_0 for the PSPA at zero damping and compare it with the critical temperature T_c of (7.33) for the LHA to real time propagation.

13

HEAVY-ION COLLISIONS AT LOW ENERGIES

The processes encountered in heavy-ion collisions at low bombarding energies have much in common with other types of large-scale collective motion like the dynamics of multipole modes or nuclear fission. All of them may be described or parameterized by a few collective degrees of freedom. The new impact experimental observations from the late 1960s and early 1970s had on theoretical descriptions was the clear evidence for the need of generalizing existing collective theories to such of non-equilibrium thermodynamics. At first one was only aiming at models based on Newtonian dynamics but it soon became clear that fluctuating forces had to be included. In the end this led to a revival of the application of Fokker–Planck equations of the type Kramers had suggested much earlier, see the discussion in the previous chapters.

From the microscopic point of view one should say, however, that such equations are much less justified for heavy-ion collisions than for fission. There are essentially two problems, both concerning the entrance phase of the reaction, the use of temperature, on the one hand, and the fact that the timescales for collective and intrinsic motion are much less separated than they are for fission. Nevertheless, since the 1970s numerous practical applications of transport equations have been employed, which at first were solely based on phenomenological input for the transport coefficients. Later the application of macroscopic models became very popular, for example the liquid drop model for the inertia (in Werner–Wheeler approximation), the wall and window formula for friction and the Einstein relation for the stochastic force. It can be said that since then not much has changed. For computational reasons one nowadays prefers the Langevin formulation over that with Fokker–Planck equations. For the transport coefficients, however, one mostly still relies on those of the macroscopic picture. Commonly these are applied even for fusion reactions in which superheavy elements are produced. This is quite peculiar, considering the well-known fact that the very existence of super-heavy elements cannot be explained without considering shell effects for the potential landscape, see e.g. the review article (Armbruster, 1999).

In the following we will restrict ourselves to presenting some of the main physical concepts and ideas developed and used in the past 30 to 40 years. It is not possible to give due credit to all of the interesting studies published in numerous papers. An exhaustive overview both of the theoretical work as well as of its application to the vast experimental data existing at that time can be found in the excellent review article by W.U. Schröder and J.R. Huizenga (1984). Our main aim will be to reflect on the physics behind the transport models and the assumptions of the underlying microscopic theories. In doing so we will confine ourselves to such models which have been in use most in the past few decades. Furthermore, we will concentrate on theories which allow for a description of the time evolution of statistical fluctuations. First, we will briefly address deep inelastic collisions and then turn to fusion reactions, in particular to those which lead to super-heavy elements. This will include a study of the importance of quantum effects both for collective dynamics as well as for microscopic transport coefficients. The chapter will be closed with some critical remarks on general questions of the applicability of transport theories to this problem.

13.1 Transport models for heavy-ion collisions

Imagine two heavy ions moving towards each other with a certain energy in their center of mass system. Before contact the relevant degrees of freedom are the components of the vector **r** of relative motion, together with the relative momentum **p**. Here, the notion "before contact" is to be understood in the sense that the mutual force does not yet excite the fragments, such that the nuclei are in their ground state, which may or may not involve a finite deformation of the ions. Because of the comparatively large reduced mass–the inertia in this channel– relative motion may be treated on the classical level of Newton's equations. The potential will be given by the mutual Coulomb energy plus an eventual beginning trace of the nucleon–nucleon force, which the nucleons in one fragment feel from those of the other.[27] As soon as the dissipative forces become sizable the fluctuating forces will begin to act as well. At that point Newtonian dynamics ceases to be correct, even if quantum features are neglected. The fluctuating force will create statistical fluctuations which manifest themselves by increasing second moments.

When the two nuclei have begun to overlap nucleons may be transferred from one fragment to the other. In (1974) W. Nörenberg suggested treating this phenomenon as a stochastic process with the time lapse over which it lasts being determined by the classical trajectory for relative motion. In a more general frame one would imagine this to be a process in a multidimensional potential landscape with the mass asymmetry being one possible degree of freedom. In this sense one is faced with the same basic problem as in all applications of models of collective dynamics, to guess which collective coordinates are necessary to parameterize certain processes and which one can one afford to consider in practical applications. But once such a model is set up one should then describe all degrees of freedom with the same transport model including statistical fluctuations.

---

[27]These interactions my lead to mutual excitations of the fragments, which in (Broglia et al., 1974) have been treated in semi-classical approximation, for more details see also (Schröder and Huizenga, 1984). In phenomenological transport models this feature is either ignored or assumed to be lumped into the ansatz for the friction force. This latter hypothesis is very questionable as the excitations of vibrational modes can hardly be expected to be proportional to the fragments relative velocity.

Which degrees of freedom are relevant in which part of the reaction is also a question of timescales. In theories of heavy-ion collisions one often distinguishes between potential landscapes of the sudden or the adiabatic limit. Suppose again the system's dynamics is characterized by a certain (sub-)set $\{Q_\mu^{\text{sud}}\}$ of parameters. For the sudden limit the system's residual degrees of freedom have no time to follow the dynamics in the Q_μ^{sud}. Rather than adjusting continuously to the $Q_\mu^{\text{sud}}(t)$, as in the adiabatic limit, the system stays in its initial configuration. For heavy-ion collisions this notion is mostly interpreted in the sense of shape degrees of freedom, see e.g. (Ngô, 1986). In the sudden approximation applied to the entrance phase of the collision the two fragments are assumed to keep their spherical shape even after the tails of the densities have started to overlap. Conversely, in the adiabatic limit the two fragments might already deform each other at the early stages of the reaction before they merge into one common system.

In a more general formulation one would picture both processes by the dynamics in a configuration space with a larger set $\{Q_\mu\}$ of parameters representing more complex deformations. Choosing this set so as to include the former one (i.e. $\{Q_\mu^{\text{sud}}\} \in \{Q_\mu\}$), motion from the Q_μ^{sud} to Q_μ^{ad} (and vice versa) becomes possible. To what extent this is realized in actual processes depends only on the energy available at decisive configurations. Indeed, the structure of the static energy $E_{\text{stat}}(Q_\mu)$ considered as function of all Q_μ may become very intricate, showing valleys separated by ridges etc. In general the static energy E_{stat} can be expected to be smaller the more freedom is allowed to optimize the shape. This implies that the "sudden valley" associated essentially with configurations of overlapping spheres lies higher in energy than the "adiabatic valley". Commonly the latter is referred to as the "fission valley", simply because in the fission process the system is claimed to have sufficient time to find its way through paths of minimal static energies. Of course, the concept of the fission valley refers to the region beyond the saddle point (for fission). Inside the latter the static energy decreases down to the minimal energy of the fused complex, the compound nucleus, provided it exists. The functional form of the static energy $E_{\text{stat}}(Q_\mu)$ depends crucially on the Coulomb energy, and, hence on the projectile–target combination. A similarly large influence exists for the total angular momentum. (Usually, as done already in Section 8.2.2 the rotational energy is counted as part of the static energy. This makes sense as long as the total spin is conserved). Large centrifugal forces may prevent the system from arriving at more compact shapes.

Besides these merely static forces there are the truly kinematical ones. Indeed, which regions of deformation space are actually reached in a given process depends on the system's dynamics. For instance, when two ions fuse to a heavier nucleus the system has to overcome the innermost saddle. This is possible only if in the dynamic stages before the system was directed on a path towards configurations of the compound nucleus and if, furthermore, not too much energy had been lost–provided the initial energy of relative motion was sufficiently high at all. Obviously the time development is governed by the transport coefficients and the way they vary with the collective degrees of freedom Q_μ as well as with the internal excitation. Various scenarios are possible which in schematic fashion may be pictured as shown in Fig. 13.1.

On the left panel different trajectories are shown in a projection on a two-dimensional space consisting of the radial distance R between the ions and the mass asymmetry degree of freedom. It is assumed that an initially asymmetric system may in the end split into two identical fragments if only the duration time of the reaction is long enough. Possible shapes, which would be determined by other collective variables, are shown by inserts. For large angular momenta (like the $l = 195\hbar$ picked for the figure) the centrifugal barrier may prevent the ions approaching each other too closely such that no dissipation of energy occurs at all. There may be processes, however, where friction forces lead to a considerable loss of energy in relative motion, and which are therefore called "deep inelastic collisions". In the event that the fragments penetrate even more deeply into each other motion in other degrees of freedom may be initiated. In the end this may lead to a sizable mass exchange, as indicated by the dotted line. Since the final stages resemble those of fission the notion "fast fission" has been coined. It is "fast" in the sense that the process has not gone through the production of a genuine compound nucleus, before the latter decays through the ordinary fission process on a much longer timescale.

How these various processes are directed by the static energy is demonstrated in the right panel. There cuts of the $E_{\text{stat}}(Q_\mu)$ are shown for projections on the relative distance; the resulting function is denoted by $V(R)$. On the top left one sees a typical trajectory for a deep inelastic collision along the potential for the "sudden configuration". On the top right a trajectory is shown which makes its way across the fusion barrier (the one of the "sudden potential") and through the potential landscape to inside the barrier of the adiabatic potential. In the end this leads to the formation of the compound nucleus. On the bottom, trajectories are shown for situations where the fragments separate again without having reached a compound nucleus but where a considerable exchange of mass has taken place. "Fast fission" and "quasi-fission" only differ in the sense that in the former case a still existing large angular momentum forces the fission barrier to disappear.

Important properties of the potential landscape are determined by the energies of the liquid drop model, in particular the combination of rotational energy, Coulomb and surface energy. For details how these energies depend on the projectile–target combination we refer to the literature, like (Schröder and Huizenga, 1984) or (Fröbrich and Lipperheide, 1996), for instance. Pioneering work on this important question has been done by W.J. Swiatecki and co-workers. Basically the physics behind this approach is similar to that illustrated above. It is only that different notions have been introduced like *extra push* and *extra-extra push*: The former refers to the energy needed to overcome the "conditional saddle" (for a static energy at frozen mass asymmetry) in addition to the energy necessary for the two ions to come into close contact at all. An extra-extra push is required for the system to overcome the inner (unconditional) barrier. In both cases energy refers to the one of the initial relative motion. Values for these pushes have been evaluated for Newtonian dynamics and on the basis of the macroscopic picture, liquid drop energy for the $E_{\text{stat}}(Q_\mu)$, wall and wall-and-window formula for friction, together with a schematic (reduced-mass) inertial force. In Section 13.3 the relevance of these extra energies for fusion reactions will be discussed for a schematic model, within which fluctuations around the classical trajectories will also be considered. Moreover, differences of macroscopic and microscopic transport coefficients in their predictions for cross sections will be explored.

As pointed out in various papers by the Los Alamos group shell effects may have considerable impact, see e.g. (Möller et al., 1997) and further references given therein. The correct barrier heights for fusion turn out to be much smaller than in the macroscopic picture. Consequently, much less energy will be needed for the extra pushes–if they are necessary at all. Such properties are of particular importance for the formation of super-heavy elements, which once produced should have as little thermal excitation as possible.

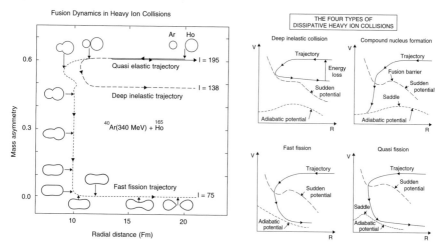

FIG. 13.1. Possible scenarios for heavy-ion collisions, see text [(Ngô, 1986)].

Let us come back once more to the difference between the sudden and the adiabatic limit. In the picture developed above nothing was said about the behavior of the intrinsic or nucleonic degrees of freedom. With the exception of the DDD model, see below, one generally assumes that these nucleonic degrees of freedom are in thermal equilibrium, independently of the shape chosen. This picture is justified as long as the relaxation time for these intrinsic degrees of freedom is still considerably shorter than the timescale for relative motion. Then, indeed, one may apply the quasi-static picture of sections 6.5 and 6.6. The latter has much in common with what elsewhere is referred to as the adiabatic model. One should be careful about this concept, however. Commonly nuclear collective motion is understood as belonging to zero thermal excitations, and many highly sophisticated theories have been developed for such a situation, as may be inferred from the many textbooks on nuclear theories like (Bohr and Mottelson, 1969), (Bohr and Mottelson, 1975), (Siemens and Jensen, 1987), (Ring and Schuck, 2000). In this context the notion "adiabatic" is to be interpreted in the sense of quantum mechanics: It simply refers to collective motion which is slow as compared to the nucleonic one. For finite thermal excitations, on the other hand, "adiabaticity" is to be understood as entropy conserving. In Section 11.2.2 some account has been given of these subtle differences, but we should like to refer once more to the discussions in (Kubo et al., 1991b), (Kubo et al., 1991a), (Balian, 1991) in the frame of statistical mechanics. In the theory of W. Nörenberg for "dissipative diabatic dynamics" *relaxation processes* from excited single particle configurations to equilibrium configurations are *taken into account explicitly*. Details are explained in Section 6.6.2.1, where it is mentioned that also the equilibrium is described within the independent particle model. It may also be said that the DDD-approach is restricted describing collective dynamics on the level of average motion. No prediction is made on fluctuating forces and their implications.

13.1.1 Commonly used inputs for transport equations

As indicated earlier, in practical applications mostly models of macroscopic nature are in use, in particular for the dynamic transport coefficients. The collective inertia, for instance, is commonly taken to be that of the Werner–Wheeler method which results in values close to those of irrotational flow. For the approach phase the choice of the inertia is commonly oriented at the reduced mass for relative motion, at times even for configurations of considerable overlap of the fragments' densities. We are now going to address more closely the evaluation of the potential landscape and two popular models for dissipation, those of surface one-body dissipation and wall-and-window friction. Later in Section 13.4 we will add some critical remarks.

Potential landscapes: As outlined before one needs to consider models with several degrees of freedom Q_μ. Therefore, realistic theoretical approaches must be able to calculate static energies for various possible shapes. The latter must include configurations of two touching spheres for the entrance phase, compact shapes for the fused system, as well as the more elongated shapes for the fission valley. To treat the approach phase correctly, special care must be taken that the energy behaves smoothly across the transition from two separated fragments over touching spheres to the more compact configurations. The problem is similar to that for fission in the scission region reported at the end of Section 5.3 referring to the work of the Berkeley–Los Alamos group. Indeed, in the calculations of P. Möller and R. Nix and others for fusion fission potentials such a smooth transition is obtained (1997). There, shell effects are taken into account by their "macroscopic-microscopic method". As mentioned previously they have considerable impact on the structure of the landscape, in particular on barrier heights for fusion and fission.

In the literature potential energies have also been evaluated on the basis of energy density functionals, be them of Skyrme type or of more phenomenological nature (see the reviews (Schröder and Huizenga, 1984) and (Ngô, 1986)). Recall the discussion in Section 2.6 and Section 3.1.1 where such densities $\mathcal{H}$ were introduced. Knowing the $\mathcal{H}$ for both the combined system with the density ρ_{comb} and the separated fragments, one may calculate the interaction potential $V(R)$ by the difference of the relevant energies: $V(R) = \int d^3r(\mathcal{H}[\rho_{\text{comb}}] - \mathcal{H}[\rho_1] - \mathcal{H}[\rho_2])$ (with R being the distance between the ion's center of mass). This procedure becomes particular appealing if for the entrance phase the "sudden limit" is supposed to hold true in the sense of using the picture of frozen spheres, where $\rho_{\text{comb}} = \rho_1 + \rho_2$. A somewhat simpler way of getting the nuclear part of interaction potential $V_N(R)$ is by folding the density ρ_1 of one ion with the mean field U_2 of the other, $V_N(R) = \int d^3r U_2(|\mathbf{R} - \mathbf{r}|)\rho_1(r)$; eventually one has add the corresponding term with 1 and 2 interchanged and to symmetrize finally. Such potentials are often used in connection with the surface friction model (Fröbrich and Lipperheide, 1996).

Finally, we should like to mention the version advocated by W.J. Swiatecki and his group which in the literature is referred to as proximity potential (Swiatecki, 1980). It is based on the liquid drop model for sharp surfaces (leptodermous systems) and takes into account interactions between the ions across gaps and crevices. This (additional) force comes from the two-body interaction of neighboring nucleons. This proximity potential can be expressed in terms of the product of two factors, a geometrical one and a universal form factor. The first accounts for the special geometry of a given shape, whereas the second represents the interaction between two parallel surfaces.

The piston model: The model was developed and used by D.H.E. Gross and co-workers to describe deep inelastic collisions (Gross, 1975), (Gross and Kalinowski, 1978), (Gross, 1979), (Gross, 1980). Like the picture behind wall friction fermions inside a "container" are subject to reflections at a moving piston. The way this system is modelled in detail leads to a strong and strongly surface peaked friction coefficient, for which reason the model is also referred to as the surface friction model (Fröbrich and Lipperheide, 1996). In many practical applications it turned out quite useful to obtain realistic cross sections for heavy-ion collisions (Gross and Kalinowski, 1978), (Schröder and Huizenga, 1984), (Fröbrich and Lipperheide, 1996). The microscopic derivation given in (Gross, 1975) has great similarities to the linear response approach described in chapters 6 and 7: (i) The nucleons are assumed to be in thermal equilibrium and may hence be considered to deliver the heat bath to which collective motion is coupled.

(ii) This coupling is treated by time-dependent perturbation theory–without exploiting the benefits of linear response theory. For the latter in (Hofmann and Siemens, 1976), (Hofmann and Siemens, 1977) the transport coefficients were treated in zero frequency limit, which for friction leads to formulas like (6.92). In (Gross and Kalinowski, 1978) this was criticized because in such a formula no room is left for a finite energy of collective motion. It is true that in the approach phase the kinetic energy of relative motion is quite large. One should not forget, however, that the quantum version of zero frequency friction requires an energy smoothing if evaluated within the independent particle model, see Section 6.4. Indeed, in (Sato et al., 1978) friction coefficients were calculated with the linear response formula which showed similar features as the surface friction model. In (Sato et al., 1979) these coefficients were applied to a numerical analysis of deep inelastic collisions (discarding fluctuating forces) in which the experimentally observed energy losses could be reproduced. Later the linear response approach was generalized (Hofmann et al., 1979), (Hofmann, 1980) to allow for considering finite values of collective energy (or frequency within the LHA). Within simple minded perturbation theory (see Section 25.6.2) it is the unperturbed energy/frequency which appears, which in (Gross and Kalinowski, 1978) comes out to be the change of the energy of relative motion. In the nonperturbative treatment of the Nakajima–Zwanzig procedure presented in Section 7.5.2 the situation is still more involved. There, the transport coefficients have to be evaluated in a self-consistent procedure, which for average motion is explained in Chapter 6. In this way the response functions which determine the transport coefficients have to be calculated at the perturbed frequency which eventually includes a contribution from damping. As demonstrated in Chapter 7 this feature has an important impact on the diffusion coefficients. In the surface friction model the latter are always taken to be given by the Einstein relation which for the approach phase is most questionable; we will return to this problem in Section 13.4. In (1979), (1980) D.H.E Gross has tried to justify his derivations from earlier work by interpreting collective motion by analogy with that of a Brownian particle. As outlined in Section 7.5.1, this analogy as well as the applicability of the Caldeira–Leggett model is most doubtful.

Wall-and-window dissipation: The physics behind this mechanism of nuclear one-body dissipation has been put forward in (Błocki et al., 1978). Based on this picture later in (Randrup, 1978) and (Feldmeier, 1987) transport equations were derived on a more formal basis in the spirit of the Kramers–Moyal expansion.[28] The nucleonic system is assumed to be represented by Fermi gases enclosed inside containers (walls) the forms of which have the typical shapes one claims relevant for nuclear collective motion. For the more compact shapes one ends up with wall friction which is in use also for fission. For the entrance phase of

---

[28] As explained in Section 25.2.3 this method requires a clear separation of timescales between intrinsic and collective motion. This property may not be given for the initial stages of the reaction, see Section 13.4 below.

heavy-ion collisions another mechanism is suggested by which energy of relative motion is lost due to the exchange of nucleons through a "window" between the two overlapping fragments. In both cases no residual interactions are considered. Rather, to justify irreversibility a "randomization hypothesis" is introduced: The nucleonic motion is assumed to be sufficiently "chaotic" such that any forced deviation of the Fermi gases from equilibrium relaxes sufficiently fast, without explicitly specifying the true origin of this randomness. For the case of the wall formula this hypothesis can be tested because this formula is nothing else but the macroscopic limit of the linear response expression applied to independent particle motion, see Section 17.1. As described in Section 6.6, any consideration of residual interactions will have great impact on the value of the friction coefficient as well as its variation with temperature. Below in Section 13.3 we are going to demonstrate for the case of fusion that such differences lead to modifications in the formation probabilities which may amount to a few orders of magnitude.

The window picture is more general in the sense that it allows for an explicit dependence of the transport coefficients on the relative velocity of the two ions, in particular for the diffusion coefficients. In (Feldmeier, 1987) it is demonstrated that this feature has some impact on the energy-angle correlations (Wilczynski plot). So far no analogous treatment has been pursued on the microscopic level of linear response theory. Discarding particle exchange a non-linear velocity dependence of the forces (for average motion) in the approach phase has been examined in (Johansen et al., 1977). There, the excitation of ^{238}U which is produced by the potential of the projectile has been studied. The outcome was less favorable for the typical friction force but consequences have not been studied any further. Conversely, in (Yamaji and Iwamoto, 1983) the linear response formalism has been applied to a reaction of ^{64}Zn on ^{196}Pt with the microscopic model based on the two-center harmonic oscillator potential. The friction coefficients have been evaluated in the zero-frequency limit of (6.92) and for two different ways of introducing irreversibility, by the frequency average described in Section 6.4.1 and by collisional damping (in a simplified version of Lorentzian smoothing, see Section 6.4.3). In the first case the result shows good agreement with the window formula,[29] for both the absolute value and for the radial dependence of the diagonal friction coefficient γ_{RR}. For this a relatively large smearing width of $\Gamma_{av} = 0.4\hbar\Omega_0$ was necessary. The temperature-dependent smoothing leads to considerably different values which furthermore vary much with T. These findings are not very surprising. On the one hand, a large (constant) averaging width Γ_{av} wipes out practically any microscopic structure. However, the damping mechanism of Section 6.4.3 cannot be expected to apply for the early stages of the reaction where the temperature concept is most questionable.

According to (Randrup, 1978) in the wall-window model the diffusion coefficients $D_{\mu\nu}$ take on the same forms which in Section 7.3.1 have been attributed

---

[29] In this publication the window formula of (Randrup, 1979) was used and called proximity friction (according to private communication by S. Yamaji).

to weak damping. As in (7.29), they are proportional to an effective temperature $T^* = (\hbar\varpi/2)\coth(\hbar\varpi/2T)$, but where the $\hbar\varpi$ has a different meaning. It represents "the change in the intrinsic excitation energy associated with the transfer of the nucleons from B to A" (from one nucleus to the other). This $\hbar\varpi$ depends linearly on the relative velocity $\Delta\mathbf{u}$ of the two nuclei. In (Feldmeier, 1987) only the high temperature limit is considered (in the sense of $|\Delta\mathbf{u}| \ll v_F$ and $2T \ll mv_F^2$ (with the Fermi velocity v_F)). Then the overall diffusion coefficient is a sum of two terms, one proportional to $|\Delta\mathbf{u}|$ one proportional to T as in the ordinary Einstein relation. However, $|\Delta\mathbf{u}|$ also appears in the other factors such that the Einstein relation proper is reached only for sufficiently small relative velocities.

13.2 Differential cross sections

In the following we are going to present briefly the calculation of cross sections on the basis of solutions of Fokker–Planck equations like that of (25.121). To begin with let us look at a solution of the type $f(\mathbf{r},\mathbf{p},t)$ with $\mathbf{r}$ being the vector of relative motion and $\mathbf{p}$ its conjugate momentum. Suppose further that this solution starts off from the sharp distribution corresponding to classical trajectories. One may then choose the coordinate system such that the initial momentum points in z direction. The initial position may be considered located within the area $bdb d\varphi$ perpendicular to the z-axis, with b being the classical impact parameter. For simplicity rotational symmetry about the z-axis is assumed. The (conditional) probability of finding *after the reaction* the system in the angle $\vartheta, \vartheta + d\vartheta$ is then given by

$$P(\vartheta;b)d\vartheta = \lim_{t\to\infty} \int_0^\infty r^2 dr \int d^3p\, f(\mathbf{r},\mathbf{p},t)\,. \tag{13.1}$$

The differential cross section is obtained by integrating over impact parameter

$$\frac{d\sigma}{d\vartheta} = \int_0^\infty 2\pi b db\, P(\vartheta;b)\,. \tag{13.2}$$

This situation portrays a beam of "particles" of unit flux with bdb particles impinging on the ring $[b, b+db]$ per unit of time. Each of them is scattered into the angle $[\vartheta, \vartheta + d\vartheta]$ with the probability given by (13.1). If one chooses to look at the differential cross section for scattering into the *solid angle* Ω one must account for the additional factor $\sin\vartheta$ in $d\Omega = \sin\vartheta d\vartheta d\varphi$, which for the cross section implies $d\sigma/d\Omega = (1/\sin\vartheta)d\sigma/d\vartheta$.

To evaluate the probability $P(\vartheta;b)$ given in (13.1) we may assume the fluctuations around the classical trajectory to be sufficiently small such that the Gaussian approximation applies. Considering the precautions discussed in Section 25.5.1 one may write $P(\vartheta;b)$ in the form

$$P(\vartheta;b)d\vartheta = \left(2\pi\Xi^{\vartheta\vartheta}(b)\right)^{-1/2} \exp\left(-\frac{(\vartheta - \vartheta^c(b))^2}{2\Xi^{\vartheta\vartheta}(b)}\right) d\vartheta\,. \tag{13.3}$$

For vanishing fluctuation $\Xi^{\vartheta\vartheta}(b)$ this distribution turns to $\lim_{\Xi\to 0} P(\vartheta;b)d\vartheta = \delta(\vartheta - \vartheta^c(b))d\vartheta$, such that the cross section becomes

$$\left.\frac{d\sigma}{d\vartheta}\right|_{\text{class}} = \int_0^\infty 2\pi b\, db\, \delta(\vartheta - \vartheta^c(b)) = \sum_m 2\pi b_m \left(\left|\frac{d\vartheta^c(b)}{db}\right|_m\right)^{-1},$$

with the b_m solving the equation $\vartheta - \vartheta^c(b) = 0$. This is the cross section for Newtonian dynamics which diverges at the so-called *rainbow angle* where $d\vartheta^c(b)/db$ vanishes. In (Hofmann and Ngô, 1976) it was demonstrated that formula (13.2) leads to finite values of the cross section beyond the rainbow angle in good agreement with experimental results. In this calculation the same transport coefficients for average motion have been used as those for trajectory calculations before where the divergence appears. The fluctuating force was determined by the Einstein relation with a time-dependent temperature. The latter is specified by the internal energy which follows the loss of energy of relative motion. In accord with the assumption of small fluctuations around the classical trajectories these energies are evaluated from average motion.

Let us look next at the double differential cross section for angle and energy. For this one may proceed like before. One only has to generalize to a two-dimensional distribution. Since (at large distances after the reaction) the kinetic energy of relative motion is given by $E = E_r = p_r^2/(2\mu)$ the cross section can be traced back to the probability

$$P(\vartheta, p_r; b)\, d\vartheta\, dp_r = (2\pi)^{-1} \Delta^{-1/2} \times$$
$$\exp\left\{-\frac{1}{2\Delta}\left[(p_r - p_r^c)^2 \Xi^{\vartheta\vartheta} - 2(p_r - p_r^c)(\vartheta - \vartheta^c)\Psi_r^\vartheta + (\vartheta - \vartheta^c)^2 \Pi_{rr}\right]\right\}. \quad (13.4)$$

leading to

$$\frac{d^2\sigma}{d\vartheta dE} = \sqrt{\frac{\mu}{2E}} \int_0^\infty 2\pi b\, db\, P(\vartheta, p_r; b) \quad (13.5)$$

Based on this expression in (Berlanger et al., 1978c) and (Berlanger et al., 1978a) so-called Wilzynski-plots of deep inelastic collisions were evaluated. They showed the large correlations between angle and energy seen in experiment. In (Berlanger et al., 1978a) the observed change in the pattern with the product $Z_1 Z_2$ of the charges of the two ions could be explained by the mutual interplay between the Coulomb interaction and the fluctuating force. The influence of the latter becomes the stronger the longer the fragments stay together.

Finally, it may be mentioned that in (Ngô and Hofmann, 1977) double differential cross sections for angle and mass exchange were evaluated. The latter mode was assumed to be statistically uncorrelated from relative motion. In (Berlanger et al., 1978b) this work was extended to get the triple differential cross sections $d^3\sigma/d\vartheta dE dZ$.

Charge equilibration as a test for quantum fluctuations: In the 1970s experimental evidence for a strong correlation between charge exchange and energy loss was found, see (Berlanger et al., 1979) or (Schröder and Huizenga, 1984). In (Hofmann et al., 1979) it was suggested that this result may be interpreted as the relaxation of the fast mode of charge excess $y = N - Z$ (of one fragment) to its quantal equilibrium. Besides y the mass asymmetry mode x was also treated explicitly, both in harmonic approximation. The relation to energy loss was parameterized as a unique relation between E_{loss} and time t in the form $E_{\text{loss}} = E_{\text{loss}}^{\max}(1 - \exp(-t/\tau_E))$, with a value of τ_E of $3 \cdot 10^{-22}$ sec. Also other parameters of the model were based on empirical estimation. The potential energy was deduced from the liquid drop model. For the x-mode a vibrational energy of $\hbar\varpi_x = 1$ MeV was assumed. The $\hbar\varpi_y$ was estimated from the longitudinal part of the isovector dipole resonance. Depending on the system considered, in this way values between $\hbar\varpi_y \simeq 3$ MeV and $\hbar\varpi_y \simeq 8$ MeV were obtained. As these values are relatively large it was perhaps safe to assume the y-mode to be under-damped. The same assumption was made for the x-mode. Under the hypothesis that the inertial tensor M_{ij} be diagonal the values of the two inertial parameters are given by the corresponding stiffnesses and the frequencies. Finally, the diagonal elements γ_{ii} of the friction tensor were obtained from the two relaxation times 10^{-20} sec and 10^{-22} sec for the x- and y-modes, respectively (interpreting these times as those for relaxation of the kinetic energy, $M_{ii}/2\gamma_{ii}$).

For the strongly under-damped y-mode its diffusion coefficient was assumed to be given by the weak damping limit of (7.29). Because of the large value of $\hbar\varpi_y$, in comparison to temperature, quantum features of the transport equation (7.34) become important. Of course, care had to be taken about a proper choice of the initial conditions. For some of them very good agreement with experimental data was obtained for the Z-variance as function of energy loss E_{loss}. Later on this was confirmed in the more extended model of (Grègoire et al., 1982). There besides x and y relative motion is also treated explicitly, with deformation degrees considered implicitly by parameterizing in some simple way the transition from the fusion to the fission valley.

From a theoretical point of view these results look very interesting, indeed, as they would be a direct hint of the importance of non-trivial quantum effects in nuclear transport. Unfortunately, however, the considerable uncertainty in the choice of both the initial conditions and of the phenomenological input of the transport equation cast some doubts about such a conclusion. Moreover, the application of the fluctuation dissipation theorem for the early stages of the reaction is uncertain, in particular in the form needed for the diffusion coefficients (7.29). A critical discussion of more experimental questions is found in the review article (Schröder and Huizenga, 1984).

13.3 Fusion reactions

The scenario for fusion reactions has been described in Section 13.1. The essential point is that the system has to overcome the barrier which separates the deformation region of the compound nucleus from the outer ones with more elongated shapes having a more or less pronounced neck. A crucial degree of freedom is the total angular momentum J with the associated rotational energy $\hbar^2 J^2/2\mathcal{J}$. In a first approximation one may look at the outer configurations only and estimate the moment of inertia by μR^2, with μ being the fragments' reduced mass and R the radial distance between their center of mass. At given incident energy E fusion is possible for angular momenta below a critical value J_{crit}. For $J > J_{\text{crit}}$ the $E_{\text{stat}}(R)$ is only repulsive whereas for smaller J a barrier exists which may be overcome (towards more compact shapes or smaller R). Let us assume that the fragments actually fuse to a compound nucleus whenever that happens. Within a description based solely on Newtonian dynamics the cross section for fusion is then simply given by $\sigma_{\text{fus}} = \pi b_{\text{crit}}^2$, with b_{crit} being the critical impact parameter; the latter relates to angular momentum and energy.[30] like $b = J/\sqrt{\mu E}$.

In such a discussion the influence of statistical fluctuations are neglected. The quality of such an assumption was checked in (Hofmann and Siemens, 1975) by developing and applying the model presented above in Section 7.4.3. The motion across the barrier was replaced by one across an inverted parabola with constant transport coefficients, with the Einstein relation taken to be valid. As one was looking at the outer barrier, motion could safely be taken to start with zero initial fluctuations. It was seen that within the simple picture adopted the influence of the dynamical fluctuations was small, but it was pointed out that this feature may change if additional degrees of freedom were taken into account. In more recent years this model has been used again to evaluate the cross section for fusion reactions, this time for those which lead to super-heavy elements (Aritomo et al., 1999), (Abe, 2002), (Shen et al., 2002), (Abe et al., 2003). The fusion reaction is written as a sum over all partial waves in the form

$$\sigma_{\text{fus}}(E) = \frac{\pi}{k^2} \sum_J (2J+1) \, P^J_{\text{fus}}(E) \equiv \frac{\pi}{k^2} \sum_J (2J+1) \, \{P^J_{\text{stick}}(E) \times P^J_{\text{form}}(E)\} \,. \tag{13.6}$$

The probability $P^J_{\text{fus}}(E)$ that a partial wave at given angular momentum J and energy E leads to fusion is supposed to split into two factors. The first one, $P^J_{\text{stick}}(E)$ defines the probability that the outer barrier is overcome, which essentially is determined by the Coulomb interaction. The "formation probability" $P^J_{\text{form}}(E)$ is evaluated by the schematic, one-dimensional model just sketched, which is to say it is identified as the limiting value $\Pi(t \to \infty) = \Pi_\infty$ defined on the right of (7.48),

$$P^J_{\text{form}}(E) \equiv \Pi_\infty = \frac{1}{2}\left(1 + \text{erf}\left[\frac{\xi^q_\infty}{\sqrt{2\xi^{qq}_\infty}}\right]\right) \,. \tag{13.7}$$

Here, the ξ^q_∞ determines the asymptotic behavior of the classical trajectory $q_c(t)$ and ξ^{qq}_∞ that of the fluctuation around it, as given by (7.44) and (7.45), respectively. As in (Hofmann and Siemens, 1975), but unlike (Rummel and Hofmann,

---

[30]To get the correct energy dependence of the cross section one may consider an energy dependence of the barrier V_B like $V_B = V(R^B) + E b_{\text{crit}}^2/R^B$, which leads to $\sigma_{\text{fus}} = \pi(R^B)^2(1 - V_B/E)$.

2003), these quantities are evaluated from Kramers' equation with the diffusion coefficients given by the Einstein relation.

In principle, the "sticking probability" $P^J_{\text{stick}}(E)$ may be evaluated by following the dynamics for step I, which is to say from the approach phase until the two fragments "stick" together at some more compact shape. Usually this is done by Langevin simulations, often using the "surface friction model". Because of its large strength almost no kinetic energy is left over at the starting point q_0. For this reason the initial conditions for the determination of Π_∞ are chosen to correspond to a Maxwell distribution in the momentum with no fluctuation in the coordinate, see e.g. (Abe, 2002) or (Shen et al., 2002). The temperature is determined from the average energy loss in the initial step. Because of inherent uncertainties in the transport coefficients the amount of this initial energy loss is not so well known.

13.3.1 Micro- and macroscopic formation probabilities

In (Rummel and Hofmann, 2003) this schematic model has been used to evaluate the Π_∞ for microscopic transport coefficients. These results were then contrasted with those of the macroscopic models and large discrepancies were noted. Moreover, the influence of the initial configurations has been examined allowing in addition for quantum effects in collective motion. In the following we will briefly describe the main features, with special focus on the difference between quantal treatments of internal and collective motion. For this purpose we need to distinguish between transport coefficients for average motion, on the one hand, and the diffusion coefficients $D_{\mu\nu}$, on the other hand, which govern the dynamics of fluctuations. To facilitate the explanations to come let us denote the former by T^k_{av}, with the upper index used to distinguish inertia, stiffness and friction. All three of them have a characteristic temperature dependence which may be parameterized by the formulas of Section 6.6.3. From (6.145), for instance, one gets the damping rate $\Gamma_{\text{kin}} = \hbar\gamma/M$ and from (6.146) the relaxation time $\tau_{\text{coll}} = \gamma/|C|$ for motion in q. The frequency $\varpi_b(T)$ is approximated by $\varpi_b(T) \approx 1$ MeV/$\hbar$. The influence of pairing is discarded, simply because in the range of temperatures relevant for our model its influence is negligible. For the T^k_{av} of the *macroscopic limit* we take those of Section 6.6.1.1, stiffness $C_{\text{LDM}}(T)$ and inertia M_{LDM} from the liquid drop model and friction from the modified wall formula, which is to say $\gamma = \gamma_{\text{wall}}/2$.

Next we turn to the diffusion coefficients $D_{\mu\nu}$ which in the quantum case may depend on the T^k_{av} in non-linear fashion. The way the $D_{\mu\nu}$ may be evaluated from the "equilibrium fluctuations" $\sum^{\text{eq,b}}_{\mu\nu}$ was presented in Section 7.3. In the results shown there the T^k_{av} were taken as given, constant parameters. But the latter themselves vary with T. This effect has to be taken into account if one wants to get the correct temperature dependence of the diffusion coefficients. Therefore, in (Rummel and Hofmann, 2003) the $|\sum^{\text{eq,b}}_{qq}|$ and $\sum^{\text{eq,b}}_{pp}$ were calculated anew, applying Drude regularization with a cutoff parameter $\omega_D \approx 10$ (for $\varpi_b(T) \approx 1$ MeV/$\hbar$). Since at low temperatures the microscopic damping strength η_b is very small, say below about $T \simeq 1$ MeV (see the lower right part of Fig. 6.11), it is no surprise that in this range the $D_{\mu\nu}$ are very well represented by the limit of weak damping. The macroscopic values of the T^k_{av}, on the other hand, imply large damping such that in this case the diffusion coefficients are close to those of the high temperature limit, for practically *all temperatures*. Actually, in this case damping is so large that one is dealing with over-damped motion, see again Fig. 6.11. At very large temperatures the microscopic and macroscopic transport coefficients approach each other, both the T^k_{av} as well as the $D_{\mu\nu}$.

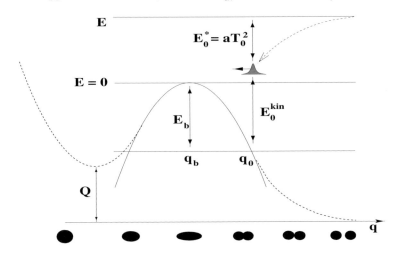

FIG. 13.2. Scenario for the formation of the compound system; see text.

Let us turn to the evaluation of Π_∞. The ingredients for such a calculation are portrayed in Fig. 13.2. If the distribution starts out with finite initial widths $\sum_{\mu\nu}(t_0)$ the argument of the error function in (13.7) becomes

$$\frac{\xi^q_\infty}{\sqrt{2\xi^{qq}_\infty}} = \frac{\sqrt{E_0^{\text{kin}}/\hbar\varpi_b} + \sqrt{E_b/\hbar\varpi_b}\,(z^-/\varpi_b)}{\sqrt{(\sum_{pp}(t_0) - \sum_{pp}^{eq,b}) + (z^-/\varpi_b)^2(\sum_{qq}(t_0) - \sum_{qq}^{eq,b}) - 2(z^-/\varpi_b)\sum_{qp}(t_0)}}.$$
(13.8)

For this expression a dimensionless coordinate q and a dimensionless momentum P is used, see Section 25.5.2. Rather than using diffusion coefficients, this form is written with the equilibrium fluctuations $\sum^{eq,b}_{\mu\nu}$. Furthermore, the ξ^q_∞ is expressed in terms of the initial kinetic energy $E_0^{\text{kin}} = \hbar\varpi_b P_0^2$ and the barrier height E_b. The latter is identified as the negative potential energy at the initial point q_0, which means to write $E_b = -V_0 = \hbar\varpi_b q_0^2/2$. As can be seen from (13.7) the average trajectory determines whether Π_∞ becomes smaller or larger than $1/2$. It is equal to $1/2$ when the trajectory just hits the top of the barrier at $\xi^q_\infty = 0$. By inspecting the numerator of (13.8), and recalling from eqn(7.31) the value of z^-, one easily recognizes that the trajectories may be classified by their initial kinetic energy E_0^{kin} in comparison with the "effective barrier height"

$$E_b^{\text{eff}} = \left(\sqrt{1+\eta_b^2} + \eta_b\right)^2 E_b \,. \qquad (13.9)$$

Trajectories with $E_0^{\text{kin}} < E_b^{\text{eff}}$ are reflected (with $\Pi_\infty < 1/2$) and those with $E_0^{\text{kin}} > E_b^{\text{eff}}$ cross the barrier ($\Pi_\infty > 1/2$); evidently this E_b^{eff} reflects directly the cause for the extra-extra push.

In (Rummel and Hofmann, 2003) a value of $E_b = 10$ MeV was chosen for the barrier height, which is in fair agreement with the typical heights of the inner barriers used in (Denisov and Hofmann, 2000). To specify the strength of dissipation in step I the heat produced by it was written as $E_0^* = aT_0^2 = (1-\mathcal{R})(E+E_b)$, with the parameter $\mathcal{R}$ being chosen to vary between 0 and 0.25. This implies that the formation process starts with the kinetic energy $E_0^{\text{kin}} = \mathcal{R}(E+E_b)$. For this stage the temperature was assumed to stay constant at its initial value $T = T_0$. The latter was evaluated with a level density parameter of $a \approx A/10$ MeV. For a system of $A = 224$ the T_0 takes on values between about 0.6 MeV and 3 MeV when E varies between 0 and 200 MeV, with only minor influence of $\mathcal{R}$. This range of temperatures is in accord with the $T_0 = 0.68 - 1.24$ MeV studied in (Aritomo et al., 1999) for the heavier system with $A = 298$ nucleons.

The initial widths in q and P were written in terms of the $\sum_{pp}^{eq,b}$ by introducing two positive numbers w and W for $\sum_{pp}(t_0) = w\sum_{pp}^{eq,b}$ and $\sum_{qq}(t_0) = W/(4w\sum_{pp}^{eq,b})$, with a vanishing cross term $\sum_{qp}(t_0) = 0$. As mentioned earlier, a strong friction force in stage I implies the momentum distribution to be almost Maxwellian, for which w is close to 1. Different to the classical case, for quantum processes the coordinate fluctuations must be finite. Requiring $W \geq 1$ warrants the uncertainty principle to be fulfilled in the form $\sum_{qq}(t_0)\sum_{pp}(t_0) \geq 1/4$. One should bear in mind, of course, that in reality finite fluctuations may have been built up which are not necessarily of Gaussian shape.

13.3.1.1 Numerical results

From (13.8) it is seen that ξ_∞^{qq} does not only depend on the $\sum_{\mu\nu}^{eq,b}$ but on the initial fluctuations $\sum_{\mu\nu}(t_0)$ as well. For the temperatures $T \gtrsim 0.6$ MeV considered the approximate versions of the diffusion coefficients $D_{\mu\nu}$ (or for the $\sum_{\mu\nu}^{eq,b}$) (high T or weak damping limit) may differ greatly from the exact results. Because of the presence of the $\sum_{\mu\nu}(t_0)$ this difference is to a large extent masked for the ξ_∞^{qq}. It should be mentioned, however, that at smaller temperatures the approximate formulas for the $D_{\mu\nu}$ would fail completely if they were evaluated for the macroscopic transport coefficients T_{av}^k for average motion.

In Fig. 13.3 the formation probabilities are plotted as functions of time. All curves are obtained for diffusion coefficients of the fully quantum calculation. For the incident energy $E = 10$ MeV two values of the initial kinetic energy were chosen, namely $E_0^{\text{kin}} = 0$ MeV ($\mathcal{R} = 0$) and $E_0^{\text{kin}} = 5$ ($\mathcal{R} = 0.25$). For the parameter w, three different values are used: $w = 1/2$, simulating a momentum distribution which is more narrow than for thermal equilibrium, $w = 1$ (as for

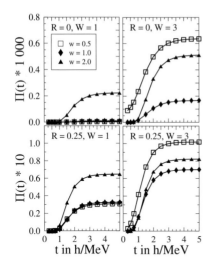 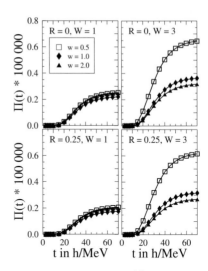

FIG. 13.3. Left: Time evolution of the formation probability $\Pi(t)$ for microscopic transport coefficients and quantal diffusion coefficients; the incident energy is $E = 10$ MeV. For more details see text and (Rummel and Hofmann, 2003). Right: Same as in left part but for macroscopic transport coefficients.

thermal equilibrium) and $w = 2$ (wider than for thermal equilibrium). For the initial width in the coordinate the two values $W = 1$ and $W = 3$ are used. They imply an initial space distribution which in most cases is almost totally located on the right of the barrier. For the choice $w = 1/2$ and $W = 3$ a small part of the distribution is already on its left initially. In this case the formation probability $\Pi(t)$ does not start from exactly zero.

The four figures in the left panel were obtained for the microscopic set of the T_{av}^k and those on the right for macroscopic values. In the microscopic case the motion is under-damped with damping strengths $\eta_b = 0.26$ and $\eta_b = 0.18$ at $T = 0.94$ MeV ($\mathcal{R} = 0$) and $T = 0.82$ MeV ($\mathcal{R} = 0.25$), respectively. Conversely, the macroscopic values of the T_{av}^k lead to strongly over-damped motion ($\eta_b = 5.9 - 4.9$ for $T = 0.6 - 3$ MeV). As a consequence, the effective barrier height (13.9) is extremely large $E_b^{eff} = 1,400 - 980$ MeV. This fact leads to ratios E_0^{kin}/E_b^{eff} of about one order of magnitude smaller than for the microscopic transport coefficients, for which one has $E_b^{eff} = 15 - 180$ MeV. This ratio crucially influences the size of the formation probability Π_∞, which for macroscopic transport coefficients T_{av}^k is several orders of magnitudes smaller than in the microscopic case. It may be noted in passing that this effect increases with decreasing E. Compared to this dramatic difference, quantum effects in the collective motion hidden in the $\sum_{\mu\nu}^{eq,b}$ are only of minor importance, for the temperatures considered. For macroscopic motion the stationary values Π_∞ are much less sensitive to the initial kinetic energy than in the case of microscopic transport coefficients. As may be seen from Fig. 13.3 in the latter case Π_∞ increases by two orders of magnitude going from $\mathcal{R} = 0$ to $\mathcal{R} = 0.25$; for macroscopic T^k_{av} the corresponding change is definitely not larger than by a factor between one and two. Finally, we should like to point to the different timescales over which $\Pi(t)$ reaches the asymptotic value. This is more or less a direct consequence of the damping strength. For over-damped motion the relevant timescale is the $\tau_{\mathrm{coll}} = \gamma/|C|$ which increases with the size of friction.

Inferences: Let us finally summarize the main results. They point not only to the great importance of using correct transport coefficients, but also to the large influence of the initial conditions. These observations have been obtained concentrating on the late stages of the whole fusion reaction. During this phase collective motion may be assumed to have slowed down considerably such that its properties resemble greatly those of fission. It may thus be assumed that the basic assumptions of the linear response approach within the LHA are given. It is for this reason that one may have confidence in the transport coefficients of Section 6.5 which served as the microscopic input T^k_{av}. Their use leads to an increase of the formation probabilities by a few orders of magnitude, as compared to those obtained for the macroscopic picture. This difference is especially large at smaller temperatures where the dynamics of nucleons in a mean field is dominated by quantum effects and where "collisional damping" is suppressed. For collective motion this implies moderate to weak damping.

Following previous work it was taken for granted that the entire fusion process occurs in two steps. For this and unlike the typical compound reactions one may not assume N. Bohr's independence hypothesis to be given in any way. For the intermediate state the system simply does not relax to a genuine equilibrium. Of course, for sufficient damping in the first step it may be safe to assume that the momentum distribution attains Maxwellian form; after all the relevant relaxation time $\tau_{\mathrm{kin}} = M/\gamma$ is inversely proportional to the friction coefficient. But this fact does not say anything about the initial distribution in the collective coordinate. From these features it is evident that for more realistic calculations of the reaction cross section it will be inevitable to follow the whole process. Attempts of this type have been undertaken accounting for shell effects in the potential landscape but staying on the purely macroscopic level with respect to the dynamical aspects otherwise, see e.g. (Aritomo and Ohta, 2005), (Zagrebaev and Greiner, 2005), (Aritomo and Ohta, 2006). However, the sensitivity of the formation probability to the initial conditions discussed above clearly shows the necessity of using the correct transport coefficients also for the first step–including the diffusion coefficients as they are responsible for the fluctuations. Unfortunately, however, it is only fair to say that a decent microscopic picture for the first stage is still lacking. As mentioned elsewhere, it should be most clear that the Einstein relation cannot be valid there.

Finally, let us return to quantum effects in collective motion, which would become relevant only at even smaller temperatures. In a sense, this is in accord with the properties presented in Section 12.5 for nuclear fission. Since at small T friction is suppressed quantal diffusion coefficients may be estimated within the so-called "zero friction limit". This feature applies only for the microscopic picture of nucleonic dynamics.

13.4 Critical remarks on theoretical approaches and their assumptions

It may be time to briefly review some of the problems one faces in descriptions of low energy heavy-ion collisions and how they are treated in different models. To begin with let us recall that during such a collision various regimes are met. Whereas in the late stages the situation greatly resembles the process of fission this is by no means so for the entrance phase. As discussed in the last chapter a description of fission as transport process was suggested by Kramers as early as 1940. On the other hand, fission can be understood as a typical example of large-scale collective motion, for which in the many years after 1940 much progress had been made. As these developments were restricted to reversible motion one of the tasks was therefore to understand generalizations to irreversible dynamics. In other words, it was necessary to discern the microscopic nature of dissipation in the light of more modern insights into nuclear physics. In doing so one must not forget, of course, the importance of the conservative forces, in particular the kinetic one which involves an effective inertia; discussions about the relevance of the latter can be found in Section 12.2 and Section 11.1. Indeed, as described in Section 6.6.1 there is an intriguing influence of the many-body effects responsible for damping on the value of the inertia and its dependence on temperature. The importance of the inertia goes along with the fact that nuclear collective motion ought to be treated self-consistently, see Section 7.5.1.

For the initial stages of the reaction the situation is very different. As noted above, experimental evidence strongly suggests that the kinetic energy of relative motion dissipates on a very short timescale of about $t = 3 \cdot 10^{-22}$ sec (see also (Aritomo and Ohta, 2005)). According to the discussion in Section 6.4 this time is comparable to the relaxation time of intrinsic excitations. For this reason one has to be very cautious not only with the Markov approximation, but with the concept of intrinsic equilibrium as well. Such assumptions are used more or less.[31] in the surface friction model (Gross, 1979), (Gross, 1980), for the proximity picture (wall-and-window) (Randrup, 1978), (Feldmeier, 1987) as well as for the early application of linear response theory in (Hofmann and Siemens, 1976), (Hofmann and Siemens, 1977).

These approaches had in common that they deliver suggestions for possible forms of the transport coefficients of average motion and the Einstein relation for the diffusion coefficients. This feature opened very appealing applications of phenomenological models. At first one may try to get information on the input

---

[31] An exception is the DDD model of W. Nörenberg et al. mentioned above, but as said there, this approach discards the dynamics of statistical fluctuations.

for Newton's equations discarding any influence of statistical fluctuations. The latter may then be incorporated through the Einstein relation which implies the diffusion coefficients to be given by the product of friction and temperature. Typically for the nuclear case the friction tensor may come out to be coordinate dependent. For Langevin equations this implies the existence of multiplicative noise, which may entail the difficulties discussed in Section 11.1.3 but which mostly have been discarded. In this context one may again refer to van Kampen's critique about the naive approaches in which no attention is paid to these subtleties.

The first dynamical descriptions of relative motion and other degrees of freedom (Hofmann and Ngô, 1976), (Ngô and Hofmann, 1977), (Berlanger et al., 1978c), (Berlanger et al., 1978a), (Berlanger et al., 1978b) (see also (Grègoire et al., 1982)) which included statistical fluctuations were based on the moment expansion of Fokker-Plank equations of Kramers' type described in Section 25.5. In more recent applications of transport models one prefers to solve Langevin equations with advanced numerical techniques. They allow for extensions to multidimensional cases with refined models for potential landscapes including shell corrections. With respect to the transport coefficients proper only those of the macroscopic, high temperature limit are used. This is inconsistent as quantum effects of nucleonic dynamics do influence not only the static energy but the transport coefficients as well. In particular this is to be expected for reactions which only lead to low thermal excitations, like those in which one wants to produce super-heavy elements.

It is true that many features of heavy-ion collisions have to some large extent successfully been described by such phenomenological models. However, this cannot be taken as a proof of the correctness either of the type of equations or of their input. There are simply too many unknown parameters involved which prohibit a clean deduction by comparing theory with experiment. One should not forget that one is not simply dealing with a problem for constant transport coefficients. Rather, the latter, as well as the potential energy, are functions not only of the dynamical variables but of the thermal excitation as well, which itself depends on the energy dissipated in the dynamical process. In addition one must also have in mind the considerable freedom one has for the choice of the collective degrees of freedom. All these points clearly indicate the need for sound theoretical approaches to the nuclear transport problem.

In the late 1970s the group of H.A. Weidenmüller claimed to have found a decent access to this goal through their Random Matrix Model. In their concluding section the authors of (Agassi et al., 1977) believe they have presented a novel transport theory tailored for small and isolated systems. They claim that more conventional approaches are bound to fail if they are based on concepts developed for macroscopic systems, see also (Weidenmüller, 1978a), (Weidenmüller, 1978b) as well as (Weidenmüller, 1980). By such concepts the authors mainly have criticized the use of (i) temperature (of a canonical ensemble instead of a microcanonical one), and (ii) perturbation theory (eventually in the form of linear response). Perhaps it may be worthwhile to look at this critique at little more closely.

Let us address first the application of temperature. The latter is claimed to require the use of a heat bath. As described in Chapter 22 such a statement does not hold true. Based on the well-known fact that temperature can be introduced even for isolated systems–provided they are in equilibrium–it is shown there that all what may happen for small systems is an uncertainty in the value of temperature. Indeed, the RMM-approach is based on the assumption that "all intrinsic states within a given small energy interval have the same a priori occupation probability" (see p. 148 of (Agassi et al., 1977)). This hypothesis implies nothing else but the application of a microcanonical picture of (local) equilibrium. However, as demonstrated in Section 23.4.4, from this property the quantal FDT for the *intrinsic* degrees of freedom can be derived within the RMM–and it is the conventional form of the FDT which involves temperature as a parameter. Moreover, for the excitations typically reached in deeply inelastic collisions of medium heavy nuclei the eventual inherent uncertainty in T does not appear big enough to destroy the applicability of this concept. Indeed, this uncertainty should be viewed in the light of the uncertainties in the parameters of the RMM as such, see (Barrett et al., 1978), together with the fact that the RMM must not be used at small intrinsic excitations.

Next we turn to question of the applicability of linear response theory. Already in its first application in (Hofmann and Siemens, 1976) it was not "the interaction between the center-of-mass distance and the intrinsic degrees of freedom" $V(x_i; R)$ (Weidenmüller, 1978a) which was treated in low order perturbation theory, but only a *local* interaction. The latter was defined in such a way that the intrinsic, nucleonic system was subject to a renormalized mean field which got adjusted continuously to the change of the center-of-mass distance,[32] the collective coordinate in this example. This feature should not be forgotten if one is to argue in favor of either a "weak" or "strong coupling limit". In the papers on the RMM cited above the latter was always favored. The estimates given in these papers were related to the assumptions of the model. There, configurations of "frozen density" were considered. This is to say that the unperturbed Hamiltonian corresponded to that of the separated fragments, even after contact had been achieved with some overlap of the densities of the individual fragments. Evidently, for such a model the interaction must turn out stronger than if rearrangements of the nucleons are allowed for. Indeed, a decisive quantity for the distinction of "weak" or "strong" turns out to be the overall strength fac-

---

[32]To achieve this adjustment the interaction $V(x_i; R)$ was rewritten as $\delta V(x_i; R, R_0) = V(x_i; R) - V(x_i; R_0)$ with the $V(x_i; R_0)$ added to the unperturbed single particle Hamiltonian. It was then argued that through a multipole expansion the $\delta V(x_i; R, R_0)$ could be written as a sum over products which factorize into "form factors" and operators for the nucleonic degrees of freedom. This form then allows for the applicability of linear response theory to describe the dynamics locally. The alert reader will notice the difference from the LHA of chapters 6 and 7 where a Taylor expansion of the interaction is used.

tor of the relevant matrix elements of the coupling, more precisely their second moments, see e.g. (Weidenmüller, 1978b). In passing it may also be noted that in the microscopic evaluation of the basic form of these second moments and their parameters given in (Barrett et al., 1978) antisymmetrization of nucleons in different fragments is discarded.

As an important issue we should finally like to point out that the application of linear response theory allows for a non-perturbative evaluation of the transport coefficients–non-perturbative with respect to the local coupling between collective and intrinsic motion. Indeed, this is the case if the coefficients of average motion are defined through the construction of the collective response function introduced in Section 6.2. For the fluctuating forces this requires a modification of the common Nakajima–Zwanzig procedure. In the usual treatment of this projection technique described in Section 25.6.2 the integral kernel is calculated to second order. Such an approximation was made in the derivation of the transport equation in (Hofmann and Siemens, 1977) as well as in (Gross and Kalinowski, 1978), (Gross, 1980). Typical for this case is the fact that the diffusion coefficients depend on friction only linearly. On the other hand, one may apply the non-perturbative Nakajima–Zwanzig method presented in Section 7.5.2–again without leaving the realm of linear response theory. This method still leaves open how one may choose the diffusion coefficients once the equations of motion are reduced to differential form. The choice favored for the LHA in Section 7.3 may not be adequate for the initial stages of a heavy-ion collisions. It requires the applicability of the *FDT for collective* motion. As mentioned below eqns(7.22) as well as in item (3) toward the end of Section 7.5.2, in (Hofmann and Jensen, 1984), (Hofmann et al., 1989) other suggestions for the diffusion coefficients have been published. They might be more adequate for representing the fluctuating forces for faster collective motion. As they depend on the frequencies for average motion in a sense they even represent non-Markovian properties. However, these suggestions never have been tested in practical applications. It might also be of interest to study to what extent they might account for the non-Markovian features found in (Agassi et al., 1977). As indicated above, this RMM approach is adapted to treat deep inelastic collisions, eventually with slight mass transfer included. To describe fusion reactions, on the other hand, one must be able to treat the development of more complex shapes including those of the compound nucleus. For such processes it is inevitable to account for inertial effects as well as for the self-consistency problem. These features are intimately connected to the fact that nuclear systems are not only isolated but self-bound. Moreover, as indicated previously, in heavy-ion collisions various phases may be passed through which belong to very different timescales. Such problems are typical for nuclear transport and are in fact very different from such in other fields. It must be said that to date no completely satisfactory answer to this general task has been reached.

14

GIANT DIPOLE EXCITATIONS

In nuclei isovector modes can be excited where neutrons move against protons with those of dipole type being the most important ones. Section 6.2.4 outlined how the mean excitation energy can be evaluated within the potential vibrating model of the Copenhagen picture of collective motion. The frequency follows the macroscopic law $\hbar\varpi_D \approx 80 A^{-1/3}$ MeV. This shows that the excitation energy is quite large implying that such modes are already damped at zero temperature–quite unlike to the low frequency modes or the slow motion of large scale encountered in nuclear fission. Typically the following values apply: the $\hbar\varpi_D$ is of order $\simeq 15$ MeV and the width is about $\Gamma_D \simeq 5$ MeV. Nevertheless, to date the fact of a sizable Γ_D is simply ignored in many theoretical descriptions. Often fancy formalisms are applied to obtain the excitation energies discarding the possibility that by way of Kramers–Kronig relations damping will also influence the position of the centroids of the resonances. In the following we will at first present some simple, but basic relations valid within a classical picture and afterwards extend this discussion to the quantum case. In the last section we will sketch possible damping mechanisms and elaborate more closely on the influence of the coupling to slow collective motion, which at finite temperature may be described by transport models.

14.1 Absorption and radiation of the classical dipole

The classical dipole is treated in many textbooks on electrodynamics as e.g. in (Jackson, 1962), (Panofsky and Phillips, 1962) and (Landau et al., 1984). The energy transferred per unit of frequency $\nu = \omega/2\pi$ from an electromagnetic wave to the dipole may be read off from (23.223) to be

$$\frac{d\mathcal{E}}{d\omega} = \omega \chi_D''(\omega)|\mathbf{E}(\omega)|^2 = \frac{\mathsf{q}^2}{m}\frac{\omega^2 \Gamma_D}{(\varpi_D^2 - \omega^2)^2 + \omega^2 \Gamma_D^2}|\mathbf{E}(\omega)|^2. \qquad (14.1)$$

The χ_D'' is the dissipative part of the electric polarizability, which in the oscillator approximation is given by $\chi_D''(\omega) = \mathsf{q}^2 \chi_{\text{osc}}''(\omega)$, as applied on the very right. Here, Γ_D is meant to represent the total damping rate of the oscillator. The *total cross section* σ_{tot} may be defined as the ratio of *all* absorbed energy to the incident energy, both measured per unit of frequency ν. As the incident flux is $(c/4\pi)|\mathbf{E}(\omega)|^2$ the σ_{tot} simply becomes

$$\sigma_{\text{tot}}(\omega) = \frac{4\pi\omega}{c}\chi_D''(\omega) = 4\pi\frac{\mathsf{q}^2}{mc}\frac{\omega^2 \Gamma_D}{(\varpi_D^2 - \omega^2)^2 + \omega^2 \Gamma_D^2}. \qquad (14.2)$$

We may note in passing that for a non-vanishing dissipative part $\chi''_{D,m}(\omega)$ of a magnetic susceptibility the $\chi''_D(\omega)$ would have to be replaced by $\chi''_D(\omega) + \chi''_{D,m}(\omega)$ (Landau et al., 1984). For the oscillator (with a constant width Γ_D) the frequency-dependent factor can be integrated over all ω to give $\pi/2$. Hence, the total absorption strength takes on the following *sum rule value*

$$\int_0^\infty \sigma_{\text{tot}}(\omega)\, d\omega/(2\pi) = \pi q^2/(mc). \tag{14.3}$$

Remember that the formula used here for the incoming flux corresponds to a plane wave for which the Poynting vector is integrated over all times. This is in accord with the derivation of (14.1) in Section 23.3.3. Indeed, one might take an alternative point of view and work with time averages over one or several cycles of the plane wave (see exercise 1).

Let us use this picture to find the scattering cross section, for which we need the energy re-emitted per unit of time. We may take the well-known formula $\overline{P} = (1/3)(\omega^4/c^3)|\mathbf{d}|^2$ which represents the angle integrated and time averaged radiative power of the oscillating dipole. The frequency-dependent dipole moment $\mathbf{d}(\omega)$ is given by (mind formula (23.222))

$$\mathbf{d}(\omega) = q\mathbf{x}(\omega) = \chi_D(\omega)\mathbf{E}(\omega) = \frac{q^2}{m} \frac{\mathbf{E}(\omega)}{\varpi_D^2 - i\omega\Gamma_D - \omega^2}. \tag{14.4}$$

After dividing by the time averaged incident energy flux $\overline{S} = (c/8\pi)|\mathbf{E}(\omega)|^2$ the *scattering cross section* becomes

$$\sigma_{\text{sca}}(\omega) = \frac{8\pi}{3}\frac{\omega^4}{c^4}|\chi_D(\omega)|^2 = \frac{8\pi}{3}\left(\frac{q^2}{mc^2}\right)^2 \frac{\omega^4}{(\varpi_D^2 - \omega^2)^2 + \omega^2\Gamma_D^2}. \tag{14.5}$$

For the dipole it can be written in the form

$$\sigma_{\text{sca}}(\omega) = \sigma_{\text{tot}}(\omega)\frac{\Gamma_{\text{rad}}}{\Gamma_D}, \qquad \text{where} \quad \Gamma_{\text{rad}} = \frac{2}{3}\frac{q^2}{mc^3}\omega^2 \tag{14.6}$$

is the *classical radiative decay width* (in units of $\hbar$). This allows one to introduce the *reaction cross section* as

$$\sigma_{\text{rea}}(\omega) \equiv \sigma_{\text{tot}} - \sigma_{\text{sca}}(\omega) = \sigma_{\text{tot}}\left(\frac{\Gamma_D - \Gamma_{\text{rad}}}{\Gamma_D}\right) \equiv \sigma_{\text{tot}}(\omega)\frac{\Gamma_{\text{rea}}}{\Gamma_D}. \tag{14.7}$$

This notation is in accord with the one used for nuclear reactions in Chapter 18. As seen from (14.6) the radiative decay width varies with frequency. For many purposes it may suffice to take it at the resonance. For instance, in deriving the sum rule (14.3) the Γ_D of (14.2) has been treated as a constant although it is to be understood as the sum $\Gamma_D = \Gamma_{\text{rad}} + \Gamma_{\text{rea}}$.

14.2 Nuclear dipole modes

Within the classical approximation the results of the previous subsection are easily applied to dipole excitations of the nucleus. It does not suffice, however, to simply account for the motion of the charged protons; the neutrons contribute as well. This comes about because only vibrations in the center of mass system are to be considered, The spurious motion of the center of mass of all nucleons has to be removed. To exhibit this feature let us use the concept of isospin and treat all A nucleons as identical particles. The dipole moment $\mathbf{D}$ of the whole nucleus then becomes:

$$\mathbf{D} = e \sum_{k=1}^{A} \left(\frac{1}{2} + t_z^{(k)} \right) (\mathbf{x}_k - \mathbf{R}) = e \sum_{k=1}^{A} t_z^{(k)} (\mathbf{x}_k - \mathbf{R}) = \frac{e}{2} \sum_{k=1}^{A} \left(t_z^{(k)} - \frac{Z-N}{A} \right) \mathbf{x}_k . \quad (14.8)$$

Here, $t_z^{(k)}$ is the z-component of the isospin for particle k and $\mathbf{R}$ stands for the vector of the center of mass. This z-component has eigenvalue $-1/2$ for the neutron and $+1/2$ for the proton (see Section 26.7). The second and third equality in (14.8) follow from the definition of $\mathbf{R}$ and from the fact that the z-component T_z of total isospin is given by $(Z-N)/2$. A short calculation shows that $\mathbf{D}$ can be written as

$$\mathbf{D} = \sum_{\text{prot}} q_p \mathbf{x}_k + \sum_{\text{neut}} q_n \mathbf{x}_k \quad \text{with} \quad q_p = eN/A, \quad q_n = -eZ/A . \quad (14.9)$$

The q_p and q_n are *effective charges* for protons and neutrons. This result enables one to use the expressions discussed before for a dipole of charge q. We simply have to sum over the contributions from neutrons and protons separately and need to consider their individual effective charges. With this precaution the formulas from the previous subsection may be taken over. In fact this implies replacing the factor q^2 by

$$\sum_{\text{prot}} q_p^2 + \sum_{\text{neut}} q_n^2 = e^2 \left\{ Z \left(\frac{N}{A} \right)^2 + N \left(\frac{Z}{A} \right)^2 \right\} = e^2 \frac{NZ}{A} . \quad (14.10)$$

Such considerations are used in the model of M. Goldhaber and E. Teller (1948). In their picture the neutron and proton clouds move against each other without changing their shape (which in the simplest case is just a sphere). In the model of H. Steinwedel and J.H.D. Jensen (1950), on the other hand, the overall shape of the nucleus is taken fixed but neutrons and protons vibrate against each other in its interior. This motion is treated within a hydrodynamic approach. Essentially, this implies that the model from before is to be applied to local volume elements. In addition to the overall factor (14.10) there appears a geometrical factor.[33] whose value is estimated to about 0.429.

---

[33] One should not get distracted by the appearance of another factor m/M which simply reflects the fact that the authors calculate *densities* of cross sections.

14.2.1 Extension to quantum mechanics

Elaborate discussions of electromagnetic transitions can be found in the classic textbooks for nuclear physics. We want to restrict ourselves to the case of dipole transitions of electric nature. The formula for the total cross section can be taken over from (the left part of) e.q.(14.2), one only has to improve on the response function $\chi''_D(\omega)$ for the dipole operator of (14.9). Let us look at the dipole moment D^z in z-direction, for which the $\chi''_D(\omega)$ then is given by the form (23.68) with $\hat{A}_\mu \equiv \hat{D}^z$. For the total cross section one gets

$$\sigma_{\text{tot}}(\omega) = \frac{4\pi}{\hbar c} \hbar\omega \, \chi''_D(\omega)$$
$$= \frac{4\pi^2}{\hbar c} \sum_{nm} \hbar\omega \, \rho(E_m) \left| \hat{D}^z_{nm} \right|^2 \left[\delta(\hbar\omega - \Omega_{nm}) - \delta(\hbar\omega + \Omega_{nm}) \right], \quad (14.11)$$

written in terms of the exact eigenstates $|m\rangle$ and energies E_m of the many-body Hamiltonian $\hat{H} = \hat{T} + \hat{V}$. For $T = 0$ and $\omega > 0$ it can be written as

$$\sigma_{\text{tot}}(\omega) = \frac{4\pi^2}{\hbar c} \sum_n (E_n - E_0) \left| \langle n|\hat{D}^z|0\rangle \right|^2 \delta(\hbar\omega - E_n + E_0). \quad (14.12)$$

To get the integrated cross section we may employ the sum rule (6.39) to obtain

$$\int_0^\infty \sigma_{\text{tot}}(\omega) \frac{d\omega}{2\pi} = \frac{\pi}{c\hbar^2} \left\langle \left[\hat{D}^z, \left[\hat{H}, \hat{D}^z \right] \right] \right\rangle \quad \Rightarrow \quad \frac{\pi e^2}{mc} \frac{NZ}{A}. \quad (14.13)$$

The expression on the very right holds true for $[\hat{V}, \hat{D}^z] = 0$ and is a clear generalization of (14.3).

14.2.2 Damping of giant dipole modes

As mentioned in the introduction to this chapter, dipole modes have a finite width even when the nucleus is not thermally excited. This width Γ_D is smaller than the modes' frequency $\hbar\varpi_D$ by a factor of about three so that the motion is under-damped. With increasing temperature the width increases by a factor of about three but seems to saturate above $T \simeq 3-4$ MeV; a detailed exposition of both experimental findings and formal developments can be found in (Bortignon et al., 1998) (see also (Bertsch and Broglia, 1994)). As mentioned at the end of Section 6.2.4 for isovector modes in general a description of the effective coupling constant in terms of the system's thermostatic energies is not possible. For this reason the characteristics for the temperature dependence described in Section 6.6 for isoscalar modes cannot be taken over. To some extent this is different for the damping mechanism. Collisional damping also plays a role for the dipole modes as well as the possibility of emission of light particles, not to mention the decay by photo emission in this context. A possible way of handling collisional damping is described in Section 6.4.3. Typically, the energy of the giant modes is larger than that of the low frequency ones such that the former may pass energy over to the latter.

For dipole modes such a coupling can be viewed from a different perspective employing a semi-classical picture (Bohr and Mottelson, 1975). Their frequency is about inversely proportional to the diameter of the nucleus, as seen from its dependence on mass number being proportional to $A^{-1/3}$ MeV. If translated to deformed nuclei the frequency for vibrations along a shorter axis is larger than that along a longer one. To first order one would have $\delta\varpi_D^k/\varpi_D \approx -\delta R^k/R$, if ϖ_D refers to a spherical nucleus of radius R. As soon as the deformation degree of freedom undergoes random motion the vibrational frequency of the dipole, too, must be treated as a stochastic variable. According to the discussion presented in chapters 6 and 7 on isoscalar modes such a situation will be given as soon as the intrinsic excitation is large enough to warrant the low frequency modes to be damped and thus by virtue of the fluctuation dissipation theorem exhibit statistical fluctuations. In this context the hint about the size of the intrinsic temperature is important because both the frictional as well as the fluctuating forces vary with T; we will return to this fact below, which has not received much attention in the literature.

To describe the physics of such processes in some transparent way the model which was invented by R. Kubo in the 1950s for nuclear magnetic resonances (for detailed descriptions see (1962) or (1991a)) is often employed (see e.g. (Bertsch and Broglia, 1994)). The model consists of a one-dimensional oscillator parameterized by one complex quantity x with its equation of motion written in the form $\dot{x} = i(\varpi_0 + \varpi_1(t))x$. Here, both frequencies are assumed to be real, but where the perturbation $\varpi_1(t)$ is a stochastic variable of a supposedly Gaussian process. This latter property allows one to calculate the strength function $S(\omega)$ as the Fourier transform of the correlation function $\langle x^*(t=0)x(t)\rangle$. To get analytic solutions one assumes the time dependence of the correlation function of the random variable $\varpi_1(t)$ to be given simply by an exponential, $\langle \varpi_1(t=0)\varpi_1(t)\rangle = \langle \varpi_1^2 \rangle \exp(-t/\tau_c)$. In this way two quantities are introduced which parameterizes the Gaussian process of the other, residual variables, the overall strength $\Delta = \langle \varpi_1^2 \rangle^{-1/2}$ and the decay time τ_c. For future reference we should like to note in passing that no further timescale is involved, as for instance the one which could describe oscillatory behavior of these "residual degrees of freedom". The strength function $S(\omega)$ is centered at the unperturbed frequency ϖ_0 and its *width decreases* with the dimensionless quantity $\alpha = \Delta\tau_c$. For two limiting cases $S(\omega)$ takes on simple analytic forms: (i) For $\alpha \ll 1$ the shape of the line (represented by $S(\omega)$) is of Lorentzian type and has the width $\Gamma_L = \Delta^2\tau_c$. (ii) For $\alpha \gg 1$, on the other hand, $S(\omega)$ becomes a Gaussian of width $\Gamma_G = \Delta$. In the first case (i) one speaks of *fast modulation* and in the second one of *slow modulation* of the line shape. Because of $\alpha \ll 1$ for (i) the width $\Gamma_L = \Delta^2\tau_c$ of the Lorentzian is much smaller than the Γ_G of the Gaussian. (Please notice that with $\Delta \to 0$ the $S(\omega)$ reduces to the δ function expected for the e.o.m. $\dot{x} = i\varpi_0 x$.) In the literature this reduction of the width due to the coupling to other dynamic and stochastic degrees of freedom is referred to as *motional narrowing*. It is important to realize that a *narrowing* of the line only occurs if these *other* degrees of freedom are *faster than the dipole*. We may note already here that for the nuclear case this condition seems hardly be fulfilled.

In a series of papers by Y. Alhassid and co-workers the coupling of dipole modes to other degrees of freedom has been treated in explicit fashion; in (1991) an extensive review is given with references to original literature. In this model quadrupole modes described by the five degrees of freedom $\alpha_{2\mu}$ are considered, which include rotations to the body fixed system, see Section 1.3.3.2. The dynamics of the $\alpha_{2\mu}$ is assumed to be a classical, stochastic process with motion being over-damped. The conservative force is derived from a free energy for which an expansion up to fourth order in β is considered. The temperature-dependent parameters entering here are obtained from a Nilsson Strutinsky procedure. The friction tensor is assumed to be diagonal, with all elements being identical to a constant friction coefficient γ (in our notation). In accord with the classic Einstein relation, the same holds true for the diffusion tensor. The transport equation for the distribution in the $\alpha_{2\mu}$ is of a structure similar to those presented in Section 11.1.2 for one collective coordinate Q. Since γ is taken to be constant, independent of both the coordinates and temperature, complications arising from an eventually multiplicative noise do not appear. In this sense the transport equation used in this work is nothing else but a direct generalization of eqn(11.11). Along the three different axes of the rotating system dipoles D^k are assumed to vibrate with frequencies ϖ_D^k taken from phenomenology, but obeying the rules mentioned above. In this way a dependence on the length R^k of the radii along these axes comes into play. A similar dependence is assumed for the inherent damping width Γ_D^k of the dipole. An ansatz is then made for the equations of motion for the the $\mathbf{D}$ and its conjugate momentum $\mathbf{P}_D$. It consists of a system of first order differential equations with dissipative forces $-(1/2)\Gamma_D\mathbf{D}$ for $\dot{\mathbf{D}}$.[34] and $-(1/2)\Gamma_D\mathbf{P}_D$ for $\dot{\mathbf{P}}_D$, as well as with centrifugal forces in both equations. It is worth noticing that for the dipoles no fluctuating forces are taken into account despite the presence of dissipation. Likewise, it is interesting to see that the motion of the $\alpha_{2\mu}$ is *independent* of that for the dipoles, the transport equation of the former is decoupled from that of the latter.

The final aim for solving the total set of equations was to determine the correlation function $\langle \mathbf{D}(t=0)\mathbf{D}(t)\rangle$ and its Fourier transform. The latter determines the double differential cross section for angle and energy. We do not want to go into further details but may refer to the case of the total cross section discussed above (where response functions were involved rather than correlation functions, but it is known that both are closely related to each other). It was assumed that the system is in complete thermal equilibrium at given temperature, with values in the range between 1 and 3 MeV. Explicit computations were performed for nuclei in the rare earth region. To introduce the concept of time,

---

[34] Remember that this construction implies that the dipoles' restoring forces depend on Γ_D^2.

fluctuating trajectories in the equilibrated ensemble were considered. They were obtained by solving for the $\alpha_{2\mu}(t)$ the associated Langevin equations (with initial conditions $\alpha_{2\mu}(t=0)$ selected from the equilibrium distribution), together with the set of equations for $\mathbf{D}$ and $\mathbf{P}_D$. The relevant question is to what extent the line shapes of the dipole modes are influenced by the motion in the $\alpha_{2\mu}$. As expected this depends on the relation of the relevant timescales of the two different different degrees of freedom. For this an "adiabaticity" parameter $\eta_{\rm ad}$ is introduced which is defined as the ratio of the relaxation rate $1/\tau_{\rm qua}$ for the quadrupoles and the spread $\Delta\varpi_D$ for the dipoles which is due to variations in the static deformation, again because of the geometric relations mentioned before. This $\Delta\varpi_D$ is then estimated from the variance in β at equilibrium. The relaxation time $\tau_{\rm qua}$, on the other hand, representing over-damped motion can easily be seen to be proportional to the friction coefficient γ. In fact, it is nothing else but the $\tau_{\rm coll} = \gamma/C$ of (6.143) (the formula used by Alhassid is obtained if the stiffness C is expressed by the equilibrium fluctuation of the coordinate(s) through the equipartition theorem).

The only parameter which was taken as open for variations was the friction coefficient γ for quadrupole motion. All the others were considered to be determined more or less by empirical experience. For $\eta_{\rm ad} \gg 1$, which is to say for large γ, during the vibration or decay of the dipole the $\alpha_{2\mu}(t)$ does not move away from the $\alpha_{2\mu}(t=0)$. Since by construction the latter is deduced from the equilibrium one may thus simply forget about the dynamics and directly employ the probability distribution $W(\alpha_{2\mu})$. Since no inertial effects are considered and as the friction coefficient is assumed to be independent of the coordinate the worries expressed in Section 11.1.4 do not apply such that this $W(\alpha_{2\mu})$ may simply be expressed through the free energy. In the literature this situation is often called the "adiabatic limit", but where this concept evidently simply refers to timescales and has nothing to do with thermodynamic properties. The case where genuine dynamical effects come into play is found for the opposite condition, namely when $\eta_{\rm ad} \ll 1$ which in this model can only be obtained by weak damping.

It has already been mentioned above that greater reservations are appropriate for accepting such a situation for actual nuclei. This shall now be looked at and in more detail.

(i) Let us begin with timescales. The whole nucleus is assumed to be in complete equilibrium. For this to make sense one needs to have a short relaxation time τ for the intrinsic degrees of freedom first of all. The present model could be classified by saying that one follows in explicit fashion different sets of collective variables, those for the dipole and those for the quadrupole, with corresponding timescales τ_D and $\tau_{\rm qua}$. Actually, as indicated above for over-damped motion the $\tau_{\rm qua}$ may well be identified with the $\tau_{\rm coll}$ used elsewhere in the book. It is evident that τ must be the shortest time of all and that the one for slow collective modes will be the longest one, and most likely this will be those for the quadrupole.

(ii) The condition $\eta_{\rm ad} \ll 1$ relevant for non-adiabatic effects is obtained only for small damping, but in such a case motion must be expected to be *under-damped*. Then inertial effects come into play and the basic equation of motion is not valid anymore. Furthermore, additional timescales would come into play, as the $\tau_{\rm kin}$ of (6.143). It can be said that the same feature would be given in Kubo's model if for the correlation function of the $\varpi_1(t)$ a more general ansatz would be used.

(iii) For $\eta_{\rm ad} \gg 1$, on the other hand, which is to say for large damping, the $\tau_{\rm coll}$ may become so large that the hypothesis of *complete* equilibrium may no longer be realistic. The heated nuclear complex may simply not stay together for a time long enough that even these slow degrees of freedom reach equilibrium.

(iv) Finally, we should like to draw the reader's attention to the microscopic estimates of timescales presented in Chapter 6. In Section 6.4.3 the intrinsic relaxation time τ was found to be of order $0.2\,\hbar/$ MeV and thus in the range of the timescales associated with the width of the dipole resonances, which is to say that $\tau \simeq \tau_{\rm D}$. The timescales for slow collective motion are discussed in Section 6.6.3, for a compact survey we may refer to Fig. 6.11. There it is seen that the transition from under- to over-damped motion occurs at $T \simeq 2$ MeV. Above that value $\tau_{\rm coll}$ is of the order of a few MeV and thus much longer than the $\tau_{\rm D}$, indeed. It would be of the order of $\tau_{\rm D}$ or smaller only at $T \simeq 1$ MeV but there the motion is under-damped. Incidentally, in this regime even the $\tau_{\rm kin}$ is hardly smaller than $\tau_{\rm D}$, as required for motional narrowing to occur.

In the literature it has been claimed that motional narrowing may be responsible for the saturation in the dipole width seen at larger thermal excitations. In the Copenhagen–Milano collaboration a theoretical approach was developed (based on early papers like (Lauritzen *et al.*, 1986)) which differs from the ones described above in the sense that for the quadrupole degrees of freedom no transport model is applied. They only appear through intrinsic multi-quasiparticle states into which the compound states are developed. The expansion coefficients are treated by applying Gaussian orthogonal ensembles. It may be said that in this way the approximations as well as the formal derivations are a bit less transparent than in the model described above.

Finally we should like to mention the work of D. Kusnezov and others (1998, 2003) in which a universal correlation between the width of the giant dipole resonance and quadrupole deformation is found. It assumes the "adiabatic" situation to be given for which the distribution of the quadrupole deformation is governed by the free energy. By comparison with experimental data a simple linear relation is found for the equilibrium value of β and the width $\Gamma_{\rm D}$ which is valid for nuclei in the range of mass number between 44 and 208, for temperatures below 4 MeV and spins below $60\hbar$.

Exercises

1. Derive formula (14.2) by employing time averages over cycles of the applied monochromatic electric field, in which case $\sigma_{\rm tot}(\omega)$ reads $\sigma_{\rm tot}(\omega) = \overline{\mathbf{E}(t)\dot{\mathbf{d}}(t)}/S$; one may make use of part (ii) of exercise 23.1.

PART III

MESOSCOPIC SYSTEMS

15
METALS AND QUANTUM WIRES

In this chapter we want to take a small detour into solid state physics and discuss conductivity in metals and semiconductor heterostructures. This is done for the following reasons: (i) Ordinary conductivity is intimately related to transport properties of electrons in the conductivity band. These properties can be described semi-classically by transport equations which very much resemble the one discussed in Part I. In particular, the relaxation time approximation also turns out to be very useful here. (ii) These properties are those of an infinite system. They may or may not survive when the metal is made smaller and smaller such that in the end only a cluster of a few to a few thousand atoms is given. One may then ask how strongly the imaginary part of the dielectric function is influenced by the conductivity of the infinite system. In this sense the problems at stake here are very similar to those in nuclear physics. How much does an understanding of nuclear matter help to comprehend the physics of finite nuclei? With respect to transport problems, a valid question is how big the influence of two-body collisions is on the dissipation mechanism of collective modes of a real nucleus. (iii) Whereas more specific properties of metal clusters will be addressed in the next chapter, here we will discuss quantum wires and see that their conductivity is quantized. This is not only an interesting feature in itself, again there is a close relation to the nuclear problem. Quantum wires or dots are systems where the electrons practically move freely inside a potential well, like the nucleons do in the mean-field approximation. The intriguing question then arises whether or not in such a system, and if so to what extent, one may speak of resistivity. Remember that it is this property which is associated with the production of heat, and thus definitely to irreversible behavior.

Clearly, our discussion must stick to the qualitative level only. We will be able to address only those topics that are directly related to the questions just mentioned. For a better understanding one will have to look into textbooks in this field, like that by N.W. Ashcroft and N.D. Mermin on solid state physics (1976) or more experimentally oriented ones like (Ibach and Lüth, 1995).

15.1 Electronic transport in metals

15.1.1 *The Drude model and basic definitions*

Let us begin with the most elementary model, commonly associated with the name of P. Drude. Ohm's law may be written as $\mathbf{j}(\mathbf{r},\omega) = \sigma(\omega)\mathbf{E}(\mathbf{r},\omega)$ with $\mathbf{j}$ being the current density induced by an external electric field $\mathbf{E}$, both calculated in frequency space. The coefficient of proportionality σ is the *conductivity*, with the *resistivity* being its inverse, $\varrho = 1/\sigma$; here both are assumed to be scalars. In the simplest version of the Drude model (refinements will be discussed in the next chapter) σ can be expressed as $\sigma(\omega) = \sigma_0/(1 - i\omega\tau)$ with $\sigma_0 = \rho q_e^2 \tau/m^*$ and where ρ is the electron's density in space, m^* their effective mass and q_e their charge. The motion of the electron is supposed to be governed by Newton's law in the form $m^* \mathbf{v}_d/\tau = q_e \mathbf{E}$, which identifies τ as a relaxation time. As m^*/τ may be interpreted as a friction coefficient, this relation reminds one of overdamped motion. Alternatively, one may speak of a stationary situation where the electrons drift with constant velocity $\mathbf{v}_d$. Without external field the electronic motion could be said to be *diffusive* with a vanishing average velocity $\langle \hat{\mathbf{v}} \rangle_{\text{eq}}$. This comes about if the electrons scatter frequently and randomly such that they essentially remain localized.

Two additional remarks are in order here. Firstly, we have tentatively assumed that the motion is governed by an effective mass m^* rather than by the electrons' bare mass. It is defined analogously to the effective mass introduced in Section 2.2.1 for nuclear matter, which is to say that $\hbar^2/m^* = \partial^2 e(k)/\partial k^2$. Secondly, in general the relaxation time will depend on momentum. The one used here is meant to correspond to that at the Fermi energy $\tau(e_F)$. Thirdly, the density should be calculated from the Fermi momentum, which according to eqn(1.22) implies $\rho = k_F^3/(3\pi^2)$ for a degeneracy factor of 2.

15.1.2 *The transport equation and electronic conductance*

The Drude model may be improved by studying the (one-body) distribution function $n(\mathbf{r}, \mathbf{p}, t)$ of the electrons. In semi-classical approximation this can be done by a Boltzmann type equation. Assuming the electrons underlie the Lorentz force $\mathbf{F}(\mathbf{r}, t) = q_e(\mathbf{E}(\mathbf{r}, t) + \mathbf{v} \times \mathbf{B}(\mathbf{r}, t)/c)$ with external electric and magnetic fields $\mathbf{E}$ and $\mathbf{H}$, respectively, this equation can be written as (Ashcroft and Mermin, 1976), (Duderstadt and Martin, 1979)

$$\frac{\partial g}{\partial t} + \mathbf{v} \frac{\partial}{\partial \mathbf{r}} g + \mathbf{F}(\mathbf{r}, \mathbf{t}) \frac{\partial}{\partial (\hbar \mathbf{k})} g = \left(\frac{\partial g}{\partial t}\right)_{\text{coll}} \approx -\frac{g - g_{\text{eq}}}{\tau(\mathbf{k})}. \tag{15.1}$$

Here, the *kinetic momentum* $\mathbf{p}_{\text{kin}} = \hbar \mathbf{k}$ appears rather than the canonical one of the Boltzmann equations of sections 2.7.1, 21.2.2 and 21; the transformation between both versions is discussed in exercise 1. For this reason the notation has been adapted to that frequently used in the literature, which means taking $g(\mathbf{r}, \mathbf{k}, t)$ instead of $n(\mathbf{r}, \mathbf{p}, t)$. On the right of (15.1) the relaxation time approximation is introduced. As can be seen from the left-hand side, in this model one assumes that in between collisions the conductance electrons move more or less freely, undisturbed by the atomic fields. The only conservative forces considered are those from the external electromagnetic fields. The equations of average motion underlying the Drude model discussed above are recovered after performing averages of velocity and position with the solutions of (15.1).

The general form of the collision term would be

$$\left(\frac{\partial g}{\partial t}\right)_{\text{coll}} = -\int \frac{d^3k'}{(2\pi)^3} \left\{ W_{\mathbf{k},\mathbf{k}'} g(\mathbf{k})\left[1 - g(\mathbf{k}')\right] - W_{\mathbf{k}',\mathbf{k}} g(\mathbf{k}')\left[1 - g(\mathbf{k})\right] \right\} \tag{15.2}$$

with $W_{\mathbf{k},\mathbf{k}'}$ representing the transition probability for the scattering in which $\mathbf{k}$ changes into $\mathbf{k}'$, see the discussion in Section 21.2. Notice that scattering into final states is allowed only to the extent that they are unoccupied, this feature being a consequence of the exclusion principle. In passing we may note that in the solid state literature the resulting transport equations are commonly referred to simply as Boltzmann equations, not Boltzmann–Uehling–Uhlenbeck equations as in nuclear physics. The form (15.2) allows one to account for various scattering processes. Whereas in the nuclear case one only has nucleon–nucleon collisions (at least as long as (i) one discards sub-nuclear degrees of freedom, like mesons etc. and (ii) the coupling to collective modes is treated explicitly), the electrons in a metal may not only scatter among themselves but also with impurities of the crystal and the phonons describing its vibrational modes.

Electron-electron collisions: This effect is completely analogous to the one encountered in the nuclear case. As discussed in Section 21.3, for low excitations of any Fermi liquid the inverse relaxation time depends quadratically on temperature and on the deviation of the single particle energy from the Fermi energy (or chemical potential μ), see (21.25) and (21.26). One may argue that for the case of a metal this can all be lumped into a term proportional to T^2, which is to say $1/\tau_{\text{e-e}} = a(e(k) - e_F)^2 + b(k_B T)^2$. Typically it can be approximated as $\simeq \mathcal{A} l(k_B T r)^2/(\hbar e_F)$, where $\mathcal{A}$ is a dimensionless number of order unity. It follows that for the resistivity of metals at room temperatures electron–electron scattering is of practically no importance.

Electron-phonon collisions: The electron-phonon collisions can be seen to vary more strongly with temperature. For T much smaller than the Debye temperature Θ_D of order of a few hundred kelvin the inverse relaxation time $1/\tau_{\text{e-ph}}$ varies like $cT^3 + dT^5$ and for $T \gg \Theta_D$ like T/Θ_D. The formulas from the Drude model imply that the resistivity then increases linearly with T, a behavior which simulates well the situation already at room temperatures.

Scattering at impurities: The scatterers may be assumed to be located at (fixed) random lattice sites. For not too high temperatures the scattering process is an elastic one. Indeed, sizable excitations (of the atoms) of the impurities can be expected to happen only if their internal excitation energies become comparable to $k_B T$. In the opposite, normally given case the scattering process is *independent* of T, for which reason this *process dominates the contributions to the resistivity* at not too large T. The transition probability can be shown to obey $W_{\mathbf{k},\mathbf{k}'} = W_{\mathbf{k}',\mathbf{k}}$, implying that the nonlinear terms of the collision term drop out, thus leading to $(\partial g/\partial t)_{\text{coll}} = -\int (d^3k'/(2\pi)^3) W_{\mathbf{k},\mathbf{k}'}[g(\mathbf{k}) - g(\mathbf{k}')]$. The $W_{\mathbf{k},\mathbf{k}'}$ is *independent* of the electron density n. These processes, too, can be simulated by a relaxation time approximation.

Matthiesen's rule: Often several statistically independent sources of scattering act simultaneously. For such a situation the total collision rate W may be taken to be the sum of the individual ones, $W = W^{(1)} + W^{(2)} + ...$ In the relaxation time approximation this implies

$$1/\tau = 1/\tau^{(1)} + 1/\tau^{(2)} + ... \qquad (15.3)$$

If the resistivity follows the Drude model where $\varrho \propto 1/\tau$ one has $\varrho = \varrho^{(1)} + \varrho^{(1)} + ...$ As described in (Ashcroft and Mermin, 1976), Matthiesen's rule should be viewed as delivering a first orientation, but one may prove that the correct ϱ cannot be smaller than the sum of the individual ones. An example where (15.3) fails are small metal clusters, which will be discussed in the next chapter. Dissipation of nuclear collective motion also falls in the same category: Despite some early claims the friction coefficient cannot be viewed as a sum of two contributions, bulk viscosity and one-body dissipation.

15.2 Quantum wires

15.2.1 *Mesoscopic systems in semiconductor heterostructures*

To define the concept "mesoscopic" for this context,[35] let us convert the relaxation time into a mean free path by $l = v_F \tau$. As the relevant scattering processes happen to take place at the Fermi surface, the velocity is estimated by the Fermi velocity. The relaxation time τ is governed by similar processes as in metals, although its temperature dependence may deviate from the estimates given above; at $T \gg \Theta_D$ one may have $\tau_{\text{e-ph}} \propto T^{-3/2}$ instead of T^{-1}. The effective mass, too, will be different from that in the bulk. Quite generally, *systems with mean free paths longer than or comparable to their size can be said to show mesoscopic behavior*. To be more precise, one needs to distinguish between mean free paths for *elastic and inelastic collisions*, with corresponding relaxation times τ_{el} and τ_{inel}, respectively. The first case is related to elastic impurity scattering. Electron-phonon collisions, on the other hand, may occur inelastically. Moreover, they may change the phase ϕ of the electron wave function. In this sense one may speak of a *phase relaxation time* τ_ϕ.

Imagine a *two-dimensional electron gas*, as given at the interface of two semiconductors like GaAs-AlGaAs. In the following this plane surface will be identified as the x,y-plane. Suppose the electrons are subject to a further constraining potential in y direction such that they are bound to move in a rectangular area of *width W and Length L*. Such a system behaves very differently from a macroscopic one if $l_\phi \gg L, W$. We may distinguish the following cases: (1) $l_\phi \gg l_e \equiv v_F \tau_{\text{el}} \gg L, W$, (2) $l_\phi \gg L \gg l_e \gg W$, (3) $l_\phi \gg L > W \gg l_e$, (4) $L, W \gg l_\phi$, discarding refinements which involve the so-called correlation

---

[35] The interested reader should again be referred to the vast literature which exists in this field, in particular to textbooks like (Datta, 1995) or (Weisbuch and Vinter, 1991), (Kelly, 1995). One may also wish to look into the more advanced theoretical discussions in (Dittrich et al., 1998), (Guhr et al., 1998), where many more references can be found.

length ξ such that one may have to distinguish between $L \gg \xi$ and $\xi \gg L \gg l_e$ (see e.g. (Guhr et al., 1998)). In what follows we shall concentrate on the ideal case (1) and shall thus neglect any scattering process; we follow essentially the discussion of Datta (1995). On a semi-classical level we might then simply use the Liouville equation, as given by (15.1) with the collision term put equal to zero. As the electrons then simply move on classical trajectories one sometimes speaks of "ballistic transport". At times this term is used even when the motion is described fully quantal by solving the Schrödinger equation.

15.2.2 Two-dimensional electron gas

In three dimensions the flow of electrons in the conduction band is determined by the following Schrödinger equation

$$\left[e_c + \frac{1}{2m^*}\left(i\hbar\nabla + q_e\mathbf{A}\right)^2 + U(\mathbf{r})\right]\Phi(\mathbf{r}) = e\Phi(\mathbf{r}). \tag{15.4}$$

Here, the e_c defines the energy corresponding to the bottom of the conduction band (at heterojunctions it may be chosen position dependent). The $\mathbf{A}$ is the vector potential, m^* the effective mass (which for GaAs, for instance, is of value $\simeq 0.07 m_e$) and $U(\mathbf{r})$ is the potential energy due to external electric fields. No interactions of the electrons with themselves are taken into account and no effort is made to account for the rapid variations of the true lattice potential, implying amongst others that no Bloch waves will show up in the solutions. All terms appearing in eqn(15.4) are smooth on the atomic scale. Microscopic features of electronic motion are simply accounted for by working with the effective mass. As in the following we will not study any magnetic effects (no quantum hall effects etc.) we are going to put $\mathbf{A} = 0$.

To get down to the *two-dimensional system* we need to use a *constraining potential* $U(z)$ *in z direction* which is decoupled from the rest. Thus the wave function takes on the form $\Phi(\mathbf{r}) = \phi(z)\Psi(x,y)$ and the total energies separate like $e = e_c + e_n + e^{(x,y)}$. This $U(z)$ may be made sufficiently narrow such that one only needs to consider the *lowest band with* $n = 1$. The Δz may take on values between 5 and 10 nm which correspond to a level spacing of about $\Delta e_n \simeq 100$ meV. In such a situation the z-mode can be said to be "frozen" to the ground state and may simply be ignored henceforth. Without any additional potential the wave function would factorize into plane waves like $\Psi(x,y) = (1/\sqrt{L_x L_y})\exp(ik_x x)\exp(ik_y y)$ with a corresponding energy $e = e_s + (\hbar^2/2m^*)(k_x^2 + k_y^2)$, where the abbreviation $e_s = e_c + e_1$ is used. Suppose we are given periodic boundary conditions. Then $k_i = n_i(2\pi/L_i)$ for $i = x, y$ and there is one state per area $4\pi^2/(L_x L_y)$ in k-space. Thus the *total number of states* with energies less than $e = e_s + \hbar^2 k^2/2m^*$ is given by

$$n_T(e) = 2\pi k^2 \frac{L_x L_y}{4\pi^2} = \frac{m^* L_x L_y}{\pi \hbar^2}(e - e_s) \tag{15.5}$$

and for the *density of states per unit area per energy* one gets

$$n_e(e) = \frac{1}{L_x L_y} \frac{d}{dE} n_T(e) = \frac{m^*}{\pi \hbar^2} \Theta(e - e_s), \tag{15.6}$$

in accord with the findings in Section 1.2.

15.2.3 Quantization of conductivity for ballistic transport

Take a piece of metal of length L in x direction and width W in y direction and apply an electric field E_x which is constant inside $-E_x = V_x/L \equiv (V_1^x - V_2^x)/L$. Integrating the current density J_x over the cross section one gets the total current $I = J_x W$, which because of $J_x = \sigma E_x = \sigma(V_1^x - V_2^x)/L = \sigma V_x/L$ becomes

$$I = \sigma \frac{W}{L} V_x \equiv G V_x \quad \text{with} \quad G = \sigma \frac{W}{L}.$$

Here $G = \sigma W/L$ is the parameter for the *conductance*, with the conductivity σ as a material parameter being independent of the sample dimension. Thus one should expect G to vary continuously with L, W. However, when both parameters are made smaller and smaller one finds deviations. This may be realized experimentally in semiconductor heterostructures in which one may speak of a two-dimensional electron gas restricted to dimensions $L, W \ll$ *mean free path*. Then experiments tell one that the *conductance is quantized* according to

$$G = \frac{2q_e^2}{h} M = \frac{q_e^2}{\pi \hbar} M \quad \text{with} \quad M = 1, 2, 3, \ldots . \tag{15.7}$$

This M defines the number of modes $\perp$ **j**. Such behavior can be understood from the following simple model for ballistic transport.

Having separated off the z direction, the Schrödinger equation (15.4) for the independent electrons reduces to:

$$\left[e_s + \frac{1}{2m^*} (i\hbar \nabla)^2 + U(y) \right] \Psi(x, y) = e \Psi(x, y). \tag{15.8}$$

As mentioned earlier the vector potential has been dropped and the potential U is supposed to vary only in the y direction. With the ansatz $\Psi(x, y) = (L)^{-1/2} \exp(ikx) \chi(y)$ for the wave function one gets:

$$\left[e_s + \frac{(\hbar k)^2}{2m^*} + \frac{p_y^2}{2m^*} + U(y) \right] \chi(y) = e \chi(y). \tag{15.9}$$

To make things simple let us assume the potential to be the one of the harmonic oscillator $U(y) = m^* \omega_0^2 y^2 / 2$, in which case the energies become

$$e(N, k) = e_s + (\hbar k)^2 / 2m^* + (N + (1/2)) \hbar \omega_0, \quad N = 0, 1, 2, 3, \ldots . \tag{15.10}$$

The quantum numbers N define the transverse modes which act as sub-bands to the conduction band. Remember that the velocity is independent of N, namely $v(N, k) = \hbar k/m^* \equiv \partial e(N, k)/\partial(\hbar k)$.

To get a finite current in the (positive) x direction one needs to apply a negative external field $-E_x = V/L$. Then electrons are pushed out of *global equilibrium* such that there are more electrons of positive momentum $k > 0$ than there are with $k < 0$. This effect may be attributed to a modified Fermi distribution. Within a semi-classical picture this feature can be described by introducing the so-called *electrochemical potential* $\mu(x) = \mu + q_e V(x)$, which in a sense describes a *local equilibrium*, where $+q_e V(x)$ is the potential energy of one electron in the external field. This procedure is well known from statistical mechanics (see e.g. Section 19 of (Brenig, 1996)) and is in use for semiconductors in general (Ashcroft and Mermin, 1976). It requires a slow variation of the external field as compared to the microscopic ones. As within our model the latter are neglected anyway, it is more than reasonable to work with this concept.

The calculation of the current may then be done as follows. With the threshold at $e_N = e(N, k = 0)$, for given energy e altogether $M(e) = \sum_N \Theta(e - e_N)$ transverse modes contribute to it. Their occupation probability is given by the Fermi distribution $n_F(e)$. To simplify matters we assume the temperature to be sufficiently small that one has $n_F(e) \approx \Theta(\mu - e)$. For a uniform electron gas with a space density ρ the current density reads as $j_x = q_e \rho v_x$. In one dimension the ρ associated with one single k-state is given by $\rho = 2/L$, with the spin degree of freedom taken into account. For the total current one thus obtains

$$I = 2\frac{q_e}{L} \sum_k v(k) M(e) \Big(\Theta(\mu_1 - e) - \Theta(\mu_2 - e) \Big). \tag{15.11}$$

The last factor explains itself from the fact that for the electrons below μ_2 ($< \mu_1$) local left and right currents average out to zero. To evaluate (15.11) we switch to a continuum of states by

$$\sum_k \longrightarrow \frac{L}{2\pi} \int dk \quad \text{where} \quad v(k) = \frac{1}{\hbar} \frac{\partial e(N, k)}{\partial k}.$$

This leads to the result

$$I = \frac{2e}{h} \int de\, M(e) \Big(\Theta(\mu_1 - e) - \Theta(\mu_2 - e) \Big) = \frac{2q_e^2}{h} M \frac{\mu_1 - \mu_2}{q_e},$$

which may be written in the form found before for ballistic transport, namely

$$I = \Big(\frac{2q_e^2}{h} M\Big) V_x \equiv G V_x. \tag{15.12}$$

As mentioned, M is the number of transverse modes $M = M(e)$ in the interval $\mu_1 \geq e \geq \mu_2$.

The picture underlying our derivation may have to be modified if, for a realistic situation, the $U(y)$ varies with x, say at the ends of the conductor. Then one may change the potential $U(y) = (1/2)m^*\omega_0^2 y^2$ to a $U(x,y) = (1/2)m^*\omega_0(x)^2 y^2$ with variable stiffness $m^*\omega_0(x)^2$. The result from above may still be valid if only an "adiabatic condition" is fulfilled. This means that the variation of $\omega_0(x)$ and thus of $U(x,y)$ with x has to be *sufficiently smooth, in which case one speaks of "reflectionless contacts"*. Otherwise, transitions to different y modes will occur with a certain probability. Effects of this type are accounted for in the so-called Landauer formula $G = (2q_e^2/h)M\mathcal{T} \equiv (2q_e^2/h)\overline{\mathcal{T}}$. The $\mathcal{T}$ acts as a *transmission coefficient* through the conduct in a given channel. It may also account for a situation in which not all electrons exit on the right of a conductor which have entered on the left. A generalization of $G = (2q_e^2/h)\overline{\mathcal{T}}$ is due to Büttiker. It applies to devices with many terminals q and specifies the transition from q to p in the form $G_{pq} = (2q_e^2/h)\overline{\mathcal{T}}_{p \leftarrow q}$ such that the current through contact p is given by

$$I_p = \sum_q G_{pq}[V_p - V_q] \quad \text{with} \quad G_{pq} = (2q_e^2/h)\overline{\mathcal{T}}_{p \leftarrow q} \quad (15.13)$$

and where V_q measures the voltage at q. Conceivably, there exist symmetry relations and sum rules etc. This method appears to be a powerful tool both for practical applications as well as for theoretical studies. Formulations from scattering theory have been taken over, which involve the "S-matrix" between channels q and p. Moreover, other theoretical tools from many-body physics have been exploited, like Green function techniques, linear response theory with the Kubo formula, see e.g. (Baranger and Stone, 1989) or W. Zwerger's review article in (Dittrich et al., 1998).

15.2.4 Physical interpretation and discussion

So far we have not elaborated on the physical differences between the conductances for a metal wire and for the ballistic conductor. Clearly, the first is determined by relaxation phenomena but the second is not, at least not by relaxation processes inside the quantum wire. The result (15.12) has been obtained discarding any irreversible behavior, which for the ordinary metal comes from stochastic collisions of the electrons with other scatterers. On the other hand, the relation (15.12) looks like Ohm's law; but can there be "resistance" without dissipation? The answer is *no*, the resistance one measures appears in the macroscopic contacts. It is the coupling to the heat bath by which irreversibility is introduced and the Joule heat is dissipated in these contacts or reservoirs. There was a "heated discussion" in the literature about ten years ago (see e.g. the review article (Guhr et al., 1998)) and Datta (1995) spends several pages illustrating and clarifying these points. It is at this place where conceptually the biggest difference to nuclear physics shows up. Indeed, in the model of the so-called "wall formula" seemingly a friction force shows up, although (in the underlying semi-classical model) the nucleons are assumed to exhibit "ballistic" motion, and although no heat bath is available. We will try to clarify this problem in the following chapters. An interesting question is whether or not irreversibility might come in because of chaotic scattering at the boundaries; recall the discussion in item (iii) in Section 6.4.1. However, it may be hard to examine this problem with those devices we discussed above. As these semiconductor heterostructures are never isolated, there will always be a heat bath present.

Another topic of great interest is the influence of electron-electron collisions (see the article by B. Kramer in (Dittrich et al., 1998)). This is expected to be important (in clean wires) even at low T, but experimental evidence still seems to be poor. One possible candidate for a theoretical model is the so-called "Luttinger liquid", which is a Fermi liquid in one dimension but where essential features of ordinary Fermi liquid theory break down.[36]

Finally, we should like to mention that work also exists on the influence of impurity scattering in "disordered wires". This mechanism leads to localization of the electron wave function, i.e. the latter no longer extends over the whole conductor. In addition to possible modifications of the value of the conductivity, the latter then shows universal fluctuations, see the article by W. Zwerger in (Dittrich et al., 1998).

Exercise

1. For a particle of mass m and charge q, which is exposed to an electromagnetic field with scalar and vector potentials $\Phi(\mathbf{r},t)$ and $\mathbf{A}(\mathbf{r},t)$, respectively, the Hamiltonian reads $\mathcal{H}(\mathbf{r},\mathbf{p},t) = (p-(q/c)\mathbf{A}(\mathbf{r},t))^2/2m + q\Phi(\mathbf{r},t)$. Show that the Liouville equation $\partial n(\mathbf{r},\mathbf{p},t)/\partial t - \{\mathcal{H}(\mathbf{r},\mathbf{p},t), n(\mathbf{r},\mathbf{p},t)\} = 0$ written with Poisson brackets transforms into

$$\frac{\partial g(\mathbf{r},\mathbf{p}_{\text{kin}},t)}{\partial t} + \mathbf{v}\frac{\partial}{\partial \mathbf{r}}g(\mathbf{r},\mathbf{p}_{\text{kin}},t) + \mathbf{F}(\mathbf{r},\mathbf{t})\frac{\partial}{\partial \mathbf{p}_{\text{kin}}}g(\mathbf{r},\mathbf{p}_{\text{kin}},t) = 0$$

if the canonical momentum $\mathbf{p}$ is replaced by the kinetic momentum $\mathbf{p}_{\text{kin}} = m\dot{\mathbf{r}} \equiv m\mathbf{v}$ and the density $g(\mathbf{r},\mathbf{p}_{\text{kin}},t)$ is defined as $g(\mathbf{r},\mathbf{p}_{\text{kin}},t) = n(\mathbf{r},\mathbf{p}_{\text{kin}}+(q/c)\mathbf{A}(\mathbf{r},t),t)$. The $\mathbf{F}(\mathbf{r},t)$ is the Lorentz force which expressed by the magnetic and electric fields $\mathbf{B} = \nabla \times \mathbf{A}$ and $\mathbf{E} = -\nabla\Phi - c^{-1}\partial \mathbf{A}/\partial t$ may be written as $\mathbf{F}(\mathbf{r},t) = q_e(\mathbf{E}(\mathbf{r},t) + \mathbf{v}\times\mathbf{B}(\mathbf{r},t)/c)$.

---

[36] One key issue is seen in the fact that the quasiparticle residue $z_\mathbf{k}$ of the one-body Green function vanishes at the Fermi surface; hence, there are no quasiparticles in the ordinary sense. Moreover, the equation for the frequency of the quasiparticle pole has no unique solution. However, taking into account only forward scattering the Hamiltonian can be diagonalized exactly using bosonization techniques. The latter is suggested because the operator of the one-body density can be made to obey boson commutation relations exactly (due to the fact that there are only left and right going particles). For this model the conductance writes $G = Kq_e^2/h$ where the coefficient K depends on the scattering amplitude.

16

METAL CLUSTERS

In the following we are going to take up the problem suggested in the introduction to the previous chapter, namely what happens to a metallic system when its size is reduced more and more. We will begin by discussing static features which are very similar to those of atomic nuclei: The stability of small clusters is governed by shell effects. Next we will turn to dynamic properties as manifested by the dielectric function. The latter specifies the behavior of the system under the influence of a time-dependent electric field. Here, again, close analogies to nuclear dynamics can be established. Specifically, we will address the question of the absorption of light at small frequencies and the intriguing mechanism behind it. Indeed, there is a very intimate relation between the conductivity of direct currents and the damping of slow collective motion.

16.1 Structure of metal clusters

Since the 1980s metal clusters have become objects of extensive studies, both experimentally as well as theoretically. A metal cluster consists of a certain number of atoms ranging from a few to a few thousand in number. It is known experimentally that such objects are more stable whenever the number of atoms reaches "magic" values, as is the case for nuclei. It is mainly because of this feature that nuclear physicists have become interested in trying to apply methods to such systems which have previously proven successful for nuclei. First of all, this concerns the description of static and dynamic properties within the spherical and deformed shell model, like the shell correction method, the Hartree–Fock approximation, density functional theories, RPA, semi-classical methods and POT. There exist excellent reviews on this subject, both in books and in articles which are recommended for further reading (Brack, 1993), (Bjørnholm, 1994), (Kreibig and Vollmer, 1995), (de Heer, 2000), (Brack, 2001), (Frauendorf and Guet, 2001).

The reason for the close correspondence between nuclear and cluster properties is as follows. As in an ordinary metal, in clusters the atoms deliver their valences electrons to the whole entity. In a certain range of temperatures the presence of the positively charged ions may be parameterized by a distribution which in the interior is almost homogeneous; in solid state physics such a picture is commonly referred to as the *jellium model*. In addition to this positive background charge the electrons are subject to their mutual but largely screened Coulomb interaction. The potentials for electronic motion found within mean-field approximations resembles very much the Woods–Saxon form used in nuclear physics. It is this feature first of all which enables one to apply many of the concepts and methods presented in the previous chapters. The jellium model works best at not too small temperatures. At very small T the ions sit on their "lattice sites" implying that the structure of the cluster is of more molecular type. With increasing T the ions start to vibrate before at even larger T they move around freely, the clusters begin to "melt". In fact, as pointed out in (Haberland, 1999) the analogy with the nuclear case is well established only for "hot" liquid clusters. Experimentally, the transition is observed in the absorption cross section: Finer structure seen in cold clusters disappear with increasing T (Schmidt et al., 1999).

The electronic structure of metallic grains may also be influenced by pairing correlations. Indeed, for such systems superconductivity has already been studied theoretically in the early 1970s, see e.g. the work of B. Mühlschlegel, D. Scalapino and R. Denton which was based on functionals of Ginzburg–Landau type (1972). In more recent years pairing interactions are often treated by the method described in Section 24.4, which uses the Hubbard–Stratonovich transformation for functional integrals. Indeed, such interactions can be assumed to be of separable nature of the type introduced in Section 3.3.2 and thus invite the application of formalisms like the SPA and PSPA; for a treatment of superconductivity in metal clusters along these lines see e.g. (Rossignoli and Canosa, 2001). Experimentally, BCS-type correlations of electron states have been reported by D. Ralph et al. in (1995, 1996, 1997). A critical review on the measurability of such and related effects in small superconductors having a discrete electronic spectrum has been given in (Braun and von Delft, 1999), see also M. Schechter et al. (2001, 2003). Of crucial importance is the relative size of the gap parameter Δ for the bulk in comparison to the mean level spacing D of the finite system.

Excited clusters may decay by emission of photons, electrons or smaller clusters. Such processes are commonly treated by the statistical model taken over from nuclear reaction theory (Frauendorf and Guet, 2001), the basic methods of which are explained in Chapter 18 in general and in Chapter 8 for decay processes in particular. One specific example is fission of a cluster into two fragments, with the decay rate being evaluated by the Bohr–Wheeler formula of Section 8.2.1. Often the potential energy surface is calculated within a macroscopic limit (liquid drop model), but more recently shell effects are also taken into account (Lyalin et al., 2002).

Because of lack of space we are not going to elaborate further on such features which mainly deal with structural effects and mostly reflect static properties of the clusters. Rather, we shall concentrate on dynamic issues being related to dissipative transport phenomena.

16.2 Optical properties

Optical properties are described by the dielectric function, which for metals is related to their conductivity. To make the reader familiar with our notation (in which we follow that of (Ashcroft and Mermin, 1976)) let us begin by recalling briefly elementary relations of electrodynamics in materials. In Gaussian units the differential form of Ampère's law reads $[\nabla \times \mathbf{H}(t)] = (4\pi/c)\mathbf{j}(t) + (1/c)\partial \mathbf{D}/\partial t$.

For the sake of simplicity let us assume first the medium to be isotropic and homogeneous and discard any space dependence of the fields. After a Fourier transformation with respect to time the resulting equation may be written in the following forms:

$$[\nabla \times \mathbf{H}(\omega)] = -i\frac{\omega}{c}\left(i4\pi\frac{\sigma(\omega)}{\omega} + \varepsilon_0(\omega)\right)\mathbf{E}(\omega) \tag{16.1}$$

$$= -i\frac{\omega}{c}\varepsilon(\omega)\mathbf{E}(\omega) = -i\frac{\omega}{c}\mathbf{D}(\omega). \tag{16.2}$$

Whereas in the first line the common definitions

$$\mathbf{j}(\omega) = \sigma(\omega)\mathbf{E}(\omega) \quad \text{and} \quad \mathbf{D}(\omega) = \varepsilon_0(\omega)\mathbf{E}(\omega), \tag{16.3}$$

for the conductivity $\sigma(\omega)$ and the "usual" dielectric function $\varepsilon_0(\omega)$ is employed, in the second line the generalized dielectric function

$$\varepsilon(\omega) = \varepsilon_0(\omega) + i4\pi\frac{\sigma(\omega)}{\omega} \tag{16.4}$$

is introduced. These two versions (16.1) and (16.2) correspond to different conventions. In the latter one free and bound charges are treated on the same footing. In this way optical properties of metals may be described like those of insulators, where all charges are bound. In general both $\varepsilon_0(\omega)$ and $\sigma(\omega)$ are complex. If they are real, the imaginary part of ε is given by $\text{Im}(\varepsilon) \equiv \varepsilon''(\omega) = 4\pi\sigma(\omega)/\omega$. Finally, let us quote the relation $\varepsilon(\omega) = 1 + 4\pi\chi_e(\omega)$ to the electric polarizability (or susceptibility) $\chi_e(\omega)$ which specifies the linear response of the polarization to the electric field, $\mathbf{P}(\omega) = \chi_e(\omega)\mathbf{E}(\omega)$.

Within the Drude–Lorentz model the dielectric function takes on the form

$$\varepsilon_D(\omega) = 1 + \frac{4\pi q_e^2 N}{m^*}\sum_{k=1}\frac{f_k}{\omega_k^2 - i\omega\Gamma_k/\hbar - \omega^2} = 1 + 4\pi\chi_e(\omega). \tag{16.5}$$

It results from treating the external electromagnetic field in dipole approximation and by parameterizing the system's excitations in terms of oscillator modes. The latter are of frequency ω_k, have a damping width Γ_k and an effective mass m^*. The response function appearing here is a generalization of that derived in exercise 23-1 for one particle (or mode) to the case that in the volume $\mathcal{V}$ there are N identical molecules with Z electrons in each. Amongst them f_k electrons are in mode k such that $\sum_k f_k = Z$. As may easily be seen the form (16.5) satisfies the "f-sum rule" (Landau et al., 1984),

$$\frac{m^*}{4\pi^2 q_e^2}\int_{-\infty}^{\infty} d\omega\, \omega\, \varepsilon''(\omega) = \frac{NZ}{\mathcal{V}} \equiv \rho_e, \tag{16.6}$$

where ρ_e is the electron density (recall the discussion below eqn(6.40)).

Conductivity in metals comes about if there are partially filled bands in which the electrons can move without experiencing any restoring force. Within our simple schematic model we may assume this situation to be represented by the mode $k = 1$ such that $m^*\omega_1^2$ vanishes. Expressing (16.5) in the form (16.4), with ε_0 accounting for terms with $k > 1$, the $\sigma_D(\omega) \equiv \sigma_D'(\omega) + i\sigma_D''(\omega)$ writes as

$$\sigma_D(\omega) = \frac{i}{4\pi}\frac{\omega_P^2}{\omega + i\Gamma_D/\hbar} = \frac{1}{4\pi}\left(\frac{\omega_P^2\Gamma_D/\hbar}{\omega^2 + (\Gamma_D/\hbar)^2} + i\frac{\omega_P^2\omega}{\omega^2 + (\Gamma_D/\hbar)^2}\right). \quad (16.7)$$

The ω_P is the plasma frequency defined by the relation $\omega_P^2 = 4\pi\mathcal{N}_P q_e^2/m_P^*$, where $\mathcal{N}_P$ is the number of conduction electrons per volume and m_P^* their effective mass. Within a semi-classical picture the damping width Γ_D may be related to the mean free path l through $\Gamma_D = \hbar v_F/l$. Here, the Fermi velocity v_F enters because only electrons in the close neighborhood of the Fermi surface are responsible for the conduction.

16.2.1 Cross sections for scattering of light

Important information on clusters is obtained by experiments with light beams. In the language of nuclear physics one would call it a scattering experiment for measuring cross sections, for instance those for scattering and absorption. We may adopt the point of view of Section 14.2 and write the total cross section as $\sigma_{\text{tot}} = \sigma_{\text{sca}} + \sigma_{\text{abs}} = \sigma_{\text{ext}}$, which in (Kreibig and Vollmer, 1995) is identified as the "extinction" cross section. The scattering of light at small metallic spheres can be described with Maxwell's equations within the long wavelength limit, as formulated by G. Mie already in 1908 (Mie, 1908). The sizes of the so-called *small clusters* with up to a few thousand particles only range up to 10 nm, which is small compared to the wavelength even of violet or ultraviolet light of about 400 nm. In (14.2) the total cross section is expressed by the dissipative part of the electric polarizability as $\sigma_{\text{tot}}(\omega) = (4\pi\omega/c)\chi_e''(\omega)$. For a dielectric sphere of radius R placed in vacuum,[37] one has (Jackson, 1962), (Landau et al., 1984)

$$\chi_e(\omega) = R^3\frac{\varepsilon(\omega) - 1}{\varepsilon(\omega) + 2} \quad (16.8)$$

such that with $\mathcal{V}_R = 4\pi R^3/3$ the $\sigma_{\text{tot}}(\omega)$ becomes

$$\sigma_{\text{tot}}(\omega) = 9\mathcal{V}_R\frac{\omega}{c}\frac{\varepsilon''(\omega)}{(\varepsilon'(\omega) + 2)^2 + (\varepsilon''(\omega))^2}. \quad (16.9)$$

As function of ω the $\chi_e(\omega)$ shows the typical behavior of a resonance, the "Mie resonance". For a smooth $\varepsilon''(\omega)$ it lies at the frequency ω_R determined by $\varepsilon'(\omega_R) + 2 = 0$.

As a first ansatz for the dielectric function $\varepsilon(\omega)$ one might assume it to be given by the bulk properties of the metal. The finite size would then only come in through the geometry which defines the boundary conditions on the metallic

---

[37] If the particle is inside a medium with dielectric constant ε_m the 1 and 2 of formula (16.8) are to be replaced by ε_m and $2\varepsilon_m$, respectively.

surface. In the following we want to address more specific intrinsic effects. Quite generally, the ε will depend on the size of the cluster, which for a spherical shape of radius R means writing $\varepsilon = \varepsilon(\omega, R)$. Evidently, with increasing R the $\varepsilon(\omega, R)$ should turn into the dielectric function of the bulk $\varepsilon_{\text{bulk}}(\omega)$. For the damping width, the inverse relaxation time, one claims the following form to hold true

$$\Gamma(R) = \Gamma_\infty + \Delta\Gamma(R) \quad \text{with} \quad \Delta\Gamma(R) = \Lambda\, v_{\text{F}}/R, \qquad (16.10)$$

with Λ being a coefficient of order unity. This ansatz may be viewed as a special version of Matthiesen's rule (15.3), which seems justified if for the electrons the surface acts like a "scatterer" with stochastic properties. In the literature this problem has been examined within the jellium model introduced above, in which electrons move freely in a mean field. For finite R special quantum features come into play, which are often referred to as *quantum size effects*, but for which it is not clear that they can solely be handled in the sense of (16.10). In other words, it is questionable that mean field dynamics alone may lead to a finite damping width Γ. In fact this problem is analogous to the one encountered in nuclear physics if an attempt is made to describe dissipation of collective motion within the model of "wall friction". The question at stake can be phrased in the following way: Suppose the system of fermions is perturbed by a time-dependent external field. Under what conditions does there exist an *irreversible* transfer of energy to the system which furthermore can be parameterized by a simple width in the one case, or a friction coefficient in the other? Below we shall at first examine this problem further for dielectric properties of metal clusters to return to specific questions about the energy transfer in the next chapter.

16.2.2 Optical properties for the jellium model

Let us assume we are given A electrons inside a volume $\mathcal{V}$ which move independently of each other in a static, spherically symmetric potential $U(r)$. For the sake of simplicity it is supposed to be zero inside $\mathcal{V}$ and to become very large outside. The perturbing Hamiltonian may be written as $\delta H = -\hat{F}_D E \sin(\omega t) \equiv (\sum_i q_e \hat{z}_i) E \sin(\omega t)$, which is valid in dipole approximation for a homogeneous electric field $E(t) = E\sin(\omega t)$ in z direction. Assuming this perturbation to be sufficiently weak we may apply linear response theory. Suppose we are interested in the power W dissipated in the volume $\mathcal{V}$ averaged over one cycle. Expressed in terms of the conductivity $\sigma(\omega)$ this quantity is given by $W = \sigma(\omega)E^2\mathcal{V}/2$ (see (Landau et al., 1984)). On the other hand, we may apply the formulas from Section 23.3.3, like (23.224) for instance, to get

$$W \equiv \overline{\frac{dE}{dt}} = \frac{\pi E^2 \chi''_D(\omega)}{(2\pi/\omega)}\mathcal{V} = \frac{1}{2}\omega \chi''_D(\omega)E^2\mathcal{V}. \qquad (16.11)$$

As already mentioned in connection with the (16.5) of the Drude–Lorentz model, the dielectric function represents the response of the charge density. For this reason, and following (6.24), the exact quantum expression for the dissipative part $\chi''_D(\omega)$ of the (dipole) response may be written as

$$\chi''_D(\omega) = \frac{\pi q_e^2}{V} \sum_{jk} (n(e_j) - n(e_k)) |\langle j|\hat{z}|k\rangle|^2 \, \delta(\hbar\omega - (e_k - e_j)) = \frac{\sigma(\omega)}{\omega} = \frac{\varepsilon''(\omega)}{4\pi}. \quad (16.12)$$

The sum is governed by the matrix elements of $\hat{z}$. It is possible to rewrite the latter in terms of such for the derivative of the potential. In this way the relevant operator can be related to those used for collective motion in nuclei or for vibrating quantum dots. One simply needs to apply the equation $(e_j - e_k)\langle j|\hat{o}|k\rangle = \langle j \mid [\hat{h}, \hat{o}] \mid k\rangle$, first for $o = z$ and then for $o = p_z$. From the commutation relations $[\hat{h}, \hat{z}] = -i\hbar \hat{p}_z/m$ (with effective inertia m) and $[\hat{h}, \hat{p}_z] = i\hbar \partial U/\partial z$ one gets

$$((e_k - e_j)/\hbar)^2 \, \langle j|e\hat{z}|k\rangle = \frac{q_e}{m} \langle j \mid \frac{\partial U}{\partial z} \mid k\rangle = -\frac{q_e}{m} \langle j|\dot{\hat{p}}_z|k\rangle \equiv -\langle j|\dot{\hat{j}}_z|k\rangle. \quad (16.13)$$

These identities may be used in (16.12) to express $\chi''_D(\omega)$, $\varepsilon''(\omega)$ or $\sigma(\omega)$ in different forms. Moreover, liberal use may be made of the fact that due to the presence of the δ function the energy difference $(e_k - e_j)$ can be replaced by $\hbar\omega$. More fundamental derivations starting from the basic equations of electrodynamics of continuous media can be found in the article (1966) by A. Kawabata and R. Kubo and in the very detailed exposition given in (1982) by D.M. Wood and N.W. Ashcroft. An early survey of existing literature was presented in (Ruppin and Yatom, 1976) and later in the textbook by U. Kreibig and M. Vollmer (1995). In (Ruppin and Yatom, 1976) the authors have demonstrated that five previously published approaches are equivalent. Actually, all of them appear to be different applications of linear response theory, although the authors reserve that name essentially only for the version of (Kawabata and Kubo, 1966). Making use of (16.13) the imaginary part of the dielectric function becomes

$$\varepsilon''(\omega) = \frac{4\pi^2 q_e^2}{m^2 \omega^4 V} \sum_{jk} (n(e_j) - n(e_k)) \left|\langle j|\frac{\partial U}{\partial z}|k\rangle\right|^2 \delta(\hbar\omega - (e_k - e_j)). \quad (16.14)$$

This expression will now be examined within a simple model.

16.2.3 The infinitely deep square well

Schematically the potential may be represented as the limit of a finite square well of radius R and depth U_0,

$$U(r) = -U_0 \Theta(R - r), \quad \text{where} \quad U_0 \to \infty. \quad (16.15)$$

For any finite U_0 the coupling operator becomes

$$\frac{\partial U}{\partial z} = -U_0 \delta(r - R) \cos\vartheta \quad (16.16)$$

if written in spherical polar coordinates. To evaluate its matrix elements we need to specify boundary conditions for the radial parts $\mathcal{R}_{nl}(r)$ of the wave functions

$\psi_{nlm}(r,\vartheta,\varphi) = \mathcal{R}_{nl}(r)Y_{lm}(\vartheta,\varphi)$. To cope with the limit $\lim_{U_0 \to \infty}$ this may be done by requiring (Koonin et al., 1977)

$$\lim_{U_0 \to \infty} \frac{\sqrt{2mU_0}}{\hbar}\mathcal{R}_{nl}(R_0) = \left.\frac{\partial \mathcal{R}_{nl}(r)}{\partial r}\right|_{r=R_0}. \qquad (16.17)$$

In (16.16) the presence of the $\cos\vartheta \propto Y_{10}(\cos\vartheta)$ reflects the fact that we are dealing with a multipole mode of order $\lambda = 1$. For possible transitions between different states it implies the selection rule that the angular momentum can only change from l to $l' = l \pm 1$. In the next chapter we are going to study quadrupole vibrations for nuclei for which the transition operator is proportional to $Y_{20}(\cos\vartheta)$. Then the corresponding selection rule is modified to permit only transitions from $l = l \pm 1$ to $l' = l, \pm 2$.

To find the solutions of the single particle Schrödinger equation it is useful to introduce the dimensionless quantity $x_{nl} = (e_{nl}/\epsilon_0)^{1/2}$, with a scale factor $\epsilon_0 = \hbar^2/2mR_0^2$. The radial part of the wave function may then be expressed by a spherical Bessel function as $\mathcal{R}_{nl}(r) = j_l(x_{nl}r/R_0)(2/R_0^3\, j_{l+1}^2(x_{nl}))^{1/2}$. The energies e_{nl} are determined by the zeros of the j_l at the boundary, which is to say by the secular equation $j_l(x_{nl}) = 0$. Its general solutions have to be found numerically. For large $k_F R_0$ one may make use of the asymptotic form of the Bessel functions (Abramowitz and Stegun, 1968). In this case their zeros lie at values x_{nl} which are determined by the two equations

$$l + 1/2 = x_{nl}\cos\phi, \quad \pi(n + 3/4) = x_{nl}(\sin\phi - \phi\cos\phi), \qquad (16.18)$$

with an "angle" ϕ satisfying $0 < \phi < \pi/2$. These equations imply the lowest energy level to correspond to $n = 0$; if one prefers to start from $n = 1$ the $(n + 3/4)$ must be replaced by $(n - 1/4)$. It can be said that the approximation (16.18) works very well even for small values of n and l, with errors of the order of 1 even for the lowest values (Barma and Subrahmanyam, 1989), (Ivanyuk, 2007). To find the variation of the energies with n and l one may differentiate (16.18) to get

$$dl = dx_{nl}\cos\phi - x_{nl}\sin\phi\, d\phi, \quad \pi dn = dx_{nl}(\sin\phi - \phi\cos\phi) + x_{nl}\phi\sin\phi\, d\phi,$$

from which one obtains

$$\left.\frac{\partial x_{nl}}{\partial n}\right|_l = \frac{\pi}{\sin\phi} = \frac{\pi}{\sqrt{1 - (l+1/2)^2/x_{nl}^2}}$$

$$\left.\frac{\partial x_{nl}}{\partial l}\right|_n = \frac{\phi}{\sin\phi} = \frac{\arccos((l+1/2)/x_{nl})}{\sqrt{1 - (l+1/2)^2/x_{nl}^2}}. \qquad (16.19)$$

The first equation specifies the average density of states $g_l(e_{nl}) \equiv (\partial n/\partial e_{nl})_l$ at fixed angular momentum. It is given by

$$g_l(e_{nl}) = \frac{1}{2\epsilon_0 x_{nl}}\left(\frac{\partial x_{nl}}{\partial n}\right)_l^{-1} = \frac{1}{2\pi e_{nl}}\sqrt{x_{nl}^2 - (l+1/2)^2}\,\Theta(x_{nl}^2 - (l+1/2)^2) \qquad (16.20)$$

and represents the semi-classical limit of $g_l(e) = \sum_n \delta(e - e_{nl})$.

Optical properties

In (1989) M. Barma and V. Subrahmanyam reported on a calculation of the $\varepsilon''(\omega)$ of (16.14), which for the square well potential at zero temperature (16.14) can be written as

$$\varepsilon''(\omega) = \frac{16\pi q_e^2}{m^2 R^5 \omega^4} S(\omega), \tag{16.21}$$

with a strength function $S(\omega)$, which for a twofold spin degeneracy is given by

$$S(\omega) = \sum_{nln'l'} \frac{1}{2}(l\delta_{l,l'+1} + l'\delta_{l',l+1})e_{nl}e_{n'l'}\Theta(e_F - e_{nl})\Theta(e_{n'l'} - e_F)\delta(\hbar\omega + e_{nl} - e_{n'l'}). \tag{16.22}$$

The Θ-functions stand for the occupation numbers, $n(e_{nl}) = \Theta(e_F - e_{nl})$, and their product represents the difference $n(e_{n'l'}) - n(e_{nl})$. It vanishes unless one state is below and one above the Fermi energy. In the limit $k_F R \to \infty$ one may try to replace the sums in nl and $n'l'$ in (16.22) by integrals. In the simplest version this means replacing $S(\omega)$ by the *quasi-continuous limit* (Wood and Ashcroft, 1982)

$$S_{qc}(\omega) = \int_0^\infty dn\, dl\, dn'\, dl'\, l\, \delta(l-l') e_{nl} e_{n'l'} \Theta(e_F - e_{nl})\Theta(e_{n'l'} - e_F)\delta(\hbar\omega + e_{nl} - e_{n'l'}). \tag{16.23}$$

Here, it was assumed that only large angular momenta contribute such that the difference between l and $l \pm 1$ become irrelevant and that $\delta_{l,l'+1}$ and $\delta_{l',l+1}$ may be substituted by $\delta(l-l')$. In (1989) M. Barma and V. Subrahmanyam succeeded in evaluating (16.23) in closed form, which as the authors claim leads to results superior to those of previous publications like those of (Kawabata and Kubo, 1966). In passing it should be mentioned that both in (16.22) and in (16.23) there is no natural cutoff. This implies that the correlation functions $S(\omega)$ and $S_{qc}(\omega)$ do not decrease with increasing ω. In (16.21) this problem is masked by the strong decrease of the pre-factor being proportional to ω^{-4}.

Because of the presence of the $\delta(l-l')$-function in (16.23) the integration in l' is trivial. Those over n and n' may then be handled by transforming them into integration in the energies e_{nl} and $e_{n'l}$ at fixed angular momentum, which requires the densities of states $g_l(e)$ introduced above. After a few trivial manipulations one ends up with

$$S_{qc}(\omega) = \int_0^\infty dl\, l \int_0^{e_F} de\, e\, g_l(e)\,(e + \hbar\omega)\, g_l(e + \hbar\omega)\, \Theta(\hbar\omega + e - e_F).$$

This leads to two different branches. Whereas for $\hbar\omega > e_F$ the remaining Θ function is of no influence, for $\hbar\omega < e_F$ it implies the form

$$S_{qc}(\omega) = \int_0^\infty dl\, l \int_{e_F-\hbar\omega}^{e_F} de\, e\, g_l(e)\,(e + \hbar\omega)\, g_l(e + \hbar\omega) \qquad \hbar\omega < e_F \tag{16.24}$$

on which we will concentrate in the following. Because of the weight l the main contributions come from large angular momenta such that one may neglect the additional terms of $1/2$ in formula (16.20) for the $g_l(e_{nl})$. Introducing the dimensionless variables $x = e/e_F$, $x_l = \epsilon_0 l^2/e_F$ and $\nu = \hbar\omega/e_F$ (16.24) turns into

$$S_{\text{qc}}(\omega) = \frac{(k_F R)^4 e_F}{8\pi^2} \int_{1-\nu}^{1} dx \int_{0}^{x} dx_l \sqrt{x - x_l}\sqrt{x + \nu - x_l}. \tag{16.25}$$

The remaining integrals can be done analytically, leading to the final result being

$$S_{\text{qc}}(\omega) = \frac{(k_F R)^4 e_F}{16\pi^2} G_{\text{qc}}(\nu) \tag{16.26}$$

with $G_{\text{qc}}(\nu) \equiv G(\nu) - G(-\nu)$, where for $\nu \leq 1$ one has

$$G(\nu) = \frac{(1+\nu)^{3/2}}{3} + \frac{\nu^2(1+\nu)^{1/2}}{4} - \frac{\nu^2(2+\nu)}{8} \ln\left|\frac{\sqrt{1+\nu}+1}{\sqrt{1+\nu}-1}\right|. \tag{16.27}$$

To understand characteristic features of the quasi-continuous limit it may suffice to study the low frequency behavior of the dielectric function. In particular it will be interesting to look at the real part of the conductivity, amongst other reasons because of possible close analogies to nuclear dissipation. Using (16.21) and (16.26) one gets

$$\sigma'_{\text{qc}}(\omega) = \frac{1}{\pi^2} \frac{q_e^2}{\hbar R} \frac{1}{\nu^2} \frac{G_{\text{qc}}(\nu)}{\nu}, \tag{16.28}$$

which is in accord with eqn(17) of (Barma and Subrahmanyam, 1989). At small frequencies, where the ratio $G_{\text{qc}}(\nu)/\nu$ turns into unity, one obtains the following behavior

$$\sigma'_{\text{qc}}(\nu \to 0) = \frac{1}{\pi^2} \frac{q_e^2}{\hbar R} \frac{1}{\nu^2} \qquad \text{for} \qquad \nu \ll e_F. \tag{16.29}$$

Although the formal divergence implied by this formula should not be taken too seriously, this expression allows for a direct comparison with the result obtained by A. Kawabata and R. Kubo (1966) (see eqn(63) of (Wood and Ashcroft, 1982) or eqn(34) of (Ruppin and Yatom, 1976)), which is smaller by a factor $8/\pi^2$. In (1982) D.M. Wood and N.W. Ashcroft also treated the example of a cubic box and presented in eqn(64) a result like (16.29) but with an additional factor of two. This difference may be attributed to the geometry of the cluster.

Next we may establish a connection with the Drude result of (16.7)–exactly in the same manner as done in (Wood and Ashcroft, 1982) for their result for the box. Suppose we look at frequencies in the range $\Gamma_D \ll \hbar\omega \ll e_F$. For such a situation one may compare the frequency dependence of the σ'_{qc} of (16.29) with the $\sigma'_D(\omega)$ given in (16.7). The damping width is easily seen to be $\Gamma_D/\hbar = (3/4)\, v_F/R$. It represents the $\Delta\Gamma(R)$ of (16.10) and clearly shows the finite size effect in being proportional to the inverse radius of the particle. It was obtained through the quasi-continuous limit without introducing a *true mechanism for irreversible behavior*, which here is to say for a vanishing collision width Γ_∞ of the bulk.[38] For this reason this mechanism is often referred to as Landau damping, like e.g. in (Wood and Ashcroft, 1982), (Yannouleas and Broglia, 1992). This kind of damping had already been introduced in 1946 by L.D. Landau (1946) in deriving dielectric functions for collisionless plasma in infinite systems; a detailed description can be found in Vol. X of the Landau, Lifshitz (and Pitajewski) series (Landau *et al.*, 1981). Landau damping is known to be entropy conserving, which is the case also for the present example because of the neglect of any genuine broadening; see also the discussion in Section 6.4.1. Indeed, within the present model the appearance of a finite $\Delta\Gamma(R)$ or Γ_D originates solely from using a quasi-continuous approximation. However, as pointed out already in (Kawabata and Kubo, 1966), the very existence of a finite width Γ_D cannot be corroborated through a quantum mechanical treatment which only takes into account mean-field dynamics. With decreasing size of the system, the discreteness of the electronic states is bound to become manifest. In this sense "... the surface no longer acts as an additional scatterer" as required by the hypothesis underlying (16.10). Indeed, had we used the $S(\omega)$ of (16.22) instead of the $S_{\rm qc}(\omega)$ of (16.23) we would have found a $\sigma'(\omega)$ which would *strictly vanish* in a finite range of small positive frequencies. This corresponds to the properties of an insulator rather than to those of a metal. For an insulator the DC-conductivity has to vanish

$$\sigma'_{\rm DC} = \lim_{\omega \to 0} \sigma'(\omega) = 0. \qquad (16.30)$$

This property is the more stringent, the larger the gap of excitations above $\hbar\omega = 0$, and hence the higher the symmetries for independent particle motion. In reality, structural details and size distributions of the clusters will remove this symmetry degeneracy for a given ensemble of clusters. In (Kawabata and Kubo, 1966) this problem was circumvented by introducing a statistical distribution of non-degenerate levels.

Let us go back to the original form (16.22)) of the strength function $S(\omega)$. It is evident that only such terms contribute to the sums where $e_{n'l'}$ is strictly larger than e_{nl}. On the one hand, this follows from the two Θ-functions which at zero temperature represent the difference in the occupation numbers, with the latter appearing in the imaginary parts of either the response function (16.12) or in the (16.14) for the dielectric function. Moreover, even at finite temperature the presence of the δ-functions ensures that for positive frequencies only transitions from lower levels to higher ones contribute to dissipation. This information is lost when sums over discrete eigenstates are replaced by integrals in the quasi-continuous limit, definitely in the final form of (16.24). Physically this implies that even at very small frequencies transitions may occur even for small systems for which the discrete level structure of the quantum result would prevent this

---

[38] It may be mentioned that there exists experimental evidence that in small clusters also electron–electron and electron–phonon scattering rates show a $1/R$ dependence, see (Muskens *et al.*, 2006) and further references given therein.

to happen. On the other hand, the transformation to a continuous limit has its own advantage. Therefore, in (Ivanyuk, 2007) a modification of the procedure exhibited above is suggested. It leads to a new function $G_{\text{mod}}(\nu)$ which vanishes below a critical value. More precisely, the ratio $G_{\text{mod}}(\nu)/\nu$, which in (16.28) is seen to determine the real part $\sigma'(\omega)$ of the conductivity, is zero for $\nu \leq 2/(k_\text{F} R)$. In this sense there is no contradiction anymore to the condition (16.30) for an insulator. Moreover, the integral in the f-sum rule (16.6) becomes finite, which for the $S_{\text{qc}}(\omega)$ of (16.25) is infinite because of the divergence of $\varepsilon''_{\text{qc}}(\omega)$ at small frequencies.

17

ENERGY TRANSFER TO A SYSTEM OF INDEPENDENT FERMIONS

In the following we are going to study a system of fermions which are enclosed in a cavity and perturbed by an external field. The particles are supposed to move in a single particle potential U and the perturbation may be understood to make this U time-dependent. As we know from previous chapters, such a model was originally exploited in nuclear physics, but in the meantime has been applied to mesoscopic systems, like quantum dots or metal clusters, as well. The main purpose of this presentation will be to exhibit close analogies to the nuclear case, but also to point out essential differences. For instance, in this model one usually ignores any trace of the self-consistency problem described in Chapter 6. In this spirit the frequency of the perturbation is being treated as a given parameter not determined by the dynamics of the isolated system through a secular equation. As an immediate consequence, there will be no need to worry about inertial effects of collective nature. In the discussion below one main issue will be to examine more closely the conditions for which the transfer of energy from the external field to the many-body system is irreversible, and we will be looking at the transition from the quantal to the classical regime. For this we shall follow in part methods which in nuclear physics have been proven successful in describing such a transition for static quantities like the system's total energy. As discussed in Chapter 5, such a transition is accompanied by the disappearance of shell effects in the many-body system. It may be expected that this feature will influence dynamical properties as well and may have an impact on the nature of irreversible behavior. Again, the external field will be assumed to be weak enough to allow for an application of linear response theory.

17.1 Forced energy transfer within the wall picture

In (Błocki et al., 1978) (see also (Jarzynski, 1993)) a formula for the energy transfer has been derived for the following classical picture. The nucleons are assumed to be put inside a container and to behave like a classical gas. The particles are allowed to "collide" with the "moving wall". Considering the transfer of momentum in each "collision" and finally summing over all events along the surface the following formula for the rate of energy transfer from the wall to the particles is obtained:

$$\frac{dE^{\text{w.f.}}}{dt} = m\rho_0 \langle v \rangle \oint u_n^2(s) ds \,. \tag{17.1}$$

Here, u_n is the normal velocity of the surface and ρ_0 the space density of particles. Their average velocity $\langle v \rangle$ is often estimated by its value at $T=0$, namely $\langle v \rangle = 3v_F/4$. For small deviations from the spherical shape one may apply formula (1.70), which for quadrupole vibrations reads $R(\vartheta, Q) = R_0(1 + QY_{20}(\cos\vartheta))$. Hence, the normal velocity is given by $u_n = \partial R(\vartheta,t)/\partial t = R_0 Y_{20}(\cos\vartheta)\dot{Q}(t)$ and the surface integral simply becomes identical to R_0^4. From (1.23) and (1.22) one gets $\rho_0 = 3A/(4\pi R_0^3) = k_F^3/(6\pi^2)$, if the degeneracy factor g is put equal to unity, as it should be for classical particles. Finally, with the abbreviation $x_F \equiv k_F R_0 = \sqrt{2me_F}R_0/\hbar$ formula (17.1) turns into

$$dE^{\text{w.f.}}/dt = \gamma_{\text{w.f.}} \dot{Q}^2, \qquad \text{where} \qquad \gamma_{\text{w.f.}} = \hbar(9A/16\pi)x_F = \hbar x_F^4/(8\pi^2) \quad (17.2)$$

is the coefficient for wall friction.

In (Koonin et al., 1977) this wall picture was tested on the basis of a linear response approach like that of (Hofmann, 1976)) described in Section 6.5.2. For the friction coefficient the zero frequency limit of (6.92) was applied and the formula $\gamma(0) = (\partial\chi''/\partial\omega)_{\omega=0}$ was analyzed semi-classically. In this way it was demonstrated that wall friction only applies in the limit of large systems. The same conclusion was reached in (Abrosimov and Hofmann, 1994) where the macroscopic limit was studied for surface modes of a slab. Inside the slab of (average) width L there was a gas of free nucleons. Their motion was treated with the Vlasov equation with boundary conditions simulating specular reflections at the walls (discarding collisions among the particles). Again, the wall formula could be reproduced in the limit of $L \to \infty$. A somewhat more exciting issue of this paper was, however, to see that this limiting value could also be obtained for a finite L when frequency smoothing to the response functions is applied. This is another example from which it can be seen that also for dynamic quantities like friction a macroscopic limit can be recovered after applying a kind of Strutinsky smoothing. Recall please that in Section 6.6.1.1 this feature was demonstrated for the case of the collective response function. Below, this issue will be studied further specifically for the friction coefficient of a spherical system treated within a quantal linear response approach.

A different formulation of the energy transfer problem was given and carried out in a series of papers by the Swiatecki group (see (Blocki et al., 1995) and references given therein). In this work the time dependence of the $Q(t)$ specifying the deformation of the mean field was assumed to be of the form $Q(t) = q_A \sin(\Omega t)$. The system was followed for one period simulating numerically the solutions of the one-body Schrödinger equation without relying on perturbation theory. In this way the total excitation energy $\Delta E(t) = E(t) - E(0)$ was calculated as a function of time. Of interest is the value the $\Delta E(t)$ has reached after one cycle $t = T$ because this may allow one to eventually evaluate the production of "heat" in the system. This delicate question encompasses two severe problems, at least, touched on elsewhere in the book at several places. In the present context they can be phrased by asking how it is possible at all (i) to get irreversible behavior by solving only the time Schrödinger equation, and (ii) to cleanly separate reversible and irreversible parts. We will return to the first question later. Fortunately, the second has a clear answer within linear response theory. Indeed, $\Delta E(T)$ can be obtained simply by integrating the rate $dE(t)/dt$ of the energy transfer over a full cycle. As demonstrated in Section 23.3.3 one obtains

$$\Delta E = -\int_0^{T/2} dt \int_{-\infty}^{\infty} ds\, \dot{Q}(t)\tilde{\chi}(t-s)Q(s) = \pi q_A^2 \chi''(\Omega)\,. \qquad (17.3)$$

Notice that only the dissipative part of the response function survives, for which reason this expression clearly measures an irreversible transfer of energy. This statement is true provided $\chi''(\Omega)$ has a meaningfully defined finite value. Assuming for the moment this to be given we may proceed and introduce a friction coefficient by looking at an *average* dissipation rate $\overline{dE/dt}$. The latter can be defined by dividing $\Delta E(T)$, the change of energy during one period of vibrations, by the length $T = 2\pi/\Omega$ of that period. In this way one gets

$$\overline{\frac{dE}{dt}} = \frac{\pi q_A^2 \chi''(\Omega)}{(2\pi/\Omega)} = \frac{\chi''(\Omega)}{\Omega}\overline{v^2} \equiv \gamma(\Omega)\overline{v^2}\,. \qquad (17.4)$$

Here, $\overline{v^2} = (1/2)(\Omega q_A)^2$ is the average squared velocity with $v = \dot{Q}$ and $\gamma(\Omega) = (\chi''(\Omega)/\Omega)$ may be identified as the friction coefficient at finite frequency. Notice that this frequency is the (real) one given by the external field, unlike the form (6.120) where ω_1 represents the actual, complex frequency the collective motion has around some Q_0 if determined self-consistently. We may note in passing that a friction coefficient of the type $\chi''(\Omega)/\Omega$ is identical to the (25.158) which appears in the transport equation for collective motion if the coupling between collective and intrinsic motion is treated perturbatively, see (25.158). Since $\chi''(\Omega)$ is an odd function of Ω this formula again turns into (6.92) in the limit of small frequencies, $\lim_{\Omega \to 0} \chi''(\Omega)/\Omega = \partial \chi''/\partial \Omega|_{\Omega=0} = \gamma(\Omega = 0)$. Whereas in (Blocki et al., 1995) the cavity was specified by an artificially deepened Woods–Saxon potential (to avoid emission of particles) we will in the following exploit the largely analytic solutions of the square-well potential, which in the previous chapter had been exploited already for the dielectric function of metal clusters.

17.1.1 Energy transfer at finite frequency

In Fig. 17.1 numerical results are presented which were obtained in (Hofmann et al., 1998) for the quantity $\Delta E(T)/E(t=0) = \Delta E/E_0$. They were calculated on the basis of formula (17.4) for the same system as used in (Blocki et al., 1995), with all parameters chosen like there. As the most striking difference from the (quantal) results of (Blocki et al., 1995) (presented there in Fig. 1), in our case we observe strong oscillations with frequency (or η), which represent nothing else but the typical strength function behavior. (These functions are smooth in ω simply because the delta functions were averaged over an interval of 0.1 MeV.) The straight line represents the result one gets from (17.2) for wall friction. The figure clearly shows that such a behavior can be obtained from the

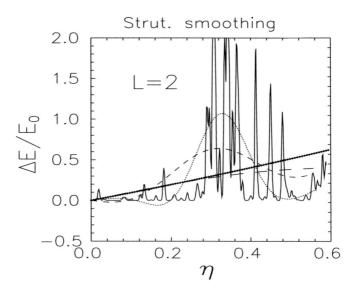

FIG. 17.1. Average energy transfer to a spherical system of independent particles by an external quadrupole perturbation, calculated for microscopic and for smoothed response functions. The η used as the abscissa is approximately given by $\eta \approx 0.02269\hbar\omega$, see text and (Hofmann et al., 1998).

quantum one only after performing averages in frequency. Such a smoothing may be represented by eqn(6.129), $\chi''_\gamma(\omega) = -\pi \sum_{jk} |F_{kj}|^2 \left(n(e_k) - n(e_j)\right) \overline{\delta_{2\gamma}}(\hbar\omega - e_{kj})$, where $\overline{\delta_{2\gamma}}$ stands for a smooth δ-function of width 2γ. For the dotted lines this averaging interval 2γ was chosen to be 5 MeV, for the short dashed ones 10 MeV and for the long dashed ones 20 MeV. It is seen that and how this smoothing leads to results similar to that of the wall formula. The example taken here are quadrupole vibrations about a spherical shape, but in (Hofmann et al., 1998) systems with sizable octupole deformations were also studied, and they show the same results; the coupling operator and its matrix elements will be specified in more detail in the next section. In (Blocki et al., 1995) only forced vibrations around the sphere were considered. As can be seen from their Fig. 1, the simulations of the Schrödinger equation indicate a straight behavior of the functional dependence of $\Delta E/E_0$ on η somewhat below that of the wall formula. This is an interesting feature considering the fact that it was obtained from a non-perturbative calculation. Indeed, starting from the (correct) formula (6.4), the $\langle \hat{F}(\hat{x}_i, \hat{p}_i, Q)\rangle_t$ need not be linear in $Q(s)$. Hence, in an integral like $\int_{-\infty}^{\infty} ds\, \tilde{\chi}(t-s)Q(s)$ the $\tilde{\chi}(t-s)$ may in principle be a complicated functional of $Q(s)$. The solution found in (Blocki et al., 1995) tells one that after performing a time average this functional leads to a linear relation with η. In a sense this time average may be considered analogous to our averaging in the spectrum. As a matter of fact, experience with the deformed shell model tells one that averaging in energy e over an interval of $\gamma_e^{av} = 10$ MeV corresponds to an average in Q over a width of order $\gamma_Q^{av} = \gamma_e^{av}/e_F \approx 1/4$. It is worth noticing that the introduction of this kind of averages has consequences for the interpretation of the friction force in a *local sense*. In applications in transport equations the $\gamma(Q(t))$ is usually taken to represent a local friction force, valid in a small region in the collective phase space. In (Blocki et al., 1995), on the other hand, a full cycle of vibrations is considered, the amplitude of which is not even considered small. Evidently, this aspect is closely related to those discussed in Section 11.2 for large-scale collective motion.

17.1.2 *Fermions inside billiards*

The transfer of energy to cavities containing independent fermions has been discussed in (Barnett et al., 2001) and in greater detail later by D. Cohen (2000, 2003). The main goal was to examine the influence of chaotic behavior in time-dependent descriptions of d-dimensional systems. Cavities or billiards for $d = 2$ and 3 are studied, which are meant to simulate (i) quantum dots whose shape is controlled by electric gates, or (ii) which are driven by time-dependent electric or magnetic fields. The starting point of this work are the formulas presented at the beginning of Section 17.1. Analogies to the nuclear case are drawn, with the latter supposedly being represented by the picture behind the wall formula. In passing we should like to mention once more that such an analogy is valid only in a very restricted sense, in that it requires us to ignore the problem of self-consistency as well as the presence of residual interactions.

As the model system one considers a "gas" of fermions whose single particle Hamiltonian is written in the form $\hat{h}(\mathbf{x},\mathbf{p};q) = \mathbf{p}^2/(2m) + U(\mathbf{x} - qD(\mathbf{x}))$ where the similarities to the nuclear case are evident. The $D(\mathbf{x})$ stands for the "deformation field" and q parameterizes its strength. Again, the confining potential U is supposed to be zero inside the cavity and to become very large outside. Unlike the nuclear case, in general the cavity's volume may not necessarily be conserved when q is varied. Evidently, the essential microscopic transition operator is the derivative of $\hat{h}$ with respect to q and is given by $-(\partial U/\partial r)D(\mathbf{r})$. The rate of energy absorption is then written in the form $\overline{dE/dt} = \mu(\Omega)\,\bar{v}^2$ and the $\mu(\Omega)$ is derived microscopically in terms of the Fourier transform $C_e(\Omega)$ of a non-symmetrized correlation function $C_e(\tau) = \langle \hat{F}(t)\hat{F}(t+\tau)\rangle_E$, but where the average is to be calculated for a microcanonical ensemble at fixed energy E. It may be said that the basic ingredients of the model as well as the general goal are quite similar to our treatment involving response functions. Bear in mind that, as outlined in Section 23.4.4, under the Random Matrix hypothesis linear response theory can also be formulated for the microcanonical ensemble including the common relation between correlation and response functions. On the other hand, in (Barnett et al., 2001) one concentrates on the classical limit to exhibit the influence of chaotic behavior of the dynamics of the particles inside the cavity. Possible extensions to the quantum case are mentioned and speculations are put forward about the origin of dissipation in the quantum case.

Finally, we should also like to briefly present the case that the external force is of magnetic nature. In (Barnett et al., 2001) a two-dimensional quantum dot in the x, y plane is considered which is perturbed by a homogeneous magnetic field $\mathbf{B}(t)$ in z direction. The corresponding Hamiltonian can be written as $\hat{h}(\mathbf{x}, \mathbf{p}; q) = (1/2m)(\mathbf{p} - e\mathbf{A}(\mathbf{x}; Q(t)))^2 + U(\mathbf{x})$, where now the confining potential $U(\mathbf{x})$ is a fixed one defining the dot. The vector potential has to be chosen according to $\mathbf{A}(\mathbf{x}; Q(t)) = (1/2)B(t)[\mathbf{e}_z \times \mathbf{x}]$, with $B(t) = Q(t)/\mathcal{A}$ being the magnitude of the magnetic field. The q represents the magnetic flux through the dot of area $\mathcal{A}$; again, it may be assumed to be $Q(t) = q_A \sin \Omega t$. Still the (average) transfer of energy to the dot is given by formulas of type (17.4), here in the form $\overline{dE/dt} = \mu(\Omega)\dot{Q}^2$ with $\dot{Q}$ representing the electromotive force and μ the conductance. The $F = \partial h/\partial q$, on the other hand, stands for the negative current $F = -I(t) = -(e/(2\mathcal{A}))[\mathbf{e}_z \times \mathbf{x}] \cdot \mathbf{v}$, which for a ring of perimeter L takes on the value $-I(t) = -ev/L$ with v being the tangential velocity.

17.2 Wall friction by Strutinsky smoothing

The Strutinsky procedure has proven a successful and practicable means of differentiating shell effects from smooth quantities. Whereas the former reflect many-body properties of single particle motion in a one-body potential, the latter represent their "macroscopic" behavior. The case where these features can be seen best is the static energy. As explained in Chapter 5, there are various ways of obtaining the macroscopic limit, be it by either averaging over particle number or by raising the internal excitation. Moreover, there exist interesting relations to semi-classical descriptions. It should therefore be of interest to use a similar technique to examine response functions. In fact, in (Magner et al., 1991) the "fluctuating part" of such functions has been calculated within periodic orbit theory. For the purpose of this chapter we are more interested in applying smoothing for getting information about the macroscopic limit of dissipative dynamics. Work along this line has been published in (Hofmann et al., 1996), in which several averaging procedures are discussed developing further earlier suggestions presented in (Hofmann and Ivanyuk, 1993), see Section 6.6.1.1.

Within the independent particle model the dissipative response function is given by (6.24), which can be rewritten in the form

$$\chi''(\omega) = \pi \int_{-\infty}^{\infty} de \sum_{jk} |F_{kj}|^2 \left(n(e - \hbar\omega/2) - n(e + \hbar\omega/2) \right) \times \qquad (17.5)$$
$$\delta(e + \hbar\omega/2 - e_k)\delta(e - \hbar\omega/2 - e_j).$$

If one were just interested in getting a smooth function of frequency one might simply apply (5.11) with x being replaced by ω. On the other hand, as the δ-functions represent the density of single particle states, one might as well average over this density (Hofmann and Ivanyuk, 1993) and (Hofmann et al., 1996). Both procedures are unsatisfactory in the following sense. In the basic expression (17.5), besides the δ-functions, there are other quantities which depend on the single particle properties and which are thus amenable to fluctuations in the spectrum, in particular the matrix elements $|F_{kj}|^2$. To account for this feature one also needs to be able to smooth over the spectral distribution. In the following this will be achieved by replacing the δ-functions in (17.5) by the smooth ones introduced in Section 5.2, as for instance in (5.16) for the single particle level density.

In the following we want to investigate properties of the friction coefficient, defined in terms of smoothed response functions, but applied to *finite systems*. The question of particular interest is whether or not, for increasing γ_{av}, the friction coefficient $\gamma(0) = (\partial \chi''/\partial \omega)_{\omega=0}$ in the zero frequency limit reaches some constant value. For the response function (17.5) $\gamma(0)$ writes as

$$\gamma(0) = -\int de \, \frac{\partial n(e)}{\partial e} \Xi(e) \approx \left(1 + \frac{\pi^2 T^2}{6} \frac{d^2}{de_F^2}\right) \overline{\gamma}, \qquad (17.6)$$

where $\overline{\gamma} \equiv \Xi(e_F)$ and the function $\Xi(e)$ is defined as

$$\Xi(e) \equiv \hbar\pi \sum_{jk} |F_{kj}|^2 \frac{1}{\gamma_{\text{av}}^2} f\left(\frac{e_k - e}{\gamma_{\text{av}}}\right) f\left(\frac{e_j - e}{\gamma_{\text{av}}}\right). \qquad (17.7)$$

The approximate form obtained in (17.6) is valid if $\Xi(e)$ is smooth around $e = e_F$ on the scale of T. Remember that the derivative $\partial n(e)/\partial e = -(4T \cosh^2((e - e_F)/2T))^{-1}$ of the Fermi function is of bell-like shape of width $2T$ centered at the Fermi energy e_F. Under such conditions the integral (17.6) can be treated by the Sommerfeld expansion described in Section 22.5.1.

Let us discuss now some details about the evaluation of the matrix elements F_{kj} for the case of spheroidal isoscalar quadrupole vibrations around a spherical equilibrium located at $Q = 0$. Following the discussion in Section 3.2 and Section 6.2.2 the single particle operator $\hat{F}(\hat{x}_i, \hat{p}_i)$ derives from the potential $U(\mathbf{r}, Q)$ like

$$\hat{F}(\mathbf{r}) = \frac{\partial U(\mathbf{r}, Q)}{\partial Q}\bigg|_{Q=0} = \frac{\partial U(r, Q)}{\partial Q}\bigg|_{Q=0} Y_{20}(\cos\vartheta). \qquad (17.8)$$

The states $|k\rangle$ correspond to a spherical system and are thus specified by the main quantum number n, the orbital angular momentum l and its projection m on the z-axis. The matrix elements of $\hat{F}$ factorize into products of radial and angular integrals. Their absolute squares can be written in the form $|F_{kj}|^2 = F^2_{nl,n'l'} h_{ll'}$, where

$$F^2_{nl,n'l'} = \left|\langle nl|\frac{\partial U(r,Q)}{\partial Q}|n'l'\rangle\right|^2 \qquad (17.9)$$

and the angular matrix elements $h_{ll'} = \sum_{mm'} |\langle Y_{lm}|Y_{20}|Y_{l'm'}\rangle|^2$ are given by

$$\begin{aligned} h_{ll'} &= \frac{1}{4\pi}\left[\frac{l(l+1)(2l+1)}{(2l+3)(2l-1)}\delta_{ll'} + \frac{3}{2}\frac{(l+1)(l+2)}{(2l+3)}\delta_{l'l+2} + \frac{3}{2}\frac{(l'+1)(l'+2)}{(2l'+3)}\delta_{ll'+2}\right] \\ &\approx \frac{1}{16\pi}\left[(2l+1)\delta_{ll'} + \frac{3}{2}(2l+3)\delta_{l'l+2} + \frac{3}{2}(2l+3)\delta_{ll'+2}\right]. \end{aligned} \qquad (17.10)$$

For the expression in the second line terms of order $1/l, 1/l'$ are discarded. Both forms fulfill the sum rule $\sum_{l'} h_{ll'} = (2l+1)/4\pi$.

A convenient model for evaluating these formulas is the square well, used already in the previous chapter. Neglecting spin-orbit forces calculations can largely be done analytically. The potential may be defined as in (16.15), with the only difference that now for any finite Q the radius $R(\vartheta, Q)$ depends on the angle ϑ. The radial part of the field operator $\hat{F}(\mathbf{r})$ of (17.8) becomes $\partial U(r,Q)/\partial Q|_{Q=0} = U_0 R \delta(r-R)$. As compared to the one of (16.16) it only contains the additional factor $R = R(Q=0)$. Otherwise the evaluation can be done as before for the dipole. We again have to consider the limit $U_0 \to \infty$, for which the boundary conditions can be chosen as in (16.17). The $F^2_{nl,n'l'}$ turn out separable in nl and $n'l'$ and are simply given by the single particle energies e_{nl}, namely $F^2_{nl,n'l'} = 4 e_{nl} e_{n'l'}$. As these matrix elements are unbound one eventually needs to introduce a parameter $E_{\rm cut}$ to cut off infinite sums in case they are not terminated by other convergence factors.

These matrix elements can be used in (17.6) and (17.7). For any given l or l' the sums over n and n' may be handled by transforming them into integrals over energies. Because the $F^2_{nl,n'l'}$ separate in the primed and unprimed quantities one gets factors like the following one:

$$\sum_n e_{nl} \frac{1}{\gamma_{\rm av}} f\left(\frac{e_{nl} - e_{\rm F}}{\gamma_{\rm av}}\right) = \int de\, g_l(e)\, e\, \frac{1}{\gamma_{\rm av}} f\left(\frac{e - e_{\rm F}}{\gamma_{\rm av}}\right) \simeq e_{\rm F} \bar{g}_l(e_{\rm F}). \quad (17.11)$$

The $g_l(e)$ were introduced in Section 16.2.3 and the $\bar{g}_l(e)$ are their smooth counterparts defined through (5.16). Next we may carry out the sum over l', which because of the angular momentum selection rules in (17.10) only consists of three terms. Hence, the friction coefficient (17.6) turns into

$$\bar{\gamma} = 4\pi \hbar e_{\rm F}^2 \sum_l \bar{g}_l(e_{\rm F}) \left(h_{ll} \bar{g}_l(e_{\rm F}) + h_{ll+2} \bar{g}_{l+2}(e_{\rm F}) + h_{ll-2} \bar{g}_{l-2}(e_{\rm F})\right). \quad (17.12)$$

The density $\bar{g}_l(e_{\rm F})$ is expected to be a smooth function both of $e_{\rm F}$ and l. Because of the degeneracy factors shown in (17.10) the sum in (17.12) will be dominated by large l. In this semi-classical regime we may assume $\bar{g}_l \approx \bar{g}_{l\pm 2}$ and thus may make use of the sum rule from above to get

$$\bar{\gamma} = \hbar e_{\rm F}^2 \sum_l (2l+1) \bar{g}_l^2(e_{\rm F}). \quad (17.13)$$

This averaged friction coefficient $\bar{\gamma}$ is solely determined by the Fermi energy $e_{\rm F}$ and the average density of levels $\bar{g}_l(e_{\rm F})$ calculated at $e_{\rm F}$ for fixed orbital momentum. It may be evaluated in various ways. One possibility is to use the Thomas–Fermi approximation, which for this case implies (Hasse, 1987)

$$\bar{g}_l^{\rm TF}(e_{\rm F}) = \sqrt{(x_{\rm F}^2 - l^2)}\, \Theta(x_{\rm F}^2 - l^2)/(2\pi e_{\rm F}). \quad (17.14)$$

It is obtained from (16.20) by neglecting $1/2$ over l. The expression (17.13) may then be evaluated by replacing the sum over l by the corresponding integral.

In this way $\overline{\gamma}$ becomes identical to the result shown in (17.2) for wall friction, namely $\overline{\gamma} = \hbar x_F^4/(8\pi^2) \equiv \gamma_{\text{w.f.}}$. There the relation $A = 2x_F^3/(9\pi)$ was used known from (1.23) for infinite matter (with the degeneracy factor $g_{\text{s.i.}}$ being one). For finite nuclei it is more meaningful to express particle number through the smooth density of single particle states $\overline{g}(e)$ as $A = \int^{e_F} \overline{g}(e)de$. This implies two different relations between particle number and the parameter x_F, and, hence, for a given nucleus different values for $\overline{\gamma}$. In Fig. 17.2 they are shown by the lines with stars. The upper one relates to a choice of the relation $A \leftrightarrow x_F$ for a finite nucleus and the lower one to that for infinite matter used in (17.2).

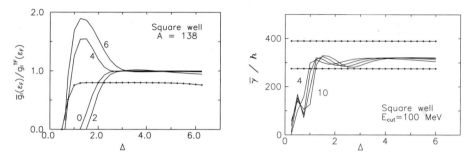

FIG. 17.2. Left panel: The averaged single particle density $\overline{g}_l(e_F)$ at fixed l (indicated by numbers) divided by the $\overline{g}_l^{\text{TF}}(e_F)$ of (17.14) as function of the averaging parameter. The solid curve with stars shows the full density $\overline{g}(e_F)$ normalized to $\overline{g}^{\text{TF}}(e_F)$, see text and (Hofmann et al., 1996). Right panel: The result for the Strutinsky averaged friction coefficient, for different degrees of the smoothing function. The two parallel lines correspond to the two versions of the wall formula explained in the text.

The value $\overline{\gamma}$ obtained above may be interpreted as a macroscopic limit of friction for finite systems. Indeed, the Thomas–Fermi approximation used for $\overline{g}_l(e_F)$ becomes exact in the limit that both R as well as l become very large. The same result may eventually be obtained by averaging over the single particle levels of the finite system, which is to say by calculating $\overline{g}_l$ starting from a microscopic description. Examples of such computations of $\overline{g}_l$ for various ls are shown in Fig. 17.2 (from (Hofmann et al., 1996)) as functions of the averaging interval γ_{av} expressed in terms of $\Delta = \gamma_{\text{av}} mR/(k_F \hbar^2)$. It is observed that the $\overline{g}_l$s reach plateaus which starts at the somewhat large values of $\Delta \approx \pi$. This feature can be traced back to properties of the zeros of Bessel functions discussed in Section 16.2.3, which asymptotically become equidistant with a spacing about π. In Fig. 17.2 we also plot, as the solid, dotted curve, the ratio obtained for the total density of levels $\overline{g}(x_F) = \sum_l (2l+1) \overline{g}_l(x_F)$. The reason that it comes out smaller than unity can be seen in the fact that the Strutinsky smoothed level density $\overline{g}$ contains corrections to that in Thomas–Fermi approximation. Usually such corrections are well represented by the extended Thomas–Fermi method discussed Section 5.2.3. For the square well the Weyl formula applies by which the single particle level density is expressed by a sum of volume, surface and curvature terms. Indeed, the surface term adds a negative correction to the volume term, with the latter corresponding to TF, see Section 4.5.2 of (Brack and Bhaduri, 1997). It is seen from the figure that $\bar{g}$ has a perfect plateau which is reached at the smaller values of $\Delta \approx 1.0$. This value corresponds to $\gamma_{\rm av} \approx$ 10 MeV, which is to say that value at which the shell structure in the single particle spectrum is washed out.

After this detour we may return to the friction coefficient. Since the matrix elements have no natural cutoff, say as a function of the single particle energy e_k, in (Hofmann et al., 1996) the summations were limited to a cutoff energy $E_{\rm cut}$ above the Fermi energy. The result of such a computation for $E_{\rm cut} = 100$ MeV are shown in Fig. 17.2, which demonstrates the existence of a plateau. Its quality depends on the degree M of the smoothing function (5.12), and is actually worse for smaller cutoff parameters. The two different values for the macroscopic limit (wall formula) mentioned above are shown by the lines with stars. The fluctuations seen in the region of small Δ have no physical meaning. They are a consequence of using the smoothing function (5.12) which can be negative due to the presence of polynomial in (5.12).

PART IV

THEORETICAL TOOLS

18

ELEMENTS OF REACTION THEORY

In this chapter we want to address basic issues of reaction theory, which are needed to understand heavy-ion collisions. More importantly, the concepts of statistical reaction theory may in a sense be considered fundamental to the main topics of the book. First we shall introduce the concept of the T-matrix for the example of potential scattering. Here, we want to restrict ourselves to a formal discussion assuming that basic elements of scattering theory are known from textbooks on quantum mechanics, like those by A. Messiah (1965) or K. Gottfried (1966). More elaborate presentations of scattering and reaction theory can be found in volume II of (Messiah, 1965), and especially in the book on scattering theory by R.G. Newton (1982) and in H. Feshbach's book on nuclear reactions (1992), which set the standard on this topic. Many formal details may also be found in (Fröbrich and Lipperheide, 1996). Finally we should like to mention the fine monograph on pre-equilibrium reactions by E. Gadioli and P.E. Hodgson (1992), where an excellent overview can also be found on the topics to be discussed in sections 18.2 and 18.4 with a nearly exhaustive list of references.

18.1 Potential scattering

18.1.1 *The T-matrix*

Suppose we are given the Hamiltonian

$$\hat{H} = \hat{H}_0 + \hat{V}, \tag{18.1}$$

and that the solutions of the Schrödinger equation

$$\hat{H}_0|\mathbf{k}, n\rangle = E|\mathbf{k}, n\rangle \equiv E|\alpha\rangle \tag{18.2}$$

for the unperturbed part are known. In the simplest case the $\hat{H}_0$ just stands for kinetic energy $-\hbar^2\Delta/2\mu$ of a free particle with the $|\alpha\rangle \equiv |\mathbf{k}\rangle$ in coordinate representation being given by the plane waves $\langle \mathbf{r}|\mathbf{k}\rangle = \exp(i\mathbf{k}\mathbf{r})$, which will be normalized according to

$$\langle \mathbf{k}|\mathbf{k}'\rangle = \int d^3\mathbf{r}\, e^{i(\mathbf{k}'-\mathbf{k})\mathbf{r}} = (2\pi)^3\delta(\mathbf{k} - \mathbf{k}'). \tag{18.3}$$

The case where the unperturbed wave functions depend on further quantum numbers n will be addressed below. Likewise we at first want to ignore that the $\hat{V}$ might depend on other degrees of freedom like spin and isospin, in addition to the coordinate vector $\mathbf{r}$. The $V(\mathbf{r})$ is assumed to drop to zero outside a certain range R.

In a reaction the "incoming" particles approach the "target", represented here by the area where $\hat{V}$ is finite, and leave it in some other direction $|\mathbf{k}'\rangle$. In the more general language one speaks of a transition from a state $|\alpha\rangle$ to a state $|\beta\rangle$. Its amplitude is parameterized by the *T-matrix*

$$T_{\alpha\beta} \equiv \langle \beta | \hat{T} | \alpha \rangle = \langle \beta | \hat{V} | \psi_\alpha^{(+)} \rangle \qquad (18.4)$$

of the transition operator $\hat{T}$. The relation on the right involves a solution $|\psi_\alpha^{(+)}\rangle$ of the Schrödinger equation $\hat{H}|\psi\rangle = E|\psi\rangle$ for the full Hamiltonian and the same energy as in the unperturbed case. The plus sign in the upper index indicates that one is dealing with *outgoing* spherical waves. Ignoring again all additional quantum numbers, the asymptotic behavior of the wave function in three-dimensional space is meant to be given by

$$\text{for} \quad r \to \infty \qquad \psi_\mathbf{k}^{(+)}(\mathbf{r}) \to e^{i\mathbf{k}\mathbf{r}} + f_\mathbf{k}^{(+)}(\Omega_{\mathbf{k}'}) \frac{e^{ikr}}{r}. \qquad (18.5)$$

The $f_\mathbf{k}^{(+)}(\Omega_{\mathbf{k}'})$, the scattering amplitude, measures the amplitude that the projectile is scattered into the solid angle $\Omega_{\mathbf{k}'}$ between $\mathbf{k}'$ and $\mathbf{k}$. In passing it may be mentioned that formally one may consider the reverse process of an incoming spherical wave and an outgoing plane wave; it will be addressed briefly below.

To ensure that the state $|\psi_\alpha^{(+)}\rangle$ has the right boundary conditions, it is to be found from the Lippmann–Schwinger equation

$$|\psi_\alpha^{(+)}\rangle = |\alpha\rangle + \hat{G}_0^{(+)}(E)\hat{V}|\psi_\alpha^{(+)}\rangle, \qquad (18.6)$$

rather than from the Schrödinger equation itself. Here, the $\hat{G}_0^{(+)}(E)$ is the *resolvent*

$$\hat{G}_0^{(+)}(E) = \lim_{\epsilon \to 0} \frac{1}{E - \hat{H}_0 + i\epsilon} \qquad (18.7)$$

associated with the Hamiltonian $\hat{H}_0$. By multiplying eqn(18.6) by $E - \hat{H}_0 + i\epsilon$ from the left it is easily seen that in the limit $\epsilon \to 0$ the $|\psi_\alpha^{(+)}\rangle$ is actually a solution of the Schrödinger equation. The proof that the wave function $\langle \mathbf{r} | \psi_\alpha^{(+)} \rangle$ in coordinate space has the desired asymptotic behavior (18.5) is left to the reader (see exercise 1). For pure potential scattering with $\langle \mathbf{r} | \hat{V} | \mathbf{s} \rangle = V(\mathbf{r})\delta(\mathbf{r} - \mathbf{s})$ the scattering amplitude is found to be given by

$$f_\mathbf{k}^{(+)}(\Omega_{\mathbf{k}'}) = -\frac{\mu}{2\pi\hbar^2} \int d^3\mathbf{s}\, e^{-i\mathbf{k}'\mathbf{s}} V(\mathbf{s}) \psi_\mathbf{k}^{(+)}(\mathbf{s}) = -\frac{\mu}{2\pi\hbar^2} \langle \mathbf{k}' | \hat{V} | \psi_\mathbf{k}^{(+)} \rangle. \qquad (18.8)$$

Let us return now to the T-matrix (18.4). The right-hand side of this definition can be expressed through (18.6) by simply multiplying this equation from the left by $\langle\beta|\hat{V}$. Then it is easy to see that the $\langle\beta|\hat{T}|\alpha\rangle$ are the matrix elements of an operator which satisfies the equation

$$\hat{T} = \hat{V} + \hat{V}\hat{G}_0^{(+)}(E)\hat{T}. \tag{18.9}$$

One needs only make use of the fact that the states $|\alpha\rangle$ form a complete set. However, one must be aware that in the general case this may involve *integration over the continuous* set of wave numbers **k** and *summation over the discrete* quantum numbers n.

Equation (18.9) allows for an easy formulation of a perturbation approach. One simply has to iterate the expression on the right-hand side to get a power series in the $\hat{V}$. One must be aware, however, that such a series may be badly convergent. The very first term in this expansion, namely $\hat{T} = \hat{V}$ defines the Born approximation. Actually, eqn(18.9) can also be written in the form $\hat{T} = \hat{V} + \hat{T}\hat{G}_0^{(+)}(E)\hat{V}$, and a formal solution can be given as

$$\hat{T} = \hat{V} + \hat{V}\frac{1}{E^{(+)} - \hat{H}}\hat{V} \quad \text{where} \quad E^{(+)} = E + i\epsilon. \tag{18.10}$$

As indicated before, the formulas involving the T-matrix are of very general nature. In the next section we are going to apply them to nuclear reactions. Often it turns out useful to relate the T-matrix to the *S-matrix*, which in operator form reads

$$\hat{S} = 1 - 2\pi i \hat{T}. \tag{18.11}$$

Next we shall cite a few further interesting and important properties of the T-matrix which will be used below.

18.1.1.1 *Properties of the T-matrix*

The cross section: The T-matrix allows one to express the transition probability per unit of time by

$$P_{\alpha\beta} = \frac{2\pi}{\hbar}|T_{\alpha\beta}|^2 \delta(E_\beta - E_\alpha), \tag{18.12}$$

written here for a transition from a state $|\alpha\rangle$ to a $|\beta\rangle$, see e.g. (Gottfried, 1966), (Landau and Lifshitz, 1982). In the event of a dense spectrum of final states this formula is to be replaced by

$$P_{\alpha\beta} = \frac{2\pi}{\hbar}|T_{\alpha\beta}|^2 \Omega_\beta(E). \tag{18.13}$$

Eventually, here the $|T_{\alpha\beta}|^2$ is to be understood as a quantity averaged over some region in energy. In lowest order perturbation theory one may replace the $\hat{T}$ by $\hat{V}$. Hence, formulas (18.12) and (18.13) reduce to those which are usually referred to as *Fermi's golden rule* derived in Section 23.2.1. To get the cross section we simply need to divide the transition rate by the initial flux $j_\alpha^{(\mathrm{in})} = \hbar k_\alpha/\mu_\alpha$:

$$d\sigma_{\alpha\beta} = P_{\alpha\beta}/j_\alpha^{(\mathrm{in})} = \frac{\mu_\alpha}{\hbar k_\alpha} P_{\alpha\beta}. \tag{18.14}$$

Here, k_α is the wave number of relative motion in the initial channel.

For the case of potential scattering discussed before the wave number for the incoming and outgoing waves are identical, $k_\alpha = k_\beta = k$. For this reason the cross section may be written as

$$d\sigma_{\alpha\beta} = \frac{2\pi\mu_\alpha}{\hbar^2 k} |T_{\alpha\beta}|^2 \, \Omega_\beta(E) \, d\Omega_\beta, \tag{18.15}$$

with the density $\Omega_\beta(E)$ of final states being given by

$$\Omega_\beta(E) \equiv \Omega_{\mathrm{kin}}(E) = (2\pi\hbar)^{-3} p^2 \frac{dp}{dE} = \frac{\mu \hbar k}{(2\pi\hbar)^3}. \tag{18.16}$$

Thus, and because of (18.4) and (18.8), the result (18.15) becomes identical to

$$\frac{d\sigma_{\mathbf{k}\to\mathbf{k}'}}{d\Omega} = \left| f_{\mathbf{k}}^{(+)}(\Omega_{\mathbf{k}'}) \right|^2. \tag{18.17}$$

Please notice that the density $\Omega_{\mathrm{kin}}(E)$ is normalized such that $\Omega_{\mathrm{kin}}(E) d\Omega dE$ measures the number of states in the energy interval $(E, E+dE)$ whose momenta point into the direction $(\Omega, \Omega + d\Omega)$. For an angle integrated cross section one would have to multiply by the factor 4π. The proof of (18.17) is readily performed by observing that the differential cross section $d\sigma_{\mathbf{k}\to\mathbf{k}'}$ may be defined through the relation

$$j_{\mathrm{in}} d\sigma_{\mathbf{k}\to\mathbf{k}'} = j_{\mathrm{r}}^{(+)} r^2 d\Omega, \tag{18.18}$$

where j_{in} is the incoming and $j_{\mathrm{r}}^{(+)}$ the radially outgoing flux. It says that the number of particles flowing through the "cross section" of value $d\sigma_{\mathbf{k}\to\mathbf{k}'}$ is identical to those flying through the surface element $r^2 d\Omega$ after the scattering. The currents involved can be calculated from the asymptotic form (18.5) of the wave function.

T-matrix for internal structure of the reacting partners: In the more general case the reacting partners may have internal structure such that the states may be written in the form $|\alpha\rangle = |\mathbf{k}, m\rangle = |\mathbf{k}\rangle|m\rangle$ introduced in (18.2). Then it might be desirable to sum over all *final internal* states $|n\rangle$. Moreover, it could happen that several or many initial states are populated with some probability $\rho(E_m)$. On the basis of (18.12) the overall transition rate would then be given by

$$\begin{aligned} P_{k_\alpha k_\beta} &= \frac{2\pi}{\hbar} \sum_{mn} \rho(E_m) \left| T_{m,k_\alpha;n,k_\beta} \right|^2 \delta\left(E_{k_\beta} + E_n - E_{k_\alpha} - E_m\right) \\ &= \frac{2\pi}{\hbar} \sum_{mn} \rho(E_m) \left| T_{m,k_\alpha;n,k_\beta} \right|^2 \delta\left(E_n - E_m - \hbar\omega\right), \end{aligned} \tag{18.19}$$

where $\hbar\omega = E_{k_\alpha} - E_{k_\beta}$ is the energy transferred to the internal system or taken away from it. For weak interactions treated in perturbation theory the form found in the second line becomes similar to that of the strength function discussed in Section 23.4.3. This formula finds application in solid state physics for scattering of neutrons, electrons or light at macroscopic bodies (see e.g. (Forster, 1975), (Chaikin and Lubensky, 1995)). There, the latter may indeed be thermally excited such that the $\rho(E_m)$ are simply given as Boltzmann factors for fixed temperature T. In nuclear physics the situation is different as there the reacting fragments start off from their ground states. However, this situation is also contained in (18.19); one simply has to use $\rho(E_m) = \delta_{0m}$.

T-matrix for incoming spherical waves: The scattering process may also be formulated in terms of *incoming spherical* waves and *outgoing plane* waves. The Lippmann–Schwinger equation (18.6) then has to be rewritten in the form

$$|\psi_\alpha^{(-)}\rangle = |\alpha\rangle + \hat{G}_0^{(-)}(E)\hat{V}|\psi_\alpha^{(-)}\rangle. \tag{18.20}$$

The operator $\hat{G}_0^{(-)}(E)$ is defined as the Hermitian conjugate to the $\hat{G}_0^{(+)}(E)$:

$$\hat{G}_0^{(-)}(E) = \left(\hat{G}_0^{(+)}(E)\right)^\dagger = \frac{1}{E^{(-)} - \hat{H}} \quad \text{with} \quad E^{(-)} = E - i\epsilon. \tag{18.21}$$

For the **k** dependence of the scattering wave function this implies the relation

$$\psi_{\mathbf{k}}^{(-)}(\mathbf{r}) = \left(\psi_{-\mathbf{k}}^{(+)}(\mathbf{r})\right)^*. \tag{18.22}$$

Hence, the asymptotic behavior at large distances changes from (18.5) (with $\mathbf{k} = \mathbf{k}_\alpha$) to

$$\psi_{\mathbf{k}_\beta}^{(-)}(\mathbf{r}) \;\rightarrow\; e^{i\mathbf{k}_\beta \mathbf{r}} + f_{\mathbf{k}_\beta}^{(-)} \frac{e^{-ikr}}{r} \quad \text{for} \quad r \to \infty. \tag{18.23}$$

The scattering amplitudes and the T-matrix elements satisfy the relations

$$f_{\mathbf{k}_\beta}^{(-)}(-\Omega_\alpha) = \left(f_{\mathbf{k}_\alpha}^{(+)}(\Omega_\beta)\right)^* \tag{18.24}$$

and

$$T_{\alpha\to\beta} \equiv \langle\beta|\hat{T}|\alpha\rangle = \langle\beta|\hat{V}|\psi_\alpha^{(+)}\rangle = \langle\psi_\beta^{(-)}|\hat{V}|\alpha\rangle, \tag{18.25}$$

respectively. The T-operator itself is still to be calculated as discussed before.

Microreversibility: The properties just discussed may be interpreted in terms of time reversal invariance. By time reversal the momenta change into their negative, with the coordinates unchanged. Denoting $-\mathbf{k}_\alpha$ and $-\mathbf{k}_\beta$ by $K\alpha$ and $K\beta$, respectively, relation (18.22) can be written as

$$\psi_\beta^{(-)} = \left(\psi_{K\beta}^{(+)}\right)^*. \tag{18.26}$$

As the unperturbed states fulfill the same relation, it is easy to convince oneself that for a Hermitian operator $\hat{H}$ one has:

$$\langle\beta|\hat{V}|\psi_\alpha^{(+)}\rangle = \langle K\alpha|\hat{V}|\psi_{K\beta}^{(+)}\rangle \quad \text{implying} \quad T_{\alpha\to\beta} = T_{K\beta\to K\alpha}. \tag{18.27}$$

This is called the *principle of microreversibility*. In terms of the antilinear operator $\hat{K}$ associated with the transformation of time reversal, the T-operator is easily seen to transform like $\hat{K}\hat{T}\hat{K}^\dagger = \hat{T}^\dagger$, provided the Hamiltonian is invariant $\hat{K}\hat{H}\hat{K}^\dagger = \hat{H}$. This property follows immediately from (18.10) regarding that for c-numbers an operation by $\hat{K}$ implies complex conjugation. Then we simply have to follow the rules for calculating matrix elements when antilinear operators are involved to finally find:

$$\begin{aligned}T_{K\beta\to K\alpha} &= \left(\langle\alpha|\hat{K}^\dagger\rangle\right)\hat{T}\left(\hat{K}|\beta\rangle\right) = \langle\alpha|\left(\hat{K}^\dagger\hat{T}\hat{K}\right)|\beta\rangle^*\\ &=\langle\beta|\left(\hat{K}^\dagger\hat{T}^\dagger\hat{K}\right)|\alpha\rangle = \langle\beta|\hat{T}|\alpha\rangle = T_{\alpha\to\beta}.\end{aligned} \tag{18.28}$$

It is worth mentioning that according to Fermi's golden rule (see Section 23.2.1) in low order perturbation theory the relation $T_{\alpha\to\beta} = T_{\beta\to\alpha}$ is warranted if only the transition operator is Hermitian, which is to say without invoking time reversal.

T-matrix for distorted waves: For

$$\hat{H} = \hat{H}_0 + \hat{V}_0 + \hat{V}_1 \equiv \hat{H}_1 + \hat{V}_1 \tag{18.29}$$

the scattering process may be described as before in terms of the eigenfunctions of $\hat{H}_1$,

$$\hat{H}_1|\chi\rangle = E|\chi\rangle, \tag{18.30}$$

which are scattering states themselves, the *distorted waves*. The T-matrix expressed in terms of the eigenstates of $\hat{H}_0$, as determined through (18.2), can be shown to be given by

$$\begin{aligned}T_{\alpha\beta} \equiv \langle\beta|\hat{T}|\alpha\rangle &= \langle\beta|\hat{V}_0|\chi_\alpha^{(+)}\rangle + \langle\chi_\beta^{(-)}|\hat{V}_1|\psi_\alpha^{(+)}\rangle\\ &= T_{\alpha\beta}^{(1)} + \langle\chi_\beta^{(-)}|\hat{V}_1|\psi_\alpha^{(+)}\rangle.\end{aligned} \tag{18.31}$$

The Born approximation with respect to the perturbation $\hat{V}_1$ is obtained by replacing $\langle\chi_\beta^{(-)}|\hat{V}_1|\psi_\alpha^{(+)}\rangle$ by $\langle\chi_\beta^{(-)}|\hat{V}_1|\chi_\alpha^{(+)}\rangle$.

18.1.2 Phase shifts for central potentials

If the potential $V(\mathbf{r})$ is rotationally invariant, the scattering problem becomes symmetric around the beam axis $\mathbf{k}$. Therefore, the cross section just depends on the *one angle* $\vartheta_{\mathbf{k}'}$. To simplify notation, let us in this case omit the indices $\mathbf{k}$ and $\mathbf{k}'$. Likewise we want to drop the suffix $^{(+)}$ and simply use $f_{\mathbf{k}}^{(+)}(\vartheta_{\mathbf{k}'}) = f(\vartheta)$. Expanding the wave function into a series of Legendre polynomials

$$\psi(r,\vartheta) = \sum_{l=0}^{\infty} \frac{u_l(r)}{r} P_l(\cos\vartheta) \qquad (18.32)$$

the u_l is seen to obey the equation

$$\left(-\frac{\hbar^2}{2\mu}\frac{d^2}{dr^2} + \frac{\hbar^2 l(l+1)}{2\mu r^2} + V(r)\right) u_l(r) = E u_l(r). \qquad (18.33)$$

As the cross section is determined by the scattering amplitude $f(\vartheta)$ for the outgoing spherical wave contained in the $u_l(r)$. An expansion for the $f(\vartheta)$ may be written as

$$f(\vartheta) = \sum_{l=0}^{\infty}(2l+1) f_l P_l(\cos\vartheta) \qquad \text{with} \qquad f_l = \frac{1}{k} e^{i\delta_l} \sin\delta_l = \frac{1}{2ik}\left(e^{2i\delta_l} - 1\right). \qquad (18.34)$$

Instead of the coefficients f_l it turns out convenient to work with the so-called phase shift δ_l introduced on the right. The plane wave, too, may be expanded into a series of Legendre polynomials. One has

$$e^{i\mathbf{k}\mathbf{r}} = \sum_l (2l+1) i^l j_l(kr) P_l(\cos\vartheta), \qquad (18.35)$$

with the spherical Bessel functions $j_l(kr)$. At large kr they take on their asymptotic forms given by

$$(2l+1) i^l j_l(kr) \sim \frac{2l+1}{2ikr}\left[(-)^{(l+1)} e^{-ikr} + e^{ikr}\right] \qquad \text{for} \quad kr \gg 1. \qquad (18.36)$$

(Notice please that $\mathbf{k}'$ and $\mathbf{r}$ are parallel vectors, see (18.175)). From (18.34) to (18.36) it follows that asymptotically the radial part of the wave function takes on the form

$$\frac{u_l}{r} \sim \frac{2l+1}{2ikr}\left[(-)^{(l+1)} e^{-ikr} + e^{2i\delta_l} e^{ikr}\right]. \qquad (18.37)$$

It is expressed in terms of the as yet unknown phase shifts δ_l. They are to be found from the solution of the Schrödinger equation (18.33) for the $u_l(r)$. Remember that outside the range of the potential spherical waves of the type found in (18.37) are indeed solutions of (18.33).

The physical meaning of the phase shift is seen by comparing the asymptotic form of the plane wave with that of the scattered wave, which are given by (18.36) and (18.37), respectively. Apparently, both consist of incoming and outgoing spherical waves, the only difference being that for the scattering state the phase of the latter is shifted by $2i\delta_l$ leading to the overall factor

$$S_l(E) = \exp(2i\delta_l) \ . \tag{18.38}$$

This is nothing else but the diagonal element of the S-matrix introduced above. As may easily be verified, the diagonal elements of the T-matrix are given by

$$T_l = -\frac{1}{\pi} e^{i\delta_l} \sin \delta_l \ , \tag{18.39}$$

such that (18.38) follows from (18.11). For a system where the currents are conserved the δ_l are real quantities implying $|S_l| = 1$. This is a special example of the more general property which states that $\hat{S}$ is a unitary operator in case the scattering process is described by a Hermitian Hamiltonian; for more details see (Messiah, 1965), (Newton, 1982), (Feshbach, 1992).

The differential cross section is found after inserting the expansion (18.34) into (18.17). It involves a double sum over angular momenta l and l'. This reduces to one summation for the total cross section, which is obtained by integrating the differential one over ϑ. Making use of the orthonormality relations $\int_{-1}^{1} dx P_l(x) P_{l'}(x) = 2\delta_{ll'}/(2l+1)$ for Legendre polynomials one finally gets

$$\sigma_{\text{tot}} = \int d\Omega \, \frac{d\sigma}{d\Omega} = 4\pi \sum_{l=0}^{\infty} (2l+1) |f_l|^2 = \frac{4\pi}{k^2} \sum_{l=0}^{\infty} (2l+1) \sin^2 \delta_l \ . \tag{18.40}$$

18.1.3 Inelastic processes

In the scattering processes looked at so far the interaction of the projectile with the target was represented by a *real potential*. In this case *all* particles which come close to the range of this potential are scattered into a certain angle. Such a situation may not always be given. It may happen that with some finite probability projectiles are "absorbed" by the target. This "absorption" may be understood literally, namely that the projectile gets stuck. Another possibility is that the reacting particles get excited through some transfer of energy from relative motion. Thus the kinetic energy of the outgoing fragments has to be associated with a different "channel". How such processes are to be described in detail will be discussed below. It is possible, however, to obtain average properties still within a potential model for a scattering process. Indeed, such an "absorption" may be accounted for by only modifying slightly the formalism from above, namely by replacing in (18.34) or (18.38)

$$e^{2i\delta_l} \quad \longrightarrow \quad S_l(E) = s_l(E) e^{2i\delta_l} \quad \text{with} \quad 0 \leq s_l(E) \leq 1 \ , \tag{18.41}$$

with the phase δ_l still being a real quantity. "Absorption" will be given whenever the $s_l(E)$ gets smaller than unity. Performing the substitution (18.41) in (18.37) the $S_l(E)$ still describes the amplitude of the outgoing spherical wave. Recalling the latter's geometrical meaning one understands that the cross section for scattering into a certain angle will still be given by the f_l which now can be expressed as (mind (18.34))

$$f_l = \frac{S_l - 1}{2ik} = \frac{1}{2k}\left[s_l \sin 2\delta_l + i(1 - s_l \cos 2\delta_l)\right]. \tag{18.42}$$

The integrated cross section then becomes

$$\sigma_{\text{el}} = 4\pi \sum_{l=0}^{\infty}(2l+1)|f_l|^2 = \frac{\pi}{k^2}\sum_{l=0}^{\infty}(2l+1)(1+s_l^2 - 2s_l\cos 2\delta_l). \tag{18.43}$$

It defines the so-called *elastic cross section*. The *inelastic* cross section is found by considering the loss of total current. Take the radial part of the wave function as given by (18.37). As one may convince oneself from (18.18), the radial flux inward is proportional to $(\hbar k/m)(4\pi/(2k)^2)$ and the one outward proportional to $(\hbar k/m)|S_l|^2(4\pi/(2k)^2)$. Hence, the difference is given by

$$j_{\text{in}}^{\text{rad}} - j_{\text{out}}^{\text{rad}} = \frac{\hbar k}{m}\left(1 - s_l^2\right)\left(\frac{\pi}{k^2}\right). \tag{18.44}$$

After summing contributions from all partial waves and by dividing by the total incoming flux j_{in} the inelastic cross section turns out to be

$$\sigma_{\text{rea}} = \frac{\pi}{k^2}\sum_{l=0}^{\infty}(2l+1)(1 - s_l^2). \tag{18.45}$$

The *total* cross section $\sigma_{\text{tot}} = \sigma_{\text{rea}} + \sigma_{\text{el}}$ for the entire process is then given by

$$\sigma_{\text{tot}} = \frac{\pi}{k^2}\sum_{l=0}^{\infty}(2l+1)2\left(1 - \text{Re}\{S_l\}\right) = \frac{2\pi}{k^2}\sum_{l=0}^{\infty}(2l+1)(1 - s_l\cos 2\delta_l). \tag{18.46}$$

The individual terms in (18.45) are proportional to the so-called *transmission coefficients*

$$T_l = 1 - |S_l|^2 = 1 - s_l^2, \tag{18.47}$$

which define the probability for a transmission into the interior of the scattering region (target).

We may note in passing that processes of this type are known from optics. A simple case of the so-called *Fraunhofer diffraction* is given by the scattering at a black sphere (see exercise 2), which also exists in nuclear physics, for details see e.g. (Fröbrich and Lipperheide, 1996).

18.2 Generalization to nuclear reactions

We are now going to present some basic features of reaction theory, mostly on a heuristic level, as a more or less natural generalization of the concepts developed in the previous section. To make the discussion as transparent as possible we will largely avoid mathematical derivations as well as lengthy formulations, in particular those which involve clumsy angular momentum couplings. In the outline of the essential steps we follow the exposition given in the textbook by H. Feshbach (1992). More detailed, formal descriptions can also be found in (Fröbrich and Lipperheide, 1996), see also (Gadioli and Hodgson, 1992).

18.2.1 Reaction channels

Suppose particle a, the projectile, impinges on particle A, the target. As we are only studying reactions on the nuclear but not sub-nuclear level, a and A will either stand for one nucleon or one nucleus. As a result of the reaction two analogous nuclear systems will be produced, which shall be called b and B. Schematically we may thus picture the reaction symbolically as

$$a + A \to B + b. \tag{18.48}$$

Traditionally, one has also used the notation $A(a,b)B$. Quantum mechanically the reaction is to be classified in terms of a set of quantum numbers, which are to be associated with the "entrance and exit channels", i.e. those configurations which characterize the system before and after the reaction. The latter is assumed to happen within a finite region of space. In shorthand notation we may use Greek letters α and β to specify the systems $a+A$ and $B+b$, respectively. Of course, as for potential scattering, the relevant system of space coordinates will be that of relative motion. An exchange of mass during the reaction will then imply that the reduced masses μ_α and μ_β in the entrance and exit "channels" will be different.

Next we need to look at the energies associated with different channels. Energy conservation requires

$$E_a^\alpha + E_A^\alpha + \epsilon_\alpha = E_B^\beta + E_b^\beta + \epsilon_\beta = E. \tag{18.49}$$

Here, the E_a^α and E_A^α are the internal energies of the fragments a and A, respectively. The

$$\epsilon_\alpha = \frac{\hbar^2 k_\alpha^2}{2\mu_\alpha} \quad \text{with} \quad \frac{1}{\mu_\alpha} = \frac{1}{M_a} + \frac{1}{M_A} \tag{18.50}$$

is the kinetic energy in the center of mass of the two partners in channel α, with μ_α being the reduced mass. The ϵ_α is referred to as the *channel energy*. Rewriting (18.49) as

$$\epsilon_\beta = \epsilon_\alpha + Q_{\alpha\beta} \quad (\geq 0) \tag{18.51}$$

allows one to introduce the "Q-value" of the reaction from channel α to channel β:

$$Q_{\alpha\beta} = E_a^\alpha + E_A^\alpha - E_B^\beta - E_b^\beta. \tag{18.52}$$

The process $\alpha \to \beta$ is possible only if the channel energy ϵ_β is positive, the channel β then is said to be "open". The "Q-value for the reaction $A(a,b)B$" is defined as in (18.52) but with the fragments being in their ground state

$$Q_{ab} = E_a^0 + E_A^0 - E_B^0 - E_b^0. \tag{18.53}$$

This Q-value may be *negative* as long as the initial channel energy ϵ_α^0 is large enough for $\epsilon_\beta^0 = \epsilon_\alpha^0 + Q_{ab}$ to be positive. For $\epsilon_\alpha^0 \leq -Q_{ab}$ the reaction $A(a,b)B$ is said to be "closed". Thus, $-Q_{ab}$ defines the *threshold of this reaction*.

18.2.2 Cross section

If we want to apply the cross section (18.15) to the reaction (18.48) we need to calculate the level density $\Omega_\beta(E)$ of the final states in the outgoing channel β. Let us look here at the simple case of spinless fragments in non-degenerate states. Then this $\Omega_\beta(E)$ is just given by (18.16) and the cross section becomes

$$\frac{d\sigma_{\alpha\beta}}{d\Omega_\beta} = \frac{1}{(2\pi)^2}\frac{1}{\hbar^4}\mu_\alpha\mu_\beta \frac{k_\beta}{k_\alpha}|T_{\alpha\beta}|^2. \tag{18.54}$$

The general case involving degeneracies is discussed below as well as in Section 8.1 in connection with the decay of the compound nucleus.

At this stage a remark is in order on the pre-factors appearing here in front of the squared T-matrix. They are intimately connected with the normalization of the wave functions which so far has been chosen to be that *corresponding to the momentum*, which for the wave function of relative motion reads

$$\int d^3x \left(\psi_{\mathbf{k}}^{(+)}(\mathbf{r})\right)^* \psi_{\mathbf{k}'}^{(+)}(\mathbf{r}) = (2\pi)^3 \delta(\mathbf{k}' - \mathbf{k}). \tag{18.55}$$

Alternatively, one may apply the so-called *energy normalization*

$$\int d^3x \left(\psi_{\mathbf{k}}^{(+)}(\mathbf{r})\right)^* \psi_{\mathbf{k}'}^{(+)}(\mathbf{r}) = \delta(E - E')\delta_\Omega(\hat{\mathbf{k}}' - \hat{\mathbf{k}}). \tag{18.56}$$

Here, δ_Ω is the delta function for the solid angle Ω. The easiest way of transforming between the two versions is to multiply the wave functions of the momentum normalization by the square root of the density $\Omega(E) = \mu\hbar k/(2\pi\hbar)^3$

$$\psi|_{\text{ener}} = \sqrt{\Omega(E)}\, \psi|_{\text{mom}}. \tag{18.57}$$

The equivalence of (18.55) and (18.56) is easily verified by accounting for the relation

$$\delta(E' - E)\delta_\Omega(\hat{\mathbf{k}}' - \hat{\mathbf{k}}) = \frac{\mu k}{\hbar^2}\delta(\mathbf{k}' - \mathbf{k}) \tag{18.58}$$

between the various δ-functions. As this manipulation has to be done both for channel α and β, formula (18.54) for the cross section turns into the simpler expression

$$\frac{d\sigma_{\alpha\beta}}{d\Omega_\beta} = \frac{(2\pi)^4}{k_\alpha^2}|T_{\alpha\beta}|^2\,. \tag{18.59}$$

In this way not only the mass factors have dropped out but even the $\hbar$ has disappeared. In what follows we shall use this latter version (18.59) for the cross section, following the customary procedure in the literature, such as (Feshbach, 1992), (Newton, 1982) or the review article by R. Lemmer (1966).

18.2.3 The T-matrix for nuclear reactions

Following H. Feshbach we exploit the formal means of orthogonal projectors, which is to say of Hermitian operators, $\hat{P} = \hat{P}^\dagger$ and $\hat{Q} = \hat{Q}^\dagger$ having the properties $\hat{P}^2 = \hat{P}$, $\hat{Q}^2 = \hat{Q}$ and

$$\hat{P} + \hat{Q} = 1 \qquad \hat{P}\hat{Q} = \hat{Q}\hat{P} = 0\,. \tag{18.60}$$

The general wave function then splits up like

$$|\Psi\rangle = \hat{P}|\Psi\rangle + \hat{Q}|\Psi\rangle\,. \tag{18.61}$$

Multiplying the Schrödinger equation $(E - \hat{H})|\Psi\rangle = 0$ from the left by $\hat{P}$ and $\hat{Q}$ yields the two separate equations

$$\left(E - \hat{H}_{PP}\right)\hat{P}|\Psi\rangle = \hat{H}_{PQ}\hat{Q}|\Psi\rangle \tag{18.62}$$

and

$$\left(E - \hat{H}_{QQ}\right)\hat{Q}|\Psi\rangle = \hat{H}_{QP}\hat{P}|\Psi\rangle\,, \tag{18.63}$$

where $\hat{H}_{PP} = \hat{P}\hat{H}\hat{P}$, $\hat{H}_{PQ} = \hat{P}\hat{H}\hat{Q}$ and so on.

Suppose the $\hat{P}$ is chosen to project on the channels one is going to study. The channel functions then satisfy the equation

$$\left(E - \hat{H}_{PP}\right)|\chi\rangle = 0\,. \tag{18.64}$$

Suppose further we take outgoing waves again. Then (18.62) can formally be solved as

$$\hat{P}|\Psi_\alpha\rangle = |\chi_\alpha^{(+)}\rangle + \frac{1}{E^{(+)} - \hat{H}_{PP}}\hat{H}_{PQ}\hat{Q}|\Psi_\alpha\rangle\,. \tag{18.65}$$

Putting this solution into (18.63) one gets

$$\left(E - \hat{H}_{QQ} - \hat{W}_{QQ}\right)\hat{Q}|\Psi_\alpha\rangle = \hat{H}_{QP}|\chi_\alpha^{(+)}\rangle\,, \tag{18.66}$$

with

$$\hat{W}_{QQ} \equiv \hat{H}_{QP}\frac{1}{E^{(+)} - \hat{H}_{PP}}\hat{H}_{PQ}\,, \tag{18.67}$$

such that the $\hat{Q}|\Psi_\alpha\rangle$ becomes

$$\hat{Q}|\Psi_\alpha\rangle = \frac{1}{E - \hat{H}_{QQ} - \hat{W}_{QQ}} \hat{H}_{QP}|\chi_\alpha^{(+)}\rangle. \quad (18.68)$$

We may turn now to the T-matrix. Equation (18.62) may be viewed as a Schrödinger equation for distorted waves, with the $\hat{H}_{PQ}\hat{Q}$ being identified as the $\hat{V}_1$ of the Hamiltonian (18.29). Applying (18.31) to the present case leads to

$$T_{\alpha\beta} = T_{\alpha\beta}^{(P)} + \langle\chi_\beta^{(-)}|\hat{H}_{PQ}\hat{Q}|\Psi_\alpha\rangle. \quad (18.69)$$

Inserting the form found in (18.68) on the right one gets the final result

$$T_{\alpha\beta} = T_{\alpha\beta}^{(P)} + \langle\chi_\beta^{(-)}|\hat{H}_{PQ}\frac{1}{E - \hat{H}_{QQ} - \hat{W}_{QQ}}\hat{H}_{QP}|\chi_\alpha^{(+)}\rangle. \quad (18.70)$$

To proceed further it is useful to introduce the eigenfunctions of the Hamiltonian $\hat{H}_{QQ}$. Let us use the following notation,

$$\begin{aligned}\hat{H}_{QQ}|\Phi_s\rangle &= \mathcal{E}_s|\Phi_s\rangle & \text{for discrete states}\\ \hat{H}_{QQ}|\Phi(\mathcal{E},\alpha)\rangle &= \mathcal{E}|\Phi(\mathcal{E},\alpha)\rangle & \text{for the continuum}.\end{aligned} \quad (18.71)$$

These states are supposed to be normalized "to the energy" like

$$\begin{aligned}\langle\Phi_s|\Phi_{s'}\rangle &= \delta_{ss'} & \langle\Phi_s|\Phi(\mathcal{E},\alpha)\rangle = 0\\ \langle\Phi(\mathcal{E}',\alpha')|\Phi(\mathcal{E},\alpha)\rangle &= \delta(\mathcal{E}'-\mathcal{E})\delta(\alpha'-\alpha).\end{aligned} \quad (18.72)$$

18.2.4 Isolated resonances

To get the essential features by which one may describe a resonance let us first make the simplifying assumption that there is just one discrete state $|\Phi_s\rangle$ of (18.71), discarding the set of continuum states. This will facilitate the analysis of the general form (18.70) of T-matrix. The latter then takes on the form

$$T_{\alpha\beta} = T_{\alpha\beta}^{(P)} + \frac{\langle\chi_\beta^{(-)}|\hat{H}_{PQ}|\Phi_s\rangle\langle\Phi_s|\hat{H}_{QP}|\chi_\alpha^{(+)}\rangle}{E - \mathcal{E}_s - \langle\Phi_s|\hat{W}_{QQ}|\Phi_s\rangle}. \quad (18.73)$$

With the notation

$$\langle\Phi_s|\hat{W}_{QQ}|\Phi_s\rangle = \Delta_s(E) - \frac{1}{2}i\Gamma_s(E) \quad (18.74)$$

and $E_s = \mathcal{E}_s + \Delta_s(E)$ this becomes

$$T_{\alpha\beta} = T_{\alpha\beta}^{(P)} + \frac{\langle\chi_\beta^{(-)}|\hat{H}_{PQ}|\Phi_s\rangle\langle\Phi_s|\hat{H}_{QP}|\chi_\alpha^{(+)}\rangle}{E - E_s + i\Gamma_s/2}. \quad (18.75)$$

As the energy denominator of (18.67) may be written as

$$\frac{1}{E^{(+)} - \hat{H}_{PP}} = \mathcal{P}\frac{1}{E - \hat{H}_{PP}} - i\pi\delta\left(E - \hat{H}_{PP}\right) \qquad (18.76)$$

the energy shift is given by

$$\Delta_s(E) = \langle \Phi_s | \hat{H}_{QP} \frac{\mathcal{P}}{E - \hat{H}_{PP}} \hat{H}_{PQ} | \Phi_s \rangle \qquad (18.77)$$

and for the width one gets

$$\Gamma_s = 2\pi \langle \Phi_s | \hat{H}_{QP} \delta(E - \hat{H}_{PP}) \hat{H}_{PQ} | \Phi_s \rangle. \qquad (18.78)$$

By inserting a complete set of states $\chi_\mu^{(+)}$ the latter expression can be written as

$$\Gamma_s = 2\pi \sum_\mu \left| \langle \Phi_s | \hat{H}_{QP} | \chi_\mu^{(+)} \rangle \right|^2 \equiv \sum_\mu \Gamma_{s\mu}(E). \qquad (18.79)$$

Here, the $\Gamma_{s\mu}(E)$ are the *partial widths* for the decay of the state $|\Phi_s\rangle$ into all channels μ which are open at the given energy E. As an illustrative example one may try to apply this projection technique to the model discussed in Section 4.3.1 for a Hamiltonian $\hat{H} = \hat{H}_0 + \hat{V}$ (see exercise 3).

At this stage a comment is in order on the assumption of using just one single state in Q-space to represent a resonance. Isolated resonances show up whenever the cross section is seen to be composed of a smooth part and a part which in some finite region exhibits a strong variation with energy. This feature may hold true even if $\hat{H}_{QQ}$ has more than just one eigenstate. This may be understood by outlining an alternative derivation of the T-matrix without going into all the details, which may be found in Section III.2 of (Feshbach, 1992). One may start solving formally eqn(18.63), instead of (18.62) as done before. The solution reads

$$\hat{Q}|\Psi\rangle = \frac{1}{E^{(+)} - \hat{H}_{QQ}} \hat{H}_{QP} \hat{P} |\Psi\rangle, \qquad (18.80)$$

assuming that in Q-space there is no incoming wave but allowing for the possibility of open channels. Inserting this expression into (18.62) the effective Schrödinger equation in P-space becomes

$$\left(E - \hat{H}_{\text{eff}}\right) \hat{P} |\Psi\rangle = 0, \qquad (18.81)$$

with

$$\hat{H}_{\text{eff}} = \hat{H}_{PP} + \hat{H}_{PQ} \frac{1}{E^{(+)} - \hat{H}_{QQ}} \hat{H}_{QP}. \qquad (18.82)$$

The resolvent $(E^{(+)} - \hat{H}_{QQ})^{-1}$ may now be sandwiched by unity operators in Q-space. Closure for the states (18.71) then requires summing over all discrete ones and integrating over the continuum. If there is a genuinely isolated resonance at an energy close to a $\mathcal{E}_s$ all the terms besides the one which involve the corresponding state $|\Phi_s\rangle$ can be expected to contribute smoothly in energy. They may therefore be lumped into an effective $\hat{\overline{H}}_{PP}$ which describes the *prompt or smooth* part of the reaction. Our derivation from before can then be repeated simply replacing the original $\hat{H}_{PP}$ by the $\hat{\overline{H}}_{PP}$.

For practical applications the following decomposition turns out to be useful. As the bound state $|\Phi_s\rangle$ may be taken to be real, one may write

$$\langle \Phi_s | \hat{H}_{QP} | \chi_\alpha^{(+)} \rangle = e^{i\delta_\alpha} \sqrt{\frac{\Gamma_{s\alpha}}{2\pi}} \equiv e^{i\delta_\alpha} \frac{g_{s\alpha}}{\sqrt{2\pi}}$$

$$\langle \chi_\beta^{(-)} | \hat{H}_{PQ} | \Phi_s \rangle = e^{i\delta_\beta} \sqrt{\frac{\Gamma_{s\beta}}{2\pi}} \equiv e^{i\delta_\beta} \frac{g_{s\beta}}{\sqrt{2\pi}} \,, \qquad (18.83)$$

where δ_α is the phase of the wave function $\chi_\alpha^{(+)} = \exp(i\delta_\alpha) \, | \, \chi_\alpha^{(+)} \, |$. Then the T-matrix (18.75) becomes

$$T_{\alpha\beta} = T_{\alpha\beta}^{(P)} + \frac{1}{2\pi} e^{i(\delta_\alpha + \delta_\beta)} \frac{g_{s\beta} g_{s\alpha}}{E - E_s + \frac{1}{2} i \Gamma_s} \,. \qquad (18.84)$$

18.2.4.1 Interference of the resonance with the background To show this interference effect between the two terms in (18.84) let us concentrate on s-wave scattering with $\alpha = \beta$. The angle integrated cross section becomes

$$\sigma_{\alpha\alpha}(E) = \frac{4\pi}{k^2} \left| \sin\delta - e^{i\delta} \frac{\Gamma_{s\alpha}/2}{E - E_s + i\Gamma_s/2} \right|^2 \,; \qquad (1.43)$$

remember that this integration absorbs a factor 4π from the pre-factor in (18.59), see also (18.98). To arrive at the form (1.43) the prompt part of the T-matrix has been taken to be $T^{(P)} = -\exp(i\delta)\sin\delta/\pi$, from (18.39). A prominent example where these features come up most clearly are neutron resonances at low energies discussed in Section 1.3; the functional behavior of $\sigma_{\alpha\alpha}(E)$ is sketched in Fig. 1.2. For a thorough treatment of this case we may recommend the beautiful article by R. Lemmer (1966).

18.2.4.2 Scattering and reaction cross section It is very instructive to analyze the resonance behavior in terms of concepts which we shall encounter again in the statistical treatment of nuclear reactions. Moreover, this may serve as a nice model in which we may study absorption of flux in channels different from the incoming one. Let us ignore the contribution from the prompt term. Concentrating on the resonance we may identify the corresponding cross section with the fluctuation cross section $\sigma_{\alpha\beta}^{(FL)}$, a notion which may become clearer below. From (18.75) one gets

$$\sigma_{\alpha\beta}^{(FL)} = \frac{\pi}{k_\alpha^2} \frac{\Gamma_{s\alpha} \Gamma_{s\beta}}{(E - E_s)^2 + (1/4)\Gamma_s^2} \,, \qquad (18.85)$$

using the notation introduced in (18.83). Summing over all final channels one gets the cross section for the formation of the "compound nucleus", which is defined by this particular state $|\Phi_s\rangle$ and which is reached by the entrance state $|\alpha\rangle$, namely

$$\sigma_\alpha^{(C)} = \sum_\beta \sigma_{\alpha\beta}^{(FL)} = \frac{\pi}{k_\alpha^2} \frac{\Gamma_{s\alpha}\Gamma_s}{(E-E_s)^2 + (1/4)\Gamma_s^2}. \tag{18.86}$$

The $\sigma_{\alpha\beta}^{(FL)}$ can thus be written as

$$\sigma_{\alpha\beta}^{(FL)} = \sigma_\alpha^{(C)} \frac{\Gamma_{s\beta}}{\Gamma_s}. \tag{18.87}$$

For $\alpha = \beta$ this becomes the cross section for elastic scattering which occurs because the compound state may decay back into the entrance channel, the so-called "compound elastic" process; one gets

$$\sigma_{el}^{(C)} \equiv \sigma_{\alpha\alpha}^{(FL)} = \sigma_\alpha^{(C)} \frac{\Gamma_{s\alpha}}{\Gamma_s}. \tag{18.88}$$

Looking only at that channel the remaining cross section may be interpreted as the reaction cross section, very much in the spirit of that concept introduced in Section 18.1.3. Indeed, introducing a Γ_s^{rea} by

$$\Gamma_s^{\text{rea}} \equiv \Gamma_s - \Gamma_{s\alpha} = \sum_{\gamma \neq \alpha} \Gamma_{s\gamma} \tag{18.89}$$

the reaction cross section may be defined as

$$\sigma_{\text{rea}}^{(C)} = \sigma_\alpha^{(C)} \frac{\Gamma_s^{\text{rea}}}{\Gamma_s}. \tag{18.90}$$

Obviously, one has

$$\sigma_{\text{rea}}^{(C)} + \sigma_{el}^{(C)} = \sigma_\alpha^{(C)} \equiv \sigma_{\text{tot}}^{(C)}, \tag{18.91}$$

in accord with the notation in Section 18.1.3. Looking only at the compound elastic process one concludes that the incoming flux is not conserved. Some part of it is flowing into the other channels, that one which in our notation makes up the reaction cross section. The total flux is conserved, of course, as reflected by the unitarity of the S-matrix calculated for all channels, $SS^\dagger = 1$. For reactions, a proof of this property would best be done involving the so-called K-matrix, for details see e.g. (Newton, 1982) or (Feshbach, 1992).

As the compound elastic scattering occurs through a delay of the emission after the incoming particle first had been "absorbed" the total cross section may be identified as the "absorption cross section" (Friedman and Weisskopf, 1955). Note that this notation is in agreement with that for absorbing an electromagnetic wave by a dipole before it gets re-emitted, see Section 16.2.

18.2.5 Overlapping resonances

If there is more than one resonance the expression (18.70) has to be calculated by evaluating the inverse of $E - \hat{H}_{QQ} - \hat{W}_{QQ}$ in the space spanned by the set of states $|\Phi_s\rangle$ of (18.71). To this end the eigenvectors $|\Theta_n\rangle$ of $\hat{H}_{QQ} - \hat{W}_{QQ}$ may be expanded in terms of $|\Phi_s\rangle$ and the complex eigenvalues E_n are given by the secular equation

$$\det\left\{E_n - \mathcal{E}_s - \langle\Phi_s|\hat{W}_{QQ}|\Phi_r\rangle\right\} = 0. \tag{18.92}$$

The form (18.73) of the T-matrix is to be replaced by

$$T_{\alpha\beta} = T_{\alpha\beta}^{(P)} + \sum_n \frac{\langle\chi_\beta^{(-)}|\hat{H}_{PQ}|\Theta_n\rangle\langle\Theta_n^{(A)}|\hat{H}_{QP}|\chi_\alpha^{(+)}\rangle}{E - E_n}, \tag{18.93}$$

where $\langle\Theta_n^{(A)}|$ is the adjoint bra to the ket $|\Theta_n\rangle$; this feature appears because $\hat{W}_{QQ}$ is not Hermitian. The invariance of the trace in linear transformations implies a simple sum rule for the imaginary parts of the eigenvalues of E_n and those of the diagonal elements W_{ss} of the $\hat{W}_{QQ}$, namely $\sum_n \text{Im}(E_n) = \text{Im}(W_{ss})$. Recalling the relation (18.78) between the $\text{Im}(W_{ss})$ and the width Γ_s for isolated resonances one gets

$$\sum_n \Gamma_n = \sum_s \Gamma_s, \tag{18.94}$$

implying that the average width is the same whether the levels overlap or not (Lemmer, 1966). This feature reflects the fact that the numerator of the resonant term in (18.93) is not independent of the eigenvalues E_n in the denominator, for more details see (Feshbach, 1992).

18.2.6 T-matrix with angular momentum coupling

The wave functions which enter the calculation of the T-matrix must be classified by the quantum numbers which specify the channel wave functions, together with those in Q space. Besides parity and isospin, perhaps, the most important ones are those associated with angular momentum. Here various coupling schemes are possible. We like to concentrate on the one which in the literature is referred to as *channel coupling* (see e.g. Section IV.1 of (Feshbach, 1992)). Suppose in the entrance channel the two fragments have spin $\mathbf{I}_a$ and $\mathbf{I}_A$. They may be coupled to the channel spin

$$\mathbf{s}_\alpha = \mathbf{I}_a + \mathbf{I}_A. \tag{18.95}$$

The total angular momentum is then given by

$$\mathbf{J} = \mathbf{s}_\alpha + \mathbf{l}_\alpha, \tag{18.96}$$

if the orbital angular momentum of the fragments in the entrance channel is denoted by $\mathbf{l}_\alpha$. Similar relations hold true for the exit channel "β" with the a, A replaced by b, B. The total angular momentum $\mathbf{J}$ is conserved, of course, but for the T-matrix elements proper angular momentum coupling has to be performed both for the entrance as well as for the exit channel. For this reason the expressions get quite involved. They simplify after integrating over angles. For our purpose it may suffice to quote results for the total cross section. Their derivation can be found in the literature; in particular we like to refer to Chapter IV of (Feshbach, 1992) and to (Lemmer, 1966). One gets

$$\sigma_{\alpha\alpha'} = \frac{\pi}{k_\alpha^2} \sum \frac{2J+1}{(2I_a+1)(2I_A+1)} |2\pi T_{\alpha\alpha'}(ls;l's';J)|^2 , \qquad (18.97)$$

with the summation being understood to extend over all angular momentum quantum numbers,[39] which are not fixed by the channel numbers α, α', like the spins of projectile and target. The channel wave functions are supposedly normalized to energy according to

$$\langle \chi_\alpha(E) | \chi_{\alpha'}(E') \rangle = 4\pi \delta_{\alpha\alpha'} \delta(E'-E) , \qquad (18.98)$$

which in relation (18.59) removes a factor 4π, mind (18.58). For spinless particles, this normalization implies the diagonal element T_l of the T-matrix for angular momentum l to take on the form (18.39). Later we may wish to rewrite (18.97) as

$$\sigma_{\alpha\alpha'} = \sum_J \sigma_{\alpha\alpha'}^{(J)} , \qquad (18.99)$$

with

$$\sigma_{\alpha\alpha'}^{(J)} = \frac{2J+1}{(2I_a+1)(2I_A+1)} \sum_{ls;l's'} \sigma_{\alpha\alpha'}(ls;l's';J) \qquad (18.100)$$

and

$$\sigma_{\alpha\alpha'}(ls;l's';J) = \frac{\pi}{k_\alpha^2} |2\pi T_{\alpha\alpha'}(ls;l's';J)|^2 . \qquad (18.101)$$

18.3 Energy averaged amplitudes

18.3.1 The optical model

The physical background of the optical model for describing nuclear reactions in some average sense is discussed in Section 1.3. The optical potential is meant to effectively portray energy averages over detailed structure of the elastic cross section. To be specific, let us speak of a reaction of a nucleon with a nucleus, although the model is of much wider use. The averaging interval ΔE is to be chosen large compared to the widths and the spacings of the individual resonances but it must be small compared to the structure of shape resonances, i.e, those resonances which are due to a scattering of the nucleon at the mean field potential of the nucleus. Indeed, both differ by orders of magnitude, the width

---

[39]Those of other relevant symmetries like parity, isospin etc. will be dropped here for the sake of simplicity.

of the compound resonances being in the range of eV whereas the shape resonances lie in the MeV range. The relationship of the optical potential to the elastic scattering amplitude was first discussed by H. Feshbach, C.E. Porter and V.F. Weisskopf (1954).

Formally the goal is to find an equation for the averaged state $\overline{\hat{P}|\Psi_\alpha\rangle}$. As shown in Section III.3 of (Feshbach, 1992) the final result is the Schrödinger equation

$$\hat{H}_{\text{opt}} \overline{\hat{P}|\Psi_\alpha\rangle} = E \overline{\hat{P}|\Psi_\alpha\rangle} \qquad (18.102)$$

with

$$\hat{H}_{\text{opt}} = \hat{H}_{PP} + \hat{H}_{PQ} \left[\left(\overline{1/e_{QQ}} \right)^{-1} + \hat{W}_{QQ} \right]^{-1} \hat{H}_{QP} \qquad (18.103)$$

and

$$\overline{1/e_{QQ}} = \int g(E - E') \frac{1}{E' - \hat{H}_{QQ} - \hat{W}_{QQ}} dE', \qquad (18.104)$$

with an averaging function $g(E - E')$. If the latter is chosen to be a Lorentzian of width $\Delta/2$ the optical model Hamiltonian turns out to be

$$\hat{H}_{\text{opt}} = \hat{H}_{PP} + \hat{H}_{PQ} \frac{1}{E - \hat{H}_{QQ} + i\Delta/2} \hat{H}_{QP}. \qquad (18.105)$$

Please observe the strikingly close relation to the effective Hamiltonian (18.82). The fact that in going from (18.103) to (18.105) the $\hat{W}_{QQ}$ has dropped out is a special feature of smoothing by a Lorentzian. Indeed, for

$$g(E - E') = \frac{1}{2\pi} \frac{\Delta}{(E - E')^2 + (\Delta/2)^2}, \qquad (18.106)$$

the result of the smoothing simply consists in shifting the energy from E to $E + i\Delta/2$. This is easily verified by calculating in (18.104) the integral by the residue theorem, observing that the energy denominator is analytic in the upper half plane (see exercise 2); remember that $\hat{W}_{QQ}$ has poles in the lower half plane.

The $\hat{H}_{\text{opt}}$ of (18.105) has a negative imaginary part accounting for a loss of the probability current. This current is lost in the channels which are effectively eliminated by the averaging procedure. The reader may notice the close analogy between this feature and the appearance of irreversible behavior in collective motion discussed in Section 6.4.1. From this analogy we may learn another important property: The energy dependence of the imaginary part of the optical potential is largely governed by the matrix elements of the operators $\hat{H}_{QP}$ and $\hat{H}_{PQ}$, together with the eigenvalues of the $\hat{H}_{QQ}$. The presence of the $i\Delta/2$ only helps to smooth out the detailed structure of the latter. For response functions this is demonstrated in Fig. 6.2 within a schematic model. This analogy might be carried even further. The case of overlapping resonances, in a sense, corresponds to the situation that the individual states which contribute to the response function are themselves strongly damped by the coupling to more complicated states (see Section 6.4.3).

18.3.2 Intermediate structure through doorway resonances

So far we have encountered two different kinds of structure in the cross section, the sharp resonances and the broad resonances associated with motion in a mean field or in the optical potential. For the case of neutron scattering at a nucleus the corresponding widths, Γ_{CN} and Γ_{sp}, respectively, would differ by several orders of magnitude, see Section 1.3. Many examples are known in which intermediate structure is seen, say in the sense of having resonances of intermediate widths

$$\Gamma_{\text{GS}} \gg \Gamma_{\text{d}} \gg \Gamma_{\text{CN}}. \tag{1.46}$$

Here the broad width has been given the index "GS" for "gross structure", rather than the Γ_{sp} mentioned before. This was done simply because the many examples of great interest in our context involve collective motion, like giant resonances or nuclear fission.

Suppose the entrance channels do not couple directly to the compound states but that there is an intermediate step in between.[40] In other words, the entrance channels may couple to states which lie in a certain part $\hat{D}|\psi\rangle$ of the total Hilbert space, and only these states are coupled to the compound states. This situation may be handled by modifying the projectors introduced in Section 18.2.3 in the following way (Feshbach et al., 1967). We add the D-part to the original P-part and accordingly change the Q-part to a Q'-part. This is to say to write

$$\begin{aligned} \hat{P}' &= \hat{P} + \hat{D} \qquad \hat{P}' + \hat{Q}' = 1 \\ \hat{P}\hat{Q}' &= \hat{P}\hat{D} = \hat{D}\hat{Q}' = 0 \,. \end{aligned} \tag{18.107}$$

For the P' plus Q' splitting the T-matrix of (18.70) may be written as

$$T_{\alpha\beta} = T_{\alpha\beta}^{(P')} + \langle \phi_\beta^{(-)}|\hat{H}_{P'Q'} \frac{1}{E - \hat{H}_{Q'Q'} - \hat{W}_{Q'Q'}} \hat{H}_{Q'P'}|\phi_\alpha^{(+)}\rangle \tag{18.108}$$

with the wave functions of the P' space being determined by $\left(E - \hat{H}_{P'P'}\right)|\phi\rangle = 0$. Looking at the decomposition $\hat{P}' = \hat{P} + \hat{D}$ it is clear from (18.62) and (18.63) that this Schrödinger equation can be split up like

$$\left(E - \hat{H}_{PP}\right)\hat{P}|\phi\rangle = \hat{H}_{PD}\hat{D}|\phi\rangle \quad \text{and} \quad \left(E - \hat{H}_{DD}\right)\hat{D}|\phi\rangle = \hat{H}_{DP}\hat{P}|\phi\rangle \,. \tag{18.109}$$

The matrix $T_{\alpha\beta}^{(P')}$ in (18.108) may exhibit resonance structure itself but without any influence of the coupling to the true compound states. The latter may be taken into account by introducing the optical potential

$$\hat{H}_{\text{opt}} = \hat{H}_{P'P'} + \hat{H}_{P'Q'} \left[\left(\overline{1/e_{Q'Q'}}\right)^{-1} + \hat{W}_{Q'Q'}\right]^{-1} \hat{H}_{Q'P'} \tag{18.110}$$

---

[40] For phenomena related to exit channels see (Feshbach, 1992).

with the associated Schrödinger equation

$$\hat{H}_{\text{opt}} \overline{\hat{P}'|\Psi\rangle} = E \overline{\hat{P}'|\Psi\rangle}. \qquad (18.111)$$

Performing these steps we have effectively averaged out the fine structure of the compound states, provided the averaging interval is in accord with the condition (1.46). Introducing the so-called *strong doorway assumption*

$$\hat{H}_{PQ'} = 0 \qquad \hat{H}_{P'Q'} = \hat{H}_{DQ'} \qquad (18.112)$$

the $\hat{H}_{\text{opt}}$ becomes

$$\hat{H}_{\text{opt}} = \hat{H}_{P'P'} + \hat{H}_{DQ'} \left[\left(\overline{1/e_{Q'Q'}} \right)^{-1} + \hat{W}_{Q'Q'} \right]^{-1} \hat{H}_{Q'D}. \qquad (18.113)$$

The final aim must be to express the T-matrix in terms of the unperturbed P-states. To this end we split the solution $\overline{\hat{P}'|\Psi\rangle}$ of (18.111) into a P and a D part,

$$\overline{\hat{P}'|\Psi\rangle} = \overline{\hat{P}|\Psi\rangle} + \overline{\hat{D}|\Psi\rangle}. \qquad (18.114)$$

In this way the Schrödinger equation (18.111) can be divided into

$$\left(E - \hat{H}_{PP} \right) \overline{\hat{P}|\Psi\rangle} = \hat{H}_{PD} \overline{\hat{D}|\Psi\rangle} \qquad (18.115)$$

and

$$\left(E - \hat{H}_{DD} - \overline{W}_{DD} \right) \overline{\hat{D}|\Psi\rangle} = \hat{H}_{DP} \overline{\hat{P}|\Psi\rangle}, \qquad (18.116)$$

with

$$\overline{W}_{DD} = \hat{H}_{DQ'} \left[\left(\overline{1/e_{Q'Q'}} \right)^{-1} + \hat{W}_{Q'Q'} \right]^{-1} \hat{H}_{Q'D}. \qquad (18.117)$$

Comparing (18.116) with (18.109) one recognizes the appearance of the potential $\overline{W}_{DD}$ in addition to the $\hat{H}_{DD}$.

Based on the Schrödinger equation (18.64), $(E - \hat{H}_{PP})|\chi\rangle = 0$, for the channel wave functions in P-space one may now repeat the steps from (18.62) to (18.70) to end up with

$$\overline{T}_{\alpha\beta} = T_{\alpha\beta}^{(P)} + \langle \chi_\beta^{(-)} | \hat{H}_{PD} \frac{1}{E - \hat{H}_{DD} - \overline{W}_{DD} - \hat{W}_{DD}} \hat{H}_{DP} | \chi_\alpha^{(+)} \rangle, \qquad (18.118)$$

where

$$\hat{W}_{DD} \equiv \hat{H}_{DP} \frac{1}{E^{(+)} - \hat{H}_{PP}} \hat{H}_{PD}. \qquad (18.119)$$

One only has to replace the $\hat{Q}$ in the original equations by the $\hat{D}$, together with identifying the $\hat{H}_{DD}$ in (18.63) by the $\hat{H}_{DD} + \overline{W}_{DD}$ of (18.116).

The resolvent appearing in (18.118) can be expressed by c-numbers if one knows the doorway states $|\psi_d\rangle$ satisfying

$$\left(E - \hat{H}_{DD}\right)|\psi_d\rangle = 0. \tag{18.120}$$

This becomes particularly simple if only one doorway state exists, in which case one gets

$$\overline{T}_{\alpha\beta} = T^{(P)}_{\alpha\beta} + \frac{\langle \chi^{(-)}_\beta|\hat{H}_{PD}|\psi_d\rangle \langle \psi_d|\hat{H}_{DP}|\chi^{(+)}_\alpha\rangle}{E - E_d + \frac{i}{2}\left(\Gamma^\uparrow_d + \Gamma^\downarrow_d\right)}, \tag{18.121}$$

where

$$E_d = \mathrm{Re}\langle \psi_d|\hat{H}_{DD} + \overline{W}_{DD} + \hat{W}_{DD}|\psi_d\rangle$$
$$\Gamma^\uparrow_d = -2\mathrm{Im}\,\langle \psi_d|\hat{W}_{DD}|\psi_d\rangle = 2\pi \sum_\beta \left|\langle \chi^{(-)}_\beta|\hat{H}_{PD}|\psi_d\rangle\right|^2 \tag{18.122}$$

and

$$\Gamma^\downarrow_d = -2\mathrm{Im}\,\langle \psi_d|\overline{W}_{DD}|\psi_d\rangle = \frac{4}{\Delta}\sum_q \left|\langle \psi_d|\hat{H}_{DQ'}|\Phi_q\rangle\right|^2. \tag{18.123}$$

The expression on the very right is valid if the average $\overline{W}_{DD}$ is calculated for Lorentz smoothing; otherwise the result would look differently. The $\Gamma^\uparrow_d$ measures the decay of the doorway state into the scattering channel, for which reason it is referred to as the "gamma up". The $\Gamma^\downarrow_d$, the "gamma down", on the other hand represents the width for the decay into the compound states.

Let us suppose further that only one channel γ is open. Then (18.121) reduces to

$$\overline{T} = T^{(P)} + \frac{1}{2\pi}e^{2i\delta}\frac{\Gamma^\uparrow_d}{E - E_d + \frac{i}{2}\left(\Gamma^\uparrow_d + \Gamma^\downarrow_d\right)}, \tag{18.124}$$

such that resonating cross section becomes

$$\overline{\sigma}(E) = \frac{4\pi}{k^2}\left|\sin\delta - e^{i\delta}\frac{\Gamma^\uparrow_d/2}{E - E_d + \frac{i}{2}\left(\Gamma^\uparrow_d + \Gamma^\downarrow_d\right)}\right|^2, \tag{1.47}$$

by analogy with the formula for one compound resonance treated in Section 18.2.4.1. Unlike this former case (see the discussion in Section 18.2.4.2) here genuine absorption takes place even if only this one channel is open. This is easily inferred after calculating the diagonal element of the S-Matrix from (18.124) and

(18.11). With the prompt part being $S^{(P)} = \exp(2i\delta)$ this element becomes the complex number

$$\overline{S} = e^{2i\delta} \frac{E - E_d + \frac{i}{2}\left(\Gamma_d^\downarrow - \Gamma_d^\downarrow\right)}{E - E_d + \frac{i}{2}\left(\Gamma_d^\uparrow + \Gamma_d^\downarrow\right)}. \tag{1.48}$$

Obviously, its absolute value is smaller than one, implying that the S-matrix is non-unitary. The spreading width $\Gamma_d^\downarrow$ acts as a source for absorption, simulating a genuine reaction, different from a mere backscattering of all flux into the entrance channel. Introducing the transmission coefficient $\mathcal{T} = 1 - \left|\overline{S}\right|^2$ of (18.47) one gets

$$\mathcal{T}(E) = \frac{\Gamma_d^\uparrow \Gamma_d^\downarrow}{(E - E_d)^2 + \frac{1}{4}\left(\Gamma_d^\uparrow + \Gamma_d^\downarrow\right)^2}. \tag{1.49}$$

Recall the discussion of Section 18.1.3 where the transmission coefficient $\mathcal{T}$ was seen to determine the absorption cross section.

18.4 Statistical theory

In this section we turn to the regime of larger energies where it becomes impossible to follow the excitations of individual resonances in detail. With increasing energies the level density takes on values which necessarily require statistical concepts for any practical description, see Section 22.2 or Chapter 10. It will be seen that after applying a few basic assumptions the famous Bohr hypothesis follows very naturally from a fully quantal description of the reactions. This hypothesis states that the decay of the compound nucleus is independent of how it is produced, see Section 1.3. This feature may well be interpreted in the sense that at an intermediate stage the nucleus is in a state of microcanonical equilibrium. The most important assumptions are:

(i) The wave functions are of complex structure, such that the elements $T_{\alpha\alpha'}$ of the T-matrix are complex numbers with *random phases*.

(ii) The necessary ensemble average can be carried out as an energy average; this hypothesis is commonly referred to as the "ergodic hypothesis" (for a critical discussion see (Feshbach, 1992)).

Averages of complex quantities with random phases can be carried out as follows. Assume the phases ϕ_n of $u_n = e^{i\phi_n}|u_n|$ to be equally probable in the interval $0 \leq \phi_n \leq 2\pi$. This implies

$$\overline{u_n} = \frac{1}{2\pi}\int_0^{2\pi} u_n d\phi_n = 0 \quad \text{and} \quad \overline{|u|^2} = \sum_n |u_n|^2. \tag{18.125}$$

The latter result is a consequence of

$$\overline{u_n^* u_m} = |u_n||u_m|\int_0^{2\pi}\frac{d\phi_n d\phi_m}{(2\pi)^2}e^{i(\phi_m - \phi_n)} = \delta_{nm}|u_n|^2. \tag{18.126}$$

These features are of course closely related to the Random Matrix Models studied in Section 4.2. To begin with we first want to look at their implications for the distribution of widths.

18.4.1 Porter–Thomas distribution for widths

This distribution may be inferred from the distribution $P(a_\mu)$ of the amplitudes a_μ of eigenvectors of random Hamiltonians given in (4.29). There, the a_μ^2 determines the probability that the compound state may be found in the μ'th model or channel state. As the corresponding width Γ_μ must be proportional to this a_μ^2 one may set

$$\frac{\Gamma_\mu}{\langle\Gamma_\mu\rangle} = N a_\mu^2 . \tag{18.127}$$

To understand this relation recall that

$$P(a_\mu) = \left(\frac{N}{2\pi}\right)^{1/2} \exp\left(-\frac{N}{2} a_\mu^2\right) \tag{4.29}$$

implies the mean square of a_μ to be proportional to $1/N$, viz $\langle a_\mu^2 \rangle = 1/N$. The (necessarily positive) width Γ_μ then follows the Porter–Thomas distribution

$$P(\Gamma_\mu) d\Gamma_\mu = \frac{1}{\sqrt{2\pi x}} e^{-x/2} dx \quad \text{with} \quad x = \frac{\Gamma_\mu}{\langle\Gamma_\mu\rangle} . \tag{18.128}$$

It is not difficult to see that this distribution $P(x)$ is quite broad. Indeed, one may readily verify that it is normalized to unity with the mean being $\langle x \rangle = 1$ but with a mean square of $\langle x^2 \rangle = 3$. The situation changes if many independent channels are open, say M altogether. The total width is then given by $\Gamma = \sum_1^M \Gamma_\mu$. If their averages are the same, $\langle \Gamma_\mu \rangle = \langle \Gamma \rangle / M$ the mean square deviation of Γ can then be seen to be given by the variance

$$\langle \Gamma^2 \rangle - \langle \Gamma \rangle^2 = \frac{2}{M} \langle \Gamma \rangle^2 \tag{18.129}$$

which decreases with increasing number of channels. In addition to the assumptions specified before it will be assumed that a large number of open channels exist such that the fluctuations in the widths become small.

18.4.2 Smooth and fluctuating parts of the cross section

The T-matrix is split again into the prompt or direct term $T^{(P)}$ and a $T^{(FL)}$ representing the resonances and thus the fluctuations

$$T = T^{(P)} + T^{(FL)} . \tag{18.130}$$

The matrix elements of $T^{(FL)}$ are assumed to be random numbers with random relative phases between $T^{(P)}$ and $T^{(FL)}$. For the average quantities this implies

$$\overline{T} = \overline{T^{(\mathrm{P})}} = T^{(\mathrm{P})} \qquad (18.131)$$
$$\overline{T^{(\mathrm{FL})}} = 0,$$

(the identification $\overline{T^{(\mathrm{P})}} = T^{(\mathrm{P})}$ is valid if the prompt channel is identified as that of the optical model). For the absolute squares, on the other hand, one gets

$$\overline{|T|^2} = \overline{|T^{(\mathrm{P})}|^2} + \overline{|T^{(\mathrm{FL})}|^2}, \qquad (18.132)$$

implying

$$\overline{\sigma} = \sigma^{(\mathrm{P})} + \overline{\sigma^{(\mathrm{FL})}} \qquad (18.133)$$

for the cross section. The fluctuating part of the T-matrix, when calculated with energy normalized wave functions attains the form

$$2\pi T^{(\mathrm{FL})}_{\alpha\beta} = e^{i(\delta_\alpha + \delta_\beta)} \sum_s e^{i\phi_s(\alpha,\beta)} \frac{g_s(\alpha) g_s(\beta)}{E - E_s + (i/2)(\Gamma_s + \Gamma'_s)}, \qquad (18.134)$$

in generalization of (18.84), allowing for a possible phase of the compound state $|\Phi_s\rangle$. The overall phase factor is a relict of the prompt channel and will drop out finally. The $g_s(\alpha)$ is determined by the coupling matrix element of the $|\Phi_s\rangle$ with the channel state $|\psi^{(+)}_\alpha\rangle$

$$g_s(\alpha) = \sqrt{2\pi} \left| \langle \Phi_s | H_{QP} | \psi^{(+)}_\alpha \rangle \right| = \sqrt{\Gamma_{s\alpha}}. \qquad (18.135)$$

The $\Gamma_{s\alpha}$ is the width the state $|\Phi_s\rangle$ gets through the channel α, and the total width is given by

$$\Gamma_s = \sum_\alpha \Gamma_{s\alpha}. \qquad (18.136)$$

The form (18.134) is written such that it is also valid for overlapping resonances for which the Γ'_s are finite but subject to the condition $\sum_s \Gamma'_s = 0$, mind (18.94).

For the sake of simplicity we discard the angular momentum quantum numbers. Of course, it is understood that in (18.134) only those compound states contribute which have the same conserved quantum numbers as the channel wave function. Dropping all quantum numbers besides the channel numbers the fluctuating cross section takes on the form (18.101), viz

$$\overline{\sigma^{(\mathrm{FL})}_{\alpha\beta}} = \frac{\pi}{k_\alpha^2} \overline{\left| 2\pi T^{(\mathrm{FL})}_{\alpha\beta} \right|^2}. \qquad (18.137)$$

After inserting the T-matrix of (18.134) one gets a double sum over s and s', with a phase factor $\phi_s - \phi_{s'}$. Assuming that the average is to be performed for an ensemble for which the phases are strict random numbers and that no correlations exist among the amplitudes $g_s(\gamma)$, only the diagonal terms survive, such that one gets

$$\overline{\sigma_{\alpha\beta}^{(FL)}} = \frac{\pi}{k_\alpha^2} \sum_s \overline{\left\{ \frac{\Gamma_{s\alpha}\Gamma_{s\beta}}{(E-E_s)^2 + (1/4)(\Gamma_s + \Gamma'_s)^2} \right\}}. \tag{18.138}$$

It may be mentioned in passing that not all of these assumptions are equally well fulfilled in actual reactions. We will return to the problem of correlations of the $g_s(\gamma)$ in Section 18.4.4.

The average appearing in (18.138) will now be evaluated when many channels are open. In this case any function of the widths Γ_k may be replaced by the same function of the average widths, $\overline{f(\Gamma_k)} = f(\overline{\Gamma_k})$. Moreover one may expect the individual ones to be uncorrelated, viz that

$$\overline{\Gamma_{s\alpha}\Gamma_{s\beta}} = \overline{(\Gamma_{s\alpha})}\,\overline{(\Gamma_{s\beta})}. \tag{18.139}$$

According to the statistical assumptions, the averages appearing here, $\overline{\Gamma_{s\alpha}}$ and $\overline{\Gamma_{s\beta}}$ may be considered independent of the states s. One may thus identify the $\overline{\Gamma_\alpha} = \overline{\Gamma_{s\alpha}}$ as the *average width of the channel* α. It is understood that these widths may depend on energy as well as on all the other conserved quantum numbers. In particular, any quantity associated with a channel α will depend on the channel energy ϵ_α. Of course, one may wish to allow for a certain finite range dE in energy the population of which being determined by the level density $\Omega(E)$. This is necessary in particular in the regime of overlapping resonances.

Under these conditions the cross section (18.138) becomes

$$\overline{\sigma_{\alpha\beta}^{(FL)}} = \frac{\pi}{k_\alpha^2} \overline{\Gamma_\alpha}\,\overline{\Gamma_\beta}\,\Sigma, \tag{18.140}$$

with

$$\Sigma = \sum_s \overline{\left[(E-E_s)^2 + (1/4)\left(\Gamma_s + \Gamma'_s\right)^2\right]^{-1}}. \tag{18.141}$$

By applying to the condition $\sum_s \Gamma'_s = 0$ the same reasoning which led us to the conclusion that $\overline{\Gamma_\alpha} = \overline{\Gamma_{s\alpha}}$ one realizes that $\overline{\Gamma'} \equiv \overline{\Gamma'_s} = 0$. Likewise the average $\overline{\Gamma_s}$ simply becomes the average width

$$\overline{\Gamma} = \sum_\beta \overline{\Gamma_\beta}. \tag{18.142}$$

The Σ appearing in (18.140) may be evaluated to

$$\Sigma = \frac{2\pi}{D_C \overline{\Gamma}} \quad \text{for} \quad \Delta E \gg D_C, \overline{\Gamma}, \tag{18.143}$$

with D_C being the average spacing of levels. This follows when in (18.141) the sum over discrete states is replaced by the continuum limit,

$$\sum_s \frac{1}{(E-E_s)^2 + (\overline{\Gamma}/2)^2} \to \frac{2}{D_C \overline{\Gamma}} \int_{E-\Delta E/2}^{E+\Delta E/2} \frac{(\overline{\Gamma}/2) dE_s}{(E-E_s)^2 + (\overline{\Gamma}/2)^2}. \tag{18.144}$$

Assuming the ΔE to be much larger than $\overline{\Gamma}$ (18.143) is seen to hold true. The fluctuating cross section of (18.140) thus turns into

$$\overline{\sigma_{\alpha\beta}^{(FL)}} = \frac{\pi}{k_\alpha^2} \frac{2\pi}{D_C} \frac{\overline{\Gamma_\alpha} \overline{\Gamma_\beta}}{\overline{\Gamma}}. \tag{18.145}$$

It should be noted once more that in deriving these relations it has always been assumed that many channels are open, such that the approximation $f(\overline{\Gamma_k}) = f(\overline{\Gamma_k})$ could be used at any step. If this is not the case, formula (18.145) has to replaced by

$$\overline{\sigma_{\alpha\beta}^{(FL)}} = \frac{\pi}{k_\alpha^2} \frac{2\pi}{D_C} \overline{\left(\frac{\Gamma_\alpha \Gamma_\beta}{\Gamma + \Gamma'}\right)}. \tag{18.146}$$

The difference between these two expressions may be phrased in terms of a *width correction factor* $W_{\alpha\beta}$, defined such that after multiplication of (18.145) by $W_{\alpha\beta}$ the latter expression turns into (18.146). This correction factor $W_{\alpha\beta}$ is smaller than unity for $\alpha \neq \beta$ but it increases the compound elastic cross section. As it is important only near thresholds and for reactions with few open channels this effect will be discarded in the following. For more details we like to refer to (Feshbach, 1992) and (Gadioli and Hodgson, 1992).

The forms (18.140) or (18.145) allows one to deduce other useful relations. The total cross section for forming the compound nucleus through the channel α may be written as

$$\overline{\sigma_\alpha^{(C)}} \equiv \sum_\beta \overline{\sigma_{\alpha\beta}^{(FL)}} = \frac{\pi}{k_\alpha^2} \overline{\Gamma_\alpha} \overline{\Gamma} \Sigma = \frac{\pi}{k_\alpha^2} \frac{2\pi}{D_C} \overline{\Gamma_\alpha}. \tag{18.147}$$

It is worth stressing that this formation cross section is solely determined by the decay rate and the level density $1/D_C$ of the compound nucleus (besides the trivial pre-factor). Such a situation is no longer given for pre-compound reactions which are discussed in Chapter 9. Inverting the last expression in (18.147) allows one to evaluate the average channel width $\overline{\Gamma_\alpha}$ in terms of the channel energy $\epsilon_\alpha = \hbar^2 k_\alpha^2 / 2\mu_\alpha$, the level density $\Omega_C = 1/D_C$ of the compound nucleus and its formation cross section,

$$\overline{\Gamma_\alpha} = \frac{\overline{\sigma_\alpha^{(C)}}}{\Omega_C} \frac{\mu_\alpha \epsilon_\alpha}{\pi^2 \hbar^2}. \tag{18.148}$$

In Section 8.1 this formula is used to calculated the decay rate of the compound nucleus. Combining (18.140) or (18.145) with (18.147) the fluctuating cross section becomes

$$\overline{\sigma_{\alpha\beta}^{(FL)}} = \overline{\sigma_\alpha^{(C)}} \frac{\overline{\Gamma_\beta}}{\overline{\Gamma}}. \tag{18.149}$$

It says that the cross section for a transition from channel α to channel β occurs in two steps, the formation of the compound nucleus through the entrance channel

α and its decay into channel β. The latter process is *independent of the entrance channel* and is governed solely by the so-called branching ratio

$$G_\beta^{(C)} = \frac{\overline{\Gamma_\beta}}{\overline{\Gamma}}. \tag{18.150}$$

Finally, we have ended up demonstrating the *Bohr hypothesis* of the compound nucleus to follow from the statistical assumptions given at the beginning of this section. Niels Bohr originally postulated this hypothesis in the 1930s to explain the sharp resonances of slow neutrons (1936). It should be realized that this situation actually corresponds to that of isolated resonances discussed in Section 18.2.4. There, this property follows more or less directly from the fact that such a resonance is governed by the quantum numbers of the compound state defining the resonance. In this sense the decay of such a state is independent of how it is produced. The situation is different for the case of many, perhaps overlapping, resonances discussed here. There this independence can only be understood if these many states are populated with equal probabilities, representing equilibration in the sense of a microcanonical ensemble. A very illuminating and profound discussion of these features is given in the article by F.L. Friedman and V.F. Weisskopf (1955), see also (Blatt and Weisskopf, 1952).

Weisskopf-Ewing relations: The formulas just derived almost immediately lead to relations associated with the names of V.F. Weisskopf and P.H. Ewing (1940) where the branching ratios are directly related to the formation probabilities of the compound nucleus. Let us first apply (18.147) to the same compound state but different entrance channels. This leads to the conclusion that the ratios

$$\frac{k_\alpha^2 \overline{\sigma_\alpha^{(C)}}}{\overline{\Gamma_\alpha}} = \frac{k_\gamma^2 \overline{\sigma_\gamma^{(C)}}}{\overline{\Gamma_\gamma}} = U(E, J, ...) \tag{18.151}$$

only depend on the energy and the conserved quantum numbers but *not on details of the channel*. Therefore the branching ratio can be brought to the form

$$G_\beta^{(C)} = \frac{k_\beta^2 \overline{\sigma_\beta^{(C)}}}{\sum_\gamma k_\gamma^2 \overline{\sigma_\gamma^{(C)}}}. \tag{18.152}$$

Hence, by way of (18.149) the average fluctuating cross section $\overline{\sigma_{\alpha\beta}^{(FL)}}$ is solely determined by the formation probabilities $\overline{\sigma_\gamma^{(C)}}$.

In practical applications of (18.152) one needs to make use of level densities both in the numerator and in the denominator. In the latter the sum over discrete quantum numbers has to be completed by an integral over energies. Such a generalization of (18.150) together with the expression (18.149) can be found in Section IV.9 of (Feshbach, 1992) where due account is also paid to angular momentum coupling, see also Chapter 3 of (Gadioli and Hodgson, 1992). We will make use of these extensions for the calculation of the decay rates in Section 8.1.1.

Reciprocity and detailed balance: Another immediate consequence of the relations (18.140) or (18.145)is

$$k_\alpha^2 \overline{\sigma_{\alpha\beta}^{(\mathrm{FL})}} = k_\beta^2 \overline{\sigma_{\beta\alpha}^{(\mathrm{FL})}}. \tag{18.153}$$

In (Blatt and Weisskopf, 1952) this is called the *reciprocity theorem* of nuclear reactions. Truth is that here it refers to the *averaged fluctuating* cross section; its derivation involves the average channel widths $\overline{\Gamma_\alpha}$. A similar relation can be shown to follow for *transitions between individual channels* whenever time reversal invariance is given (Blatt and Weisskopf, 1952), (Fröbrich and Lipperheide, 1996). In this sense the principles of reciprocity and microreversibility discussed in Section 18.1.1 are identical. Notice, please, that the latter principle was not invoked to get the form (18.153), neither did we use any kind of perturbation theory. As pointed out by Feshbach (1992) microreversibility need not survive the averaging procedures applied. Incidentally, in this connection he refers to *detailed balance*. As a matter of fact, at times these concepts are used synonymously. Indeed, applying the relation (18.28) to the transition probability $P_{\alpha\beta}$ of (18.13) immediately gives

$$\Omega_\alpha P_{\alpha\beta} = \Omega_\beta P_{\beta\alpha} \tag{18.154}$$

when $\beta \to \alpha$ is the time reversed process of $\alpha \to \beta$. This form identifies the relation of detailed balance in closer sense, as used also in statistical mechanics, see e.g. (van Kampen, 2001). For the microcanonical ensemble, specified within a certain range ΔE of energy, the probability $w(\epsilon_\alpha)$ of finding a state of energy ϵ_α is given by $w(\epsilon_\alpha) = \rho \Omega_\alpha$ with ρ being constant in the interval at stake, see (22.63) and (22.64). Hence, if both α and β are states of this type relation (18.154) implies

$$w_\alpha P_{\alpha\beta} = w_\beta P_{\beta\alpha}. \tag{18.155}$$

A variant of this form for a canonical ensemble is given by equation (23.93) from which finally the fluctuation dissipation theorem of linear response theory follows.

18.4.3 Hauser–Feshbach theory

The Weisskopf relations show that the fluctuating cross section can be related to the formation probabilities, which in a sense reflect more gross properties of the compound states. It is still open as to how these quantities may be calculated theoretically. This important step is delivered by the Hauser–Feshbach theory (1952). The clue is to identify first the prompt channels as those determined by a properly chosen optical model, and hence to define $T^{(\mathrm{P})}$ as the $T^{(\mathrm{opt})}$. Next one generalizes the definition (18.47) of the transmission coefficient to

$$\mathcal{T}_\alpha = 1 - \sum_\beta \left| S_{\alpha\beta}^{(\mathrm{opt})} \right|^2. \tag{18.156}$$

Using unitarity of the S-matrix, $SS^\dagger = 1$ for $S = S^{(\mathrm{opt})} + S^{(\mathrm{fl})}$, it is not difficult to prove that $\mathcal{T}_\alpha = \sum_\beta \overline{\left|S^{(\mathrm{fl})}_{\alpha\beta}\right|^2}$. Under the same statistical assumptions as used above it can be shown that

$$\mathcal{T}_\alpha = \overline{\Gamma_\alpha}\,\overline{\Gamma}\,\Sigma \equiv \frac{2\pi}{D_C}\overline{\Gamma_\alpha}. \tag{18.157}$$

Hence, according to (18.147) the formation probabilities $\overline{\sigma_\alpha^{(C)}}$ can be calculated as

$$\overline{\sigma_\alpha^{(C)}} = \frac{\pi}{k_\alpha^2}\mathcal{T}_\alpha. \tag{18.158}$$

For fluctuating cross section this together with (18.140) implies

$$\overline{\sigma_{\alpha\beta}^{(\mathrm{FL})}} = \frac{\pi}{k_\alpha^2}\frac{\mathcal{T}_\alpha \mathcal{T}_\beta}{\sum_\gamma \mathcal{T}_\gamma}. \tag{18.159}$$

As said before, the importance of this formula is to be seen in the fact that the transmission coefficients can be calculated from the optical model. In practical applications, formula (18.159) has of course to be completed by adding detailed information about the channels like angular momenta, energy, etc.

18.4.3.1 *Average and fluctuating cross sections for s-waves* The basic features of the Hauser–Feshbach theory may be illuminated by a simple model described in (Friedman and Weisskopf, 1955). Take the splitting of the total cross section into one for elastic scattering and one for the reaction (or absorption), $\sigma_{\mathrm{tot}} = \sigma_{\mathrm{rea}} + \sigma_{\mathrm{el}}$, introduced in Section 18.1.3 for a partial wave analysis. Concentrating on s-waves the cross sections may be expressed by the corresponding element $S \equiv S_0$ of the S-matrix as

$$\sigma_{\mathrm{el}} = \frac{\pi}{k^2}|1 - S|^2, \quad \sigma_{\mathrm{rea}} = \frac{\pi}{k^2}\left(1 - |S|^2\right), \quad \text{and} \quad \sigma_{\mathrm{tot}} = \frac{\pi}{k^2}2\left(1 - \mathrm{Re}\{S\}\right). \tag{18.160}$$

Suppose we perform an energy average of the type associated with the optical model. For the total cross section this leads to $\overline{\sigma_{\mathrm{tot}}} = (\pi/k^2)2(1 - \mathrm{Re}\{\overline{S}\})$ and for the reaction cross section one may write $\overline{\sigma_{\mathrm{rea}}} = (\pi/k^2)(1 - |\overline{S}|^2) - \sigma_{\mathrm{fl}}$, with a fluctuation part defined as $\sigma_{\mathrm{fl}} = (\pi/k^2)(\overline{|S|^2} - |\overline{S}|^2)$. Likewise the average elastic cross section becomes $\overline{\sigma_{\mathrm{el}}} = (\pi/k^2)\left|1 - \overline{S}\right|^2 + \sigma_{\mathrm{fl}}$. Requiring $S^{\mathrm{opt}} \equiv \overline{S}$, the optical model may be defined through relations like those in (18.160),

$$\sigma_{\mathrm{el}}^{\mathrm{opt}} = \frac{\pi}{k^2}\left|1 - S^{\mathrm{opt}}\right|^2 \tag{18.161}$$

$$\sigma_{\mathrm{rea}}^{\mathrm{opt}} = \frac{\pi}{k^2}\left(1 - \left|S^{\mathrm{opt}}\right|^2\right) \tag{18.162}$$

$$\sigma_{\mathrm{tot}}^{\mathrm{opt}} = \frac{\pi}{k^2}2\left(1 - \mathrm{Re}\{S^{\mathrm{opt}}\}\right), \tag{18.163}$$

where again $\sigma_{\mathrm{tot}}^{\mathrm{opt}} = \sigma_{\mathrm{rea}}^{\mathrm{opt}} + \sigma_{\mathrm{el}}^{\mathrm{opt}}$. To establish a connection with the statistical model let us take over the analog decomposition already introduced in Section

18.2.4, by which the total compound cross section is split into a reaction (absorption) part and one for compound elastic scattering,

$$\sigma_{\text{tot}}^{(C)} = \sigma_{\text{rea}}^{(C)} + \sigma_{\text{el}}^{(C)} . \tag{18.164}$$

Following the picture of Hauser–Feshbach we should identify the average total cross section as the absorption cross of the optical model, see (18.158) together with (18.156). Hence, from (18.162) one gets

$$\overline{\sigma_{\text{tot}}^{(C)}} = \sigma_{\text{abs}}^{\text{opt}} \equiv \sigma_{\text{rea}}^{\text{opt}} = \frac{\pi}{k^2} \left(1 - \left| S^{\text{opt}} \right|^2 \right) . \tag{18.165}$$

On the other hand, the decomposition (18.164) suggests

$$\overline{\sigma_{\text{tot}}^{(C)}} = \frac{\pi}{k^2} \left(1 - \left| S^{\text{opt}} \right|^2 \right) - \sigma_{\text{fl}} + \overline{\sigma_{\text{el}}^{(C)}} . \tag{18.166}$$

Both expressions become identical only for $\sigma_{\text{fl}} = \overline{\sigma_{\text{el}}^{(C)}}$. Indeed, this relation is very much in the spirit of the statistical model, in particular when the elastic cross section consists of many resonances. Looking at the averaged elastic cross section, one gets

$$\overline{\sigma_{\text{el}}} = \sigma_{\text{el}}^{\text{opt}} + \sigma_{\text{fl}} . \tag{18.167}$$

The $\sigma_{\text{el}}^{\text{opt}}$, which behaves as a smooth, non-fluctuating function of energy, is sometimes referred to as the shape elastic cross section. Summarizing, one has

$$\sigma_{\text{el}}^{\text{opt}} = \overline{\sigma_{\text{el}}} - \overline{\sigma_{\text{el}}^{(C)}} \quad \text{and} \quad \sigma_{\text{abs}}^{\text{opt}} = \overline{\sigma_{\text{rea}}} + \overline{\sigma_{\text{el}}^{(C)}} . \tag{18.168}$$

Here, in the relation on the left, only cross sections for elastic scattering occur. The interpretation for the second relation is as follows. The $\overline{\sigma_{\text{rea}}}$ represents the contribution from *genuine reactions* or real inelastic processes in the prompt channel. The $\overline{\sigma_{\text{el}}^{(C)}}$, on the other hand, accounts for the loss of flux which simply occurs because of the very long time it takes for the re-emission from the compound states back into the incident channel, see also the discussion in Section 1.3.1.2.

18.4.4 Critique of the statistical model

Bohr's independence hypothesis is intimately related to a practically complete relaxation of all relevant degrees of freedom to equilibrium. Deviations may thus be expected in the event that this equilibrium cannot be reached because of emission processes taking place at earlier times. For instance, there may be correlations between direct and delayed processes. Indeed, in the derivation given before it was assumed that the amplitudes $g_s(\alpha)$ are random, totally uncorrelated numbers. However, the constraint of unitarity of the (full) S-matrix may require special measures, to allow for *correlations among the* $g_s(\alpha)$, for instance. Assuming that the latter appear only pairwise, in the sense of having

$$\overline{g_s^2(\alpha)g_s^2(\beta)} = \left(\overline{g_s^2(\alpha)}\right)\left(\overline{g_s^2(\beta)}\right) + \left(\overline{g_s(\alpha)g_s(\beta)}\right)\left(\overline{g_s(\beta)g_s(\alpha)}\right), \quad (18.169)$$

the Hauser–Feshbach relations (18.159) changes to

$$\overline{\sigma_{\alpha\beta}^{(\text{FL})}} = \frac{\pi}{k_\alpha^2} \frac{\mathcal{T}_\alpha \mathcal{T}_\beta + \mathcal{T}_{\alpha\beta}^2}{\sum_\gamma \mathcal{T}_\gamma}, \quad (18.170)$$

where

$$\mathcal{T}_{\alpha\beta} = X_{\alpha\beta} \operatorname{tr} X + (X)_{\alpha\beta}^2 \quad \text{with} \quad X_{\alpha\beta} = \overline{g_s(\alpha)g_s(\beta)} \quad (18.171)$$

(and $\mathcal{T}_\alpha = \mathcal{T}_{\alpha\alpha}$). For compound elastic scattering ($\alpha = \beta$) the result (18.170) is larger than (18.159) by exactly a factor of 2, which is established experimentally. For an overview of theories dealing with this topic see the review article by C. Mahaux and H.A. Weidenmüller (1979) or Section IV.8 of (Feshbach, 1992).

The decay widths increase with energy. For heavier nuclei already at a few tens of MeV one reaches the so-called "regime of overlapping resonances" where $\overline{\Gamma} \gg D$. Thus the average time $\overline{\tau}$ it takes for the decay of the nucleus by evaporation of light particles becomes much smaller than the Poincaré recurrence time τ_{rec},

$$\overline{\tau} = \frac{\hbar}{\overline{\Gamma}} \ll \tau_{\text{rec}} \sim \frac{2\pi\hbar}{D}. \quad (18.172)$$

At larger excitations there cannot be complete equilibrium simply because the nucleus may decay before the latter is reached. This is the realm of "pre-equilibrium" or "pre-compound" reactions which is addressed in Chapter 9.

Exercises

1. (i) Convince yourself that for a free particle the resolvent $\hat{G}_0^{(+)}(E)$ of (18.7) has the following coordinate representation

$$\langle \mathbf{r} \mid \hat{G}_0^{(+)}(E) \mid \mathbf{s} \rangle = -\frac{\mu}{2\pi\hbar^2} \frac{e^{ik|\mathbf{r}-\mathbf{s}|}}{|\mathbf{r}-\mathbf{s}|} = G_0^{(+)}(E; \mathbf{r}, \mathbf{s}) \quad (18.173)$$

and satisfies the equation

$$\frac{\hbar^2}{2\mu} \left(\Delta + k^2\right) G_0^{(+)}(E; \mathbf{r}, \mathbf{s}) = \delta(\mathbf{r} - \mathbf{s}). \quad (18.174)$$

(ii) Make use of

$$\lim_{r\to\infty} \frac{e^{ik|\mathbf{r}-\mathbf{s}|}}{|\mathbf{r}-\mathbf{s}|} \sim \frac{e^{ikr}}{r} e^{-i\mathbf{k}'\cdot\mathbf{s}} + O(s/r) \quad \text{with} \quad \mathbf{k}' = k\mathbf{r}/r, \quad (18.175)$$

and prove (18.5) with the scattering amplitude in the form (18.8).

2. Diffraction at the black sphere (or disk) of radius R at large energies: Idealize the S-matrix by

$$S_l = \begin{cases} 1 & \text{for} \quad l > L \equiv kR \gg 1 \\ 0 & \text{for} \quad l \leq L \equiv kR \gg 1 \end{cases} \tag{18.176}$$

such that all partial waves with an "impact parameter" $b = l/k \leq R$ are supposed to be *totally absorbed*. Convince yourself that the integrated elastic cross section from (18.43) is identical to the reaction cross section (or inelastic one) defined in (18.45),

$$\sigma_{\text{el}} = \sigma_{\text{rea}} = \frac{\pi}{k^2} \sum_{l=0}^{L} (2l+1) \approx \frac{\pi}{k^2} l^2 = \pi R^2. \tag{18.177}$$

Why is the total cross section larger than the area of the disk seen by the beam particles, $\sigma_{\text{tot}} = 2\pi R^2$?

3. Prove the relations found in Section 4.3.2 for width and energy shift of a strength function. For a $\hat{H} = \hat{H}_0 + \hat{V}$ take the unperturbed eigenstates of (4.30), $\hat{H}_0|s\rangle = E_s^0|s\rangle$ and $\hat{H}_0|\mu\rangle = E_\mu^0|\mu\rangle$ and assume the coupling matrix elements to be $V_{s\mu} = \langle s|V|\mu\rangle = \langle \mu|V|s\rangle^* = V_{\mu s}^*$ and $V_{ss} = V_{\mu\mu'} = 0$. Interpret the $|\mu\rangle$ as the "scattering states" and $|s\rangle$ as the special state Φ_s. Define projectors by $\hat{P} = \sum_\mu |\mu\rangle\langle\mu|$ and $\hat{Q} = |s\rangle\langle s|$ such that the various parts of the Hamiltonian read $\hat{H}_{PP} = \sum_\mu E_\mu^0 |\mu\rangle\langle\mu|$ and $\hat{H}_{QQ} = E_s^0|s\rangle\langle s|$ as well as $\hat{H}_{QP} = \sum_\mu V_{s\mu}|s\rangle\langle\mu| = \hat{H}_{PQ}^\dagger$. Calculate the partial widths of (18.79) to $\Gamma_{s\mu} = 2\pi|V_{s\mu}|^2 \delta(E - E_\mu^0)$. Show that after introducing average matrix elements $V_{s\mu} \to v(E)$ the total width takes on the simple form

$$\Gamma_s = 2\pi \sum_\mu |V_{s\mu}|^2 \delta(E - E_\mu^0) \equiv 2\pi \frac{v^2(E)}{D(E)}$$

with $1/D(E)$ being the density of μ-states. Compare this result with (4.49) and show that the energy shift (18.77) becomes

$$\Delta_s(E) = \sum_\mu |V_{s\mu}|^2 \frac{\mathcal{P}}{E - E_\mu^0}$$

19
DENSITY OPERATORS AND WIGNER FUNCTIONS

19.1 The many-body system

19.1.1 Hilbert states of the many-body system

We will use the shorthand notation $|m\rangle \equiv |m_1, m_2, m_3, \dots\rangle$ to classify many-body states which depend on a whole set of quantum numbers by a single letter m. The normalization condition then reads

$$\delta_{nm} = \langle n|m\rangle \equiv \langle n_1, n_2 \dots |m_1, m_2 \dots\rangle = \prod_\mu \delta_{n_\mu m_\mu} \tag{19.1}$$

In coordinate representation such a state is given by the wave function

$$\Phi_m(\mathbf{x}_1, \mathbf{x}_2 \dots \mathbf{x}_A) = \langle \mathbf{x}_1, \mathbf{x}_2 \dots \mathbf{x}_A | m \rangle. \tag{19.2}$$

With volume element $d^A\mathbf{x} = \prod_{i=1}^A d\mathbf{x}_i$ the scalar product is defined as:

$$\langle n|m\rangle = \int d^A\mathbf{x}\, \Phi_n^*(\mathbf{x}_1, \mathbf{x}_2 \dots \mathbf{x}_A) \Phi_m(\mathbf{x}_1, \mathbf{x}_2 \dots \mathbf{x}_A) \tag{19.3}$$

The completeness relations for the states $|m\rangle$ may be written as

$$\sum_m |m\rangle\langle m| \equiv \sum_{m_1, m_2 \dots} |m_1 m_2 \dots\rangle\langle m_1, m_2 \dots| = \hat{1}_A, \tag{19.4}$$

and

$$\int d^A\mathbf{x}\, |\mathbf{x}_1, \mathbf{x}_2 \dots \mathbf{x}_A\rangle\langle \mathbf{x}_1, \mathbf{x}_2 \dots \mathbf{x}_A| = \hat{1}_A \tag{19.5}$$

in coordinate space. The symbol $\hat{1}_A$ represents the unity operator in the Hilbert space for A particles. If spin and isospin need not appear explicitly, one has for wave functions:

$$\sum_m \Phi_m^*(\mathbf{x}_1, \mathbf{x}_2 \dots \mathbf{x}_A) \Phi_m(\mathbf{x}'_1, \mathbf{x}'_2 \dots \mathbf{x}'_A) = \prod_{i=1}^A \delta(\mathbf{x}_i - \mathbf{x}'_i) \tag{19.6}$$

19.1.2 Density operators and matrices

Given a (pure) state $|\psi\rangle$ the expectation value of any observable $\hat{O}$ can be expressed through the Hermitian density operator $\hat{\rho} = |\psi\rangle\langle\psi|$ in the form

$$\langle \hat{O} \rangle = \langle \psi | \hat{O} | \psi \rangle = \operatorname{tr} \hat{\rho} \hat{O}, \tag{19.7}$$

if only $|\psi\rangle$ is normalized to unity. The trace may be defined with respect to any complete orthonormal set as

$$\operatorname{tr} \hat{X} = \sum_m \langle m|\hat{X}|m\rangle. \tag{19.8}$$

(Needless to say that with the convention set up above these formulas are valid independently of particle number.) For the space part of an operator the trace may also be performed with the states $|\mathbf{x}_1, \mathbf{x}_2 \ldots \mathbf{x}_A\rangle$ in which case it reads (for A particles)

$$\operatorname{tr} \hat{X} = \int d^A\mathbf{x}\, \langle \mathbf{x}_1, \mathbf{x}_2 \ldots \mathbf{x}_A|\hat{X}|\mathbf{x}_1, \mathbf{x}_2 \ldots \mathbf{x}_A\rangle. \tag{19.9}$$

Generalizations to the spin and isospin degrees of freedom are more or less straightforward.

This density operator $\hat{\rho}$ should be normalizable and Hermitian, which means to require

$$\operatorname{tr} \hat{\rho} = 1 \quad \text{and} \quad \hat{\rho} = \hat{\rho}^\dagger. \tag{19.10}$$

For a pure state it fulfills the condition

$$\hat{\rho}^2 = \hat{\rho}. \tag{19.11}$$

Evidently, the normalization to unity is in accord with (19.7). Given any basis $|m\rangle$ the $\hat{\rho}$ can be written as

$$\hat{\rho} = \sum_{nm} |m\rangle\langle m|\hat{\rho}|n\rangle\langle n| = \sum_{nm} |m\rangle \rho_{mn} \langle n|, \tag{19.12}$$

with the *density matrix* $\rho_{mn} = \langle m|\hat{\rho}|n\rangle$. According to (19.10) and (19.11) it fulfills the conditions $\sum_m \rho_{mm} = 1$ and $\rho_{mn} = (\rho_{nm})^*$ as well as

$$\sum_{n'} \rho_{mn'} \rho_{n'n} = \rho_{mn} \quad \text{for a pure state}. \tag{19.13}$$

If the $\hat{\rho}$ is a functional of the Hamilitonian $\hat{H}$ the density matrix ρ_{mn} is diagonal in the basis of eigenstates:

$$\rho_{mn} = \rho_m \delta_{mn} \quad \text{for} \quad \hat{\rho} = \hat{\rho}[\hat{H}] \quad \text{with} \quad \hat{H}|m\rangle = E_m|m\rangle. \tag{19.14}$$

Notice that because of (19.11) the diagonal matrix elements are positive numbers $\rho_m = \langle m|\hat{\rho}|m\rangle = \langle m|\hat{\rho}^2|m\rangle \geq 0$, and because of the normalization they must be smaller than one. Hence, they represent the probabilities for the system to be in the eigenstate $|m\rangle$.

Density operators may also be defined for *statistical mixtures* for which property (19.11) is no longer required, and hence (19.13) for the corresponding density matrix. Prime examples are the density operators associated with the equilibrium distributions of statistical mechanics, for which one has

$$\hat{\rho}_{\rm ST} = \sum_m \rho_m^{\rm ST} |m\rangle\langle m| \quad \text{with} \quad 0 \le \rho_m^{\rm ST} \le 1. \tag{19.15}$$

The $\hat{\rho}_{\rm ST}$ is still Hermitian and normalized to unity but (19.11) and (19.13) are replaced by:

$$\hat{\rho}_{\rm ST}^2 \ne \hat{\rho}_{\rm ST} \quad \text{and} \quad \operatorname{tr} \hat{\rho}_{\rm ST}^2 = \sum_m \left(\rho_m^{\rm ST}\right)^2 \le 1. \tag{19.16}$$

Let us finally address explicitly the density matrix in coordinate space. In complete accordance with the general expression $\rho_{mn} = \langle m|\hat{\rho}|n\rangle$ we define the following (complex) function

$$\rho(\mathbf{x}_1, \mathbf{x}_2, \ldots; \mathbf{x'}_1, \mathbf{x'}_2, \ldots) = \langle \mathbf{x}_1, \mathbf{x}_2, \ldots |\hat{\rho}| \mathbf{x'}_1, \mathbf{x'}_2, \ldots \rangle. \tag{19.17}$$

As before, its diagonal elements $\rho(\mathbf{x}_1, \mathbf{x}_2, \ldots; \mathbf{x}_1, \mathbf{x}_2, \ldots)$ represent the probability density of finding particles at $\mathbf{x}_1, \mathbf{x}_2, \ldots$, or in a small volume element around it, rather. In the basis $|m\rangle$ in which the density matrix is diagonal one has

$$\rho(\mathbf{x}_1, \mathbf{x}_2, \ldots; \mathbf{x}_1, \mathbf{x}_2, \ldots) = \sum_m \rho_m \left| \Phi_m(\mathbf{x}_1, \mathbf{x}_2, \ldots) \right|^2. \tag{19.18}$$

19.1.3 Reduction to one- and two-body densities

Starting from a density operator $\hat{\rho}(A)$ of the A particle system one may construct a density operator $\hat{\rho}^{(a)}$ for the a-particle system by performing the trace over the remaining $A - a$ degrees of freedom. This could be done such that the new operator is normalized to unity again. For the one-body operator in nuclear physics one often adopts a slightly different normalization by writing

$$\hat{\rho}^{(1)} = A \operatorname{tr}|_{2\ldots A} \hat{\rho}(A) \quad \text{with} \quad \operatorname{tr} \hat{\rho}^{(1)} = A, \tag{19.19}$$

which we will take over for our purposes. The definition (19.19) is naturally generalized to the two-body case by writing

$$\hat{\rho}^{(2)} = A(A-1) \operatorname{tr}|_{3\ldots A} \hat{\rho}(A) \quad \text{with} \quad \operatorname{tr} \hat{\rho}^{(2)} = A(A-1). \tag{19.20}$$

The benefit we gain with this definition is seen when one wants to calculate averages of one and two-body operators. One speaks of a one-body operator $\hat{O}^{(1)}$ when it is of the form

$$\hat{O}^{(1)} = \sum_{i=1}^{A} \hat{o}_i^{(1)}, \tag{19.21}$$

where the $\hat{o}_i^{(1)}$ only acts on the coordinates of particle i. Likewise, a two-body operator $\hat{O}^{(2)}$ consists of a sum of terms like

$$\hat{O}^{(2)} = \sum_{i<j}^{A} \hat{o}_{ij}^{(2)} \tag{19.22}$$

where the $\hat{o}_{ij}^{(2)}$ is an interaction or coupling between particles i and j. The average of a one-body operator with the A-body density $\hat{\rho}(A)$ then turns out to be

$$\langle \Phi | \hat{O}^{(1)} | \Phi \rangle = \operatorname{tr} \hat{\rho}(A) \hat{O}^{(1)} = \operatorname{tr} \hat{\rho}^{(1)} \hat{o}^{(1)} \tag{19.23}$$

where in the expression on the very right only one particle is involved. Likewise, for a two-body operator one gets

$$\langle \Phi | \hat{O}^{(2)} | \Phi \rangle = \operatorname{tr} \hat{\rho}(A) \hat{O}^{(2)} = \frac{1}{2} \operatorname{tr} \hat{\rho}^{(2)} \hat{o}_{12}^{(2)} . \tag{19.24}$$

These are consequences of the fact that the wave functions involved are totally antisymmetric in a permutation of their arguments. As they appear twice, an interchange of the order of particles does not modify the averages. Since the sum $\sum_{i=1}^{A} \hat{o}_i^{(1)}$ is symmetric in the index i one gets the result

$$\begin{aligned}
\langle \Phi | \hat{O}^{(1)} | \Phi \rangle &= \sum_{i=1}^{A} \int d^3x_1 \ldots d^3x_A \Phi^*(\mathbf{x}_1, \mathbf{x}_2, \ldots, \mathbf{x}_A) \hat{o}_i^{(1)} \Phi(\mathbf{x}_1, \mathbf{x}_2, \ldots, \mathbf{x}_A) \\
&= A \int d^3x_1 \ldots d^3x_A \Phi^*(\mathbf{x}_1, \mathbf{x}_2, \ldots, \mathbf{x}_A) \hat{o}_1^{(1)} \Phi(\mathbf{x}_1, \mathbf{x}_2, \ldots, \mathbf{x}_A) ,
\end{aligned} \tag{19.25}$$

which can be recast in the form (19.23). A similar process of reasoning may be performed for two-body operators to prove (19.24).

Let us discuss a few examples in coordinate representation. For a general one-body operator one gets $\operatorname{tr} \hat{\rho}^{(1)} \hat{o}^{(1)} = \int d^3x \langle \mathbf{x} | \hat{\rho}^{(1)} \hat{o}^{(1)} | \mathbf{x} \rangle$. This form becomes particularly simple if the $\hat{o}_i^{(1)}$ does not depend on momentum,

$$\operatorname{tr} \hat{\rho}^{(1)} \hat{o}^{(1)} = \int d^3x \, \rho^{(1)}(\mathbf{x}; \mathbf{x}) o^{(1)}(\mathbf{x}) \qquad \text{for} \qquad \hat{o}^{(1)} = \hat{o}^{(1)}(\hat{\mathbf{x}}) . \tag{19.26}$$

For a $\hat{o}^{(1)} = \hat{o}^{(1)}(\hat{\mathbf{p}})$ one needs to take care of the coordinate representation of the momentum operator $\hat{\mathbf{p}}$:

$$\langle \mathbf{x} | \hat{\mathbf{p}} | \mathbf{x}' \rangle = \frac{\hbar}{i} \nabla_{\mathbf{x}} \delta(\mathbf{x} - \mathbf{x}') = -\frac{\hbar}{i} \nabla_{\mathbf{x}'} \delta(\mathbf{x} - \mathbf{x}') . \tag{19.27}$$

It follows that the expectation value of the kinetic energy can be written as

$$\left\langle \frac{(\hat{\mathbf{p}})^2}{2m} \right\rangle = \frac{\hbar^2}{2m} \int d^3x \, d^3x' \, \delta(\mathbf{x}' - \mathbf{x}) \nabla_{\mathbf{x}} \nabla_{\mathbf{x}'} \rho^{(1)}(\mathbf{x}; \mathbf{x}') . \tag{19.28}$$

The average of any two-body interaction of the form $\hat{V} = \sum_{i<j} V(\mathbf{x}_i - \mathbf{x}_j) = (1/2) \sum_{i \neq j} V(\mathbf{x}_i - \mathbf{x}_j)$ is given by

$$\operatorname{tr} \hat{\rho}^{(2)} \hat{V} = \int d^3x \, d^3x' \, \langle \mathbf{x}, \mathbf{x}' | \hat{\rho}^{(2)} | \mathbf{x}, \mathbf{x}' \rangle V(\mathbf{x} - \mathbf{x}') . \tag{19.29}$$

19.2 Many-body functions from one-body functions

With the help of four dimensional spinors $\chi(m_s, m_\tau)$ for spin and isospin the one-body wave function $\varphi_\alpha(\mathbf{x})$ can be written as:

$$\langle \mathbf{x} | \varphi_\alpha \rangle \equiv \varphi_\alpha(\mathbf{x}) = \psi_n(\mathbf{x}) \chi(m_s, m_\tau) \qquad \alpha \equiv \{\{n\}, m_s, m_\tau\}. \qquad (19.30)$$

The $\{n\}$ represents the set of quantum numbers which is necessary to classify completely the part $\psi_n(\mathbf{x})$ of the wave function in three-dimensional space. In this notation, the $\varphi_\alpha(\mathbf{x})$ still has the spinor components, but the $\psi_n(\mathbf{x})$ is simply a function of $\mathbf{x}$. For the corresponding "bra" one has

$$\langle \varphi_\alpha | \mathbf{x} \rangle \equiv \varphi_\alpha^\dagger(\mathbf{x}) = \chi^\dagger(m_s, m_\tau) \psi_n^*(\mathbf{x}). \qquad (19.31)$$

The overlap of two such generalized wave functions is then given by

$$\begin{aligned} \langle \varphi_\alpha | \varphi_{\alpha'} \rangle &= \left(\int d^3x \, \psi_n^*(\mathbf{x}) \psi_{n'}(\mathbf{x}) \right) \chi^\dagger(m_s, m_\tau) \chi(m_s', m_\tau') \\ &= \delta_{nn'} \delta_{m_s m_s'} \delta_{m_\tau m_\tau'} \equiv \delta_{\alpha \alpha'}, \end{aligned} \qquad (19.32)$$

where in the last line we have assumed orthonormalized states.

In the following, for the single particle functions introduced in (19.30) we will use the shorthand notation $\varphi_{\alpha_i}(i)$. The i inside the bracket just specifies that we look at the coordinates and spins of particle number i. These functions will be assumed to deliver an orthonormal, complete set in the one-body space. A possible A-body function is obtained through the *direct product*:

$$\Phi_{\alpha_1 \ldots \alpha_A}^{\mathrm{DP}}(1, \ldots, A) = \varphi_{\alpha_1}(1) \varphi_{\alpha_2}(2) \cdots \varphi_{\alpha_A}(A). \qquad (19.33)$$

Such a function does not obey any symmetry with respect to the permutation $\mathcal{P}$ of the particles. For fermions we need totally *antisymmetric ones*, a symmetry which is obviously achieved by the *Slater determinant*

$$\begin{aligned} \Phi_{\alpha_1 \ldots \alpha_A}^{\mathrm{SL}}(1, \cdots, A) &= \frac{1}{\sqrt{A!}} \sum_{\mathcal{P}} (-1)^{\mathcal{P}} \mathcal{P} \varphi_{\alpha_1}(1) \varphi_{\alpha_2}(2) \cdots \varphi_{\alpha_A}(A) \\ &= \frac{1}{\sqrt{A!}} \begin{vmatrix} \varphi_{\alpha_1}(1) & \cdots & \varphi_{\alpha_1}(A) \\ \vdots & \ddots & \vdots \\ \varphi_{\alpha_A}(1) & \cdots & \varphi_{\alpha_A}(A) \end{vmatrix}. \end{aligned} \qquad (19.34)$$

The states given in (19.33) and (19.34) form complete sets of orthonormal states in Hilbert spaces associated with A particles (see exercise 1). In passing it may be mentioned that a convenient way of representing many-body states of independent particle motion is delivered by the formalism of second quantization, which may be found in Section 26.8.

19.2.1 One- and two-body densities

Applying (19.19) to the Slater determinant (19.34) and expanding the latter, one readily derives the following simple relation for the one-body density:

$$\hat{\rho}_{\text{SL}}^{(1)} = A \, \text{tr}|_{2\ldots A} \, |\Phi_{\alpha_1\ldots\alpha_A}^{\text{SL}}(1,\cdots,A)\rangle\langle\Phi_{\alpha_1\ldots\alpha_A}^{\text{SL}}(1,\cdots,A)|$$

$$= \sum_{k,l=1}^{A} |\varphi_{\alpha_l}(1)\rangle\langle\varphi_{\alpha_k}(1)|(-1)^{k+l} \left[\text{tr}|_{2\ldots A} \, |\Phi_{\alpha_1\ldots\alpha_{l-1}\alpha_{l+1}\ldots\alpha_A}^{\text{SL}}\rangle\langle\Phi_{\alpha_1\ldots\alpha_{k-1}\alpha_{k+1}\ldots\alpha_A}^{\text{SL}}| \right]$$

$$= \sum_{l=1}^{A} |\varphi_{\alpha_l}(1)\rangle\langle\varphi_{\alpha_l}(1)| \equiv \sum_{l=1}^{A} |\alpha_l\rangle\langle\alpha_l|. \qquad (19.35)$$

Notice that in the last expression no reference to a specific particle is found. With this form it becomes particularly easy (i) to see that the trace over $\hat{\rho}^{(1)}$ gives A (as it should be according to (19.19)), and (ii) to prove that the condition (19.10) for a pure state is fulfilled. The two-body density is seen to be given by

$$\hat{\rho}_{\text{SL}}^{(2)} = A(A-1) \, \text{tr}|_{3\ldots A} \, |\Phi_{\alpha_1\ldots\alpha_A}^{\text{SL}}(1,\cdots,A)\rangle\langle\Phi_{\alpha_1\ldots\alpha_A}^{\text{SL}}(1,\cdots,A)|. \qquad (19.36)$$

Again using expansions of determinants one obtains (see exercise 2)

$$\hat{\rho}_{\text{SL}}^{(2)} = \sum_{m,l=1}^{A} |\varphi_{\alpha_l}(1)\rangle|\varphi_{\alpha_m}(2)\rangle \Big[\langle\varphi_{\alpha_m}(2)|\langle\varphi_{\alpha_l}(1)| - \langle\varphi_{\alpha_l}(2)|\langle\varphi_{\alpha_m}(1)| \Big]$$

$$= \sum_{m,l=1}^{A} \Big[|\varphi_{\alpha_l}(1)\rangle|\varphi_{\alpha_m}(2)\rangle - |\varphi_{\alpha_m}(1)\rangle|\varphi_{\alpha_l}(2)\rangle \Big] \langle\varphi_{\alpha_m}(2)|\langle\varphi_{\alpha_l}(1)|. \qquad (19.37)$$

Notice that the second form is obtained from the first one by interchanging $m \leftrightarrow l$ in the second term. With a similar manipulation more symmetric forms can be derived, as for instance the very plausible generalization of the form for the one-body density,

$$\hat{\rho}_{\text{SL}}^{(2)} = \sum_{m,l=1}^{A} |\Phi_{\alpha_l\alpha_m}^{\text{SL}}(1,2)\rangle\langle\Phi_{\alpha_l\alpha_m}^{\text{SL}}(1,2)|, \qquad (19.38)$$

which involves the two-dimensional Slater determinant

$$|\Phi_{\alpha_l\alpha_m}^{\text{SL}}(1,2)\rangle = \frac{1}{\sqrt{2}} \begin{vmatrix} \varphi_{\alpha_l}(1) & \varphi_{\alpha_l}(2) \\ \varphi_{\alpha_m}(1) & \varphi_{\alpha_m}(2) \end{vmatrix}. \qquad (19.39)$$

Let us turn now to density matrices and address first those in coordinate representation. The one-body density related to the Slater determinant may be written as

$$\rho_{\text{SL}}^{(1)}(\mathbf{x};\mathbf{x}') = \sum_{l=1}^{A} \langle\mathbf{x}|\alpha_l\rangle\langle\alpha_l|\mathbf{x}'\rangle = \sum_{l=1}^{A} \varphi_{\alpha_l}^*(\mathbf{x}')\varphi_{\alpha_l}(\mathbf{x}), \qquad (19.40)$$

with the probability density in coordinate space given by $\rho_{\text{SL}}^{(1)}(\mathbf{x};\mathbf{x})$. Notice that these expressions are still operators in spin-isospin space. After averaging over the latter and integrating over space one gets $\int d^3x \, \rho_{\text{SL}}^{(1)}(\mathbf{x};\mathbf{x}) = A$. For the two-body density matrix

$$\rho_{\text{SL}}^{(2)}(\mathbf{x}_1,\mathbf{x}_2;\mathbf{x'}_1,\mathbf{x'}_2) = \sum_{m,l=1}^{A} \langle \mathbf{x}_1,\mathbf{x}_2 | \Phi_{\alpha_l \alpha_m}^{\text{SL}} \rangle \langle \Phi_{\alpha_l \alpha_m}^{\text{SL}} | \mathbf{x'}_1,\mathbf{x'}_2 \rangle. \quad (19.41)$$

the first line of (19.37) leads to

$$\rho_{\text{SL}}^{(2)}(\mathbf{x}_1,\mathbf{x}_2;\mathbf{x'}_1,\mathbf{x'}_2) = \rho_{\text{SL}}^{(1)}(\mathbf{x}_1,\mathbf{x'}_1)\rho_{\text{SL}}^{(1)}(\mathbf{x}_2,\mathbf{x'}_2) - \rho_{\text{SL}}^{(1)}(\mathbf{x}_1,\mathbf{x'}_2)\rho_{\text{SL}}^{(1)}(\mathbf{x}_2,\mathbf{x'}_1). \quad (19.42)$$

Equations (19.40) and (19.42) can be generalized to any basis for the two particle system which factorizes like into orthonormal one particle functions like $|\mu,\nu\rangle = |\mu;1\rangle|\nu;2\rangle$. In this way the one-body density matrix becomes

$$\rho_{\mu\nu}^{(1,\text{SL})} = \langle \mu | \hat{\rho}_{\text{SL}}^{(1)} | \nu \rangle = \sum_{l=1}^{A} \langle \mu | \alpha_l \rangle \langle \alpha_l | \nu \rangle \quad (19.43)$$

and, as in (19.42), the two-body density matrix can be written in terms of the one-body densities as

$$\langle \mu\nu | \hat{\rho}_{\text{SL}}^{(2)} | \mu'\nu' \rangle = \rho_{\mu\mu'}^{(1,\text{SL})} \rho_{\nu\nu'}^{(1,\text{SL})} - \rho_{\mu\nu'}^{(1,\text{SL})} \rho_{\nu\mu'}^{(1,\text{SL})} \quad (19.44)$$

This feature is typical of an *uncorrelated* system, meaning that there are no correlations beyond those which come from the Pauli principle.

19.3 The Wigner transformation

The Wigner transformation is an important tool for describing features on the border between classical and quantum physics. We shall begin by introducing its basic relations for the special case of three dimensions. Extensions to other cases are straightforward.

19.3.1 *The Wigner transform in three dimensions*

Suppose we are given some one-body operator $\hat{O}(t)$ whose matrix elements in coordinate representation may be written as $O(\mathbf{r}_1,\mathbf{r}_2,t) = \langle \mathbf{r}_1 | \hat{O}(t) | \mathbf{r}_2 \rangle$. Likewise, in momentum space one has $O(\mathbf{p}_1,\mathbf{p}_2,t) = \langle \mathbf{p}_1 | \hat{O}(t) | \mathbf{p}_2 \rangle$. The transformation between both representations is given by the Fourier transformation

$$|\mathbf{p}\rangle = \int d^3r \, |\mathbf{r}\rangle\langle \mathbf{r} | \mathbf{p} \rangle = \int d^3r \, \exp(\frac{i}{\hbar}\mathbf{p}\,\mathbf{r})|\mathbf{r}\rangle. \quad (19.45)$$

Here, for the sake of simplicity, we have chosen to work with genuine plane waves $\langle \mathbf{r} | k \rangle = \exp((i/\hbar)\mathbf{p}\,\mathbf{r})$ in an infinite system. In other words, our momenta are continuous variables. If necessary, one might go back to a finite volume and employ periodic boundary conditions. In addition, we shall work with the momentum itself, rather than the wave vector $\mathbf{k}$. This means that the completeness relation has to be written as

$$\int \frac{d^3p}{(2\pi\hbar)^3} \langle \mathbf{r}_1|\mathbf{p}\rangle\langle\mathbf{p}|\mathbf{r}_2\rangle = \int \frac{d^3p}{(2\pi\hbar)^3} e^{i\mathbf{p}(\mathbf{r}_1-\mathbf{r}_2)} = \delta(\mathbf{r}_1 - \mathbf{r}_2). \qquad (19.46)$$

The only difference is the new normalization factor $(2\pi\hbar)^3$, which is also needed to get the inverse of the transformation (19.45). Following E. Wigner (and H. Weyl) it is possible to define a mixed transformation by

$$O^{\mathrm{W}}(\mathbf{r},\mathbf{p},t) = \int d^3s\, e^{-\frac{i}{\hbar}\mathbf{p}\mathbf{s}} O(\mathbf{r}+\frac{\mathbf{s}}{2}, \mathbf{r}-\frac{\mathbf{s}}{2}, t), \qquad (19.47)$$

where $\mathbf{r}$ and $\mathbf{s}$ are the center of mass and the relative coordinates given by

$$\mathbf{r} = \frac{1}{2}(\mathbf{r}_1 + \mathbf{r}_2) \quad \mathbf{s} = \mathbf{r}_1 - \mathbf{r}_2 \quad \text{or} \quad \mathbf{r}_1 = \mathbf{r} + \frac{\mathbf{s}}{2} \quad \mathbf{r}_2 = \mathbf{r} - \frac{\mathbf{s}}{2}. \qquad (19.48)$$

The inverse of (19.47) is defined as

$$O(\mathbf{r}_1,\mathbf{r}_2,t) = \int \frac{d^3p}{(2\pi\hbar)^3} e^{\frac{i}{\hbar}\mathbf{p}\mathbf{s}} O^{\mathrm{W}}(\mathbf{r},\mathbf{p},t). \qquad (19.49)$$

Of particular interest is the Wigner transform of the density matrix (which may or may not depend on time) and which we want to call $n(\mathbf{r},\mathbf{p},t)$ in what follows:

$$n(\mathbf{r},\mathbf{p},t) = \int d^3s\, \exp(-\frac{i}{\hbar}\mathbf{p}\mathbf{s})\, \rho(\mathbf{r}+\frac{\mathbf{s}}{2}, \mathbf{r}-\frac{\mathbf{s}}{2}, t). \qquad (19.50)$$

It may be considered the quantum analog to the density distribution in classical phase space. However, in the quantum case positive definiteness is no longer guaranteed. It is only for its integrals over either momenta or coordinates that they can be interpreted as probabilities. Evidently, for the integral over momenta one obtains

$$\int \frac{d^3p}{(2\pi\hbar)^3} n(\mathbf{r},\mathbf{p},t) = \rho(\mathbf{r},\mathbf{r},t) = \langle \mathbf{r}|\hat{\rho}(t)|\mathbf{r}\rangle \qquad (19.51)$$

which is the probability density for finding the system at $\mathbf{r}$. Integration over the coordinates specifies that in momentum space:

$$\int d^3r\, n(\mathbf{r},\mathbf{p},t) = \int d^3r_1 d^3r_2 \exp(-\frac{i}{\hbar}\mathbf{p}(\mathbf{r}_1 - \mathbf{r}_2))\, \rho(\mathbf{r}_1,\mathbf{r}_2,t) \qquad (19.52)$$

$$= \int d^3r_1 d^3r_2 \exp(-\frac{i}{\hbar}\mathbf{p}(\mathbf{r}_1 - \mathbf{r}_2))\, \langle \mathbf{r}_1|\hat{\rho}(t)|\mathbf{r}_2\rangle = \langle \mathbf{p}|\hat{\rho}(t)|\mathbf{p}\rangle = \rho(\mathbf{p},\mathbf{p},t).$$

Finally, we need to look at the evaluation of average values. Given the operator $\hat{O}(t)$ with its Wigner transform $O^{\mathrm{W}}(\mathbf{r},\mathbf{p},t)$. As is easily verified, the expectation value associated with the density operator $\hat{\rho}(t)$ may then be calculated as

$$\langle \hat{O}\rangle_t = \mathrm{tr}\, \hat{O}(t)\hat{\rho}(t) = \int \frac{d^3r\, d^3p}{(2\pi\hbar)^3} O^{\mathrm{W}}(\mathbf{r},\mathbf{p},t)\, n(\mathbf{r},\mathbf{p},t). \qquad (19.53)$$

19.3.2 Many-body systems of indistinguishable particles

Most of the formulas derived above, like (19.53) for the evaluation of averages, may trivially be generalized to any dimensions, for instance to the case of N distinguishable particles in three-dimensional space. The situation becomes more subtle when the many-body system consists of indistinguishable particles, say of identical bosons or fermions. In this case one must work with either totally symmetrized or totally antisymmetrized Wigner functions for the many-body densities. As the first step one needs to introduce density matrices obeying such symmetries, for instance as described in Section 2.8 of (Feynman, 1988)). The construction may be considered a generalization of the discussion in Section 19.2 for a system of independent fermions. To derive the corresponding Wigner functions from such density matrices is beyond the scope of our presentation. It is of interest, however, to address the simpler task of getting the classical limit, for which all exchange terms have to be discarded. Actually, as shown in Section 2.3 of (Balian, 1991), for instance, such terms can be seen to be of higher order in $\hbar$, as compared with the leading, direct terms. But even the classical macro-state must be unaffected when the position or the momentum of two identical particles are exchanged. Hence, the classical density distribution $n(\mathbf{r}_1, \mathbf{p}_1, \ldots \mathbf{r}_N, \mathbf{p}_N; t)$ must remain identical when the indices for two particles are interchanged. This feature may be accounted for by modifying the $6N$-dimensional volume element according to

$$\prod_{i=1}^{N}\left(\frac{d^3 r_i\, d^3 p_i}{(2\pi\hbar)^3}\right) \implies \frac{1}{N!}\prod_{i=1}^{N}\left(\frac{d^3 r_i\, d^3 p_i}{(2\pi\hbar)^3}\right) \equiv d\tau_N\,. \qquad (19.54)$$

Applying this to the calculation of a trace one may write

$$\operatorname{tr} \hat{F} \implies \int d\tau_N\ F^{\mathrm{cl}}(\mathbf{r}_1, \mathbf{p}_1, \ldots \mathbf{r}_N, \mathbf{p}_N) \qquad (19.55)$$

with F^{cl} being the classical analog of the operator $\hat{F}$. The F^{cl} might be obtained from the Wigner transform of $\hat{F}$ by neglecting all terms which involve commutators. A prime example for an application of (19.53) is the calculation of the partition function $Z = \operatorname{tr} \exp(-\beta \hat{H})$ of an ideal, classical gas. Suppose the particles experience the potential $V(\mathbf{r}_i)$ without any mutual interaction. Then the Z is easily verified to be

$$Z(\beta) = \int d\tau_N\ \exp\!\left(-\beta \sum_{1}^{N}\left(\frac{\mathbf{p}_i^2}{2m} + V(\mathbf{r}_i)\right)\right) = \frac{z^N}{N!} \qquad (19.56)$$

with

$$z(\beta) = \int \frac{d^3 r\, d^3 p}{(2\pi\hbar)^3}\ \exp\!\left(-\beta\!\left(\frac{\mathbf{p}^2}{2m} + V(\mathbf{r})\right)\right). \qquad (19.57)$$

This result becomes particularly simple if the potential just defines the boundaries of the container of volume $\mathcal{V}$ inside which the gas is free. Then the z simply reduces to $z = \mathcal{V}/\lambda$, with the thermal wavelength

$$\lambda = \sqrt{2\pi\hbar^2\beta/m}, \qquad (19.58)$$

such that the partition function for the N particles becomes

$$Z(\beta) = \frac{V^N}{N!\lambda^{3N}} \simeq \left(\frac{V}{N\lambda^3}\right)^N. \qquad (19.59)$$

In the last step Stirling's formula has been applied, but only to leading order in N, which is to say in the form $\ln N! \simeq N(\ln N - e)$. This result may be used to deduce

$$\mathcal{F} = NT\left(\ln\frac{N\lambda^3}{V} - 1\right) \qquad (19.60)$$

for the free energy $\mathcal{F} = -T\ln Z(\beta)$ and the Sackur–Tetrode formula

$$\mathcal{S} = N\left(\ln\frac{V}{N\lambda^3} + \frac{5}{2}\right) \qquad (19.61)$$

for the entropy from $\mathcal{S} = -(\partial\mathcal{F}/\partial T)_Q$. In applying these formulas one should not forget that the classical limit is valid only for low densities and large temperatures, which generally implies $V/(N\lambda^3) \gg 1$ such that $\mathcal{S}/N \gg 1$.

19.3.3 Propagation of wave packets

For the sake of simplicity we restrict ourselves to a one-dimensional system; a generalization should be straightforward. In quantum mechanics a wave function $\psi(Q,t) = \langle Q|\psi(t)\rangle$ propagates according to the law

$$\langle Q|\psi(t)\rangle = \int dQ_0\, k(Q,Q_0,t)\,\psi(Q_0,0) = \int dQ_0\,\langle Q|e^{-\frac{i}{\hbar}\hat{H}t}|Q_0\rangle\langle Q_0|\psi(0)\rangle. \qquad (19.62)$$

There are various ways to calculate the propagator $k(Q,Q_0,t)$, the most interesting one being through Feynman's path integrals (Feynman and Hibbs, 1965), which are introduced in Chapter 24. We want to study how such formulas or results can be translated to Wigner functions. For the matrix elements $\langle Q_1|\hat{d}(t)|Q_2\rangle = \psi(Q_1,t)\psi^*(Q_2,t)$ of the density operator $\hat{d}(t) = |\psi(t)\rangle\langle\psi(t)|$ the Wigner transform reads

$$d^W(Q,P,t) = \int_{-\infty}^{\infty} ds\, e^{-\frac{i}{\hbar}Ps}\psi(Q+\frac{s}{2},t)\psi^*(Q-\frac{s}{2},t) \qquad (19.63)$$

The time evolution of the wave functions is given by (19.62). A short calculation, which amongst others makes use of the transformation (19.48), shows that one may introduce a propagator for the Wigner function, namely

$$K^W(Q,P;Q_0,P_0;t) \qquad (19.64)$$
$$= \int_{-\infty}^{\infty} ds\, ds_0\, e^{-\frac{i}{\hbar}(Ps-P_0 s_0)}\, k(Q+\frac{s}{2},Q_0+\frac{s_0}{2},t)\, k^*(Q-\frac{s}{2},Q_0-\frac{s_0}{2},t),$$

to get

$$d^W(Q, P, t) = \int dQ_0 \, \frac{dP_0}{2\pi\hbar} K^W(Q, P; Q_0, P_0; t) \, d^W(Q_0, P_0, t = 0). \quad (19.65)$$

Obviously the $K^W(Q, P; Q_0, P_0; t)$ must fulfill the following initial condition

$$K^W(Q, P; Q_0, P_0; t \to 0) = 2\pi\hbar \, \delta(Q - Q_0)) \, \delta(P - P_0)). \quad (19.66)$$

This may indeed be verified explicitly starting from the definition of the propagators of wave functions given in (19.62). Actually, for the oscillator the K^W is entirely determined by classical trajectories, see (Hofmann, 1997).

19.3.4 Correspondence rules

Let us turn now to discuss features which become relevant away from equilibrium. In Chapter 7 equations of motion for the collective density operator $\hat{d}(t)$ are encountered whose classical, Markovian limit is expected to have a structure known from the Fokker–Planck equation. Both are related by the correspondence rules given in eqn(7.75). Let us see how such rules come about through the Wigner transformation, which shows them to be correct identities. There appear terms which involve operators like $[\hat{A}, \hat{B}]_\pm$ with $\hat{A}$ being either $\hat{Q}$ or $\hat{P}$ and $\hat{B}$ being the $\hat{d}(t)$ itself or some combination of it with $\hat{Q}$ or $\hat{P}$. To get the Wigner transform of $[\hat{A}, \hat{B}]_\pm$ we first have to evaluate their basic matrix elements in coordinate representation. The one for $\hat{Q}$ is trivial, namely

$$\langle Q_1 | \left[\hat{Q}, \hat{B}\right]_\pm | Q_2 \rangle = (Q_1 \pm Q_2) \langle Q_1 | \hat{B} | Q_2 \rangle. \quad (19.67)$$

To get the one for P one may insert a complete set of coordinate eigenstates and obtains:

$$\langle Q_1 | \left[\hat{P}, \hat{B}\right]_\pm | Q_2 \rangle = \frac{\hbar}{i} \left\{ \frac{\partial}{\partial Q_1} \mp \frac{\partial}{\partial Q_2} \right\} \langle Q_1 | \hat{B} | Q_2 \rangle. \quad (19.68)$$

To deduce their Wigner transforms it is convenient to first rewrite (19.47) as

$$\mathcal{T}^W \left[\hat{O}(t)\right] = \int_{-\infty}^{\infty} dQ_1 dQ_1 \, \delta(Q - \frac{Q_1 + Q_2}{2}) e^{-\frac{i}{\hbar} P(Q_1 - Q_2)} \langle Q_1 | \hat{O}(t) | Q_2 \rangle. \quad (19.69)$$

The equivalence of both expressions is readily verified from elementary properties of the δ-function observing (19.48). From (19.67) we thus obtain

$$\mathcal{T}^W \left[[\hat{Q}, \hat{B}]_\pm\right] = \int_{-\infty}^{\infty} dQ_1 dQ_1 \, \delta(Q - \frac{Q_1 + Q_2}{2}) e^{-\frac{i}{\hbar} P(Q_1 - Q_2)} (Q_1 \pm Q_2) \langle Q_1 | \hat{B} | Q_2 \rangle \quad (19.70)$$

from which, again with the help of (19.48), it is easily seen that for $\mathcal{T}^W \left[\hat{Q}, \hat{B}\right]_\pm$ the final expressions can be written as:

$$T^{\mathrm{W}}\left[\left[\hat{Q},\hat{B}\right]\right]=i\hbar\frac{\partial}{\partial P}B^{\mathrm{W}} \qquad T^{\mathrm{W}}\left[\left[\hat{Q},\hat{B}\right]_{+}\right]=2QB^{\mathrm{W}} \qquad (19.71)$$

by using for the Wigner transform of $\hat{B}$ the notation B^{W} as suggested by (19.47). To derive the results for $\hat{Q}$ replaced by $\hat{P}$ it is convenient to perform partial integrations, once (19.68) has been inserted into (19.69). One gets:

$$T^{\mathrm{W}}\left[\left[\hat{P},\hat{B}\right]_{\pm}\right]=-\frac{\hbar}{i}\int_{-\infty}^{\infty}\mathrm{d}Q_1\mathrm{d}Q_2\,\langle Q_1|\hat{B}|Q_2\rangle\times \\ \left(\frac{\partial}{\partial Q_1}\mp\frac{\partial}{\partial Q_2}\right)\delta(Q-Q_1+Q_22)e^{-\frac{i}{\hbar}P(Q_1-Q_2)}. \qquad (19.72)$$

For the derivatives appearing here eqns(19.48) imply the following set of relations

$$\frac{\partial}{\partial Q}=\frac{\partial}{\partial Q_1}+\frac{\partial}{\partial Q_2} \qquad 2\frac{\partial}{\partial s}=\frac{\partial}{\partial Q_1}-\frac{\partial}{\partial Q_2} \qquad (19.73)$$

which can be used to find for (19.72):

$$T^{\mathrm{W}}\left[\left[\hat{P},\hat{B}\right]\right]=\frac{\hbar}{i}\frac{\partial}{\partial Q}B^{\mathrm{W}}. \qquad (19.74)$$

All these results have been obtained more or less by straightforward calculations. They can be summarized by:

$$T^{\mathrm{W}}\left[-\frac{i}{\hbar}\left[\hat{A},\hat{B}\right]\right]=\{A^{\mathrm{W}},B^{\mathrm{W}}\}=\frac{\partial A^{\mathrm{W}}}{\partial Q}\frac{\partial B^{\mathrm{W}}}{\partial P}-\frac{\partial A^{\mathrm{W}}}{\partial P}\frac{\partial B^{\mathrm{W}}}{\partial Q} \qquad (19.75)$$

and

$$T^{\mathrm{W}}\left[\frac{1}{2}\left[\hat{A},\hat{B}\right]_{+}\right]=A^{\mathrm{W}}\,B^{\mathrm{W}}, \qquad (19.76)$$

and make up the correspondence rules (7.75) used in Chapter 7. It is important to recall that these results are correct only for the special case that $\hat{A}$ is either the coordinate or the momentum operator. For any other operator (19.75) or (19.76) is correct only to zero order in $\hbar$ (Balian, 1991); for a proof of this statement one may simply apply the same tricks as described in Section 21.1 for the von Neumann equation.

19.3.5 *The equilibrium distribution of the oscillator*

In one dimension the Hamiltonian may be written as $\hat{H}=\hat{P}^2/2M+M\varpi^2\hat{Q}^2/2$ (with a supposedly positive stiffness $C=M\varpi^2$). Hence, the density operator of the canonical equilibrium with temperature $T=1/\beta$ is given by

$$\hat{d}_{\mathrm{eq}}(\hat{q},\hat{P};\beta)=\frac{1}{Z}\exp\left(-\beta\left[\frac{\hat{P}^2}{2M}+\frac{C\hat{Q}^2}{2}\right]\right), \qquad (19.77)$$

with the partition function $Z = \int_{-\infty}^{+\infty} dQ \langle Q|\hat{k}_{\text{eq}}(\beta)|Q\rangle$ where

$$\hat{k}_{\text{eq}}(\hat{q},\hat{P};\beta) = \exp\left(-\beta\left[\frac{\hat{P}^2}{2M} + \frac{C\hat{Q}^2}{2}\right]\right). \qquad (19.78)$$

This operator satisfies the Bloch equation $\partial \hat{k}(\beta)/\partial\beta = -\hat{H}\,\hat{k}(\beta)$. It allows one to construct matrix elements of $\hat{k}_{\text{eq}}(\beta)$ (Feynman, 1988), but we may also make use of expression (24.57) obtained in Section 24.2 within the path integral formalism. From this expression one readily calculates the Wigner function

$$k_{\text{eq}}^{\text{W}}(Q,P;\beta) = \zeta \exp\left[-\left(\frac{P^2}{2mT^*} + \frac{m\varpi^2 Q^2}{2T^*}\right)\right] \qquad \zeta = \frac{1}{\cosh(\hbar\varpi\beta/2)} \qquad (19.79)$$

from (19.47) and (19.48). Here, the effective temperature

$$T^*(\varpi,T) = \frac{\hbar\varpi}{2}\coth\frac{\hbar\varpi}{2T} \quad \Longrightarrow \quad \begin{cases} \hbar\varpi/2 & \text{for } \hbar\varpi \gg T \\ T & \text{for } \hbar\varpi \ll T. \end{cases} \qquad (19.80)$$

was introduced, with its limits for small and high temperatures specified on the right. With a normalization factor of $\mathcal{N} = (1/2)\coth(\hbar\varpi/2T)$ the Wigner form of the density distribution is given by

$$d_{\text{eq}}^{\text{W}}(Q,P;\beta) = \frac{1}{\mathcal{N}}\exp\left[-\left(\frac{P^2}{2mT^*} + \frac{m\varpi^2 Q^2}{2T^*}\right)\right], \qquad (19.81)$$

and the partition function is easily seen to become $Z = (2\sinh(\hbar\varpi/(2T)))^{-1}$.

The Wigner function of the density allows one to calculate expectation values of some operator $\hat{O}$ through

$$\langle \hat{O} \rangle = \int \frac{dQ\,dP}{2\pi\hbar}\, O(Q,P)\, d_{\text{W}}(Q,P;\beta). \qquad (19.82)$$

This is straightforward if O is only a function either of momentum or coordinate. For mixed functions one eventually has to consider ordered combinations. Equation (19.82) immediately leads to the *quantal version of the equipartition theorem*

$$\langle P^2/2M \rangle = T^*(\varpi,T)/2 = \langle M\varpi^2 Q^2/2 \rangle. \qquad (19.83)$$

The mixed expectation value has to vanish $\langle QP \rangle = \langle PQ \rangle = 0$. This follows already from the fact that Q is even under time reversal whereas P is odd with $\hat{d}$ being time reversal invariant. For very small and very large temperatures one obtains from (19.80) the results expected from quantum mechanics and from classical statistical physics: In the first case one gets the value known for the quantal ground state. The high temperature limit leads to the classical equipartition theorem. Note that for both limits the scale is set by the $\hbar\varpi$. Let us look next at the uncertainty relation in equilibrium. As $\langle Q \rangle = \langle P \rangle = 0$ and because of the vanishing mixed fluctuation one has

$$\Delta Q^2 \Delta P^2 = \frac{\hbar^2}{4} \coth^2\left(\frac{\hbar \varpi}{2T}\right) \geq \frac{\hbar^2}{4} \qquad (19.84)$$

(with the precise definition of $\Delta u^2 = \langle u^2 \rangle - \langle u \rangle^2$). On the very right of (19.84) the equality sign holds true for $T = 0$; in the high temperature limit one gets $(T/\varpi)^2$. In terms of the equilibrium fluctuations $\sum_{qq}^{\mathrm{eq}} = \Delta Q^2$ and $\sum_{pp}^{\mathrm{eq}} = \Delta P^2$ the Wigner functions for the density and for the operator $\hat{k}_{\mathrm{eq}}$ fulfill the following identities

$$d_{\mathrm{eq}}^{\mathrm{W}}(Q, P; \beta) = \frac{1}{\sqrt{\sum_{qq}^{\mathrm{eq}} \sum_{pp}^{\mathrm{eq}}}} \exp\left(-\frac{P^2}{2\sum_{pp}^{\mathrm{eq}}} - \frac{Q^2}{2\sum_{qq}^{\mathrm{eq}}}\right) \equiv \frac{1}{Z} k_{\mathrm{eq}}^{\mathrm{W}}(Q, P; \beta); \quad (19.85)$$

the ζ of (19.79) and the partition function Z relate to each other as $\zeta = Z/\sqrt{\sum_{qq}^{\mathrm{eq}} \sum_{pp}^{\mathrm{eq}}}$.

Exercises

1. Prove that the $\Phi_{\alpha_1 \ldots \alpha_A}^{\mathrm{SL}}$ of (19.34) are normalized to unity, for instance by expanding the Slater determinant to its first column.
2. Derive (19.37) from the definition (19.36) for the two-body density for independent particles.

20
THE HARTREE–FOCK APPROXIMATION

In many fields a good approximation to the many-body problem may be obtained assuming the particles to move independently of each other. Certainly, like it is in the case of the two-body density (recall eqn(19.42) and (19.44)), the notion "independent" is to be understood in the sense that any correlations beyond those given by the Pauli principle are discarded. The Hartree–Fock method allows one to find an optimal wave function for fermions, optimal in the sense of calculating the total energy of the system, starting from an appropriate two-body interaction. It is based on the variational principle which relies on the fact that for any state $|\Psi\rangle$ the normalized expectation value of the Hamiltonian cannot be smaller than the exact ground state energy E_0, viz $E[\Psi] = \langle\Psi|\hat{H}|\Psi\rangle/\langle\Psi|\Psi\rangle \geq E_0$. As the functional $E[\Psi]$ is bound from below one may introduce trial functions Ψ_{MOD}, which span a subspace of the total Hilbert space. The Ψ_{MOD} may then be found by minimizing $E[\Psi_{\text{MOD}}]$.

20.1 Hartree–Fock with density operators

For the Hartree–Fock procedure one chooses the model functions Ψ_{MOD} to be Slater determinants $\Psi_{\text{MOD}} \equiv \Psi^{\text{HF}}$. The latter are build out of single particle states $\varphi_{\alpha_l}(l)$ for the A particles with $l = 1\ldots A$. The variational procedure then leads to Hartree–Fock equations for these states $\varphi_{\alpha_l}(l)$ and the corresponding single particle energies e_l. Rather than following this conventional procedure we shall formulate the problem in terms of density matrices.

As the starting point we assume the Hamiltonian of the A-particle system to be given by

$$\hat{H} = \hat{T} + \hat{V} = \sum_i^A \hat{t}_i + \frac{1}{2}\sum_{i \neq j} \hat{V}_{ij}. \tag{20.1}$$

The interaction V_{ij} is meant to act between particles i and j, but it might depend on the one-body density effectively simulating some kind of three-body forces (see Section 2.6.2). In the Hartree–Fock approximation the energy functional $E[\Psi^{\text{HF}}]$ may be expressed by the one-body density $\hat{\rho}^{(1)}_{\text{HF}}$, simply to be identified as $\hat{\rho}$ in the following. For the one-body operator of the kinetic energy this is trivial and leads to

$$\langle\Psi^{\text{HF}}|\hat{T}|\Psi^{\text{HF}}\rangle = \operatorname{tr}\hat{t}\hat{\rho} = \sum_{kl}\langle k|\hat{t}|l\rangle \rho_{lk}. \tag{20.2}$$

The calculation of the average of the two-body operator $\hat{V}$ is a bit more complicated. The general expression is given by the two-body density according to formula (19.24) which here reads

$$\langle \Psi^{\mathrm{HF}}|\hat{V}|\Psi^{\mathrm{HF}}\rangle = \frac{1}{2}\operatorname{tr}\hat{V}_{12}\hat{\rho}^{(2)}_{\mathrm{HF}}\,. \tag{20.3}$$

This expression may be rewritten as

$$\langle \Psi^{\mathrm{HF}}|\hat{V}|\Psi^{\mathrm{HF}}\rangle = \frac{1}{2}\sum_{kl}\langle kl|\hat{V}_{12}\hat{\rho}^{(2)}_{\mathrm{HF}}|kl\rangle = \frac{1}{2}\sum_{klmn}\langle kl|\hat{V}_{12}|mn\rangle\langle mn|\hat{\rho}^{(2)}|kl\rangle$$

$$= \frac{1}{2}\sum_{klmn}\langle kl|\hat{V}_{12}|mn\rangle(\rho_{mk}\rho_{nl} - \rho_{ml}\rho_{nk}) \tag{20.4}$$

$$\equiv \frac{1}{2}\sum_{klmn}\langle kl|\overline{V_{12}}|mn\rangle\rho_{mk}\rho_{nl}\,.$$

The second line was obtained because for Slater determinants the two-body density can be expressed by the one-body density as $\langle mn|\hat{\rho}^{(2)}|kl\rangle = \rho_{mk}\rho_{nl} - \rho_{ml}\rho_{kn}$ (remember eqn(19.44)). The matrix element $\overline{V_{12}}$ is defined through the third line. It uses the convention

$$\langle kl|\overline{V_{12}}|ij\rangle \equiv \langle kl|V_{12}\bigl(|ij\rangle - |ji\rangle\bigr)\,, \tag{20.5}$$

which may be taken as a definition of the operator $\overline{V_{12}}$ in the Hilbert space of those two particle wave functions which are simply products of one-body functions $|ij\rangle = |i\rangle|j\rangle$. The necessary antisymmetrization is taken care of automatically by the definition (20.5) (see exercise 1).

For the energy functional one thus gets

$$E[\hat{\rho}] = \langle \Psi^{\mathrm{HF}}|\hat{H}|\Psi^{\mathrm{HF}}\rangle = \sum_{kl}\langle k|\hat{t}|l\rangle\rho_{lk} + \frac{1}{2}\sum_{klmn}\langle kl|\overline{V_{12}}|mn\rangle\rho_{mk}\rho_{nl}\,. \tag{20.6}$$

the variation of which will lead to the optimal one-body density operator. As the latter is to represent pure states one has to make sure that the constraint $\hat{\rho}^2 = \hat{\rho}$ is fulfilled, which can be done with the help of Lagrange multipliers. In matrix notation this may be expressed as

$$\delta\Bigl[E[\rho] - \operatorname{tr}\Lambda(\rho^2 - \rho)\Bigr] = 0\,, \tag{20.7}$$

with some matrix Λ. A first order variation of $E[\rho]$ leads to the one-body operator

$$\hat{h} = \frac{\delta E[\rho]}{\delta\hat{\rho}}\,, \tag{20.8}$$

whose matrix elements $h_{ij} \equiv \langle i|\hat{h}|j\rangle$ are given by

$$h_{ij} = \frac{\partial E[\rho]}{\partial \rho_{ji}} = \langle i|\hat{t}|j\rangle + \sum_{kl}\langle ik|\overline{V_{12}}|jl\rangle \rho_{lk} + \frac{1}{2}\sum_{rskl}\langle rs|\frac{\partial \overline{V_{12}}}{\partial \rho_{ji}}|kl\rangle \rho_{kr}\rho_{ls}. \quad (20.9)$$

The last term on the right, the so-called *rearrangement term*, is present as soon as the interaction depends on the density. It will be ignored for the rest of this chapter.

With the $\hat{h}$ of (20.8), and because of tr $\Lambda\rho^2 = $ tr $\rho\Lambda\rho$ the variational equations can be written as tr $[(h - \rho\Lambda - \Lambda\rho + \Lambda)\,\delta\rho] = 0$ As they must be valid for *any* $\delta\rho$ one must have
$$h - \rho\Lambda - \Lambda\rho + \Lambda = 0\,.$$
Let us multiply this equation once from left and once from right by ρ and subtract the resulting equations from each other. After taking into account the identity $\rho^2 = \rho$ once more, the final result may be written as

$$[h,\,\rho] = 0, \qquad \text{or in operator form:} \quad [\hat{h},\,\hat{\rho}] = 0\,. \qquad (20.10)$$

Solving these equations amounts to diagonalizing the Hamilton matrix (20.9). Indeed, two operators commute if and only if they have a common basis $|\lambda_\nu\rangle$ of eigenvectors, in which one thus has

$$\hat{h} = \sum_\nu e_\nu |\lambda_\nu\rangle\langle\lambda_\nu| \qquad (20.11)$$

$$\hat{\rho} = \sum_\nu |\lambda_\nu\rangle\langle\lambda_\nu|\,. \qquad (20.12)$$

In coordinate representation the one-body Hamilton operator $\hat{h}$ of (20.11) is given by

$$\langle \mathbf{r}_1|\hat{h}|\mathbf{r}_2\rangle = \sum_\nu e_\nu \varphi_{\lambda_\nu}(\mathbf{r}_1)\varphi^*_{\lambda_\nu}(\mathbf{r}_2)\,. \qquad (20.13)$$

Its diagonal elements represent an energy density of dimension energy per volume.

20.1.1 The Hartree–Fock equations

To obtain the common version of the Hartree–Fock equations from
$$\hat{h}|\lambda_\nu\rangle = e_\nu|\lambda_\nu\rangle \qquad (20.14)$$
we return to the coordinate representation and assume a two-body interaction of the form $\hat{V} = (1/2)\sum_{i\neq j}\hat{V}(\hat{\mathbf{x}}_i - \hat{\mathbf{x}}_j)$. Then (20.14) may be written as

$$\left[-\frac{\hbar^2}{2m}\Delta_\mathbf{x} + \hat{U}^{\rm HF}\otimes\right]\varphi_\nu(\mathbf{x}) = e_\nu\varphi_\nu(\mathbf{x})\,. \qquad (20.15)$$

Here, the operator $\hat{U}^{\rm HF}$ acts on the wave function in a *non-local way*, for which reason we have used the symbol $\otimes$. Its meaning is defined by writing the set of equations for $\nu = 1\ldots A$ in more explicit fashion like

$$\left[-\frac{\hbar^2}{2m}\Delta_{\mathbf{x}} + U^D(\mathbf{x})\right]\varphi_\nu(\mathbf{x}) - \int d^3x' U^X(\mathbf{x};\mathbf{x}')\varphi_\nu(\mathbf{x}') = e_\nu\varphi_\nu(\mathbf{x}), \quad (20.16)$$

with

$$U^D(\mathbf{x}) = \int d^3x'\, V(\mathbf{x}-\mathbf{x}')\rho(\mathbf{x}',\mathbf{x}') \quad (20.17)$$

and

$$U^X(\mathbf{x};\mathbf{x}') = V(\mathbf{x}-\mathbf{x}')\rho(\mathbf{x}',\mathbf{x}). \quad (20.18)$$

The first, *direct term* U^D just has the common form of a *local* potential. Often this is called the "mean field" (although one must say that this notion is not uniquely defined, as sometimes it is used for the full operator as well). Through the *exchange term* U^X (20.18) the mean field operator becomes non-local: In eqn(20.16) it acts on the single particle states $\varphi_\nu(\mathbf{x})$ through the integral over d^3x'. Formally, this equation resembles a (non-local) Schrödinger equation. It is important to realize, however, that both the U^D and the U^X depend on the wave functions $\varphi_i(\mathbf{x})$ implying the Hartree–Fock equations (20.16) to be *nonlinear*. They actually have to be solved iteratively, in the sense of starting with a guess for the wave functions, and thus for the potentials, and finding improved ones after solving the equations repeatedly. After convergence of this procedure the potentials are determined in *self-consistent* fashion.

Finally, we should like to add two remarks. (i) In the *Hartree approximation* the *product wave function* Φ^{DP} given in (19.33) are used for the variational procedure. In this case only the term U^D survives. It is the antisymmetrization, and thus the Pauli principle which in the end leads to the exchange term. (ii) The variational principle only involves a *variation to first order*. Whether or not the energy functional takes on a maximum or a minimum (or is indifferent) is immaterial at first. However, the Hartree–Fock (or Hartree) solutions become unstable when $E^{\mathrm{HF}}[\Psi^{\mathrm{HF}}]$ turns out to be maximal.

20.1.2 *The ground state energy in HF-approximation*

According to the basic concept of the variational principle we know that it is the *ground state energy* E_0^{HF} which one aims at by approximating the true wave function by the trial function. The ground state wave function is that Slater determinant in which all single particle states are filled up to the *Fermi energy* e_F. The total energy E_0^{HF}, however, *is not* given by simply summing up all single particle energies up to (and including) e_F. The reason can be traced back to the fact that in the e_ν the interaction is counted twice! Indeed, a simple calculation shows

$$\begin{aligned}E_0^{\mathrm{HF}} &= \langle\Psi^{\mathrm{HF}}|\hat{H}|\Psi^{\mathrm{HF}}\rangle_0 = \sum_\nu^A e_\nu - \frac{1}{2}E_{\mathrm{pot}}^{\mathrm{HF}} \\ &\equiv \sum_\nu^A e_\nu - \frac{1}{2}\int d^3x\, d^3x'\, V(\mathbf{x}-\mathbf{x}')\Big[\rho(\mathbf{x};\mathbf{x})\rho(\mathbf{x}',\mathbf{x}') - \rho(\mathbf{x};\mathbf{x}')\rho(\mathbf{x}',\mathbf{x})\Big]\end{aligned}$$

$$(20.19)$$

To obtain this result one needs only multiply equation (20.16) by $\varphi_\nu^*(\mathbf{x})$, integrate over $\mathbf{x}$, sum over ν and take into account that the functional of the total energy splits up like

$$\begin{aligned}E^{\rm HF}[\Psi^{\rm HF}] &= \langle \Psi^{\rm HF}|\hat{H}|\Psi^{\rm HF}\rangle = E_{\rm kin}^{\rm HF} + E_{\rm pot}^{\rm HF} \\ &= \frac{\hbar^2}{2m}\int d^3x\, d^3x'\,\delta(\mathbf{x}'-\mathbf{x})\nabla_{\mathbf{x}}\nabla_{\mathbf{x}'}\rho(\mathbf{x};\mathbf{x}')\\ &\quad + \frac{1}{2}\int d^3x\, d^3x'\,V(\mathbf{x}-\mathbf{x}')\Big[\rho(\mathbf{x};\mathbf{x})\rho(\mathbf{x}',\mathbf{x}') - \rho(\mathbf{x};\mathbf{x}')\rho(\mathbf{x}',\mathbf{x})\Big]\end{aligned} \quad (20.20)$$

into contributions from the kinetic and the potential energy.

20.2 Hartree–Fock at finite temperature

A natural extension of the variational equation (20.7) to finite thermal excitation is performed by varying the grand canonical potential $\Omega_{\rm gce}$ instead of the energy functional $\mathcal{E}_{\rm HF}[\hat{\rho}]$. In Section 22.1.2.2 the $\Omega_{\rm gce}$ is defined in terms of the many-body density operator, see e.g. (22.52). For independent particles described by the one-body operator $\hat{\rho}$ one has

$$\Omega_{\rm HF}[\hat{\rho}] = -T S_{\rm HF}[\hat{\rho}] + \mathcal{E}_{\rm HF}[\hat{\rho}] - \mu\hat{\rho}\,. \qquad (20.21)$$

The functional of the energy $\mathcal{E}_{\rm HF} = \mathcal{E}_{\rm HF}[\hat{\rho}]$ still has the form given in (20.6),

$$\mathcal{E}_{\rm HF}[\hat{\rho}] = {\rm tr}\,\hat{t}\hat{\rho} + \frac{1}{2}{\rm tr\,tr}\,\overline{V_{12}}\,\hat{\rho}\hat{\rho}\,. \qquad (20.22)$$

As here the trace is understood to be performed in the single particle model, in the term for the two-body interaction the symbol tr has to appear twice. The entropy is expressed as

$$S_{\rm HF}[\hat{\rho}] = -{\rm tr}\left\{[1-\hat{\rho}]\ln[1-\hat{\rho}] + \hat{\rho}\ln\hat{\rho}\right\}, \qquad (20.23)$$

see (22.137). The Hartree–Fock equations may then be written as

$$\left.\frac{\delta\Omega_{\rm HF}[\hat{\rho}]}{\delta\hat{\rho}}\right|_{\hat{\rho}_{\rm HF}} = 0. \qquad (20.24)$$

Here, the variation of the energy is still given by (20.8), which is to say $\delta\mathcal{E}_{\rm HF}[\hat{\rho}] = \delta\hat{\rho}\hat{h}\,[\hat{\rho}]$. The variation for the entropy (20.23) reads

$$\delta S_{\rm HF}[\hat{\rho}] = {\rm tr}\,\delta\hat{\rho}\ln[1-\hat{\rho}] - {\rm tr}\,\delta\hat{\rho}\ln\hat{\rho}\,, \qquad (20.25)$$

which may be proven analogously to that for the general relation (22.13) of thermostatics.

Without proof we state that this more general procedure includes correctly the zero temperature case. It then reduces to the standard form $[\hat{h},\hat{\rho}_0]=0$ found in (20.10). The condition $\hat{\rho}_0^2=\hat{\rho}_0$ is taken care of by the entropy term in (20.21). This is possible because in the limit $T\to 0$ the $T\partial S_{HF}/\partial \hat{\rho}_0$ remains finite and approaches $\hat{h}(\hat{\rho}_0)-\mu$. An elaborate discussion of the whole subject can be found in (des Cloizeaux, 1968) and a more concise one in (Balian and Vénéroni, 1989).

20.2.1 TDHF at finite T

In Section 2.7 the equation of time-dependent Hartree–Fock was written in the form

$$\frac{d}{dt}\hat{\rho}=\frac{1}{i\hbar}[\hat{h},\hat{\rho}]. \tag{2.77}$$

It remains valid even at finite thermal excitations. As a consequence, and because of (20.8) and (20.25) both the total energy $\mathcal{E}_{HF}$ as well as the entropy S_{HF} are constants of motion. Although in Section 22.5 the proof of (22.137) and hence of (20.23) is given for equilibrium it can be said to be valid also out of equilibrium, see e.g. §55 of (Landau and Lifshitz, 1980).

An essential difference from the case at $T=0$ is seen in the linearized EOM, not in its form but in its content. Writing the time-dependent density as

$$\hat{\rho}(t)=\hat{\rho}_{\text{eq}}+\Delta\hat{\rho}(t) \tag{20.26}$$

and evaluating (2.77) to first order in $\Delta\hat{\rho}(t)$ one gets

$$\frac{d}{dt}\Delta\hat{\rho}=\frac{1}{i\hbar}\left[\hat{h}(\hat{\rho}_{\text{eq}}),\Delta\hat{\rho}(t)\right]+\frac{1}{i\hbar}\left[\operatorname{tr}\overline{V_{12}}\Delta\hat{\rho}(t),\hat{\rho}_{\text{eq}}\right]. \tag{20.27}$$

Here, the $\hat{\rho}_{\text{eq}}$ is that of equilibrium which at finite T is to be determined from the variation of the Ω_{HF} given in (20.21). Actually, with the help of the latter the linearized equation (20.27) can be written in the more concise form (Balian and Vénéroni, 1989)

$$\frac{d}{dt}\Delta\hat{\rho}=\frac{1}{i\hbar}\left[\operatorname{tr}\frac{\partial^2\Delta\Omega_{HF}}{\partial\hat{\rho}_{\text{eq}}^2}\Delta\hat{\rho}(t),\hat{\rho}_{\text{eq}}\right]. \tag{20.28}$$

Exercises

1. Prove these relations by making use of the technique of second quantization described in Section 26.8.

21

TRANSPORT EQUATIONS FOR THE ONE-BODY DENSITY

Let us assume the equation of motion for the one-body density to have the form

$$\frac{d}{dt}\hat{\rho}(t) - \frac{1}{i\hbar}[\hat{h}, \hat{\rho}] = \hat{I}[\hat{\rho}(t)]. \qquad (21.1)$$

In principle, the right-hand side is meant to represent all effects of the interaction beyond those given through the "mean field" in the one-body operator $\hat{h}$. In practice, however, one often only accounts for the effects of real collisions among the particles; we are going to adhere to the same approximation later on. This term will then be treated in some statistical way, for which reason one should no longer expect the relation $\rho^2 = \rho$ to be fulfilled, as it would be given for Hartree–Fock at zero thermal excitations. As a matter of fact, for the following discussion it is immaterial how the Hamiltonian $\hat{h}$ is defined; it may or may not be a functional of the one-body density $\hat{\rho}$, as it would be the case in both Landau theory and for an extended version of time-dependent Hartree–Fock. In the present context, the term on the right-hand side is meant to express the presence of non-conservative forces, and will thus be called "collision term". It should be stressed that this form of the equation does not represent the most general reduction to the one-body picture. This goal could be achieved by applying the technique of Green functions, as described by L.P. Kadanoff and G. Baym in (1962), or by the formalism developed by Keldish, L.V. (1964). In the following we want to study the semi-classical limit of (21.1) by applying Wigner transforms. This will first be done in explicit fashion for the left-hand side of this equation. An adequate form for the collision term will be discussed later.

21.1 The Wigner transform of the von Neumann equation

To be specific we will stick to the one-body system we are addressing in this chapter. It can be said, however, that the discussion of this subsection is of more general nature as an extension to other systems is more or less straightforward. For a system with a Hamiltonian $\hat{h}$ the von Neumann equation may be written as

$$\frac{d}{dt}\hat{\rho}(t) - \frac{1}{i\hbar}[\hat{h}, \hat{\rho}(t)] = 0. \qquad (21.2)$$

In coordinate representation it becomes

$$\frac{\partial}{\partial t}\rho(\mathbf{r}_1, \mathbf{r}_2, t) - \frac{1}{i\hbar}\int d^3r' \left(h(\mathbf{r}_1, \mathbf{r}', t)\rho(\mathbf{r}', \mathbf{r}_2, t) - \rho(\mathbf{r}_1, \mathbf{r}', t)h(\mathbf{r}', \mathbf{r}_2, t) \right) = 0, \qquad (21.3)$$

as is easily verified making use of the completeness relation $\int d^3r |\mathbf{r}\rangle\langle\mathbf{r}| = \hat{1}$. Applying (19.47) and its inverse (19.49) allows one to derive the equation of motion for the Wigner transform. After a lengthy calculation one obtains

$$\frac{\partial}{\partial t}n(\mathbf{r},\mathbf{p},t) - \frac{1}{i\hbar}\int \frac{d^3p'}{(2\pi\hbar)^3}\frac{d^3q'}{(2\pi\hbar)^3}\int d^3r'\, d^3s'\, \exp\left(\frac{i}{\hbar}\left[(\mathbf{p}'-\mathbf{p})\mathbf{s}' + \mathbf{q}'(\mathbf{r}-\mathbf{r}')\right]\right) \times$$
$$\left[h(\mathbf{r}'+\mathbf{s}'/2,\mathbf{p}'+\mathbf{q}'/2,t) - h(\mathbf{r}'-\mathbf{s}'/2,\mathbf{p}'-\mathbf{q}'/2,t)\right]n(\mathbf{r}',\mathbf{p}',t) = 0. \quad (21.4)$$

To get the semi-classical limit we need to collect terms of given power in $\hbar$. Let us abbreviate the integrals by $\mathcal{I}(\mathbf{r},\mathbf{p},t)/i\hbar$. Performing the following transformation $\mathbf{q}' \to \hbar\mathbf{x}$ and $\mathbf{s}' \to \hbar\mathbf{y}$, the $\mathcal{I}(\mathbf{r},\mathbf{p},t)$ can be written as

$$\mathcal{I}(\mathbf{r},\mathbf{p},t) = \int \frac{d^3p'}{(2\pi)^3} d^3r' \int \frac{d^3x}{(2\pi)^3} d^3y\, \exp\left(i\left[(\mathbf{p}'-\mathbf{p})\mathbf{y} + \mathbf{x}(\mathbf{r}-\mathbf{r}')\right]\right) \times$$
$$\left[h(\mathbf{r}'+(\hbar/2)\mathbf{y},\mathbf{p}'+(\hbar/2)\mathbf{x},t) - h(\mathbf{r}'-(\hbar/2)\mathbf{y},\mathbf{p}'-(\hbar/2)\mathbf{x},t)\right]n(\mathbf{r}',\mathbf{p}',t). \quad (21.5)$$

(Please notice that $\hbar$ no longer appears in the "volume" elements of the integrals.) This form enables one to derive a power series in $\hbar$. We just have to expand the Hamilton functions appearing in the square brackets with respect to $(\hbar/2)\mathbf{y}$ and $(\hbar/2)\mathbf{y}$. As a result one gets a sum over terms like:

$$\delta\mathcal{I}^m(\mathbf{r},\mathbf{p},t) = \hbar^m \int \frac{d^3p'}{(2\pi)^3} d^3r'\, \mathcal{J}_m(\mathbf{r}',\mathbf{p}',t) n(\mathbf{r}',\mathbf{p}',t), \quad (21.6)$$

where the dependence on $\hbar$ appears only as given by the pre-factors. The integrals over $\mathbf{y}$ and $\mathbf{x}$ are hidden in the $\mathcal{J}_m(\mathbf{r}',\mathbf{p}',t)$. The integrand there consists of terms involving the following types of factors: exponentials, powers in $\mathbf{y}$ and $\mathbf{x}$, and differential operators of order m in $\partial/\partial\mathbf{r}'$ and $\partial/\partial\mathbf{p}'$ which act on $h(\mathbf{r}',\mathbf{p}',t)$ (at $\mathbf{y}=0$ and $\mathbf{x}=0$). The final integrations over $d^3p'\, d^3r'$ may be carried out relating to simple identities of Fourier transformations, which in the end will involve δ-functions in $(\mathbf{r}-\mathbf{r}')$ and $(\mathbf{p}-\mathbf{p}')$. After performing appropriate partial integrations the powers in $\mathbf{y}$ and $\mathbf{x}$ finally show up in the corresponding powers of the differential operators $\partial/\partial\mathbf{r}$ and $\partial/\partial\mathbf{p}$ acting on $n(\mathbf{r},\mathbf{p},t)$. The leading order is found for $m=1$, which is to say by just considering $\delta\mathcal{I}^{m=1}(\mathbf{r},\mathbf{p},t)$ (see exercise 1). One obtains

$$\frac{\partial}{\partial t}n(\mathbf{r},\mathbf{p},t) - \left\{h(\mathbf{r},\mathbf{p},t),\, n(\mathbf{r},\mathbf{p},t)\right\} = 0, \quad (21.7)$$

with the curly brackets being the Poisson brackets (see (2.82)). Evidently, the form of this equation is identical to that of the Liouville equation.

At this stage a few comments may be appropriate.

(1) First one should note that just introducing the Wigner transform does not imply anything about a semi-classical limit: Equation (21.4) is *totally equivalent* to the original form (21.2) of the von Neumann equation.

(2) All manipulations which lead us from (21.4) to (21.7) *only* involve the *structure* of the equation, but *not the function* $n(\mathbf{r}, \mathbf{p}, t)$ we are looking for. There may be solutions to (21.7) which still contain genuine quantum effects. This feature is most readily seen from the special case of harmonic oscillators, see the next item. Truly, the equation itself does not contain any $\hbar$, but these quantum effects may come in through the additional conditions which one needs to pose for this differential equation in order to render the solutions unique.

(3) For any Hamiltonian which *depends on coordinates and momenta at most to order two*, *eqn(21.7) is equivalent to (21.4)*: In such a case all terms $\delta\mathcal{I}^m(\mathbf{r}, \mathbf{p}, t)$ with $m > 1$ vanish identically. (The reader is recommended to check this statement, for a simple example at least, say for one dimension.)

(4) For the general case, it is not too difficult to actually calculate the $\delta\mathcal{I}^m(\mathbf{r}, \mathbf{p}, t)$ to all orders. One then finds that the Wigner transform of the commutator can be written in the following concise form

$$\left(\frac{1}{i\hbar}[\hat{h}, \hat{\rho}]\right)^{\mathrm{W}} = \frac{2}{\hbar} h(\mathbf{r}, \mathbf{p}, t) \cdot \sin\left(\frac{\hbar}{2}\left[\overleftarrow{\nabla_\mathbf{r}}\overrightarrow{\nabla_\mathbf{p}} - \overleftarrow{\nabla_\mathbf{p}}\overrightarrow{\nabla_\mathbf{r}}\right]\right) \cdot n(\mathbf{r}, \mathbf{p}, t), \qquad (21.8)$$

which unfortunately is not very useful for practical applications.

(5) Finally, a warning regarding the expansion in powers of $\hbar$, which is a mere formal one. So far nothing is said about the convergence of this series. To ensure the latter, one may way wish to do some averaging over the spectrum (see e.g. (Strutinsky and Magner, 1976)).

21.2 Collision terms in semi-classical approximations

In this section we want to look at the collision term. In principle, its form should be derived in genuinely microscopic fashion starting from a description of the many-body system. This is possible within the formalisms mentioned at the beginning of this chapter; in the following we again want to refer mainly to the Kadanoff–Baym version. As such theories are beyond the scope of the present volume we will have to confine the discussion to a more heuristic derivation. It will be based mainly on semi-classical arguments, where amongst others we may treat subsequent collisions as being independent of each other. To simplify matters we will discard spin degrees of freedom and will finally apply perturbation theory, after having indicated a more correct procedure; for elaborate discussions of the general case we may refer to monographs on Landau theory, like that of D. Pines and P. Nozières, (1966) or of G. Baym and C.J. Pethick (1991)as well as of E.M. Lifshitz and I.P. Pitajewski (Landau et al., 1981).

The collision term is meant to portray how at some given point $(\mathbf{r}, \mathbf{p})$ in phase space the density $n(\mathbf{r}, \mathbf{p}, t)$ changes in time due to the collisions among particles, for which reason it is often written as

$$I[n(\mathbf{r}, \mathbf{p}, t)] \equiv \left.\frac{\partial}{\partial t} n(\mathbf{r}, \mathbf{p}, t)\right|_{\mathrm{coll}} \qquad (21.9)$$

There are two ways in which such a change may come about. Firstly, it may happen because the particle represented by the density $n(\mathbf{r}, \mathbf{p}, t)$ meets another one with momentum $\mathbf{p}_2$. Due to the mutual interaction they scatter to a different region in momentum space. Within a semi-classical picture we may assume this to happen at one given point $\mathbf{r} = \mathbf{r}_1 = \mathbf{r}_2 = \mathbf{r}_3 = \mathbf{r}_4$ in space. This process may thus simply be classified by the change of the momenta from $\mathbf{p} \equiv \mathbf{p}_1$ and $\mathbf{p}_2$ before the interaction to $\mathbf{p}_3$ and $\mathbf{p}_4$ after. It leads to a *depletion* of the volume element in phase space around $\mathbf{r}, \mathbf{p}$, for which reason the corresponding term in $\partial n(\mathbf{r}, \mathbf{p}, t)/\partial t|_{\text{coll}}$ is commonly called *"loss term"*. The *"gain term"* comes about because the inverse process is possible as well, where particles having momenta $\mathbf{p}_3$ and $\mathbf{p}_4$ scatter to $\mathbf{p}$ and $\mathbf{p}_2$ and thus *increase* the density inside the volume element at $\mathbf{r}, \mathbf{p}$.

Suppose the transition or the scattering from $\mathbf{p}, \mathbf{p}_2$ to $\mathbf{p}_3$ and $\mathbf{p}_4$ happens with probability $W(\mathbf{p}, \mathbf{p}_2; \mathbf{p}_3, \mathbf{p}_4)dt$. Because of the general invariance with respect to time and space reversal (parity) this transition probability (per unit time) must satisfy $W(\mathbf{p}, \mathbf{p}_2; \mathbf{p}_3, \mathbf{p}_4) = W(\mathbf{p}_3, \mathbf{p}_4; \mathbf{p}, \mathbf{p}_2)$. To get the loss of $n(\mathbf{r}, \mathbf{p}, t)$ per unit of time we have to consider three more points. First of all, we need to multiply the $W(\mathbf{p}, \mathbf{p}_2; \mathbf{p}_3, \mathbf{p}_4)$ by the probabilities $n(1)$ and $n(2)$ that the two single particle states are occupied. Evidently, these probabilities are simply given by the corresponding densities themselves $n(i) \equiv n(\mathbf{r}, \mathbf{p}_i, t)$ (all evaluated at $\mathbf{r}$). Secondly, as we are dealing with fermions not all final states are possible. This may be accounted for simply by multiplication with the probabilities of having the single particle states $|\mathbf{p}_3\rangle$ and $|\mathbf{p}_4\rangle$ *unoccupied*, which is to say by $1 - n(3)$ and $1 - n(4)$. Thirdly, we need to integrate over all states of the partners of particle 1. This means to integrate over all momenta $\mathbf{p}_2, \mathbf{p}_3, \mathbf{p}_4$, which leads to

$$\frac{\partial}{\partial t} n(\mathbf{r}, \mathbf{p}, t)\bigg|_{\text{coll}} = -\int \frac{d^3 p_2}{(2\pi\hbar)^3} \int \frac{d^3 p_3}{(2\pi\hbar)^3} \int \frac{d^3 p_4}{(2\pi\hbar)^3} \, W(\mathbf{p}, \mathbf{p}_2; \mathbf{p}_3, \mathbf{p}_4) \times$$
$$\left[n(\mathbf{r}, \mathbf{p}, t) n(2) \left(1 - n(3)\right) \left(1 - n(4)\right) - \left(1 - n(\mathbf{r}, \mathbf{p}, t)\right) \left(1 - n(2)\right) n(3) n(4) \right]. \quad (21.10)$$

Here, the "gain term" has already been included. It is obtained by interchanging in the integrand $(1, 2)$ by $(3, 4)$ and by accounting for an additional overall minus sign. It may be noted that for the scattering processes we are discussing here, the transition must obey conservation of total energy and momentum. Often this is written out explicitly using a slightly different notation for the $W(\mathbf{p}, \mathbf{p}_2; \mathbf{p}_3, \mathbf{p}_4)$.

It is worthwhile to reconsider the structure of (21.10) by simply looking explicitly at particle "1" before and after the collision. Identifying "particle 3" from above as "particle $1'$" one gets

$$\frac{\partial}{\partial t} n(1)\bigg|_{\text{coll}} = -\int \frac{d^3 p'}{(2\pi\hbar)^3} \left[W_{\mathbf{p},\mathbf{p}'} n(1) \left(1 - n(1')\right) - W_{\mathbf{p}',\mathbf{p}} \left(1 - n(1)\right) n(1') \right]. \quad (21.11)$$

Starting from (21.10) the probability $W_{\mathbf{p},\mathbf{p}'}$ for the transition of the momentum $\mathbf{p}$ to $\mathbf{p}'$ is easily seen to contain the two remaining integrals over $d^3 p_2 d^3 p_4$. The form

(21.11), however, is of more general nature. It also applies when the transition of the momentum of the particle we consider does not come from collisions with other particles of the same nature, as would be the case in mixtures of different constituents. Moreover, the transition might be caused by interactions with other degrees of freedom. One prime example is found for conduction electrons in a metal or semiconductor which may interact with impurities in the crystal or with the vibrational modes of the latter (the "phonons"). This association is needed in Chapter 15.

21.2.1 The collision term in the Born approximation

In principle, the transition rate $W(\mathbf{p}, \mathbf{p}_2; \mathbf{p}_3, \mathbf{p}_4)$ is to be calculated from the T-matrix governing the scattering process involved, see Section 2.7.1.3. In the Born approximation we may simply apply *Fermi's golden rule* (see Section 23.2.1):

$$W(\mathbf{p}, \mathbf{p}_2; \mathbf{p}_3, \mathbf{p}_4) = \frac{2\pi}{\hbar} \left| \langle 12|V|[34]\rangle \right|^2 \delta(e_1 + e_2 - e_3 - e_4). \qquad (21.12)$$

Here, V is the two-body interaction behind the collision and $|[\mathbf{p}_3\mathbf{p}_4]\rangle = (|\mathbf{p}_3\mathbf{p}_4\rangle - |\mathbf{p}_4\mathbf{p}_3\rangle)/\sqrt{2}$ is the antisymmetrized wave function for the final state. For a $V = V(\mathbf{r}_1 - \mathbf{r}_2)$ the matrix element $\langle 12|V|34\rangle$ can be expressed through the Fourier transform of $V(\mathbf{s})$

$$v(\mathbf{q}) = \int d^3s \, e^{-\frac{i}{\hbar}\mathbf{q}\mathbf{s}} V(\mathbf{s}). \qquad (21.13)$$

(Notice that because of $V(\mathbf{s}) = V(-\mathbf{s})$ and $V^*(\mathbf{s}) = V(\mathbf{s})$ one has $v(\mathbf{q}) = v(-\mathbf{q})$ as well as $v^*(\mathbf{q}) = v(\mathbf{q})$ (for real $\mathbf{q}$)). One gets

$$\langle \mathbf{p}_1\mathbf{p}_2|V|\mathbf{p}_3\mathbf{p}_4\rangle = v(\mathbf{p}_3 - \mathbf{p}_1)(2\pi\hbar)^3 \delta(\mathbf{p}_4 + \mathbf{p}_3 - \mathbf{p}_2 - \mathbf{p}_1), \qquad (21.14)$$

where the delta function stands for momentum conservation in the two-body collision. This form is verified by the following short calculation, exploiting again the transformation (19.48) to relative and center of mass coordinates,

$$\langle \mathbf{p}_1\mathbf{p}_2|V|\mathbf{p}_3\mathbf{p}_4\rangle =$$
$$\int d^3r \, d^3s \, \exp\!\left(\frac{i}{\hbar}(\mathbf{p}_3 - \mathbf{p}_1)(\mathbf{r} + \frac{\mathbf{s}}{2})\right) V(\mathbf{s}) \exp\!\left(\frac{i}{\hbar}(\mathbf{p}_4 - \mathbf{p}_2)(\mathbf{r} - \frac{\mathbf{s}}{2})\right) = \qquad (21.15)$$
$$(2\pi\hbar)^3 \delta(\mathbf{p}_4 + \mathbf{p}_3 - \mathbf{p}_2 - \mathbf{p}_1) \int d^3s \, \exp\!\left(\frac{i}{\hbar}(\mathbf{p}_3 - \mathbf{p}_1)\mathbf{s}\right) V(\mathbf{s}).$$

Piecing together all ingredients the collision term attains the form:

$$\frac{\partial}{\partial t} n(\mathbf{r}, \mathbf{p}, t)\bigg|_{\text{coll}} = -\frac{2\pi}{\hbar} \int \frac{d^3p_2}{(2\pi\hbar)^3} \int \frac{d^3p_3}{(2\pi\hbar)^3} \int \frac{d^3p_4}{(2\pi\hbar)^3}$$
$$\frac{1}{2}\Big(v(\mathbf{p} - \mathbf{p}_3) - v(\mathbf{p} - \mathbf{p}_4)\Big)^2 \delta(e + e_2 - e_3 - e_4)(2\pi\hbar)^3 \delta(\mathbf{p} + \mathbf{p}_2 - \mathbf{p}_3 - \mathbf{p}_4) \times \qquad (21.16)$$
$$\Big[n(\mathbf{r}, \mathbf{p}, t) n(2)(1 - n(3))(1 - n(4)) - (1 - n(\mathbf{r}, \mathbf{p}, t))(1 - n(2)) n(3) n(4) \Big].$$

Before we discuss some general properties of this functional, we should like to contemplate some aspects of validity. We have to go back to our starting point,

Fermi's golden rule for the transition probability. Quite generally, formulas like (2.88) cannot be correct for very small and very large time lapses Δt. More precisely, it can be argued that the following relation must be satisfied (see e.g. the books on quantum mechanics by G. Baym (1969) and F. Schwabl (1995))

$$\frac{2\pi\hbar}{\Delta e} \ll \Delta t \ll \frac{2\pi\hbar}{\delta e}. \qquad (21.17)$$

Here, Δe is the spread in energy over which the interaction or coupling acts, which causes the transition, and δe is the average distance between neighboring levels. In more pictorial language suitable to the present problem, one may say that the "duration" of one collision must be shorter than the time interval to the next one. Let us not try to relate these two formulations in a more quantitative sense. For the moment it may suffice to know of the existence of such conditions. In any case it should be clear from this discussion that our approximation has to break down for the more dense systems.

21.2.2 The BUU and the Landau–Vlasov equation

Putting the expression found in (21.16) on the right-hand side of (21.7) one obtains the equation

$$\frac{\partial}{\partial t}n(\mathbf{r},\mathbf{p},t) - \left\{h(\mathbf{r},\mathbf{p},t),\, n(\mathbf{r},\mathbf{p},t)\right\} = \left.\frac{\partial}{\partial t}n(\mathbf{r},\mathbf{p},t)\right|_{\text{coll}}. \qquad (21.18)$$

As will be explained shortly, it encompasses several versions of transport theories, of which each one plays an immense role in different fields of physics. Unfortunately, this equation is not easy to solve. It is of integro-differential nature and *nonlinear* in the function $n(\mathbf{r},\mathbf{p},t)$ one has to solve for. A word of warning may be in order concerning the handling of perturbation theory. Defining the collision term through (21.16) in second order in the interaction v does not mean that the whole equation should be solved to the same order. It is totally legitimate to seek general solutions in which the $I[n]$ is taken into account in full, non-perturbative sense. We will demonstrate this feature in Section 21.3 for the example where (21.18) is solved explicitly for a simple situation.

Let us finally note the connection with the original Boltzmann equation. If the system behaves classically the occupation numbers $n(i)$ are much smaller than unity. Then the $1-n(i)$ may be approximated by 1, and the collision term described above turns into the one of the Boltzmann equation. Such a situation is given, for instance, close to an equilibrium of sufficiently high temperatures. An equation like (21.18) was first suggested by E.A. Uehling and G.E. Uhlenbeck (1933) for a system where the potential in $h(\mathbf{r},\mathbf{p},t)$ is given by an "external" field. In nuclear physics literature this equation is commonly referred to as the BUU-equation. It represents an extension of the classic Boltzmann equation to the realm of quantum mechanics. As indicated previously, the form of Boltzmann's equation may be obtained formally by putting the factors $(1-n(i))$ equal to unity, neglecting in this way all relics of the Pauli principle. (An extensive discussion of the Boltzmann equation and its properties, together with applications, is found in the book by J.J. Duderstadt and W.R. Martin (1979)).

Conservation laws: It can be shown that the transport equation satisfies the fundamental *conservation laws* for *total particle number, total energy and total momentum*. They are represented by differential relations for the corresponding densities. With respect to particle number it is just the common continuity relation between the space density and its associated current density $\mathbf{j}(\mathbf{r}, t)$, namely

$$\frac{\partial}{\partial t}\rho(\mathbf{r}, t) + \nabla_\mathbf{r} \mathbf{j}(\mathbf{r}, t) = 0, \qquad (21.19)$$

with the $\mathbf{j}(\mathbf{r}, t)$ being defined as

$$\mathbf{j}(\mathbf{r}, t) = \int \frac{d^3 p_2}{(2\pi\hbar)^3} \frac{\mathbf{p}}{m} n(\mathbf{r}, \mathbf{p}, t). \qquad (21.20)$$

Similar relations hold true for the other laws. They are obtained by first multiplying (21.18) by:

(1) 1 for particle number,
(2) $e(\mathbf{r}, \mathbf{p}, t)$ for energy and
(3) $\mathbf{p}$ for momentum

and then integrating the resulting equations over all $\mathbf{p}$. In all these cases the integrals of the collision term vanish. Notice, please, that for the proof of energy conservation (2) one needs to take that energy $e(\mathbf{r}, \mathbf{p}, t)$ which also appears in the δ-function of the collision term and which ought to correspond to the Hamilton function $h(\mathbf{r}, \mathbf{p}, t)$ of the Liouville part of the transport equation. Only in the case of free particles will this be the kinetic energy $p^2/2m$, as is appropriate for dilute gases. We refrain from writing down the explicit forms of the conservation laws for energy and momentum. They may be found, for instance, in the books by Kadanoff–Baym (1962) or Landau and Lifshitz (1981).

Let us add two remarks on the content of quantum effects. Firstly, according to the comments made at the end of Section 21.1, it may be clear that they do not appear simply because of the Pauli factors. Secondly, to avoid any misunderstanding, even when the latter are included this does not mean that the medium effects are taken into account correctly.

21.3 Relaxation to equilibrium

For a finite collision term a transport equation like (21.1) or its various semi-classical versions describe relaxation to equilibrium. Let us clarify this concept by studying a specific but simple process. Suppose we take our (many-body) system at global equilibrium at some finite temperature T. Assuming homogeneity, the $n(\mathbf{r}, \mathbf{p}_i, t)$ becomes independent of $\mathbf{r}$ and is simply given by the Fermi distribution

$$n(\mathbf{r}, \mathbf{p}_i, t) = n_{\text{eq}}(p_i) = \frac{1}{1 + \exp((e(p_i) - \mu)/T))}, \qquad (21.21)$$

with the chemical potential μ. Notice, that the Poisson bracket of the Hamiltonian with the equilibrium density vanishes. The same can be seen to hold true for the collision term, implying that the Fermi distribution is a solution of the basic equation (21.18) (not only within the Born approximation). Suppose further that at time $t = 0$ we add one particle of momentum $\mathbf{p}$ to this system. Whereas the distribution $n(\mathbf{r}, \mathbf{p}, t)$ of this particle will start with a $n(\mathbf{r}, \mathbf{p}, t = 0) \neq n_{\text{eq}}(p)$ and will remain so for some time, those of all the other particles may safely be assumed always to stay very close to $n_{\text{eq}}(p_i)$. How close it is will actually depend on the size of this initial disturbance in relation to the size of the total system, such that in a "Gedanken experiment" we may assume our condition to be perfectly fulfilled. Then the influence of the collision term can be represented by the following simple relation

$$\frac{\partial}{\partial t} n(\mathbf{r}, \mathbf{p}, t) \bigg|_{\text{coll}} = -n(\mathbf{r}, \mathbf{p}, t) \frac{\Sigma^>(p)}{\hbar} + \left(1 - n(\mathbf{r}, \mathbf{p}, t)\right) \frac{\Sigma^<(p)}{\hbar}. \qquad (21.22)$$

This is obtained by singling out the factors $n(\mathbf{r}, \mathbf{p}, t)$ and $(1 - n(\mathbf{r}, \mathbf{p}, t))$ on the right-hand side of (21.16). Because of the special situation just mentioned the remaining factors, which are called $\Sigma^>(p)/\hbar$ and $\Sigma^<(p)/\hbar$, respectively, do not depend anymore on the unknown function $n(\mathbf{r}, \mathbf{p}, t)$, but on the equilibrium distribution instead. It is for this reason that they just appear as mere factors to the unknown functions $n(\mathbf{r}, \mathbf{p}, t)$ and $(1 - n(\mathbf{r}, \mathbf{p}, t))$, and thus simply as coefficients in a differential equation, which in this way has become a *linear one*. To simplify matters further, let us assume that the "transport" in $\mathbf{r}$ and $\mathbf{p}$ can be neglected, which in (21.18) would be determined by the Poisson brackets on the left-hand side. In this case, the $\mathbf{r}, \mathbf{p}$ may be treated as fixed quantities such that we simply obtain an ordinary differential equation in time for the quantity $n(t) = n(\mathbf{r}, \mathbf{p}, t)$, namely

$$\frac{d}{dt} n(t) = -n(t) \frac{1}{\hbar} \left(\Sigma^>(p) + \Sigma^<(p)\right) + \frac{\Sigma^<(p)}{\hbar}. \qquad (21.23)$$

This form suggests introducing the following object

$$\Gamma(p) = \Sigma^>(p) + \Sigma^<(p) \qquad (21.24)$$

Obviously, (21.23) implies that the *time-in*dependent solution has to satisfy the relation $\Sigma^<(p) = \Gamma(p) n_{\text{eq}}(p)$. Thus the time-dependent solution $n(t)$ will be given by

$$n(p, t) = n_{\text{eq}}(p) + e^{-\Gamma(p) t/\hbar} \left(n(p, t = 0) - n_{\text{eq}}(p)\right) \qquad (21.25)$$

showing that the Γ, as given by (21.24), determines the relaxation to equilibrium, with the *relaxation time* τ being given by $\tau = \hbar/\Gamma$. More precisely, as we have discarded the effects of the left-hand side of (21.16), we ought to call it the *collisional* relaxation time.

Fortunately, according to P. Morel and P. Nozières (1962) the integrals over the momenta which appear in the expressions for $\Sigma^{\gtrless}$ and thus for Γ, and which are given by (21.16), can be calculated under quite general conditions. To lowest order in both the temperature as well as in the deviation of the particles' energy from the chemical potential one finds

$$\Gamma(p) = \frac{1}{\Gamma_0}\left[(e(p)-\mu)^2 + \pi^2 T^2\right]. \tag{21.26}$$

Now the interaction is hidden in the coefficient $1/\Gamma_0$. It is written in such a way that the Γ_0 has the dimension of energy. A typical value for nuclear physics is $\Gamma_0 \approx 33$ MeV or $1/\Gamma_0 \approx 0.03(\text{MeV})^{-1}$ (Hofmann, 1997). Notice, please that the discussion here is not restricted to using perturbation theory when defining the collision term as in (21.16). In general, the self-energies, and thus the $\Gamma(p)$ are to be calculated from the T (or G) matrix. However, if (21.16) had been used, the $1/\Gamma_0$ would of course be of second order in v. Notice also that even in such a case the solution (21.25) contains v to all orders. This is a simple demonstration of the remarks made below eqn(21.18).

It is perhaps worthwhile to examine the form of the $\Gamma(p)$ given in (21.26). For zero temperature it vanishes at the Fermi surface and it will be small in its immediate neighborhood. Evidently, this is an effect of the Pauli principle. Recall that in the present case the $\Gamma(p)$ corresponds to a particle added to the system in equilibrium. At $T=0$, this is possible only for a $e(p)$ which is slightly larger than the chemical potential μ. For very small excitations the Pauli principle will still be effective and it will become weaker the larger the excitation will be. The same effect is responsible for the temperature dependence. Essentially, the only difference is seen in the fact that $e(p) - \mu$ measures a specific excitation of an individual particle, whereas T stands for an excitation in the statistical sense of the system as a whole. Clearly, for a $T>0$ a particle may also be added for $e(p) < \mu$, as then the Fermi sea gets depleted to an extent as given by the Fermi distribution (21.21). Finally, we should like to refer to Section 4.3 where such aspects are looked at from a different point of view, with the aim of direct applications to finite systems. Actually, in Section 4.3.4.4 it is seen that at $T=0$ the form (21.26) does well represent the average trend for the widths of particles and holes in finite nuclei.

21.3.1 *Relaxation time approximation to the collision term*

The physical meaning of the relaxation time described in the previous section may serve as justification for approximating the collision term by the following expression

$$\left.\frac{\partial}{\partial t}n(\mathbf{r},\mathbf{p},t)\right|_{\text{coll}} \approx -\frac{n(\mathbf{r},\mathbf{p},t) - n_{\text{eq}}}{\tau} \equiv -\frac{\delta n(\mathbf{r},\mathbf{p},t)}{\tau}. \tag{21.27}$$

This appears to be meaningful for situations where one is interested in describing the dynamics of a system close to thermal equilibrium. Indeed, forgetting once more the Poisson bracket on the left-hand side of the equation of motion (21.18), the result (21.25) obtained above follows more or less directly. One only needs to replace the left-hand side of (21.27) by $\partial n(t)/\partial t \equiv \partial \delta n(t)/\partial t$, which is possible because n_{eq} is constant in time.

Evidently, the transport equation following here from,

$$\frac{\partial}{\partial t} n(\mathbf{r},\mathbf{p},t) - \left\{ h(\mathbf{r},\mathbf{p},t)\,,\, n(\mathbf{r},\mathbf{p},t) \right\} = -\frac{n(\mathbf{r},\mathbf{p},t) - n_{eq}}{\tau}, \qquad (21.28)$$

is much easier to solve than the BUU equation with the full collision term, simply because (21.28) is *linear in the unknown function* $n(\mathbf{r},\mathbf{p},t)$. It is important to note that the equilibrium distribution n_{eq} appearing here is that of *global equilibrium*: It does not depend on the position $\mathbf{r}$ anymore. An improved version of the relaxation time approximation is possible by introducing the concept of *local equilibrium*. There, both the temperature and the chemical potential may change with $\mathbf{r}$. Moreover, the Fermi function (21.21) will get an additional term in the exponent if the fluid of particles moves with a finite local velocity. To account for such effects the right-hand side of (21.28) has to be modified. Then special care must be taken to ensure that (21.28) still obeys the conservation laws. For more details the reader is referred to the literature given above.

21.3.2 A few remarks on the concept of self-energies

The features found in the discussion of the relaxation phenomenon allows one to establish connection with the concept of self-energies and to understand a few general properties of the latter. To be specific let us assume the $h(\mathbf{r},\mathbf{p},t)$ in the transport equation to be that of Hartree–Fock (for the Skyrme type of interaction). The energy $e(p)$ of our particle which corresponds to this h is then given by $e(p) = p^2/2m + U^D$ (recall eqn(2.83), for instance). Suppose we are dealing with homogeneous matter once more, then the U^D is either a constant or a function of p. The form of the solution (21.25) suggests the interpretation that the collisions with the other particles modify this energy to a complex one:

$$e(p) = p^2/2m + U^D \quad \Longrightarrow \quad e(p) = p^2/2m + U^D - i\Gamma/2. \qquad (21.29)$$

It has become customary to call the $-\Gamma/2$ the imaginary part of the *self-energy*. More generally, one may say that in the present approximation the total self-energy Σ_p is given by

$$\Sigma_p = U^D - i\Gamma/2 \equiv \Sigma'_p + i\Sigma''_p, \qquad (21.30)$$

with Σ'_p and Σ''_p being the real and imaginary parts and of the self-energy to the particle which is classified by its momentum p (in general it eventually would have to be the vector $\mathbf{p}$).

Our introduction of self-energies may not sound convincing to readers who have never seen this concept before. They may be assured, however, that the relations deduced here are valid in general. This is true not only in connection with one-body Green functions but with the transport equation for the one-body density as well. The $\Sigma^{\gtrless}$ are different branches of the Σ Those found here represent that self-energy which is due to collisions; in more general terms, they represent effects of the interaction beyond Hartree–Fock. The general Σ will depend not only on p but on frequency (or time) as well. The notation $\gtrless$ comes from the time-dependent picture: The $\Sigma^{\gtrless}(p, t - t')$ represent the $\Sigma(p, t - t')$ for times $(t - t') \gtrless 0$, respectively. In fact, in the Kadanoff–Baym theory the quantum mechanically correct transport equation is expressed entirely in terms of self-energies.

In the Σ_p deduced above the Σ'_p is of order one in the interaction whereas Σ''_p is of order two. For the bare interaction we have been using here, the second property is correct and of general nature. The feature mentioned first, however, is incorrect and due to our approximate way of handling things. The contribution to Σ'_p which is of second order in the interaction has simply been discarded. Quite generally, real and imaginary parts ought to be connected to each other by dispersion relations.

In Brueckner–Hartree–Fock theory things are a bit more subtle, in the sense that there an imaginary part of the self-energy shows up already in *first order* in the G-matrix. The latter only has to be calculated along the lines indicated in Section 2.2.4 in connection with the optical model. This may be taken as another hint at deficiencies of the common applications of BUU in connection with the Skyrme type interactions used to define $h(\mathbf{r}, \mathbf{p}, t)$.

Exercises

1. Evaluate the $\mathcal{I}(\mathbf{r}, \mathbf{p}, t)$ of (21.5) up to first order in $\hbar$ and show that to this order the transport equation eqn(21.4) turns into (21.7).

22

NUCLEAR THERMOSTATICS

Nuclear excitations are often visualized by thermal properties like temperature, heat or entropy. However, as the nucleus is a small and isolated system it must be clarified to what extent these concepts make sense at all. Of crucial issue are the relative thermodynamic fluctuations which (for small systems) may become quite large. We shall return to such questions later in this chapter. At first we shall present the most relevant features of thermostatics. In their selection we shall be guided largely by a pragmatic point of view, namely as to discuss those topics needed for the understanding of nuclear collective dynamics at finite thermal excitations. As in this section we will only study static situations we prefer to use the title thermo*statics* rather than the more common notion thermodynamics. Evidently, as finally we shall address the question of applicability to nuclear physics, a few points of more fundamental nature are unavoidable. As far as common textbook knowledge is concerned we will heavily benefit from the expositions given in the following texts (Landau and Lifshitz, 1980), (Balian, 1991), (Brenig, 1996), (Schwabl, 2000).

22.1 Elements of statistical mechanics

For our purpose it is of interest to start from quantum statistical mechanics. One of the basic tools is the density operator $\hat{\rho}$ normalized to unity by tr $\hat{\rho} = 1$. It is meant to contain all information on the system which possibly is available. At equilibrium it is supposedly specified by the Hamiltonian $\hat{H}$. If the latter is precisely known, the dynamics would be governed by the von Neumann equation

$$i\hbar \frac{d\hat{\rho}}{dt} = [\,\hat{H}\,,\,\hat{\rho}(t)\,]\,. \qquad (22.1)$$

Provided it is strictly valid, any density operator which is a functional of this Hamiltonian, $\hat{\rho} = \hat{\rho}[\hat{H}]$, is a stationary solution of this equation. In preparation for the second law it is vital to realize that such a situation is not given for irreversible systems, to which we will come below. To begin with let us look at the two most common ensembles for which the static density operator $\hat{\rho}$ is a functional of the Hamiltonian $\hat{H}$ alone with no other observables involved.

(i) Microcanonical ensemble: In terms of the energies E_m and eigenstates $|m\rangle$ of $\hat{H}$ the density operator of this ensemble may be written as

$$\hat{\rho}(E) = \begin{cases} (\Delta G(E))^{-1} \sum_m |m\rangle\langle m| & \text{if } E - \Delta \leq E_m \leq E, \\ 0 & \text{otherwise}. \end{cases} \qquad (22.2)$$

where $\Delta G(E)$ is the number of levels in the given interval. The width is tacitly assumed to be much smaller than the energy E itself,

$$\Delta/E \ll 1, \tag{22.3}$$

but large enough to include sufficiently many states such that a statistical treatment makes sense.

(ii) Canonical ensemble: It is defined with respect to the following density operator (in units where the Boltzmann constant k_B is equal to one).

$$\hat{\rho}(\beta) = \frac{1}{Z(\beta)} e^{-\beta \hat{H}} \equiv \frac{1}{Z(\beta)} \sum_m e^{-\beta E_m} |m\rangle\langle m|, \tag{22.4}$$

and involves the parameter β which can be shown to be the inverse temperature $\beta = 1/T$, see below. The $Z(\beta)$ is the partition function given by:

$$Z(\beta) = \operatorname{tr} e^{-\beta \hat{H}} = \sum_m e^{-\beta E_m}. \tag{22.5}$$

Notice that the operators (22.2) and (22.4) of both ensembles are of the form (19.15), namely $\hat{\rho} = \sum_m \rho_m |m\rangle\langle m|$ where $\operatorname{tr} \hat{\rho} = 1$ implies $\sum_m \rho_m = 1$.

One defines as internal energy $\mathcal{E}$ the average of the Hamiltonian,

$$\mathcal{E} = \operatorname{tr} \hat{\rho} \hat{H}, \tag{22.6}$$

and the entropy is given by

$$\mathcal{S}[\hat{\rho}] = -\langle \ln \hat{\rho} \rangle = -\operatorname{tr}(\hat{\rho} \ln \hat{\rho}). \tag{22.7}$$

In energy representation (19.15) both expressions write

$$\mathcal{E} = \sum_m \rho_m E_m \qquad \mathcal{S} = -\sum_m \rho_m \ln \rho_m. \tag{22.8}$$

In the microcanonical ensemble the entropy takes on L. Boltzmann's form,

$$\mathcal{S} = \ln \Delta G(E). \tag{22.9}$$

The two ensembles can be seen to maximize the entropy under the conditions of (i) fixed energy E (to precision Δ), or (ii) fixed temperature (adjusted to the constraint of given average energy $\mathcal{E}$, see the discussion below in Section 22.1.2).

22.1.1 Thermostatics for deformed nuclei

Let us address the situation typical for low energy nuclear physics, in which the density operator may depend on some additional parameters Q_μ specifying a possible deformation of the nucleus. Evidently, such a dependence comes in through a Hamiltonian of the type $\hat{H} = \hat{H}(\hat{x}_i, \hat{p}_i, Q_\mu)$ discussed in Chapter 6. To keep the notation simple we will often just take one collective variable and will refrain from writing explicitly the nucleonic coordinates $\hat{x}_i$ and momenta $\hat{p}_i$. To begin with, we are going to study the first law by analogy with the situation known from textbooks, where the Q typically takes the role of the system's volume, for instance.

22.1.1.1 The first law
The first law involves variations of energy and entropy in relation to the dQ. For $Q + dQ$ we may write

$$\hat{H}(Q + dQ) = \hat{H}(Q) + \hat{F} dQ \equiv \hat{H}(Q) + d\hat{H} \quad \text{with} \quad \hat{F} = \left.\frac{\partial \hat{H}}{\partial Q}\right|_Q \quad (22.10)$$

for the Hamiltonian and

$$\hat{\rho}(Q + dQ, X + dX) = \hat{\rho}(Q, X) + d\hat{\rho} \equiv \hat{\rho} + d\hat{\rho} \quad (22.11)$$

for the density operator, if X is the additional parameter necessary to specify the thermal excitation, like entropy or temperature. To first order in $d\hat{H}$ and $d\hat{\rho}$ the variation of energy is given by

$$\begin{aligned} d\mathcal{E} &\equiv \operatorname{tr} \hat{\rho}(Q + dQ, X + dX)\hat{H}(Q + dQ) - \operatorname{tr} \hat{\rho}(Q, X)\hat{H}(Q) \\ &\simeq \operatorname{tr} \hat{\rho}(Q, X) d\hat{H} + \operatorname{tr} d\hat{\rho} \hat{H}(Q) \,. \end{aligned} \quad (22.12)$$

The variation of entropy can be proven to be (see exercise 1)

$$d\mathcal{S}[\hat{\rho}] = -\operatorname{tr}\left(d\hat{\rho} \ln \hat{\rho}(Q, X)\right). \quad (22.13)$$

Let us assume the $\hat{\rho}(Q, X)$ to represent the canonical distribution (22.4) of an equilibrium specified by the Hamiltonian $\hat{H}(Q)$ and by a temperature $X = T$. Accounting for the normalization of the density one easily gets

$$d\mathcal{S} = \beta \operatorname{tr} d\hat{\rho} \, \hat{H}(Q) \,. \quad (22.14)$$

Note that β comes in through the $\ln \hat{\rho}$ which appears in (22.13). No assumption needs to be made about the temperature being constant. A possible variation still may be accounted for in $d\hat{\rho}$. Thus the relation (22.12) attains the common form of the first law

$$d\mathcal{E} = \delta \mathcal{W} + T d\mathcal{S} \quad \text{with} \quad \delta \mathcal{W} = \langle \hat{F} \rangle_Q dQ \quad (22.15)$$

being the *work* exchanged with "external forces". In practical applications to macroscopic systems one needs to introduce the concept of *heat* $\mathcal{Q}_h$ through the relation

$$\delta \mathcal{Q}_h = \operatorname{tr} d\hat{\rho} \, \hat{H} = T \, d\mathcal{S} \,. \quad (22.16)$$

As one knows, unlike $d\mathcal{S}$ the $\delta \mathcal{Q}_h$ is not a complete differential. In case of a vanishing $\delta \mathcal{Q}_h$ one speaks of a *thermally* isolated system (Landau and Lifshitz, 1980). A change in energy is then only possible through a variation of the parameter Q, which in general represents an external field, but see the discussion in Section 22.1.1.3 for an interpretation of this concept in the nuclear case. For a genuinely *isolated or closed system* no coupling to the external world exists at all. Even for a thermally isolated system, i.e. for $d\mathcal{S} = 0$, the temperature may change. Below in eqn(22.31) a formula is given for the variation of T with the parameter Q, which in Section 23.6.2 is generalized to the multidimensional case. Often it is useful to rewrite (22.12) in the representation of the eigenstates $|m\rangle$ of the Hamiltoinan,

$$d\mathcal{E} = \sum_m \rho_m dE_m + \sum_m d\rho_m E_m . \tag{22.17}$$

The $d\rho_m$ measures a change in the occupation number representing either a warming up or a cooling of the system.

At fixed entropy $d\mathcal{S} = 0$ the variation of the internal energy is given solely by the first term of (22.12), for which reason one gets

$$\left(\partial \mathcal{E}/\partial Q\right)_\mathcal{S} = \langle \partial \hat{H}/\partial Q \rangle \equiv \text{tr}\,[\hat{\rho}\,\partial \hat{H}/\partial Q], \tag{22.18}$$

as a direct generalization of the case of zero temperature. Introducing the free energy $\mathcal{F}(Q,T)$ through the partition function by

$$Z(Q,T) = \text{tr}\left[\exp(-\hat{H}(Q)/T)\right] = \exp(-\mathcal{F}(Q,T)/T), \tag{22.19}$$

the quasi-static force (22.18) may be rewritten in the form

$$\left(\partial \mathcal{F}/\partial Q\right)_T = -T\left(\partial \ln Z/\partial Q\right)_T = \langle \partial \hat{H}/\partial Q \rangle = \text{tr}\,[\hat{\rho}\,\partial \hat{H}/\partial Q]. \tag{22.20}$$

Both expressions are seen to be generalizations of the well-known formula which relates pressure to derivatives of energies with respect to volume (at fixed particle number $\mathcal{N}$, see below), namely

$$\mathcal{P} = -\left(\partial \mathcal{E}/\partial \mathcal{V}\right)_\mathcal{S} = -\left(\partial \mathcal{F}/\partial \mathcal{V}\right)_T . \tag{22.21}$$

Indeed, the volume $\mathcal{V}$ may easily be related to parameters specifying the Hamiltonian, say for the motion of a gas or liquid inside a container. Because of this analogy this quasi-static force may be denoted by the symbol $\mathcal{P}_Q$, which means identifying

$$-\mathcal{P}_Q \equiv \langle \partial \hat{H}/\partial Q \rangle = \left(\partial \mathcal{F}/\partial Q\right)_T = \left(\partial \mathcal{E}/\partial Q\right)_\mathcal{S} . \tag{22.22}$$

From the microscopic definition of entropy (22.7), together with that of the density operator (22.4) in the canonical ensemble one obtains

$$\mathcal{E} = \mathcal{F} + T\mathcal{S}. \tag{22.23}$$

This equation defines the Legendre transformation from temperature to entropy, which in differential form reads

$$d\mathcal{E} = T d\mathcal{S} - \mathcal{P}_Q dQ \quad \text{or} \quad d\mathcal{F} = -\mathcal{S} dT - \mathcal{P}_Q dQ, \tag{22.24}$$

from which the formula $\mathcal{S} = -(\partial \mathcal{F}/\partial T)_Q$ becomes evident. These relations allow one to transform from the one combination (T,Q) to the other one $(\mathcal{S},Q)$, without involving the density operator itself.

22.1.1.2 Second derivatives and susceptibilities
Within a harmonic approximation to collective motion one needs to consider derivatives of second order. Starting from (22.20) (or (22.18)) one gets for $\mathcal{X} = \mathcal{F}$ (or $\mathcal{E}$) and $X = T$ (or S)

$$(\partial^2 \mathcal{X}/\partial Q^2)_X = \langle \partial^2 \hat{H}/\partial Q^2 \rangle + \mathrm{tr}\,[(\partial \hat{H}/\partial Q)(\partial \hat{\rho}/\partial Q)_X].$$

For fixed $Q = Q_0$ the derivative of first order, $\partial \hat{H}/\partial Q$, becomes identical to the generator $\hat{F}$ of collective motion around the Q_0 (see Section (6.1.1)). Using the definition

$$\chi^X = -\mathrm{tr}\,\left[\hat{F}\left(\frac{\partial \hat{\rho}}{\partial Q}\right)_X\right]_{Q_0, X_0} = -\mathrm{tr}\,\left[\frac{\partial \hat{H}}{\partial Q}\left(\frac{\partial \hat{\rho}}{\partial Q}\right)_X\right]_{Q_0, X_0} \quad (22.25)$$

for either the adiabatic ($X = S$) or isothermal susceptibility ($X = T$), one may thus write:

$$\left\langle \partial^2 \hat{H}/\partial Q^2 \right\rangle_{Q_0, X_0} = (\partial^2 \mathcal{E}/\partial Q^2)_{Q_0, S_0} + \chi^{\mathrm{ad}} = (\partial^2 \mathcal{F}/\partial Q^2)_{Q_0, T_0} + \chi^{\mathrm{T}}. \quad (22.26)$$

Relations of this type are commonly derived for cases where the Hamiltonian depends only linearly on Q, in which case the left-hand side vanishes. We will come back to this situation later in Section 22.1.2.1 (see eqn(22.48)). Of interest is the difference of both susceptibilities,

$$\chi^{\mathrm{T}} - \chi^{\mathrm{ad}} = \left(\frac{\partial^2 \mathcal{E}}{\partial Q^2}\right)_{Q_0, S_0} - \left(\frac{\partial^2 \mathcal{F}}{\partial Q^2}\right)_{Q_0, T_0} = -\left(\frac{\partial S}{\partial Q}\right)_T \frac{\partial^2 \mathcal{E}}{\partial Q \partial S}. \quad (22.27)$$

Here, for the final expression on the very right the relation

$$(\partial^2 \mathcal{F}/\partial Q^2)_T = (\partial^2 \mathcal{E}/\partial Q^2)_S + (\partial^2 \mathcal{E}/\partial Q \partial S)(\partial S/\partial Q)_T \quad (22.28)$$

was used. The latter is obtained by differentiating the force $\mathcal{P}_Q$ of (22.22) with respect to Q obeying the relation

$$(\partial \mathcal{P}_Q/\partial Q)_T = (\partial \mathcal{P}_Q/\partial Q)_S + (\partial \mathcal{P}_Q/\partial S)_Q (\partial S/\partial Q)_T, \quad (22.29)$$

for the functional form $\mathcal{P}_Q = \mathcal{P}_Q(Q, S(T, Q))$. Had we interchanged S and T, i.e. used the form $\mathcal{P}_Q = \mathcal{P}_Q(Q, T(S, Q))$, we would have arrived at

$$\chi^{\mathrm{T}} - \chi^{\mathrm{ad}} = \left(\frac{\partial T}{\partial Q}\right)_S \frac{\partial^2 \mathcal{F}}{\partial Q \partial T}. \quad (22.30)$$

The first factor on the right can be written as

$$\left(\frac{\partial T}{\partial Q}\right)_S = -\left(\left(\frac{\partial S}{\partial T}\right)_Q\right)^{-1} \left(\frac{\partial S}{\partial Q}\right)_T = -\left(\left(\frac{\partial^2 \mathcal{F}}{\partial T^2}\right)_Q\right)^{-1} \frac{\partial^2 \mathcal{F}}{\partial Q \partial T}. \quad (22.31)$$

This follows from $dS = (\partial S/\partial Q)_T \, dQ + (\partial S/\partial T)_Q \, dT$ after taking $dS = 0$ and using twice $S = -(\partial \mathcal{F}/\partial T)_Q$. Putting these results together one arrives at

$$\chi^{\text{T}} - \chi^{\text{ad}} = -\left[\left(\left(\frac{\partial^2 \mathcal{F}}{\partial T^2}\right)_Q\right)^{-1}\left(\frac{\partial^2 \mathcal{F}}{\partial T \partial Q}\right)^2\right]_{Q_0, T_0}. \tag{22.32}$$

In Section 23.6.2 it is shown how the difference of the two susceptibilities can be expressed in terms of fluctuations. For the present case formula (23.159) reduces to (see exercise 2)

$$\chi^{\text{T}} - \chi^{\text{ad}} = \frac{1}{T_0}\left(\frac{\langle \Delta \hat{F} \Delta \hat{H} \rangle \langle \Delta \hat{H} \Delta \hat{F} \rangle}{\langle \Delta \hat{H}^2 \rangle}\right)_{Q_0 T_0}, \tag{22.33}$$

where $\Delta \hat{O} = \hat{O} - \langle \hat{O} \rangle$. It is seen that $\chi^{\text{T}} - \chi^{\text{ad}} \geq 0$. Equation (22.30) implies the following variation of temperature with Q at constant entropy

$$dT = \left(\frac{\partial^2 \mathcal{F}}{\partial Q \partial T}\right)^{-1}\left(\chi^{\text{T}} - \chi^{\text{ad}}\right) dQ \quad \text{for} \quad dS = 0, \tag{22.34}$$

which is in accord with the result derived in Section 23.6.2 (see exercise 3).

22.1.1.3 *Comments on applications to finite nuclei* Before we continue some remarks are in order on the notion "external" and "internal" as used for nuclear physics. As indicated above, the Q may define the shape of the mean field in which the nucleons move. In this case this Q is "external" to the system of nucleons, like the volume is to the motion of a gas or liquid inside a container. Within a thermo-*static* picture one likes to treat the value of Q as a free parameter. In the nuclear case this is *not possible* in a real sense, as there is no genuine external field of *static* nature. On the other hand, time-dependent perturbations do exist, but they will excite the nucleus as a whole–including excitations of the shape degrees of freedom, which imply dynamics in the Q itself. Such excitations are then to be understood as being "internal" with respect to the truly external (time-dependent) fields which produce the excitation. Indeed, this interpretation of $Q(t)$ being an internal degree of freedom shows up more or less naturally as soon as *one considers self-consistency* between the density and the mean field. As demonstrated in Chapter 6, the $Q(t)$ then is related to an average of a one-body operator of the system of nucleons. Nevertheless, developing a theory for the time evolution of the full system, the quasi-static picture may well be used as a valid theoretical means. In this spirit one may also wish to exploit a concept of a "local" equilibrium–defined here in the sense that the equilibrium distribution of the whole set of nucleons may explicitly depend on the collective variable Q. The notion of "local" equilibrium is borrowed from hydrodynamics. There, however, "locality" is to be understood with respect to ordinary three-dimensional space. This type of equilibrium requires the mean free path of the constituents to be much smaller than the diameter of the system as a whole. Such a condition is not necessary in our case.

Another question of debate concerns the applicability of the concept of a "heat bath". Definitely, for a finite, isolated nucleus the latter does not exist. To what extent one may work with temperature, nevertheless, will be clarified below. In the past people have adopted a hypothetical model, in which the nucleus is put inside a gas of nucleons. This picture may well be applied to a drop of liquid inside a closed container, but not in any real sense to an isolated nucleus. If heated up the latter may evaporate light particles, but they move away and are not bounced back in any way. The situation is different, of course, for nuclei inside stellar objects.

22.1.2 *Generalized ensembles*

Let us suppose we are given a Hamiltonian of the type $\hat{H}_{\rm sc} = \hat{H} + \hat{H}_{\rm cpl}$ with a coupling $\hat{H}_{\rm cpl} = \sum_\nu \hat{A}_\nu a_\nu$ to external fields, similar to that of (23.18) used in Section 23.3.2 for linear response theory, but where now the a_ν are static quantities; to simplify notation the superscript "ext" is dropped. Evidently, this structure includes a form like that given in (22.10) which results from an explicit Q-dependence of the system's Hamiltonian $\hat{H}(Q)$. In addition there might be genuine external forces. To some extent the difference is a question of notation and in the end depends on the distinction between ex- and internal quantities. One only must be aware that the choice one prefers will influence the definition of the internal energy, which amongst other places enters the first law; an example which elucidates this question will be given below in the discussion of the grand canonical ensemble. Of course, there may be cases where terms like those in $\hat{H}_{\rm sc}$ are added which commonly one would not associate with parts of the Hamiltonian. A prime example is the number of particles. Although being a conserved quantity in the strict sense, it may be impossible to fulfill this constraint exactly. Its average, however, can be accounted for with the help of Lagrange multipliers.

Whatever the nature of the forces is, let us assume that there is a set of observables $\hat{A}_i$ whose average values $\langle \hat{A}_i \rangle$ can be controlled. Suppose further this to be the only information we have on the system whose density operator we want to find. This is to say we are looking for a $\hat{\rho}$ which satisfies the constraints

$$\operatorname{tr} \hat{\rho} \hat{A}_i = \langle \hat{A}_i \rangle , \qquad (22.35)$$

with the $\langle \hat{A}_i \rangle$ on the right having given values. This problem can be solved with the help of the "maximum statistical entropy principle". It states that the equilibrium is associated with that statistical ensemble (represented here by the $\hat{\rho}$) which has the largest statistical entropy $\mathcal{S}[\hat{\rho}]$–among all possible distributions satisfying the constraints (22.35) In passing it may be mentioned that there is a close relation to information theory, details of which may be found in chapters 3 and 4 of (Balian, 1991), see also (Brenig, 1996). Considering the constraints through *Lagrange multipliers* λ_i the variation of the entropy leads to the following density operator

$$\hat{\rho} = e^{-\sum_i \lambda_i \hat{A}_i}/Z \quad \text{with} \quad Z = \operatorname{tr} e^{-\sum_i \lambda_i \hat{A}_i}. \tag{22.36}$$

The λ_i are to be fixed by requiring the density to have the averages given on the right-hand side of (22.35). In Section 22.1.3 we will set up inequalities which demonstrate the extremal feature of the entropy associated with the equilibrium operator (22.36). The reader may have noticed that in writing eqns(22.35) and (22.36) the Hamiltonian itself is meant to be one of the $\hat{A}_i$, say $\hat{H} = \hat{A}_0$, with $\lambda_0 = \beta$. For $i \geq 1$ one would thus have $\lambda_i = \beta a_i$, or in summary

$$\lambda_0 = \beta \quad \text{and} \quad \lambda_i = \beta a_i \quad (i > 0). \tag{22.37}$$

The partition function Z of (22.36) may easily be differentiated with respect to the λ_i. The derivative of the trace of an exponential can be calculated as if the operators involved would commute, see (26.27). As the result one gets the set of equations.[41]

$$\frac{\partial}{\partial \lambda_i} \ln Z(\lambda_0, \cdots \lambda_n) = -\langle \hat{A}_i \rangle \quad i = 0, 1, \cdots n, \tag{22.38}$$

which may be used to determine the λ_i. Equation (22.38) implies the differential form

$$d \ln Z = -\sum_i \langle \hat{A}_i \rangle d\lambda_i. \tag{22.39}$$

Conversely, from the differential of the entropy (22.7) one gets

$$d\mathcal{S} = \sum_i \lambda_i \, d\langle \hat{A}_i \rangle \quad \Longleftrightarrow \quad \partial \mathcal{S}/\partial \langle \hat{A}_i \rangle = \lambda_i. \tag{22.40}$$

Indeed, inserting the form (22.36) into the definition (22.7) one finds

$$\mathcal{S} = \ln Z + \sum_i \lambda_i \langle \hat{A}_i \rangle = \ln Z - \sum_i \lambda_i \partial \ln Z/\partial \lambda_i, \tag{22.41}$$

such that (22.40) follows with the help of (22.39).

The differential forms (22.39) and (22.40) exhibit the meaning of the notion of *natural variables*, which are the λ_i for $\ln Z$ and the $\langle \hat{A}_i \rangle$ for the entropy. The *Legendre transformation* (22.41) allows one to transform between the two representations. Often the quantities $\langle \hat{A}_i \rangle$ and λ_i are called *conjugate variables*, the first one being *extensive* in nature and the second one *intensive*. As $\lambda_0 \equiv \beta$ and $\langle \hat{A} \rangle_0 \equiv \langle \hat{H} \rangle = \mathcal{E}$ relation (22.19) is found again after using the abbreviation $\ln Z = -\beta \mathcal{F}$. (see exercise 4) The entropy may then be calculated as $\mathcal{S} = -(\partial \mathcal{F}/\partial T)_{\lambda_i}$ for $i \geq 1$.

---

[41] Following the very definition of a partial derivative, say with respect to λ_i, the other λ_js are to be kept fixed. In our notation this is understood implicitly and will not be indicated explicitly, as often done in thermostatics.

At this stage we may add a clarifying remark on the concepts of partial and complete equilibria. Suppose we are given an isolated system, in the sense that all quantities $\hat{A}_i$ considered in the density operator $\hat{\rho}$ of (22.36) are internal ones. This is to say we discard any coupling to the outside, besides those behind the constraints in the λ_i which are necessary to warrant the averages $\langle \hat{A}_i \rangle$ to take on their presumed values. The equilibrium associated with this $\hat{\rho} = \hat{\rho}(\lambda_0, \cdots \lambda_n)$ can be said to be of *partial nature*. It does not correspond to the system's complete or total equilibrium, at given energy or temperature, which in the course of time will be reached after all constraints have been released. The latter process will take some time, of course, and in this sense defines the typical relaxation process of non-equilibrium thermodynamics; examples of such processes are addressed in Section 23.7.1.

22.1.2.1 Equilibrium fluctuations for commuting observables

If all observables $\hat{A}_i$ that specify the partition function (22.36) commute with each other, there is no problem in differentiating Z twice.[42] to get:

$$\frac{\partial^2 Z}{\partial \lambda_i \partial \lambda_j} = \operatorname{tr} e^{-\sum_k \lambda_k \hat{A}_k} \hat{A}_i \hat{A}_j = \frac{\partial^2 Z}{\partial \lambda_j \partial \lambda_i}. \tag{22.42}$$

Subtracting the product of the averages one gets

$$\frac{\partial^2 \ln Z}{\partial \lambda_i \partial \lambda_j} = \langle \hat{A}_i \hat{A}_j \rangle - \langle \hat{A}_i \rangle \langle \hat{A}_j \rangle. \tag{22.43}$$

As the $\langle \hat{A}_i \rangle$ themselves are given by the first derivatives through (22.38) one finally gets the interesting result

$$\left\langle \left(\hat{A}_i - \langle \hat{A}_i \rangle \right) \left(\hat{A}_j - \langle \hat{A}_j \rangle \right) \right\rangle = -\frac{\partial \langle \hat{A}_i \rangle}{\partial \lambda_j} = -\frac{\partial \langle \hat{A}_j \rangle}{\partial \lambda_i}. \tag{22.44}$$

It says that static fluctuations (correlations) can be obtained by varying averages with respect to their conjugate variables, which are often represented by "external fields". Evidently, as the diagonal elements are positive, the average $\langle \hat{A}_i \rangle$ *decreases with its conjugate variable* λ_i. In fact even the full matrix on the right-hand side of (22.43) is positive (see exercise 5). Thus $\ln Z$ is a *convex* function of its variables λ_i. The entropy, on the other hand, is easily seen to be given by the negative inverse of this matrix. Indeed, differentiating twice the relation on the right of (22.40) one obtains

$$\frac{\partial^2 S}{\partial \langle \hat{A}_i \rangle \partial \langle \hat{A}_j \rangle} = \frac{\partial \lambda_j}{\partial \langle \hat{A}_i \rangle} = \frac{\partial \lambda_i}{\partial \langle \hat{A}_j \rangle} = -\left(\frac{\partial^2 \ln Z}{\partial \lambda_i \partial \lambda_j} \right)^{-1}, \tag{22.45}$$

implying S to be a *concave* function of its (natural) variables $\langle \hat{A}_i \rangle$.

---

[42] The calculation of fluctuations of non-commuting observables is addressed in Section 23.6.2.

Let us look at some simple examples. For the canonical ensemble there is only one variable, the energy $\mathcal{E} = \langle \hat{H} \rangle$. Differentiating with respect to the one Lagrange multiplier β one gets from (22.43)–(22.45):

$$-\left(\frac{\partial^2 \ln Z}{\partial \beta^2}\right)^{-1} = \frac{\partial^2 \mathcal{S}}{\partial \mathcal{E}^2} = \frac{\partial \beta}{\partial \mathcal{E}} = \left(\frac{\partial \mathcal{E}}{\partial \beta}\right)^{-1} = -\left\langle \left(\hat{H} - \mathcal{E}\right)^2 \right\rangle^{-1} = -\frac{1}{\Delta \mathcal{E}^2}. \quad (22.46)$$

Equation (22.44) relates the energy fluctuation to the specific heat $C = (\partial \mathcal{E}/\partial T)$,

$$C = \frac{\partial \mathcal{E}}{\partial T} = -\frac{1}{T^2}\frac{\partial \mathcal{E}}{\partial \beta} \equiv \frac{1}{T^2}\left\langle \left(\hat{H} - \mathcal{E}\right)^2 \right\rangle = \frac{1}{T^2}\Delta\mathcal{E}^2 > 0. \quad (22.47)$$

An important question is the extent to which this positivity may be ensured for finite nuclei; it will be addressed in Section 22.4. The variation of the average $\langle \hat{A}_i \rangle$ with the external variable a_j (at fixed T) is commonly referred to as the isothermal susceptibility. From (22.44) and (22.37) one gets (see exercise 6)

$$T\chi^T_{ij} \equiv -T\partial\langle \hat{A}_i \rangle/\partial a_j = \left\langle \left(\hat{A}_i - \langle \hat{A}_i \rangle\right)\left(\hat{A}_j - \langle \hat{A}_j \rangle\right) \right\rangle. \quad (22.48)$$

It is important to recall that this relation has been derived assuming that this set $\hat{A}_i$ commutes. In Section 23.6 a more general definition of χ^T_{ij} will be presented which is also valid for non-commuting observables.

22.1.2.2 Grand canonical ensemble The notion of a grand canonical ensemble is applied if one of the controlled observables $\langle \hat{A}_i \rangle$ of (22.35) represents the number of particles $\langle \hat{A} \rangle = \mathcal{A}$. Its conjugate variable is the chemical potential μ, or more precisely, according to (22.37) the $\lambda_\mu \equiv -\alpha = -\beta\mu$. The (grand) partition function thus reads

$$Z_{\text{gce}} = \text{tr} \exp\left(-\beta\left(\hat{H} - \mu\hat{A}\right)\right) = \text{tr} \exp\left(-\beta\hat{H} + \alpha\hat{A}\right). \quad (22.49)$$

Introducing the grand potential by $\Omega_{\text{gce}} = -T \ln Z_{\text{gce}}$ the probability distribution $w(E_n^{(A)}, A)$ for finding a system of A particles at energy $E_n^{(A)}$ is given by

$$w(E_n^{(A)}, A) = \exp\left(\frac{\Omega_{\text{gce}} + \mu A - E_n^{(A)}}{T}\right) \quad (22.50)$$

where $\sum_A \sum_n w(E_n^{(A)}, A) = 1$ such that the partition function itself takes on the form

$$\Omega_{\text{gce}} = -T \ln Z_{\text{gce}} \quad \text{and} \quad Z_{\text{gce}} = \sum_A \left(e^{\mu A/T} \sum_n e^{-E_n^{(A)}/T}\right). \quad (22.51)$$

The chemical potential is to be found from the condition that the average particle number equals a given value, $\sum_A \sum_n w(E_n^{(A)}, A)A = \mathcal{A}$.

Often, besides $\mathcal{A}$, additional quantities need to be considered explicitly, the most prominent one being that of the volume $\mathcal{V}$. In applications to the deformed shell model it will be one (or several) collective variable(s); it is even conceivable that both cases may have to be considered simultaneously. For the moment let us simply work with one additional "degree of freedom" q. Following the notation introduced in Section 22.1.1.1 one may be inclined to replace in (22.49) the $\hat{H}$ by $\hat{H}_0 + \hat{F}q$ (with a q instead of the dQ in (22.10), and $\hat{H}_0 = \hat{H}(Q_0)$). Taking up the notation of (22.22), the average of this additional term is to be written as $-\mathcal{P}_Q q$ and the corresponding intensity is given by $\lambda_q = -\beta q$. The energy would then be given by $\mathcal{E}_0 = \langle \hat{H}_0 \rangle$. Applying (22.41) to the present case one gets

$$\Omega_{\text{gce}} = -T\mathcal{S} + \mathcal{E}_0 - \mathcal{P}_Q q - \mu \mathcal{A}. \tag{22.52}$$

Its total differential turns out to be

$$d\Omega_{\text{gce}} = -\mathcal{S}dT - \mathcal{P}_Q dq - \mathcal{A}d\mu, \tag{22.53}$$

as may easily verified with the help of (22.39). Likewise, from (22.40) the first law of thermostatics may be written as $d\mathcal{E}_0 = Td\mathcal{S} + qd\mathcal{P}_Q + \mu d\mathcal{A}$. The reader may have expected the form

$$d\mathcal{E} = Td\mathcal{S} - \mathcal{P}_Q dq + \mu d\mathcal{A}, \tag{22.54}$$

simply by analogy with the common case for which one has $-\mathcal{P}d\mathcal{V}$ instead of $-\mathcal{P}_Q dq$. Indeed, both forms are correct, as the two energies are connected by a Legendre transformation, namely $\mathcal{E} = \mathcal{E}_0 - \mathcal{P}_Q q$. Recall, that in the version of (22.15) the $d\mathcal{E}$ was involved as can be seen from the definition (22.13). Defining the free energy as in (22.23), $\mathcal{F} = \mathcal{E} - T\mathcal{S}$ one gets from (22.52) $\Omega_{\text{gce}} = \mathcal{F} - \mu\mathcal{A}$.

If the external variable q refers to the volume $\mathcal{V}$ of an extensive system the grand potential Ω_{gce} satisfies the *Gibbs–Duhem relation* (see e.g. Section 5.6.5 of (Balian, 1991))

$$\Omega_{\text{gce}} = -P\mathcal{V}. \tag{22.55}$$

Indeed, replacing in (22.52) the $-\mathcal{P}_Q q$ by $-\mathcal{P}\mathcal{V}$ it is seen from (22.53) that Ω_{gce} depends on the two "intensive" variables T and μ but only on the one "extensive" variable $\mathcal{V}$. Hence, Ω_{gce} must be proportional to $\mathcal{V}$ such that $-\mathcal{P} \equiv (\partial \Omega_{\text{gce}}/\partial \mathcal{V})_{T,\mu} = \Omega_{\text{gce}}/\mathcal{V}$. After differentiating (22.55) and by obeying (22.54) one may prove the differential version $\mathcal{S}dT - \mathcal{V}d\mathcal{P} + \mathcal{A}d\mu = 0$.

22.1.2.3 Equilibrium between two systems Suppose there are two systems between which some conserved quantity A may be exchanged, which in the sub-system has values of $\langle \hat{A}_1 \rangle$ and $\langle \hat{A}_2 \rangle$. If the total system is in equilibrium the entropy has to be maximal. We may thus use relation (22.40) with $d\mathcal{S} = 0$. As the sum $A = A_1 + A_2$ is conserved one must have $dA_1 + dA_2 = 0$. Consequently, the corresponding *intensive quantities* have to be equal $\lambda_1 = \lambda_2$. Prime examples are transitions between different phases of one material. For pressure, temperature and chemical potential of two phases one thus has $\mathcal{P}_1 = \mathcal{P}_2$, $T_1 = T_2$ and $\mu_1 = \mu_2$, respectively.

22.1.3 Extremal properties

Extremal properties of thermal quantities derive from the fact of the entropy $\mathcal{S}[\hat{\rho}] = -\mathrm{tr}\,(\hat{\rho}\ln\hat{\rho})$ being a concave function of the density operator. This latter feature may be put into the form

$$\mathrm{tr}\,\{\hat{\rho}'\,(\ln\hat{\rho} - \ln\hat{\rho}')\} \leq 0 \qquad \text{with} \qquad \mathrm{tr}\,\hat{\rho}' = \mathrm{tr}\,\hat{\rho} = 1\,. \tag{22.56}$$

Besides being normalizable the two operators appearing here only need to be Hermitian. The proof is quite simple and is left as an exercise (see exercise 7). Suppose the $\hat{\rho}$ of (22.56) corresponds to the density operator (22.36) of that equilibrium which is specified by the averages $\langle\hat{A}_i\rangle$ of all the $\hat{A}_i$ (which determine the λ_i uniquely). Inserting its logarithm $\ln\hat{\rho} = -\ln Z - \sum_i \lambda_i \hat{A}_i$ into (22.56) averages $\langle\hat{A}_i\rangle'$ of the $\hat{A}_i$ appear which are to be calculated with the density operator $\hat{\rho}'$. The $\ln Z$, on the other hand, survives the averaging and can be expressed by the entropy $\mathcal{S}$ and the equilibrium values of the $\langle\hat{A}_i\rangle$. As the entropy associated with the $\hat{\rho}'$ is given by $\mathcal{S}' = \mathcal{S}[\hat{\rho}'] = -\mathrm{tr}\,(\hat{\rho}'\ln\hat{\rho}')$ one arrives at

$$\mathcal{S} \geq \mathcal{S}' + \sum_i \lambda_i \left(\langle\hat{A}_i\rangle - \langle\hat{A}_i\rangle'\right)\,. \tag{22.57}$$

This inequality reflects the extremal property of entropy we spoke about in Section 22.1.2: Among all densities $\hat{\rho}'$ leading to the same average values, $\langle\hat{A}_i\rangle = \langle\hat{A}_i\rangle'$, the $\hat{\rho}$ of the generalized ensemble (22.36) has maximal entropy. Later, in Section 23.6.1 this property will be studied further within linear response theory by explicitly constructing a quadratic form for the deviation of entropy from its equilibrium value.

The inequality (22.57) can be applied to various other situations, too. This flexibility stems from the fact that the $\hat{\rho}'$ can be chosen in many ways. Let us discuss first another application to equilibrated systems to address the increase of entropy in irreversible processes below. Suppose a free energy $\mathcal{F}'$ is associated with the density operator $\hat{\rho}'$ in the same way as done by (22.19): $\ln Z' = -\beta\mathcal{F}'$. To the $\mathcal{S}'$ and the averages $\langle\hat{A}_i\rangle'$ it will be in the same relation as found before in (22.41), namely $\mathcal{S}' = \ln Z' + \sum_i \lambda_i \langle\hat{A}_i\rangle'$. The inequality (22.57) then implies $\ln Z \geq \ln Z'$, or

$$\mathcal{F}(T) \leq \mathcal{F}'(T) \qquad \text{at fixed} \qquad T\,. \tag{22.58}$$

Hence, for given temperature, amongst all possible distributions the canonical one of equilibrium has the smallest free energy.

The inequality (22.58) may serve as the starting point of a variational procedure, designed to optimize approximations to the true density operator, provided one takes the same temperature. Indeed, in many cases, like for almost all many-body systems, it is impossible to work with the correct, full Hamiltonian $\hat{H}$, not to speak of the density operator itself. However, physical intuition may guide one to some approximate form $\hat{H}_{\mathrm{app}}$, with a corresponding density $\hat{\rho}' = \hat{\rho}_{\mathrm{app}} = \exp(-\beta\hat{H}_{\mathrm{app}})/Z_{\mathrm{app}}$ and thus a free energy $\mathcal{F}' = \mathcal{F}_{\mathrm{app}}$. Suppose, one wants to specify total energy only on average. In this case (22.58) may be written as

$$\mathcal{F} \leq \mathcal{E}_{\text{app}} - T\mathcal{S}_{\text{app}} \quad \text{with} \quad \mathcal{E}_{\text{app}} = \operatorname{tr} \hat{\rho}_{\text{app}} \hat{H} \quad \text{and} \quad \mathcal{S}_{\text{app}} = -\operatorname{tr} \hat{\rho}_{\text{app}} \ln \hat{\rho}_{\text{app}}. \tag{22.59}$$

It says, whatever the approximate Hamiltonian will be the difference $\mathcal{E}_{\text{app}} - T\mathcal{S}_{\text{app}}$ is bound from below by the system's correct free energy. It is worth noticing that the energy $\mathcal{E}_{\text{app}}$ is calculated for the true Hamiltonian and not for $\hat{H}_{\text{app}}$. Formula (22.59) finds many applications. For instance, it is used to extend quantal variational procedures to finite thermal excitations, e.g. the Jastrow method discussed in Section 2.4.

22.1.3.1 *The second law and quasi-static processes* Let us first look again at (22.14), which relates the variation of entropy by $Td\mathcal{S} = \operatorname{tr} d\hat{\rho}\hat{H}(Q)$ to a variation of the density. We may note first that even for a finite $d\hat{\rho}$ the entropy may well stay constant. Indeed, this may be the case if $d\hat{\rho}$ is a functional of $d\hat{H}$. Take for $\hat{\rho}(t)$ a solution of the von Neumann equation (22.1) for a Hamiltonian which depends on the time-dependent parameter $Q(t)$. Without any further manipulation, the $d\mathcal{S} = \dot{\mathcal{S}}dt$ is easily seen to vanish identically (see exercise 8).

Above it was argued that for systems that are not thermally isolated the variation $d\mathcal{S}$ in entropy gets a contribution from the heat $\delta\mathcal{Q}_h$ transferred to or from the system which is given by $Td\mathcal{S} = \delta\mathcal{Q}_h$. On top of this amount, which may have any sign, there is the part $d\mathcal{S}_{\text{irr}}$ which is due to *irreversible processes* and which cannot be negative. This statement may be put in the following version of the *second law*:

$$d\mathcal{S} = \frac{\delta\mathcal{Q}_h}{T} + d\mathcal{S}_{\text{irr}} \quad \text{where} \quad d\mathcal{S}_{\text{irr}} \geq 0. \tag{22.60}$$

In the end this is a consequence of the concavity of the functional $\mathcal{S}[\hat{\rho}]$, and thus of the inequalities (22.56) and (22.57), in particular. The latter may be rewritten as $\Delta\mathcal{S} \geq \sum_i \lambda_i \Delta\langle\hat{A}_i\rangle$, where the $\Delta\langle\hat{A}_i\rangle$ stand for the differences $\langle\hat{A}_i\rangle - \langle\hat{A}_i\rangle'$ and the $\Delta\mathcal{S}$ for $\mathcal{S} - \mathcal{S}'$. In fact the $\hat{\rho}'$ may even be associated with a state of non-equilibrium. The inequality then says that *any approach to equilibrium is associated with an increase in entropy*. Of particular interest, also in nuclear physics, are *quasi-static processes*. There the changes in the $\langle\hat{A}_i\rangle$ are assumed to be sufficiently slow such that the system always passes through states of partial equilibria. The latter are parameterized by the form (22.36) with variable parameters λ_i; details are worked out in Section 23.7 for linear processes.

Let us turn to possible origins of the increase in entropy. As mentioned above, if time evolution is governed by a *Hamiltonian which is completely known* the entropy $S[\hat{\rho}] = -\operatorname{tr}(\hat{\rho}\ln\hat{\rho})$ remains *constant in time*. However, there may be various situations where the *microscopic Hamiltonian is not* known. One possibility is encountered in the statistical treatment of nuclear reactions described in Section 18.4. In Section 3.2.3 of his book (1991), R. Balian studies explicitly how and why in such a case the entropy necessarily increases in time. Random Matrix

Theory supplies one realization of such Hamiltonians. On the other hand, even if the Hamiltonian is known in one way or other macroscopic equations of motion are necessarily associated with *coarse graining*. One possibility is simply given by suitably energy averaging relevant quantities. In reaction theory this leads to the appearance of an imaginary part in the optical potential, as explained in Section 18.3. Applying such energy averages to microscopic quantities which define the effective forces for collective motion leads to dissipative forces, and hence to irreversible behavior. These features are described in detail in Section 6.4, and in a different context in Section 23.7. Eventually, they may be influenced by chaotic properties of the underlying dynamics. A question of great debate is whether or not chaotic motion alone may imply irreversibility and the production of entropy, see e.g. (Kubo *et al.*, 1991*b*), (Dorfman, 1999).

In any case, nuclear collective motion shares this property with truly macroscopic systems: the von Neumann equation is not valid, the transport exhibits irreversible features and hence the entropy does in general not stay constant.

22.2 Level densities and energy distributions

The discussion in the previous section was kept on a general level without any reference to specific properties of a given system, if only its Hamiltonian was known. In this spirit, all averages were expressed by traces over density operators which were expressed as functionals of the basic Hamiltonian. In most cases where thermodynamic concepts are applicable to many-body systems their energy levels are so densely spaced that introducing the concept of the level density $\Omega(\epsilon)$ is inevitable. The latter is defined such that

$$dG = \Omega(\epsilon)d\epsilon \equiv d\epsilon/D(\epsilon) \qquad (22.61)$$

measures the number of levels in the energy interval $d\epsilon$. Here, $D(\epsilon)$ is the average spacing of neighboring levels. The $G(\epsilon)$ is meant to represent the number of all states up to ϵ,

$$G(E) = \int_0^E d\epsilon\, \Omega(\epsilon)\,; \qquad (22.62)$$

here, the ground state energy E_0 has been put equal to zero.

In the development of statistical mechanics the concept of the level density has played a major role. It may be viewed as completely determining the mechanical structure of a given system, for which reason $\Omega(\epsilon)$ is often called the "structure function". In fact, one may develop a statistical theory of thermal excitations and its relation to thermostatics without any reference to a Hamiltonian (Tisza, 1963), (Tisza and Quay, 1963), (Mandelbrot, 1964), see also (Lavenda, 1991). All what is needed is this structure function $\Omega(\epsilon)$, together with the occupation numbers $\rho(\epsilon)$ as an additional factor. It may be said that the $\Omega = \Omega(\epsilon)$ need not even be continuous, in this way one may also handle a system with discrete eigenvalues (using a form like (22.98)). Within such language, the energy distribution is then determined by

$$dW(\epsilon) = w(\epsilon)d\epsilon = \rho(\epsilon)dG(\epsilon) = \rho(\epsilon)\Omega(\epsilon)d\epsilon. \tag{22.63}$$

Here, $w(\epsilon)$ is the probability density for the energy to lie in the interval $\epsilon, \epsilon + d\epsilon$. The properties specific to the ensemble one wants to study are solely contained in the $\rho(\epsilon)$. For the microcanonical ensemble one simply has

$$\rho(\epsilon, E) = \begin{cases} 1/\Delta G(E) & \text{for } E - \Delta \leq \epsilon \leq E \\ 0 & \text{otherwise.} \end{cases} \tag{22.64}$$

Evidently, with a Hamiltonian at once's disposal the connection with (22.2) is readily performed, although this association is not really needed.

As stated in (22.3) the width Δ of (22.64) is assumed to be much smaller than the energy E itself, $\Delta \ll E$. In such a case the $\Delta G(E)$ appearing in (22.64) may be estimated to

$$\Delta G(E) = G(E) - G(E - \Delta) \approx \Omega(E)\Delta. \tag{22.65}$$

As we may see from tables 22.1 and 22.2, the condition $\Delta \ll E$ is fairly well fulfilled even for a nucleus–provided the latter is not too small and that the excitation energy E (with the ground state energy $E_0 = 0$) is not much lower than say 10 MeV. The value of the ratio Δ/E turns out less than 1%, i.e. much smaller than the corresponding quantity of the canonical distribution. Again, a typical example will be useful: For a nucleus like ^{208}Pb having an excitation energy of only 5 MeV there would already be more than a million states within an interval of only 10 keV making this ratio only 0.2%.

The very definition of G implies proper normalization of the energy distribution $w(\epsilon)$, namely $\int d\epsilon w(\epsilon) = 1$. For the thermal entropy we may use its microscopic definition (22.7) to find from (22.63) and (22.64)

$$S = -\operatorname{tr} \hat{\rho} \ln \hat{\rho} \equiv -\int d\epsilon\, w(\epsilon) \ln \rho(\epsilon, E) = \ln \Delta G(E), \tag{22.66}$$

in accord with (22.9). Evidently, since for the microcanonical distribution the energy is sharply defined, the average energy

$$\mathcal{E} = \int_0^\infty d\epsilon\, \epsilon w(\epsilon) \tag{22.67}$$

is practically identical to the value of E–to the extent expressed by the relation (22.3). In principle, the E itself may, however, be understood as a fluctuating variable, a feature which will be exploited further below. In this sense the entropy, too, becomes a fluctuating quantity, say if written as $S(E) = \ln \Delta G(E)$.

The canonical ensemble, parameterized as

$$\rho(\epsilon, \beta) = \frac{1}{Z} e^{-\beta \epsilon} \quad \Longleftrightarrow \quad w(\epsilon; \beta) = \frac{\Omega(\epsilon)}{Z(\beta)} \exp(-\beta \epsilon), \tag{22.68}$$

is specified by the temperature $T = 1/\beta$, which through (22.67) (with $w(\epsilon) = w(\epsilon;\beta)$) fixes the average energy. The average energy $\mathcal{E}$ can be expressed as $\mathcal{E} = -\partial \ln Z(\beta)/\partial \beta$, as is easily seen from the form

$$Z(\beta) = \int_0^\infty d\epsilon \, e^{-\beta\epsilon} \, \Omega(\epsilon) \qquad (22.69)$$

for the partition function. The *most likely energy* ϵ_m, on the other hand, is determined through the maximum of the energy distribution, which implies

$$0 = \left.\frac{\partial w(\epsilon;\beta)}{\partial \epsilon}\right|_{\epsilon_m} = -\beta + \left.\frac{\partial \ln \Omega(\epsilon)}{\partial \epsilon}\right|_{\epsilon_m}. \qquad (22.70)$$

It becomes identical to the $\mathcal{E}$ only in the thermostatic limit, see below.

22.2.1 Composite systems

Let us for the moment go back to a description in terms of density operators, although the following derivation can also be done without such a reference. The density operator of the microcanonical ensemble was given in (22.2). Suppose the total system consists of two coupled systems "1" and "2" where the coupling is so weak that it may be neglected in the energy balance $E_m = E_\mu^{(1)} + E_\nu^{(2)}$. The eigenstates $|m\rangle$ are then given by the direct product $|m\rangle = |1;\mu\rangle|2;\nu\rangle$, such that the density operator for subsystem "1" turns into

$$\hat{\rho}_1(E_\mu^{(1)}) = \mathrm{tr}\,_2\,\hat{\rho}(E) \equiv \sum_\mu \frac{\Delta G_2(E - E_\mu^{(1)})}{\Delta G(E)} |1;\mu\rangle\langle 1;\mu|. \qquad (22.71)$$

Here, $\Delta G_2(E - E_\mu^{(1)})$ is the number of states of system "2" the energies of which satisfy the constraint $E - \Delta - E_\mu^{(1)} \leq E_\nu^{(2)} \leq E - E_\mu^{(1)}$. Since $\hat{\rho}(E)$ is normalized to unity the same must be true for the density $\hat{\rho}_1$, $\mathrm{tr}\,_1\hat{\rho}_1(E_\mu^{(1)}) = 1$. From the expression on the very right of (22.71) it thus follows that $\Delta G(E) = \sum_\mu \Delta G_2(E - E_\mu^{(1)})$.

Suppose the distribution of levels of system "1" may again be classified by a continuous level density $\Omega_1(\epsilon)$. Obviously, the probability density for the energy of system "1" must then be given by

$$w_1(\epsilon; E) = \frac{\Delta G_2(E - \epsilon)}{\Delta G(E)} \Omega_1(\epsilon). \qquad (22.72)$$

Let us assume the interval $E, E + \Delta$ to be sufficiently small that one may replace the ΔGs by $\Delta G \approx \Omega(\epsilon)\Delta$, which is to say replacing in $dG = \Omega(\epsilon)d\epsilon$ the differential by the finite difference Δ (mind (22.65)). Then one obtains

$$w_1(\epsilon; E) = \frac{\Omega_1(\epsilon)\Omega_2(E - \epsilon)}{\Omega(E)}. \qquad (22.73)$$

Normalization then leads to

$$\Omega(E) = \int_0^E d\epsilon\, \Omega_2(E-\epsilon)\Omega_1(\epsilon)\,, \tag{22.74}$$

which says that the level density of the composite system is given by the folding integral over the individual densities.

The discussion so far has only involved microcanonical distributions, but the results have important implications for the corresponding canonical ensembles. Since the partition function is the Laplace transform of the level density, see (22.69), it follows from (22.74) and the convolution theorem that the partition function of a composite system simply factorizes:

$$Z(\beta) = Z_1(\beta)Z_2(\beta)\,. \tag{22.75}$$

Given this fact it is easily recognized from (22.68) that the $w_1(\epsilon; E)$ of (22.73) can be written as

$$w_1(\epsilon; E) = \frac{w_1(\epsilon; \beta)w_2(E-\epsilon; \beta)}{w(E; \beta)}\,. \tag{22.76}$$

On the right-hand side the energy distributions of the three systems appear, where each one is described by the (22.68) of the canonical ensemble at the same β. The intriguing feature of (22.76) is that the left-hand side represents a microcanonical quantity, fixed only by the energy E, whereas the distributions on the right-hand side are all determined by the parameter β–although, as we have seen, this dependence drops out entirely. This latter observation has been used in (Mandelbrot, 1964) and (Tisza and Quay, 1963), taking up an idea by L. Szilard (1925), to prove that the $w(E; \beta)$ of any canonical distribution must necessarily be of the form (22.68). Indeed, eqn(22.76), which then serves as the starting point, has its own physical interpretation. The joint probability $w_1(\epsilon; \beta)w_2(E-\epsilon; \beta)$ of finding the two systems at energies ϵ and $E-\epsilon$ must be equal to the probability $w(E; \beta)$ of finding the composite system at energy E multiplied by the conditional probability $w_1(\epsilon; E)$ for the correct splitting of the total energy.

Next we look at the situation of a small system "1" coupled weakly to a large one. Suppose system "2" is big enough such that its energy fluctuations may be considered small–which in a strict sense would only be given for a macroscopically large system, see below. Replacing E by its average $\mathcal{E} = \mathcal{E}_1 + \mathcal{E}_2$, which makes sense for sufficiently small Δ, one may calculate $\Omega_2(E-\epsilon)$ as

$$\Omega_2(E-\epsilon) \simeq \Omega_2(\mathcal{E}_2)e^{\beta\mathcal{E}_1}e^{-\beta\epsilon} \quad \text{with} \quad \beta = \frac{\partial \ln \Omega_2(\mathcal{E}_2)}{\partial \mathcal{E}_2}\,. \tag{22.77}$$

Then the energy distribution of system "1" turns out to be that of the canonical ensemble

$$w_1(\epsilon; \beta) = \frac{e^{-\beta\epsilon}}{Z_1(\beta)}\Omega_1(\epsilon)\,, \tag{22.78}$$

with the partition function

$$Z_1(\beta) = \frac{\Omega(E)}{\Omega_2(\mathcal{E}_2)} e^{-\beta \mathcal{E}_1} = \int_0^\infty d\epsilon \, e^{-\beta \epsilon} \Omega_1(\epsilon) \,. \qquad (22.79)$$

This follows because $w_1(\epsilon)$ is properly normalized, as can be seen from (22.73) and (22.74). Notice, please, that according to (22.77) the temperature is determined by the large system "2" which may thus be considered the heat bath for system "1".

Finally, it should be noted that the properties discussed here are of great relevance in nuclear physics. This can be seen in Chapter 8 where the decay of excited systems is treated by statistical means. In principle such a process ought to be handled within a microcanonical ensemble. Indeed, a formula like (22.73) appears in that for the decay rate given by (8.9). A temperature of a canonical ensemble may then be deduced from the energy spectrum of evaporated light particles, very much in the spirit of a formula like (22.78); for details we may refer to Chapter 10.3.

22.2.2 A Gaussian approximation

The aim of the discussion to come is to find possible ways of defining canonical ensembles for isolated systems. Such derivations are based on the observation that the level density is a strongly growing exponential function of energy. As suggested by (22.66) this dependence is governed by entropy, $\Omega(\epsilon) \propto \exp(S(\epsilon))$. One possibility consists in simply expanding this entropy to second order (see exercise 9). It is more instructive, however, to follow the argumentation of (Tisza and Quay, 1963).

To begin with, let us introduce the function

$$S(\epsilon; \beta) = \ln Z(\beta) + \beta \epsilon \,, \qquad (22.80)$$

considered here as a *random function of the random variable* ϵ, with β being a fixed numerical parameter. Next we use (22.63) for a finite energy difference $d\epsilon = \Delta$ to obtain for the logarithm of the number of levels

$$\ln \Delta G(\epsilon) = S(\epsilon; \beta) + \ln \Delta \mathcal{W}(\epsilon; \beta) \,, \qquad (22.81)$$

provided we assume the $\rho(\epsilon; \beta)$ to be given by (22.68). Notice that the last term on the right is negative, since the probability $\Delta \mathcal{W}(\epsilon; \beta)$ cannot be larger than unity. Therefore, the $\ln \Delta G(\epsilon)$ is bound from above by $S(\epsilon; \beta)$. Suppose we approximate $\Delta \mathcal{W}$ by the Gaussian distribution

$$\Delta \mathcal{W}(\epsilon; \beta) = \frac{\Delta}{\sqrt{2\pi}\Delta\mathcal{E}} \exp\left(-\frac{(\epsilon - \mathcal{E})^2}{2\Delta\mathcal{E}^2}\right) \equiv w(\epsilon)\Delta \,, \qquad (22.82)$$

with a mean energy $\mathcal{E}$ and fluctuation $\Delta\mathcal{E}$. We may note in passing that in this approximation the most likely value ϵ_m of the energy coincides with its average value. The Δ on the right-hand side of (22.82) may be interpreted as the "width" over which the distribution function $\Delta W(\epsilon; \beta)$ will be examined. To warrant proper normalization of $w(\epsilon)$ one must have

$$\int_0^\infty w(\epsilon)d\epsilon = 1 \quad \Longleftrightarrow \quad \Delta \mathcal{E} \ll \mathcal{E}. \tag{22.83}$$

At $\epsilon = \mathcal{E}$ we get by inserting (22.82) into (22.81)

$$\ln \Delta G(\mathcal{E}) = \mathcal{S}(\mathcal{E}; \beta) + \ln\left(\frac{\Delta}{\sqrt{2\pi}\Delta\mathcal{E}}\right). \tag{22.84}$$

Provided the second term on the right can be neglected, the entropy $\mathcal{S}(\mathcal{E}; \beta)$ becomes identical to that of the form (22.9), $\mathcal{S}(\mathcal{E}; \beta) = S(\epsilon = \mathcal{E}; \beta) = \ln \Delta G(\mathcal{E})$. One possible condition for this to hold true is to choose the width Δ to be of the order of the standard deviation of ΔW, namely $\Delta = \sqrt{2\pi}\Delta\mathcal{E}$. In passing we may note that for extensive systems this requirement is immaterial in the thermostatic limit, where the quantities $\mathcal{E}$, $\mathcal{S}$ and $\Delta \mathcal{E}$ are proportional to particle number A, such that for fixed Δ one has $\ln(\Delta/(\sqrt{2\pi}\Delta\mathcal{E})) = O(\ln A/A)$.

In the Gaussian approximation introduced in (22.82) the level density can be written as

$$\Omega_G(\epsilon) = \frac{1}{\sqrt{2\pi}\Delta\mathcal{E}} \exp\left(\mathcal{S}(\mathcal{E}) + \beta(\epsilon - \mathcal{E}) - \frac{(\epsilon - \mathcal{E})^2}{2\Delta\mathcal{E}^2}\right). \tag{22.85}$$

This follows from (22.81) by using $\Delta G = \Omega \Delta$ (mind (22.65) and after expanding $S(\epsilon, \beta)$ to first order around the average energy $\mathcal{E}$ putting $\beta = \partial S/\partial \mathcal{E}$ in accord with (22.80). Inserting this level density $\Omega_G(\epsilon)$ into formula (22.69) for the partition function $Z_G(\beta)$ one ends up with (remember the normalization of $w(\epsilon)$ according to (22.83))

$$Z_G(\beta) = \exp\left(-\beta\mathcal{E} + \mathcal{S}(\mathcal{E})\right) \equiv \exp(-\beta\mathcal{F}). \tag{22.86}$$

Thus one gets the common relations of thermostatics $\mathcal{F} = \mathcal{E} - T\mathcal{S}$, with $\mathcal{S}$ (calculated at $\mathcal{E}$) being the (thermal) entropy, $\mathcal{E}$ the internal and $\mathcal{F}$ the free energy.

Through the linear relation (22.80) between ϵ and S the probability distribution $w(\epsilon)$ may be transformed into one for entropy $S \equiv S(\epsilon; \beta)$. One simply has to identify

$$w(\epsilon)d\epsilon = w(\epsilon(S))\frac{dS}{\beta} = \frac{1}{\sqrt{2\pi}\Delta S} \exp\left(-\frac{(S - \mathcal{S}(\mathcal{E}))^2}{2\Delta S^2}\right) dS \equiv w_S(S)dS, \tag{22.87}$$

with the fluctuation in entropy being given by the specific heat (see (22.47) and (Landau and Lifshitz, 1980))

$$\Delta S^2 = \beta^2 \Delta \mathcal{E}^2 \equiv C. \tag{22.88}$$

Equation (22.87) reflects in direct fashion the principle that in equilibrium the entropy becomes maximal. Finally, we may wish to refer to Section 23.6.1 where the fluctuations in the entropy will be studied in more detail.

The identification of the entropy with the logarithm of the level density has an interesting consequence for composite systems. Suppose we have two systems at fixed β. Then the entropy introduced in (22.80) is an additive quantity:

$$S(\epsilon) = S_1(\epsilon_1) + S_2(\epsilon_2) \quad \text{for} \quad \epsilon = \epsilon_1 + \epsilon_2. \tag{22.89}$$

Thus $\mathcal{S}(\mathcal{E}; \beta) = \ln \Delta G(\mathcal{E})$ implies

$$\Delta G(\mathcal{E}) = \Delta G_1(\mathcal{E}_1) \, \Delta G_2(\mathcal{E}_2), \tag{22.90}$$

or, with $\Delta G = \Omega \Delta$ again,

$$\Omega(\mathcal{E}) = \Omega_1(\mathcal{E}_1) \, \Omega_2(\mathcal{E}_2) \, \Delta. \tag{22.91}$$

Obviously, this is an approximate version only of the correct result (22.74), which involves a folding integral. The former would follow from the latter in case the integrand of (22.74) concentrated around the upper boundary $E = \mathcal{E}$, inside an interval of order Δ.

As mentioned earlier the Gaussian approximation may also be introduced on the basis of the hypothesis that

$$\Omega(\epsilon) = \mathcal{N} \exp(S(\epsilon)), \tag{22.92}$$

One need only assume that the energy dependence of the pre-factor $\mathcal{N}$ is much weaker than that of the exponential. The level density (22.92) may then be combined with the occupation number $\rho(\epsilon, T)$ of a (still hypothetical) canonical distribution (22.68). The resulting probability for the energy distribution then reads $w(\epsilon) = (\mathcal{N}/Z) \exp(S(\epsilon) - \beta \epsilon)$, which invites one to perform an expansion to second order in the exponent. Finally one ends up with the formulas just found for the quantities of interest like $w(\epsilon)$ or $\Omega_G(\epsilon)$ etc (see exercise 9).

Estimates for the Fermi gas model: For not too high thermal excitations the total energy $\mathcal{E}$ of a Fermi system is known to depend quadratically on temperature, $\mathcal{E} = aT^2$ (we again put the ground state energy equal to zero). This simple law is valid provided the level density is a sufficiently smooth function of energy, and in this sense for the nucleus corresponds to the "macroscopic limit", see Section 22.5.

TABLE 22.1. Dependence on A for $\mathcal{E} = 20$ MeV, $\Delta = 0.01$ MeV

A	50	100	150	200
T	1.8	1.3	1.0	0.9
$\Delta \mathcal{E}/\mathcal{E}$	0.42	0.36	0.32	0.30
$\Delta/\mathcal{E}$	$5.0 \cdot 10^{-4}$	$5.0 \cdot 10^{-4}$	$5.0 \cdot 10^{-4}$	$5.0 \cdot 10^{-4}$
$G(\mathcal{E})$	$3.7 \cdot 10^5$	$3.9 \cdot 10^9$	$4.8 \cdot 10^{12}$	$1.9 \cdot 10^{15}$

It is straightforward to apply the formulas derived above, in particular as besides the energy no other quantities $\langle \hat{A}_i \rangle$ are involved. It may suffice to quote

TABLE 22.2. Dependence on $\mathcal{E}$ for $A = 100, \Delta = 0.001$

$\mathcal{E}$	25	75	125	175
T	1.4	2.5	3.2	3.7
$\Delta\mathcal{E}/\mathcal{E}$	0.34	0.26	0.22	0.21
$\Delta/\mathcal{E}$	$4.0 \cdot 10^{-5}$	$1.3 \cdot 10^{-5}$	$8.0 \cdot 10^{-6}$	$5.7 \cdot 10^{-6}$
$G(\mathcal{E})$	$1.3 \cdot 10^{10}$	$7.5 \cdot 10^{20}$	$2.5 \cdot 10^{28}$	$3.4 \cdot 10^{34}$

the results for the entropy, the specific heat C as well as the relative fluctuations for the energy and the entropy. One has

$$S(\mathcal{E}) = 2\sqrt{a\mathcal{E}} = 2aT \qquad (22.93)$$

$$C = T\frac{\partial S}{\partial T} = \frac{\partial \mathcal{E}}{\partial T} = 2aT = S \qquad (22.94)$$

$$\frac{\Delta\mathcal{E}^2}{\mathcal{E}^2} = \frac{2}{\sqrt{a\mathcal{E}}} = \frac{4}{C} = 4\frac{\Delta S^2}{S^2} \qquad (22.95)$$

Using the value $a = A/8$ MeV^{-1} for the parameter a, one gets a reasonable estimate for the dependence of the relative energy fluctuations on particle number,

$$\frac{1}{2}\frac{\Delta\mathcal{E}}{\mathcal{E}} = \frac{2}{\sqrt{AT}} = \left(\frac{2\text{MeV}}{A\mathcal{E}}\right)^{1/4}. \qquad (22.96)$$

Anticipating the result (22.129) for temperature fluctuations in a microcanonical ensemble one may draw an intriguing connection with entropy fluctuations of the canonical distribution,

$$\left.\frac{\Delta S^2}{S^2}\right|_{\text{canon}} = \frac{1}{C} = \left.\frac{\Delta T^2}{T^2}\right|_{\text{micro}}. \qquad (22.97)$$

In Section 22.3.2 the right part will be shown to hold true if the temperature is introduced for an isolated system, albeit the notion of fluctuation has a slightly different meaning.

To illustrate how large the relative fluctuations may be we exhibit in tables 22.1 and 22.2 those of the energy $\Delta\mathcal{E}/\mathcal{E}$ as a function of mass number A. Also shown are the temperature T, the number of levels $G(E)$ in a given interval Δ of the microcanonical distribution and the relative width Δ/E of this interval. The $G(E)$ is evaluated to first order in Δ (remember (22.65)) with the level density given by (22.92). Evidently, for the typical nuclear case the energy fluctuations of the canonical distribution are too large for the thermostatic limit to apply: There is no way that they might become as small as the Δ of the microcanonical distribution. This question will be taken up again in Section 22.4.

22.2.3 Darwin–Fowler method for the level density

The Darwin–Fowler method allows for an approximate way of calculating the level density which in nuclear physics is of wide use. We will begin developing some basic features restricting ourselves to the energy as the only observable.

The method is based on the observation that the level density and the partition function are Laplace transforms of one another. The microscopic expression for the level density reads

$$\Omega(\epsilon) = \sum_n \delta(\epsilon - E_n), \qquad (22.98)$$

where n subsumes a whole set of quantum numbers. The δ-functions may be replaced by their Fourier representation

$$\delta(\epsilon - E_n) = \frac{1}{2\pi} \int_{-\infty}^{\infty} d\tau\, e^{i\tau(\epsilon - E_n)}, \qquad (22.99)$$

where integration is to be performed along the real axis. Inserting this into (22.98) and after transforming the integration variable to $i\tau = \beta = \beta^*$ the level density can be written as the inverse Laplace transform (see (26.25))

$$\Omega(\epsilon) = \mathcal{L}^{-1}[Z(\beta)] = \frac{1}{2\pi i} \int_{-i\infty}^{i\infty} d\beta\, e^{\beta\epsilon}\, Z(\beta) \qquad (22.100)$$

of the partition function $Z(\beta) = \sum_n e^{-\beta E_n} = \int_0^\infty d\epsilon\, e^{-\beta\epsilon}\, \Omega(\epsilon)$; the application to the one-dimensional oscillator is treated in exercise 10.

The Darwin–Fowler approximation now consists in calculating this integral by the method of steepest descent (see Section 26.2). The essential idea is to deform the original integration contour such that the integrand finally becomes proportional to a real Gaussian. In the present case it is convenient to shift the line of integration from the imaginary axis to the one from $\beta_0 - i\infty$ to $\beta_0 + i\infty$, where the $\beta_0 > 0$ is chosen to satisfy the following relation to the energy ϵ:

$$\epsilon + \left(\frac{d\ln Z}{d\beta}\right)_{\beta=\beta_0} = 0. \qquad (22.101)$$

Expanding the exponent of the integrand to second order around this β_0 the integrand takes on the form

$$e^{\beta\epsilon}\, Z(\beta) = e^{\beta\epsilon + \ln Z(\beta)} \simeq \exp\left(\beta_0\epsilon + \ln Z(\beta_0) + \frac{1}{2}(\beta - \beta_0)^2 \left(\frac{d^2 \ln Z}{d\beta^2}\right)_{\beta_0}\right). \qquad (22.102)$$

Introducing the real quantity $\beta'' = -i(\beta - \beta_0)$ the integral in (22.100) turns into

$$\Omega_{\rm DF}(\epsilon) \simeq \frac{1}{2\pi} e^{\beta_0\epsilon + \ln Z(\beta_0)} \int_{-\infty}^{\infty} d\beta''\, \exp\left(-\frac{1}{2}(\beta'')^2 \left(\frac{d^2 \ln Z}{d\beta^2}\right)_{\beta_0}\right), \qquad (22.103)$$

such that within this approximation the level density becomes

$$\Omega_{\rm DF}(\epsilon) = \frac{1}{\sqrt{2\pi\Delta\epsilon^2}}\, e^{\beta_0\epsilon + \ln Z(\beta_0)} \equiv \frac{1}{\sqrt{2\pi\Delta\epsilon^2}}\, e^{S(\beta_0(\epsilon))}, \qquad (22.104)$$

with
$$\Delta\epsilon^2 = \left(\frac{d^2 \ln Z}{d\beta^2}\right)_{\beta_0} \geq 0. \qquad (22.105)$$

Evidently, the continuous function of energy $\Omega_{\text{DF}}(\epsilon)$ of (22.104) cannot be considered to represent the $\Omega(\epsilon)$ of (22.98) in a literal sense. Rather, one should compare $\Omega_{\text{DF}}(\epsilon)$ with an averaged density $\widetilde{\Omega}(\epsilon)$. The latter might be defined through the relation $\Delta G(\epsilon) = \widetilde{\Omega}(\epsilon)\Delta$ where $\Delta G(\epsilon)$ is the number of levels in the interval from $\epsilon - \Delta/2$ to $\epsilon - \Delta/2$. The same number may be obtained by integrating $\Omega_{\text{DF}}(\epsilon)$ over the same interval. Two conditions must then be fulfilled if one wants to safely identify $\widetilde{\Omega}(\epsilon)$ by the $\Omega_{\text{DF}}(\epsilon)$. On the one hand, the number $\Delta G(\epsilon)$ must be big enough such that one may apply statistical concepts at all. On the other hand, the interval must be small enough to allow for the approximation $\int_{\epsilon-\Delta/2}^{\epsilon+\Delta/2} \Omega_{\text{DF}}(\epsilon)d\epsilon \approx \Omega_{\text{DF}}(\epsilon)\Delta$. Of course, this problem only reflects again the difficulty one may have to represent a microcanonical ensemble by the canonical one.

Formally, the $\mathcal{S}(\beta) = \beta\epsilon + \ln Z(\beta)$ satisfies a relation similar to (22.80) used above, with the essential difference, however, that ϵ and β are not independent of each other. The manipulations which lead us from the microscopic definition (22.98) of the level density to the form (22.104) might be considered to be of purely formal nature without attributing any physical content to the variable β. Indeed, such a point of view is often taken in the literature. However, the relations involved almost ask for an interpretation in terms of thermostatics, especially if we compare them with those discussed in the previous section. In this spirit one should then like to associate the following quantities with each other:

$$\begin{array}{ccc} \epsilon & \longleftrightarrow & \mathcal{E} \\ \mathcal{S}(\beta_0) & \longleftrightarrow & \mathcal{S}(\mathcal{E}) \\ \Delta\epsilon^2 & \longleftrightarrow & \Delta\mathcal{E}^2. \end{array} \qquad (22.106)$$

In this way the level density (22.104) becomes identical to that obtained in (22.85), if only both are calculated at the same mean energy: $\Omega_{\text{DF}}(\beta(\mathcal{E})) = \Omega_{\text{G}}(\epsilon = \mathcal{E})$.

The essential difference of both derivations may be seen in the following. In some sense both treatments involve a probability law which is proportional to $\exp(S)$ and in both cases a Gaussian approximation has been involved. In the first case we started from the distribution in energy such that the expansion was performed in ϵ. In (22.102), on the other hand, the expansion was done with respect to temperature. Effectively, this led to a width $\Delta\beta^2$ in β. Most interestingly, this width turned out to be given by the inverse of that in energy, such that one may write

$$\Delta\beta\,\Delta\mathcal{E} = 1. \qquad (22.107)$$

In the literature this is called the *uncertainty relation of thermostatics*. The deduction we adopted here is due to H. Feshbach (1987). We shall return to this problem in the next section, where a brief account will be given of the interest paid to this relation in the literature. Before that, however, we shall briefly generalize the Darwin–Fowler approximation to more than one parameter.

22.2.3.1 Level density with constraints Suppose we are given a system for which there exists a set of n commuting observables $\hat{A}_i$ with the eigenvalues in their common set of eigenstates being $A_{\nu_i}^{(i)}$. The partition function of (22.36) may then be written in the form

$$Z(\lambda_0, \cdots \lambda_n) = \mathrm{tr}\, \exp\bigl(-\sum_i \lambda_i \hat{A}_i\bigr) = \sum_{\nu_0 \ldots \nu_n} \exp\bigl(-\sum_i \lambda_i A_{\nu_i}^{(i)}\bigr). \qquad (22.108)$$

In generalization of (22.100) the multidimensional level density becomes

$$\Omega(A_0, \ldots A_n) = \frac{1}{(2\pi i)^n} \int_{-i\infty}^{i\infty} \mathrm{d}\lambda_0 \cdots \mathrm{d}\lambda_n \, Z(\lambda_0, \cdots \lambda_n) \exp\bigl(\sum_i \lambda_i A_i\bigr). \qquad (22.109)$$

The integrand can be written as an exponential of the function

$$S(\lambda_0, \cdots \lambda_n; A_0, \ldots A_n) = \ln Z(\lambda_0, \cdots \lambda_n) + \sum_i \lambda_i A_i, \qquad (22.110)$$

such that the stationary points are given by

$$0 = \frac{\partial S}{\partial \lambda_i} = A_i + \frac{\partial}{\partial \lambda_i} \ln Z(\lambda_0, \cdots \lambda_n) \equiv A_i - \langle \hat{A}_i \rangle \qquad i = 0, 1, \cdots n; \qquad (22.111)$$

here for the last equation (22.38) was used. Obeying this condition implies replacing in $S(\lambda_0, \cdots \lambda_n; A_0, \ldots A_n)$ the A_i by those averages $\langle \hat{A}_i \rangle$ which are calculated with that density operator which corresponds to the partition function (22.108). In this way the S of (22.110) can be identified as the thermal entropy $\mathcal{S}(\lambda_0, \cdots \lambda_n)$, in the same spirit as done before in (22.106). Expanding the exponent to second order the integrand of (22.109) becomes a multidimensional Gaussian, which allows for analytic integration such that the result becomes (see (26.7))

$$\Omega_{\mathrm{DF}}(A_0, \ldots A_n) = \frac{\exp\bigl(\mathcal{S}(\lambda_0, \cdots \lambda_n)\bigr)}{\sqrt{(2\pi)^n |\mathcal{D}(\lambda_0, \cdots \lambda_n)|}}. \qquad (22.112)$$

Here, $\mathcal{D}(\lambda_0, \cdots \lambda_n)$ is the determinant of the coefficient matrix of the quadratic form found in the exponent. With the help of (22.43) and (22.44) it can be expressed by either the fluctuations of the $\hat{A}_i$ or by the derivatives of their averages with respect to the λ_i,

$$\mathcal{D} = \det\left\{\frac{\partial^2}{\partial \lambda_i \partial \lambda_j} \ln Z\right\} = \det\left\{\langle \Delta \hat{A}_i \Delta \hat{A}_i \rangle\right\} = (-)^n \det\left\{\frac{\partial \langle \hat{A}_i \rangle}{\partial \lambda_j}\right\}. \qquad (22.113)$$

In passing we note that in Section 23.6.1 the evaluation of fluctuations in the entropy will be viewed from a different angle.

22.3 Uncertainty of temperature for isolated systems

The question of the extent to which a temperature may be defined precisely for an isolated system is still controversial–despite the fact that it was answered in the 1960s in pioneering papers by B.B. Mandelbrot (1964) and by L. Tisza (1963) and others thereafter. The concepts used there are similar to those previously developed in a special branch of mathematical statistics, and which are nicely described in the book by B.H. Lavenda *Statistical physics, a probabilistic approach* (1991). Incidentally, similar ideas have been put forward in a similar context by J. Lindhard (1974). As we shall see below, they are related to the theory of measurement. As described in (1986) by J. Lindhard, associations with the problem of complementarity between energy and temperature were raised a long time ago.

In the context of nuclear physics, it was H. Feshbach (1987) who pointed to the problem one faces when temperature is introduced. Indeed, if estimated on the basis of the Fermi gas model its uncertainties may not be negligible; the relevant formula was given above in eqn(22.97), see also (Hofmann, 1997). Later this assertion of temperature being subject to such uncertainties was questioned in an opinion piece in *Physics Today* by C.Kittel in (1988), with a reply being given in the same journal by B.B. Mandelbrot (1989) in which he refers to his own papers and to the appropriate mathematical background.

22.3.1 The physical background

In the following discussion we want to restrict ourselves to systems with temperature and energy being the only conjugate variables. As outlined in (Tisza and Quay, 1963) and (Tisza, 1963), which serve as the basis of the discussion here, generalizations to a set of variables $(A_i;\ \lambda_i)$ are more or less straightforward. Generalizing (22.63) one would have to introduce a level density $\Omega(A_i)$ which depends on all A_i. The latter are to be treated as fluctuating variables replacing –in the language used above–the operators $\hat{A}_i$ appearing in the density operator $\hat{\rho}$ of (22.36). As mentioned before, the authors of (Mandelbrot, 1964), (Tisza and Quay, 1963) and (Tisza, 1963) stress that all the derivations need is the structure function $\Omega(\epsilon)$ without any reference to a Hamiltonian.

For the relation of entropy to the partition function we may start with the relation $S(\epsilon; \beta) = \ln Z(\beta) + \beta\epsilon$ introduced above in (22.80). Next we use the ansatz

$$w(\epsilon; \beta) = \frac{\Omega(\epsilon)}{Z(\beta)} \exp(-\beta\epsilon) \qquad (22.114)$$

for the probability density of energy. The dependence of the function $S(\epsilon; \beta)$ on the fluctuating variable ϵ is the same as that of the thermal entropy $\mathcal{S}$ on the average energy $\mathcal{E}$. The parameter β could be fixed in the way standard to thermostatics, namely by exploiting the relation $\mathcal{E} = -\partial \ln Z/\partial \beta$. However, one might also take the most likely value ϵ_m of the energy, which according to (22.70) relates to β as

$$\beta = \left.\frac{\partial S(\epsilon;\beta)}{\partial \epsilon}\right|_{\epsilon_m}. \tag{22.115}$$

For finite systems both procedures may not lead to the same answer. Indeed, one of the problems already lies in the determination of the average $\mathcal{E}$. This has to be done by measuring the, in principle, fluctuating quantity ϵ. Each measurement will give a certain value ϵ' from which an estimate β' of the unknown quantity β can be obtained. Evidently, since the ϵ is fluctuating no unique answer will be found in any finite number of measurements. Hence, β can be expected to exhibit fluctuations, too.

To treat the latter, Mandelbrot (1964) suggested considering the $w(\epsilon;\beta)$ as a *function of β*, called the *likelihood function* in mathematical statistics.[43] (see the literature mentioned above as well as the book by B.H. Lavenda (1991)) and maximizing this function with respect to β, for given ϵ. In other words, the estimate β' of the unknown intensity β is associated with the most "probable" value, more precisely the one with largest "likelihood".[44] The maximum of the (positive) quantity $w(\epsilon;\beta)$ of (22.114), now as function of β, is determined through

$$\left(\frac{\partial S(\epsilon;\beta)}{\partial \beta}\right)_{\beta=\beta'} = \left(\frac{\partial \ln Z}{\partial \beta}\right)_{\beta=\beta'} + \epsilon' = 0. \tag{22.116}$$

This equation might also be written as

$$-\left.\frac{\partial \ln Z}{\partial \beta}\right|_{\beta'} \equiv \mathcal{E}'(\beta') = \epsilon'. \tag{22.117}$$

Here, on the right there appear the empirically determined values ϵ' which thus equal the canonical averages as determined by the β'. Notice that the second derivative of the $S(\epsilon;\beta)$ with respect to β (at fixed ϵ) is positive (semi) definite. This follows because of the relation

$$\frac{\partial^2 S(\epsilon;\beta)}{\partial \beta^2} = \frac{\partial^2 \ln Z(\beta)}{\partial \beta^2} \tag{22.118}$$

and because the second derivative of the logarithm of the partition function has this property, see e.g. (22.43) and the discussion after. Therefore, the extremal value of the likelihood function $w(\epsilon;\beta)$ corresponds to a maximum, indeed.

Finally, two additional remarks are in order. (i) Evidently, the procedure just outlined leads to the canonical distribution. This follows because eqn(22.116) has a unique solution. If the ϵ' is chosen to be the average $\mathcal{E}$ of all observations one gets the average value of β. (ii) It is to be expected that in the limit of $T=0$ it will become difficult to deduce information on T. There, the fluctuation in energy will vanish, which in turn is represented by the $\partial^2 \ln Z(\beta) / \partial \beta^2$. Therefore the

---

[43] This notion is sometimes used for $w(\epsilon;\beta)/\Omega(\epsilon)$.

[44] It should be noted that the likelihood function does not define a probability law in a true sense. As a function of β it may not even be normalizable.

$\mathcal{E}'(\beta')$ of (22.117) will be insensitive to variations in β. We shall return to this problem below.

Let us add a word on notation. In mathematical statistics the $\Omega(\epsilon)$ of (22.114) is called *structure function* or *prior density*. As mentioned earlier, it is this quantity which contains all "mechanical information" of the system, and which solely defines the microcanonical distribution. In this terminology the canonical distribution $w(\epsilon; \beta)$ of (22.114) is called *posterior density*.

22.3.2 The thermal uncertainty relation

The main goal is to find a proper "measure" for estimating the "fluctuations" in β'. Taking (22.114) as our basis let us define the *posterior probability density* $\widetilde{w}(\beta'; \epsilon')$ for the fluctuating variable β'

$$\widetilde{w}(\beta'; \epsilon') = \left(\widetilde{Z}(\epsilon')\right)^{-1} e^{\mathcal{L}(\beta';\epsilon')} \widetilde{w}(\beta'), \qquad (22.119)$$

with a *prior density* $\widetilde{w}(\beta')$. Here, the *likelihood function* $\mathcal{L}(\beta'; \epsilon')$ is defined as

$$\mathcal{L}(\beta'; \epsilon') = \ln\left(\frac{w(\epsilon'; \beta')}{\Omega(\epsilon')}\right) = -\ln Z(\beta') - \beta'\epsilon' = -S(\epsilon'; \beta') \qquad (22.120)$$

(remember (22.80)) and the normalization factor is given by

$$\widetilde{Z}(\epsilon') = \int d\beta'\, e^{\mathcal{L}(\beta';\epsilon')} \widetilde{w}(\beta'). \qquad (22.121)$$

Differentiating with respect to ϵ' leads to an average value $\overline{\beta}$ by

$$\frac{\partial \ln \widetilde{Z}}{\partial \epsilon'} = -\overline{\beta}. \qquad (22.122)$$

The most probable value $\hat{\beta}$, on the other hand, is determined by the stationary point of $\mathcal{L}(\beta'; \epsilon') + \ln \widetilde{w}(\beta')$. To find a Gaussian approximation to the $\widetilde{w}(\beta'; \epsilon')$ we may proceed as done before in Section 22.2.2 for the energy distribution and expand to second order:

$$\mathcal{L}(\beta'; \epsilon') + \ln \widetilde{w}(\beta') \simeq \mathcal{L}(\hat{\beta}; \epsilon') + \ln \widetilde{w}(\hat{\beta}) + \frac{1}{2}\left(\beta' - \hat{\beta}\right)^2 \frac{\partial^2}{\partial \hat{\beta}^2}\left(\mathcal{L}(\hat{\beta}; \epsilon') + \ln \widetilde{w}(\hat{\beta})\right). \qquad (22.123)$$

Finally, we will be interested in evaluating $\widetilde{Z}$ at the average energy $\epsilon' = \mathcal{E}$. This will be possible after $\mathcal{E}$ has been determined in a number r of measurements. The $\mathcal{L}(\hat{\beta}; \epsilon')$ will involve a term which grows with r and will for large r dominate the $\ln \widetilde{w}(\hat{\beta})$, for which reason the latter contribution may be neglected such that we get

$$\widetilde{Z}(\mathcal{E}) = \left[2\pi \left(\frac{\partial^2 \ln Z}{\partial \hat{\beta}^2}\right)^{-1}\right]^{1/2} \exp\left(-\hat{\beta}\mathcal{E} - \ln Z(\hat{\beta})\right) \widetilde{w}(\hat{\beta}). \qquad (22.124)$$

With these approximations, which will be valid for a sufficiently large sample size, we found $\hat{\beta} \simeq \bar{\beta} \simeq \beta(\mathcal{E})$, which allows us to identify $-\ln \widetilde{Z}(\mathcal{E})$ with the thermal entropy $\mathcal{S}(\mathcal{E})$

$$\frac{\partial \mathcal{S}(\mathcal{E})}{\partial \mathcal{E}} = \hat{\beta} = -\frac{\partial \ln \widetilde{Z}(\mathcal{E})}{\partial \mathcal{E}}. \tag{22.125}$$

A second differentiation leads to

$$\frac{\partial^2 \mathcal{S}(\mathcal{E})}{\partial \mathcal{E}^2} = \frac{\partial \hat{\beta}}{\partial \mathcal{E}} = -\frac{\partial^2 \ln \widetilde{Z}(\mathcal{E})}{\partial \mathcal{E}^2} = -\Delta \beta^2. \tag{22.126}$$

Since the expression on the very left is nothing else but the inverse energy fluctuation (see (22.46)) we finally get

$$\Delta \mathcal{E} \Delta \beta = 1. \tag{22.127}$$

As we know from (22.47) the energy fluctuation is related to the specific heat by $\Delta \mathcal{E}^2 = C/\beta^2$. Thus the relative fluctuation of β becomes

$$\frac{\Delta \beta^2}{\beta^2} = \frac{1}{C}. \tag{22.128}$$

Since $\beta = 1/T$ one has (see exercise 11)

$$\left| \frac{d\beta}{\beta} \right| = \frac{dT}{T} \quad \text{and hence} \quad \frac{\Delta T^2}{T^2} = \frac{1}{C}, \tag{22.129}$$

a formula which is also given in (Landau and Lifshitz, 1980), as well as in (Lindhard, 1986) and (Feshbach, 1987).

To proceed further or to get deeper insight into the problem one needs to invoke the mathematical concept of the "Fisher information". In this way it becomes possible to derive the *uncertainty* relation proper (Lavenda, 1991), (Mandelbrot, 1989)

$$\Delta \mathcal{E}^2 \Delta \beta^2 \geq 1. \tag{22.130}$$

22.4 The lack of extensivity and negative specific heats

So far in our discussion we have not touched on the problem of the existence of the thermostatic limit in a proper sense. It only became clear that relative fluctuations of thermal quantities may decrease with increasing particle number. Moreover, for a canonical ensemble a close relation between specific heats and energy fluctuations has been seen to exist, as given by (22.47) or $C_Q = (\partial E/\partial T)_Q \equiv \beta^2 \Delta E_Q^2 \equiv \beta^2 \langle \Delta \hat{H}^2 \rangle_Q$. This relation clearly implies that the canonical specific heat is definitely *positive*. From the inequality $C_Q > 0$ other stability conditions can be deduced. The $C_\mathcal{V} > 0$ at constant volume $\mathcal{V}$, for instance, may be shown to imply that pressure must decrease with increasing volume, see e.g. (Landau and Lifshitz, 1980) and (Balian, 1991). In such proofs the system is assumed to be *homogeneous and extensive*. Extensivity may be defined through the following property of entropy studied as function of (extensive) variables like volume, energy and particle number, $\mathcal{S}(x\mathcal{V}, xE, xN) = x\mathcal{S}(\mathcal{V}, E, N)$. In Section 5.5 of (1991) R. Balian puts this condition at the beginning of his discussion of the thermostatic limit. The point is, however, that for a truly macroscopic body this condition suffices to be fulfilled *locally*: It must be possible to choose regions inside the body which are macroscopically small but microscopically large, and inside which matter is homogeneous, (see also (Landau and Lifshitz, 1980)). We may note in passing that this condition implies that any surface effects must be negligibly small. In such a case the positivity of $C_\mathcal{V} > 0$ can be shown to be given locally.

A specific heat C can also be defined for microcanonical ensembles. In this case it turns out that C may be negative. In fact this may be so even for truly macroscopic systems if only C is calculated for the body as a whole. This is the case whenever long range forces play a crucial role, e.g. for gravitation inside stellar objects or for the Coulomb forces for a charged metal. For astrophysical applications this property was clarified long ago, see e.g. (Hertel and Thirring, 1971), (Lynden-Bell, 1999) and references to early work cited therein. More recently in (Thirring et al., 2003) the connection of a negative specific heat with non-ergodic behavior has been demonstrated within a simple model.

In the past decade negative specific heats have been subject of much interest in connection with micro-systems like atomic nuclei and metal clusters. Amongst the vast literature on this field we may refer to reports like (Gross, 2001) and (Chomaz and Gulminelli, 2005) for nuclear aspects and (Haberland, 1999) for metal clusters (for the original experimental work see (Schmidt et al., 2001)). In nuclei this phenomenon is claimed to be seen in a phase transition similar to that from a liquid to a gas. Such a transition may occur in the very last stage of heavy-ion collisions at energies larger than those described in Chapter 13. In such reactions the nuclear complex is heated up so much that it begins to expand until it has reached a sufficiently low density at which it is thermodynamically more stable that homogeneous matter breaks up into small fragments. Such a reaction cannot be parameterized in terms of a few shape degrees of freedom and thus is not amenable to that kind of transport theory we mainly address in this book. Therefore, it must suffice to refer to the papers mentioned above, as well as to a few more recent ones like (Campi et al., 2005) and (Tõke et al., 2005).

It is evident that for self-bound micro-systems like finite nuclei or metal clusters the *conditions for extensivity are not fulfilled*. Such systems have to be taken as entire, quantum objects. Shell effects, for instance, show up only for motion inside given boundaries and get lost if the system is split into smaller pieces, or if its volume increases indefinitely. But even when shell effects are discarded–after all they disappear with increasing thermal excitation–one must keep in mind that in a realistic nucleus many of its nucleons occupy the surface layer. Usually, this feature is neglected in the liquid drop model. However, this drop is charged and the long-range Coulomb interaction then has similar implications as those mentioned before in the context of the specific heat of macroscopic systems.

Often it is argued that because of this lack of extensivity, and the therefore obvious problems with the thermostatic limit, the use of the canonical ensemble is not trustworthy. This argument is simply wrong, or at least not very precise. Indeed, as explained in Section 22.1.2, even for a small and isolated system one may apply the *maximum entropy principle*–provided the system is in equilibrium at all, but this assumption underlies all theoretical models mentioned in this present context. The crucial question is that of the interpretation of the measured energies. If only their *average* value $\langle E \rangle$ were known one may well introduce the canonical density operator and deduce the temperature from the usual relation to the $\langle E \rangle$, with obvious generalizations to the inclusion of other constraints. For this it is immaterial whether or not the system is isolated and how big it is. The size of the system plays a role only with respect to the size of the fluctuations of the thermal quantities around their averages. As explained in Section 22.3, for the nuclear case one should indeed worry about *fluctuations in temperature*, which to the best of our knowledge has been discarded in all the theories published in the literature; recall please the discussion in Section 10.1.3 on structure in the specific heat associated with the disappearance of pairing correlations. In this respect experiments with metal clusters are different as they are produced in such a way that they can be considered to be in thermal equilibrium with their surroundings (Schmidt *et al.*, 1999), (Haberland, 1999).

Let us come back to the relation between the measured energy E and temperature. If this energy (or a set of energies, rather) can really be associated with a microcanonical ensemble then temperature has to be defined by the level density $\Omega(E)$ through the formula $1/T = \partial \ln \Omega(E)/\partial E$. In the common procedure one would evaluate this derivative at the *most likely* value E_m of the energy distribution $w(E)$, see Section 22.2. Whereas for large systems this E_m is practically identical to $\langle E \rangle$ this is not the case for small systems. This fact leads to some ambiguity in the definition of temperature and hence is one of the causes for the different expressions for the specific heat (Behringer *et al.*, 2005). In models where the existence of a negative microcanonical C is demonstrated one often takes a "bi-modal" distribution $w(E)$, consisting typically of two separated Gaussians of the type employed in Section 22.2.2 (Schmidt *et al.*, 2001), (Chomaz and Gulminelli, 2005). It is then easy to see that as a function of energy the entropy $S(E) \propto \ln \Omega(E)$ attains a dip, which lies in the region between the two maxima of $w(E)$–similar to the behavior expected for phase transitions (see e.g. Fig. 6.2 of (Balian, 1991)). In this region temperature exhibits "back bending" with the implication that the microcanonical specific heat $C = T\partial S/\partial E$ becomes negative.

22.5 Thermostatics of independent particles

In second quantized form the Hamiltonian of independent particle motion reads (see Section 26.8)

$$\hat{H} \equiv \hat{H}_{\text{ipm}} = \sum_k e_k \hat{c}_k^\dagger \hat{c}_k \equiv \sum_k e_k \hat{n}_k \,. \tag{22.131}$$

The creation and annihilation operators $\hat{c}_k^\dagger, \hat{c}_k$ are associated with the single particle states $|k\rangle$. The latter are eigenstates of the $\hat{n}_l = \hat{c}_l^\dagger \hat{c}_l$ with eigenvalues $0, 1$, viz $\hat{n}_l|k\rangle = \delta_{kl}|k\rangle$, such that the number operator reads $\hat{A} = \sum_k \hat{n}_k$. The eigenvalues of $\hat{H}$ and $\hat{A}$ in the A-body states $|m, A\rangle$ may be expressed by

$$E_{m,A} = \sum_k n_k^{(m,A)} e_k \quad \text{and} \quad A = \sum_k n_k^{(m,A)} \,. \tag{22.132}$$

The $n_k^{(m,A)}$ are either 1 or 0 depending on whether in the Slater determinant $|m, A\rangle$ the single particle state $|k\rangle$ is occupied or empty. The partition function given in (22.51) may therefore be expressed as

$$\begin{aligned} Z_{\text{gce}}^{\text{ipm}} &= \sum_A \sum_m \exp\left(-\beta E_{mA} + \alpha A\right) \quad \text{with} \quad \alpha = \beta \mu \\ &= \prod_k \sum_{n_k} \exp\left(-n_k\left(\beta e_k - \alpha\right)\right) = \prod_k \left(1 + e^{-(\beta e_k - \alpha)}\right) \,. \end{aligned} \tag{22.133}$$

According to (22.51) the grand potential is given by

$$\Omega_{\text{gce}}^{\text{ipm}} = -T \ln Z_{\text{gce}}^{\text{ipm}} = -T \sum_k \ln\left(1 + e^{(\alpha - \beta e_k)}\right) \quad \text{with} \quad \alpha = \beta \mu \,. \tag{22.134}$$

The internal energy $\mathcal{E}$ and the average particle number $\mathcal{A}$ are defined by the thermal average over the expressions given above. As we know from (22.38) they may be evaluated directly by differentiating the $\ln Z$ with respect to the associated intensities. For the particle number one therefore gets

$$\mathcal{A} = \left(\frac{\partial \ln Z_{\text{gce}}^{\text{ipm}}}{\partial \alpha}\right)_\beta = \sum_k \frac{\exp\left(\alpha - \beta e_k\right)}{1 + \exp\left(\alpha - \beta e_k\right)} = \sum_k n(e_k) \,, \tag{22.135}$$

with the average occupation number for fermions $n(e_k) = (\exp(\beta e_k - \alpha) + 1)^{-1}$ being the averages of the operators $\hat{n}_k$ introduced above. Likewise, the internal energy becomes

$$\mathcal{E}_{\text{ipm}} = -\left(\frac{\partial \ln Z_{\text{gce}}^{\text{ipm}}}{\partial \beta}\right)_\alpha = \sum_k e_k n(e_k) \,. \tag{22.136}$$

In principle, the sum extends over all single particle states found in a given mean field. Evidently, for a potential of finite depth the continuum of scattering states ought to be included as well. In principle, special care would then be required to deal with this problem. However, for small temperatures it is only the level density around the Fermi energy e_F which matters, and for the common nucleus this e_F is situated well below the energy where the continuum starts.

As may be seen from (22.53) the entropy is given by the negative derivative of the grand potential with respect to temperature, with the other possible variables fixed. In this way one gets the expression

$$S_{\text{ipm}} = -\sum_k \left\{ [1 - n(e_k)] \ln[1 - n(e_k)] + n(e_k) \ln n(e_k) \right\}. \tag{22.137}$$

After introducing the single particle level density $g(e) = \sum_k \delta(e - e_k)$ of (5.15) the expressions above can be cast into the following integral form

$$I = \int_0^\infty \phi(e) n(e)\, de, \tag{22.138}$$

with $\phi(e) = \varphi(e) g(e)$. For $\varphi(e) = 1$ or $\varphi(e) = e$ the I represents particle number or internal energy, respectively. The $\ln Z_{\text{gce}}^{\text{ipm}}$ is obtained by replacing $\varphi(e) n(e)$ by $\ln(1 + \exp(\alpha - \beta e))$. Because of $(d/de) \ln(1 + \exp(\alpha - \beta e)) = -\beta n(e)$ the grand potential of (22.134) becomes

$$\Omega_{\text{gce}}^{\text{ipm}}(T) = -\int_0^\infty n(e) G(e)\, de \quad \text{with} \quad G(e) = \int_0^e d\epsilon\, g(\epsilon), \tag{22.139}$$

following the definition of $G(e)$ given in (22.62). This form of $\Omega_{\text{gce}}^{\text{ipm}}$ involves a partial integration, the only condition being that $\lim_{e \to \infty} G(e) \ln(1 + \exp(\alpha - \beta e)) = 0$. We may note in passing that for zero temperature the grand potential simply turns into $\Omega_{\text{gce}}^{\text{ipm}}(T = 0) = \Omega_0(e = e_F)$ with

$$\Omega_0(e) = -\int_0^e G(\epsilon)\, d\epsilon. \tag{22.140}$$

To proceed further let us suppose that at $e \to \infty$ the integrated function $\Phi(e)$,

$$\Phi(e) = \int_0^e \phi(e')\, de', \quad \text{i.e.} \quad \phi(e) = \frac{d\Phi(e)}{de}, \tag{22.141}$$

grows less than exponential (and is well behaved at $e = 0$). Then partial integration leads to the form

$$I = -\int_0^\infty \Phi(e) \left(\frac{\partial n(e)}{\partial e} \right) de, \tag{22.142}$$

where the factor

$$\frac{\partial n(e)}{\partial e} = -\frac{\beta}{4} \frac{1}{\cosh^2(\beta e - \alpha)/2} \equiv n'(e) \tag{22.143}$$

of the integrand has a pronounced peak at $e = \mu$ of width $4T$. Since we restrict ourselves to cases where $T \ll \mu$ this property may lead us to good approximations to the integral (22.142).

Before we come to that let us quote the result for the *Fermi gas* proper where $g(e) \propto \sqrt{e}$ such that $G(e) \propto (2/3)e\sqrt{e}$. Thus the grand potential,

$$\frac{\Omega_{\text{gce}}}{\mathcal{V}} = -T g_{\text{eff}} \int_0^\infty \sqrt{e} \ln\left(1 + e^{\frac{\mu-e}{T}}\right) de \quad \text{with} \quad g_{\text{eff}} = g_{\text{s.i.}} \frac{m^{3/2}}{\sqrt{2\pi^2 \hbar^3}}, \tag{22.144}$$

may be expressed through the internal energy, which in this case simply represents the kinetic energy $\mathcal{E}_{\text{ipm}} \equiv \mathcal{E}_{\text{kin}}$:

$$\frac{\Omega_{\text{gce}}}{\mathcal{V}} = -\frac{2}{3}\frac{\mathcal{E}_{\text{kin}}}{\mathcal{V}} \quad \text{implying} \quad \mathcal{PV} = \frac{2}{3}\mathcal{E}_{\text{kin}}. \tag{22.145}$$

In (22.144) the factor $g_{\text{s.i.}}$ stands for the spin-isospin degeneracy; it is equal to 2 if protons and neutrons are treated separately, or to 4 otherwise. The expression on the very right of (22.145) follows from (22.55) and represents a special form of the equation of state (EOS)(see exercise 12).

22.5.1 *Sommerfeld expansion for smooth level densities*

Sommerfeld has developed a method for deducing the temperature dependence of integrals like those given in (22.138) provided that the function $\phi(e)$ is sufficiently smooth in the neighborhood of $\mu = \alpha/\beta$. Since the derivative of the Fermi function is peaked at $e = \mu$, the I of (22.142) may be evaluated by expanding $\Phi(e)$ around this point. For $T \ll \mu$ the lower limit of the integral in (22.142) may be replaced by $-\infty$. In this way it becomes obvious that only even powers in $e - \mu$ survive, which finally implies an expansion into even powers of T. The coefficients may then be expressed by the Riemann zeta function multiplied by the odd derivatives of the original ϕ calculated at μ. We save ourselves from going into more details of the derivation, which may be found in the literature (Landau and Lifshitz, 1980), (Reif, 1984), (Ashcroft and Mermin, 1976). To leading order in T one finds

$$I = \int_0^\mu \phi(e') \, de' + \frac{\pi^2}{6} T^2 \left[\frac{d}{de} \phi(e)\right]_\mu. \tag{22.146}$$

A direct application of this method to a finite nucleus is not possible without taking further precautions, simply because the original density (5.15) of (bound) single particle levels does not meet the condition of being a smooth function. To benefit from this expansion one needs to introduce a *smooth level density* by averaging the correct one (5.15) over some finite energy interval γ_{av}, say as performed in Section 5.2 in connection with the Strutinsky method,

$$\bar{g}(e) = \frac{1}{\gamma_{\text{av}}} \int f\left(\frac{e'-e}{\gamma_{\text{av}}}\right) g(e') de' = \frac{1}{\gamma_{\text{av}}} \sum_k f\left(\frac{e_k - e}{\gamma_{\text{av}}}\right). \tag{22.147}$$

Of course, the averaging interval γ_{av} need not necessarily be the same as for the Strutinsky procedure. It must only be larger than the average level spacing D. Only in this case the extreme variation of g with e, which is present for any finite system, gets washed out. For a $\gamma_{\rm av}$ smaller than the average shell spacing $\hbar\Omega_0^{\rm sh}$ an oscillatory behavior of $g(e)$ would remain. Eventually, one may wish to smooth out even the latter, in which case $\gamma_{\rm av}$ would have to be chosen like in the Strutinsky method. Applications are discussed in Section 5.5.

For a level density $\overline{g}(e)$ which is sufficiently smooth the I of (22.138) or (22.142) may be approximated as sketched above. For internal energy, grand potential and particle number one gets results typical of the "Fermi gas", namely

$$\overline{\mathcal{E}}_{\rm ipm}(T,\mu) = \int_0^\mu e\overline{g}(e)de + \frac{\pi^2}{6}\left(\overline{g}(\mu) + \mu\overline{g}'(\mu)\right)T^2 \qquad (22.148)$$

$$\overline{\Omega}_{\rm gce}^{\rm ipm}(T,\mu) = -\int_0^\mu de \int_0^e de'\,\overline{g}(e') - \frac{\pi^2}{6}\overline{g}(\mu)T^2 \qquad (22.149)$$

$$\mathcal{A}(T,\mu) = \int_0^\mu \overline{g}(e)de + \frac{\pi^2}{6}\overline{g}'(\mu)T^2 \,. \qquad (22.150)$$

Here, $\overline{g}'(\mu)$ is the derivative of the level density calculated at the chemical potential. Notice, please that the expression for particle number is in accord with (22.53) which requires $\mathcal{A}(T,\mu) = -(\partial\overline{\Omega}_{\rm gce}^{\rm ipm}/\partial T)_\mu$. Likewise, from the same relation one gets for the entropy

$$\overline{\mathcal{S}}_{\rm gce}^{\rm ipm}(T) = -\left(\frac{\partial\overline{\Omega}_{\rm gce}^{\rm ipm}}{\partial T}\right)_\mu = \frac{\pi^2}{3}\overline{g}(e_{\rm F})\,T\,. \qquad (22.151)$$

These expressions are correct to order T^2 for given μ. Since the latter will vary with T, too, one may make further expansions around $\mu = e_{\rm F}$. To lowest order one gets for the grand potential

$$\overline{\Omega}_{\rm gce}^{\rm ipm}(T,\mu) = \overline{\Omega}_{\rm gce}^{\rm ipm}(T=0) - \overline{g}(e_{\rm F})\frac{(\mu - e_{\rm F})^2}{2} - \frac{\pi^2}{6}\left(\overline{g}(e_{\rm F}) + (\mu - e_{\rm F})\overline{g}'(e_{\rm F})\right)T^2 \qquad (22.152)$$

If the particle number $\mathcal{A}(T,\mu)$ may be considered independent of temperature one may identify it with $\mathcal{A} = \int_0^{e_{\rm F}} \overline{g}(e)de$. The relation (22.150) then implies that a finite value of $\overline{g}'(\mu)$ goes along with a change in the chemical potential, $\mu - e_{\rm F} = -(\pi^2/6)T^2\overline{g}'(e_{\rm F})/\overline{g}(e_{\rm F})$. Notice that for a positive $\overline{g}'(e_{\rm F})$ the chemical potential decreases with T below $e_{\rm F}$. Often, the terms which involve $\overline{g}'$ are discarded. Then formula (22.148) reduces to $\overline{\mathcal{E}}_{\rm ipm} = \overline{E}_0^{\rm ipm} + (\pi^2/6)\overline{g}(e_{\rm F})T^2$, where $\overline{E}_0^{\rm ipm} = \int_0^{e_{\rm F}} de\,e\overline{g}(e)$ is the *smoothed* ground state energy.

22.5.2 Thermostatics for oscillating level densities

To study the influence of the oscillating density on the thermostatic potentials let us write it in the form

$$\delta g(e) = g(e) - \overline{g}(e) = \sum_{n=1}^\infty A_n(e)\cos\left(W_n(e)/\hbar\right)\,. \qquad (22.153)$$

For constant amplitudes A_n and the special choice $W_n(e)/\hbar = n\omega e - \phi_n$ (with constant phases ϕ_n) this is nothing else but a Fourier series.[45] The more general version (22.153), on the other hand, is compatible with the Gutzwiller trace formula (5.28). Even for the general case, One may assume the amplitudes $A_n(e)$ to be smooth functions of energy such that integral of the type (22.142) are dominated by the variation of the phases $W_n(e)$. Hence, it suffices to know the following integral:

$$J(T) = -\int_0^\infty A(e) \exp\left(\frac{i}{\hbar}W(e)\right) \frac{\partial n(e)}{\partial e} de \qquad (22.154)$$

As shown in (Richter et al., 1996) to leading order in $\hbar$ and T this $J(T)$ factorizes into $J(T) = J(T=0)R(T)$, with the value at $T=0$ being given by

$$J(T=0) = A(\mu)\exp(iW(\mu)/\hbar) \qquad (22.155)$$

and the temperature-dependent factor by $R(T = 1/\beta)$

$$R(T) = \frac{\pi T^W/\hbar\beta}{\sinh(\pi T^W/\hbar\beta)} \equiv \frac{\tau}{\sinh\tau} \quad \text{where} \quad T^W = \left(\frac{dW(e)}{de}\bigg|_{e=\mu}\right). \qquad (22.156)$$

This result is obtained by an astute application of the residue theorem. One first closes the contour by adding to the path $0 \to \infty$ the parallel one which stretches from $\infty + i2\pi T$ to $0 + i2\pi T$ and by including the corresponding two short vertical lines. For small T/μ the contributions from the latter are negligibly small; this is due to the factor $n'(e)$ which as we know from (22.143) differs from zero only in the neighborhood of μ. This contour encloses a (double) pole of $n'(e)$ located at $e = \mu + i\pi T$. It is assumed that the energy dependence of the amplitude $A(e)$ can be neglected and that the energy dependence of the $W(e)$ may be considered to first order, that is $W(e) = W(\mu) + (e-\mu)/\widetilde{\omega}(\mu)$.

With this outcome one may evaluate the integrals (22.138) or (22.142). Let us look at the grand potential, in which case the Φ of (22.142) is given by the variation $\delta\Omega_0$ of the Ω_0 at zero temperature introduced in (22.140). According to (22.155) we need to know it at the chemical potential. Integrating (22.153) twice over energy one gets

$$\delta\Omega_0^n(\mu) = A_n(\mu)\left(\hbar/T_n^W\right)^2 \cos\left(W_n(\mu)/\hbar\right) \qquad (22.157)$$

as the contribution from the nth term. Noting that the damping factor, too, may in general vary from term to term, the sum for the variation of the grand potential is seen to become

---

[45] For such a form calculation of thermal quantities had already been done in 1968 by A. Gilbert in an unpublished report of the University of California Radiation Laboratory (UCRL-18095) of which use was made in (Kataria et al., 1978).

$$\delta\Omega_{\text{gce}}^{\text{ipm}}(T,\mu) = \sum_{n=1}^{\infty} R_n(T)\delta\Omega_0^n(\mu). \tag{22.158}$$

For given T and μ one may proceed in calculating the various thermostatic quantities one is interested in. For the entropy, for instance, one gets from (22.53)

$$\delta\mathcal{S}_{\text{gce}}^{\text{ipm}}(T,\mu) = -\sum_{n=1}^{\infty} \frac{dR_n(T)}{dT}\delta\Omega_0^n(\mu). \tag{22.159}$$

One must be aware, however, that the two parameters T and μ are influenced not only by the oscillating part of the level density but by the smooth part as well.

In application to the Gutzwiller trace formula (5.28) the T_n^W are seen to become the periods of the periodic orbits $T_{p.o.} = kT_{p.p.o.}$ with the $T_{p.p.o.} = dW_{p.p.o.}/de$ (calculated at the chemical potential) being that of the principal ones, see (5.29). The form of the damping factors $R_n(T)$ implies that only the shortest periodic orbits are important, more precisely the larger the temperature the shorter the orbits which contribute. In practice this amounts to estimating the $\tau_{p.p.o.}$ from $T_{p.p.o.} \simeq 2\pi/\hbar\Omega_0^{\text{sh}}$, with $\hbar\Omega_0^{\text{sh}}$ being the typical shell spacing parameter. In Section 5.5.2 these features are elucidated for the example of a schematic but realistic nuclear model.

22.5.3 Influence of angular momentum

The nuclear total angular momentum J is a conserved quantum number. It is difficult to trace it back to the angular momenta of the individual particles. For the projection M on a given axis, however, one may simply sum up the corresponding magnetic quantum numbers m_k of the occupied single particle states $|k\rangle$. This may be accounted for by adding to the exponent in (22.133) for the grand partition function $Z_{\text{gce}}^{\text{ipm}}$ a term $-\gamma M$, or in that for the grand potential $\Omega_{\text{gce}}^{\text{ipm}}$ in (22.134) a term $-\gamma m_k$. The single particle level density $g(e) = \sum_k \delta(e-e_k)$ has to be modified to $g(e,m) = \sum_k \delta(e-e_k)\delta_{m,m_k}$. Without going into details we just like to quote the main results one gets for the Fermi gas model, where e and m are treated as continuous parameters, with corresponding level densities $\bar{g}(e)$ and $\bar{g}(e,m)$ and where $\bar{g}(e) = \int \bar{g}(e,m)dm$. Discarding derivatives of $\bar{g}(\mu)$ one gets (Bethe, 1937), (Bloch, 1954) for energy, grand potential and average angular momentum

$$\overline{\mathcal{E}}_{\text{ipm}}(\alpha,\beta,\gamma) = \overline{\mathcal{E}}_{\text{ipm}}(\alpha,\beta) + \frac{1}{2}\left(\frac{\gamma}{\beta}\right)^2 \bar{g}(\mu)\langle m^2\rangle \tag{22.160}$$

$$\overline{\Omega}_{\text{gce}}^{\text{ipm}}(\alpha,\beta,\gamma) = \overline{\Omega}_{\text{gce}}^{\text{ipm}}(\alpha,\beta) - \frac{1}{2}\left(\frac{\gamma}{\beta}\right)^2 \bar{g}(\mu)\langle m^2\rangle \tag{22.161}$$

$$\overline{\mathcal{M}}_{\text{ipm}}(\alpha,\beta,\gamma) = -\frac{\gamma}{\beta}\bar{g}(\mu)\langle m^2\rangle. \tag{22.162}$$

The expression for particle number $\mathcal{A}$ is unchanged. The $\langle m^2\rangle$ is an average of the squared magnetic quantum number defined as

Thermostatics of independent particles   473

$$\bar{g}(\mu)\langle m^2\rangle = \int_{-\infty}^{\infty} \bar{g}(e,m)\, m^2\, dm \approx \mathcal{J}^{\text{rig}}/\hbar^2\,. \qquad (22.163)$$

The estimate through the rigid body value $\mathcal{J}^{\text{rig}}$ of the moment of inertia follows in a semi-classical approximation. The formulas presented here are derived for the symmetry $g(e,m) = g(e,-m)$. The more general case is treated in (Bloch, 1968) and (Feshbach, 1992) which allows one to include isospin.

Exercises

1. For any $d\hat{\rho}$ the variation of the entropy (22.7) reads $d\mathcal{S} = -\text{tr}\,(\hat{\rho}+d\hat{\rho})\ln[\hat{\rho}+d\hat{\rho}] + \text{tr}\,\hat{\rho}\ln\hat{\rho}$. Show that to lowest order $d\mathcal{S}$ is given by (22.13) by using $\ln(1+x) \simeq x$ and $\text{tr}\,d\hat{\rho} = 0$.
2. Prove formula (22.33) by deriving from (22.19) and (22.20)

$$\left(\frac{\partial^2 \mathcal{F}}{\partial T^2}\right)_Q = -\frac{1}{T^3}\left(\langle \hat{H}^2\rangle - \langle \hat{H}\rangle^2\right)_Q$$

$$\left(\frac{\partial^2 \mathcal{F}}{\partial T \partial Q}\right)_Q = \frac{1}{T^2}\left(\langle \hat{H}\hat{F}\rangle - \langle \hat{H}\rangle\langle \hat{F}\rangle\right)_Q.$$

3. Convince yourself that (22.34) is equivalent to (23.156) (if applied to the present one-dimensional case).
4. (a) Compare (22.24) with (22.40) and show $\mathcal{S} = -(\partial \mathcal{F}/\partial T)_{\lambda_i}$. (b) Prove relation (22.23), namely $\mathcal{E} = \mathcal{F} + T\mathcal{S}$, from (22.38). (c) Show that relation (22.21) between pressure and free energy is a special case of (22.38).
5. Show that $\sum_{ij} C_i C_j^* (\langle \hat{A}_i \hat{A}_j \rangle - \langle \hat{A}_i \rangle \langle \hat{A}_j \rangle) = \langle |\sum_i C_i (\hat{A}_i - \langle \hat{A}_i \rangle)|^2 \rangle \geq 0$.
6. Check the relation of eqn(22.48) to formula (22.26).
7. With the help of the eigenfunctions of both operators, prove the inequality (22.56), namely $\text{tr}\,\{\hat{\rho}'(\ln \hat{\rho} - \ln \hat{\rho}')\} \leq 0$ with $\text{tr}\,\hat{\rho}' = \text{tr}\,\hat{\rho} = 1$, by making use of the inequality $\ln x \leq x - 1$.
8. Assume the density $\hat{\rho}(t)$ satisfies the von Neumann equation (22.1). Convince yourself that the entropy $\mathcal{S}[\hat{\rho}]$ given by (22.7) stays constant in time by making use of (22.13) and the invariance of the trace with respect to cyclic permutations.
9. Assume the level density relates to entropy as given in (22.92). Derive formula (22.86) for the partition function, together with (22.82) for $w(\epsilon)$ and (22.85) for the level density $\Omega_G(\epsilon)$ with the energy fluctuation being given by

$$\Delta\mathcal{E}^{-2} = -\left.\frac{\partial^2 \ln \Omega(\epsilon)}{\partial \epsilon^2}\right|_{\mathcal{E}} \simeq -\left.\frac{\partial^2 S(\epsilon)}{\partial \epsilon^2}\right|_{\mathcal{E}} = -\left.\frac{\partial}{\partial \epsilon}\frac{1}{T}\right|_{\mathcal{E}} = \frac{1}{T^2}\frac{\partial T}{\partial \mathcal{E}}\,.$$

10. Given a one-dimensional oscillator of frequency ϖ, the quantal level density of which reads $\Omega(\epsilon) = \sum_{n=0}^{\infty} \delta(\epsilon - (n+1/2)\hbar\varpi))$. Convince yourself that the partition function is given by $Z(\beta) = (2\sinh(\hbar\varpi\beta/2))^{-1}$ and that, hence,

the mean energy becomes $\mathcal{E} = -\partial \ln Z/\partial \beta = (\hbar\varpi/2)\coth(\hbar\varpi\beta/2)$ and its fluctuation $\Delta\mathcal{E}^2 = \partial^2 \ln Z/\partial \beta^2 = (\hbar\varpi/2)^2(\sinh(\hbar\varpi\beta/2))^{-2} = (\hbar\varpi)^2 Z^2$. Show that in Darwin–Fowler approximation the level density reads

$$\Omega_{\mathrm{DF}}(\mathcal{E}) = \frac{1}{\sqrt{2\pi\Delta\mathcal{E}^2}} Z(\beta) e^{\beta\mathcal{E}} = \frac{1}{\sqrt{2\pi\hbar\varpi}} \exp(\frac{\hbar\varpi\beta}{2} \coth(\frac{\hbar\varpi\beta}{2})).$$

Compare this Ω_{DF} with the exact expression given in (24.95) and interpret the difference.

11. Suppose the prior density $\tilde{w}(\beta)$ is given by the normal distribution (25.29) with a sufficiently narrow width $\Delta\beta$. Find the normal distribution for $\tilde{w}_T(T)$, where $\tilde{w}(\beta)d\beta = \tilde{w}_T(T)dT$, and show that $\Delta T = \overline{T}^2 \Delta\beta$.

12. The EOS for the Fermi gas at small T:
    (i) Show that for $T = 0$ eqn(22.145) implies that the so-called *adiabatic index* $\Gamma \equiv (d\ln P/d\ln \rho)_{\mathcal{S}=0}$ is equal to $5/3$ (with ρ being the density of (1.23)).
    (ii) With the help of (22.148) prove that to lowest order in T

$$P(T,\rho) = P(T=0,\rho)\left(1 + \frac{5\pi^2}{12}\left(\frac{T}{e_F}\right)^2\right)$$

    (iii) Convince yourself that for the classical ideal gas the value $\Gamma = 5/3$ is valid at any $T > 0$ with finite entropy. It suffices to apply the canonical ensemble and make use of formulas (19.60) for the free energy and (19.61) for the entropy, respectively.

23

LINEAR RESPONSE THEORY

In this introduction to linear response theory special emphasis is put on properties that are relevant for our purpose, following the more detailed expositions given in (Kadanoff and Martin, 1963), (Pines and Nozières, 1966), (Forster, 1975). Generally one addresses the problem of evaluating averages $\langle \hat{A}_\mu \rangle_t$ of some operators $\hat{A}_\mu$ when the latter couple to a given system in the following factorized form

$$\hat{H}_{\mathrm{cpl}}(t) = \sum_\nu \hat{A}_\nu a_\nu^{\mathrm{ext}}(t). \tag{23.1}$$

The *linear* response shall then be defined as

$$\chi_{\mu\nu}(\omega) = \left. \frac{\delta(\langle \hat{A}_\mu \rangle_\omega)}{\delta(-a_\nu^{\mathrm{ext}}(\omega))} \right|_{\{a_\nu^{\mathrm{ext}}=0\}}, \tag{23.2}$$

with $\langle \hat{A}_\mu \rangle_\omega$ and $a_\nu^{\mathrm{ext}}(\omega)$ being the Fourier transforms (see Section 26.4) of the average $\langle \hat{A}_\mu \rangle_t$ and the $a_\nu^{\mathrm{ext}}(t)$, respectively.

23.1 The model of the damped oscillator

Let us first address the simple case of the harmonic oscillator, which is actually more than just an example for studying generic features. Within the locally harmonic approximation to collective motion it serves as one of the basic building blocks for the development of the transport equations.

For this system there is just one quantity $\langle \hat{A} \rangle_t = q(t)$ to be studied. The $q(t)$ is the coordinate measuring the elongation from to the stationary point at $q=0$. Under the presence of an external force $-q^{\mathrm{ext}}(t)$ the basic equation of motion reads

$$M\frac{d^2 q}{dt^2} + \gamma \frac{dq}{dt} + Cq(t) = -q^{ext}(t). \tag{23.3}$$

The Fourier transform $q(\omega)$ satisfies the equation

$$\chi_{\mathrm{osc}}^{-1}(\omega) q(\omega) = -q^{ext}(\omega), \quad \text{with} \quad \chi_{\mathrm{osc}}(\omega) = \frac{1}{-\omega^2 M - i\omega\gamma + C}. \tag{23.4}$$

The function $\chi_{\mathrm{osc}}(\omega)$ is in accord with the definition (23.2) of the linear response. Its poles lie at the solutions $\omega^\pm$ of the secular equation $-\omega^2 M - i\omega\gamma + C = 0$, which are given by

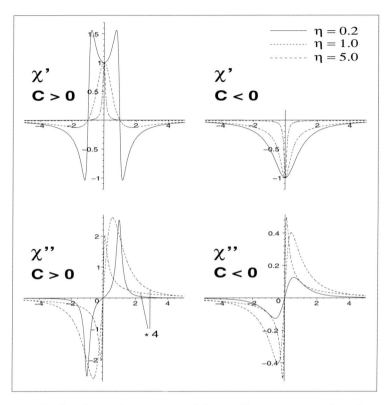

FIG. 23.1. Real and imaginary parts of the oscillator response function as functions of ω/ϖ given by eqn(23.7) for stable and unstable modes at various damping strengths η; mind the different scales used in the lower left part: the dashed curves are stretched by a factor of 4.

$$\omega^\pm = \pm \mathcal{E} - i\frac{\Gamma}{2} = \varpi\left(\pm\sqrt{\text{sign}C - \eta^2} - i\eta\right) \quad \text{where} \quad \begin{cases} \Gamma &= \gamma/M \\ \varpi^2 &= |C|/M \\ 2\eta &= \gamma/\sqrt{M|C|}. \end{cases} \quad (23.5)$$

The $\chi_{\text{osc}}(\omega)$ of (23.4) may thus be rewritten in the form

$$\chi_{\text{osc}}(\omega) = -\frac{1}{2M\mathcal{E}}\left(\frac{1}{\omega - \omega^+} - \frac{1}{\omega - \omega^-}\right). \tag{23.6}$$

For many purposes it is convenient to split this response function into real and imaginary parts when calculated along the real axis: For $\omega = \omega^*$ one gets

$$\chi_{\text{osc}}(\omega) = \chi'_{\text{osc}}(\omega) + i\chi''_{\text{osc}}(\omega) = \frac{1}{M\varpi^2}\left(\frac{\text{sign}C - (\omega/\varpi)^2}{N(\omega/\varpi)} + i2\eta\frac{\omega/\varpi}{N(\omega/\varpi)}\right), \tag{23.7}$$

with the denominator being defined as $N(x) = x^4 + 2(-\text{sign}C + 2\eta^2)x^2 + 1$. These functions are plotted in Fig. 23.1. One recognizes that $\chi'_{\text{osc}}(\omega)$ and $\chi''_{\text{osc}}(\omega)$ are odd and even functions of ω, respectively. The $\chi''_{\text{osc}}(\omega)$ has the typical Lorentz or Breit–Wigner form. For $\omega > 0$ it is positive having just one maximum at $\omega = \omega_m$ where

$$\omega_m^2 = \frac{\varpi^2}{3}\left(2\sqrt{1 - \text{sign}C\eta^2 + \eta^4} + \text{sign}C - 2\eta^2\right), \quad (23.8)$$

with the following limiting values for small and large damping:

$$\lim_{\eta \to 0} \omega_m^2 = \frac{\varpi^2}{3}(2 + \text{sign}C) \qquad \lim_{\eta \to \infty} \omega_m^2 = 0. \quad (23.9)$$

For large damping the concentration of the strength distribution $\chi''_{\text{osc}}(\omega)$ at very small frequencies can be seen directly from the figure. Let us discuss separately the two cases of negative and positive stiffness.

Positive stiffness, $(C > 0)$: Both for undercritical damping $(\eta < 1)$, as well as for overcritical damping $(\eta > 1)$ the poles lie in the lower part of the complex plane. For $\eta < 1$ the $\mathcal{E}$ is real and the poles are situated symmetrically with respect to the imaginary axis at $\omega^\pm = \varpi(\pm\sqrt{1-\eta^2} - i\eta)$. For $\eta > 1$, $\mathcal{E}$ is purely imaginary such that the poles come to lie on the imaginary axis at $\omega^\pm = -i\varpi(\eta \mp \sqrt{\eta^2 - 1})$. Notice, that in both cases the real part of the response function vanishes at $\omega = \varpi$ and is positive at $\omega = 0$, with the value at $\omega = 0$ being independent of η.

Let us briefly look at the limiting case of vanishing friction. It would not be wise to put Γ strictly equal to zero. Rather we should identify $\Gamma/2$ as an infinitely small but positive quantity ϵ, to get from (23.6)

$$\chi_{\text{osc}}(\omega) = -\frac{1}{2M\varpi}\left(\frac{1}{\omega - \varpi + i\epsilon} - \frac{1}{\omega + \varpi + i\epsilon}\right) \quad (23.10)$$

Applying the Dirac formula (26.16) the real and imaginary parts turn into

$$\chi'_{\text{osc}}(\omega) = \frac{1}{2M\varpi}\mathcal{P}\frac{2\varpi}{\varpi^2 - \omega^2}, \quad (23.11)$$

and

$$\chi''_{\text{osc}}(\omega) = \frac{\pi}{2M\varpi}\left(\delta(\omega - \varpi) - \delta(\omega + \varpi)\right), \quad (23.12)$$

respectively.

Negative stiffness $C < 0$: In this case the system is unstable. Therefore it is not surprising that the response function $\chi_{\text{osc}}(\omega)$ has some peculiar features. For instance, the $\mathcal{E}$ now is purely imaginary no matter what the size of η will be: $\omega^\pm = i\varpi(\pm\sqrt{1+\eta^2} - \eta)$. One of the poles (ω^+) lies in the upper half of the complex plane. As a consequence, when applied to the response function $\chi_{\text{osc}}(\omega)$, the contour in (23.123) must lie above (ω^+). Otherwise, the $\tilde{\chi}_{\text{osc}}(t)$

would not vanish for negative times. Nevertheless, along the real axis $\chi_{\text{osc}}(\omega)$ can be split like in (23.7) with the $\chi'_{\text{osc}}(\omega)$ and $\chi''_{\mu\nu}(\omega)$ still having the general symmetry properties discussed above. The time-dependent counterparts of these functions have to be calculated from (26.20). Now, however, the real part $\chi'_{\text{osc}}(\omega)$ is negative at the origin $\omega = 0$, increases monotonically with ω and has no zeroes along the real axis. The imaginary part, on the other hand still has the same behavior: $\chi''_{\mu\nu}(\omega) > 0$ for $\omega > 0$. This may be inferred directly from Fig. 23.1.

Let us finally turn to moments of the dissipative response defined as

$$2 S^{(n)} = \int_C \frac{d\omega}{\pi} \omega^n \chi''_{\mu\nu}(\omega). \tag{23.13}$$

For the oscillator those for $n = -1, 0, 1$ can easily be seen to be

$$2 S^{(-1)}_{\text{osc}} = 1/C \qquad S^{(0)}_{\text{osc}} = 0 \qquad 2 S^{(1)}_{\text{osc}} = 1/M. \tag{23.14}$$

Over-damped motion: In nuclear physics the case of large damping is quite important. Formally, this limit may be obtained by putting in (23.3) the inertia equal to zero and by evaluating (23.5) for $\eta \to \infty$. For $\eta > 1$ both frequencies are purely imaginary, and for $\eta \to \infty$ the ω^+ turns into $-iC/\gamma$ whereas the ω^- moves along the negative imaginary axis to $-i\infty$. More precisely for $\eta \to \infty$ one gets to leading order,

$$\mathcal{E} = \varpi\sqrt{\text{sign}C - \eta^2} \quad \stackrel{\eta \to \infty}{\longrightarrow} \quad i\varpi\eta = i\frac{\gamma}{2M} = i\frac{1}{2\tau_{\text{kin}}}, \tag{23.15}$$

such that the frequencies turn into

$$i\omega^+ \stackrel{\eta \to \infty}{\longrightarrow} \text{sign}C\frac{|C|}{\gamma} \equiv \text{sign}C\frac{1}{\tau_{\text{coll}}} \qquad i\omega^- \stackrel{\eta \to \infty}{\longrightarrow} \frac{1}{\tau_{\text{kin}}}. \tag{23.16}$$

Because of these features we may neglect the second term of the response function $\chi_{\text{osc}}(\omega)$ given by (23.6). Since $2M\mathcal{E}$ becomes $i\gamma$ the $\chi_{\text{osc}}(\omega)$ turns into

$$\chi_{\text{ovd}}(\omega) = \frac{i}{\gamma}\frac{1}{\omega + iC/\gamma} = \frac{C + i\gamma\omega}{(\gamma\omega)^2 + C^2} \equiv \chi'_{\text{ovd}}(\omega) + i\chi''_{\text{ovd}}(\omega). \tag{23.17}$$

23.2 A brief reminder of perturbation theory

We imagine the perturbation $\hat{H}_{\text{cpl}}(t)$ of (23.1) added to the Hamiltonian H of our system to give:

$$\hat{H}_{\text{sc}} = \hat{H} + \hat{H}_{\text{cpl}}(t) \tag{23.18}$$

The calculation of the averages $\langle \hat{A}_\mu \rangle_t$ may be done within the Heisenberg picture, which means evaluating the operators $\hat{A}_\mu$ at time t by $\hat{A}_\mu(t) = \hat{\mathcal{U}}^\dagger_{\text{sc}}(t)\hat{A}_\mu \hat{\mathcal{U}}_{\text{sc}}(t)$ with the time evolution operator $\hat{\mathcal{U}}_{\text{sc}}$ satisfying the basic equation of motion

A brief reminder of perturbation theory

$$i\hbar \frac{d}{dt}\hat{\mathcal{U}}_{\text{sc}}(t) = \hat{H}_{\text{sc}}\,\hat{\mathcal{U}}_{\text{sc}}(t) \qquad \hat{\mathcal{U}}_{\text{sc}}(t_0) = \hat{1}\,. \qquad (23.19)$$

Here and in the following it is assumed that the perturbation starts at some initial time t_0. For later purposes it will be convenient to work with the interaction picture defined through $\hat{\mathcal{U}}_I(t) = \exp(i\hat{H}t/\hbar)\hat{\mathcal{U}}_{\text{sc}}(t)$ with the corresponding equation of motion

$$i\hbar \frac{d}{dt}\hat{\mathcal{U}}_I(t) = \hat{H}^I_{\text{cpl}}(t)\,\hat{\mathcal{U}}_I(t) \qquad \text{where} \qquad \hat{H}^I_{\text{cpl}}(t) = e^{i\hat{H}t/\hbar}\hat{H}_{\text{cpl}}e^{-i\hat{H}t/\hbar}\,. \qquad (23.20)$$

This differential equation may be integrated to give:

$$\hat{\mathcal{U}}_I(t) = e^{i\hat{H}t_0/\hbar} - \frac{i}{\hbar}\int_{t_0}^{t} \hat{H}^I_{\text{cpl}}(s)\,\hat{\mathcal{U}}_I(s)\,ds\,. \qquad (23.21)$$

The first term on the right accounts for the proper initial condition as determined by (23.19). For the special form (23.1) one gets to *first order* in a_ν^{ext}:

$$\hat{A}_\mu(t) = \hat{A}^I_\mu(t - t_0) - \frac{i}{\hbar}\sum_\nu \int_{t_0}^{t} \left[\hat{A}^I_\mu(t - t_0)\,,\,\hat{A}^I_\nu(s - t_0)\right] a_\nu^{\text{ext}}(s)\,ds \qquad (23.22)$$

To obtain this result one needs to iterate (23.21) *once*, which means calculating a $\hat{\mathcal{U}}_I^{(1)}(t)$ by replacing in (23.21) the $\hat{\mathcal{U}}_I(s)$ in the integrand by $\exp(i\hat{H}t_0/\hbar)$.

A full formal solution for the time evolution operator can be found by taking into account all iterations of eqn(23.21). To simplify matters let us put $t_0 = 0$. Then the term of nth order would read

$$I^{(n)} = \left(\frac{-i}{\hbar}\right)^n \int_0^t ds_1 \int_0^{s_1} ds_2 \ldots \int_0^{s_{n-1}} ds_n\, \hat{H}^I_{\text{cpl}}(s_1)\hat{H}^I_{\text{cpl}}(s_2)\ldots\hat{H}^I_{\text{cpl}}(s_n)\,, \qquad (23.23)$$

where naturally the product of the $\hat{H}^I_{\text{cpl}}(s_k)$ evaluated at different times s_k is ordered according to the sequence $s_1 \geq s_2 \geq \cdots \geq s_{n-1} \geq s_n$. It can be shown that this n-dimensional integral can be rewritten as

$$I^{(n)} = \left(\frac{-i}{\hbar}\right)^n \frac{1}{n!} \int_0^t dt_1 \int_0^t dt_2 \ldots \int_0^t dt_n\, \mathcal{T}\left[\hat{H}^I_{\text{cpl}}(t_1)\hat{H}^I_{\text{cpl}}(t_2)\ldots\hat{H}^I_{\text{cpl}}(t_n)\right]\,. \qquad (23.24)$$

The so-called *time-ordering operator* $\mathcal{T}$ is defined such as to make sure that the operators $\hat{H}^I_{\text{cpl}}(t_k)$ are again ordered in the same sequence as before. For any two of them this means writing

$$\mathcal{T}\left[\hat{H}^I_{\text{cpl}}(t_i)\hat{H}^I_{\text{cpl}}(t_k)\right] = \hat{H}^I_{\text{cpl}}(t_k)\hat{H}^I_{\text{cpl}}(t_i) \qquad \text{if} \qquad t_k \geq t_i\,. \qquad (23.25)$$

We will refrain from discussing the formal proof of this statement, which may for instance be found in Chapter 3 of (Fetter and Walecka, 1971). The reader may convince himself that the identity of (23.23) and (23.24) would indeed follow for commuting operators. The result of the correct proof may then be interpreted in the following sense: In any practical manipulations the $\hat{H}^I_{\rm cpl}(t_i)$ may indeed be treated as commuting operators provided care is taken of the correct order by putting at the end the $\mathcal{T}$ at the right place. Having version (23.24) at one's disposal the $\hat{\mathcal{U}}_I(t)$ of (23.21) can be written as an exponential such that the full evolution operator attains the compact form

$$\hat{\mathcal{U}}_{\rm sc}(t) = \exp\left(-\frac{i}{\hbar}\hat{H}t\right)\left\{\mathcal{T}\exp\left(-\frac{i}{\hbar}\int_0^t ds\hat{H}^I_{\rm cpl}(s)\right)\right\} \qquad \text{(for} \quad t_0 = 0). \quad (23.26)$$

23.2.1 Transition rate in lowest order

As a first application let us derive Fermi's golden rule for the rate of transitions between two unperturbed states, defined as the eigenstates $|m\rangle$ of the Hamiltonian $\hat{H}$ with the corresponding Schrödinger equation $\hat{H}|m\rangle = E_m|m\rangle$. The amplitude $\mathcal{A}_{mn}$ for a transition between an initial state $|m\rangle$ to a final state $|n\rangle$ is given by $\mathcal{A}_{mn}(t) = \langle n|\hat{\mathcal{U}}_{\rm sc}(t)|m\rangle$. To first order and for $n \neq m$ this amplitude is given by

$$\begin{aligned}\mathcal{A}^{(1)}_{mn}(t) &= -\frac{i}{\hbar}\int_{t_0}^t \langle n|e^{-i\hat{H}t/\hbar}\hat{H}^I_{\rm cpl}(s)e^{i\hat{H}t_0/\hbar}|m\rangle \, ds \\ &= -\frac{i}{\hbar}\int_{t_0}^t e^{-iE_n(t-s)/\hbar}\langle n|\hat{H}_{\rm cpl}(s)|m\rangle e^{iE_m(s-t_0)/\hbar} \, ds\,.\end{aligned} \quad (23.27)$$

For the transition probability one gets

$$w^{(1)}_{mn}(t) = \left|\mathcal{A}^{(1)}_{mn}(t)\right|^2 = \frac{1}{\hbar^2}\left|\int_{t_0}^t e^{i(E_n-E_m)s/\hbar}\langle n|\hat{H}_{\rm cpl}(s)|m\rangle \, ds\right|^2. \quad (23.28)$$

For a periodic external perturbation of the form (with $\hat{A}^\dagger = \hat{A}$)

$$\hat{H}_{\rm cpl} = a(t)\hat{A} = \frac{a_0}{2}\left(e^{i\omega t} + e^{-i\omega t}\right)\hat{A} \qquad \text{for} \quad t > t_0\,. \quad (23.29)$$

one gets

$$w^{(1)}_{mn}(t) = \frac{1}{\hbar^2}\left|\int_{t_0}^t a(s)e^{i\omega_{nm}s} \, ds\right|^2 |\langle m|A|n\rangle|^2\,, \quad (23.30)$$

where we have introduced the shorthand notation $\hbar\omega_{nm} = E_n - E_m$. To evaluate the time-dependent first factor we modify the $a(t)$ of (23.29) by introducing a convergence factor in replacing.[46]

$$a(t) \quad \longrightarrow \quad e^{\epsilon t}a(t) \qquad \epsilon > 0\,. \quad (23.31)$$

Then the lower limit of the integral can be put equal to $-\infty$. Effectively, the external perturbation acts on a time lapse of the order of ϵ^{-1}. Finally we may

---

[46] Evidently, this form is meaningful only around $t \simeq 0$. At large positive times the perturbation would simply grow too large to allow for treating it in low order.

wish to turn to the limit $\epsilon \to 0$, which may be interpreted physically as the field being switched on *adiabatically slowly*. The integral at stake is easily seen to become

$$I(t) \equiv -\frac{i}{\hbar}\int_{-\infty}^{t} a(s)e^{i\omega_{nm}s}\,ds = \frac{a_0}{2\hbar}\left(\frac{e^{i(\omega_{nm}+\omega-i\epsilon)t}}{\omega_{nm}+\omega-i\epsilon} + \frac{e^{i(\omega_{nm}-\omega-i\epsilon)t}}{\omega_{nm}-\omega-i\epsilon}\right). \tag{23.32}$$

For the squared absolute value one gets

$$|I(t)|^2 = \frac{a_0^2}{4\hbar^2}\left\{\frac{e^{2\epsilon t}}{(\omega_{nm}+\omega)^2+\epsilon^2} + \frac{e^{2\epsilon t}}{(\omega_{nm}-\omega)^2+\epsilon^2} \right.$$
$$\left. \frac{e^{2(i\omega+\epsilon)t}}{(\omega_{nm}+\omega-i\epsilon)(\omega_{nm}-\omega+i\epsilon)} + \text{c.c.}\right\}. \tag{23.33}$$

Evidently, the terms in the second line result from the interference between the two terms of the cosine in $a(t)$. Actually, in most cases we are not interested in the transition probability as such but in the rate $P_{mn} = dw_{mn}^{(1)}/dt$ at which this transition occurs. After differentiating (23.33) the terms in the first line become proportional to Lorentzians of the type

$$\frac{2\epsilon}{(\omega_{nm}\pm\omega)^2+\epsilon^2} \equiv 2\pi\bar{\delta}_\epsilon(\omega_{nm}\pm\omega) \quad\to\quad 2\pi\delta(\omega_{nm}\pm\omega) \quad \text{for}\quad \epsilon\to 0. \tag{23.34}$$

The *smooth delta function* $\bar{\delta}_\epsilon$ introduced here has a width ϵ. For any time t satisfying the condition

$$\epsilon|t| \ll 1 \tag{23.35}$$

the only time dependence of the transition rate would be given through the interference terms. They vanish if one applies a time average over one (or several) cycle(s) of length $T = 2\pi/\omega$ of the external perturbation, with the ω and ϵ fulfilling the condition

$$(2\pi/\omega)\epsilon \ll 1. \tag{23.36}$$

For this time averaged transition rate one gets

$$\overline{P}_{mn}^T(\omega) = \frac{a_0^2}{4}\frac{2\pi}{\hbar}\left|\langle m|A|n\rangle\right|^2 \left(\bar{\delta}_\epsilon(E_n-E_m-\hbar\omega) + \bar{\delta}_\epsilon(E_n-E_m+\hbar\omega)\right). \tag{23.37}$$

The ω_{nm} has been replaced by the energy difference considering that $\delta(x)/\hbar = \delta(\hbar x)$. Evidently, the same result would be obtained if in (23.29) the cosine were replaced by a sine function.

In the literature forms like (23.37) are commonly derived not for a real external field but for $a(t) = \exp(-i\omega t)$ instead. In such cases the time average is not required, and, as one may also see from our derivation, the result is

$$P_{mn}(\omega) = \frac{2\pi}{\hbar}\left|\langle m|A|n\rangle\right|^2 \bar{\delta}_\epsilon(E_n-E_m-\hbar\omega) \quad\text{for}\quad \hat{H}_{\text{cpl}} = e^{-i\omega t}\hat{A}. \tag{23.38}$$

This form allows one to perform the static limit $\omega \to 0$, which represents a time-independent perturbation $\hat{A}$. One gets

$$P_{mn}(\omega) = \frac{2\pi}{\hbar} \left|\langle m|A|n\rangle\right|^2 \bar{\delta}_\epsilon (E_n - E_m) \,. \tag{23.39}$$

Because of the condition (23.36) this limit cannot be applied to the form (23.37). Indeed, for $\omega = 0$ the "interference terms" do not cancel. As can be seen from (23.33) the terms in the second line become identical to those in the first line; actually *all four* terms reduce to the same expression. The reader may recall that our result was obtained for a situation where the perturbation was switch on adiabatically, mind (23.31). If one is to work with an $a(t)$ which acts only over a finite interval of time the smooth delta function is to be replaced by another one (defined through a $\sin^2$ function) with a "width" ΔE of the order of $2\pi/t$.

Finally, we need to discuss the choice of the width $\delta E = \hbar \epsilon$ of the smooth delta function measured in energy. To be specific we concentrate on the form (23.39), anticipating that the case of (23.38) can be treated analogously, but leave aside the condition (23.36) on the frequency for the example of a real periodic perturbation. Obviously, this width δE must be larger than the average spacing D of the levels such that (23.35) can be written as

$$|t| \ll 2\pi\hbar/D \equiv \tau_{\rm rec} \,, \tag{23.40}$$

with $\tau_{\rm rec}$ being the recurrence time, mind eqn(4.56). For shorter times the width may be larger but still satisfying (23.35). In this case the smooth delta function may encompass several or many (final) levels $|n\rangle$. Suppose within a range of ΔE the matrix elements $\langle m|A|n\rangle$ do not vary much with n. Then one may simply add up their contribution which in the end means to rewrite (23.39) as

$$\overline{P}_{mn}(\omega) = \frac{2\pi}{\hbar} \overline{|A_{mn}|^2} \, \Omega(E_n)|_{E_n = E_m} \,. \tag{23.41}$$

Here, the abbreviation $\overline{|A_{mn}|^2} = \overline{|\langle m|A|n\rangle|^2}$ is used and $\Omega(E_n)$ stands for the density of levels in the range at stake. Often this form is referred to as *Fermi's golden rule*. Besides (23.40) it requires the time lapse to be of the order of $\hbar/\Delta E$.

23.3 General properties of response functions

In this section we want to derive properties of the response function assuming that our microscopic system follows Hamiltonian dynamics, as described by the H introduced above.

23.3.1 Basic definitions

To get the response functions from eqn(23.22) we need to perform an average with a time-independent density operator $\hat{\rho}$ to get $\langle \hat{A}_\mu\rangle_t = {\rm tr}\hat{\rho}\hat{A}_\mu(t)$. For most of the properties to be discussed below it is only necessary that the Hamiltonian commutes with this density, $[\hat{\rho}, \hat{H}] = 0$, a condition which would be satisfied even within the microcanonical ensemble. Restricting ourselves to the canonical distribution $\hat{\rho} = (1/Z)\exp(-\hat{H}/T)$ will be necessary only for a few properties. The generalization to the grand canonical ensemble will be of no concern here.

Let us introduce the deviation

$$\Delta \langle \hat{A}_\mu \rangle_t = \langle \hat{A}_\mu \rangle_t - \langle \hat{A}_\mu \rangle \equiv \operatorname{tr} \hat{\rho} \, \hat{A}_\mu(t) - \operatorname{tr} \hat{\rho} \, \hat{A}_\mu \qquad (23.42)$$

of the average from its unperturbed value. (The latter comes from the first term on the right-hand side of (23.22)). From (23.22) we obtain:

$$\Delta \langle \hat{A}_\mu \rangle_t = -2i \sum_\nu \int_{t_0}^t \tilde{\chi}''_{\mu\nu}(t-s) a_\nu^{\mathrm{ext}}(s) ds \qquad (23.43)$$

with the (purely imaginary) function

$$\tilde{\chi}''_{\mu\nu}(t-s) = \frac{1}{2\hbar} \operatorname{tr} \left(\hat{\rho} \, [\hat{A}_\mu^I(t), \hat{A}_\nu^I(s)] \right). \qquad (23.44)$$

Here, the time evolution of the operators is as specified in (23.20).[47] Notice, that the t_0 introduced before, which was still present in (23.22), has dropped out both in (23.44) as well as in the second term on the right-hand side of (23.42). This is for the same reason that the function defined in (23.44) depends only on the time *difference*. This very fact is a consequence of the cyclic invariance of the trace and of our choosing the density operator to commute with the Hamiltonian, which implies $\operatorname{tr} \hat{\rho} \exp(-i\hat{H}t_0/\hbar)\hat{A}\exp(i\hat{H}t_0/\hbar) = \operatorname{tr} \hat{\rho}\hat{A} \equiv \langle \hat{A} \rangle$. For practical applications we need to perform two changes, with the first one being entirely of physical origin. Usually, functions like the one in (23.44) are different from zero only within some finite time interval τ. If we now assume the difference $t - t_0$ to be larger than τ we may safely replace the lower limit of the integral by $-\infty$. If the system does not allow one to rely on such a physical condition from the start one may introduce a convergence factor in the external field $a_\nu^{\mathrm{ext}}(t) \longrightarrow \exp(\epsilon t) a_\nu^{\mathrm{ext}}(t)$ like before in (23.31).

The second change is of purely mathematical origin and completely rigorous. Introducing the real function

$$\tilde{\chi}_{\mu\nu}(t-s) = 2i\Theta(t-s)\tilde{\chi}''_{\mu\nu}(t-s) \qquad (23.45)$$

(23.43) can be rewritten as:

$$\Delta \langle \hat{A}_\mu \rangle_t = -\sum_\nu \int_{-\infty}^\infty \tilde{\chi}_{\mu\nu}(t-s)\, a_\nu^{\mathrm{ext}}(s)\, ds. \qquad (23.46)$$

---

[47] Henceforth, the time dependence of these operators will always be referring to that of the system defined with $\hat{H}$. Therefore, we will drop the upper index "I", which was introduced above for the interaction picture with respect to the total Hamiltonian (23.18) including the time-dependent coupling.

According to the definition (23.2) it is easily recognized as the response function associated with the two operators $\hat{A}_\mu$ and $\hat{A}_\nu$. We only have to apply the convolution theorem for Fourier transforms (26.21) and to differentiate with respect to the external fields.

The form (23.45) makes apparent the causal behavior of the physical response. After a Fourier transformation, observing the convolution theorem (26.22) and the Fourier transform (26.23) of the Θ-function one gets the relation

$$\chi_{\mu\nu}(\omega) = \int_{-\infty}^{\infty} \frac{d\Omega}{\pi} \frac{\chi''_{\mu\nu}(\Omega)}{\Omega - \omega - i\epsilon} \,. \tag{23.47}$$

For real ω we can write

$$\chi_{\mu\nu}(\omega) = \chi'_{\mu\nu}(\omega) + i\chi''_{\mu\nu}(\omega) \,, \tag{23.48}$$

with $\chi'_{\mu\nu}(\omega)$ being given by the Kramers–Kronig relation:

$$\chi'_{\mu\nu}(\omega) = \mathcal{P} \int_{-\infty}^{\infty} \frac{d\Omega}{\pi} \frac{\chi''_{\mu\nu}(\Omega)}{\Omega - \omega} \,. \tag{23.49}$$

This follows easily by applying to (23.47) Dirac's formula (26.16). The $\chi'_{\mu\nu}(\omega)$ and $\chi''_{\mu\nu}(\omega)$ are called reactive and dissipative parts of the response function, for reasons which will become more clear below.

23.3.2 Basic properties

Symmetry relations Due to basic properties of the commutator in (23.44), and because the operators $\hat{A}_\mu$ were chosen to be Hermitian, these functions have simple symmetry properties. For instance, for time inversion from $t - s$ to $s - t$ one gets from (23.44)

$$\widetilde{\chi}''_{\mu\nu}(s-t) = -\widetilde{\chi}''_{\nu\mu}(t-s) \,. \tag{23.50}$$

Applying the Fourier transformation and obeying the rules for complex conjugation one obtains:

$$\begin{aligned}(\chi'_{\mu\nu}(\omega))^* &= \chi'_{\mu\nu}(-\omega) = \chi'_{\nu\mu}(\omega) & (23.51)\\ (\chi''_{\mu\nu}(\omega))^* &= -\chi''_{\mu\nu}(-\omega) = \chi''_{\nu\mu}(\omega) \,. & (23.52)\end{aligned}$$

If the A_μ have a definite behavior under time reversal, $\mathcal{T}\hat{A}_\mu(t)\mathcal{T}^{-1} = \varepsilon_\mu^{\mathcal{T}}\hat{A}(-t)$ one has two additional symmetry relations, valid for both the frequency and the time-dependent functions,

$$\begin{aligned}\chi'_{\mu\nu}(\omega) &= \varepsilon_\mu^{\mathcal{T}}\varepsilon_\nu^{\mathcal{T}}\chi'_{\mu\nu}(-\omega) & \chi''_{\mu\nu}(\omega) &= -\varepsilon_\mu^{\mathcal{T}}\varepsilon_\nu^{\mathcal{T}}\chi''_{\mu\nu}(-\omega) & (23.53)\\ \widetilde{\chi}'_{\mu\nu}(t) &= -\varepsilon_\mu^{\mathcal{T}}\varepsilon_\nu^{\mathcal{T}}\widetilde{\chi}'_{\mu\nu}(-t) & \widetilde{\chi}''_{\mu\nu}(t) &= -\varepsilon_\mu^{\mathcal{T}}\varepsilon_\nu^{\mathcal{T}}\widetilde{\chi}''_{\mu\nu}(-t) \,. & (23.54)\end{aligned}$$

For $\varepsilon_\mu^{\mathcal{T}}\varepsilon_\nu^{\mathcal{T}} = 1$ it is seen that in (23.48) the separation into reactive and dissipative parts just means splitting the response into its real and imaginary parts (for real

ω). For all diagonal elements this is true irrespective of a special behavior under time reversal. Therefore, the primes introduced for the oscillator response in Section 23.1 have the same meaning as those used here for the more general definition. It may be said that in (Kubo et al., 1991a) the same notation is used for the susceptibilities as in (23.48) but the $\chi'_{\mu\nu}(\omega)$ and $\chi''_{\mu\nu}(\omega)$ are strictly defined as the real and imaginary parts of $\chi_{\mu\nu}(\omega)$.

Behavior at large frequencies: The form (23.47) allows one to deduce information on the decay of the response function at large frequencies. The following relation is easily seen to hold true

$$\lim_{\omega\to\infty}\omega^2\chi_{\mu\nu}(\omega) = \lim_{\omega\to\infty}(-\omega)\int_{-\infty}^{+\infty}\frac{d\Omega}{\pi}\frac{\chi''_{\mu\nu}(\Omega)}{1-\frac{\Omega-i\eta}{\omega}} = -\int_{-\infty}^{+\infty}\frac{d\Omega}{\pi}\chi''_{\mu\nu}(\Omega)\Omega, \tag{23.55}$$

provided the moments of the dissipative parts are finite:

$$\int_{-\infty}^{+\infty}\frac{d\Omega}{\pi}\chi''_{\mu\nu}(\Omega) = 0 \qquad \int_{-\infty}^{+\infty}\frac{d\Omega}{\pi}\chi''_{\mu\nu}(\Omega)\Omega^n < \infty \qquad (n>0).$$

Whereas the first condition is usually fulfilled because of symmetries (see (23.53)), the latter should be so for physical reasons. In practical applications it may have to be imposed by cut off procedures.

Special functions: So far the nature of the operators was not specified any further. One often faces a situation where one operator is the derivative of another one. Suppose we are given the response function $\tilde\chi_{AA}(t-s) = (i/\hbar)\Theta(t-s)\langle[\hat A(t),\hat A(s)]\rangle$. From the operator identity $\dot{\hat A}(t) = (i/\hbar)[\hat H,\hat A(t)] = d\hat A(t)/dt$ one easily derives the following relations:

$$\frac{d}{dt}\tilde\chi_{AA}(t-s) = \tilde\chi_{\dot A A}(t-s) = \frac{i}{\hbar}\Theta(t-s)\langle[\dot{\hat A}(t),\hat A(s)]\rangle \tag{23.56}$$

$$\frac{d}{ds}\tilde\chi_{AA}(t-s) = \tilde\chi_{A\dot A}(t-s) = -\tilde\chi_{\dot A A}(t-s). \tag{23.57}$$

Here, use has been made of the fact that the commutator in $\tilde\chi_{AA}(t-s)$ vanishes when calculated at equal times. This is no longer true for the one appearing in (23.56). Differentiating this equation once more one therefore gets:

$$\frac{d^2}{dt^2}\tilde\chi_{AA}(t-s) = -\frac{d}{ds}\tilde\chi_{\dot A A}(t-s)$$

$$= -\frac{i}{\hbar}\Theta(t-s)\langle[\ddot{\hat A}(t),\hat A(s)]\rangle + \frac{i}{\hbar}\delta(t-s)\langle[\dot{\hat A}(t),\hat A(s)]\rangle \tag{23.58}$$

$$= -\tilde\chi_{\ddot A A}(t-s) + \frac{i}{\hbar}\delta(t-s)\langle[\dot{\hat A}(t-s),\hat A]\rangle.$$

Finally, we note that the Fourier transforms obey the following relations:

$$\chi_{\dot{A}A}(\omega) = -\frac{i}{\hbar}\omega\chi_{AA}(\omega) = -\chi_{A\dot{A}}(\omega) \qquad \chi''_{\dot{A}A}(\omega) = -\frac{i}{\hbar}\omega\chi''_{AA}(\omega), \qquad (23.59)$$

$$\chi_{\dot{A}\dot{A}}(\omega) = \frac{\omega^2}{\hbar^2}\chi_{AA}(\omega) + \frac{i}{\hbar}\langle[\hat{A},\hat{A}]\rangle \qquad \chi''_{\dot{A}\dot{A}}(\omega) = \frac{\omega^2}{\hbar^2}\chi''_{AA}(\omega). \qquad (23.60)$$

23.3.3 Dissipation of energy

Let us look at the change of the *total* energy $\langle\hat{H}_{\rm sc}\rangle_t$ given by:

$$\frac{dE_{\rm sc}}{dt} = \left\langle\frac{\partial\hat{H}_{\rm cpl}}{\partial t}\right\rangle_t = \sum_\mu \langle\hat{A}_\mu\rangle_t \dot{a}_\mu^{\rm ext}(t).$$

Applying (23.46) one obtains:

$$\frac{dE_{\rm sc}}{dt} = \sum_\mu \langle\hat{A}_\mu\rangle \dot{a}_\mu^{\rm ext}(t) - \sum_{\mu\nu}\dot{a}_\mu^{\rm ext}(t)\int_{-\infty}^\infty ds\, \tilde{\chi}_{\mu\nu}(t-s)\, a_\nu^{\rm ext}(s)\,.$$

This expression still contains reactive and dissipative parts. To separate both we integrate over some interval of time $\overline{\Delta E_{\rm sc}} = \int_{-t_1}^{t_1} dt\, dE_{\rm sc}/dt$ to get:

$$\overline{\Delta E_{\rm sc}} = \sum_\mu \langle\hat{A}_\mu\rangle \left(a_\mu^{\rm ext}(t_1) - a_\mu^{\rm ext}(-t_1)\right) - \\ \int_{-t_1}^{t_1} dt \int_{-\infty}^\infty ds \sum_{\mu\nu}\dot{a}_\mu^{\rm ext}(t)\,\tilde{\chi}_{\mu\nu}(t-s)\,a_\nu^{\rm ext}(s)\,. \qquad (23.61)$$

To simplify matters, we will assume a situation where the first term vanishes identically. (Otherwise we would easily recognize its associated energy change as the one typically for a potential.) This will be the case for oscillating external fields with t_1 being some integer multiple of the fundamental period. Another possibility would be a field which vanishes at some large initial and final time. Apparently we may cover both cases without much loss of generality by letting t_1 go to infinity, eventually introducing a convergence factor $\exp(-|t|\epsilon/\hbar)$ if necessary. From the convolution theorem (26.21) for Fourier transformations we get by identifying $i\omega a_\mu^{\rm ext}(-\omega)$ as the transform of the $\dot{a}_\mu^{\rm ext}(t)$ at $-\omega$:

$$\overline{\Delta E_{\rm sc}} = -\int_{-\infty}^\infty \frac{d\omega}{2\pi}\, i\omega \sum_{\mu\nu} a_\mu^{\rm ext}(-\omega)\,\chi_{\mu\nu}(\omega)\,a_\nu^{\rm ext}(\omega)$$

$$= \int_{-\infty}^\infty \frac{d\omega}{2\pi} \sum_{\mu\nu} a_\mu^{\rm ext}(-\omega)\,a_\nu^{\rm ext}(\omega)\,\omega\,\chi''_{\mu\nu}(\omega) \qquad (23.62)$$

$$= \int_{-\infty}^\infty \frac{d\omega}{2\pi} \sum_{\mu\nu} \dot{a}_\mu^{\rm ext}(-\omega)\,\dot{a}_\nu^{\rm ext}(\omega)\,\frac{\chi''_{\mu\nu}(\omega)}{\omega}\,. \qquad (23.63)$$

The second line follows because $\sum_{\mu\nu} a_\mu^{\rm ext}(-\omega)\,a_\nu^{\rm ext}(\omega)\,\omega\,\chi'_{\mu\nu}(\omega)$ is an odd function of ω due to the symmetry (23.51). In the third line the Fourier transforms of the velocities of the external fields have been introduced. In both cases the remaining integrand is even in ω. The whole expressions measure the energy that is lost to the system *irreversibly*, for which reason $\chi''_{\mu\nu}(\omega)$ is called the *dissipative* part of the response function. Indeed, the integrand of (23.62) can be shown to be positive (semi-) definite (see exercise 2)

$$\sum_{\mu\nu} a_\mu^{\text{ext}}(-\omega)\, \omega\, \chi''_{\mu\nu}(\omega)\, a_\nu^{\text{ext}}(\omega) \geq 0. \tag{23.64}$$

Below in eqn(23.189) it is demonstrated that and how this property reflects the increase of entropy. Conversely, the so-called *reactive* part $\chi'_{\mu\nu}(\omega)$ represents that energy which at some time flows into the system but is restored later. Let us look at the case of the oscillator. There, it is seen from (23.7) that the dissipative part is proportional to the friction coefficient γ. (In exercise 1 the example of an oscillator that is coupled to an electric field in dipole approximation is studied.) Realize, please, that this separation does not lose its physical meaning for the inverted oscillator with $C < 0$. Indeed, the step from the first to the second line of (23.62) just involves the symmetry properties under $\omega \to -\omega$, which are independent of the sign of C.

23.3.4 Spectral representations

Beginning with the dissipative part (23.44), we may evaluate the response functions in terms of energies E_m and eigenstates of the Hamiltonian, $\hat{H}|m\rangle = E_m|m\rangle$. We will assume the $|m\rangle$ to be complete, in the sense of $\sum_m |m\rangle\langle m| = 1$, without bothering explicitly about a continuous part of the spectrum. With the shorthand notations $\langle m|\hat{A}^\mu|n\rangle = A^\mu_{mn}$ and $E_n - E_m = \hbar\Omega_{nm}$ one obtains for $\tilde{\chi}''_{\mu\nu}(t)$ the form

$$\tilde{\chi}''_{\mu\nu}(t) = \frac{1}{2\hbar} \sum_{nm} \rho(E_m) \left(e^{-i\Omega_{nm}t} A^\mu_{mn} A^\nu_{nm} - e^{+i\Omega_{nm}t} A^\nu_{mn} A^\mu_{nm} \right) \tag{23.65}$$

$$= -\frac{i}{\hbar} \sum_{nm} \rho(E_m) A^\mu_{mn} A^\nu_{nm} \sin(\Omega_{nm}t) \quad (A^\mu_{mn} A^\nu_{nm} = A^\nu_{mn} A^\mu_{nm}). \tag{23.66}$$

For the Fourier transformed functions this implies:

$$\chi''_{\mu\nu}(\omega) = \frac{\pi}{\hbar} \sum_{nm} \Big[\rho(E_m) - \rho(E_n)\Big] A^\mu_{mn} A^\nu_{nm}\, \delta(\omega - \Omega_{nm}) \tag{23.67}$$

for the first line and

$$\chi''_{\mu\nu}(\omega) = \frac{\pi}{\hbar} \sum_{nm} \rho(E_m) A^\mu_{mn} A^\nu_{nm} \Big[\delta(\omega - \Omega_{nm}) - \delta(\omega + \Omega_{nm})\Big] \tag{23.68}$$

for the second one, which is to say if the matrix elements are either symmetric or antisymmetric $A^\eta_{mn} = \pm A^\eta_{nm}$ both for $\eta = \mu$ and ν. If one is symmetric and the other antisymmetric the $-i\sin(\Omega_{nm}t)$ in (23.66) has to be replaced by $\cos(\Omega_{nm}t)$ and the minus sign between the δ-functions turns into a plus.

For most practical applications it will be easiest to describe the excitation in terms of temperature, in which case the spectral weight $\rho(E_m)$ will be given by Boltzmann factors:

$$\rho(E_m) = \frac{e^{-E_m/T}}{\sum_n e^{-E_n/T}} = \frac{e^{-(E_m-E_0)/T}}{G_0 + \sum_{E_n>E_0} e^{-(E_n-E_0)/T}}, \quad (23.69)$$

with G_0 measuring the degeneracy of the ground state. From this form it is easy to perform the transition to zero temperature, for which the $\rho(E_m)$ vanishes for $E_m > E_0$. Assuming the ground state to be non-degenerate, the form (23.68) turns into

$$\chi''_{\mu\nu}(\omega) = \frac{\pi}{\hbar} \sum_n A^\mu_{0n} A^\nu_{n0} \left[\delta(\omega - \Omega_{n0}) - \delta(\omega + \Omega_{n0}) \right]. \quad (23.70)$$

For the last expression we may recognize more easily than for the case of finite excitation–although there the physics is exactly the same–a relation to Fermi's golden rule: $\chi''_{\mu\nu}(\omega)$ measures the probability to excite the system to some energies determined by the spectrum. Conversely, energy from the external fields is absorbed only at those frequencies which correspond to real transitions. Apparently, the $\chi''_{\mu\nu}$ is positive for positive frequencies and it represents *absorption* of energy only. This feature is clearly due to the fact that we look at the response of a system which is either in its ground state or in statistical equilibrium.

Finally, the function $\chi_{\mu\nu}(\omega)$ is obtained by inserting (23.67) into (23.47). One gets

$$\chi_{\mu\nu}(\omega) = \sum_{nm} \frac{\rho(E_n) - \rho(E_m)}{\hbar\omega - (E_n - E_m) + i\epsilon} A^\mu_{mn} A^\nu_{nm}, \quad (23.71)$$

which for symmetric or antisymmetric matrix elements can be written in the form

$$\chi_{\mu\nu}(\omega) = -\sum_{nm} \rho(E_m) A^\mu_{mn} A^\nu_{nm} \left[\frac{1}{\hbar\omega - (E_n - E_m) + i\epsilon} - (E_n \leftrightarrow E_m) \right]. \quad (23.72)$$

In such a case the reactive part $\chi'_{\mu\nu}(\omega)$ becomes

$$\chi'_{\mu\nu}(\omega) = \sum_{nm} \rho(E_m) A^\mu_{mn} A^\nu_{nm} \left[\mathcal{P} \frac{2(E_n - E_m)}{(E_n - E_m)^2 - (\hbar\omega)^2} \right]. \quad (23.73)$$

Comparing these expressions with the (23.10) for the undamped oscillator it is seen that the squared transition matrix elements of the operator $\hat{q}$ must be identified as $(2M\varpi)^{-1}$ and that the response in this case has *no explicit* dependence on temperature (for given transport coefficients).

Finally, we shall restrict ourselves to independent particle motion concentrating on the case where the $\hat{A}_\mu$ are the one-body operators $\hat{F}_\mu$ that appear in the microscopic derivation of the equations of motion. The Hamiltonian then reads $\hat{H} \equiv \hat{H}_0 = \sum_k e_k \hat{c}_k^\dagger \hat{c}_k$, if expressed in second quantized form with the creation and annihilation operators $\hat{c}_k^\dagger, \hat{c}_k$ associated with the single particle state $|k\rangle$. The time dependence of $\hat{F}_\mu^I(t)$ can then be calculated explicitly to

$$\hat{F}_\mu^I(t) = \sum_{jk} \langle j | \hat{F}_\mu | k \rangle \hat{c}_j^\dagger(t) \hat{c}_k(t) = \sum_{jk} \langle j | \hat{F}_\mu | k \rangle e^{\frac{i}{\hbar}(e_j - e_k)t} \hat{c}_j^\dagger \hat{c}_k . \quad (23.74)$$

The dissipative part of the response function $\tilde{\chi}_{\mu\nu}''(t) = (1/(2\hbar)) \langle [\hat{F}_\mu^I(t), \hat{F}_\nu] \rangle$ thus may be written as

$$\tilde{\chi}_{\mu\nu}''(t) = \sum_{jkj'k'} \langle j | \hat{F}_\mu | k \rangle \langle k' | \hat{F}_\nu | j' \rangle \frac{1}{2\hbar} e^{\frac{i}{\hbar}(e_j - e_k)t} \langle \left[\hat{c}_j^\dagger \hat{c}_k, \hat{c}_{k'}^\dagger \hat{c}_{j'} \right] \rangle \quad (23.75)$$

The expectation value on the right-hand side is easily calculated using the anti-commutation relations for the fermion operators; using the Fermi function $n(e)$ one gets

$$\langle \left[\hat{c}_j^\dagger \hat{c}_k, \hat{c}_{k'}^\dagger \hat{c}_{j'} \right] \rangle = \langle \hat{c}_j^\dagger \hat{c}_{j'} \rangle \delta_{kk'} - \langle \hat{c}_{k'}^\dagger \hat{c}_k \rangle \delta_{jj'} \equiv (n(e_j) - n(e_k)) \delta_{jj'} \delta_{kk'} . \quad (23.76)$$

Repeating all the steps performed for the many-body states, the following result for the full response function is obtained:

$$\chi_{\mu\nu}(\omega) = \sum_{jk} \langle j | \hat{F}_\mu | k \rangle \langle k | \hat{F}_\nu | j \rangle \frac{n(e_k) - n(e_j)}{\hbar\omega - (e_k - e_j) + i\epsilon} . \quad (23.77)$$

23.4 Correlation functions and the fluctuation dissipation theorem

23.4.1 Basic definitions

Like the response functions one may define correlation functions, essentially by replacing in (23.44) the commutator by an anti-commutator. One only has to care about non-vanishing expectation values $\langle \hat{A}_\mu \rangle = tr \hat{\rho} \hat{A}_\mu$ of the operators involved to have

$$\tilde{\psi}_{\mu\nu}''(t-s) = \frac{1}{2} \text{tr} \, \hat{\rho} \left[\hat{A}_\mu(t) - \langle \hat{A}_\mu \rangle, \hat{A}_\nu(s) - \langle \hat{A}_\nu \rangle \right]_+ ; \quad (23.78)$$

for conventional reasons the factor $1/\hbar$ in (23.44) has been dropped. For the sake of brevity we left out the upper index "I", but it is understood that the time dependence is given by the same Hamiltonian $\hat{H}$ as in the case of the response function. Again, one may generalize to obtain:

$$\tilde{\psi}_{\mu\nu}(t-s) = 2i\Theta(t-s)\tilde{\psi}_{\mu\nu}''(t-s) = \tilde{\psi}_{\mu\nu}'(t-s) + i\tilde{\psi}_{\mu\nu}''(t-s), \quad (23.79)$$

with the Fourier transform given by:

$$\psi_{\mu\nu}(\omega) = \int \frac{d\Omega}{\pi} \frac{\psi_{\mu\nu}''(\Omega)}{\Omega - \omega - i\epsilon} = \mathcal{P} \int \frac{d\Omega}{\pi} \frac{\psi_{\mu\nu}''(\Omega)}{\Omega - \omega} + i\psi_{\mu\nu}''(\omega) \equiv \psi_{\mu\nu}'(\omega) + i\psi_{\mu\nu}''(\omega), \quad (23.80)$$

where the last two equations are valid for real ω. The form for the spectral density appearing in (23.80), the Fourier transform of (23.78), can be guessed from the corresponding results for the response function. Observing the differences mentioned before we only have to multiply (23.68) by $\hbar$, replace the $\hat{A}_\lambda$ by

$$\Delta \hat{A}_\lambda = \hat{A}_\lambda - \langle \hat{A}_\lambda \rangle = \hat{A}_\lambda - \operatorname{tr} \rho A_\lambda \qquad (23.81)$$

and change the minus sign in between the two delta functions by a plus sign. Assuming the validity of (23.68) the $\psi''_{\mu\nu}(\omega)$ becomes

$$\psi''_{\mu\nu}(\omega) = \pi\hbar \sum_{nm} \rho(E_m) \langle m|\Delta \hat{A}_\mu|n\rangle \langle n|\Delta \hat{A}_\nu|m\rangle \qquad (23.82)$$
$$[\delta(\hbar\omega - (E_n - E_m)) + \delta(\hbar\omega + (E_n - E_m))] \;.$$

Unlike the case of the response function, here the terms with $\omega = 0$ do not vanish. The best way to express this fact is to split off from the sum over all states $|m\rangle, |n\rangle$ the terms for which both energies are identical. Introducing the abbreviation

$$\psi^0_{\mu\nu} = \sum_{E_n = E_m} \rho(E_m) \langle m|\Delta \hat{A}_\mu|n\rangle \langle n|\Delta \hat{A}_\nu|m\rangle \qquad (23.83)$$

(with summation over states n and m of the same energy) one may write (23.82) in the form

$$\psi''_{\mu\nu}(\omega) = \psi^0_{\mu\nu} 2\pi\hbar\delta(\hbar\omega) + {}_R\psi''_{\mu\nu}(\omega) \;, \qquad (23.84)$$

remember that $2\pi\hbar\delta(\hbar\omega) = 2\pi\delta(\omega)$. Here the ${}_R\psi''_{\mu\nu}(\omega)$ just stands for the residual parts, which is to say it is given by a sum like the one in (23.82) but with the restriction $E_n \neq E_m$. Hence, this ${}_R\psi''_{\mu\nu}(\omega)$ does not get contributions from the unperturbed average values $\langle \hat{A}_\lambda \rangle$.

At this stage it is useful to introduce the notion of the "zero frequency component" $\hat{A}^0_\mu$ of the operators $\hat{A}_\mu$ (Brenig, 1989), (Kubo et al., 1991a). In our notation these operators are obtained by evaluating the Fourier transform of

$$\hat{A}_\mu(t) = \exp(i\hat{H}t/\hbar)\hat{A}_\mu \exp(-i\hat{H}t/\hbar) \qquad \text{at} \qquad \omega = 0 \;. \qquad (23.85)$$

In terms of the eigenstates of the Hamiltonian $\hat{H}$ we may first write:

$$\hat{A}_\mu(t) = \sum_{nm} |m\rangle\langle m|\hat{A}_\mu(t)|n\rangle\langle n| \qquad (23.86)$$

As the matrix elements are given by $\langle m|\hat{A}_\mu(t)|n\rangle = \langle m|\hat{A}_\mu|n\rangle \exp(i(E_m - E_n)t/\hbar)$ the zero frequency component can be written as

$$\hat{A}^0_\mu = (2\pi\hbar)^{-1} \hat{A}_\mu(\hbar\omega = 0)$$
$$= \sum_{nm} \langle m|\hat{A}_\mu|n\rangle |m\rangle\langle n| \, \delta(E_n - E_m) \equiv \sum_{E_n = E_m} \langle m|\hat{A}_\mu|n\rangle |m\rangle\langle n| \;. \qquad (23.87)$$

It commutes with the Hamiltonian and its average with the (static) density operator $\hat{\rho} = \hat{\rho}(\hat{H})$ is identical to the one of the original operator: $\langle \hat{A}^0_\mu \rangle = \langle \hat{A}_\mu \rangle$.

More generally, one readily verifies that $\langle m|\hat{A}_\mu^0|n\rangle\delta(E_n-E_m) = \langle m|\hat{A}_\mu|n\rangle\delta(E_n-E_m)$. This means that in (23.83) the operators appearing on the right-hand sides can be replaced by their zero frequency parts. This feature may be summarized in writing

$$\psi_{\mu\nu}^0 = \langle\Delta\hat{A}_\mu^0\Delta\hat{A}_\nu^0\rangle = \psi_{\nu\mu}^0. \tag{23.88}$$

Actually, as we will show below, this quantity is related to the difference between the isothermal susceptibility and the static response.

As for response functions there exist symmetry relations, such as

$$\psi_{\mu\nu}^{''*}(\omega) = \psi_{\mu\nu}^{''}(-\omega) = \psi_{\nu\mu}^{''}(\omega) \qquad \psi_{\mu\nu}^{'*}(\omega) = -\psi_{\mu\nu}^{'}(-\omega) = \psi_{\nu\mu}^{'}(\omega) \tag{23.89}$$
$$\psi_{\mu\nu}^{''}(\omega) = \varepsilon_\mu^T\varepsilon_\nu^T\psi_{\mu\nu}^{''}(-\omega) \qquad \psi_{\mu\nu}^{'}(\omega) = -\varepsilon_\mu^T\varepsilon_\nu^T\psi_{\mu\nu}^{'}(-\omega). \tag{23.90}$$

23.4.2 The fluctuation dissipation theorem

For the correlation functions the same can be said as for the response functions: The properties derived so far are valid for any density operator which commutes with the Hamiltonian $\hat{H}$. But now restricting to the canonical distribution leads to a real simplification: The Fourier transform of (23.78) can then be shown to become related to $\chi_{\mu\nu}^{''}(\omega)$ by the celebrated fluctuation dissipation theorem (FDT) (see e.g. (Forster, 1975)):

$$\psi_{\mu\nu}^{''}(\omega) = \hbar\coth\frac{\hbar\omega}{2T}\chi_{\mu\nu}^{''}(\omega). \tag{23.91}$$

The proof is based solely on the fact that the operator $\exp(-\beta\hat{H})$ may be considered an analytical continuation of the time evolution operator. Notice first that

$$\mathrm{tr}\, e^{-\beta\hat{H}}\hat{A}(t)\hat{B}(s) = \mathrm{tr}\,\hat{A}(t+i\beta)e^{-\beta\hat{H}}\hat{B}(s) = \mathrm{tr}\, e^{-\beta\hat{H}}\hat{B}(s)\hat{A}(t+i\beta)$$

which implies

$$\frac{1}{2}\mathrm{tr}\,\hat{\rho}\,\Delta\hat{A}_\mu(t)\Delta\hat{A}_\nu(s) = \int_{-\infty}^\infty\frac{d\omega}{2\pi}e^{-i\omega(t-s)}s_{\mu\nu}(\omega), \tag{23.92}$$

with the $\Delta\hat{A}_\lambda$ defined in (23.81). From the identity

$$\int_{-\infty}^\infty\frac{d\omega}{2\pi}e^{-i\omega(t-s)}s_{\mu\nu}(\omega) = \int_{-\infty}^\infty\frac{d\omega}{2\pi}e^{-i\omega(s-t-i\beta)}s_{\nu\mu}(\omega)$$
$$= \int_{-\infty}^\infty\frac{d\omega}{2\pi}e^{-i\omega(t-s)}s_{\nu\mu}(-\omega)e^{\beta\omega}$$

one gets the relations

$$s_{\mu\nu}(\omega) = s_{\nu\mu}(-\omega)e^{\beta\omega} \tag{23.93}$$
$$\hbar\chi_{\mu\nu}^{''}(\omega) = s_{\mu\nu}(\omega) - s_{\nu\mu}(-\omega) = s_{\mu\nu}(\omega)(1-e^{-\beta\omega}) \tag{23.94}$$
$$\psi_{\mu\nu}^{''}(\omega) = s_{\mu\nu}(\omega) - s_{\nu\mu}(-\omega) = s_{\mu\nu}(\omega)(1+e^{-\beta\omega}) \tag{23.95}$$

and thus
$$\chi''_{\mu\nu}(\omega) = \frac{1}{\hbar}\tanh\frac{\hbar\omega}{2T}\psi''_{\mu\nu}(\omega). \qquad (23.96)$$

The form given in (23.91) follows immediately if the $\psi''_{\mu\nu}(\omega)$ does not have a pole at $\omega = 0$; otherwise one needs to generalize to (23.99).

One great advantage of (23.91) is the possibility of evaluating equilibrium fluctuations
$$\Sigma^{eq}_{\mu\nu} = \frac{1}{2}\langle\left[\Delta\hat{A}_\mu, \Delta\hat{A}_\nu\right]_+\rangle \qquad (23.97)$$
from knowledge of response functions according to the formula:
$$\frac{1}{2}\langle\left[\Delta\hat{A}_\mu, \Delta\hat{A}_\nu\right]_+\rangle = \tilde{\psi}''_{\mu\nu}(t=0) = \int\frac{d\hbar\omega}{2\pi}\coth\left(\frac{\hbar\omega}{2T}\right)\chi''_{\mu\nu}(\omega) \qquad (23.98)$$

The expression on the very right is obtained from (23.91). Actually, as outlined before, some caution is appropriate when applying this form of the FDT, in case the correlation function has the δ-type contribution at $\omega = 0$, seen in (23.84). This feature comes about for the spectral representations (23.68) and (23.82) of the response and correlation functions valid for a genuinely discrete spectrum. To cover such situations, the FDT of (23.91) would have to be replaced by

$$\psi''_{\mu\nu}(\omega) = \psi^0_{\mu\nu}2\pi\delta(\omega) + \hbar\coth\frac{\hbar\omega}{2T}\chi''_{\mu\nu}(\omega). \qquad (23.99)$$

Ifs which are realistic for quantal dissipation, however, the spectrum cannot be handled as being truly discrete. If the dynamics is bound to exhibit irreversible features it cannot be associated with a Hamiltonian. Then the pole of the correlation function at $\omega = 0$ will acquire a width, such that the $\delta(\omega)$ will become a smooth peak, the strength of which being proportional to $\psi^0_{\mu\nu}$, and the appropriate version of the FDT is the form (23.91); these features are demonstrated in Section 6.6.3.

At this stage we may return to the conjecture raised in (23.64) about the energy transfer (23.62) showing irreversible features. Indeed, it is easy to prove that the overall strength function

$$S(\omega) = \sum_{\mu\nu} a^{\text{ext}}_\mu(-\omega)\,\mathbf{s}_{\mu\nu}(\omega)\,a^{\text{ext}}_\nu(\omega) \geq 0 \qquad (23.100)$$

cannot be negative (see exercise 2). From this fact, and with the help of (23.94) together with $\omega(1 - e^{-\beta\omega}) \geq 0$ the statement (23.64) follows immediately.

23.4.3 Strength functions for periodic perturbations

For a periodic perturbation, say like the $a_0\cos(\omega t)\hat{A}$ of (23.29), the average rate $\overline{P}_{mn}(\omega)$ for a transition from states $|m\rangle$ to $|n\rangle$ is given by (23.37). In the spirit of the present section let us take this formula in the limit of $\epsilon \to 0$, which is to say that we replace the smooth delta function by real ones. Suppose the final states

$|n\rangle$ are not specified, or, in other words, that we look at transitions to all of them. Then the $\overline{P}_{mn}(\omega)$ have to be summed over all n. Suppose further that the states $|m\rangle$ are occupied with probability $\rho(E_m)$, in which case the total rate is given by $\overline{P}(\omega) = \sum_{mn} \rho(E_m) \overline{P}_{mn}(\omega)$. For $\omega > 0$ this rate becomes proportional to the correlation function given in (23.82) and (23.84):

$$\overline{P}(\omega) = \frac{a_0^2}{4} \frac{2\pi}{\hbar} \sum_{mn} \rho(E_m) |\langle m|A|n\rangle|^2 \left(\delta(E_n - E_m - \hbar\omega) + \delta(E_n - E_m + \hbar\omega) \right)$$

$$= \frac{a_0^2}{2\hbar^2} \psi''_{AA}(\omega) \qquad \text{for} \qquad \omega > 0. \tag{23.101}$$

The pre-factor depends on the precise choice of the external field. Moreover, if instead of the $a_0 \cos(\omega t)$ one simply chose $a_0 \exp(-i\omega t)$ the result would be

$$\overline{P}(\omega) = \frac{2a_0^2 \pi}{\hbar} \sum_{mn} \rho(E_m) |\langle m|A|n\rangle|^2 \delta(E_n - E_m - \hbar\omega) = \frac{2a_0^2}{\hbar} s_{AA}(\omega), \tag{23.102}$$

with the $s_{AA}(\omega)$ defined through (23.92). As can be inferred from these relations the correlation functions specify the *strength* with which the excitations of the system occur at a given external field. Notice that here the correlation functions at finite T are involved. This feature must be considered when the transition rate is to be expressed in terms of response functions through the FDT. Of course, formulas (23.101) and (23.102) are easily reduced to zero temperature simply by taking $\rho(E_m) = \delta_{m0}$, as explained in Section 23.3.4 for response functions. In solid state physics transitions at finite T play a role for scattering of electrons, neutrons or light at crystals or liquids. The cross section for these reactions can then be expressed in terms of density-density correlation functions, see e.g. (Pines and Nozières, 1966), (Forster, 1975), (Chaikin and Lubensky, 1995). In typical nuclear reactions in the lab, on the other hand, the two partners react with each other starting from the ground state. This situation is different, however, in stellar objects where finite nuclei are in equilibrium with their surroundings, see Section 2.6.3. Then formulas like (23.101) and (23.102) may be used for the description of nuclear reactions as well, not only for the decay of heated nuclei.

23.4.4 Linear response for a Random Matrix Model

It is instructive to calculate response and correlation functions for an ansatz of the matrix elements which has been used in applications of Random Matrix Theory to nuclear transport. To simplify the discussion we restrict ourselves to the one-dimensional case, following the exposition in (Hofmann, 1977). This is to say that the $\hat{H}_{\rm cpl}$ of (23.1) is simply given by $q(t)\hat{F}$. The corresponding response and correlation functions, $\chi''(\omega)$ and $\psi''(\omega)$, respectively, may then be written in the form

$$\Xi(\omega) = \sum_{nm} \rho(E_m) |\langle m|F|n\rangle|^2 \xi(\omega, \Omega_{nm}), \tag{23.103}$$

if only the $\Xi(\omega)$ and $\xi(\omega, \Omega_{nm})$ are given by

$$\Xi(\omega) = \begin{cases} \chi''(\omega) \\ \psi''(\omega)/\hbar \end{cases} \quad \text{for} \quad \xi(\omega, \Omega_{nm}) = \pi \begin{cases} \delta(\hbar\omega - \Omega_{nm}) - \delta(\hbar\omega + \Omega_{nm}) \\ \delta(\hbar\omega - \Omega_{nm}) + \delta(\hbar\omega + \Omega_{nm}). \end{cases} \quad (23.104)$$

Suppose the Hamiltonian $\hat{H}$ is too complex to allow for an explicit calculation of its eigenstates $|m\rangle$ and energies E_m. Eventually one may assume the matrix elements $\langle m|F|n\rangle$ to be random numbers. As was done in Section 4.2 for the energies E_m, one may apply Gaussian ensembles with zero average, $\overline{\langle m|F|n\rangle} = 0$. Following (Weidenmüller, 1980) (see also (Bulgac et al., 1996)) the squared matrix elements may be approximated by the following smooth function of the two energies E_n and E_m:

$$\overline{\langle m|F|n\rangle\langle n|F|m\rangle} = \overline{F_R^2} \sqrt{D(E_n)\,D(E_m)} \, \exp\left(-\frac{(E_n - E_m)^2}{2\Gamma_R^2}\right) \equiv \mathcal{F}_R^2(E_n, E_m). \quad (6.71)$$

Notice, that in distinction to (Weidenmüller, 1980), (Bulgac et al., 1996) but in accord with the spirit of linear response, no dependence of the matrix elements on q is taken into account here. The very form of (6.71) suggests using a continuous spectrum with the $D(E)$ being the average level spacing. The Γ_R represents the "bandwidth" over which the operator $\hat{F}$ couples. Along these lines of arguments the double sum in (23.103) may be replaced by a twofold integral like

$$\sum_{nm} \quad \to \quad \int_0^\infty \frac{dE_n}{D(E_n)} \int_0^\infty \frac{dE_m}{D(E_m)} \quad (23.105)$$

such that the $\Xi(\omega)$ becomes

$$\Xi(\omega) = \int_0^\infty \frac{dE}{D(E)} \rho(E) \int_0^\infty \frac{dE'}{D(E')} \mathcal{F}_R^2(E', E) \, \xi(\omega, E' - E). \quad (23.106)$$

So far the occupation probability $\rho(E)$ has not been specified; it might still represent a microcanonical ensemble. This is of particular interest as both the response and the correlation functions are determined by almost the same expression, giving hope for finding realized an eventually generalized fluctuation dissipation theorem. In fact, it will soon be seen that the conventional form (23.91) follows, which involves the concept of temperature, if only a few natural conditions are fulfilled. Please observe that in (23.106) the first integral is of the same structure as those encountered in Section 22.2, namely

$$\Xi(\omega) = \int_0^\infty dE \, w(E) \Upsilon(\omega, E) \quad \text{with} \quad w(E) = \frac{\rho(E)}{D(E)} \quad (23.107)$$

being the probability density for the energy E. The $\Upsilon(\omega, E)$ introduced here may be rewritten as

$$\Upsilon(\omega, E) = \overline{F_{\rm R}^2} \int_{-E}^{\infty} de \sqrt{\frac{D(E)}{D(E+e)}} \exp\left(-\frac{e^2}{2\,\Gamma_{\rm R}^2}\right) \xi(\omega, e). \qquad (23.108)$$

For negative e the integrand is bound at least by the width $\Gamma_{\rm R}$ of the Gaussian. In (Weidenmüller, 1980) its value is estimated to $\Gamma_{\rm R} \approx 5 - 9$ MeV. This number may safely be assumed to be small compared to the typical excitation E, which is determined by the distribution $w(E)$ governing the integral in (23.107). Hence, in (23.108) the lower limit of the integration over e may be replaced by $-\infty$. If the average level spacing $D(E+e)$ decreases sufficiently strongly with e the following logarithmic expansion is appropriate: $-\ln D(E)/D(E+e) \approx +\partial \ln D(E)/\partial E\, e + \partial^2 \ln D(E)/\partial E^2 e^2/2$. Inserting this into (23.108) the $\Upsilon(\omega, E)$ becomes

$$\Upsilon(\omega, E) = \overline{F_{\rm R}^2} \int_{-\infty}^{\infty} de \exp\left(-\frac{e^2}{2\,(I(E))^2} + \frac{e}{2}\beta(E)\right) \xi(\omega, e), \qquad (23.109)$$

with $\quad \beta(E) = -\dfrac{\partial \ln D(E)}{\partial E} \quad$ and $\quad \dfrac{1}{(I(E))^2} = \dfrac{1}{\Gamma_{\rm R}^2} + \dfrac{1}{2}\dfrac{\partial^2 \ln D}{\partial E^2} \equiv \dfrac{1}{\Gamma_{\rm R}^2} + \dfrac{1}{2\Delta\mathcal{E}^2}.$

These relations are meaningful as long as the width of the energy distribution $w(E)$ is sufficiently narrow. In the ideal case we might take the $w(E)$ to represent a microcanonical ensemble of narrow width around the average $\mathcal{E}$. Then $\beta(E)$ may be replaced by $\beta(\mathcal{E})$ and thus becomes the usual inverse of temperature (see the discussion in Section 22.2). Notice that the $\Delta\mathcal{E}$ is the width of the energy distribution of the canonical ensemble. Typically this width is much larger than the $\Gamma_{\rm R}$, such that $I(E) \approx \Gamma_{\rm R}$.

The response and correlation functions turn into

$$\begin{aligned}\chi''(\omega) &= 2\pi \overline{F_{\rm R}^2}\, \exp\left(-\frac{(\hbar\omega)^2}{2\,\Gamma_{\rm R}^2}\right) \sinh\left(\frac{\hbar\omega}{2T}\right) \\ \psi''(\omega) &= 2\pi \overline{F_{\rm R}^2}\, \hbar\, \exp\left(-\frac{(\hbar\omega)^2}{2\,\Gamma_{\rm R}^2}\right) \cosh\left(\frac{\hbar\omega}{2T}\right)\end{aligned} \qquad (23.110)$$

It is seen that they fulfill the fluctuation dissipation theorem in the classic form (23.91). In (Trost, 1994) integrals like that in (23.109) have been evaluated numerically applying the Fermi gas model for the level densities. Writing the relation between $\psi''_{\mu\nu}(\omega)$ and $\chi''_{\mu\nu}(\omega)$ as $\psi''_{\mu\nu}(\omega) = \hbar \coth(\hbar\omega/2T)\chi''_{\mu\nu}(\omega)(1 + C_{\rm FDT})$ the correction $C_{\rm FDT}$ was seen to be less than 0.01 for temperatures above 0.5 MeV and frequencies larger than 0.1 MeV/$\hbar$. Of course, one must be aware that the very definition of temperature is subject to the common uncertainties, which in particular show up at small excitation energies (see Section 22.3). On the other hand, however, the RMM used here is supposed to be applicable only for heavier nuclei and above an excitation energy of about $15 - 20$ MeV, which roughly corresponds to $T \gtrsim 0.5$ MeV. Finally, we should like to draw the reader's attention to the intimate connection between the behavior of $\psi''(\omega)$ at small frequencies, on the one hand, and the property of ergodicity in the sense defined in Section 23.6.3. Indeed, since the FDT comes out in the form (23.91) it means that the first term shown in (23.99) cannot be very large, if different from zero at all. (It should be evident that for the smooth (squared) matrix elements of the RMM an eventual peak at $\omega = 0$ can at best be a smooth one.) This implies that the $\psi^0_{\mu\nu}$ must be small and because of (23.161) also $\chi^T_{\mu\nu} - \chi_{\mu\nu}(0)$.

23.5 Linear response at complex frequencies

So far we have tacitly assumed the frequency ω to be real. Clearly, functions like (23.4) or (23.6) for the oscillator response are analytic everywhere, except at the position of the poles, which lie apart from the real axis. Actually, this feature is true quite generally for all physical modes of macroscopic motion. It will therefore be necessary to understand how the fundamental expressions we have encountered in this subsection can be continued analytically to complex frequencies. We will begin looking at stable modes whose poles lie in the lower half plane.

The spectral representation (23.47) was constructed so as to guarantee causal behavior for the time-dependent function (23.45), often called the *retarded* response function. Its spectral representation can be continued into the complex plane to become one branch of the function

$$X_{\mu\nu}(z) = \int_{-\infty}^{\infty} \frac{d\Omega}{\pi} \frac{\chi''_{\mu\nu}(\Omega)}{\Omega - z} \qquad \text{Im}\, z \neq 0 \qquad (23.111)$$

namely $\chi_{\mu\nu}(z) \equiv \chi^R_{\mu\nu}(z) = X_{\mu\nu}(z)$ for $\text{Im}\, z > 0$. The other one, $\chi^A_{\mu\nu}(z) = X_{\mu\nu}(z)$ for $\text{Im}\, z < 0$, defines the *advanced* function, whose Fourier transform would vanish for positive time arguments.

The form (23.111) implies symmetry relations similar to those seen before for real frequencies, like $(X_{\mu\nu}(z))^* = X_{\nu\mu}(z^*)$ and $X_{\mu\nu}(-z) = X_{\nu\mu}(z)$, which are immediate consequences of (23.52). For real frequencies one gets from (23.111) and the Dirac formula (26.16)

$$2\chi'_{\mu\nu}(\omega) = (X_{\mu\nu}(\omega + i\epsilon) + X_{\mu\nu}(\omega - i\epsilon)) \qquad (23.112)$$

$$2i\chi''_{\mu\nu}(\omega) = (X_{\mu\nu}(\omega + i\epsilon) - X_{\mu\nu}(\omega - i\epsilon))\,, \qquad (23.113)$$

for the reactive and dissipative parts, respectively. The property $(X_{\mu\nu}(z))^* = X_{\nu\mu}(z^*)$ may be used to introduce for $2i\chi''_{\mu\nu}(\omega)$ in the last term on the right the same argument as in the first one, namely $\omega + i\epsilon$, a frequency slightly above the real axis. The resulting expression may then be generalized to any z in the upper half plane. In this way we get.[48]

$$2i\chi''_{\mu\nu}(z) = \left(\chi_{\mu\nu}(z) - \chi^*_{\nu\mu}(z^*)\right) \qquad (23.114)$$

as the analytical continuation of $\chi''_{\mu\nu}(\omega)$ into the upper half of the complex plane, expressed in terms of the *physical* response function $\chi_{\mu\nu}(z) \equiv \chi^R_{\mu\nu}(z)$.

---

[48] In (Kubo et al., 1991a) the $\chi''_{\mu\nu}(\omega)$ is defined as representing the imaginary part of $\chi_{\mu\nu}(\omega)$ for $\omega = \omega^*$; in this case this relation changes to $2i\chi''_{\mu\nu}(z) = (\chi_{\mu\nu}(z) - \chi^*_{\mu\nu}(z))$.

One will often need to evaluate this function $\chi_{\mu\nu}(z)$ in the lower half plane as well, which is possible under certain circumstances. For instance, if $\chi''_{\mu\nu}(t)$ decays fast enough, say like

$$\lim_{|t|\to\infty} e^{\frac{1}{2\hbar}\Gamma|t|}\tilde{\chi}''_{\mu\nu}(t) = 0, \tag{23.115}$$

then $\chi_{\mu\nu}(z)$ will be analytic in the lower half-plane *at least* for $\operatorname{Im} z > -\Gamma/2$.

Let us illustrate these features for the example of the damped oscillator (with $C > 0$) presented in Section 23.1. The function (23.4) or (23.6) is analytic not only in the *full upper* half plane, but also for *all* z with $\operatorname{Im} z > -\Gamma/2$; it is even analytic for all z besides at the two poles. For real ω, the form given in (23.7) for the imaginary part $\chi''_{\mathrm{osc}}(\omega)$ is in accord with (23.114). It can be used in (23.111) to evaluate the retarded or advanced functions, respectively. In the first case one recovers the $\chi_{\mathrm{osc}}(\omega)$, of course; the result of the second one is:

$$\chi^{\mathrm{A}}_{\mathrm{osc}}(\omega) \equiv \chi^*_{\mathrm{osc}}(\omega) = -\frac{1}{2M\mathcal{E}}\left(\frac{1}{\omega-(\omega^+)^*} - \frac{1}{\omega-(\omega^-)^*}\right) \qquad \omega^* = \omega. \tag{23.116}$$

This is easily understood by conveniently closing the integral of (23.111) in the appropriate half plane. We see that for this one-dimensional case and for real ω the advanced function is obtained just by complex conjugation. Looking back at the basic form (23.4) we realize that the corresponding secular equation is obtained by replacing γ by $-\gamma$. Of course, there is the following generalization to complex frequencies $\left(\chi^{\mathrm{A}}_{\mathrm{osc}}(z)\right)^* = \chi^{\mathrm{R}}_{\mathrm{osc}}(z^*)$.

23.5.1 Relation to thermal Green functions

To describe equilibrium properties of many-body systems it is often useful to introduce thermal Green function (Negele and Orland, 1987), also-called temperature Green function (Fetter and Walecka, 1971) or Matsubara functions (Mahan, 1983). In Section 24.4 we will need a special type of them, namely those that may be associated with response functions. Let us outline this property for the example of functions for one Hermitian operator $\hat{F}$. For our purpose it will be convenient to define the thermal Green function as

$$\widetilde{\mathcal{G}}(\tau - \sigma) = -\frac{1}{\hbar}\langle \mathcal{T}\hat{F}(\tau)\hat{F}(\sigma)\rangle. \tag{23.117}$$

The $\hat{F}(\tau)$ represents the operator $\hat{F}$ in the interaction picture for the imaginary time $t = -i\tau$ (where τ is real; see Section 24.2); thus one has $\hat{F}(\tau) = \exp(\hat{H}\tau/\hbar)\hat{F}\exp(-\hat{H}\tau/\hbar)$. The $\mathcal{T}$ is the time-ordering operator introduced at the beginning of Section 23.2. Please notice that all equations discussed there retain their validity if everywhere t is replaced by $-i\tau$.

The spectral representation of $\widetilde{\mathcal{G}}(\tau)$ is easily seen to be

$$\widetilde{\mathcal{G}}(\tau) = -\frac{1}{\hbar}\sum_{nm}\rho(E_m)\left|\langle m|F|n\rangle\right|^2 e^{\tau(E_m-E_n)/\hbar}, \qquad \text{for}\quad \tau > 0. \tag{23.118}$$

In applications one often needs its frequency transform $\mathcal{G}(i\nu_r)$. Both are connected to each other by the relations

$$\mathcal{G}(i\nu_r) = \int_0^{\hbar\beta} d\tau\, \widetilde{\mathcal{G}}(\tau)\, e^{i\tau\nu_r} \qquad \nu_r = 2\pi r/\hbar\beta$$

$$\widetilde{\mathcal{G}}(\tau) = \frac{1}{\hbar\beta} \sum_r \mathcal{G}(i\nu_r) e^{-i\tau\nu_r}$$
(23.119)

For the spectral representation of $\mathcal{G}(i\nu_r)$ one gets

$$\mathcal{G}(i\nu_r) = \sum_{nm} |\langle m|F|n\rangle|^2 \, \frac{\rho(E_m) - \rho(E_n)}{i\hbar\nu_r + (E_m - E_n)} = \mathcal{G}(-i\nu_r), \qquad (23.120)$$

observing that $\rho(E_m) = \exp(-\beta E_m)/Z$ and $\exp(i\nu_r\hbar\beta) = 1$; the symmetry in $i\nu_r \to -i\nu_r$ follows from the invariance of the $|\langle m|F|n\rangle|^2$ to an exchange of the summation indices. The form (23.120) reminds one of the Fourier transform $\chi(\omega)$ of the (retarded) response function $\widetilde{\chi}(t)$, which according to (23.72) reads

$$\chi(\omega) = -\sum_{nm} |\langle m|F|n\rangle|^2 \, \frac{\rho(E_m) - \rho(E_n)}{\hbar\omega + (E_m - E_n) + i\epsilon}. \qquad (23.121)$$

Hence, both functions are connected by the following analytic continuation (see also (Mahan, 1983))

$$\mathcal{G}(i\nu_r) \quad \overleftrightarrow{i\nu_r \leftrightarrow \omega + i\epsilon} \quad -\chi(\omega). \qquad (23.122)$$

23.5.2 Response functions for unstable modes

As for stable modes, in the time-dependent picture the physical response function describing an instability should vanish for negative times. On the other hand, we know that in such a case at least one of the characteristic frequencies lies in the upper half plane. To reconcile both features, one needs to choose carefully the contour of integration in the frequency plane, when one wants to Fourier-transform from frequency back to time. This physical response, showing causal behavior, may be identified again by the *retarded function*, viz

$$\widetilde{\chi}^{\mathrm{R}}_{\mu\nu}(t) = \int_{\mathcal{C}} \frac{d\omega}{2\pi}\, e^{-i\omega t} \chi^{\mathrm{R}}_{\mu\nu}(\omega) \qquad (23.123)$$

for which the contour $\mathcal{C} = \mathcal{C}^{\mathrm{R}}$ needs to be defined to cross the imaginary axis *above* that pole in the *upper* half plane which is farthest away from the real axis. For the oscillator this is the $\omega_{\mathrm{R}}^+ = iz^+$, with the z^+ given by (7.31). For the advanced function $\widetilde{\chi}^{\mathrm{A}}_{\mu\nu}(t)$ the same form of the transformation applies but with the contour $\mathcal{C} = \mathcal{C}^{\mathrm{A}}$ chosen to cross the imaginary axis *below* $\omega_{\mathrm{A}}^- = (\omega_{\mathrm{R}}^+)^*$ in the lower half plane. The position of these two poles follows from the fact that for the damped oscillator the transformation from "retarded" to "advanced" is to be performed by changing the sign of the damping constant.

As for the stable case, the Fourier transformed functions $\chi^{\text{R}}_{\mu\nu}(\omega)$ and $\chi^{\text{A}}_{\mu\nu}(\omega)$ are branches of an $X(\omega)$ defined by

$$X_{\mu\nu}(\omega) = \int_{\mathcal{C}} \frac{d\Omega}{\pi} \frac{\chi''_{\mu\nu}(\Omega)}{\Omega - \omega} \qquad \text{Im}\,\omega \neq \text{Im}\,\mathcal{C} \equiv \text{Im}\,\Omega \quad \Omega \in \mathcal{C}, \qquad (23.124)$$

in complete analogy with (23.111). Specifically, for the retarded function one has

$$\chi_{\mu\nu}(\omega) \equiv \chi^{\text{R}}_{\mu\nu}(\omega) = \int_{\mathcal{C}} \frac{d\Omega}{\pi} \frac{\chi''_{\mu\nu}(\Omega)}{\Omega - \omega} \qquad \text{Im}\,\omega > \text{Im}\,\mathcal{C}. \qquad (23.125)$$

(The advanced one is obtained by the same expression if only the contour is chosen to obey $\text{Im}\,\omega < \text{Im}\,\mathcal{C}$.) For complex conjugation one gets $\left(\chi^{\text{A}}_{\mu\nu}(\omega)\right)^* = \chi^{\text{R}}_{\nu\mu}(\omega^*)$. This may be used to calculate the dissipative part of the response function through a relation like (23.114). Some care is necessary if this form of $\chi''(\omega)$ appears in integral representations such as (23.124) and (23.125) or in the fluctuation dissipation theorem

$$\Sigma^{eq,b}_{\mu\nu} = \int_{\mathcal{C}} \frac{d\omega}{2\pi} \coth\left(\frac{\omega}{2T}\right) \chi''_{\mu\nu}(\omega). \qquad (23.126)$$

For stable modes the integration is performed along the real axis. This contour has to be changed to a $\mathcal{C}$, which must be defined to lie *above* the poles of the *retarded* part, i.e. $\chi(\omega)$, and to lie *below* those of the *advanced* part, which is to say below those of $\chi^*(\omega^*)$. In other words, the $\mathcal{C}$ should leave the $\equiv \omega^+_{\text{R}}$ below and the ω^-_{A} above itself. For the contour $\mathcal{C}$ in (23.126) there is a further restriction as the integrand contains another factor with poles in the complex plane, the coth which diverges at the Matsubara frequencies $\omega^n_{\text{M}} = \pm n 2\pi T/\hbar$. As in the stable case, the contour $\mathcal{C}$ has to cross the imaginary axis in between those which lie closest to the real axis, namely $\pm 2\pi T/\hbar$. This condition actually puts a lower limit on the range of temperatures for which such a construction may work, namely $T > T^0 \equiv \hbar\omega^+_{\text{R}}$. All these relations may be tested again for the case of the oscillator. In fact, for the application discussed in this book instabilities only occur for collective motion. As the latter is treated within the locally harmonic motion all we need to do about the extension of linear response theory is contained in the oscillator model.

23.5.3 Equilibrium fluctuations of the oscillator

For coordinate and momentum the FDT in the form (23.98) leads to

$$\Sigma^{eq}_{qq} = \int (d\omega/2\pi)\,\hbar \coth(\hbar\omega/2T)\,\chi''_{qq}(\omega) \qquad (7.12)$$

$$\Sigma^{eq}_{pp} = M^2 \int (d\omega/2\pi)\,\hbar \coth(\hbar\omega/2T)\,\omega^2 \chi''_{qq}(\omega). \qquad (7.14)$$

If for $\chi''_{qq}(\omega)$ the response function of the damped oscillator (23.7) is used the integral for Σ^{eq}_{qq} is well defined as for large ω the dissipative part of the oscillator response drops like $\sim \omega^{-3}$. Because of the additional factor ω^2 this feature is no longer given for the momentum, the integral appearing there diverges logarithmically. This deficiency can be cured at the expense of introducing one more parameter. For instance, one might simply use a cutoff for the integral as done in (Hofmann et al., 1989). A more elegant way is to exploit the Drude regularization of (Grabert et al., 1984b) (see also (Grabert et al., 1988) or (Weiss, 1993)). In this method one modifies the integrand by replacing the oscillator response by a function which has one more pole. This can be achieved, for instance, by rewriting the response function (23.4) as

$$\chi_{\rm osc}^{\rm dru}(\omega) = \frac{1}{-\omega^2 M - i\omega\gamma(\omega) + C} \quad \text{with} \quad \gamma(\omega) = \frac{\gamma}{1 - i\omega/\varpi_D} \qquad (23.127)$$

where friction has become a simple frequency-dependent function. With increasing ω the imaginary part of $\chi_{\rm osc}^{\rm dru}(\omega)$ decreases sufficiently fast. In the ideal case the parameter ϖ_D can be chosen in such a way that the difference from the mere oscillator response function is small in regions of maximal strength. Applications to the nuclear case were presented in (Hofmann and Kiderlen, 1998), where a discussion of problems related to the choice of ϖ_D can also be found, see also (Hofmann, 1997) as well as (Kiderlen, 1994) and (Kiderlen and Hofmann, 1994) where instabilities in Fermi liquids have been considered. In the following we want to show some basic features only and refer to these publications for more details.

To evaluate the integrals appearing in the expressions for $\sum_{qq}^{\rm eq}$ and $\sum_{pp}^{\rm eq}$ one expands the hyperbolic cotangent into a uniformly convergent series, which by using the Matsubara frequencies $\nu_n = n2\pi T/\hbar$ becomes

$$\coth\left(\frac{\hbar\omega}{2T}\right) = \frac{2T}{\hbar}\left[\frac{1}{\omega} + \sum_{n=1}^{\infty}\left(\frac{1}{\omega + i\nu_n} + \frac{1}{\omega - i\nu_n}\right)\right]. \qquad (23.128)$$

From here on one may apply the residue theorem taking care of the analytic structure of the response functions. The fluctuations of the coordinate turn out to be given by

$$\sum_{qq}^{\rm eq} = T\left(1/C + 2\sum_{n=1}^{\infty}\chi_{qq}(i\nu_n)\right). \qquad (23.129)$$

For $\sum_{pp}^{\rm eq}$ the Drude regularization leads to

$$\sum_{pp}^{\rm osc}(\varpi_D) = TM\left(+2\sum_{n=1}^{\infty}[C + \nu_n\gamma(i\nu_n)]\chi_{\rm osc}^{\rm dru}(i\nu_n)\right), \qquad (23.130)$$

after having replaced $\chi_{qq}''(\omega)$ by the function $\chi_{\rm osc}''^{\rm dru}(\omega)$.

23.6 Susceptibilities and the static response

In the previous sections we looked at the variation of expectation values $\langle \hat{A}_\mu \rangle_t$ which are induced by time-dependent "external" fields $a_\nu^{\text{ext}}(t)$. Within such a picture one may define the *static response* as the limit of vanishing frequency:

$$\chi_{\mu\nu}(0) \equiv \lim_{\omega \to 0} \chi_{\mu\nu}(\omega) = \chi'_{\mu\nu}(0). \qquad (23.131)$$

The last equation follows because the dissipative part of the response function vanishes at $\omega = 0$. In this procedure no constraint is imposed on thermostatic quantities like temperature T or entropy S. However, one may have processes where one of them is kept constant. In linear response one then has to evaluate $\Delta\langle \hat{A}_\mu \rangle$ under appropriate constraints. This will allow one to define *isothermal* or *isentropic (adiabatic)* susceptibilities, χ^T and χ^{ad}, respectively. For this purpose we need to replace the a_ν^{ext} by time *in*dependent quantities a_ν. The total Hamiltonian can then be written as

$$\hat{H}_{\text{sc}}(a) = \hat{H} + \sum_{\nu=1} a_\nu \hat{A}_\nu = \hat{H}_{\text{sc}}(a^0) + \sum_{\nu=1} \xi_\nu \hat{A}_\nu. \qquad (23.132)$$

The ξ_ν are meant to represent the deviations $\xi_\nu = a_\nu - a_\nu^0$ of the a_ν from some fixed values a_ν^0. The latter may for instance be associated with the equilibrium of a generalized ensemble, which may also described by the Hamiltonian $\hat{H}_{\text{sc}}(a^0) = \hat{H} + \sum_{\nu=1} a_\nu^0 \hat{A}_\nu$. Later we may wish to replace the ξ_ν by differentials $d\xi_\nu$.

23.6.1 Static perturbations of the local equilibrium

In the following we will employ the density operator $\hat{\rho}(\lambda) = \exp(-\sum_i \lambda_i \hat{A}_i)/Z(\lambda)$ of the generalized ensemble introduced in Section 22.1.2. The $\hat{A}_i$ are to be constructed from the Hamiltonian $\hat{H}_{\text{sc}}(a)$. To keep the notation as simple as possible we will replace the $\hat{H}_{\text{sc}}(a^0)$ by an $\hat{H}$, again. As we want to allow for a variation of the temperature we may write

$$\beta \hat{H}_{\text{sc}} = \beta_0 \Big(\hat{H} + \sum_{j=0} \xi_j \hat{A}_j \Big) \equiv \sum_{j=0} \lambda_j \hat{A}_j \qquad (23.133)$$

identifying $\hat{H} = \hat{A}_0$ and as in (22.37)) $\lambda_0 = \beta = \beta_0 + \Delta\beta$ with $\Delta\beta = \beta_0 \xi_0$. With the λ_ν being equal to $\beta_0 \xi_\nu$ for $\nu \geq 1$ the density operator takes on the form

$$\hat{\rho}_{\text{qs}}(\lambda) = \frac{1}{Z(\lambda)} \exp\Big(-\beta_0 \Big(\hat{H} + \sum_{j=0} \xi_j \hat{A}_j\Big)\Big). \qquad (23.134)$$

In this notation the unperturbed density may symbolically be associated with a set λ^0 where $\lambda_0^0 = \beta_0$ and $\lambda_\nu^0 = 0$. For differential forms, where ξ_i is to be replaced by $d\xi_i$, the $d\lambda_i$ are given by $d\lambda_i = \beta_0 d\xi_i$ (with $d\beta = \beta_0 d\xi_0$). In this sense the relations (22.39) and (22.40) become

$$d\ln Z(\lambda) = -\beta_0 \sum_{i=0} \langle \hat{A}_i \rangle_\lambda \, d\xi_i \tag{23.135}$$

$$d\mathcal{S} = (\beta_0 + \Delta\beta)\, d\langle \hat{H} \rangle_\lambda + \beta_0 \sum_{\nu=1} \xi_\nu \, d\langle \hat{A}_\nu \rangle_\lambda . \tag{23.136}$$

At this stage we need to fix β_0. This can be done by requiring that for $\Delta\beta = \xi_\nu = 0$ the system is in equilibrium with respect to the Hamiltonian $\hat{H} \equiv \hat{H}_{\text{sc}}(a^0)$. Hence, β_0 is defined as

$$\beta_0 = \left. \frac{\partial \mathcal{S}}{\partial \langle \hat{H} \rangle} \right|_{\xi=0} . \tag{23.137}$$

To zero order in the ξ_i the variations in entropy and energy (about this equilibrium) vanish, $d\mathcal{S}|_{\xi=0} = \beta_0 \, d\langle \hat{H} \rangle_{\xi=0} = 0$ and the first order terms become

$$d\mathcal{S} = \beta_0 \sum_{j=0} \xi_j \, d\langle \hat{A}_j \rangle_\lambda . \tag{23.138}$$

Pursuing our lines of argument we will be able to obtain an expression for $\Delta\mathcal{S}$ which is valid to second order in the ξ_i.

First we need to evaluate the density operator $\hat{\rho}_{\text{qs}}(\lambda)$ to linear order (including $\Delta\beta = \beta_0 \xi_0$). The difference from the unperturbed operator $\hat{\rho}_{\text{qs}}(\lambda^0)$ becomes

$$\Delta\hat{\rho}_{\text{qs}} = \hat{\rho}_{\text{qs}}(\lambda) - \hat{\rho}_{\text{qs}}(\lambda^0) = \hat{\rho}_{\text{qs}}(\lambda^0)\left(-\Delta\beta\Delta\hat{H} - \beta_0 \sum_{\nu=1} \xi_\nu \int_0^{\beta_0} d\zeta \, \Delta\hat{A}_\nu^{\text{Im}}(\zeta)\right). \tag{23.139}$$

Here the abbreviation $\Delta\hat{A}_i = \hat{A}_i - \langle \hat{A}_i \rangle_{\lambda^0}^{\text{qs}}$ is used (with the average calculated from $(\hat{\rho}_{\text{qs}}(\lambda^0))$) and $\Delta\hat{A}_\nu^{\text{Im}}(\zeta)$ is defined as

$$\Delta\hat{A}_\nu^{\text{Im}}(\zeta) = e^{\zeta\hat{H}} \Delta\hat{A}_\nu e^{-\zeta\hat{H}} . \tag{23.140}$$

This result is easily obtained from (26.26) and (26.27) (see exercise 3). Evidently, any other term in the sum for which the associated $\hat{A}_\nu$ commutes with the $\hat{H}$ reduces to the simpler form $\beta\Delta\hat{A}_\nu\xi_\nu$. One such candidate could be particle number. In this case the ξ_ν would be the deviation of the chemical potential from its value in equilibrium. The corresponding term would represent a *thermal force*. Here, we shall mainly be interested in that one which comes from the variation of temperature. Restricting ourselves to this simplification, the sum over ν represents mechanical forces, the latter may be of "internal" or "external" nature.

With the $\Delta\hat{\rho}_{\text{qs}}$ given through (23.139) one may evaluate the variation of the averages of the operators A_i. Employing the Mori product defined in (26.28), which in the present notation reads

$$(\hat{A}|\hat{B}) = \frac{1}{\beta_0} \int_0^{\beta_0} d\zeta \, \text{tr}\, \hat{\rho}_{\text{qs}}(\lambda^0) \hat{A}^{\text{Im}}(\zeta)\hat{B} = \frac{1}{\beta_0} \int_0^{\beta_0} d\zeta \, \text{tr}\, \hat{\rho}_{\text{qs}}(\lambda^0) \hat{A}\hat{B}^{\text{Im}}(-\zeta), \tag{23.141}$$

one gets:

$$\langle \hat{A}_i \rangle^{qs}_\lambda - \langle \hat{A}_i \rangle^{qs}_{\lambda^0} = -\Delta\beta \langle \Delta\hat{H} \Delta\hat{A}_i \rangle - \beta_0 \sum_{\nu=1} \xi_\nu \left(\Delta\hat{A}_\nu | \Delta\ddot{A}_i \right) \qquad (23.142)$$

or $\quad \Delta\langle \hat{A}_i \rangle^{qs}_\lambda = -\beta_0 \sum_{j=0} \left(\Delta\hat{A}_i | \Delta\hat{A}_j \right) \xi_j$. $\qquad (23.143)$

For the infinitesimal variations needed in (23.138) we may replace the ξ_i of (23.143) by $d\xi_i$ and evaluate the $\left(\Delta\hat{A}_i | \Delta\hat{A}_j \right)$ at $\xi_i = 0$. In this way one gets the differential

$$d\langle \hat{A}_i \rangle^{qs}_{\xi=0} = -\sum_{j=0} \mathcal{A}_{ij} \, d\xi_j, \qquad \text{where} \qquad \mathcal{A}_{ij} = \beta_0 \left(\Delta\hat{A}_i | \Delta\hat{A}_j \right) = \mathcal{A}_{ji},$$
$$(23.144)$$

which for the variation of the entropy (23.138) leads to

$$d\mathcal{S} = -\beta_0 \sum_{i,j} \xi_i \, \mathcal{A}_{ij} \, d\xi_j \ . \qquad (23.145)$$

At this point it may be worthwhile to reconsider the variation of entropy by starting from the basic formula (22.13), which for the quasi-static density operator $\hat{\rho}_{qs}(\lambda)$ of (23.134) reads $d\mathcal{S}|_{qs} = -\mathrm{tr}\left(d\hat{\rho}_{qs} \ln \hat{\rho}_{qs}(\lambda) \right)$. As the quantities to be varied are the ξ_j, the $d\hat{\rho}_{qs}$ may be taken from (23.139) if only the ξ_j are replaced by $d\xi_j$, which is to say

$$d\hat{\rho}_{qs} = -\beta_0 \hat{\rho}_{qs}(\lambda^0) \sum_{j=0} d\xi_j \int_0^{\beta_0} d\zeta \, \Delta\hat{A}^{\mathrm{Im}}_j(\zeta) . \qquad (23.146)$$

Evaluating $\ln \hat{\rho}_{qs}(\lambda)$ from (23.134) one gets:

$$d\mathcal{S}|_{qs} = -\beta_0^2 \sum_{j=0} \langle \Delta\hat{H} | \Delta\hat{A}_j \rangle d\xi_j - \beta_0 \sum_{i,j} \xi_i \, \mathcal{A}_{ij} \, d\xi_j . \qquad (23.147)$$

With the help of (23.144) the first term is easily recognized to be identical to $\beta_0 \, d\langle \hat{H} \rangle_{\xi=0}$ such that we may write

$$T d\mathcal{S}|_{\xi=0} = \mathrm{tr}\, d\hat{\rho}_{qs} \hat{H} = -\beta_0 \sum_{j=0} \langle \Delta\hat{H} | \Delta\hat{A}_j \rangle d\xi_j \equiv d\langle \hat{H} \rangle_{\xi=0} . \qquad (23.148)$$

Earlier these expressions were put equal to zero, thus specifying the condition for *global or total equilibrium*: The entropy $\mathcal{S}(\lambda^0)$ associated with the density operator $\hat{\rho}_{qs}(\lambda^0)$ is extremal with respect to variations in all ξ_j. It must be *larger* than the entropy $\mathcal{S}(\lambda)$ for the *conditional or partial equilibrium* given for *finite values* of the ξ_j. This fact may be proven as follows by evaluating the finite difference $\Delta\mathcal{S} \equiv \mathcal{S}(\lambda^0) - \mathcal{S}(\lambda)$. Remember that we want to look at the variation measured from finite ξ_i down to $\xi_i = 0$. This convention requires a change of sign for eqn(23.145) such that we get the following quadratic form:

$$\Delta \mathcal{S} \equiv \mathcal{S}(\lambda^0) - \mathcal{S}(\lambda) = \frac{1}{2}\beta_0^2 \sum_{i,j=0} \xi_i \left(\Delta\hat{A}_i | \Delta\hat{A}_j\right) \xi_j = \frac{1}{2}\beta_0^2 \sum_{i,j=0} \xi_i \mathcal{A}_{ij} \xi_j. \quad (23.149)$$

By inverting (23.143) to

$$\xi_i = -\sum_{j=0} \frac{1}{\beta_0} (\mathcal{A})_{ij}^{-1} \Delta\langle\hat{A}_j\rangle_\lambda^{\text{qs}} \quad (23.150)$$

this result can be written as

$$\Delta\mathcal{S} = \frac{1}{2} \sum_{i,j=0} \Delta\langle\hat{A}_i\rangle_\lambda^{\text{qs}} (\mathcal{A})_{ij}^{-1} \Delta\langle\hat{A}_j\rangle_\lambda^{\text{qs}}. \quad (23.151)$$

Furthermore, by invoking (23.133) for (23.149) $\Delta\mathcal{S}$ is seen to be identical to

$$\Delta\mathcal{S} = \frac{1}{2}\left(\beta\Delta\hat{H}_{\text{sc}} - \beta_0\Delta\hat{H} \middle| \beta\Delta\hat{H}_{\text{sc}} - \beta_0\Delta\hat{H}\right) \geq 0. \quad (23.152)$$

The fact of $\Delta\mathcal{S} = \mathcal{S}(\lambda^0) - \mathcal{S}(\lambda)$ being positive definite follows because the Mori product of two identical operators must be larger or equal to zero.

23.6.2 Isothermal and adiabatic susceptibilities

As mentioned earlier, the isothermal and adiabatic susceptibilities measure the variation of expectation values with external parameters where either temperature or entropy are kept constant. The infinitesimal deviations $d\langle\hat{A}_\mu\rangle_\lambda^{\text{qs}}$ of the averages from their equilibrium values are given by (23.144). (To simplify the notation we will replace β_0 by β). These relations invite us to apply the following definition of the *isothermal susceptibilities*:

$$\chi_{ij}^{\text{T}} = \left(\frac{\partial\langle\hat{A}_i\rangle_\lambda^{\text{qs}}}{\partial(-\xi_j)}\right) = \beta\left(\Delta\hat{A}_i | \Delta\hat{A}_j\right) = \chi_{ji}^{\text{T}}. \quad (23.153)$$

Remember, please, that for the evaluation of the partial derivative all other variables are to be kept constant, such that $d\xi_{k\neq j} = 0$. Hence, for all $j = \nu \geq 1$ the χ_{ij}^{T} are calculated at *constant temperature*, indeed. The χ_{00}^{T}, on the other hand, simply stands for the *specific heat* at constant ξ_ν for all $\nu \geq 1$ (recall that $\hat{A}_0 = \hat{H}$). Remember that if either $\hat{A}_i$ or $\hat{A}_j$ commute with $\hat{H}$, expression (23.153) reduces to the form given in (22.48). In terms of the eigenstates and energies of $\hat{H}$, the χ_{ij}^{T} can be given the following spectral representation,

$$\chi_{ij}^{\text{T}} = \sum_{nm} \langle m|\Delta\hat{A}_i|n\rangle\langle n|\Delta\hat{A}_j|m\rangle \frac{\rho(E_m) - \rho(E_n)}{E_n - E_m}. \quad (23.154)$$

For degeneracies as well as for all diagonal matrix elements one has to apply the relation

$$\lim_{E_n \to E_m} \frac{\rho(E_m) - \rho(E_n)}{E_n - E_m} = \lim_{E_n \to E_m} \rho(E_m) \frac{1 - \exp[-\beta(E_n - E_m)]}{E_n - E_m} = \rho(E_m)\beta. \quad (23.155)$$

It is easily recognized that the diagonal elements of $\chi_{ij}^{\rm T}$ are positive, $\chi_{ii}^{\rm T} > 0$.

We may turn now to the *adiabatic* susceptibilities, for which we will use the symbol $\chi_{\mu\nu}^{\rm ad}$ rather than $\chi_{\mu\nu}^{\rm S}$. To calculate them in the canonical ensemble we may make use of (23.148) for $d\mathcal{S} = 0$. This implies that $d\beta$ and $d\xi_\nu$ are related to each other by (with $\Delta\hat{H} \equiv \Delta\hat{H}_{\rm sc}(a^0)$)

$$\frac{d\beta}{\beta} = -\sum_{\nu=1} \frac{\langle \Delta\hat{H}\Delta\hat{A}_\nu \rangle}{\langle \Delta\hat{H}\Delta\hat{H} \rangle} d\xi_\nu \equiv -\frac{\langle \Delta\hat{H}\Delta\hat{H}_{\rm cpl} \rangle}{\langle \Delta\hat{H}\Delta\hat{H} \rangle}. \quad (23.156)$$

We may note in passing that this property reflects the "adiabatic theorem" of statistical mechanics, see Section 11.2.2. The $d\beta$ of (23.156) is to be inserted into (23.144), which for the variation of the mean value $\langle \hat{A}_i \rangle_\lambda^{\rm qs}$ leads to

$$d\langle \hat{A}_i \rangle_\lambda^{\rm qs}\Big|_{\mathcal{S}} = -\beta \sum_\nu \left(-\langle \Delta\hat{A}_i \Delta\hat{H} \rangle \frac{\langle \Delta\hat{H}\Delta\hat{A}_\nu \rangle}{\langle \Delta\hat{H}\Delta\hat{H} \rangle} + (\Delta\hat{A}_i | \Delta\hat{A}_\nu) \right) d\xi_\nu$$

$$\equiv -\sum_{\nu=1} \chi_{\mu\nu}^{\rm ad} d\xi_\nu. \quad (23.157)$$

Observe, please, that the term with $i = 0$ vanishes identically, such that we may only look at terms with $i = \mu \geq 1$. Together with the factor β the second term on the right is identical to $\chi_{\mu\nu}^{\rm T}$ such that one has

$$\chi_{\mu\nu}^{\rm ad} = \chi_{\mu\nu}^{\rm T} - \beta \frac{\langle \Delta\hat{A}_\nu \Delta\hat{H} \rangle \langle \Delta\hat{H}\Delta\hat{A}_\mu \rangle}{\langle \Delta\hat{H}\Delta\hat{H} \rangle}. \quad (23.158)$$

From the identity $\partial \langle \hat{A}_\mu \rangle / \partial \beta = -\langle \Delta\hat{H}\Delta\hat{A}_\mu \rangle = -\langle \Delta\hat{A}_\mu \Delta\hat{H} \rangle$ the difference between the two susceptibilities may be written as

$$\chi_{\mu\nu}^{\rm T} - \chi_{\mu\nu}^{\rm ad} = \beta \frac{\langle \Delta\hat{A}_\nu \Delta\hat{H} \rangle \langle \Delta\hat{H}\Delta\hat{A}_\mu \rangle}{\langle \Delta\hat{H}\Delta\hat{H} \rangle} = \frac{T}{C}\left(\frac{\partial \langle \hat{A}_\nu \rangle}{\partial T} \right)\left(\frac{\partial \langle \hat{A}_\mu \rangle}{\partial T} \right). \quad (23.159)$$

Here, the specific heat $C = \partial \langle \hat{H} \rangle / \partial T = \beta^2 \langle \Delta\hat{H}\Delta\hat{H} \rangle$ has been introduced, see (22.47). It is seen that for diagonal elements the difference (23.159) cannot be negative, viz $\chi_{\mu\mu}^{\rm T} \geq \chi_{\mu\mu}^{\rm ad}$.

23.6.3 Relations to the static response

The isothermal case: The form (23.154) of the isothermal susceptibility allows for an easy connection with the static response $\chi_{\mu\nu}(0)$. The spectral representation for the latter can be traced back to that for the reactive part of the dynamic response (remember (23.131)), which simply means using (23.73) at $\omega = 0$. Recalling that this form (23.73) has been derived assuming the matrix elements to be real, one readily verifies that in such a case the static response can be written as

$$\chi_{\mu\nu}(0) = \sum_{nm} \langle m|\hat{A}_\mu|n\rangle \langle n|\hat{A}_\nu|m\rangle \, \mathcal{P} \frac{\rho(E_m) - \rho(E_n)}{(E_n - E_m)}. \qquad (23.160)$$

At times $\chi_{\mu\nu}(0)$ is also referred to as *isolated susceptibility* (Kubo et al., 1991a). This is due to the fact that the linear response functions are calculated as if the system were isolated from any reservoir or heat bath. In contrast to $\chi_{\mu\nu}^{\mathrm{T}}$, the $\chi_{\mu\nu}(0)$ gets contributions *only* from states of *different energies*. This restriction is left over from the fact that the reactive part of the response function is governed by a principal value integral. Thus the difference between the isothermal susceptibility and the static response is just given by the sum over contributions from all states having the same energy. From (23.155) one gets obeying (23.83) and (23.88)

$$\chi_{\mu\nu}^{\mathrm{T}} - \chi_{\mu\nu}(0) = \beta \sum_{E_n = E_m} \langle m|\Delta\hat{A}_\mu|n\rangle \langle n|\Delta\hat{A}_\nu|m\rangle \, \rho(E_m) = \beta \psi_{\mu\nu}^0 = \beta \langle \Delta\hat{A}_\mu^0 \Delta\hat{A}_\nu^0 \rangle. \qquad (23.161)$$

For $\mu = \nu$ this expression is positive definite, showing that $\chi_{\mu\mu}^{\mathrm{T}} \geq \chi_{\mu\mu}(0)$. At zero excitation this difference vanishes in case the ground state is non-degenerate or if the $\hat{A}_\nu$ do not couple between the possible states with $E_n = E_m = E_0$.

The isentropic case For the difference between the adiabatic and the isolated susceptibilities we may exploit the results from above, namely (23.158) and (23.161):

$$\chi_{\mu\nu}^{\mathrm{ad}} - \chi_{\mu\nu}(0) = \chi_{\mu\nu}^{\mathrm{ad}} - \chi_{\mu\nu}^{\mathrm{T}} + \chi_{\mu\nu}^{\mathrm{T}} - \chi_{\mu\nu}(0)$$
$$= -\beta \frac{\langle \Delta\hat{A}_\nu \Delta\hat{H}\rangle \langle \Delta\hat{H}\Delta\hat{A}_\mu\rangle}{\langle \Delta\hat{H}\Delta\hat{H}\rangle} + \beta\langle \Delta\hat{A}_\mu^0 \Delta\hat{A}_\nu^0\rangle.$$

The expression on the very right may be evaluated further with the help of basic properties of the Mori product. After some lengthy calculation one gets (Kubo et al., 1991a) (see also (Fick and Sauermann, 1986), (Brenig, 1989), (Hofmann, 1997))

$$\chi_{\mu\nu}^{\mathrm{ad}} - \chi_{\mu\nu}(0) = \beta \sum_{\gamma \geq 1} \frac{\langle \Delta\hat{A}_\mu \Delta\hat{H}_\gamma\rangle \langle \Delta\hat{H}_\gamma \Delta\hat{A}_\nu\rangle}{\langle \Delta\hat{H}_\gamma^2\rangle}. \qquad (23.162)$$

The sum extends over a set $\gamma = 0, 1, 2, 3...$ involving all operators $\hat{H}_\gamma$ which commute with $\hat{H}$ (with $\hat{H} \equiv \hat{H}_{\gamma=0}$). From this result it is easily seen that

$$\chi_{\mu\nu}^{\mathrm{ad}} - \chi_{\mu\nu}(0) \geq 0 \qquad \text{for} \qquad \mu = \nu. \qquad (23.163)$$

Indeed, because of $\mathrm{tr}\rho\Delta\hat{H}_\gamma\Delta\hat{A}_\mu = \mathrm{tr}\Delta\hat{H}_\gamma\rho\Delta\hat{A}_\mu = \mathrm{tr}\rho\Delta\hat{A}_\mu\Delta\hat{H}_\gamma$ in this case the sum on the right-hand side of (23.162) extends over non-negative numbers.

Evidently, $\chi^{\text{ad}}_{\mu\nu} - \chi_{\mu\nu}(0)$ will be zero in case that there are no constants of motion $\hat{H}_\gamma$ which contribute to the sum in (23.162). In the ideal case the expectation values $\langle\Delta\hat{H}_\gamma\Delta\hat{A}_\mu\rangle$ vanish, for all $\hat{H}_\gamma$ besides the Hamiltonian itself. In general this condition will hardly be fulfilled, as in (23.162) powers of the Hamiltonian are also to be considered, i.e. operators like $\hat{H}_\gamma = (\hat{H})^k$. However, in this case another argument may be employed. Contributions of this type can be expected to be *small if the fluctuations in energy are small*, as it would be the case in the thermostatic limit. The two conditions under which we may expect the adiabatic susceptibility to become identical to the isolated one, the static response, i.e. $\chi^{\text{ad}}_{\mu\nu} = \chi_{\mu\nu}(0)$ may be summarized as: It is sufficient to have (i) a non-degenerate spectrum E_m, and (ii) a narrow distribution of the occupied states. In (Brenig, 1989) a system having this property is identified as "ergodic". Often, this notion is in use for relaxation processes. A possible connection between the two concepts shall be discussed now.

23.7 Linear irreversible processes

23.7.1 *Relaxation functions*

We return to the discussion of Section 23.6.1. The density operator (23.134) will now be interpreted as to represent the system at time $t = 0$, brought into that state by external forces, which then are released. For $t > 0$ the system may evolve freely which may be described by either the von Neumann equation for the density operator or Heisenberg's equation of motion for the operators. In any case, the evolution operator is determined by the Hamiltonian $\hat{H} = \hat{H}_{\text{sc}}(a^0)$ and the ξ_i are time-independent quantities which specify the initial conditions. Within the Heisenberg picture the expectation values can be written as

$$\langle\hat{A}_i\rangle^{\text{qs}}_t = \text{tr}\left\{\hat{\rho}_{\text{qs}}(\lambda)\hat{A}_i(t)\right\} \simeq \langle\hat{A}_i\rangle^{\text{qs}}_{\lambda^0} + \text{tr}\,\Delta\hat{\rho}_{\text{qs}}\hat{A}_i(t), \quad (23.164)$$

with $\Delta\hat{\rho}_{\text{qs}}$ given by (23.139) and $\hat{A}_i(t) = \exp(i\hat{H}t/\hbar)\hat{A}_i\exp(i\hat{H}t/\hbar)$. Evidently, the time dependence drops out for any operator which commutes with the Hamiltonian $\hat{H}$. In the following we will assume this not to be the case for all $\hat{A}_\mu$ with $\mu \geq 1$. As we want to concentrate on time-dependent quantities we discard the case of $i = 0$. Employing the Mori product (23.141) the very last term in (23.164) can be written as

$$\text{tr}\,\Delta\hat{\rho}_{\text{qs}}\hat{A}_\mu(t) = -\Delta\beta\langle\Delta\hat{H}\Delta\hat{A}_\mu\rangle - \beta_0\sum_{\nu=1}\xi_\nu\left(\Delta\hat{A}_\nu|\Delta\hat{A}_\mu(t)\right) \quad (23.165)$$

(remember that $\text{tr}\,\hat{\rho}_{\text{qs}}(\lambda^0)\Delta\hat{A}^{\text{Im}}_\nu(\zeta) = 0$). For the deviation of $\langle\hat{A}_\mu\rangle^{\text{qs}}_t$ from the unperturbed value this implies

$$\Delta\langle\hat{A}_\mu\rangle^{\text{qs}}_t = -\beta_0\sum_{j=0}\left(\Delta\hat{A}_\mu(t)|\Delta\hat{A}_j\right)\xi_j. \quad (23.166)$$

As expected by construction, at $t = 0$ these expressions are identical to (23.143). We may note in passing that the term $j = 0$ does not change in time at all.

Let us first restrict ourselves to the case where *the elongation* to the static value $\langle \hat{A}_\mu \rangle_\lambda^{qs}$ was performed *at constant temperature*. In this case $\Delta\beta = \beta_0 \xi_0$ vanishes such that the summation in (23.166) only extends over $j = \nu \geq 1$.

The value the $\Delta\langle \hat{A}_\mu \rangle_t^{qs}$ take on when time progresses to $+\infty$ may be obtained by applying the final value theorem of Laplace transforms, which says

$$\lim_{t\to\infty} \hat{A}_\mu(t) = \lim_{\epsilon\to 0^+} \epsilon \int_0^\infty d\tau\, e^{-\epsilon\tau} \hat{A}_\mu(\tau). \tag{23.167}$$

Because of

$$\lim_{\epsilon\to 0^+} \epsilon \int_0^\infty d\tau\, e^{-\epsilon\tau - i\omega\tau} = \lim_{\epsilon\to 0^+} \epsilon \int_{-\infty}^0 d\tau\, e^{\epsilon\tau + i\omega\tau} = \lim_{\epsilon\to 0^+} \frac{\epsilon}{\epsilon + i\omega}$$
$$= \begin{cases} 1 & \text{for} \quad \omega = 0 \\ 0 & \text{for} \quad \omega \neq 0 \end{cases} \tag{23.168}$$

one has

$$\lim_{\epsilon\to 0^+} \epsilon \int_0^\infty d\tau\, e^{-\epsilon\tau} \hat{A}_\mu(\tau) = \frac{1}{2\pi} \hat{A}_\mu(\omega = 0) = \hat{A}_\mu^0, \tag{23.169}$$

where $\hat{A}_\nu(\omega)$ is the frequency transform of $\hat{A}_\nu(\tau)$ and $\hat{A}_\nu^0$ the associated zero frequency component introduced in (23.87). Since this component commutes with the Hamiltonian the average of $\Delta\langle \hat{A}_\mu \rangle_t^{qs}$ turns to

$$\lim_{t\to\infty} \Delta\langle \hat{A}_\mu \rangle_t^{qs} = -\beta_0 \sum_{\nu=1} \langle \Delta\hat{A}_\mu^0 \Delta\hat{A}_\nu^0 \rangle \xi_\nu = -\sum_{\nu=1} \left(\chi_{\mu\nu}^T - \chi_{\mu\nu}(0)\right) \xi_\nu, \tag{23.170}$$

where through (23.161) the isothermal and the isolated susceptibilities show up.

In physical terms the results may be interpreted as follows: Before $t = 0$ the averages are shifted from their unperturbed values by the finite amount $\Delta\langle \hat{A}_\mu \rangle_\lambda^{qs}$; the unperturbed state is described by the density operator (23.134) for $\xi_\mu = 0$. Once the system moves freely it *does not* "relax" back to the unperturbed state *unless the* $\chi_{\mu\nu}^T - \chi_{\mu\nu}(0)$ vanish identically. Only in the latter case may it be said to behave ergodically (under the circumstances that the initial state was reached at constant temperature rather than constant entropy, see below). These features may be described concisely by introducing relaxation functions, for which different definitions exist. Adapting that of (Brenig, 1989) we introduce the functions

$$\tilde{\Psi}_{\mu\nu}^T(t) = \beta_0 \left(\Delta\hat{A}_\mu(t) | \Delta\hat{A}_\nu\right) \Theta(t), \tag{23.171}$$

which have the limiting values

$$\lim_{t\to 0^+} \tilde{\Psi}_{\mu\nu}^T(t) = \chi_{\mu\nu}^T \quad \text{and} \quad \lim_{t\to\infty} \tilde{\Psi}_{\mu\nu}^T(t) = \chi_{\mu\nu}^T - \chi_{\mu\nu}(0). \tag{23.172}$$

With these functions the time-dependent expectation values can be written as

$$\Delta\langle\hat{A}_\mu\rangle^{qs}_t = -\sum_{\nu=1}\widetilde{\Psi}^T_{\mu\nu}(t)\,\xi_\nu \qquad \text{for} \qquad t \geq 0. \tag{23.173}$$

Let us see how the $\mathring{\Psi}^T_{\mu\nu}(t)$ relate to the response functions $\widetilde{\chi}_{\mu\nu}(t)$. With the current $\hat{J}_\mu(t) = d\hat{A}_\mu(t)/dt$ the latter can be written as $\widetilde{\chi}_{\mu\nu}(t) = -\beta\big(\hat{J}_\mu(t)|\hat{A}_\nu\big)$, see (23.216).

Because of $\widetilde{\chi}_{\mu\nu}(t) \equiv \Theta(t)\widetilde{\chi}_{\mu\nu}(t))$ this implies the identity

$$\frac{d}{dt}\widetilde{\Psi}^T_{\mu\nu}(t) = \chi^T_{\mu\nu}\delta(t) - \widetilde{\chi}_{\mu\nu}(t) \qquad \text{and} \qquad -i\omega\Psi^T_{\mu\nu}(\omega) = \chi^T_{\mu\nu} - \chi_{\mu\nu}(\omega) \tag{23.174}$$

for the Fourier transforms. After integrating the first equation over time one gets the form

$$\widetilde{\Psi}^T_{\mu\nu}(t) = \left(\chi^T_{\mu\nu} - \int_0^t \widetilde{\chi}_{\mu\nu}(s)\,ds\right)\Theta(t) \tag{23.175}$$

which correctly represents the limiting values given in (23.172). We may use these relations to get the differential law (see exercise 4)

$$\frac{d}{dt}\langle\hat{A}_\mu\rangle^{qs}_t = -\beta_0\sum_{\nu=1}(\hat{J}_\mu(t)|\hat{A}_\nu)\xi_\nu = \sum_{\nu=1}\widetilde{\chi}_{\mu\nu}(t)\xi_\nu\,. \tag{23.176}$$

Let us finally address another situation in which the elongation $\Delta\langle\hat{A}_i\rangle^{qs}_\lambda|_S$ is performed at constant entropy. For this one may introduce the relaxation functions (see exercise 5)

$$\widetilde{\Psi}^{ad}_{\mu\nu}(t) = \chi^{ad}_{\mu\nu} - \int_0^t \widetilde{\chi}_{\mu\nu}(s)\,ds \equiv \widetilde{\Psi}^T_{\mu\nu}(t) - \chi^T_{\mu\nu} + \chi^{ad}_{\mu\nu}\,. \tag{23.177}$$

The connection with the response function is given by $\widetilde{\chi}_{\mu\nu}(t) = -d\widetilde{\Psi}^{ad}_{\mu\nu}(t)/dt$, but for the limiting values one has

$$\lim_{t\to 0}\widetilde{\Psi}^{ad}_{\mu\nu}(t) = \chi^{ad}_{\mu\nu} \qquad \text{and} \qquad \lim_{t\to\infty}\widetilde{\Psi}^{ad}_{\mu\nu}(t) = \chi^{ad}_{\mu\nu} - \chi_{\mu\nu}(0)\,. \tag{23.178}$$

The time-dependent expectation values would now be given by

$$\Delta\langle\hat{A}_\mu\rangle^{qs}_t = -\sum_{\nu=1}\widetilde{\Psi}^{ad}_{\mu\nu}(t)\,\xi_\nu \qquad \text{for} \qquad t \geq 0. \tag{23.179}$$

Indeed, this relation satisfies the right initial conditions, but the time dependence as such is governed by the Hamiltonian, which is to say by the response functions $\widetilde{\chi}_{\mu\nu}(t)$. Now "ergodicity" is defined in accord with the discussion in Section 23.6.3: The system does not relax to zero unless $\chi^{ad}_{\mu\nu} = \chi_{\mu\nu}(0)$.

23.7.2 Variation of entropy in time

For static processes the variation of entropy with averages of observables have been given in (22.40) or (23.138). These relations may actually be used for time-dependent processes as well, provided they are sufficiently slow. More precisely, the time lapse Δt over which the dynamics may be followed and for which such a concept is valid must be macroscopically small but microscopically large:

$$\tau_{\text{micro}} \ll \Delta t \ll \tau_{\text{macro}}. \tag{23.180}$$

In this spirit the derivative of entropy with respect to time may be written as

$$\frac{dS}{dt} = \beta_0 \sum_\mu \xi_\mu \frac{d}{dt} \langle \hat{A}_\mu \rangle_t^{\text{qs}} \equiv \beta_0 \sum_\mu \xi_\mu \langle \hat{J}_\mu \rangle_t^{\text{qs}}. \tag{23.181}$$

The expressions on the right still contain reversible as well as irreversible parts. In Section 4.6 of (Kubo et al., 1991a) the authors present a method of how these terms may be separated which includes a discussion about the conditions for macroscopic laws to hold true. Rather than following their approach in greater detail we shall present a shorter version which essentially leads to the same results.

Our starting point will be eqn(23.176) which involves the response function $\widetilde{\chi}_{\mu\nu}(t)$. In the following we will need it at the microscopically large time Δt, such that we may write

$$\begin{aligned}\widetilde{\chi}_{\mu\nu}(\Delta t) &= \widetilde{\chi}_{\mu\nu}(t=0) + \int_0^{\Delta t} dt \left(\frac{d}{dt}\widetilde{\chi}_{\mu\nu}(t)\right) \\ &\simeq \widetilde{\chi}_{\mu\nu}(t=0) + \lim_{\epsilon \to 0^+}\int_0^\infty dt\, e^{-\epsilon \tau}\left(\frac{d}{dt}\widetilde{\chi}_{\mu\nu}(t)\right).\end{aligned} \tag{23.182}$$

Whereas the first line reflects a mere identity the approximation in *the second line is valid for time lapses Δt much larger than the microscopic relaxation time τ_{micro} of the function $d\widetilde{\chi}_{\mu\nu}(t)/dt$*. The limit $\epsilon \to 0^+$ has to be understood in the light of condition (23.180) rather than in a genuine mathematical sense. Actually, this modification is in complete agreement with the spirit of the discussion of Section 6.4.1. There it was demonstrated how the introduction of a cutoff factor in time may be understood in the sense of a frequency average, which in the end is directly responsible for the appearance of a finite friction coefficient. This discussion is completed here by showing that such features lead to an increase of entropy. Before proceeding it may be worth noting that after using the second line of (23.182) the character of eqn(23.176) changes considerably. In this way the coefficients in front of the ξ_ν become constants–if considered on the microscopic scale. Macroscopically, however, both these coefficients as well as the ξ_ν themselves do vary in time.

For comparisons with the formulation in (Kubo et al., 1991a) we should express the response functions in terms of Mori products involving currents. Indeed, as will be proven in detail in Section 23.8 there exist the following identities

$$d\widetilde{\chi}_{\mu\nu}(t)/dt = \widetilde{\chi}_{J_\mu A_\nu}(t) = \beta_0\big(\hat{J}_\mu(t)|\hat{J}_\nu\big) = \beta\big(\hat{J}_\nu(-t)|\hat{J}_\mu\big) \qquad t>0$$
$$\widetilde{\chi}_{\mu\nu}(t=0) = -\beta_0\big(\hat{J}_\mu|\hat{A}_\nu\big) = \beta_0\big(\hat{J}_\nu|\hat{A}_\mu\big). \tag{23.183}$$

With these results inserted into (23.182) equation (23.176) turns into

$$\frac{d}{dt}\langle\hat{A}_\mu\rangle_t^{\text{macro}} = \beta_0 \sum_{\nu=1} \Big(-\big(\hat{J}_\mu|\hat{A}_\nu\big) + G_{\mu\nu}\Big)\xi_\nu = \beta_0\sum_{\nu=1} L_{\mu\nu}\xi_\nu, \tag{23.184}$$

with the definitions:

$$G_{\mu\nu} = \lim_{\epsilon\to 0^+}\int_0^\infty e^{-\epsilon\tau}\big(\hat{J}_\mu(\tau)|\hat{J}_\nu\big)d\tau \equiv L_{\mu\nu} + \big(\hat{J}_\mu|\hat{A}_\nu\big). \tag{23.185}$$

The reason for this peculiar construction becomes clear after finding the symmetries of the individual terms with respect to an exchange of the indices μ and ν. In (23.183) it was indicated already that $\big(\hat{J}_\mu|\hat{A}_\nu\big)$ is *anti*symmetric. To demonstrate that $L_{\mu\nu}$ is *symmetric* we had best look at its spectral representation. The latter as well as that for $\big(\hat{J}_\mu|\hat{A}_\nu\big)$ may easily be gained from those of the response functions derived in Section 23.3.4. From (23.65) one has

$$-\beta\big(\hat{J}_\mu|\hat{A}_\nu\big) \equiv 2i\widetilde{\chi}''_{\mu\nu}(t=0) = -\frac{i}{\hbar}\sum_{nm}\big(\rho(E_n) - \rho(E_m)\big)A^\mu_{mn}A^\nu_{nm}. \tag{23.186}$$

The identity $\beta_0\big(\hat{J}_\mu(\tau)|\hat{J}_\nu\big) \equiv \widetilde{\chi}_{J_\mu A_\nu}(t)$ implies $\beta_0 G_{\mu\nu} = \chi_{J_\mu A_\nu}(\omega=0)$ such that one gets from (23.71)

$$\beta_0 G_{\mu\nu} = \chi_{J_\mu A_\nu}(\omega=0) = \frac{i}{\hbar}\sum_{nm}\frac{\rho(E_n)-\rho(E_m)}{\Omega_{mn}+i\epsilon}\Omega_{mn}A^\mu_{mn}A^\nu_{nm}, \tag{23.187}$$

simply by replacing there A^μ_{mn} by $J^\mu_{mn} = i\Omega_{mn}A^\mu_{mn}$. Combining both expressions one gets

$$\begin{aligned}\beta_0 L_{\mu\nu} &= \frac{1}{\hbar}\sum_{nm}\big(\rho(E_n)-\rho(E_m)\big)\frac{\epsilon}{\Omega_{mn}^2+\epsilon^2}\Omega_{mn}A^\mu_{mn}A^\nu_{nm}\\ &= \frac{1}{\hbar}\sum_{nm}\big(\rho(E_n)-\rho(E_m)\big)\frac{\epsilon}{\Omega_{mn}^2+\epsilon^2}\frac{1}{\Omega_{mn}}J^\mu_{mn}J^\nu_{nm},\end{aligned} \tag{23.188}$$

where terms of order ϵ^2 have been dropped. Despite involving the spectral representations of response functions the results (23.186)-(23.188) may of course be obtained directly by using the definitions of these quantities in terms of Mori products. Remember that in both versions (23.188) the overall factors which contain the energy-dependent quantities are positive–provided both energies are different. Evidently, contributions from states of the same energy drop out of the expression in the first line and they must be excluded from the summation in the second line. Moreover, one may easily convince oneself that the $L_{\mu\nu}$ is symmetric with respect to an exchange in μ and ν. Therefore, after putting (23.184) into (23.181) one gets a finite value for the energy production

$$T_0 \frac{dS}{dt}\bigg|_{\mathrm{irr}} = \beta_0 \sum_{\mu\nu=1} \xi_\mu L_{\mu\nu} \xi_\nu \geq 0. \qquad (23.189)$$

The property of being positive semi-definite can be proven analogously to (23.64). Notice please the close analogy to formula (23.62) for energy dissipation.

Let us finally briefly comment on the difference of our derivation from that performed in (Kubo et al., 1991a). Rather than starting from the microscopic expression (23.176) for the time derivative of the averages the authors of (Kubo et al., 1991a) define its macroscopic counterpart as the ratio of differences,

$$\frac{d}{dt}\langle \hat{A}_\mu \rangle_t^{\mathrm{macro}} \simeq \frac{\langle \hat{A}_\mu \rangle_t^{\mathrm{qs}} - \langle \hat{A}_\mu \rangle_\lambda^{\mathrm{qs}}}{\Delta t}. \qquad (23.190)$$

The right-hand side is evaluated by calculating from the formulas found in the previous section the numerator $\langle \hat{A}_i \rangle_t^{\mathrm{qs}} - \langle \hat{A}_i \rangle_\lambda^{\mathrm{qs}}$ to first in Δt.

23.7.3 Time variation of the density operator

It is instructive to study the variation of the density operator and to identify those parts which are responsible for the variation of the entropy. To this end we will look at solutions of the von Neumann equation to re-derive from them formula (23.189); again our procedure is inspired by the discussion presented in (Kubo et al., 1991a). As before, the time evolution will be defined by the Hamiltonian $\hat{H}$, but where measures are taken to get irreversible behavior. For the von Neumann equation it is convenient to make use of the notion of the Liouville operator $\hat{L}$. It is defined through the commutator of $\hat{H}$ with the density operator, see e.g. (25.140). To comply with the previous discussion we like to introduce in explicit fashion a damping term with which the (modified) von Neumann equation reads

$$\frac{d\hat{\rho}(t)}{dt} + i\hat{L}\,\hat{\rho}(t) = -\epsilon\big(\hat{\rho}(t) - \hat{\rho}_{\mathrm{qs}}(\lambda(t))\big). \qquad (23.191)$$

In other words one may speak of introducing a relaxation term with a "relaxation time" $\tau_\epsilon = \epsilon^{-1}$, where ϵ has a similar meaning as before. Like in (23.185) we will in the end look for the limit $\epsilon \to 0$. A formal solution of (23.191) is given by

$$\hat{\rho}(t) = \epsilon \int_{-\infty}^{t} dt'\, \exp\left(-(\epsilon + i\hat{L})(t-t')\right) \hat{\rho}_{\mathrm{qs}}(\lambda(t')). \qquad (23.192)$$

We may note in passing that this form satisfies (23.191) independently of the precise choice of $\hat{\rho}_{\mathrm{qs}}(\lambda(t))$. In the spirit of the relaxation time approximation we should like to take an expression like (23.134) but *now* with *time-dependent* parameters $\xi_j(t)$. This means using the quasi-static density operator $\hat{\rho}_{\text{qs}}(\lambda(t)) = \exp(-\beta_0(\hat{H} + \sum_{j=0} \xi_j(t)\hat{A}_j))/Z(t)$ to represent the *local* density. In addition to this inherent time dependence there is the explicit one given by the action of the evolution operator $\exp(-i\hat{L}(t-t'))$. Before we interpret the solution (23.192) in more physical terms, let us perform a few manipulations. At first let us change the integration variable from t' to $\tau = t - t'$ to get

$$\hat{\rho}(t) = \epsilon \int_0^\infty d\tau\, e^{-\epsilon\tau} \left\{ e^{-i\hat{L}\tau} \hat{\rho}_{\text{qs}}(\lambda(t-\tau)) \right\}. \tag{23.193}$$

For small ϵ this form may again be understood as representing the expression

$$\hat{\rho}(t) \simeq e^{-i\hat{L}\Delta t} \hat{\rho}_{\text{qs}}(\lambda(t-\Delta t)). \tag{23.194}$$

The Δt is meant to fulfill condition (23.180) on the separation of timescale, implying it to be (infinitely) large on the microscopic scale. In this sense the final value theorem of Laplace transform may be applied, which has already been used in (23.167). The form (23.194) may then be interpreted such that $\hat{\rho}_{\text{qs}}(t_0)$ specifies the distribution at some initial time $t_0 = t - \Delta t$ before it evolves to its final value at time $t = t_0 + \Delta t$.

Since we are not really interested in the strict limit $\epsilon = 0$ we need to study (23.193) a bit more closely. As may easily be checked, this expression can be written in the form

$$\hat{\rho}(t) = \hat{\rho}_{\text{qs}}(\lambda(t)) + \int_0^\infty d\tau\, e^{-\epsilon\tau} \frac{d}{d\tau} \left\{ e^{-i\hat{L}\tau} \hat{\rho}_{\text{qs}}(\lambda(t-\tau)) \right\} \equiv \hat{\rho}_{\text{qs}}(\lambda(t)) + \Delta\hat{\rho}(t) \tag{23.195}$$

$$\simeq \hat{\rho}_{\text{qs}}(\lambda(t)) + \int_0^\infty d\tau\, e^{-\epsilon\tau} \frac{d}{d\tau} \left\{ e^{-i\hat{L}\tau} \hat{\rho}_{\text{qs}}(\lambda(t)) \right\} \equiv \hat{\rho}_{\text{qs}}(\lambda(t)) + \Delta\hat{\rho}_{\text{qs}}^{\text{st}}(\lambda(t)) \tag{23.196}$$

Whereas the expression in the first line is exact, in the second line the dependence of the $\hat{\rho}_{\text{qs}}(\lambda(t-\tau))$ on τ has been discarded. This approximation is valid again under the assumption of (23.180), now with respect to the change of the λ in time being small on the microscopic scale. This may be called a "stationary" situation, which explains the use of the upper index "st". Evidently, this limit concurs with the Markov approximation to the density operator $\hat{\rho}(t)$ and hence to the averages of any operators calculated from it, see e.g. eqn(23.200) below.

Next we are going to apply perturbation theory and evaluate $\hat{\rho}(t)$ to first order in the ξ_i. The quasi-static part $\hat{\rho}_{\text{qs}}(\lambda(t))$ is easily seen to be

$$\hat{\rho}_{\text{qs}}(\lambda(t)) = \hat{\rho}_{\text{qs}}(\lambda^0)\left(1 - \sum_{j=0} \xi_j(t) \int_0^{\beta_0} d\zeta\, e^{\zeta\hat{H}} \Delta\hat{A}_j\, e^{-\zeta\hat{H}}\right). \tag{23.197}$$

The same expression is to be used for the calculation of $\Delta\hat{\rho}_{\text{qs}}^{\text{st}}(\lambda(t))$, for which one needs to know the action of $\exp(-i\hat{L}(\tau))$ on the operators $\hat{A}_j$. This leads to nothing else but the time evolution in Heisenberg sense although in reverse direction, namely $\exp(-i\hat{L}(\tau))\hat{A}_j = \hat{A}_j(-\tau)$. Because of $\exp(-i\hat{L}(\tau))\hat{\rho}_{\text{qs}}(\lambda^0) = \hat{\rho}_{\text{qs}}(\lambda^0)$ only the second term of (23.197) contributes to $\Delta\hat{\rho}_{\text{qs}}^{\text{st}}(\lambda(t))$ such that one gets

$$\Delta\hat{\rho}_{\text{qs}}^{\text{st}}(\lambda(t)) = -\sum_{\nu=1}\hat{\rho}_{\text{qs}}(\lambda^0)\,\xi_\nu(t)\int_0^\infty d\tau\, e^{-\epsilon\tau}\int_0^{\beta_0}d\zeta\, e^{\zeta\hat{H}}\Delta\dot{\hat{A}}_\nu(-\tau)e^{-\zeta\hat{H}}. \qquad (23.198)$$

Evidently, the term with $j=0$ vanishes; the $\hat{H}$ itself does not vary in time. Summarizing the results one may say that the density operator $\hat{\rho}(t)$ is being replaced by the following quasi-stationary operator

$$\hat{\rho}_{\text{qs}}^{\text{st}}(\lambda(t)) = \hat{\rho}_{\text{qs}}(\lambda^0) + \Delta\hat{\rho}_{\text{qs}}(\lambda(t)) + \Delta\hat{\rho}_{\text{qs}}^{\text{st}}(\lambda(t)). \qquad (23.199)$$

As all terms depend on time *only* through the parameters $\xi_\nu(t)$ the density operator can be considered *stationary* on the *microscopic scale*, indeed. Hence, the time derivative of $\langle\hat{A}_\mu\rangle$ may be defined as the average of $d\hat{A}_\mu/dt$ calculated with $\hat{\rho}_{\text{qs}}^{\text{st}}(\lambda(t))$, which is to say as (see exercise 6)

$$\frac{d}{dt}\langle\hat{A}_\mu\rangle_t^{\text{macro}} \equiv \operatorname{tr}\hat{\rho}_{\text{qs}}^{\text{st}}(\lambda(t))\,\dot{\hat{A}}_\mu. \qquad (23.200)$$

The contribution of the first term of (23.199) vanishes. The next one, the quasi-static change $\Delta\hat{\rho}_{\text{qs}}(\lambda(t))$ of the density operator leads to those terms which in (23.184) are given by the matrix $(\hat{J}_\mu|\hat{A}_\nu)$. This is an interesting observation as these terms definitely do not contribute to the time variation of the entropy. It is only the $\Delta\hat{\rho}_{\text{qs}}^{\text{st}}(\lambda(t))$ which leads to a finite value for dS/dt. Indeed, if used in (23.181) one obtains

$$\left.\frac{dS}{dt}\right|_{\text{irr}} = \beta_0\sum_\mu \xi_\mu \left.\frac{d}{dt}\langle\hat{A}_\mu\rangle_t^{\text{macro}}\right|_{\text{irr}} = \beta_0\sum_\mu \xi_\mu\,\operatorname{tr}\Delta\hat{\rho}_{\text{qs}}^{\text{st}}(\lambda(t))\dot{\hat{A}}_\mu. \qquad (23.201)$$

Inserting here the $\Delta\hat{\rho}_{\text{qs}}^{\text{st}}$ of (23.198), the $G_{\mu\nu}$ defined in (23.185) shows up of which only its symmetric part $L_{\mu\nu}$ survives such that one gets back to (23.189). This follows from the symmetry property of Mori products $(\hat{A}(-\tau)|\hat{B}) = (\hat{A}|\hat{B}(\tau)) = (\hat{B}(\tau)|\hat{A})$ as a consequence of (23.225) and (26.29).

It is worth looking at these results from a different angle. So far the derivation of (23.189) has been based on formulas like (23.138) or (23.201) which employ the variation of entropy with that for the averages of the $\hat{A}_j$. In (23.195) and (23.196) the density was split into the quasi-static density operator $\hat{\rho}_{\text{qs}}(\lambda(t))$ plus its correction, say the $\Delta\hat{\rho}(t)$ of (23.195). For such a form a ΔS may be evaluated on the basis of eqn(22.13), which to first order in $\Delta\hat{\rho}(t)$ may be written as

$$\Delta \mathcal{S} = -\text{tr}\left(\Delta\hat{\rho}(t)\ln\hat{\rho}_{\text{qs}}(\lambda(t))\right) = \beta_0 \text{tr}\,\Delta\hat{\rho}(t) \sum_{j=0} \xi_j\,\hat{A}_j\,. \qquad (23.202)$$

Evaluating it to first order in Δt and approximating $d\mathcal{S}/dt$ by $\Delta\mathcal{S}/\Delta t$ it is easily seen that one gets back to (23.181) with the time derivatives of the averages being replaced by their "macroscopic" estimates given in (23.190). This is most easily verified by simply taking.[49] $\Delta\hat{\rho}(t) = \hat{\rho}(t) - \hat{\rho}_{\text{qs}}(\lambda(t))$ and noticing that $\langle\hat{A}_\mu\rangle_t^{\text{qs}} = \text{tr}\,\hat{\rho}(t)\hat{A}_\mu$ (this average now being calculated within the Schrödinger picture instead of the Heisenberg picture from before) and $\langle\hat{A}_\mu\rangle_\lambda^{\text{qs}} = \text{tr}\,\hat{\rho}_{\text{qs}}(\lambda(t))\hat{A}_\mu$.

23.7.4 Onsager relations for macroscopic motion

Previously in this section linear relations were derived between the time derivatives of the averages of a set of observables $\hat{A}_\mu$ and the parameters ξ_μ which define the density operator $\hat{\rho}_{\text{qs}}(\lambda)$ of (23.134) and thus in the present context the initial conditions for the averages themselves. The $\hat{\rho}_{\text{qs}}(\lambda)$ is fixed by the Hamiltonian $\hat{H}$ together with the coupling terms between the ξ_ν and the $\hat{A}_\nu$. The time evolution itself was solely determined by the $\hat{H}$, which led to the result (23.176) in which the coefficients of the linear relations are given by the response functions $\widetilde{\chi}_{\mu\nu}(t)$. As expected the energy associated with the $\hat{H}$ stays constant in time. In (23.176) the ' 'generalized forces" $-\widetilde{\chi}_{\mu\nu}(t)\xi_\nu$ still contain reversible as well as irreversible parts. To separate both we concentrated on slow motion and we introduced fluxes, which through (23.190) may be defined by ratios of differences. This led to the set of equations (23.184). In their derivation the notion "slow motion" was understood in the sense specified by condition (23.180). This allows one to apply basic relations of thermostatics whenever one looks at microscopically large times. For instance, through (23.138) one may express the ξ_ν by the derivatives of the entropy with respect to the averages $\langle\hat{A}_\mu\rangle$. In this way the set (23.184) can be written in the form.[50]

$$\frac{d}{dt}\langle\hat{A}_\mu\rangle_t = \beta_0 \sum_{\nu=1} L_{\mu\nu}\xi_\nu = \beta_0 \sum_{\nu=1} L_{\mu\nu} \frac{\partial \mathcal{S}}{\partial \langle\hat{A}_\nu\rangle}\,. \qquad (23.203)$$

These relations are commonly attributed to Onsager who suggested that "fluxes" $d\langle\hat{A}_\mu\rangle_t/dt$ should be in linear relations to the "generalized forces" expressed by the derivatives of the entropy with respect to the averages $\langle\hat{A}_\nu\rangle_t$ (see e.g. (Landau and Lifshitz, 1980), (Brenig, 1989)), (Kubo et al., 1991a), (Balian, 1992). Whenever the entropy $\mathcal{S}$ is known as a function of the averages $\langle\hat{A}_\nu\rangle$ the set of equations (23.203) can be considered a closed one of differential equations for

---

[49] As a matter of fact, for $\Delta\hat{\rho}(t)$ one may also take the integral form given in (23.195). There one may replace $\int_0^\infty d\tau \exp(-\epsilon\tau)...$ by $\int_0^{\Delta t} d\tau...$ (remember (23.185)), which may be evaluated to be $\exp(-i\hat{L}\Delta t)\hat{\rho}_{\text{qs}}(\lambda(t-\Delta t)) - \hat{\rho}_{\text{qs}}(\lambda(t)) \approx \exp(-i\hat{L}\Delta t)\hat{\rho}_{\text{qs}}(\lambda(t)) - \hat{\rho}_{\text{qs}}(\lambda(t))$. Evidently, this last expression would be obtained directly if the integral expression of (23.196) had been used.

[50] To simplify the notation we leave out the index "macro".

the unknown quantities $\langle \hat{A}_\mu \rangle_t$. Considering the $L_{\mu\nu}$ as functionals of the $\langle \hat{A}_\mu \rangle_t$ this set of equations might even be used for non-linear motion.

Let us concentrate on the linear case and try to derive closed sets of linear differential equations, first for the averages themselves and then for the $\xi_\mu(t)$. In (23.203) a mixture of $\langle \hat{A}_\mu \rangle_t$ and ξ_ν appears. By the same arguments employed before one may use eqns(23.150) to express the parameters ξ_i directly by the averages $\Delta \langle \hat{A}_j \rangle_\lambda$. Unfortunately, the set (23.203) only involves terms and parameters for $i > 0$ whereas in (23.150) or (23.143) all entries of the matrix $\mathcal{A}_{ij}$ appear. The first feature simply reflects energy conservation. Indeed, if the matrix L were formally extended to include terms with $i = 0$ (keeping the same structure as given on the very right of (23.203)) all elements L_{0j} would vanish identically. To remedy the situation we may restrict the set (23.143) to the subset $i > 0$. However, as on the right-hand side the sums involve *all* ξ_j we first need a subsidiary condition to obtain a set of equations which is invertible. Here, again, the two possibilities come to mind which we encountered in Section 23.7.1 for the definition of relaxation functions: constant temperature or constant entropy. In the first case we simply have to put $\xi_0 = 0$ and the desired set of differential equations is given by

$$\frac{d}{dt}\langle \hat{A}_\mu \rangle_t = -\sum_{\nu=1} L^X_{\mu\nu} \Delta\langle \hat{A}_\nu \rangle_t \qquad \text{for} \quad \mu = 1, 2 \ldots \qquad (23.204)$$

where here X stands for temperature with the matrix $L^T_{\mu\nu}$ being defined.[51] as $L^T = (L) \times ((\chi^T)^{-1})$. The case of constant entropy is more subtle. Evidently, this may at best be required on the quasi-static level and to order one in the ξ_j. In this case the condition to be posed can be written as

$$\Delta\beta = -\beta_0 \sum_{\nu=1} \frac{\langle \Delta\hat{H}\Delta\hat{A}_\nu \rangle}{\langle \Delta\hat{H}\Delta\hat{H} \rangle} \xi_\nu. \qquad (23.205)$$

It is identical to (23.156) if only differentials are replaced by differences, more precisely $d\beta$ by $\Delta\beta = \beta_0\xi_0$ and the $d\xi_\nu$ by ξ_ν; this feature may also be seen by inspecting the integral version of (23.147). Employing (23.205) for the set (23.143) it is seen that the linear relations between the $\Delta\langle \hat{A}_\mu \rangle^{qs}_\lambda$ and the ξ_ν are determined by the matrix elements $\chi^{ad}_{\mu\nu}$, mind (23.157) for the differential version. As a consequence, the final set of equations is found from (23.204) if only the matrix L^X is interpreted as $L^{ad} = (L) \times ((\chi^{ad})^{-1})$.

In the relations discussed so far in this subsection no time derivatives of the ξ_ν appeared. In Section 23.7.3, on the other hand, the quasi-static density operator $\hat{\rho}_{qs}(\lambda(t))$ was introduced, where the $\lambda(t)$ stands as representative for the $\xi_\mu(t)$ which are to be evaluated at time t. At any time t, the values of the $\xi_\mu(t)$ are then

---

[51] Actually, rather than (23.143) one might think to invert the time-dependent version which results after imposing the same constraints to (23.166). This would involve a time-dependent matrix $((\chi^T_{\mu\nu})^{-1})$. But to lowest order it may be replaced by the one of (23.143).

fixed by requiring the quasi-static relations to the averages $\langle \hat{A}_\nu \rangle_{\lambda(t)}$ to hold true, say in the form (23.143). Evidently, here we again have to rely on the assumption of a small microscopic timescale, or equivalently, on a sufficiently fast relaxation of the microscopic degrees of freedom. Taking that for granted, we may then relate the fluxes $d\langle \hat{A}_\mu \rangle_{\lambda(t)}/dt$ to the time derivatives of the $\xi_\mu(t)$. In principle, one may start from (23.144) to get (see exercise 7)

$$\frac{d\langle \hat{A}_i \rangle^{qs}_{\lambda(t)}}{dt} = -\sum_{j=0} \mathcal{A}_{ij} \frac{d\xi_j}{dt}, \qquad (23.206)$$

encountering the same problems as before. Therefore, in these relations the $\mathcal{A}_{ij}$ ought to be replaced by the $\chi^X_{\mu\nu}$ to get

$$\frac{d\langle \hat{A}_\mu \rangle^X_t}{dt} = -\sum_{\nu=1} \chi^X_{\mu\nu} \frac{d\xi_\nu}{dt} \qquad \text{with} \qquad X = T \text{ or "ad"}. \qquad (23.207)$$

This set may be used to rewrite (23.203) as a set of differential equations for the $\xi_\mu(t)$, namely

$$\frac{d\xi_\mu(t)}{dt} = -\sum_{\nu=1} \Lambda^X_{\mu\nu} \xi_\nu(t) \qquad \text{with} \qquad \Lambda^X = ((\chi^X)^{-1}) \times (L). \qquad (23.208)$$

Let us return to the variation of entropy for which the result (23.189) was derived above. It may be rewritten in various forms by exploiting the different linear relations obtained above, see also (Landau and Lifshitz, 1980). Firstly one may convince oneself that (23.189) is regained from (23.181) by making use of (23.207) and (23.208). These reassuring features can be summarized as

$$T_0 \frac{dS}{dt} = -\beta_0 \sum_{\mu\nu} \xi_\mu \chi^X_{\mu\nu} \frac{d\xi_\nu}{dt} = \beta_0 \sum_{\mu\nu} \xi_\mu L_{\mu\nu} \xi_\nu = T_0 \frac{dS}{dt}\bigg|_{\text{irr}}.$$

One may also write this bilinear form in terms of the $\Delta\langle \hat{A}_\nu \rangle_t$ by again invoking the set (23.143), or the integral version of (23.206), rather, which is to say $\Delta\langle \hat{A}_\mu \rangle^X = -\sum_{\nu=1} \chi^X_{\mu\nu} \xi_\nu$. In this way one gets

$$T_0 \frac{dS}{dt}\bigg|_{\text{irr}} = \beta_0 \sum_{\mu\nu} \Delta\langle \hat{A}_\mu \rangle^X \left((\chi^X)^{-1} L (\chi^X)^{-1}\right)_{\mu\nu} \Delta\langle \hat{A}_\nu \rangle^X. \qquad (23.209)$$

Of interest also are forms that are bilinear in the fluxes or the velocities $\dot{\xi}_j$. Because of the symmetries $L_{\nu\mu} \equiv L_{\mu\nu}$ one gets the identities

$$T_0 \frac{dS}{dt}\bigg|_{\text{irr}} = \beta_0 \sum_{\mu\nu} \frac{d\langle \hat{A}_\mu \rangle_t}{dt} L_{\mu\nu} \frac{d\langle \hat{A}_\nu \rangle_t}{dt} = \beta_0 \sum_{\mu\nu} \frac{d\xi_\mu}{dt} \left(\chi^X L \chi^X\right)_{\mu\nu} \frac{d\xi_\nu}{dt}.$$

$$(23.210)$$

23.8 Kubo formula for transport coefficients

Let us begin by expressing the response functions in terms of the Mori products (26.28). This is possible by simply rewriting them in the following way

$$\widetilde{\chi}_{\mu\nu}(t) = \frac{i}{\hbar}\Theta(t)\,\mathrm{tr}\,\left(\hat{\rho}\left[\hat{A}_\mu(t),\hat{A}_\nu\right]\right) = \frac{i}{\hbar}\Theta(t)\,\mathrm{tr}\,\left(\hat{A}_\mu(t)\left[\hat{A}_\nu,\hat{\rho}\right]\right). \quad (23.211)$$

Next we observe that for any operator $\hat{A}$ and its propagation in imaginary time $\hat{A}^{\mathrm{Im}}(\lambda) = \exp(\lambda\hat{H})\hat{A}\exp(-\lambda\hat{H})$ the following identities hold true:

$$\hat{A}^{\mathrm{Im}}(\beta) = \hat{A} + \int_0^\beta d\lambda\, \frac{d\hat{A}^{\mathrm{Im}}}{d\lambda}(\lambda) = \hat{A} + \int_0^\beta d\lambda\, \left[\hat{H},\hat{A}^{\mathrm{Im}}(\lambda)\right] \quad (23.212)$$

Multiplication by $e^{-\beta\hat{H}}$ from the left and rearranging the equation leads to

$$\left[\hat{A}, e^{-\beta\hat{H}}\right] = e^{-\beta\hat{H}} \int_0^\beta d\lambda\, \frac{d\hat{A}^{\mathrm{Im}}}{d\lambda}(\lambda) = e^{-\beta\hat{H}} \int_0^\beta d\lambda\, \left[\hat{H},\hat{A}^{\mathrm{Im}}(\lambda)\right] \quad (23.213)$$

The commutator on the very right is proportional to a "current", but the latter has to be defined in real time. Hence one has

$$\hat{J}_A(t) = \frac{d}{dt}\hat{A}(t) \equiv \frac{i}{\hbar}\left[\hat{H},\hat{A}(t)\right] \equiv \frac{i}{\hbar}\frac{d}{d\lambda}\hat{A}^{\mathrm{Im}}(\lambda), \quad (23.214)$$

where $\hat{A}^{\mathrm{Im}}(\lambda) = \hat{A}(t=-i\hbar\lambda)$ such that (23.213) becomes

$$\left[\hat{A}, e^{-\beta\hat{H}}\right] = \exp(-\beta\hat{H}) \int_0^\beta d\lambda\, \left(-i\hbar\hat{J}_A(t=-i\hbar\lambda)\right). \quad (23.215)$$

Using this relation in (23.211) for $t > 0$ the response function can be written as (Kubo et al., 1991a)

$$\widetilde{\chi}_{\mu\nu}(t) = \mathrm{tr}\,\hat{\rho} \int_0^\beta d\lambda\, \hat{J}_\nu(-i\hbar\lambda)\hat{A}_\mu(t) = \beta\big(\hat{J}_\nu|\hat{A}_\mu(t)\big) = -\beta\big(\hat{A}_\nu|\hat{J}_\mu(t)\big) \quad (23.216)$$

with the Mori product (26.28). The expression on the very right follows from (23.226) (see exercise 8); remember also the symmetry relation (26.29).

Let us apply these formulae to averages of currents $\hat{J}_\mu(t) = d\hat{A}_\mu(t)/dt$ where the $\hat{A}_\mu(t)$ themselves appear in the coupling operator $\hat{H}_{\mathrm{cpl}} = \sum_\nu \hat{A}_\nu a_\nu^{\mathrm{ext}}(t)$ of (23.1). As the unperturbed average of the current vanishes $\langle\hat{J}_\mu\rangle = \mathrm{tr}\,\hat{\rho}\hat{H} = 0$ eqn(23.46) reduces to

$$\langle\hat{J}_\mu\rangle_t = -\sum_\nu \int_{-\infty}^\infty \widetilde{\chi}_{J_\mu A_\nu}(t-s) a_\nu^{\mathrm{ext}}(s)ds \quad (23.217)$$

with the response function $\widetilde{\chi}_{J_\mu A_\nu}(t) = (i/\hbar)\Theta(t)\mathrm{tr}\,(\hat{\rho}[\hat{J}_\mu(t),\hat{A}_\nu])$. By differentiation with respect to time one finds after considering (23.216) and (23.225)

$$d\widetilde{\chi}_{\mu\nu}(t)/dt - 2i\delta(t)\widetilde{\chi}''_{\mu\nu}(t) = \widetilde{\chi}_{J_\mu A_\nu}(t) = \beta\big(\hat{J}_\mu(t)|\hat{J}_\nu\big) = \beta\big(\hat{J}_\nu(-t)|\hat{J}_\mu\big). \quad (23.218)$$

The example of electric conductivity: Suppose there are given a set of electric charges e_i sitting at positions $\mathbf{r}_i$. An external electric field $\mathbf{E}(t)$ then couples to the system through

$$\hat{H}_{\text{cpl}} = -\sum_i e_i \left(\hat{\mathbf{r}}_i \cdot \widehat{\mathbf{E}(t)}\right) = -\sum_\nu \sum_i e_i \hat{x}^i_\nu E_\nu(t). \tag{23.219}$$

For the averages of the components $\hat{J}_\nu = \sum_i e_i \widehat{\dot{x}^i}_\nu$ of (the operators for) the total current we may use (23.217) with $a_\nu^{\text{ext}}(s) = E_\nu(s)$. Comparing the Fourier transforms $J_\mu(\omega) = \sum_\nu \chi_{J_\mu x_\nu}(\omega) E_\nu(\omega)$ of $J_\mu(t) \equiv \langle \hat{J}_\mu \rangle_t$ with Ohm's law

$$J_\mu(\omega) = \sum_\nu \sigma_{\mu\nu}(\omega) E_\nu(\omega) \tag{23.220}$$

one realizes that the complex electric conductivity may be written in the Kubo form:

$$\sigma_{\mu\nu}(\omega) = \Phi_{\mu\nu}(\omega) = \beta \int_0^\infty dt\, e^{i\omega t} \left(\hat{J}_\mu(t)|\hat{J}_\nu\right). \tag{23.221}$$

The conductivity for direct currents is obtained as the zero frequency limit $\lim_{\omega \to 0+}$; for its evaluation see the discussion in Section 23.7.2 and recall that for this case the $-\left(\hat{J}_\mu|\hat{A}_\nu\right)$ of (23.186) vanish as the $\hat{A}_\nu$ simply stand for the electrons' coordinates. Formula (23.221) may easily be generalized to the continuous charge densities and currents as well as to electric fields which vary in space. One then applies Fourier transforms also to the coordinate, but must take care about the right order of limits, in the sense of performing $\lim_{\omega \to 0+}$ after $\lim_{\mathbf{q} \to 0}$, if necessary.

Exercises

1. Energy transfer to charged dipole: In dipole approximation (which is to say for small velocities with $|\dot{\mathbf{x}}|/c \ll 1$) the force exerted by an electromagnetic wave on a particle of charge q and mass m is given by $q\mathbf{E}(t)$. Suppose the particle moves in a harmonic potential $m\omega_0^2 \mathbf{x}^2/2$ and underlies a friction force $-(\Gamma/m)\dot{\mathbf{x}}$. Show that the Fourier transform $\mathbf{x}(\omega)$ can be expressed as

$$\mathbf{x}(\omega) = \frac{q}{m} \frac{\mathbf{E}(\omega)}{\omega_0^2 - i\omega\Gamma - \omega^2} \equiv \chi_{\text{osc}}(\omega) \left[q\mathbf{E}(\omega)\right], \tag{23.222}$$

where $\chi_{\text{osc}}(\omega)$ is the oscillator response function of (23.4).

(i) The energy transferred from the field to the oscillator per unit of time is given by $d\mathcal{E}/dt = \int \mathbf{E}\mathbf{j}\, d\mathbf{x}'$. With the dipole moment being $\mathbf{d}(t) = q\mathbf{x}(t)$ the current will becomes $\mathbf{j} = \dot{\mathbf{d}}(t)\delta(\mathbf{x}' - \mathbf{x}(t))$ such that the total energy transfer can be written as $\overline{\Delta \mathcal{E}} = \int_{-\infty}^{\infty} dt \int \mathbf{E}\mathbf{j}\, dx' \equiv \int_{-\infty}^{\infty} dt\, \mathbf{E}(t)\, \dot{\mathbf{d}}(t)$. Applying the folding integral for Fourier transforms, prove that one finally gets

$$\overline{\Delta \mathcal{E}} = \mathsf{q}^2 \int_{-\infty}^{\infty} \frac{d\omega}{2\pi}\, |\mathbf{E}(\omega)|^2\, (-i\omega)\, \chi_{\rm osc}(\omega) = \mathsf{q}^2 \int_{-\infty}^{\infty} \frac{d\omega}{2\pi}\, |\mathbf{E}(\omega)|^2\, \omega \chi''_{\rm osc}(\omega)$$

$$= \frac{\mathsf{q}^2}{m} \int_{-\infty}^{\infty} \frac{d\omega}{2\pi}\, |\mathbf{E}(\omega)|^2\, \frac{\omega^2 \Gamma}{(\omega_0^2 - \omega^2)^2 + \omega^2 \Gamma^2}. \qquad (23.223)$$

(ii) Assume to be given a strictly harmonic external field, say $\mathbf{E}(t) = \mathbf{E}_0 \cos \Omega t$. Show that the time-dependent dipole moment is given by

$$\mathbf{d}(t) = \mathsf{q}^2 \mathbf{E}_0 (\chi'_{\rm osc}(\Omega) \cos \Omega t + \chi''_{\rm osc}(\Omega) \sin \Omega t).$$

Calculate the rate of energy transfer $\overline{d\mathcal{E}/dt}$ averaged over one cycle with $T = 2\pi/\Omega$ and prove that with $\chi''_{\rm e}(\Omega) = \mathsf{q}^2 \chi''_{\rm osc}(\Omega)$ it can be written as

$$\overline{\frac{d\mathcal{E}}{dt}} \equiv \frac{1}{T} \int_t^{t+T} dt\, \mathbf{E}(t)\, \dot{\mathbf{d}}(t) = \frac{1}{2} \Omega \chi''_{\rm e}(\Omega)\, \mathbf{E}_0^2. \qquad (23.224)$$

(iii) Interpret this result in terms of the dielectric function $\varepsilon(\omega)$ of (16.5) of the Drude–Lorentz model, which follows from (23.222) and the relation of the polarization to the electric field. Use Poynting's law (Landau et al., 1984) to show that this relation between $\overline{d\mathcal{E}/dt}$ and $\varepsilon''(\Omega)$ does not only apply to the Drude–Lorentz model.

2. Proof of the validity of (23.100): Convince yourself that the spectral representation of $s_{\mu\nu}(\omega)$ is given by

$$s_{\mu\nu}(\omega) = \pi \sum_{nm} \rho(E_m)(\Delta \hat{A}^\mu)_{mn}(\Delta \hat{A}^\nu)_{nm}\, \delta(\omega - \Omega_{nm})$$

and show that $S(\omega)$ can be written as

$$S(\omega) = \pi \sum_{nm} \rho(E_m) \left| \langle m | \Delta \hat{H}_{\rm cpl}(\omega) | n \rangle \right|^2 \delta(\omega - \Omega_{nm}) \geq 0$$

where $\hat{H}_{\rm cpl}(\omega) = \sum_\nu \hat{A}_\nu a_\nu^{\rm ext}(\omega)$ (one may assume real external fields and Hermitian operators $\hat{A}_\nu$).

3. Derive (23.139) for the case of $\Delta \beta = 0$. To this end write the density operator (23.134) in the form $\hat{\rho}_{\rm qs}(a_\nu, \beta) = \hat{k}_{\rm qs}(a_\nu, \beta)/Z_{\rm qs}(a_\nu, \beta)$ and identify $\hat{k}_{\rm qs} = \hat{\mathcal{U}}_{\rm qs}(t - t_0 = -i\hbar\beta)$. The $\hat{\mathcal{U}}_{\rm qs}(t) = \exp(-i\hat{H}t/\hbar)\, \hat{\mathcal{U}}_I(t)$ can be found from (23.21). Changing there the integration variable to $\zeta = (s - t_0)i/\hbar$ one will find $\hat{k}_{\rm qs}(a_\nu, \beta) = e^{-\beta \hat{H}}(1 - \int_0^\beta e^{\zeta \hat{H}} \hat{H}_{\rm cpl} e^{-\zeta \hat{H}}\, d\zeta)$ and, hence, with $\Delta \hat{H}_{\rm cpl} = \hat{H}_{\rm cpl} - \langle \hat{H}_{\rm cpl} \rangle$

$$\hat{\rho}_{\rm qs}(a_\nu, \beta) = \hat{\rho}_{\rm qs}(a_\nu^0, \beta)(1 - \int_0^\beta e^{\zeta \hat{H}} \Delta \hat{H}_{\rm cpl} e^{-\zeta \hat{H}}\, d\zeta).$$

4. (i) Show that (23.176) can be derived from Ehrenfest's equations in the form
$$\langle \hat{J}_\mu \rangle^{qs}_t \equiv \frac{d}{dt} \langle \hat{A}_\mu \rangle^{qs}_t = \operatorname{tr} \hat{\rho}_{qs}(\lambda) \frac{i}{\hbar} \left[\hat{H}, \hat{A}_\mu(t) \right].$$
Make use of the relation $\left[\hat{\rho}_{qs}(\lambda), \hat{H} + \sum_{j=0} \xi_j \hat{A}_j \right] = 0$ to prove
$$\langle \hat{J}_\mu \rangle^{qs}_t = \sum_{j=0} \tilde{\chi}^\lambda_{\mu j}(t) \xi_j \simeq \sum_{\nu=1} \tilde{\chi}_{\mu\nu}(t) \xi_\nu \quad in\ first\ order,$$
where the function $\tilde{\chi}^\lambda_{\mu\nu}(t)$ is defined like the $\tilde{\chi}_{\mu\nu}(t)$ but with the unperturbed density operator $\hat{\rho}_{qs}(\lambda^0)$ being replaced by $\hat{\rho}_{qs}(\lambda)$.

(ii) Show that this result is in accord with the $\langle \hat{J}_\mu \rangle_t = -\sum_\nu \int_{-\infty}^\infty \tilde{\chi}_{J_\mu\nu}(t-s) a^{\text{ext}}_\nu(s) ds$ of (23.217) by choosing the "external fields" to be $a^{\text{ext}}_\nu(s) = \Theta(-s)\xi_\nu$.

5. Show that (23.177) follows naturally from the basic relation (23.166). One only has to realize that there the time-independent parameters ξ_i may be chosen such as to fulfill the condition of $\Delta \langle \hat{A}_i \rangle^{qs}_\lambda |_S$ being evaluated at constant S.

6. Show that for $\epsilon \to 0$ one gets $i\hat{L} \hat{\rho}^{\text{st}}_{qs}(\lambda(t)) = 0$. Convince yourself that in this limit (23.200) follows from $\operatorname{tr}(d\hat{\rho}(t)/dt) \hat{A}_\mu$ with $\hat{\rho}(t)$ being the solution (23.199) of (23.191).

7. Convince yourself that (23.206) may be obtained directly from
$$\frac{d\langle \hat{A}_\mu \rangle^{qs}_{\lambda(t)}}{dt} = \frac{d}{dt} \operatorname{tr}\left(\hat{A}_\mu \hat{\rho}_{qs}(\lambda(t))\right) = \sum_{j=0} \operatorname{tr}\left(\hat{A}_\mu \frac{\partial \hat{\rho}_{qs}(\lambda(t))}{\partial \xi_j}\right) \frac{d\xi_j}{dt}.$$

8. Apply the Mori product (26.28) to operators $\hat{A}(t)$ and $\hat{B}(t)$ with the time dependence being given by $\hat{A}(t) = \exp(i\hat{H}t/\hbar) \hat{A} \exp(-i\hat{H}t/\hbar)$ (and likewise for $\hat{B}(t)$) and show:
$$\left(\hat{A}(t_1) | \hat{B}(t_2)\right) = \left(\hat{A}(t_1+\tau) | \hat{B}(t_2+\tau)\right) \qquad (23.225)$$
$$\left(\dot{\hat{A}}(t_1) | \hat{B}(t_2)\right) + \left(\hat{A}(t_1) | \dot{\hat{B}}(t_2)\right) = 0 \qquad (23.226)$$

24

FUNCTIONAL INTEGRALS

In the past few decades functional integrals have become more and more important in various fields of modern theoretical physics. Indeed, many developments would have been impossible without them. For instance, this is true for the understanding of the interplay between quantum features, on the one hand, and dissipative phenomena on the other. In this chapter we want to present the main ideas and some of the important techniques which we may make use of at other places in the book. We shall begin with quantum mechanics to then pass on to statistical mechanics and discuss important issues of many-body systems. There exist several excellent textbooks and monographs in these fields, like (Schulman, 1981), (Negele and Orland, 1987), (Kleinert, 1990), beginners are strongly recommended to consult the classic one by R.P. Feynman and A.R. Hibbs (1965).

24.1 Path integrals in quantum mechanics

24.1.1 *Time propagation in quantum mechanics*

Suppose we are given the time evolution operator $\hat{\mathcal{U}}(t_b, t_a)$ of a quantal system, which for the sake of simplicity will be assumed to be one-dimensional. Lets us look at the propagation in coordinate space. The probability amplitude that in some time lapse $\Delta t = t_b - t_a$ the system moves from $|x_a\rangle$ to $|x_b\rangle$ is given by the propagator

$$k(x_b, t_b; x_a, t_a) = \begin{cases} \langle x_b | \hat{\mathcal{U}}(t_b, t_a) | x_a \rangle & \text{for } t_b \geq t_a, \\ 0 & \text{for } t_b < t_a. \end{cases} \quad (24.1)$$

It satisfies the initial condition

$$\lim_{t_b \to t_a} k(x_b, t_b; x_a, t_a) = \delta(x_b - x_a). \quad (24.2)$$

With the help of $k(x_b, t_b; x_a, t_a)$ the wave function at time t_b can be expressed through its values at t_a by

$$\Psi(x_b, t_b) = \int dx_a \, k(x_b, t_b; x_a, t_a) \, \Psi(x_a, t_a). \quad (24.3)$$

The corresponding equation for the propagator implies the "group property"

$$k(x_b, t_b; x_a, t_a) = \int dx \, k(x_b, t_b; x, t) \, k(x, t; x_a, t_a) \qquad t_b > t > t_a. \quad (24.4)$$

Here, t may be any intermediate time and the integration has to occur over the full x-space, which is to say over all possible values of x.

In passing we should like to mention that these properties are very similar to those of Markovian transport processes. There the equation which is analogous to (24.4) is referred to as the Chapman–Kolmogorov equation. The basic difference is that the latter (see (25.48)) deals with transition *probabilities* whereas (24.4) involves transition *amplitudes*.

Given a Hamiltonian $\hat{H}$ the time evolution operator $\mathcal{U}(t_b, t_a)$ is commonly derived from the corresponding Schrödinger equation (see Section 23.2). If the latter is independent of time one has $\mathcal{U}(t_b, t_a) = \exp(-i\hat{H}(t_b - t_a)/\hbar)$. Then the propagator satisfies the following differential equation

$$\left(-i\hbar \partial_{t_b} + \hat{H}_{x_b}\right) k(x_b, t_b; x_a, t_a) = -i\hbar \delta(x_b - x_a) \delta(t_b - t_a), \qquad (24.5)$$

where the operators in the bracket are to act on the propagator as function of the final values x_b and t_b (with $\partial_{t_b} \equiv \partial/\partial_{t_b}$). Equation (24.5) can easily be proved by making use of the definition (24.1).

There may be classical systems which cannot be described within a Hamiltonian picture but are described well by the Lagrange formalism. One such case is an electrically charged particle coupled to electromagnetic fields. The latter may be integrated out so that only the particle's degrees of freedom survive. The particle then undergoes complicated time-dependent interactions which cannot be put in the frame of Hamilton's equations. This problem was the starting point for R.P. Feynman to develop his access to quantum mechanics which is based on knowledge of a Lagrangian $\mathcal{L}(x, \dot{x}, t)$ rather than a Hamiltonian. Indeed, perhaps the most general formulation of time evolution for classical systems is given through canonical transformations which make use of the action integral

$$S[x(t)] = \int_{t_a}^{t_b} dt\, \mathcal{L}(x, \dot{x}, t) \qquad \text{with fixed} \qquad x_b = x(t_b) \text{ and } x_a = x(t_a). \quad (24.6)$$

Eventually, this action might be used to define the evolution operator of a quantum system, again through canonical transformations. The essential step forward along these lines was that Feynman realized that this concept works for infinitesimally short times, whereas it was known to fail for global propagation. The interested reader is strongly recommended to read R.P. Feynman's Nobel Lecture (1972) where he beautifully explains the story of how he found this clue to improve on previous attempts by P.A.M. Dirac. Suppose we are given a particle of mass M which moves in a potential $V(x)$ such that the Lagrangian reads

$$\mathcal{L}(x, \dot{x}, t) = M\dot{x}(t)^2/2 - V(x(t), t). \qquad (24.7)$$

If then (24.3) is applied for *infinitesimally short times* $\varepsilon = t_b - t_a$, together with the *short-time propagator*

$$k_\varepsilon(x_b, x_a, \varepsilon) = \sqrt{\frac{M}{2\pi i \hbar \varepsilon}} \exp\left(\frac{i}{\hbar} \varepsilon \mathcal{L}\left(\frac{x_b - x_a}{\varepsilon}, \frac{x_b + x_a}{2}\right)\right) \qquad (24.8)$$

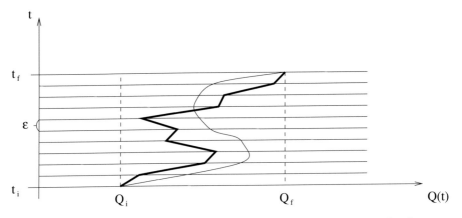

FIG. 24.1. An example for a polygonal path for real time propagation between $x(t_i \equiv t_a = 0) = Q_i$ and $x(t_f \equiv t_b = N\varepsilon) = Q_f$ with nine intermediate time steps (fully drawn line), together with the average trajectory.

the wave function $\Psi(x_b, t_b)$ can be shown to satisfy the Schrödinger equation. As noted in Section 25.2.3 the proof given in (Feynman and Hibbs, 1965) was copied to derive the Fokker–Planck equation from the Chapman–Kolmogorov equation.

To get the propagator for finite time lapses one simply has to make use of (24.4). Indeed, this property may be applied repeatedly, dividing the interval $t_b - t_a$ in N intervals of length $\varepsilon = t_n - t_{n-1}$, where n runs from $n = 1\ldots N$. With $A = \sqrt{2\pi i\hbar\varepsilon/M}$ and identifying $t_b = t_N$ and $t_a = t_0$ one ends up with

$$k(x_b, t_b; x_a, t_a) = \lim_{\substack{N\to\infty \\ N\varepsilon = t_b - t_a}} \frac{1}{A} \int_{-\infty}^{\infty} \prod_{n=1}^{N-1} \frac{dx_n}{A} \times \\ \exp\left(\frac{i}{\hbar}\varepsilon \sum_{n=1}^{N} \mathcal{L}\left(\frac{x_n - x_{n-1}}{\varepsilon}, \frac{x_n + x_{n-1}}{2}, n\varepsilon\right)\right). \quad (24.9)$$

A pictorial representation for a finite polygonal path is shown in Fig. 24.1. For $N \to \infty$ any continuous path may be represented this way. In this limit the "sum" over paths turns into the *path integral* for which one commonly adopts the following shorthand notation using the action integral of (24.6):

$$k(b; a) \equiv k(x_b, t_b; x_a, t_a) = \int_a^b \mathcal{D}[x(t)] \, \exp\left(\frac{i}{\hbar} S[x(t)]\right). \quad (24.10)$$

All these representations are also correct when the Lagrangian depends on time in explicit fashion. For a system which has a time-*independent* Hamiltonian $\hat{H}$ the path integral may also be obtained through the following form which is valid for any finite time lapse $\Delta t_n = t_n - t_{n-1}$:

$$k(x_b, t_b; x_a, t_a) = \langle x_b | \exp\bigl(-\tfrac{i}{\hbar}\hat{H}(t_b - t_a)\bigr) | x_a \rangle \equiv \langle x_b | \mathcal{K}(t_b - t_a) | x_a \rangle$$
$$\mathcal{K}(t_b - t_a) = \int \prod_{n=1}^{N} dx_n \, |x_n\rangle\langle x_n| e^{-i\hat{H}\Delta t_n/\hbar} |x_{n-1}\rangle\langle x_{n-1}|, \tag{24.11}$$

where $\langle x_b| = \langle x_N|$ and $|x_{n=0}\rangle = |x_a\rangle$. At any individual step the completeness relation $\int dx_n |x_n\rangle\langle x_n| = 1$. For $\Delta t = \varepsilon$ the $\langle x_n| \exp(-i\hat{H}\varepsilon/\hbar)|x_{n-1}\rangle$ can be seen to be identical to the form (24.8) of the short-time propagator to leading order in ε (Negele and Orland, 1987). To this order in ε one has

$$\exp(-i\hat{H}\varepsilon/\hbar) = \exp\left(-i\hat{T}\varepsilon/\hbar\right) \exp\left(-i\hat{V}\varepsilon/\hbar\right) \times + \mathcal{O}(\varepsilon^2). \tag{24.12}$$

Unfortunately, it is only for simple examples that the propagator $k(b;a)$ can be calculated analytically. The simplest example is the free particle, of course. In this case the integrals in (24.9) can simply be carried out. One finds the result

$$k(b;a) = \left[\frac{2\pi i\hbar(t_b - t_a)}{M}\right]^{-1/2} \exp\left(\frac{iM(x_b - x_a)^2}{2\hbar(t_b - t_a)}\right), \tag{24.13}$$

which says that in this case the propagator for a finite time lapse is identical in form to that of the corresponding short-time propagator. Let us state already here the result: for the oscillator with $\mathcal{L} = M\dot{x}(t)^2/2 - M\varpi^2 x^2/2$ the propagator turns out to be

$$k(x_b, x_a, t) = \sqrt{\frac{M\varpi}{2\pi\hbar i \sin(\varpi t)}} \exp\left\{\frac{iM\varpi}{2\hbar \sin(\varpi t)}\left[(x_b^2 + x_a^2)\cos(\varpi t) - 2x_b x_a\right]\right\} \tag{24.14}$$

although a proof will only be given in the next section. Notice that in this case the exponential is determined by the *classical* action (see exercise 1).

24.1.2 Semi-classical approximation to the propagator

We are now going to examine how much information about quantum physics one can find in the classical action $S[x_c(t)]$. The latter is given by (24.6) where the $x_c(t)$ represents the classical trajectory with the appropriate boundary conditions $x_a = x(t_a)$ and $x_b = x(t_b)$. Let us change to the new variable y defined through

$$x(t) = x_c(t) + y(t) \quad \text{with} \quad y(t_a) = y(t_b) = 0 \tag{24.15}$$

Expanding the action to second order in $y(t)$ one gets

$$S[x(t)] = S[x_c(t)] + \Delta S[x(t)] \approx S[x_c(t)] + \delta S[y(t); x_c(t)] + \tfrac{1}{2}\delta^2 S[y(t); x_c(t)]$$
$$= S_c + \int_{t_a}^{t_b}\left(\frac{\partial \mathcal{L}}{\partial \dot{x}_c}\dot{y} + \frac{\partial \mathcal{L}}{\partial x_c}y\right)dt + \frac{1}{2}\int_{t_a}^{t_b}\left(\frac{\partial^2 \mathcal{L}}{\partial \dot{x}_c^2}\dot{y}^2 + 2\frac{\partial^2 \mathcal{L}}{\partial \dot{x}_c \partial x_c}\dot{y}y + \frac{\partial^2 \mathcal{L}}{\partial x_c^2}y^2\right)dt,$$

where $S_c \equiv S[x_c(t)]$, δS and $\delta^2 S/2$ define the terms of order zero, one and two. For the partial derivatives of the Lagrangian we have used a shorthanded notation. The precise meaning of the cross term of second order, for instance, is

$$\frac{\partial^2 \mathcal{L}}{\partial x_c \partial \dot{x}_c} \equiv \left. \frac{\partial^2 \mathcal{L}}{\partial x \partial \dot{x}} \right|_{x=x_c(t)}.$$

The S_c is the classical action associated with $\delta S = 0$ and, hence, to the Euler-Lagrange equation

$$\delta S = 0 \quad \Longleftrightarrow \quad \frac{d}{dt}\frac{\partial \mathcal{L}}{\partial \dot{x}_c} - \frac{\partial \mathcal{L}}{\partial x_c} = 0, \tag{24.16}$$

obtained by partially integrating the term which involves $\dot{y}$. To second order $S[x(t)]$ is determined by the classical action $S_c = S[x_c(t)]$ together with the fluctuation $\delta^2 S$. For a Lagrangian which is at *most quadratic* in coordinate and velocity this expansion represents represents the *exact result*. The terms of order zero and one are completely absorbed in the classical action $S_c = S[x_c(t)]$.

Let us proceed further and evaluate the path integral (24.9). For the integrals involved the transformation (24.15) implies

$$x_i = x_c(t_i) + y_i \quad \Longrightarrow \quad \int_{-\infty}^{\infty} dx_i \quad \to \quad \int_{-\infty}^{\infty} dy_i$$

For each integral the measure simply transforms from dx_i to dy_i and the boundaries are unchanged. Likewise there is no reason to change the measure in the path integral (24.10). However, as we need to specify the initial and final points, this must be accounted for in the transformation of the entire integral by writing:

$$\mathcal{D}[x(t)] \equiv \mathcal{D}[y(t)] \quad \text{and} \quad \int_{x_a,t_a}^{x_b,t_b} \mathcal{D}[x(t)] \quad \to \quad \int_{0,t_a}^{0,t_b} \mathcal{D}[y(t)] \tag{24.17}$$

Thus to second order one gets

$$k(b;a) \equiv k(x_b,t_b;x_a,t_a) = \int_{0,t_a}^{0,t_b} \mathcal{D}[y(t)] \exp\left[\frac{i}{\hbar}\left(S_c + \frac{1}{2}\delta^2 S[y(t);x_c(t)]\right)\right].$$

This form can be rewritten in the following way

$$k(b;a) = k(x_b,t_b;x_a,t_a) = K(0,t_b;0,t_a) e^{\frac{i}{\hbar} S_c(x_b,t_b;x_a,t_a)} \tag{24.18}$$

where the first factor may be calculated from the form

$$K(0,t_b;0,t_a) = \lim_{\substack{N \to \infty \\ N\varepsilon = t_b - t_a}} \frac{1}{A} \int \prod_{k=1}^{N-1} \frac{dy_k}{A} \exp\left\{\frac{i}{2\hbar} \delta^2 S[y(t);x_c(t)]\right\}. \tag{24.19}$$

To perform these last steps one has to recognize the different nature of the two exponential factors. As the S_c does not depend on the paths the first one can be taken out of the integrals. Recall, please, that the integrand of a path integral consists of c-numbers only, so that there is no problem of non-commuting quantities. An interesting feature of (24.18) is seen in the fact that the *coordinate dependence of $k(b;a)$ is entirely given by the classical action*. The pre-factor is determined by propagating from point zero back to point zero, no matter what the x_a and x_b are. It is this pre-factor, however, which contains quantum effects.

Let us return to the general case. By applying partial integrations on may see that the fluctuation of the action can be brought to the form

$$\delta^2 S[y(t)] \equiv \int_{t_a}^{t_b} \left[\frac{\partial^2 \mathcal{L}}{\partial \dot{x}_c^2} \dot{y}^2 + 2\frac{\partial^2 \mathcal{L}}{\partial \dot{x}_c \partial x_c} \dot{y} y + \frac{\partial^2 \mathcal{L}}{\partial x_c^2} y^2 \right] dt$$
$$= \int_{t_a}^{t_b} y(t) \left[-\frac{d}{dt} \frac{\partial^2 \mathcal{L}}{\partial \dot{x}_c^2} \frac{d}{dt} - \left(\frac{d}{dt} \frac{\partial^2 \mathcal{L}}{\partial \dot{x}_c \partial x_c} \right) + \frac{\partial^2 \mathcal{L}}{\partial x_c^2} \right] y(t) \, dt. \quad (24.20)$$

Hence, by solving the differential equation

$$\mathbf{D}\, y(t) \equiv \left[-\frac{d}{dt} \frac{\partial^2 \mathcal{L}}{\partial \dot{x}_c^2} \frac{d}{dt} - \left(\frac{d}{dt} \frac{\partial^2 \mathcal{L}}{\partial x_c \partial \dot{x}_c} \right) + \frac{\partial^2 \mathcal{L}}{\partial x_c^2} \right] y(t) = \lambda y(t) \quad (24.21)$$

for the boundary conditions (24.15), $y(t_a) = y(t_b) = 0$, the fluctuation can be expressed by the eigenvalue λ as $\delta^2 S[y(t)] = \lambda \int_{t_a}^{t_b} y^2(t)\, dt$. The differential equation (24.21) is of Sturm–Liouville type, which is well studied in the mathematical literature, see e.g. (Courant and Hilbert, 1953).[52]

Suppose a set $\varphi_n(t)$ of eigenfunctions of the Sturm–Liouville equation is at our disposal, then we may expand the solution $y(t)$ to find for the $\delta^2 S$ the diagonal form expressed in terms of the eigenvalues λ_n, namely

$$y(t) = \sum_{n=1}^{\infty} a_n \varphi_n(t) \qquad \delta^2 S[y(t)] = \sum_{n=1}^{\infty} \lambda_n a_n^2 \quad (24.22)$$

This form makes it obvious that the *classical action S_c becomes a minimum if and only if there are no negative eigenvalues*.

24.1.2.1 Summation over continuous paths

We are now going to evaluate the $K(0, t_b; 0, t_a)$ by applying (24.22). Notice that this expansion defines an exact representation of the general path $y(t)$. The latter is *uniquely* determined by the

---

[52] Essential properties of the Sturm–Liouville equation may be summarized as follows: Given the differential equation $\frac{d}{dt} p(t) \frac{d}{dt} y(t) - q(t) y(t) + \lambda \rho(t) y(t) = 0$ with $p(t), \rho(t) > 0$ one may prove:
(1) The eigenvalues λ_n form a *denumerable* sequence $\lambda_1, \lambda_2, \lambda_3, \ldots$
(2) The eigenvalues are *non-degenerate*.
(3) There is at most a *finite number of negative* eigenvalues,
(3.1) for $\lambda_n < 0$ the corresponding solution φ_n is that of an *aperiodic motion*;
(3.2) for $q(t) \geq 0$ all eigenvalues are positive, $\lambda_n > 0$.
(4) The set of eigenfunctions φ_n is *complete* and may be normalized as: $\int_{t_a}^{t_b} \varphi_n(t) \varphi_m(t) \rho(t) dt = \delta_{nm}$

(in general) *infinite* set a_n. Therefore it must be possible to transform the path integral from the one over $y(t)$ to the one over the a_ns. Of course, in practice we will again have to define the infinitely dimensional path integral by first introducing time slices. For the N-fold integral this transformation then reads

$$\int \prod_{k=1}^{N-1} dy_k \exp\left\{\frac{i}{2\hbar}\delta^2 S[y(t)]\right\} = \int \prod_{k=1}^{N-1} dy_k \exp\left\{\frac{i}{2\hbar}\sum_{k=1}^{N} \lambda_k a_k^2\right\} \quad (24.23)$$

For the sake of simplicity the normalization constants A have been dropped. For any finite N the path $y(t)$ will then be approximated by an $y^{(N)}(t)$ with

$$y(t) = \lim_{N\to\infty} y^{(N)}(t) \equiv \lim_{N\to\infty} \sum_{k=1}^{N} a_k\, \varphi_k(t)$$

Please notice that the $y^{(N)}(t)$ is a *smooth* function everywhere, even for finite N, for which the first derivative of $y(t)$ jumps at the grid points. There the $y^{(N)}(t)$ takes on the values given by

$$y^{(N)}(t_l) \equiv y_l = \sum_{k=1}^{N} a_k\, \varphi_k(t_l) \quad (24.24)$$

This defines a *linear* transformation between the y_l and the a_k, with the transformation matrix being given by the $\varphi_k(t_l)$. Thus the measure of the integral transforms according to $dy_1 \cdots dy_N = \det(\varphi_k(t_l))da_1 \cdots da_N$, where $\det(\varphi_k(t_l))$ is the Jacobian $\det(\delta y/\delta a)$. It can be evaluated after approximating the integral by the corresponding Riemannian sum:

$$\int_{t_a}^{t_b} \varphi_n(t)\varphi_m(t)\,dt = \delta_{nm} \quad \Longrightarrow \quad \sum_l \varphi_n(t_l)\varphi_m(t_l)\Delta t = \delta_{nm}$$

(where $\Delta t = \epsilon$). The expression on the very right represents a multiplication of two matrices such that their determinant is given by $\det(\varphi_k(t_l)) = 1/\sqrt{(\Delta t)^N}$. For the Jacobian, on the other hand, this implies that it is simply a function of $T = t_b - t_a$ and N. Let us go back to the integral on the right-hand side of (24.23). For its essential part we may write

$$\prod_{k=1}^{N-1} \int \frac{da_k}{\sqrt{2\pi i\hbar}} \exp\left(\frac{i}{2\hbar}\lambda_k a_k^2\right) = \prod_{k=1}^{N-1} \frac{1}{\sqrt{\lambda_k}} \quad (24.25)$$

omitting for the moment the Jacobian.

Now we are ready to put together all factors to finally obtain condensed expressions for the path integral (24.19). Let us lump together the factors $1/A$ from (24.19) as well as the Jacobi determinant into a function $\Delta(N,T)$ of (N,T). As far as specific properties of our system are concerned, as represented by the

Lagrangian, this function may at most depend on the inertia M, namely through the As, but not on the potential $V(x)$. Recalling that the λ_k are nothing else but the eigenvalues of the differential operator $\mathbf{D}$ given in (24.21) the result of the path integral may finally be written as

$$\int_{0,t_a}^{0,t_b} \mathcal{D}[y(t)] \exp\left\{ \frac{i}{2\hbar} \int_{t_a}^{t_b} y(t)\,\mathbf{D}\,y(t)\,dt \right\}$$

$$\equiv \lim_{N\to\infty} \Delta(N,T) \int \prod_{l=1}^{N-1} \frac{da_l}{\sqrt{2\pi i\hbar}} \exp\left\{ \frac{i}{2\hbar} \sum_{k=1}^{N} \lambda_k a_k^2 \right\} \quad (24.26)$$

$$= \lim_{N\to\infty} \Delta(N,T) \prod_{k=1}^{N-1} \frac{1}{\sqrt{\lambda_k}} = \Delta\,[\det(\mathbf{D})]^{-1/2}$$

The Δ appearing in the last form is defined through the limiting process. The essential point is that it does *not depend on any* properties that are *specific* to our system–provided we look at systems having the same mass M. Essentially, it depends only on the basic definition or construction of the path integral itself. Of course, we already know from previous experience that the limit does exist. For simple cases we even know how to calculate it explicitly, say for a system with some given potential $W(x)$. Thus the integral for the system with the potential $V(x)$ may be obtained by exploiting the ratio

$$\frac{K_V(0,t_b;0,t_a)}{K_W(0,t_b;0,t_a)} = \frac{\int_{0,t_a}^{0,t_b} \mathcal{D}[y(t)] \exp\left\{ \frac{i}{2\hbar} \int_{t_a}^{t_b} y(t)\,\mathbf{D}_V\,y(t)\,dt \right\}}{\int_{0,t_a}^{0,t_b} \mathcal{D}[y(t)] \exp\left\{ \frac{i}{2\hbar} \int_{t_a}^{t_b} y(t)\,\mathbf{D}_W\,y(t)\,dt \right\}} = \sqrt{\frac{\det(\mathbf{D}_W)}{\det(\mathbf{D}_V)}}$$

(24.27)

In this way it is not even necessary to calculate the factor Δ itself.

Illustrative examples: For a particle in a potential $V(x)$ the differential operator $\mathbf{D}_V$ is given by

$$\mathbf{D}_V = -M\frac{d^2}{dt^2} - \left.\frac{\partial^2 V(x)}{\partial x^2}\right|_{x=x_c(t)} \quad (24.28)$$

and the Sturm–Liouville equation (24.21) becomes

$$M\ddot{y}(t) + C(t)y(t) + \lambda y(t) = 0 \quad \text{with} \quad C(t) \equiv \left.\partial^2 V(x)/\partial x^2\right|_{x=x_c(t)}.$$

(24.29)

For the free particle this reduces to $M\ddot{y}(t) + \lambda y(t) = 0$ and the solutions are

$$\varphi_n(t) = \sqrt{2/T}\sin\left(n\pi(t-t_a)/T\right), \quad \lambda_n(T) = M\left(n\pi/T\right)^2 > 0. \quad (24.30)$$

All eigenvalues are positive such that the classical trajectory $x(t) = x_a + (x_b - x_a/T)(t-t_a)$ minimizes the action. For the oscillator with, $C(t) = M\varpi^2 = \text{const}$ (24.29) becomes

$$M\ddot{\varphi}(t) + (M\varpi^2 + \lambda_n)\varphi(t) = 0. \quad (24.31)$$

Its solutions are similar to those of the free particle, only the eigenvalues change:

$$\varphi_n(t) \propto \sin\left((n\pi(t-t_a)/T)\right) \quad \text{with} \quad \lambda_n = M\left((n\pi/T)^2 - \varpi^2\right) \quad (24.32)$$

These λ_n depend on M, C and T. For given $C > 0$ (and M) there exists a minimal time $T_{min} = \pi/\varpi$ such that for $T < T_{min}$ all eigenvalues are positive, $\lambda_n(T) > 0$. Conversely, for $T > T_{min}$ there exist a finite number of negative eigenvalues. These features may to some extent be considered generic for any potential $V(x)$; we will come back to them later. We may recall from the exercise the special property the point x_b has: All trajectories which at $t = 0$ start at x_a will go through x_b at time $T = \pi/\varpi$; $x_b = x(T)$ is called *focal or conjugate point* of x_a. To simulate motion across a potential barrier let us look at the *inverted oscillator* for which (24.31) and (24.32) apply but with C changed to $C < 0$. Here, *all* eigenvalues are *positive*, no matter how large the T is. In this case *all classical trajectories minimize the action* $S[x(t)]$.

24.1.2.2 *The quantal fluctuations for the oscillator* Let us study (24.27) for the following cases, the free particle with $W = 0$ and the oscillator with $V(x) = \varpi^2 x^2/2$ (for the sake of simplicity we put the inertia equal to unity). With the eigenvalues found before, one gets

$$\det\{-d^2/dt^2\} = \prod_{n=1}^{\infty} \lambda_n = \prod_{n=1}^{\infty} (n\pi/T)^2$$

and from (24.32)

$$\det\left\{-\frac{d^2}{dt^2} - \varpi^2\right\} = \prod_{n=1}^{\infty}\left(\left(\frac{n\pi}{T}\right)^2 - \varpi^2\right) = \prod_{n=1}^{\infty}\left(\frac{n\pi}{T}\right)^2\left(1 - \frac{\varpi^2 T^2}{n^2\pi^2}\right) \quad (24.33)$$

For the ratios this implies

$$\frac{\det\{-d^2/dt^2 - \varpi^2\}}{\det\{-d^2/dt^2\}} = \frac{\prod_{n=1}^{\infty}\left(\frac{n\pi}{T}\right)^2 \prod_{n=1}^{\infty}\left(1 - \frac{\varpi^2 T^2}{n^2\pi^2}\right)}{\prod_{n=1}^{\infty}\left(\frac{n\pi}{T}\right)^2} = \prod_{n=1}^{\infty}\left(1 - \frac{\varpi^2 T^2}{n^2\pi^2}\right) \quad (24.34)$$

provided we are keen enough to divide out factors which themselves diverge. Of course we might again go back to finite N and discuss the limit $N \to \infty$ more fully. According to (24.26) this would require taking into account the factor $\Delta(N, T)$. It is only in combination with the latter that the product of eigenvalues (more precisely the square root of it) leads to finite results.

To interpret (24.34) further we make use of Euler's product formula

$$\prod_{n=1}^{\infty}\left(1 - z^2/(n^2\pi^2)\right) = \sin z/z \quad (24.35)$$

which is valid for *all* z of the complex plain. The result for the path integral for the free particle is known to be given by (24.13), which implies

$$\int_{0,t_a}^{0,t_b} \mathcal{D}[y(t)]\, \exp\left\{\frac{i}{2\hbar}\int_{t_a}^{t_b} y(t)\, \mathbf{D}_{W=0}\, y(t)\, dt\right\} = \sqrt{\frac{M}{2\pi i\hbar T}} \qquad (24.36)$$

So for the oscillator we get from (24.27) and (24.34)

$$\int_{0,t_a}^{0,t_b} \mathcal{D}[y(t)]\, \exp\left\{\frac{i}{2\hbar}\int_{t_a}^{t_b} y(t) \left\{-\frac{d^2}{dt^2} - \varpi^2\right\} y(t)\, dt\right\} = \sqrt{\frac{M\varpi}{2\pi i\hbar \sin \varpi T}}; \qquad (24.37)$$

this formula is valid for *all* ϖT (see exercise 2).

24.1.2.3 Counting $\hbar$ for the WKB approximation The propagator found by expanding the action to second order describes quantum features on a semi-classical level. It is related to the WKB approximation of quantum mechanics. In general there may be more than one classical trajectory which combines a given starting point with a fixed end point. In this case one has to sum up the amplitudes for all these possibilities, to finally get

$$K_{\text{WKB}}(x_b, t_b; x_a, t_a) \equiv \sum_{\text{class.traj.}} K(0, t_b; 0, t_a)\Big|_{\text{2nd}} e^{\frac{i}{\hbar} S_c(x_b, t_b; x_a, t_a)}. \qquad (24.38)$$

Ordering the $\hbar$s one may see the following rules to hold true:

$$\begin{aligned} K(0, t_b; 0, t_a)\Big|_{\text{2nd}} &\propto 1/\sqrt{\hbar} \\ k(x_b, t_b; x_a, t_a) &= K_{\text{WKB}}(x_b, t_b; x_a, t_a)\Big(1 + O(\hbar)\Big). \end{aligned} \qquad (24.39)$$

Proofs of these statements may for instance by found in (Schulman, 1981).

24.1.3 *The path integral as a functional*

Suppose there is a two-dimensional system with some coupling $V(q, x)$ between the two degrees of freedom. One might think, for instance of two particles of mass M and m which move along two axes, q and x. The action will be given by

$$S[q(t), x(t)] = \int_{t_a}^{t_b} \left[\frac{M}{2}\dot{q}^2 + \frac{m}{2}\dot{x}^2 - V(q, x)\right] dt \qquad (24.40)$$

and the path integral for the propagator may be written as:

$$K(q_b, x_b, t_b; q_a, x_a, t_a) = \int_a^b \mathcal{D}[q(t)] \int_a^b \mathcal{D}[x(t)]\, \exp\left[\frac{i}{\hbar} S[q(t), x(t)]\right] \qquad (24.41)$$

The latter may be evaluated in two steps. For instance, we might integrate over $x(t)$ first. In this way we would introduce an amplitude $\mathcal{T}$ by

$$\mathcal{T}[q(t)] = \int_a^b \mathcal{D}[x(t)] \left[\exp\left\{\frac{i}{\hbar}\int_{t_a}^{t_b}\left(\frac{m}{2}\dot{x}^2 - V(q,x)\right)dt\right\}\right] \qquad (24.42)$$

which is a *functional* of the path $q(t)$. The final result then becomes

$$K(q_b, x_b, t_b; q_a, x_a, t_a) = \int_a^b \mathcal{D}[q(t)] \left[\exp\left(\frac{i}{\hbar}\int_{t_a}^{t_b}\frac{M}{2}\dot{q}^2\, dt\right)\right] \mathcal{T}[q(t)]. \qquad (24.43)$$

The steps performed here might look very formal at first sight. However, they not only reflect an extremely important physical situation, but also open up a wide class of possible applications of path integrals.

One such case is the Caldeira–Leggett model (Caldeira and Leggett, 1983) which has played an immense role in the past 20 years and found application in various fields of physics. The Lagrangian of this model reads

$$\mathcal{L} = \frac{M}{2}\dot{q}^2 - V(q) + \sum_{k=1}^{N}\lambda_k\, x_k\, g(q) + \sum_{k=1}^{N}\left(\frac{m_k}{2}\dot{x}_k^2 - \frac{m_k\varpi_k^2}{2}x_k^2\right) \qquad (24.44)$$

and is of the form $\mathcal{L} = \mathcal{L}_{\text{sys}} + \mathcal{L}_{\text{coupl}} + \mathcal{L}_{\text{bath}}$. One is interested in studying the dynamics in the system which is described by the coordinate q. This system is coupled to a "heat bath" modelled by a set of N oscillators. Since this coupling is chosen to be linear in the x_k the bath variables can be "integrated out" to get an effective action for the system (see exercise 3).

24.2 Path integrals for statistical mechanics

An application of path integration to statistical mechanics is easiest whenever one deals with distributions at given temperature $T \equiv \beta^{-1}$, as for the canonical or grand canonical ensembles. To be specific we shall concentrate on the first example; a generalization is almost trivial. The statistical operator $\hat{\rho}(\beta) = \exp(-\beta\hat{H})/Z$ can be traced back to the operator $\hat{k}(\beta) = \exp(-\beta\hat{H})$ which satisfies the *Bloch equation*

$$\partial\hat{k}(\beta)/\partial\beta = -\hat{H}\,\hat{k}(\beta) \qquad \text{with} \qquad \hat{k}(0) = \hat{1},$$

where $\hat{1}$ stands for the unit operator. In terms of this $\hat{k}$ the partition function can be written as

$$Z(\beta) = \operatorname{tr}\hat{k}(\beta) = \int_{-\infty}^{\infty}\langle x|\hat{k}(\beta)|x\rangle\, dx = \int_{-\infty}^{\infty}k(x,x;\beta)\, dx. \qquad (24.45)$$

Here, on the very right the coordinate representation of the $\hat{k}(\beta)$ was introduced:

$$k(x_b, x_a; \beta) \equiv \langle x_b|\hat{k}(\beta)|x_a\rangle = \langle x_b|e^{-\beta\hat{H}}|x_a\rangle. \qquad (24.46)$$

It allows one to establish connection to path integration, simply by way of the association

$$\langle x_b|\hat{k}(\beta)|x_a\rangle \Leftrightarrow \langle x_b|\exp\left[-\frac{i}{\hbar}\hat{H}t\right]|x_a\rangle \quad \text{for} \quad t \Leftrightarrow -i\hbar\beta = -i\tau. \quad (24.47)$$

This implies that the $k(x_b, x_a; \beta)$ can be expressed by the *Euclidean path integral*

$$k(x_b, x_a; \beta) \equiv k(x_b, -i\hbar\beta; x_a, 0) \equiv K_E(x_b, \tau = \hbar\beta; x_a, 0)$$
$$= \int_{x(0)=x_a}^{x(\hbar\beta)=x_b} \mathcal{D}[x(\tau)] \exp\left\{-\frac{S_E[x(\tau)]}{\hbar}\right\} \quad (24.48)$$

Path integrals of this type involve the so-called *Euclidean action* $S_E[x(\tau)]$. Let us first see where this name comes from. In field-theoretical applications the transformation $t = -i\tau$ (with $\tau = \tau^*$) is called "Wick rotation" or transformation to the "Euclidean metric". The latter notion is related to the transformation of the *Minkowski metric* $ds^2 = -dt^2 + d\mathbf{x}^2$ of special relativity to one which is positive definite, namely $ds^2 = d\tau^2 + d\mathbf{x}^2$. Here we shall discuss the implication of such a transformation for the path integral for the example of a particle in a potential $V(x)$, which is to say for the Lagrange function $\mathcal{L}(\dot{x}, x, t) = M\dot{x}^2/2 - V(x)$. The action transforms according to

$$S[x(t)] = \int_{t_a}^{t_b} \left(\frac{M}{2}\left(\frac{dx}{dt}\right)^2 - V(x)\right) dt = i\, S_E[x(\tau)] \quad (24.49)$$

with the (real) Euclidean action

$$S_E[x(\tau)] = \int_{\tau_a}^{\tau_b} \left(\frac{M}{2}\left(\frac{dx}{d\tau}\right)^2 + V(x)\right) d\tau \quad (24.50)$$

It is observed that $S[x(t)]$ *effectively* turns into $S_E[x(\tau)]$ by replacing t by τ and by changing $V(x(t))$ simultaneously into $-V(x(\tau))$. Associated to this Euclidean action is a "classical path" $\bar{x}(\tau)$ being defined by

$$\delta S_E = 0 \quad \Longleftrightarrow \quad M\frac{d^2x}{d\tau^2} = V'(x) \equiv \frac{dV(x)}{dx} \quad (24.51)$$

The notion "classical path" simply is a shorthand notation used by analogy with the traditional case of real time propagation. It has not much to do with "classical mechanics".

With $a = (2\pi\hbar\eta/M)^{-1}$ the Euclidean path integral is defined as

$$K_E(x_b, \tau_b; x_a, \tau_a) = \int_{x(\tau_a)=x_a}^{x(\tau_b)=x_b} \mathcal{D}[x(\tau)] \exp\{-S_E[x(\tau)]/\hbar\} \quad (24.52)$$
$$= \lim_{\substack{N\to\infty \\ N\eta=\tau_b-\tau_a}} \frac{1}{a}\int \prod_{k=1}^{N-1} \frac{dx_k}{a} \exp\left\{-\frac{\eta}{\hbar}\sum_{k=0}^{N}\left[\frac{M}{2}\left(\frac{x_{k+1}-x_k}{\eta}\right)^2 + V\left(\frac{x_{k+1}+x_k}{2}\right)\right]\right\}$$

As can be seen the ϵ, which in the normal path integral defines the infinitesimal time step, has been replaced by $-i\eta$. Thus the η is defined as $\eta = (\tau_b - \tau_a)/N$ and the normalization factor $1/A$ has been replaced by $1/a$.

Let us look how the final results of Section 24.1.2.1 transform to the Euclidean case. From the form of the Euclidean action (24.50) it follows that the differential operator (24.28) is to be changed into

$$\mathbf{D}_V^E = -M\frac{d^2}{d\tau^2} + \left.\frac{\partial^2 V(x)}{\partial x^2}\right|_{x=x_c(\tau)}. \tag{24.53}$$

Hence, the eigenvalue equation (24.29) becomes

$$-M\ddot{y}(\tau) + C(\tau)y(\tau) = \lambda y(\tau) \quad \text{with} \quad C(\tau) \equiv \left.\frac{\partial^2 V(x)}{\partial x^2}\right|_{x=x_c(\tau)} \tag{24.54}$$

and $y(\tau_a) = y(\tau_b) = 0$.[53] The eigenvalues λ_k may be used in

$$\prod_{k=1}^{N} \int \frac{da_k}{\sqrt{2\pi\hbar}} \exp\left(-\frac{1}{2\hbar}\lambda_k a_k^2\right) = \prod_{k=1}^{N} \frac{1}{\sqrt{\lambda_k}}, \tag{24.55}$$

which is the transform of formula (24.25). It is observed that by these manipulations we have succeeded in throwing out the imaginary unit i everywhere (provided one does not try for pathological potentials). As the most important consequence, the individual integrals now converge in a pure, absolute sense. As a matter of fact, functional integrals of this type had already been introduced by N. Wiener in the 1920s while studying the physics of Brownian motion. For this reason, the measure involved here at times is called the *Wiener measure*.

Evidently, it is much easier to compute a functional integral of the type (24.52) than those described in previous sections, for instance in the form given by (24.9). The essential difference is the fact that we are now dealing with *real Gaussians*, as they appear from the kinetic energy terms, for instance. This holds true already for the simpler cases of bilinear Lagrangians, which can be done analytically, now without having to define especial convergence procedures. However, the real virtue of this distinct formulation comes into play when one wants to evaluate the integrals numerically. In practice, one will of course not actually be able to carry out the limit $N \to \infty$. However, for many applications this may not really be necessary in a strict sense. It is conceivable that it may be possible to approximate the correct limit by a *grid* of some *finite step size* η. How small the η may be chosen depends on the computer available and on the time one wants to spend on the computation. The finer the grid, the better will be the approximation. Suppose such a computation can be done and that in this way one gets the τ dependence of the $K_E(x_b, \tau_b; x_a, \tau_a)$ (in its approximate form, of course). The corresponding propagator for real time $k(x_b, t_b; x_a, t_a)$ may then be obtained by again making use of the analyticity properties, for the transformation $t = -i\tau$, but now in backward form.

---

[53] We may remark that the Sturm–Liouville equation retains its validity also for boundary conditions $y(\tau_a) = y(\tau_b)$ even when these values do not vanish.

Let us see what happens for the simple cases of the free particle and the harmonic oscillator. For real time propagation the propagators are given in (24.13) and (24.14), respectively. The thermal equivalents are

$$k_{\text{eq}}(x_b, x_a; \beta) = \sqrt{\frac{M}{2\pi\hbar^2\beta}} \exp\left[-\frac{M(x_b - x_a)^2}{2\hbar^2\beta}\right] \tag{24.56}$$

and

$$k_{\text{eq}}(x_b, x_a; \beta) = \langle x_b | \hat{k}(\beta) | x_a \rangle = \tag{24.57}$$

$$\sqrt{\frac{M\varpi}{2\pi\hbar\sinh(\hbar\varpi\beta)}} \exp\left\{-\frac{M\varpi}{2\hbar\sinh(\hbar\varpi\beta)}\left[(x_b^2 + x_a^2)\cosh(\hbar\varpi\beta) - 2x_b x_a\right]\right\}.$$

Notice that the structure of $\hat{k}(\beta)$ requires to have $\tau_a = 0$ and $\tau_b = \hbar\beta$ if used for the statistical operator $\hat{\rho}(\beta)$ of a physical system. Nevertheless, in a formal way we may make use of the group property (24.4), $k(b; a) = \int k(b; c)k(c; a)dx_c$, where τ_c would be some "imaginary time" between τ_b and τ_a, for instance between $\tau_a = 0$ and $\tau_b = \hbar\beta$.

For the partition function the path integral takes on a characteristic form. It comes up because the trace requires an additional integral. Indeed, from (24.45) it follows that $Z(\beta)$ can be expressed by closed integrals

$$Z(\beta) = \oint_{x(0)=x(\hbar\beta)} \mathcal{D}[x(\tau)] \exp\left\{-S_{\text{E}}\left[x(\tau)\right]/\hbar\right\} = e^{-\beta\mathcal{F}} \tag{24.58}$$

On the very right the free energy $\mathcal{F}$ has been introduced through the common relation to Z. A related type of path integral has been encountered already in the pre-factor of (24.18). In both cases the initial and final points are the same $x(0) = x(\hbar\beta)$. Whereas in the former case these points are fixed to $x(0) = x(\hbar\beta) = 0$, now they define a variable over which one needs to integrate. The structure of this functional integral becomes even clearer if we look at its representation in terms of time slices:

$$Z(\beta) = \lim_{\substack{N\to\infty \\ N\varepsilon=\tau_b-\tau_a}} \int \frac{dx}{a} \prod_{k=1}^{N-1} \frac{dx_k}{a} \exp\left\{-\frac{\eta}{\hbar}\sum_{k=0}^{N}\mathcal{L}\left(\frac{x_k - x_{k-1}}{\eta}, \frac{x_k + x_{k-1}}{2}, \tau_k\right)\right\}$$

with $\quad x = x_0 \equiv x(0) = x_N \equiv x(\hbar\beta)$

$$\tag{24.59}$$

Notice that *no additional* factor $1/a$ is left over in front of the integrals. In exercise 4 this formula is applied to the one-dimensional oscillator.

24.2.1 The classical limit of statistical mechanics

One may expect that the classical limit is reached for high temperatures, for instance in view of the result for the harmonic oscillator. For the Euclidean propagator this implies that we ought to look for "short times" $\hbar\beta$. To lowest order we would just take the *short-time propagator*, which is written as

$$k(y, x; \beta) = \sqrt{\frac{M}{2\pi\hbar^2\beta}} \exp\left[-\frac{M(y-x)^2}{2\hbar^2\beta}\right] e^{-\beta V\left(\frac{y+x}{2}\right)} \qquad (24.60)$$

For small $\hbar\beta$ this function differs from zero only for $|y - x| \ll \Delta x \equiv \hbar/\sqrt{MT}$, where Δx is the width of the Gaussian factor in (24.60), which is proportional to the *thermal wavelength* $\lambda = \sqrt{2\pi\hbar^2\beta/M}$. For a potential $V(x)$) that is smooth on the scale of this Δx we may approximate the $V((y+x)/2)$ by its value at the initial point x, which is to say

$$V\left(\frac{y+x}{2}\right) \approx V(x) \qquad \text{or} \qquad V[x(\tau)] \approx V(x), \qquad (24.61)$$

to get

$$k_c(y, x; \beta) \approx \left(\sqrt{\frac{M}{2\pi\hbar^2\beta}} \exp\left[-\frac{M(y-x)^2}{2\hbar^2\beta}\right]\right) e^{-\beta V(x)} \qquad (24.62)$$

This function corresponds to the limit of *classical statistical mechanics*. Indeed, the corresponding partition function reads

$$Z_c(\beta) = \sqrt{\frac{M}{2\pi\hbar^2\beta}} \int_{-\infty}^{\infty} dx \, e^{-\beta V(x)}. \qquad (24.63)$$

The $\hbar$ appears there because of the normalization of the phase space volume. Without knowledge of the latter, which comes from quantum mechanics, the classical partition function could not be determined absolutely. The result (24.63) was reached simply by exploiting the short-time propagator (24.60). A more correct derivation would have to begin with the full expression (24.48) of the path integral. Obviously, the final result is obtained by requiring that the *actual paths never move away in any essential means from the initial point* $x_a = x$. The precise formulation of this condition, and hence of (24.61) can be formulated as

$$\frac{1}{V}\left\langle\frac{\partial V}{\partial x}\right\rangle \Delta x \ll 1. \qquad (24.64)$$

Actually, for typical solids and liquids at room temperature this condition is satisfied (Feynman and Hibbs, 1965), implying that for this temperature solid state physics is largely governed by classical statistical mechanics.

24.2.2 Quantum corrections

Next we want to get the low order corrections to the classical partition function. For a first orientation let us look at the harmonic oscillator, again. Expanding the $\sinh(\hbar\varpi\beta/2)$ to third order in $\hbar\varpi\beta/2$ we get

$$Z_{\text{osc}} = \frac{1}{2\sinh(\hbar\varpi\beta/2)} \approx \frac{1}{\hbar\varpi\beta}\left(1 - \frac{1}{6}\left(\frac{\hbar\varpi\beta}{2}\right)^2\right). \qquad (24.65)$$

The first term determines the classical partition function, in accord with the result one gets from (24.63) for $V(x) = M\varpi^2/2$. The leading term of the quantum correction is given by the second term. The form which determines its relative magnitude will be encountered again below in the general case; we need only relate the ϖ^2 to the second derivative of the potential energy $V''(x) = M\varpi^2$. However, it is not yet clear how this expression will appear and at what value of the coordinate this second derivative is to be calculated.

In the previous subsection the classical limit was obtained by replacing in the path integral the potential by its initial value, see (24.61). Also for this reason it is meaningful to search for the corrections by expanding the potential, for instance about the "classical path". Following Feynman and Hibbs it may be more promising to expand about the mean position

$$\bar{x} = \frac{1}{\hbar\beta}\int_0^{\hbar\beta} x(\tau)\,d\tau \qquad \text{with} \qquad x(0) = x(\hbar\beta) = x \qquad (24.66)$$

of each path. To get the partition function we need to integrate over x. As every path (which starts and ends at x) may be classified by this average $\bar{x}$ the integral over x may be replaced by the one over $\bar{x}$, such that

$$Z = \int d\bar{x} \oint_{\substack{x(0)=x \\ x(\hbar\beta)=x}} \mathcal{D}[x(\tau)]\,\exp\left(-\frac{1}{\hbar}\int_0^{\hbar\beta}\left[\frac{M}{2}\left(\frac{dx}{d\tau}\right)^2 + V[x(\tau)]\right]d\tau\right) \qquad (24.67)$$

Expanding $V[x(\tau)]$ about $\bar{x}$ to second order the integral over τ becomes

$$\int_0^{\hbar\beta} V[x(\tau)]\,d\tau = \hbar\beta V(\bar{x}) + \int_0^{\hbar\beta}[x(\tau) - \bar{x}]V'(\bar{x})\,d\tau + \frac{1}{2}\int_0^{\hbar\beta}[x(\tau) - \bar{x}]^2 V''(\bar{x})\,d\tau \qquad (24.68)$$

Evidently, the second term vanishes. For the partition function this implies

$$Z(\beta) = \sqrt{\frac{M}{2\pi\hbar^2\beta}}\int d\bar{x}\,e^{-\beta V(\bar{x})}\,\mathcal{C}_{q.c.}(\bar{x}) \qquad (24.69)$$

with a *quantum correction factor* defined as

$$\mathcal{C}_{q.c.}(\bar{x}) = \sqrt{\frac{2\pi\hbar^2\beta}{M}}\oint_{\substack{x(0)=x \\ x(\hbar\beta)=x}} \mathcal{D}[x(\tau)] \times \exp\left(-\frac{1}{\hbar}\int_0^{\hbar\beta}\left[\frac{M}{2}\left(\frac{dx}{d\tau}\right)^2 + \frac{1}{2}[x(\tau) - \bar{x}]^2 V''(\bar{x})\right]d\tau\right). \qquad (24.70)$$

Indeed, the classical result (24.63) is regained for $\mathcal{C}_{q.c.} = 1$.

To proceed further one might introduce a variable $y(\tau) = x(\tau) - \bar{x}$ which underlies the constraint that $\int_0^{\hbar\beta} y(\tau)\, d\tau = 0$. Actually this constraint can be fulfilled automatically by exploiting the Fourier series

$$x(\tau) = \bar{x} + \sum_{r\neq 0} x_r \exp(i\nu_r \tau) \equiv \bar{x} + \sum_{r\neq 0} x_r \exp\left(i\frac{2\pi r}{\hbar\beta}\tau\right), \qquad (24.71)$$

if the ν_r are chosen to be the Matsubara frequencies $\nu_r = 2\pi r/\hbar\beta$. Obviously, the $\bar{x}$ may be identified as x_0. As the $x(\tau)$ is real the coefficients obey the relation $x_{-r} = x_r^*$. One readily verifies that one gets (with $\lambda_r = M\nu_r^2 + V''(\bar{x}) = \lambda_{-r}$)

$$S_E[x(\tau)] = \int_0^{\hbar\beta} \left[\frac{M}{2}\left(\frac{dx}{d\tau}\right)^2 + \frac{1}{2}[x(\tau) - \bar{x}]^2 V''(\bar{x})\right] d\tau = \hbar\beta \sum_{r>0} \lambda_r |x_r|^2 d. \qquad (24.72)$$

Notice, please, that these λ_r are nothing else but the eigenvalues of the Sturm–Liouville equation (24.54) and eigenfunctions $\varphi_r(\tau) \propto \exp(i\nu_r \tau)$. The latter satisfy the boundary conditions $\varphi_r(0) = \varphi_r(\hbar\beta)$. These values do not and need not vanish as the expansion has been performed around the average $\bar{x}$. Let us first assume that all eigenvalues λ_r are positive such that the resulting Gaussian integrals converge.

To actually carry out the path integral we need to know the integral measure for the variables x_r which define the Fourier series (24.71). Since in (24.72) we have restricted ourselves to positive r-values the following measure is appropriate (Feynman and Hibbs, 1965), (Kleinert, 1990):

$$\sqrt{\frac{2\pi\hbar^2\beta}{M}} \oint_{\substack{x(0)=x\\x(\hbar\beta)=x}} \mathcal{D}[x(\tau)] \cdots \equiv \lim_{N\to\infty} \int_{-\infty}^{\infty} \prod_{r=1}^{(N-1)/2} \frac{M\hbar\beta\nu_r^2}{\pi}\, dx'_r\, dx''_r \cdots . \qquad (24.73)$$

Here x'_r and x''_r are the real and imaginary parts of x_r. For each $|x_r|^2 = (x'_r)^2 + (x''_r)^2$ the integrals over x'_r and x''_r can be carried out separately. Introducing $\varpi^2(\bar{x}) = V''(\bar{x})/M$ one gets the result

$$C_{q.c.}(\bar{x}) = \lim_{\substack{N\to\infty\\N\eta=\hbar\beta}} \prod_{r=1}^{(N-1)/2} \frac{\nu_r^2}{\nu_r^2 + \varpi^2(\bar{x})} = \frac{\hbar\beta\varpi(\bar{x})}{2\sinh(\hbar\beta\varpi(\bar{x})/2)}, \qquad (24.74)$$

if for the last step Euler's product formula (24.35) is used again, now for $z \to iz$. Notice that the integral measure was written so as to contain the factors ν_r^2. In this way those divergences are removed which before in Section 24.1.2.2 were encountered for the oscillator itself. Indeed, for the latter the result (24.74) together with (24.69) becomes exact, as one would then have $V(\bar{x}) = M\varpi^2\bar{x}^2/2$ with the constant frequency $\varpi(\bar{x}) \equiv \varpi$.

Applying the expansion used in (24.65) one may evaluate the quantum corrections to second order. In this way one gets a formula known as the leading term in the *Wigner–Kirkwood expansion* for the partition function $Z(\beta)$, see e.g. (Brack and Bhaduri, 1997). Generalizing to the d-dimensional case the formula reads

$$Z_{\text{WK}}(\beta) = \left(\frac{1}{\lambda}\right)^d \int d^d x \, e^{-\beta V(\mathbf{x})} \left\{ 1 - \frac{\beta^2}{12} \frac{\hbar^2}{2m} \nabla^2 V + \mathcal{O}(\hbar^4) \right\}, \qquad (24.75)$$

with the thermal wavelength introduced in (19.58). Another approximation to (24.69) is to evaluate the $\bar{x}$ integral by the saddle point approximation. Assuming just one saddle point, one gets from (26.13) with (24.74)

$$Z_{\text{s.p.}}(\beta) = \frac{\exp(-\beta V(\bar{x}_{\text{s.p.}}))}{\hbar \beta \varpi(\bar{x}_{\text{s.p.}})} \mathcal{C}_{q.c.}(\bar{x}_{\text{s.p.}}) = \frac{\exp(-\beta V(\bar{x}_{\text{s.p.}}))}{2 \sinh(\hbar \beta \varpi(\bar{x}_{\text{s.p.}})/2)}. \qquad (24.76)$$

The condition $\lambda_r > 0$ may break down in barrier regions where the local stiffness $V''(\bar{x})$ of the potential becomes negative, mind (24.72). Evidently, the critical eigenvalue is the λ_1. Hence, the condition $\lambda_r > 0$ will be fulfilled for all r if only the temperature is larger than the *local critical temperature*

$$T_0(\bar{x}) = \frac{\hbar \sqrt{-V''(\bar{x})/M}}{2\pi} \qquad \text{for} \qquad V''(\bar{x}) < 0. \qquad (24.77)$$

Correspondingly one may introduce the *global critical temperature* T_0 as the maximal value of all $T_0(\bar{x})$. The meaning of these critical temperatures can easily be seen by inspecting the functional form on the very right of (24.74). For an unstable mode the $\varpi(\bar{x})$ becomes purely imaginary, $\varpi(\bar{x}) = i|\varpi(\bar{x})|$. Then the $\mathcal{C}_{q.c.}(\bar{x})$ of (24.74) turns into

$$\mathcal{C}_{q.c.}(\bar{x}) = \frac{\hbar \beta |\varpi(\bar{x})|}{2 \sin(\hbar \beta |\varpi(\bar{x})|/2)} \qquad \text{for} \qquad V''(\bar{x}) < 0. \qquad (24.78)$$

This form diverges at the local critical temperature $T_0(\bar{x})$ where the argument of the sine approaches π. On the other hand, with increasing temperature $\mathcal{C}_{q.c.}(\bar{x})$ turns to unity. This limit is reached whenever $\hbar \beta |\varpi(\bar{x})|/2 \ll 1$ no matter whether the local modes are stable or not. In the first case, where the $\varpi(\bar{x})$ are real, the $\mathcal{C}_{q.c.}(\bar{x})$ increases like $\hbar \beta \varpi(\bar{x})/2$ with decreasing T. For given T it is the larger the larger the frequency.

24.3 Green functions and level densities

To have a specific (and important) example in mind let us look at the motion of one particle in three-dimensional space governed by a Hamiltonian $\hat{H}$ which is assumed to be independent of time. In generalization of the one-dimensional case the propagator of (24.1) may be written as as

$$k(\mathbf{x}', \mathbf{x}; t'-t) = \begin{cases} \langle \mathbf{x}'| \exp\left(-\frac{i}{\hbar}\hat{H}(t'-t)\right) |\mathbf{x}\rangle & \text{for } t' \geq t, \\ 0 & \text{for } t' < t. \end{cases} \qquad (24.79)$$

This $k(\mathbf{x}', \mathbf{x}; t)$ may of course be expressed through path integrals, but one may as well exploit the set $|n\rangle$ of (supposedly orthonormalized) eigenstates of $\hat{H}$, which in coordinate representation will be given as $\langle \mathbf{x} | n \rangle = \phi_n(\mathbf{x})$. This allows for the form

$$k(\mathbf{x}', \mathbf{x}; t) = \sum_n \langle \mathbf{x}' | n \rangle \langle n | e^{-i\hat{H}t/\hbar} | \mathbf{x} \rangle = \sum_n \phi_n(\mathbf{x}') \phi_n^*(\mathbf{x}) e^{-i e_n t/\hbar}, \qquad (24.80)$$

with e_n being the energies. A (half-sided) Fourier transform leads to the energy-dependent Green function

$$G(\mathbf{x}', \mathbf{x}; z) = \frac{-i}{\hbar} \int_0^\infty dt\, k(\mathbf{x}', \mathbf{x}; t)\, e^{\frac{i}{\hbar} z t} = \sum_n \phi_n(\mathbf{x}') \phi_n^*(\mathbf{x}) \frac{1}{z - e_n}, \qquad (24.81)$$

defined this way for Im$\{z\} > 0$. The propagator obeys an equation like (24.5) such that the equation for the Green function becomes

$$\left(z - \hat{H}_{\mathbf{x}'} \right) G(\mathbf{x}', \mathbf{x}; z) = \delta(\mathbf{x}' - \mathbf{x}). \qquad (24.82)$$

Normalization of the eigenstates implies that the integral over $G(\mathbf{x}' = \mathbf{x}, \mathbf{x}; z)$ attains the form

$$\int d^3 x\, G(\mathbf{x}, \mathbf{x}; z) = \sum_n \frac{1}{z - e_n} \equiv \mathcal{G}(z). \qquad (24.83)$$

The function $\mathcal{G}(z)$ represents the trace of the matrix $G(\mathbf{x}', \mathbf{x}; z)$. For values of z just above the real axis one may use the Dirac formula (26.16) to find the following relation to the level density

$$g(e) = \sum_n \delta(e - e_n) = -\frac{1}{\pi} \text{Im}\,\{\mathcal{G}(e + i\eta)\} \qquad \eta > 0, \qquad (24.84)$$

which was seen before to be also given by the inverse Laplace transform of the partition function $Z(\beta)$,

$$g(e) = \mathcal{L}_e^{-1}[Z(\beta)] \equiv \frac{1}{2\pi i} \int_{\eta - i\infty}^{\eta + i\infty} d\beta\, e^{\beta e} Z(\beta) \qquad (24.85)$$

(see Section 22.2.3). Actually, the analytic structure of both expressions is quite similar as in both cases a trace formula is involved. Before we continue to study properties of the level density we shall address a semi-classical approximation to the $G(\mathbf{x}', \mathbf{x}; e)$.

24.3.1 Periodic orbit theory

As just seen the level density is obtained by summing up the amplitudes for all events where the particle starts at some arbitrary point and returns to it at some later time. Furthermore one needs to integrate over all times after introducing some definite weighting factor as defined by a half-sided Fourier or

Green functions and level densities

Laplace transform. So far, however, no "orbits" are visible. They come into play as soon as one evaluates the propagator or the Green function within the semi-classical approximation. In generalization of (24.39) we may write $k(\mathbf{y}, \mathbf{x}; t) = K_{\text{WKB}}(\mathbf{y}, \mathbf{x}; t)(1 + O(\hbar))$, where the WKB-propagator is given by:

$$K_{\text{WKB}}(\mathbf{y}, \mathbf{x}; t) = \sum_\alpha \sqrt{\det\left(\frac{i}{2\pi\hbar}\frac{\partial^2 S_\alpha}{\partial \mathbf{y} \partial \mathbf{x}}\right)} \exp\left(\frac{i}{\hbar} S_\alpha(\mathbf{y}, \mathbf{x}; t)\right)$$

$$= \sum_\alpha \left(\frac{i}{2\pi\hbar}\right)^{d/2} \sqrt{\left|\det\frac{\partial^2 S_\alpha}{\partial \mathbf{y} \partial \mathbf{x}}\right|} \exp\left(\frac{i}{\hbar} S_\alpha(\mathbf{y}, \mathbf{x}; t) - i\kappa\frac{\pi}{2}\right). \tag{24.86}$$

The action $S_\alpha(\mathbf{y}, \mathbf{x}; t)$ is a solution of the Hamilton–Jacobi equation

$$\frac{\partial S(\mathbf{y}, \mathbf{x}; t)}{\partial t} + H\left(\mathbf{y}, \frac{\partial S(\mathbf{y}, \mathbf{x}; t)}{\partial \mathbf{y}}\right) = 0. \tag{24.87}$$

In general, there may be many such S_α which correspond to classical trajectories α which start at $\mathbf{x}$ at $t = 0$ and end up at $\mathbf{y}$ at time t. The equations have been written here for a d-dimensional space for the vectors $\mathbf{x}$. The κ appearing in the second line of (24.86) is the number of negative eigenvalues of the matrix $\partial^2 S_\alpha / \partial \mathbf{y} \partial \mathbf{x}$ (which are actually determined by the orders of the so-called "focal points" along the trajectory α). The determinant itself is often referred to as the *van Vleck determinant*.

To obtain the Green function $G_{\text{WKB}}(\mathbf{y}, \mathbf{x}; e + i\eta)$ one needs to evaluate the Fourier integral (24.81) on the same level of approximation as used before. As this goes along with $\hbar \to 0$ it amounts to applying the method of *stationary phase* (see Section 26.2) which, in the limit $\eta \to 0$, leads to the following condition

$$\frac{\partial}{\partial t}\Big(et + S_\alpha(\mathbf{y}, \mathbf{x}; t)\Big) = 0. \tag{24.88}$$

By way of the Hamilton–Jacobi equation (24.87) this relates the *quantal energy* e to the *classical Hamilton function* $H(\mathbf{x}, \mathbf{p})$ by

$$e = -\frac{\partial S_\alpha(\mathbf{y}, \mathbf{x}; t_{\alpha\beta})}{\partial t}. \tag{24.89}$$

Here, a second index β appears simply because for given e and α several values of t may satisfy this equation. Carrying out the stationary phase integral one gets (see e.g. Chapter 18 of (Schulman, 1981); remember, however, that there the Fourier transform (24.81) is defined without the $-i$)

$$G_{\text{WKB}}(\mathbf{y}, \mathbf{x}; e) = \sum_{\alpha\beta} \left(\frac{i}{2\pi\hbar}\right)^{(d/2-1)} \sqrt{|\widetilde{D}_{\alpha\beta}|} \exp\left(\frac{i}{\hbar} W_{\alpha\beta} - i\kappa\frac{\pi}{2}\right), \tag{24.90}$$

with

$$\widetilde{D}_{\alpha\beta} = \det \begin{pmatrix} \frac{\partial^2 W}{\partial y \partial x} & \frac{\partial^2 W}{\partial y \partial e} \\ \frac{\partial^2 W}{\partial x \partial e} & \frac{\partial^2 W}{\partial e^2} \end{pmatrix}_{\alpha\beta}. \quad (24.91)$$

In these final steps the action $S(\mathbf{y}, \mathbf{x}; t)$ was replaced by its analogue $W(\mathbf{y}, \mathbf{x}; e)$, which describes classical motion at fixed energy e. Both are related to each other by a Legendre transformation

$$S(\mathbf{y}, \mathbf{x}; t) = W(\mathbf{y}, \mathbf{x}; e) - et. \quad (24.92)$$

Finally, to get the $\mathcal{G}_{\text{WKB}}(e)$ required for the level density in (24.84) one needs to calculate the integral $\int d^d x\, G_{\text{WKB}}(\mathbf{x}, \mathbf{x}; e)$ applying the method of stationary phase, once more. The phase appearing here becomes stationary when the following condition is fulfilled

$$0 = \left(\frac{\partial W(\mathbf{y}, \mathbf{x}; e)}{\partial \mathbf{y}} + \frac{\partial W(\mathbf{y}, \mathbf{x}; e)}{\partial \mathbf{x}} \right)_{\mathbf{x}=\mathbf{y}} \equiv \mathbf{p}_y - \mathbf{p}. \quad (24.93)$$

On the very right we just identified the derivatives of the action with respect to initial and final coordinates as the corresponding momenta, with a minus sign in the first case, as demanded by Hamilton–Jacobi theory. It is seen that, finally, only *periodic orbits* contribute. This feature is reached after performing the following two steps: (i) *Closed orbits* come up when calculating the trace, for which one needs to put $(\mathbf{x}' \equiv)\mathbf{y} = \mathbf{x}$. (ii) *Periodicity* is then guaranteed because the momenta, too, become identical at the initial and final time as soon as one evaluates the trace in semi-classical fashion, namely by way of the method of stationary phase.

24.3.2 The level density for regular and chaotic motion

In Chapter 5 a general feature of the level density was exploited which says that it may be split into a smooth and an oscillating part, $g(e) = \bar{g}(e) + \delta g(e)$. By the very definition, the $\bar{g}(e)$ is that part which survives averaging over an energy interval big enough to wash out typical quantum effects. Commonly it is identified as the level density in Thomas–Fermi approximation, or its extended version, see eqs. (5.25) and (5.26). Within periodic orbit theory the $\delta g(e)$, on the other hand is given by

$$\delta g(e) = \sum_{p.p.o} \sum_{k=1}^{\infty} \mathcal{A}_{p.p.o}^{(k)}(e) \cos\left(\frac{k}{\hbar} W_{p.p.o}(e) - \sigma_{p.p.o}^{(k)} \frac{\pi}{2} \right), \quad (5.28)$$

which is usually referred to as the *Gutzwiller trace formula*. Here, the summation extends over all periodic orbits. They consist of the "primitive" orbits, classified as "p.p.o", and those where the former are traversed k times. Characteristic details of these trajectories enter the amplitudes $\mathcal{A}_{p.p.o}^{(k)}(e)$, the actions $W_{p.p.o}(e)$

as well as the phase indices $\sigma_{p.p.o}^{(k)}$, which can all be deduced from the *classical* actions of the trajectories. Note, however, that the phase indices in (5.28), sometimes called "Maslov indices", are different from the n of (24.90), simply because formula (5.28) involves further integrations within the stationary phase method which come in building the trace in (24.83). For more details of the derivation of (5.28) as well as for its general interpretation, the reader is referred to the literature, in particular to (Gutzwiller, 1990) and (Brack and Bhaduri, 1997). In this latter reference much care is taken to elaborate on the differences which come about when the orbits are isolated or non-isolated. The latter feature is given in case that there are symmetries, which in the quantum case lead to the well-known degeneracies. One of the first generalizations of the original trace formula to this latter case, which for nuclear physics is quite important was worked out by V.M. Strutinsky and A.G. Magner in (1987). Whereas this work dealt with spherical shapes, in more recent times one has been able to study deformed systems, see e.g. (Magner et al., 2002) and references given therein.

For simple regular systems a form like (5.17) may be derived by application of the Poisson summation formula (see e.g. (Lighthill, 1962))

$$\sum_{k=0}^{\infty} \delta(e-k) = 1 + 2\sum_{n=1}^{\infty} \cos(2\pi n e) \qquad e \geq 0. \qquad (24.94)$$

For the oscillator of frequency ϖ in one and three dimensions (of spherical symmetry) one finds (see exercise 5)

$$g(e) = \begin{cases} (\hbar\varpi)^{-1}\left\{1 + 2\sum_{k=1}^{\infty}(-1)^k \cos\left(2\pi k \frac{e}{\hbar\varpi}\right)\right\} & d=1 \\ (\hbar\varpi)^{-3}\left(e^2/2 - (\hbar\varpi)^2/8\right)\left\{1 + 2\sum_{k=1}^{\infty}(-1)^k \cos\left(2\pi k \frac{e}{\hbar\varpi}\right)\right\} & d=3. \end{cases} \qquad (24.95)$$

Both are of the structure of (5.17) with the smooth parts being given by the factors in front of the curly brackets. They agree with the level density (5.27) in *extended Thomas–Fermi approximation (ETF)*, mind exercise 5.1.

An interesting example is the infinite square well in one dimension for which both the $k(x, y; t)$ as well as the $G(x, y; e)$ can be calculated analytically in direct fashion (see exercise 6). This allows one to get an exact expression for $g(e)$ including both the smooth as well as the fluctuating part.

24.4 Functional integrals for many-body systems

24.4.1 The Hubbard–Stratonovich transformation

Many-body systems are essentially governed by two-body interactions, such that in general no exact solution is possible. For this reason one has to rely on approximations. Within the functional integral approach an elegant starting point is delivered by the Hubbard–Stratonovich transformation. It allows one to remove this two-body interaction at the expense of the introduction of new degrees of freedom. In this sense it shows similarities to the Bohm–Pines method of Section 7.5.1.1, but there are also essential differences to which we shall come to below. The essential trick is first to rewrite the general two-body interaction (3.10) in one of the separable forms (3.11) to (3.15), depending on the application one aims at, see e.g. (Negele and Orland, 1987) where references to original literature can be found. The eventually resulting one-body part may be added to the always present $\hat{H}_0$. For our purpose it may suffice to start with the simpler form given in (3.16). Actually, since we are going to work in the grand canonical ensemble, we may write

$$\hat{H} = \hat{K}_0 + (k/2)\hat{F}\hat{F} \quad \text{with} \quad \hat{K}_0 \equiv \hat{H}_0 - \mu\hat{N}, \qquad (24.96)$$

with the chemical potential μ and the particle number operator $\hat{N}$. The two-body part is meant to involve the generator $\hat{F}$ for collective motion.

For such a system we want to calculate the partition function Z as the trace of the operator $\hat{k}(\beta) = \exp(-\beta\hat{H}) = (\exp(-\eta\hat{H}/\hbar))^N$. The product form will again help to express Z through functional integrals in the limit $\eta \to 0$. Although $\hat{H}_0$ and $\hat{F}^2$ will not commute with each other, the associated "evolution operators" will do to lowest order in η. This property was already encountered in (24.12) for real time propagation. For the present case the relevant formula reads: $\exp(-\eta\hat{H}/\hbar) = \exp(-\eta\hat{K}_0/\hbar)\exp(-\eta k\hat{F}^2/(2\hbar)) + \mathcal{O}(\eta^2)$. The second factor on the right can be expressed as an integral over an auxiliary variable q simply by recalling the following relation among Gaussians:

$$\exp\left(-\frac{\eta}{\hbar}\frac{k}{2}\hat{F}^2\right) = \sqrt{\frac{\eta}{2\pi\hbar|k|}} \int_{-\infty}^{\infty} dq \, \exp\left(\frac{\eta}{\hbar}\left(\frac{1}{2k}q^2 - q\hat{F}\right)\right). \qquad (24.97)$$

Here, it is used for operators but this does not hurt at all as those which appear here commute with each other. Moreover, as we want to concentrate on isoscalar modes for which $k < 0$ there is no convergence problem with the q-integration. Such a replacement may be done for every time slice to get

$$\hat{k}(\beta) = \prod_{n=1}^{N} \left[\exp\left(-\frac{\eta}{\hbar}\hat{K}_0\right) \sqrt{\frac{\eta}{2\pi\hbar|k|}} \int_{-\infty}^{\infty} dq_n \, \exp\left(\frac{\eta}{\hbar}\left(\frac{1}{2k}q_n^2 - q_n\hat{F}\right)\right) \right]$$
$$= \prod_{n=1}^{N} \left[\int_{-\infty}^{\infty} \sqrt{\frac{\eta}{2\pi\hbar|k|}} dq_n \, \exp\left(\frac{\eta}{\hbar}\left(\frac{1}{2k}q_n^2 - \hat{K}_0 - q_n\hat{F}\right)\right) \right]. \qquad (24.98)$$

Strictly speaking this relation is correct only in the limit of infinitesimal η. For $N \to \infty$ or $\eta \to 0$ (with $N\eta = \hbar\beta$) we may switch to the continuous limit identifying the discrete values q_n as representations of a continuous function $q(\tau)$ with $q_n = q(\tau = n\eta)$. In this way sums over n turn into integrals over the imaginary time τ. Finally, (24.98) may be written as

$$\hat{k}(\beta) = \int \mathcal{D}[q(\tau)] \, \exp\left(\frac{1}{2\hbar k}\int_0^{\hbar\beta} d\tau \, q(\tau)^2\right) \hat{U}_{\text{HS}}[q(\tau)] \qquad (24.99)$$

where the integral measure is defined as (see exercise 7)

$$\int_{-\infty}^{\infty} \mathcal{D}[q(\tau)] \cdots = \lim_{N\to\infty} \int_{-\infty}^{\infty} \prod_{n=1}^{N} \sqrt{\frac{\eta}{2\pi\hbar|k|}} \, dq_n \cdots \qquad (24.100)$$

and where the imaginary time propagator $\hat{U}_{\mathrm{HS}}[q(\tau)]$ stands for

$$\hat{U}_{\mathrm{HS}}[q(\tau)] = \mathcal{T} \exp\left(-\frac{1}{\hbar}\int_0^{\hbar\beta} d\tau\, \hat{H}_{\mathrm{HS}}[q(\tau)]\right) \equiv \lim_{\eta\to 0} \exp\left(-\frac{\eta}{\hbar}\sum_n\left(\hat{K}_0 + q_n\hat{F}\right)\right) \qquad (24.101)$$

with the one-body Hamiltonian

$$\hat{H}_{\mathrm{HS}}[q(\tau)] = \hat{K}_0 + q(\tau)\,\hat{F}. \qquad (24.102)$$

In formula (24.101) the time-ordering operator $\mathcal{T}$ is introduced. Its role is simply to make sure that the sequence of time slices follows those of the original version (24.98). Please notice that so far no path integral representation is given for the nucleonic degrees of freedom. Their evolution operator $\hat{U}_{\mathrm{HS}}[q(\tau)]$ is still formulated with the operators $\hat{K}_0$ and $\hat{F}$. This will change as soon as we seek to evaluate average quantities like the partition function $Z(\beta)$. One must bear in mind that finally the intrinsic operators will have to be evaluated at different intermediate times. This follows from the definition of $\hat{U}_{\mathrm{HS}}[q(\tau)]$ as a product of operators at different time slices. In fact this feature is already encountered for the simple case discussed in Section 24.1.1, namely in using (24.12) in (24.11) and performing the continuous limit finally.

Formally, the $Z(\beta)$ may be written in the form (24.58), namely

$$Z(\beta) = \oint_{q(0)=q(\hbar\beta)} \mathcal{D}[q(\tau)] \, \exp\left[-S_{\mathrm{E}}[q(\tau)]/\hbar\right] \qquad (24.103)$$

if we only introduce the effective Euclidean action

$$S_{\mathrm{E}}[q(\tau)] = -\frac{1}{2k}\int_0^{\hbar\beta} d\tau \left[q^2(\tau) - \hbar \ln \operatorname{tr} \hat{U}_{\mathrm{HS}}[q(\tau)]\right]. \qquad (24.104)$$

Now in $\operatorname{tr} \hat{U}_{\mathrm{HS}}$ the trace is to be built over all nucleonic degrees of freedom which appear in the Hamiltonian $\hat{\mathcal{H}}_{\mathrm{HS}}$ and thus in the $\hat{U}_{\mathrm{HS}}$. So far nothing has been said about how this trace may be evaluated. It is clear, however, that one is dealing only with a *one-body problem*, as the two-body interaction has successfully been removed through (24.97). Instead of the two-body interaction there now appears a *"time"-dependent mean field* and an effective potential energy $q(\tau)^2/2k$; this is the essence of the *Hubbard–Stratonovich transformation*. The q may be interpreted as an additional but not superfluous variable which represents collective motion. In this sense this procedure can be said to be similar to the Bohm–Pines method described in Section 7.5.1.1. There, however, a *collective inertia* is also introduced. This goes along with the fact that in the Bohm–Pines Hamiltonian (7.50) besides the qF a second coupling operator appears which is linear in the momentum or velocity. Finally, we should like to draw the readers' attention to the special form of the integral measure (24.100) which is different from the one we got acquainted with above. Firstly, no inertia appears, a fact which reflects the properties just mentioned. Secondly, the infinitesimal quantity η shows up in the *numerator* rather than in the denominator of the square root.

24.4.2 The high temperature limit and quantum corrections

To evaluate the contributions to $Z(\beta)$ from the $\hat{U}_{\mathrm{HS}}[q(\tau)]$ one proceeds like in the case of a particle in a potential described in sections 24.2.1 and 24.2.2. One introduces the average $\bar{q}$ defined by an equation like (24.66) and expands $q(\tau)$ about it by a Fourier series like (24.71), with the same Matsubara frequencies $\nu_r = 2\pi r/\hbar\beta$ and again with $q_r^* = q_{-r}$. With $q_r' = \mathrm{Re}(q_r)$ and $q_r'' = \mathrm{Im}(q_r)$ the appropriate integral measure is then given by (see exercise 7)

$$\mathcal{D}[q(\tau)] = \sqrt{\frac{\beta}{2\pi|k|}}\, d\bar{q} \cdot \lim_{N\to\infty} \prod_{r=1}^{(N-1)/2} \frac{\beta}{\pi|k|}\, dq_r'\, dq_r'' \equiv \sqrt{\frac{\beta}{2\pi|k|}}\, d\bar{q} \cdot \overline{\mathcal{D}}[q(\tau)]. \tag{24.105}$$

The key element for getting the partition function (24.103) is to know the trace over the evolution operator $\hat{U}_{\mathrm{HS}}$. The separation of the path $q(\tau) = \bar{q} + \Delta q(\tau)$ into a static and a τ-dependent term implies a similar splitting for the Hamiltonian of (24.102), namely $\hat{H}_{\mathrm{HS}}[q(\tau)] = \hat{H}_{\mathrm{HS}}(\bar{q}) + \Delta q(\tau)\hat{F}$. This invites one to work with the interaction picture and to introduce the function

$$Z(\beta, [q(\tau)]) = \mathrm{tr}\left[\exp\left(-\beta\left(\hat{H}_{\mathrm{HS}}(\bar{q}) - \bar{q}^2/2k\right)\right) \times \left\{\mathcal{T}\exp\left(-\frac{1}{\hbar}\int_0^{\hbar\beta} d\tau\, \Delta q(\tau)\hat{F}(\tau)\right)\right\}\right] \tag{24.106}$$

where $\hat{F}(\tau) = \exp(\hat{H}_{\mathrm{HS}}(\bar{q})\tau/\hbar)\hat{F}\exp(-\hat{H}_{\mathrm{HS}}(\bar{q})\tau/\hbar)$. This follows immediately from the form (23.26) derived in Section 23.2 for real time propagation by simply applying the "Wick rotation" to imaginary time.

For convenience the term $\bar{q}^2/2k$ was included in the exponent of the first factor. The trace over this factor alone defines the partition function

$$Z^{\mathrm{SPA}}(\beta, \bar{q}) = \mathrm{tr}\, \exp\left(-\beta\left(\hat{H}_{\mathrm{HS}}(\bar{q}) - \frac{1}{2k}\bar{q}^2\right)\right) \equiv \exp\left(-\beta\mathcal{F}^{\mathrm{SPA}}(\beta, \bar{q})\right). \tag{24.107}$$

for the system if all contributions from the dynamic part $\Delta q(\tau)$ are discarded. In the literature,[54] this approximation is called the "static path approximation" (SPA) (Arve et al., 1988). A similar method had been applied before

---

[54] Our notation may differ slightly from others. Whereas the $Z^{\mathrm{SPA}}(\beta, \bar{q})$ of (24.107) still depends on $\bar{q}$, the name SPA-partition function is also used for the $Z(\beta)$ which is obtained after the dependence on $\bar{q}$ is integrated out according to (24.108).

in (Mühlschlegel et al., 1972) to small superconducting particles. Actually, the $\hat{H}_{\text{HS}}(\bar{q})$ is the Hamiltonian of the *static* mean-field (or Hartree) approximation to the two-body interaction present in (24.96). The c-number term $-\bar{q}^2/(2k)$ is nothing else but the correction needed in this approximation to get the correct total energy of the system. The function $\mathcal{F}^{\text{SPA}}(\beta,\bar{q})$ introduced in (24.107) represents the free energy of the system at fixed $\bar{q}$. It might thus be interpreted as the system's potential energy (at fixed temperature) for the collective degree of freedom $q(\tau)$; recall the discussion in Chapter 6 and in Section 6.2.3, in particular. Taking this for granted, there is, however, the *important* difference from (24.63) seen in the pre-factor of the integral. Whereas (24.63) corresponds to a real *dynamical* system, which has an inertia M, any trace of the latter is missing in the present case, see e.g. (24.108).

All τ-dependent contributions, which are functionals of the fluctuation $\Delta q(\tau) = q(\tau) - \bar{q} = \sum_{r \neq 0} q_r \exp(i\nu_r \tau)$ may again be put into a quantum correction factor $\mathcal{C}_{q.c.}(\beta,\bar{q})$. Hence, the $Z(\beta)$ gets a form analogous to (24.69), namely,

$$Z(\beta) = \sqrt{\frac{\beta}{2\pi |k|}} \int d\bar{q} \, \exp\left(-\beta \mathcal{F}^{\text{SPA}}(\beta,\bar{q})\right) \mathcal{C}_{q.c.}(\beta,\bar{q}) \,. \tag{24.108}$$

Unlike (24.70) here the factor $\mathcal{C}_{q.c.}(\beta,\bar{q})$ will be defined without implying any approximation, namely as

$$\mathcal{C}_{q.c.}(\beta,\bar{q}) = \oint_{\substack{q(0)=q \\ q(\hbar\beta)=q}} \overline{\mathcal{D}}[q(\tau)] \, \exp(-\Delta S_{\text{E}}/\hbar) \,, \tag{24.109}$$

with the fluctuation of the action being given by

$$\Delta S_{\text{E}}[q(\tau)] = -\hbar \left(\frac{\beta}{2k} \sum_{r \neq 0} |q_r|^2 + \ln \frac{Z(\beta,[q(\tau)])}{Z^{\text{SPA}}(\beta,\bar{q})} \right) \,. \tag{24.110}$$

The $\Delta S_{\text{E}}[q(\tau)]$ may be expanded into powers of the $\Delta q(\tau)$. In this way the expansion

$$\ln \frac{Z(\beta,[q(\tau)])}{Z^{\text{SPA}}(\beta,\bar{q})} = \sum_{n=1}^{\infty} \frac{1}{n!} \int_0^{\hbar\beta} d\tau_1 \ldots d\tau_n \Delta q(\tau_1) \ldots \Delta q(\tau_n) \, \Xi_n(\tau_1,\ldots \tau_n) \tag{24.111}$$

can be used where

$$\Xi_n(\tau_1,\ldots \tau_n) \equiv \left. \frac{\delta^n \ln Z(\beta,[q(\tau)])}{\delta q(\tau_1) \delta q(\tau_2) \ldots \delta q(\tau_n)} \right|_{\Delta q(\tau)=0} \,. \tag{24.112}$$

A closer look shows one that it is nothing else but a cumulant expansion for the functional $Z(\beta,[q(\tau)])$. It has similar properties to the one described in Section 25.1.3 for exponential functions of stochastic variables. Here, the cumulants Ξ_n

are the expectation values of products of $\hat{F}(\tau_k)$ which are to be evaluated with the density operator of the SPA,

$$\hat{\rho}_{\rm HS}(\beta,\overline{q}) = \exp\bigl(-\beta(\hat{H}_{\rm HS}(\overline{q}) - \overline{q}^2/2k)\bigr)/Z^{\rm SPA}(\beta,\overline{q}). \tag{24.113}$$

To get these cumulants one may benefit from the presence of the time-ordering operator. The $\hat{F}(\tau_k)$ defined at different times may be treated as commuting objects if finally the $\mathcal{T}$ is placed at the left of the product whose expectation value is to be calculated (in this sense the situation differs from that discussed in Section 22.1.2.1). For the first cumulant one simply gets $\Xi_1 = \langle \hat{\mathcal{T}} \hat{F}(\tau_1) \rangle_{\overline{q}} = \langle \hat{F} \rangle_{\overline{q}}$, the time-independent average of $\hat{F}$. A straightforward calculation shows the second and third ones to be given by:

$$\Xi_2 = \langle \hat{\mathcal{T}} \hat{F}(\tau_1)\hat{F}(\tau_2) \rangle_{\overline{q}} - \langle \hat{F} \rangle_{\overline{q}}^2, \tag{24.114}$$

$$\Xi_3 = \langle \hat{\mathcal{T}} \hat{F}(\tau_1)\hat{F}(\tau_2)\hat{F}(\tau_3) \rangle_{\overline{q}} - 3\langle \hat{\mathcal{T}} \hat{F}(\tau_1)\hat{F}(\tau_2) \rangle_{\overline{q}} \langle \hat{F} \rangle_{\overline{q}} + 2\langle \hat{F} \rangle_{\overline{q}}^3. \tag{24.115}$$

Once these expressions are inserted into (24.111) only those terms contribute which do not have the constant factor $\langle \hat{F} \rangle_{\overline{q}}$. Indeed, the latter could be pulled out of the corresponding integral over τ, but $\int_0^{\hbar\beta} \Delta q(\tau) d\tau = 0$. The same feature holds true for the higher cumulants. We are now going to discuss in somewhat greater detail the approximation where the sum in (24.111) is terminated at $n = 2$.

Before we move on a remark is in order about the general nature of the expansion (24.111), see e.g. (Negele and Orland, 1987). If one were to use creation and annihilation operators, $\hat{c}_\alpha^\dagger$ and $\hat{c}_\alpha$, respectively, one would introduce couplings of a given system to external "currents" like $\hat{H}_{\rm cpl}(t) = \sum_\alpha (\hat{c}_\alpha^\dagger \hat{J}_\alpha(\tau) + \hat{c}_\alpha \hat{J}_\alpha^\dagger(\tau))$. In this case the $Z(\beta,[q(\tau)])/Z^{\rm SPA}(\beta,\overline{q})$ would be the generating functional for thermal Green functions defined for the $\hat{c}_\alpha^\dagger(\tau)$ and $\hat{c}_\alpha(\tau)$, whereas the analog of the expansion (24.111) would lead to the so-called *connected* (thermal) Green functions, thus replacing the Ξ_k. In this sense we may identify the $\langle \hat{\mathcal{T}} \hat{F}(\tau_1) \ldots \hat{F}(\tau_n) \rangle_{\overline{q}}$ as the Green function $\widetilde{\mathcal{G}}^{(n)}(\tau_1 \ldots \tau_n)$ of order n for the operator $\hat{F}$.

24.4.3 The perturbed static path approximation (PSPA)

As in the case of a particle in a potential discussed in Section 24.2.2, here, too, the leading quantum corrections are those which result from the inclusion of quadratic fluctuations.[55] To this order (24.111) becomes

$$\ln \frac{Z(\beta,[q(\tau)])}{Z^{\rm SPA}(\beta,\overline{q})}\bigg|_{\rm PSPA} = -\frac{1}{2\hbar} \sum_{r,s \neq 0} q_r q_s \int_0^{\hbar\beta} d\tau \int_0^{\hbar\beta} d\sigma\, e^{i\nu_r \tau} e^{i\nu_s \sigma} \widetilde{\mathcal{G}}(\overline{q}, \tau - \sigma) \tag{24.116}$$

with the thermal Green function $\widetilde{\mathcal{G}}(\overline{q}, \tau - \sigma) = -\langle \mathcal{T} \hat{F}(\tau)\hat{F}(\sigma) \rangle_{\overline{q}}/\hbar$, see Section 23.5.1. The integrals on the right of (24.116) can easily be calculated if

---

[55] In the literature different names have been used for this approximation, PSPA (for "perturbed SPA") in (Attias and Alhassid, 1997) (which we take over), RPA-SPA in (Puddu et al., 1991) and "correlated SPA" in (Rossignoli et al., 1999)).

$\tilde{\mathcal{G}}(\bar{q}, \tau - \sigma)$ is expressed by its frequency transform $\mathcal{G}(i\nu_r)$ through $\tilde{\mathcal{G}}(\tau) = \sum_r \mathcal{G}(i\nu_r) e^{-i\tau\nu_r}/(\hbar\beta)$. One finds

$$\ln \frac{Z(\beta, [q(\tau)])}{Z^{\text{SPA}}(\beta, \bar{q})}\bigg|_{\text{PSPA}} = -\frac{\beta}{2} \sum_{r \neq 0} |q_r|^2 \mathcal{G}(\bar{q}, i\nu_r) = \frac{\beta}{2} \sum_{r \neq 0} |q_r|^2 \chi(\bar{q}, i\nu_r). \quad (24.117)$$

As elsewhere in this book nuclear dynamics is described by response functions, the latter has been introduced on the very right. Both functions are connected to each other by analytic continuation, in the sense $\mathcal{G}(\bar{q}, i\nu_r) \leftrightarrow -\chi^R(\bar{q}, \omega)$ when $i\nu_r$ is replaced by $\omega + i\epsilon$ and vice versa, see Section 23.5.1. Using the result (24.117) for (24.110) one finally arrives at a simple form for the $\Delta S_{\text{E}}[q(\tau)]$ of the PSPA. With $\lambda_r(\beta, \bar{q}) = 1 + k\chi(\bar{q}, i\nu_r)$ it reads

$$\Delta S_{\text{E}}[q(\tau)] = -\frac{\hbar\beta}{k} \sum_{r>0} \lambda_r(\beta, \bar{q}) |q_r|^2 . \quad (24.118)$$

Notice that the summation now only extends over positive r, which is possible because $\mathcal{G}(\bar{q}, i\nu_r)$ is even in r, see (23.120). As we stick to the case $k < 0$, the remaining Gaussian integrals hidden in $\bar{\mathcal{D}}[q(\tau)]$ cause no problem as long as $\lambda_r(\beta, \bar{q}) = 1 + k\chi(\bar{q}, i\nu_r) > 0$ for all $r > 0$. One gets

$$\mathcal{C}^{\text{PSPA}}(\beta, \bar{q}) = \prod_{r>0} \left(\lambda_r(\beta, \bar{q})\right)^{-1} . \quad (24.119)$$

To evaluate the λ_r one takes advantage of the fact that any meromorphic function $f(z)$ can be expressed through the ratio $H(z)/G(z)$ of two entire functions, where $H(z)$ accounts for the zeros and $G(z)$ for the poles of $f(z)$ (Hurwitz and Courant, 1964). For real ω the $\chi(\omega)$ of the independent particle model is given by (6.22). It has poles at the finite differences $e_{kj} = e_k - e_j \neq 0$ of the single particle energies, but no one at $\omega = 0$ because the static response $\chi(0)$ ought to be finite. Hence, the same feature is given for the function $\lambda(\omega) = 1 + k\chi(\omega)$. On the other hand, the relation $\lambda(\omega) = 0$ is identical to the secular equation (6.18) such that the zeros of $\lambda(\omega)$ are determined by the frequencies ϖ_μ which in Chapter 6 have been identified as those of the collective modes (although only a few of them will show the typical collective behavior). Let us first concentrate on stable modes for which all of the ϖ_μ^2 are positive. From the form (6.22) of the response function one obtains the following expression for λ_r,

$$\lambda_r(\beta, \bar{q}) = \frac{\prod_\mu (\nu_r^2 + \varpi_\mu^2(\beta, \bar{q}))}{\prod'_{k>l}(\nu_r^2 + e_{kl}^2(\bar{q}))} = \prod_\mu \frac{(\nu_r^2 + \varpi_\mu^2(\beta, \bar{q}))}{(\nu_r^2 + \Omega_\mu^2(\beta, \bar{q}))} . \quad (24.120)$$

The prime in the product in the denominator of the first formula is to indicate that only finite excitations e_{kl} are to be considered. As we may recall from the discussion in Section 6.1.3 the collective modes ϖ_μ are in one to one correspondence to these intrinsic excitations such that we may count the latter by the same index μ and denote the associated "frequencies" by Ω_μ. For stable modes the ϖ_μ is smaller than the corresponding Ω_μ, which implies the $\lambda_r(\beta,\bar{q})$ to be smaller than unity. In the ideal case almost all strength is concentrated in the lowest mode ϖ_1, which in this sense is "most collective". Whereas this ϖ_1 is well separated from the corresponding Ω_1, all the other ϖ_μ lie close to their counterpart Ω_μ. For this situation one may estimate the $\lambda_r(\beta,\bar{q})$ by

$$\lambda_r(\beta,\bar{q}) \lesssim \frac{\nu_r^2 + \varpi_1^2(\beta,\bar{q})}{\nu_r^2 + \Omega_1^2(\beta,\bar{q})}. \qquad (24.121)$$

Hence, the correction factor (24.119) is bound from below by

$$\mathcal{C}^{\text{PSPA}}(\beta,\bar{q}) \gtrsim \prod_{r>0} \frac{\nu_r^2 + \Omega_1^2(\beta,\bar{q})}{\nu_r^2 + \varpi_1^2(\beta,\bar{q})} = \frac{\sinh(\hbar\beta\Omega_1/2)}{\hbar\beta\Omega_1/2} \frac{\hbar\beta\varpi_1/2}{\sinh(\hbar\beta\varpi_1/2)}. \qquad (24.122)$$

The form on the very right comes up if for each r the fraction is multiplied and divided by ν_r^2, which then allows one to apply formula (24.74) twice. The appearance of the factor $\hbar\beta\varpi_1/2/\sinh(\hbar\beta\varpi_1/2)$ is expected from the discussion of the simple model of a particle in a potential. After all, the role of the particle's local mode is now taken up by the collective one ϖ_1. However, as we see from (24.122), in addition there is a factor that depends on the (most relevant) intrinsic excitation $\hbar\Omega_1$, and its variation with $\hbar\beta\Omega_1/2$ is just opposite to the other one. Here an interesting property may come up, which is not untypical for nuclear dynamics: Eventually *collective* motion can be treated *classically* but *not that of the nucleons*. Indeed, as $\varpi_1 \ll \Omega_1$ one may have $\hbar\beta\varpi_1/2 \ll 1$ but not the analogous relation for Ω_1.

The breakdown of the PSPA In Section 24.2.2 it was seen that the Gaussian integrals may not converge for unstable local modes. This happens at or below a certain local critical temperature $T_0(\bar{x})$ given in (24.77). Exactly the same feature also becomes true for the PSPA, as is easily seen from the form (24.120). In Section 6.1.3 it is shown that a negative stiffness $\partial^2 \mathcal{F}^{\text{SPA}}(\beta,\bar{q})/\partial \bar{q}^2$ of the free energy implies exactly *one* unstable collective mode. The free energy meant here is that of the SPA given in (24.107). Like before, the critical eigenvalue λ_r is that for $r=1$. This leads to exactly the same form for the *local* critical temperature, namely

$$T_0(\bar{q}) = \frac{\hbar|\varpi_1(\beta,\bar{q})|}{2\pi} \qquad \text{where} \qquad \varpi_1^2 < 0, \qquad (24.123)$$

and, likewise, for the *global* critical temperature T_0 as the maximum of $T_0(\bar{q})$. Notice that for the model discussed before, in which most strength is concentrated in the collective mode, one gets from (24.122)

$$\mathcal{C}^{\text{PSPA}}(\beta,\bar{q}) \gtrsim \frac{\sinh(\hbar\beta\Omega_1/2)}{\hbar\beta\Omega_1/2} \frac{\hbar\beta|\varpi_1|/2}{\sin(\hbar\beta|\varpi_1|/2)}. \qquad (24.124)$$

The existence of the critical temperature T_0 was recognized in the works on PSPA mentioned above. Before that a similar problem was already seen to occur in the treatment of dissipative tunneling within the Caldeira–Leggett model, see the discussion in Section 12.5.2.

Extension to damped motion Within the PSPA all microscopic information about nucleonic motion was seen to be contained in single particle Green functions. As described in Section 24.4.3 they may be replaced by nucleonic response functions which by (24.118) define the eigenvalues λ_r. The latter specify the coefficients of the expansion of the Euclidean action to order two in the deviation from the average $\bar{q}$. Actually, even if this expansion is carried further, the coefficients can always be expressed in terms of one-body Green functions (Rummel, 2004). This feature allows one to apply the microscopic damping mechanism explained in Section 6.4.3 which implies simply using these Green functions with complex self-energies. Hence, while sticking to the Gaussian approximation the $\Delta S_E[q(\tau)]$ may still be expressed through (24.118), but now with the response function for damped motion discussed in Section 6.5. For those neither the forms (24.120) nor that of (24.121) is valid anymore. However, we may pursue the line of arguments which lead us from (24.120) to (24.121). Actually, this is in complete accord with the procedure of Section 6.5.

To proceed further it is best first to express the intrinsic response function by the collective one using the identity $\chi_{\text{coll}} = \chi(\omega)/(1 + k\,\chi(\omega))$ of (6.33). This leads to

$$\lambda_r(\beta, \bar{q}) = (1 + k\chi(i\nu_r)) = \left(1 - k\,\chi_{\text{coll}}(i\nu_r)\right)^{-1}. \qquad (24.125)$$

The $\chi_{\text{coll}}(\omega)$ may now be replaced by $\chi_{\text{osc}}(\omega) = -M^{-1}(\omega^2 + i\omega\Gamma - \varpi^2)^{-1}$, the response function of the damped oscillator (see Section 23.1). Please, bear in mind that both the effective inertia $M = M(\beta, \bar{q})$ and the width $\Gamma = \Gamma(\beta, \bar{q})$ may vary with temperature and with the collective coordinate. Evaluating the $\chi_{\text{osc}}(\omega)$ at $\omega = i\nu_r$ after insertion into (24.125) one gets

$$\lambda_r(\beta, \bar{q}) = \frac{\nu_r^2 + \nu_r\Gamma + \varpi^2}{\nu_r^2 + \nu_r\Gamma + \Omega^2} \quad \text{with} \quad \Omega^2 = \varpi^2 - k/M. \qquad (24.126)$$

Again, because of the negative k the $\lambda_r(\beta, \bar{q})$ is smaller than one. As a consequence, the (still appropriate) quantum correction factor (24.119) is larger than unity. For unstable modes (with a negative ϖ^2) the condition for a positive λ_1 has now to be determined from the equation $\nu_1^2 + \nu_1\Gamma - |\varpi|^2 = 0$, which is easily seen to be related to the secular equation of the damped oscillator. As this equation has two solutions $\nu_1^{\pm}$ one must take the larger one, $\nu_1^{+} = |\varpi|\,(\sqrt{1 + \eta^2} - \eta)$ where $\eta = \Gamma/(2\,|\varpi|)$. Translated into temperature this leads to

$$T_0(\bar{q}) = \frac{\hbar\,|\varpi(\beta, \bar{q})|}{2\pi}\left(\sqrt{1 + \eta^2} - \eta\right). \qquad (24.127)$$

By comparing with (24.123) it is seen that damping *reduces* the value of $T_0(\bar{q})$.

Exercises

1. Calculate the classical action for the oscillator which is coupled to a time-dependent external field by the potential $\delta V = -e(t)x$ and show that one gets

$$S_c(x_b, t_b; x_a, t_a) = \frac{M\varpi}{2\sin\varpi(t_b-t_a)}\left[(x_b^2+x_a^2)\cos\varpi(t_b-t_a) - 2x_a x_b\right.$$

$$+ \frac{2x_b}{M\varpi}\int_{t_a}^{t_b} e(t)\sin\varpi(t-t_a)dt + \frac{2x_a}{M\varpi}\int_{t_a}^{t_b} e(t)\sin\varpi(t_b-t)dt \quad (24.128)$$

$$\left. - \frac{2}{M^2\varpi^2}\int_{t_a}^{t_b}\int_{t_a}^{t} e(t)e(s)\sin\varpi(t_b-t)\sin\varpi(s-t_a)dt\,ds\right].$$

2. What does the propagator look like for the driven oscillator with $\delta V = -e(t)x$?

3. Calculate the transition amplitude $\mathcal{T}[q(t)]$ for the Lagrangian (24.44) by integrating out the bath variables analogously to (24.42); one may put $g(q) = q$. Rewrite the remaining path integral in $q(t)$ in terms of an effective action.

4. Derive $Z = (2\sinh\hbar\varpi/2T)^{-1}$ for the partition function of the oscillator.

5. Suppose the single particle energy e_k is a unique function $e_k = \mathcal{I}^{-1}(k)$ of k such that the inverse function $k = \mathcal{I}(e)$ exists. (i) Convince yourself that the level density $g(e)$ can be written as (Brack and Bhaduri, 1997)

$$g(e) = \sum_k \delta(e - e_k) = \mathcal{N}(e)\,|\mathcal{I}'(e)|\sum_k \delta(k - \mathcal{I}(e))$$

$$= \mathcal{N}(e)\,|\mathcal{I}'(e)|\left\{1 + 2\sum_{k=1}^{\infty}\cos(2\pi k\mathcal{I}(e))\right\} \quad e \geq 0, \quad (24.129)$$

where $\mathcal{N}(k)$ represents the degeneracy of level k. Prove (24.95) for the oscillator (with $\mathcal{N}(k) = (k+1)(k+2)/2$ for $d = 3$).

6. Take a particle moving in a one-dimensional infinite square well, say with walls located at $x = 0, a$. As in between the walls the particle moves freely, one may exploit the propagator for free motion provided one takes into account the (infinitely many) reflections at the walls, each of which is to be attributed a phase factor of $\exp(i\pi) = -1$. One may or may not exploit the method of mirror charges. Show that the formula for $k(x, y; t)$ is identical to the (24.80) which involves the quantal eigenfunctions of the square well. This may be done with the help of the Jacobi theta function $d_3(z\mid t) \equiv \sum_{n=-\infty}^{\infty}\exp(2niz)\exp(\pi itn^2) = 1 + 2\sum_{n=1}^{\infty}\cos(2nz)\exp(\pi itn^2)$, about which more properties can be found in Section 23 of (Schulman, 1981) or Section 21.1 of (Whittaker and Watson, 1992). Perform the Fourier transform to the $G(x, y; e)$ by making use of the formula $\int_0^{\infty} du\,\exp(-a/u^2 - bu^2) = \sqrt{\pi/4b}\exp(-2\sqrt{ab})$. Calculate the level density from (24.84) and compare it with (24.129).

7. Convince yourself about the validity of the integral measures (24.100) and (24.105): (i) For (24.100) it suffices to examine the formula

$$\int \mathcal{D}[q(\tau)] \exp\left(-\frac{1}{2\hbar |k|} \int_0^{\hbar\beta} d\tau\, q(\tau)^2\right) = 1, \quad (a)$$

as the continuum expression of the form which may be found from (24.98),

$$\lim_{N\to\infty} \int_{-\infty}^{\infty} \left(\prod_{n=1}^{N} \sqrt{\frac{\eta}{2\pi\hbar |k|}}\, dq_n\right) \exp \sum_{n=1}^{N} \left(\frac{\eta}{\hbar}\left(-\frac{1}{2|k|} q_n^2\right)\right) = 1. \quad (b)$$

(ii) For (24.105) one needs to consider the relation

$$\lim_{N\to\infty} \int_{-\infty}^{\infty} \left(\prod_{r=1}^{(N-1)/2} \frac{\beta}{\pi |k|}\, dq'_r\, dq''_r\right) \exp\left\{-\sum_{r=1}^{(N-1)/2} \frac{\beta}{|k|}\left((q'_r)^2 + (q''_r)^2\right)\right\} = 1,$$

together with the other factor $\sqrt{\beta/2\pi |k|} \int_{-\infty}^{\infty} d\bar{q}\, \exp(\beta/2 |k| \bar{q}^2) = 1$. Show first that they come up when the exponent of eqn(a) is evaluated with the Fourier series $q(\tau) = \sum_r q_r \exp(i\tau\, 2\pi r/(\hbar\beta))$ for the real $q(\tau)$. Notice that the last equation could also be written in a form where negative rs are still involved, one would only have to replace $|k|$ by $2|k|$. Is there any relation to the transformation (24.24)?

25

PROPERTIES OF LANGEVIN AND FOKKER–PLANCK EQUATIONS

In this chapter we are going to present some important features of transport theory for the dynamics of one or a few degrees of freedom, which in nuclear physics represent collective motion. In many respects the latter resembles that of a Brownian particle. Indeed, both examples may be described by the same type of equations. Their derivation, however, must be different as soon as collective motion is treated in self-consistent fashion. These problems are addressed specifically in Section 7.5 based on the discussion in Section 25.6 about the more general aspects of microscopic derivations. At first we want to get acquainted with basic properties, which will be introduced for the example of the typical Brownian particle.

25.1 The Brownian particle, a heuristic approach

25.1.1 Langevin equation

For a particle of mass M dispersed in a fluid of other particles of much smaller masses m_i, Newton's law may be written as $M\dot{\vec{v}} = \vec{K} + \vec{F}_{B-F}$, with $\vec{v}$ being its velocity, $\vec{K}$ some applied conservative force and $\vec{F}_{B-F}$ the force which is exerted on the particle by the fluid. Typically, $m_i \ll M$ such that timescales $\tau_i = \tau_{\text{micro}}$ and $\tau_M = \tau_{\text{macro}}$ will be well separated, with $\tau_{\text{micro}} \ll \tau_{\text{macro}}$. Indeed, such a situation is also given in nuclear physics as in a typical collective motion many nucleons participate. As it is very difficult (if not impossible) to know the $\vec{F}_{B-F}$ in detail one often relies on phenomenology. In a first approximation this $\vec{F}_{B-F}$ may be parameterized by a force linear in the velocity, $\vec{F}_{frict} = -\gamma \vec{v}$. (For our purpose it may be sufficient to work with one friction coefficient γ rather than a friction tensor.) The presence of a *friction force alone* still implies *deterministic behavior*. This changes as soon as stochastic properties must be taken into account, in which case the particle performs a "random walk". According to Langevin such a motion can be described by generalizing the force $\vec{F}_{B-F}$ to

$$\vec{F}_{B-F} = -\gamma \vec{v} + \vec{L}(t), \qquad (25.1)$$

where $\vec{L}(t)$ represents the *fluctuating or stochastic* part. It is interesting to note that van Kampen (2001) has referred to such systems as "mesoscopic" ones. Unlike the present use of this notion it was meant to distinguish from the "macroscopic" systems we know from everyday life and for which only average quantities matter.

The presence of a fluctuating force asks for averaging procedures with appropriately defined ensembles of identical systems, similar to the procedure in equilibrium statistical mechanics. It is clear that the notion "identical" only refers to the mesoscopic level. Since we are unable to describe the fluid in detail, we cannot say more about its actual state than what will be reflected by the properties of the underlying "stochastic process". The latter will be assumed to follow the basic laws

$$\langle L(t) \rangle = 0, \tag{25.2}$$

$$\langle L(t) L(t+s) \rangle = \Lambda \delta(s). \tag{25.3}$$

For the sake of simplicity we have restricted ourselves to one-dimensional motion, and will do so henceforth in this section. Actually, these two conditions have to be complemented by a third one which involves correlations of higher order: for Gaussian processes it is assumed that all of them factorize and can be traced back to the first two. In more technical terms one may say that the cumulants of $L(t)$ of order three and larger vanish (see Section 25.1.3).

In (25.2) the average of $L(t)$ has been put equal to zero, simply because its contribution to $\vec{F}_{B-F}$ is assumed to be already parameterized by the friction force. On the right-hand side of (25.3) a δ-function in the time difference s appears. So when the force $L(t)$ acts twice any consequence of this action can be felt only within a *mesoscopically small* time interval. At this point a condition like $\tau_{\text{micro}} \ll \tau_{\text{meso}}$ on the different timescales is taken to its extreme. Somewhat more realistic would be to assume that the "auto-correlation function" $\langle L(t) L(t+s) \rangle$ has a *finite memory*, perhaps parameterized by a function $\varphi(s)$ which decays to zero on a small, but finite timescale of the order of τ_{micro}. After such a time the "collisions" with the molecules of the fluid become *statistically uncorrelated*. For a τ_{micro} which is negligibly small on the mesoscopic scale, and for which we may thus put $\tau_{\text{micro}} \approx 0$, this implies

$$\langle L(t) L(t+s) \rangle \approx \langle L(t) \rangle \langle L(t+s) \rangle = 0 \qquad \text{for} \qquad s > 0, \tag{25.4}$$

where the zero comes about because of (25.2).

The presence of this irregular behavior of $L(t)$ implies that in strict mathematical sense the equation of motion

$$\dot{v} + \Gamma v = L(t) \qquad \text{with} \qquad \Gamma = \gamma/M \tag{25.5}$$

has pathological features.[56] For this reason an argument is needed to clarify how one wants to interpret this force physically. We know that it will act on an (in the idealistic case of (25.3)) infinitely short time interval. Therefore we had better look for its contribution on some small but *finite* time interval Δt.

---

[56] In mathematics, equations of that type are called "stochastic differential equations"; details may be found in (Gardiner, 2002) as well as in (van Kampen, 2001) and (Risken, 1989).

For a Langevin force $L(t)$ like the one given here, which is independent of the dynamical variable v, we may simply introduce the "Kraftstoß"

$$\Delta w = \int_t^{t+\Delta t} L(s) ds \tag{25.6}$$

and integrate (25.5) over Δt to get

$$\Delta v(t) = -\Gamma v(t) \Delta t + \Delta w. \tag{25.7}$$

Notice, please, that for a truly continuous function $L(t)$, and according to the mean value theorem of integration, the Δw could be written as $\Delta w = L(\widetilde{t}) \Delta t$, where $\widetilde{t}$ is some time with $t < \widetilde{t} < t + \Delta t$. Such a situation would indeed be given if one were working with some finite value of the τ_{micro}, which then may be put equal to zero after all necessary formal steps have been undertaken (see (van Kampen, 2001) and (Risken, 1989)). Mathematically, however, one may define the Δw by the Stieltjes integral in (25.6) and proceed by solving (25.5) just as we would do if all terms therein were well defined. In this sense the $L(t)$ may be considered the inhomogeneous term to the differential equation of first order for $v(t)$. Its solution can thus be written as

$$v(t) = a e^{-\Gamma t} + e^{-\Gamma t} \int_0^t ds\, e^{\Gamma s} L(s) \tag{25.8}$$

Averaging over the ensemble we find

$$\langle v(t) \rangle = a e^{-\Gamma t} + e^{-\Gamma t} \int_0^t ds\, e^{\Gamma s} \langle L(s) \rangle = \langle v \rangle_0 e^{-\Gamma t} \tag{25.9}$$

after deliberately interchanging the ensemble average with time integration. The last equality is obtained because of (25.2). Due to the final form written on the very right the integration constant a has been identified as the average initial velocity. To get the dynamics for $\langle v^2(t) \rangle$, the second moment of the velocity, we may apply (25.8) twice to get

$$\langle v^2(t) \rangle = \langle v \rangle_0^2 e^{-2\Gamma t} + e^{-2\Gamma t} \int_0^t ds \int_0^t ds'\, e^{\Gamma(s+s')} \langle L(s) L(s') \rangle, \tag{25.10}$$

after dropping the vanishing term $\langle L(s) \rangle$. For the double integral one may make use of (25.3) to find

$$\int_0^t ds \int_0^t ds'\, e^{\Gamma(s+s')} \langle L(s) L(s') \rangle = \int_0^t ds \int_0^t ds'\, e^{\Gamma(s+s')} \Lambda \delta(s' - s) = \frac{\Lambda}{2\Gamma} \left(e^{2\Gamma t} - 1 \right) \tag{25.11}$$

such that one gets:

$$\langle v^2(t) \rangle = \langle v \rangle_0^2 e^{-2(\gamma/M)t} + \frac{M\Lambda}{2\gamma} \left(1 - e^{-2(\gamma/M)t} \right). \tag{25.12}$$

From (25.9) the first term on the right is seen to become $\langle v(t)\rangle^2$. Hence, the fluctuation in the velocity, or its mean square deviation, can be written as

$$\Sigma_{vv}(t) \equiv \langle v^2(t)\rangle - \langle v(t)\rangle^2 = \frac{M\Lambda}{2\gamma}\left(1 - e^{-2(\gamma/M)t}\right) \qquad (25.13)$$

It vanishes identically for $\Lambda = 0$, which is to say for a vanishing "Langevin" force. In this way we get back to a *deterministic* description.

The solution (25.12) shows that for $t \gg M/\gamma$ the $\langle v^2(t)\rangle$ approaches a constant value, namely

$$\langle v^2(t)\rangle \longrightarrow \frac{M\Lambda}{2\gamma} \equiv \langle v^2\rangle_{\text{eq}} \qquad (25.14)$$

This observation can be used to actually fix the parameter Λ. We only have to remember the physical situation our equation of motion is meant to describe. The fluid in which the Brownian particle moves acts like a "heat bath". So once equilibrium is reached we may exploit the virial theorem, which here reads $M\langle v^2\rangle_{\text{eq}}/2 = T/2$. Therefore we must have

$$\frac{\Lambda}{2} = \frac{\gamma}{M}\langle v^2\rangle_{\text{eq}} = \frac{\gamma}{M^2}T. \qquad (25.15)$$

25.1.2 Fokker–Planck equations

In the previous discussion a finite fluctuation of the velocity has been encountered. For finite stochastic forces this feature is to be expected for all dynamical quantities. It is therefore most natural to try describing the time evolution of density distributions. One possibility is by employing partial differential equations, which in this context are generally called Fokker–Planck equations. In this section we want to establish their basic nature for simple examples, starting from Kramers' version which plays a special role in nuclear physics.[57]

Kramers' equation: It was originally set up in (Kramers, 1940) to derive a formula for the fission rate which is addressed in Chapter 12 and reads

$$\frac{\partial}{\partial t}f(Q,P,t) = \left[-\frac{\partial}{\partial Q}\frac{P}{M} + \frac{\partial}{\partial P}\left(\frac{dV(Q)}{dQ}\right) + \frac{\partial}{\partial P}\frac{P}{M}\gamma + D_{pp}\frac{\partial^2}{\partial P\partial P}\right]f(Q,P,t). \qquad (25.16)$$

Here, inertia M and friction γ have the same meaning as before, the quantity $V(Q)$ is the system's potential and the *diffusion coefficient* D_{pp} is given by the so-called Einstein relation $D_{pp} = \gamma T$. As is easily recognized, for vanishing friction (25.16) reduces to the Liouville equation. The density distribution is supposed

---

[57]For a discussion of historical developments of transport equations the reader may be referred to the book by van Kampen (2001).

to be normalized to unity like $\int_{-\infty}^{\infty} dQ dP \ f(Q,P,t) = 1$. By simply integrating (25.16) over momentum one obtains the continuity equation in the form

$$\frac{\partial}{\partial t} n(Q,t) + \frac{\partial}{\partial Q} j(Q,t) = 0, \qquad (25.17)$$

with the density and the current density defined as

$$n(Q,t) = \int_{-\infty}^{\infty} dP \ f(Q,P,t) \quad \text{and} \quad j(Q,t) = \int_{-\infty}^{\infty} dP \ \frac{P}{M} f(Q,P,t), \qquad (25.18)$$

respectively. This is easily inferred after realizing that terms involving derivatives with respect to momentum vanish, like

$$\int_{-\infty}^{\infty} dP \ \frac{\partial}{\partial P} G(Q,P,t) \ f(Q,P,t) = \lim_{P \to \infty} G(Q,P,t) \ f(Q,P,t) = 0. \qquad (25.19)$$

This is true for any function $G(Q,P,t)$ which is sufficiently well behaved in P, meaning that it does not counter the strong decrease of $f(Q,P,t)$ with P at large values of P.

Rayleigh equation: For a vanishing conservative force $dV(Q)/dQ = 0$ integration of (25.16) over the coordinate leads to the following equation associated with Lord Rayleigh (van Kampen, 2001)

$$\frac{\partial}{\partial t} m(P,t) = \left[\frac{\partial}{\partial P} \frac{P}{M} \gamma + D_{pp} \frac{\partial^2}{\partial P \partial P} \right] m(P,t), \qquad (25.20)$$

where $m(P,t) = \int_{-\infty}^{\infty} dQ \ f(Q,P,t)$ is the reduced density in momentum. Let us take this simple example to describe how one may deduce equations for first and second moments in P from the transport equations. They are to be defined as

$$P^c(t) \equiv \langle P \rangle_t = \int_{-\infty}^{\infty} dP \ P \ m(P,t), \qquad (25.21)$$

$$\Sigma_{pp}(t) \equiv \langle (P - P^c(t))^2 \rangle_t = \int_{-\infty}^{\infty} dP \ (P - P^c(t))^2 \ m(P,t), \qquad (25.22)$$

respectively. The $dP^c(t)/dt$ can be expressed by $\partial m(P,t)/\partial t$, the partial derivative of the density which satisfies eqn(25.20). The derivatives with respect to P can be handled by partial integration, which becomes particularly simple if we assume the coefficients γ and D_{pp} to be independent of P. For the first moment no contribution is obtained from the derivative of second order, such that one gets Newton's equation in the form $dP^c(t)/dt = -\gamma P^c(t)/M$, reflecting "classical" motion, hence the index c. The equation for the second moment becomes $d\Sigma_{pp}(t)/dt + 2(\gamma/M)\Sigma_{pp}(t) = 2D_{pp}$, which by the equipartition theorem of classical statistical mechanics implies $D_{pp} = \gamma T$. Applying the relation $v = P/M$

one easily establishes a connection with the solutions found in Section 25.1.1. The density distribution itself is given by the Gaussian (see exercise 1)

$$m(P,t) = \left(2\pi \sum_{pp}(t)\right)^{-1/2} \exp\left(-\frac{(P - P^c(t))^2}{2\sum_{pp}(t)}\right). \tag{25.23}$$

Smoluchowski equation: A simple equation for the density $n(Q,t)$ of (25.18) is given by the Smoluchowski equation:

$$\frac{\partial}{\partial t} n(Q,t) = \frac{\partial}{\partial Q}\left(\frac{1}{\gamma}\frac{\partial V(Q)}{\partial Q} + \frac{T}{\gamma}\frac{\partial}{\partial Q}\right) n(Q,t), \tag{25.24}$$

valid for over-damped motion only. Important formal properties of this equation are discussed in Section 25.3, even for variable friction, for applications to physical problems see Section 11.1.2 and Chapter 12.

25.1.3 *Cumulant expansion and Gaussian distributions*

In the previous discussion it was seen that Gaussian functions may be used to represent probability distributions for stochastic variables. Before this latter concept is described in more detail in the next section we shall discuss characteristic features of Gaussian distributions. To this end we first introduce the *characteristic function* $G(k)$ associated with a distribution $P(y)$ which defines the probability with which the stochastic variable Y takes on the value y. This $P(y)$ should be positive definite and normalized to unity $\int dy P(y) = 1$ (where the integral is to extend over the range in which y is defined). The $G(k)$ is defined by the average of an exponential, namely

$$G(k) = \langle e^{ikY} \rangle = \int dy\, e^{iky} P(y). \tag{25.25}$$

Expanding the exponential, one gets

$$G(k) = \sum_{r=0}^{\infty} \frac{(ik)^r}{r!} \langle Y^r \rangle \quad \text{with} \quad \langle Y^r \rangle = \int dy\, y^r P(y) \tag{25.26}$$

provided all moments exist. Notice that here (and in the next section) we distinguish between the stochastic variable Y and its realization y. In the text we often use the sloppy notation and write averages simply as $\langle y \rangle$. Equation (25.26) tells us that the moments can be calculated from the derivatives of the (supposedly known) generating function. By taking the logarithm one gets an expansion in terms of the *cumulants* Ξ_r

$$\ln G(k) = \sum_{r=1}^{\infty} \frac{(ik)^r}{r!} \Xi_r. \tag{25.27}$$

The Ξ_r are combination of the moments, for the first three one has

$$\Xi_1 = \langle Y \rangle$$
$$\Xi_2 = \langle Y^2 \rangle - \langle Y \rangle^2 = \Sigma_{yy} \qquad (25.28)$$
$$\Xi_3 = \langle Y^3 \rangle - 3\langle Y^2 \rangle \langle Y \rangle + 2\langle Y \rangle^3.$$

Here, in the second line the variance Σ_{yy} has been introduced. For the Gaussian distribution

$$P_{\text{Gauss}}(y) = \left(2\pi \Sigma_{yy}\right)^{-1/2} \exp\left(-\frac{(y - \langle y \rangle)^2}{2\Sigma_{yy}}\right) \qquad (25.29)$$

the characteristic function becomes

$$G(k) = \exp\left(i\langle y \rangle k - \frac{1}{2}\Sigma_{yy} k^2\right). \qquad (25.30)$$

Both functions are determined uniquely by the first and second moment. Such a distribution is also-called *normal*.

These results are easily generalized to n dimensions.

$$P_{\text{Gauss}}(\mathbf{y}) = (2\pi)^{-n/2} \left(\det \Sigma_{\mu\nu}\right)^{-1/2} \times \qquad (25.31)$$
$$\exp\left\{-\frac{1}{2} \sum_{\mu,\nu=1}^{n} (y_\mu - \langle Y_\mu \rangle)(y_\nu - \langle Y_\nu \rangle)\left(\Sigma^{-1}\right)_{\mu\nu}\right\}.$$

The $\left(\Sigma^{-1}\right)_{\mu\nu}$ is the inverse of the *covariance matrix*

$$\Sigma_{\mu\nu} = \langle (Y_\mu - \langle Y_\mu \rangle)(Y_\nu - \langle Y_\nu \rangle) \rangle = \langle Y_\mu Y_\nu \rangle - \langle Y_\mu \rangle \langle Y_\nu \rangle. \qquad (25.32)$$

For $\mu = \nu$ the $\Sigma_{\mu\mu}$ is referred to as *variance*, but we will often simply call the $\Sigma_{\mu\nu}$ a fluctuation of the corresponding variables. The generalization of (25.28) becomes

$$\Xi_\mu^{(1)} = \langle Y_\mu \rangle$$
$$\Xi_{\mu\nu}^{(2)} = \Sigma_{\mu\nu} \qquad (25.33)$$
$$\Xi_{\mu\nu\rho}^{(3)} = \langle Y_\mu Y_\nu Y_\rho \rangle - \langle Y_\mu Y_\nu \rangle \langle Y_\rho \rangle - \langle Y_\nu Y_\rho \rangle \langle Y_\mu \rangle - \langle Y_\mu Y_\rho \rangle \langle Y_\nu \rangle + 2\langle Y_\mu \rangle \langle Y_\nu \rangle \langle Y_\rho \rangle.$$

Coming back to the Langevin approach described in Section 25.1.1, it may be said that the stochastic force $L(t)$ is given by a normal distribution. The first two cumulants are determined by (25.2) and (25.3) and all higher ones vanish.

25.2 General properties of stochastic processes

Having set up elementary physical aspects we like to put the discussion of stochastic processes on more general grounds. This will include processes where the existence of the fluctuating force need not necessarily be traced back to the fluctuation dissipation theorem.

25.2.1 Basic concepts

Suppose we are given a stochastic variable X which may take on numbers x with probability $P_X(x)$. Be f some mapping from X to another variable Y. Suppose this mapping depends on time t, then the assignment $Y_X(t) = f(X,t)$ is called a "stochastic process". Taking for X one of its possible values x we get an ordinary function of t, namely

$$Y_x(t) = f(x,t) \tag{25.34}$$

which is called the "sample function" or "realization" of the process. Knowing the probability distribution $P_X(x)$ one may calculate averages like that of the n'th moment of the distribution

$$\langle Y(t_n)\cdots Y(t_2)Y(t_1)\rangle = \int Y_x(t_n)\cdots Y_x(t_2)Y_x(t_1) P_X(x) dx. \tag{25.35}$$

Often the values of these moments do not change when *all* times are shifted by the same amount,

$$\langle Y(t_n+\tau)\cdots Y(t_2+\tau)Y(t_1+\tau)\rangle = \langle Y(t_n)\cdots Y(t_2)Y(t_1)\rangle \tag{25.36}$$

Processes of this type are called *stationary*. For $n=1$ it implies $\langle Y(t+\tau)\rangle = \langle Y(t)\rangle$ and, hence, that the average of $Y(t)$ is *independent* of t. Combining these averages with moments for $n=2$ one may define the *auto-correlation function* as

$$\psi(t_2,t_1) = \langle Y(t_2)Y(t_1)\rangle - \langle Y(t_2)\rangle\langle Y(t_1)\rangle \equiv \langle\langle Y(t_2)Y(t_1)\rangle\rangle. \tag{25.37}$$

For $t_2 = t_1 = t$ it becomes identical to what elsewhere in the text is called *fluctuation* of the quantity $Y(t)$. For a stationary process the $\psi(t_2,t_1)$ depends on time only through *time difference*: $\psi(t_2,t_1) = \psi(t = t_2 - t_1)$.

So far we have been looking only at averages of the $Y_X(t)$. Their probability density may be traced back to the $P_X(x)$ through the relation $P_1(y,t) = \int \delta(y - Y_x(t)) P_X(x) dx$. More generally we may write

$$P_n(y_n,t_n;\cdots;y_2,t_2;y_1,t_1) = \int \delta(y_n - Y_x(t_n))\cdots\delta(y_2 - Y_x(t_2))\delta(y_1 - Y_x(t_1)) P_X(x) dx. \tag{25.38}$$

This $P_n(y_n,t_n;\cdots;y_2,t_2;y_1,t_1)$ is called the *joint probability density*; it is a measure of the process to take on the values y_i at time t_i for *all* $i = 1$ to n, with the n being some arbitrary number and $t_i \neq t_j$ for $i \neq j$. The averages introduced in (25.35) may now be calculated as

$$\langle Y(t_n)\cdots Y(t_2)Y(t_1)\rangle = \int y_n\cdots y_2 y_1 P_n(y_n,t_n;\cdots;y_2,t_2;y_1,t_1) dy_n\cdots dy_2 dy_1. \tag{25.39}$$

The P_n need to fulfill a few more or less evident conditions. First of all they must be positive (semi-) definite and normalized like $\int P_1(y_1,t)dy_1 = 1$. As they should fulfill the relation

$$\int P_n(y_n,t_n;y_{n-1},t_{n-1};\cdots;y_1,t_1)dy_n = P_{n-1}(y_{n-1},t_{n-1};\cdots;y_1,t_1) \quad (25.40)$$

the normalization is then guaranteed for all n. This equation may be interpreted in the following way. Enlarging n means specifying the stochastic process in greater detail. Integration over an y_n implies giving up information such that one naturally comes back on the level of information contained in P_{n-1}. Evidently, this integration could also be done over any other of the y_i. In this sense, the P_n should stay constant when two pairs (y_k,t_k) and (y_l,t_l) are interchanged. Given these conditions, it can be shown that this "hierarchy" of joint probabilities P_n may be taken as one possibility of defining a stochastic process. For more details see van Kampen's book (2001), with references to the original literature.

The joint probability P_n defines the measure of finding the system in the "state" y_i at the time t_i for all $i = 1\ldots n$, no matter what the values of y are at times not contained in the set t_i ($i = 1\ldots n$). One may introduce the *conditional probability* $P_{1|1}(y_2,t_2|y_1,t_1)$ as the probability of finding for Y the value y_2 at time t_2 given that its value at t_1 is y_1. Again; the $P_{1|1}$ is non-negative and it should be normalized like

$$\int P_{1|1}(y_2,t_2|y_1,t_1)dy_2 = 1 \quad (25.41)$$

In a natural way the concept of the conditional probability can be generalized to a $P_{l|k}$ for which Y is fixed at k different times $t_1,\ldots,t_k$ and which defines the joint probability that l values are found at other times $t_{k+1},\ldots,t_{k+l}$. Apparently the following relation must then be fulfilled

$$P_{k+l}(y_{k+l},t_{k+l};\cdots y_{k+1},t_{k+1};y_k,t_k;\cdots;y_1,t_1)$$
$$= P_{l|k}(y_{k+l},t_{k+l};\cdots y_{k+1},t_{k+1}\,|\,y_k,t_k;\cdots;y_1,t_1)P_k(y_k,t_k;\cdots;y_1,t_1) \quad (25.42)$$

The $P_{l|k}$ is symmetric in the set of k pairs of variables, and also in the set of the l pairs. For "stationary processes" one has, in accord with (25.36),

$$P_n(y_n,t_n+\tau;\cdots;y_2,t_2+\tau;y_1,t_1+\tau) = P_n(y_n,t_n;\cdots;y_2,t_2;y_1,t_1). \quad (25.43)$$

For $n = 1$ this means that the $P_1(y,t)$ does not depend on time:

$$P_1(y,t+\tau) = P_1(y,t) = P_1(y). \quad (25.44)$$

25.2.2 Markov processes and the Chapman–Kolmogorov equation

The relations given so far in this chapter get a special meaning when we assume the times $t_1,\ldots,t_n$ to be ordered in the sense $t_1 < t_2 < \cdots t_n$. Then the $P_{1|n-1}(y_n,t_n|y_{n-1},t_{n-1};\cdots y_1,t_1)$ defines the probability that the system reaches the value y_n at the *final* time t_n *provided* it has passed through the values $y_1, \ldots, y_{n-1}$ at *previous* times. As the n may be arbitrary this will in general not only involve lots of information but implies a large "memory" of the system on the history of the motion. A very important class of processes is of simpler nature and fulfills the property

$$P_{1|n-1}(y_n, t_n | y_{n-1}, t_{n-1}; \cdots y_1, t_1) = P_{1|1}(y_n, t_n | y_{n-1}, t_{n-1}). \qquad (25.45)$$

It is seen that for these *Markov processes* the conditional probability density at t_n is *uniquely defined* by the values y_{n-1} at t_{n-1} and does not depend on any knowledge of the values at earlier times. A closer look tells one that the processes treated in the previous section by Langevin or Fokker Planck equations are all Markovian.

As a Markov process is entirely determined by $P_1(y_1, t_1)$ and $P_{1|1}(y_2, t_2 | y_1, t_1)$ all other joint probabilities P_n can be constructed from these two distributions. Let us see this for the case of $n = 3$, for which one may write

$$\begin{aligned} P_3(y_3, t_3; y_2, t_2; y_1, t_1) &= P_{1|2}(y_3, t_3 | y_2, t_2; y_1, t_1) P_2(y_2, t_2; y_1, t_1) \\ &= P_{1|1}(y_3, t_3 | y_2, t_2) P_{1|1}(y_2, t_2 | y_1, t_1) P_1(y_1, t_1). \end{aligned} \qquad (25.46)$$

Integrating this relation over y_2 one obtains for $t_1 < t_2 < t_3$

$$P_2(y_3, t_3; y_1, t_1) = \left[\int dy_2 P_{1|1}(y_3, t_3 | y_2, t_2) P_{1|1}(y_2, t_2 | y_1, t_1) \right] P_1(y_1, t_1). \qquad (25.47)$$

On dividing both sides by $P_1(y_1, t_1)$ one gets (observing (25.42))

$$P_{1|1}(y_3, t_3 | y_1, t_1) = \int dy_2 P_{1|1}(y_3, t_3 | y_2, t_2) P_{1|1}(y_2, t_2 | y_1, t_1) \qquad (25.48)$$

This relation is called the *Chapman–Kolmogorov equation*.

In applications of these concepts to nuclear physics we will be concerned with conditional probabilities of the type $P_{1|1}(y_2, t_2 | y_1, t_1)$. Sometimes it is also-called the *transition probability* for the transition from the "state" y_1 at time t_1 to the "state" y_2 at time t_2. For the joint probabilities (25.48) implies

$$P_1(y_2, t_2) = \int dy_1 P_{1|1}(y_2, t_2 | y_1, t_1) P_1(y_1, t_1). \qquad (25.49)$$

In this sense the $P_{1|1}(y_2, t_2 | y_1, t_1)$ *propagates* the initial distribution $P_1(y_1, t_1)$ to the final one. Often in the text we will identify $P_{1|1}(y_2, t_2 | y_1, t_1)$ as the *propagator* $K(y_2, t_2 | y_1, t_1) \equiv P_{1|1}(y_2, t_2 | y_1, t_1)$. Evidently, these conditional probabilities must satisfy the initial condition

$$\lim_{t_2 \to t_1} K(y_2, t_2 | y_1, t_1) = \lim_{t_2 \to t_1} P_{1|1}(y_2, t_2 | y_1, t_1) = \delta(y_2 - y_1). \qquad (25.50)$$

Physically this may be interpreted as saying that no transitions take place in a vanishing time interval.

Of course, all relations discussed so far in this section may easily be generalized to n dimensions. One simply has to replace the ys by the n-dimensional vector $\mathbf{y}$ (and the stochastic variable Y accordingly). The same generalization has to be done for the integrals, with the infinitesimal dy replaced by the corresponding volume element $d^n y$.

Stationary Markov processes: Evidently, a *necessary condition* for a Markov process to become stationary is that the *transition probability* $P_{1|1}(y_2, t_2 | y_1, t_1)$ depends *only* on the time *difference*, in which case one may write

$$P_{1|1}(y_2, t_2 | y_1, t_1) = P_{1|1}(y_2 | y_1; \tau) \quad \text{with} \quad \tau = t_2 - t_1. \quad (25.51)$$

A process of this type is called *homogeneous*. Actually this property of the transition probability does *not suffice* to make the *process stationary*. Indeed, by its very definition (25.43), this feature refers to the *joint* probability. Therefore, for Markov processes, besides K, one needs special properties of P_1 to be given as well. A sufficient condition is found when the P_1 is chosen to represent the equilibrium of the process at stake. For more details see (van Kampen, 2001).

For a genuinely stationary process it makes sense to define the $P_{1|1}(y_2 | y_1; \tau)$ of (25.51) for negative τ. Using (25.42) one may write

$$P_2(y_2, t_2; y_1, t_1) = P_{1|1}(y_2 | y_1; \tau) P_1(y_1). \quad (25.52)$$

As P_2 is symmetric one must have the identity

$$P_{1|1}(y_2 | y_1; \tau) P_1(y_1) = P_{1|1}(y_2 | y_1; -\tau) P_1(y_2) \quad (25.53)$$

for all stationary Markov processes. Realize, please, that integrating eqn(25.53) over y_1 leads to $P_1(y_2)$, in consistency with (25.49) and (25.44) for the left-hand side, and with (25.41) (written for $P_{1|1}(y_2 | y_1; \pm \tau)$) for the right-hand side. Here, two remarks may be in order. (1) Equation (25.51) to (25.53) are also valid for *reversible* dynamics, for which the $P_{1|1}(y_2 | y_1; \tau)$ can simply be expressed through Liouville's equation. In this case all the statements just made are more or less evident. (2) For *irreversible* motion the handling of a $P_{1|1}(y_2 | y_1; -\tau)$ may not always be easy, for which reason one mostly refers to the transition probability for *positive* τs.

25.2.3 Fokker–Planck equations from the Kramers–Moyal expansion

The Fokker–Planck equation is a differential equation of first order in time. For this reason one may expect that only properties of the short-time propagator $K(\mathbf{y}_2, t + \tau | \mathbf{y}_1, t)$ matter. This feature is exploited in the Kramers–Moyal expansion, which allows for a systematic derivation of the Fokker–Planck equation from the Chapman–Kolmogorov equation (25.48). It requires the $K(\mathbf{y}_2, t + \tau | \mathbf{y}_1, t)$ to satisfy the following three properties:

$$\int d^n y_2 \, (\mathbf{y}_2 - \mathbf{y}_1)_\mu K(\mathbf{y}_2, t + \tau | \mathbf{y}_1, t) = A_\mu(\mathbf{y}_1, t) \tau + O(\tau^2) \quad (25.54)$$

$$\frac{1}{2}\int d^n y_2\, (\mathbf{y}_2-\mathbf{y}_1)_\mu (\mathbf{y}_2-\mathbf{y}_1)_\nu K(\mathbf{y}_2,t+\tau|\mathbf{y}_1,t) = D_{\mu\nu}^{\text{n.s.}}(\mathbf{y}_1,t)\tau + O(\tau^2)$$
(25.55)

$$\frac{1}{m!}\int d^n y_2\, (\mathbf{y}_2-\mathbf{y}_1)_{\mu_1}\cdots(\mathbf{y}_2-\mathbf{y}_1)_{\mu_m} K(\mathbf{y}_2,t+\tau|\mathbf{y}_1,t) = O(\tau^2) \quad \text{for} \quad m>2.$$
(25.56)

Here, $(\mathbf{y}_2-\mathbf{y}_1)_\mu$ is the μth component of the difference vector. These conditions can be seen to imply the Fokker–Planck equation

$$\frac{\partial}{\partial t_2}K(\mathbf{y}_2,t_2|\mathbf{y}_1,t_1) = \left[-\frac{\partial}{\partial y_2^\mu}A_\mu(\mathbf{y}_2,t_2) + \frac{\partial^2}{\partial y_2^\mu \partial y_2^\nu}D_{\mu\nu}^{\text{n.s.}}(\mathbf{y}_2,t_2)\right] K(\mathbf{y}_2,t_2|\mathbf{y}_1,t_1).$$
(25.57)

Here, use is made of the common summation convention (by summing over repeated indices). It is essential to notice that in both terms the coefficients for drift and diffusion, $A_\mu(\mathbf{y}_2,t_2)$ and $D_{\mu\nu}^{\text{n.s.}}(\mathbf{y}_2,t_2)$, respectively, stand to the right of the differential operators. If the diffusion coefficients and the distribution functions are smoothly differentiable the differentials of second order can be interchanged. The corresponding two diffusion terms can then be written as one, with a symmetric diffusion coefficient $D_{\nu\mu}$ defined through

$$\frac{\partial^2}{\partial y_2^\mu \partial y_2^\nu} D_{\mu\nu}^{\text{n.s.}} + \frac{\partial^2}{\partial y_2^\nu \partial y_2^\mu} D_{\nu\mu}^{\text{n.s.}} = \frac{\partial^2}{\partial y_2^\mu \partial y_2^\nu}\left(D_{\mu\nu}^{\text{n.s.}} + D_{\nu\mu}^{\text{n.s.}}\right) \equiv 2\frac{\partial^2}{\partial y_2^\mu \partial y_2^\nu} D_{\mu\nu}.$$
(25.58)

The propagator $K(\mathbf{y}_2,t_2|\mathbf{y}_1,t_1)$ must satisfy the initial condition (25.50) which here reads

$$\lim_{t_2\to t_1} K(\mathbf{y}_2,t_2|\mathbf{y}_1,t_1) = \delta(\mathbf{y}_2-\mathbf{y}_1).$$
(25.59)

After multiplying (25.57) by some given initial distribution $d(\mathbf{y}_1,t_1)$, integrating over the initial coordinates y_1^μ and observing the Chapman–Kolmogorov equation in the form (25.49), $d(y_2,t_2) = \int dy_1 K(y_2,t_2|y_1,t_1)d(y_1,t_1)$, one obtains the corresponding equation for the joint probability distribution, which we write as

$$\frac{\partial}{\partial t}d(\mathbf{y},t) = \left[-\frac{\partial}{\partial y^\mu}A_\mu(\mathbf{y},t) + \frac{\partial^2}{\partial y^\mu \partial y^\nu}D_{\mu\nu}^{\text{n.s.}}(\mathbf{y},t)\right] d(\mathbf{y},t).$$
(25.60)

The reader who wonders why the differential equations stops at second order in $\partial/\partial y^\mu$ is reminded of the Pawula theorem. It states that for any positive definite propagator either this is the case or one must take into account derivatives to all orders, see e.g. Section 4.3 of (Risken, 1989).

We want to save ourselves from going through a rigorous derivation of (25.57) and refer to textbooks like (Risken, 1989), (Honerkamp, 1994), (van Kampen, 2001) and (Gardiner, 2002) instead. Rather, we want to present a more pedestrian proof for slightly simplified conditions. It is based on the observation that to lowest non-trivial order the short-time propagator is of Gaussian form. Actually, our proof is analogous to R.P. Feynman's derivation of the Schrödinger equation from the short-time propagator of path integrals (Feynman and Hibbs, 1965). Indeed, relations (25.54)–(25.56) tell one that the functional form of $K(\mathbf{y}_2, t+\tau|\mathbf{y}_1, t)$ (with respect to the final variables $\mathbf{y}_2$) is determined by the first and second moments, as is the case for the Gaussian of (25.31). From eqn(25.54) it is observed that to lowest order in τ the first moments are given by

$$\langle \mathbf{y}_2 \rangle_\tau \equiv \int d^n y_2 \, \mathbf{y}_2 \, K(\mathbf{y}_2, t+\tau|\mathbf{y}_1, t) = \mathbf{y}_1 + \mathbf{A}(\mathbf{y}_1, t)\,\tau. \tag{25.61}$$

For the second moments one gets

$$\Sigma_{\mu\nu}(\tau) \equiv \int d^n y_2 \left(y_2^\mu - \langle y_2^\mu \rangle_\tau\right)\left(y_2^\nu - \langle y_2^\nu \rangle_\tau\right) K(\mathbf{y}_2, t+\tau|\mathbf{y}_1, t) = 2D_{\mu\nu}^{\text{n.s.}}(\mathbf{y}_1, t)\,\tau. \tag{25.62}$$

The easiest way to see this is to rewrite the first two factors of the integrand of (25.55) as

$$(\mathbf{y}_2 - \mathbf{y}_1)_\mu (\mathbf{y}_2 - \mathbf{y}_1)_\nu \equiv \left(y_2^\mu - \langle y_2^\mu \rangle_\tau + \langle y_2^\mu \rangle_\tau - y_1^\mu\right)\left(y_2^\nu - \langle y_2^\nu \rangle_\tau + \langle y_2^\nu \rangle_\tau - y_1^\nu\right), \tag{25.63}$$

to observe that $\int d^n y_2 \, (\mathbf{y}_2 - \langle \mathbf{y}_2 \rangle_\tau) K(\mathbf{y}_2, t+\tau|\mathbf{y}_1, t) = 0$ and to discard terms of order two in τ. The Gaussian finally becomes

$$K(\mathbf{y}_2, t+\tau|\mathbf{y}_1, t) = (2\pi)^{-n/2} \left[\det(2\tau\mathbf{D})\right]^{-1/2} \exp\left(-\frac{A}{4\tau}\right) \tag{25.64}$$

with

$$A = [\mathbf{y}_2 - \mathbf{y}_1 - \mathbf{A}(\mathbf{y}_1, t_1)\tau]_\mu \left[\mathbf{D}^{-1}(\mathbf{y}_1, t_1)\right]_{\mu\nu} [\mathbf{y}_2 - \mathbf{y}_1 - \mathbf{A}(\mathbf{y}_1, t)\tau]_\nu. \tag{25.65}$$

As expected, for $\tau \to 0$ this $K(\mathbf{y}_2, t+\tau|\mathbf{y}_1, t)$ satisfies the initial condition (25.59).

To simplify the problem let us stick to the one-dimensional case. Furthermore, we shall assume a constant diffusion coefficient D and a drift term which depends only on y. Let us start by inserting into the Chapman–Kolmogorov equation the Gaussian form of the short-time propagator:

$$d(y, t+\tau) = \int_{-\infty}^{+\infty} dy_0 \, K(y, t+\tau|y_0, t) d(y_0, t)$$
$$\simeq \int_{-\infty}^{+\infty} \frac{dy_0}{\sqrt{4\pi\tau D}} \exp\left[-\frac{1}{4\tau D}\left(y - y_0 - \tau A(y_0)\right)^2\right] d(y_0, t) \tag{25.66}$$

Let us introduce the coordinate $\xi = y_0 - y$ and transform the integral to

$$d(y, t+\tau) \simeq \int_{-\infty}^{+\infty} \frac{d\xi}{\sqrt{4\pi\tau D}} \times$$
$$\exp\left[-\frac{1}{4\tau D}\left(\xi^2 + 2\tau\xi A(y+\xi) + \tau^2 (A(y+\xi))^2\right)\right] d(y+\xi, t) \tag{25.67}$$

The integrand is dominated by the normalized Gaussian

$$G(\xi) = (4\pi\tau D)^{-1/2} \exp[-\xi^2/4\tau D] \qquad (25.68)$$

whose width increases with τ. To lowest order we may treat the fluctuations in ξ as $\xi^2 \sim \tau D$. Besides the ξ^2 itself all other terms in the exponent of (25.67) are of higher order such that we may apply the expansion

$$\frac{1}{\sqrt{4\pi\tau D}} \exp\left[-\frac{1}{4\tau D}\left(\xi^2 + 2\tau\xi A(y+\xi) + \tau^2 A(y+\xi)^2\right)\right]$$
$$\simeq G(\xi) \exp\left[-\frac{\xi}{2D} A(y+\xi) - \frac{\tau}{4D} A(y+\xi)^2\right] \qquad (25.69)$$
$$\simeq G(\xi)\left[1 - \frac{\xi A(y)}{2D} - \left(\xi^2\left\{-\frac{1}{2D}\frac{\partial A(y)}{\partial y} - \frac{1}{2}\frac{A(y)^2}{(2D)^2}\right\} - \frac{\tau A(y)^2}{4D}\right)\right]$$

Likewise we may expand the $d(y_0 + \xi, t)$

$$d(y+\xi, t) \simeq d(y,t) + \xi\frac{\partial d(y,t)}{\partial y} + \frac{1}{2}\xi^2\frac{\partial^2 d(y,t)}{\partial y^2}$$

Combining both and observing that our normalized Gaussian has a vanishing first moment and that the second one gives $\int_{-\infty}^{+\infty} d\xi\, G(\xi)\xi^2 = \tau d$ we get for the $d(y, t+\tau)$ of (25.67) the form

$$d(y, t+\tau) \simeq d(y,t) - \tau\frac{\partial}{\partial y} A(y) d(y,t) + \tau D \frac{\partial^2}{\partial y^2} d(y,t) \qquad (25.70)$$

which for infinitesimal τ turns into the expected Fokker–Planck equation

$$\frac{\partial}{\partial t} d(y,t) = \left[-\frac{\partial}{\partial y} A(y) + D\frac{\partial^2}{\partial y^2}\right] d(y,t). \qquad (25.71)$$

25.2.3.1 *The corresponding Langevin equations* The transport coefficients A_μ and $D_{\mu\nu}^{\text{n.s.}}$ of (25.57) are uniquely specified by the first and second moments of the short-time propagator. If a particular process is defined by Langevin equations this feature allows one to deduce the corresponding Fokker–Planck equation. For an n-dimensional system the set of the general Langevin equations (with $\mu = 1 \cdots n$) may be written as

$$\dot{Y}_\mu(t) = a_\mu(\mathbf{Y}(t), t) + c_{\mu\nu}(\mathbf{Y}(t), t) L_\nu(t), \qquad \mathbf{Y}(t_1) = \mathbf{y}_1. \qquad (25.72)$$

The $\mathbf{Y}(t)$ is the vector of the stochastic variables we like to study. The $L_\mu(t)$, which determine the stochastic forces, are meant to be given by the normal Gaussian distribution of zero average and width Λ:

$$\langle L_\nu(t) \rangle = 0 \qquad \langle L_\mu(t) L_\nu(t+s) \rangle = \Lambda \delta_{\mu\nu} \delta(s). \qquad (25.73)$$

According to (25.54) and (25.55) the transport coefficients are found from

$$A_\mu(\mathbf{y}, t) = \lim_{\tau \to 0} \frac{1}{\tau} \langle Y_\mu(t+\tau) - y_\mu \rangle |_{\mathbf{Y}(t)=\mathbf{y}} \tag{25.74}$$

$$D_{\mu\nu}^{\text{n.s.}}(\mathbf{y}, t) = \frac{1}{2} \lim_{\tau \to 0} \frac{1}{\tau} \langle (Y_\mu(t+\tau) - y_\mu)(Y_\nu(t+\tau) - y_\nu) \rangle |_{\mathbf{Y}(t)=\mathbf{y}}. \tag{25.75}$$

The average values may be calculated from (25.72) in the same way as explained in Section 25.1.1 for the one-dimensional case. It had already been indicated there that a problem appears whenever the coefficient $c_{\mu\nu}(\mathbf{Y}(t), t)$ of the stochastic force varies with the stochastic variable $\mathbf{Y}$. In the literature such a case is called *multiplicative noise*; otherwise the noise is referred to as *additive*. For multiplicative noise one must first define precisely how the equations are to be integrated. Two suggestions are known in the literature, one by K. Itô and one by R.L. Stratonovich. We shall discuss this difficulty and its physically motivated solution in Section 25.3 for the example of a system with one coordinate. It will be argued that for all systems in which the noise comes from internal motion the Stratonovich prescription is the right one. In this case the drift and diffusion coefficients of the Fokker–Planck equation are given by

$$A_\mu(\mathbf{y}, t) = a_\mu(\mathbf{y}, t) + \frac{\Lambda}{2} c_{\nu\rho}(\mathbf{y}, t) \frac{\partial}{\partial y_\nu} c_{\mu\rho}(\mathbf{y}, t) \tag{25.76}$$

$$D_{\mu\nu}^{\text{n.s.}}(\mathbf{y}, t) = \frac{\Lambda}{2} c_{\mu\rho}(\mathbf{y}, t) c_{\nu\rho}(\mathbf{y}, t), \tag{25.77}$$

respectively. It is seen that the drift term gets a *noise induced* contribution.

25.2.4 The master equation

Like the Fokker–Planck equation, the master equation too can be derived from the Chapman–Kolmogorov equation if only the short-time propagator $K(y_2, t + \tau | y_1, t)$ is known. We restrict ourselves to homogeneous processes for the one-dimensional case. To first order in τ one may write

$$K(y_2, t + \tau | y_1, t) = [1 - a_0(y_2)\tau] \delta(y_2 - y_1) + \tau W(y_2|y_1) + O(\tau^2). \tag{25.78}$$

Here, $W(y_2|y_1)$ is the *transition rate*, which is to say the *transition probability per unit of time* from y_1 to y_2, and $[1 - a_0(y_2)\tau]$ is the probability that within the time lapse τ no transition occurs at all. As both must add up to one when summing over all final states, one must have $a_0(y_1) = \int dy_2 \, W(y_2|y_1)$. Please notice that the system is allowed to make finite jumps, in contrast to the case of the Fokker–Planck equation where the individual trajectories are continuous (without necessarily having a continuous derivative). The form (25.78) may now be inserted into the Chapman–Kolmogorov equation for the propagator of the form (25.48). After performing the limit $\tau \to 0$ one gets a master equation for the propagator $K(y_3|y_1; \tau) = K(y_3, t + \tau|y_1, t)$:

$$\frac{\partial}{\partial \tau} K(y_3|y_1;\tau) = \int dy_2 \Big(W(y_3|y_2) K(y_2|y_1;\tau) - W(y_2|y_3) K(y_3|y_1;\tau) \Big) \quad (25.79)$$

Multiplying this equation by $P(y_1,t)$ and integrating over y_1 one arrives at

$$\frac{\partial}{\partial t} P(y,t) = \int \Big[W(y|y') P(y',t) - W(y'|y) P(y,t) \Big] dy'. \quad (25.80)$$

Such equations are also valid for cases where Y takes on only *discrete* values labelled by n, which then reads

$$\frac{d}{dt} P_n(t) = \sum_{n'} \Big[W_{nn'} P_{n'}(t) - W_{n'n} P_n(t) \Big]. \quad (25.81)$$

It is easily recognized that these equations are of the form of *balance equations*: The first term measures the *gain*, the second one the *loss* of probability. In (25.81), for example, the first term accounts for all processes by which the state n is "fed" from all other states with the transition rate $W_{nn'}$. By the same mechanism, the state n *loses* probability to all the others. This terminology indicates that such equations may as well be applied to *quantal processes*. For instance, we might calculate the transition probability from Fermi's golden rule

$$W_{nn'} = \frac{2\pi}{\hbar} |V_{nn'}|^2 \Omega_\beta(E_n) \quad (25.82)$$

for a transition between quantal states $|n\rangle$ of energy E_n with level density $\Omega_\beta(E_n)$ and the transition matrix element $V_{nn'}$. In a non-perturbative approach one may replace this $V_{nn'}$ by the T-matrix $T_{nn'}$, see the discussion in Section 18.1.1.

The master equation and its ingredients have a few very important properties (van Kampen, 2001) which we would not like to miss mentioning. These remarks are directed mainly at common physical systems, where the microscopic dynamics is determined by a Hamiltonian. We will refer to the discrete form (25.81), but a transcription to the continuous forms shown before should not cause any problems.

(1) The solutions become stationary for

$$\sum_{n'} \Big[W_{nn'} P_{n'}(t) - W_{n'n} P_n(t) \Big] = 0. \quad (25.83)$$

This solution is reached for $t \to \infty$. (The proof requires a finite number of discrete states, but the statement is valid for the majority of other systems as well).
(2) This limit is related to thermal equilibrium for which entropy reaches a maximal value. Defining entropy by $S = -k_B P_n(t) \ln P_n(t)$ it can be shown to *increase* with t. This is the essential content or a consequence of the famous "H-theorem".

(3) In equilibrium there exists the relation of *detailed balance*,

$$W_{nn'} P^{eq}_{n'} = W_{n'n} P^{eq}_{n}, \qquad (25.84)$$

which is much stronger than (25.83). In Section 18.4.2 it is discussed in connection with reaction theory and used in Section 9.1 for pre-equilibrium reactions. Remember also the similar relation (25.53) shown in the previous section. However, there is an essential difference in that now the $K(-\tau)$ appearing on the right-hand side of (25.53) is replaced by $K(\tau)$. Indeed, (25.84) is valid only for systems with time reversal invariance, which is given, for instance, when the microscopic Hamiltonian is an even function in the momenta.

25.3 Non-linear equations in one dimension

We now take up the discussion of the problems one has with the Langevin equation mentioned above for non-linear forces. As the main nuclear application is found for Smoluchowski type equations we choose as variable the collective coordinate Q. This discussion will be accomplished by a presentation of further properties of the Fokker–Planck equation which will be needed in the subsequent section.

25.3.1 Transport equations for multiplicative noise

In one dimension the Langevin equation for non-linear forces has the form

$$\dot{Q} = a(Q) + c(Q) L(t). \qquad (25.85)$$

Any explicit time dependence of the coefficients will be ignored. Like before, the L is supposed to vanish on the ensemble average $\langle L(t) \rangle = 0$ and to be correlated according to eqn(25.73). The "Kraftstoß" Δw cannot be defined simply as in (25.6). One needs to specify how the function $c(Q(t))$ behaves within the time interval Δt. For this problem two prescriptions are known in the mathematical literature. The problem does not show up if one replaces the sharp delta function by a smooth function $\bar{\delta}(s)$ having a finite width associated with the microscopic timescale τ_{micro}. This changes (25.3) or (25.73) into $\langle L(t) L(t+s) \rangle = \Lambda \bar{\delta}(s)$, and, more importantly, equation (25.85) behaves smoothly in time. This allows one to apply the common transformation to a new variable $\widetilde{Q}$ for which the Langevin equation becomes

$$\dot{\widetilde{Q}} = \widetilde{a}(\widetilde{Q}) + L(t). \qquad (25.86)$$

It is obtained by dividing (25.85) by $c(Q)$ and by defining the $\widetilde{Q}$ through $d\widetilde{Q} = dQ/c(Q)$ with $\widetilde{a}(\widetilde{Q}) = a(Q)/c(Q)$. In this new equation the dangerous term has disappeared and the corresponding Fokker–Planck equation reads

$$\frac{\partial}{\partial t} \widetilde{f}(\widetilde{Q}, t) = -\frac{\partial}{\partial \widetilde{Q}} \widetilde{a}(\widetilde{Q}) \widetilde{f}(\widetilde{Q}, t) + \frac{\Lambda}{2} \frac{\partial^2}{\partial \widetilde{Q}^2} \widetilde{f}(\widetilde{Q}, t). \qquad (25.87)$$

Recall that for a linear $\widetilde{a}$ this equivalence had been demonstrated in Section 25.1, but it also holds true for this *quasi-linear* case (van Kampen, 2001) in which the non-linearity appears only in the first term (the *drift term*). To get the Fokker–Planck equation associated with the original Langevin equation (25.85) we only have to transform back to the Q coordinate. In doing so we must also transform the density $\tilde{f}(\tilde{Q}, t)$ to the $f(Q, t)$ obeying $\tilde{f}(\tilde{Q}, t) = f(Q, t)c(Q)$, such that both are normalized properly in each system. In this way one obtains

$$\frac{\partial}{\partial t} f(Q, t) = -\frac{\partial}{\partial Q} a(Q) f(Q, t) + \frac{\Lambda}{2} \frac{\partial}{\partial Q} c(Q) \frac{\partial}{\partial Q} c(Q) f(Q, t), \qquad (25.88)$$

where the differential operators act on all factors on their right. For future reference let us rewrite this Fokker–Planck equation in the general form

$$\frac{\partial}{\partial t} f(Q, t) = -\frac{\partial}{\partial Q} \Big(A(Q, t) f(Q, t) \Big) + \frac{\partial^2}{\partial Q^2} \Big(D(Q, t) f(Q, t) \Big). \qquad (25.89)$$

It is equivalent to (25.88) if only the coefficients are chosen as

$$A(Q) = a(Q) + \frac{\Lambda}{2} c(Q) \frac{\partial c(Q)}{\partial Q} \quad \text{and} \quad D(Q) = \frac{\Lambda}{2} [c(Q)]^2. \qquad (25.90)$$

This form makes evident that the non-linear stochastic force implies a ' 'noise induced" contribution to the drift term.

Let us return now to the question of how one may treat the original Langevin equation (25.85) in a mathematically proper way. As suggested by Stratonovich this is possible after defining the integral in (25.6) in the following way:

$$\Delta w = c \left(\frac{Q(t) + Q(t + \Delta t)}{2} \right) \int_t^{t+\Delta t} L(s) ds. \qquad (25.91)$$

For a "smooth delta function" $\bar{\delta}(s)$ this result is very plausible. It only requires that the $c(Q(t))$ vary slowly on the scale of the $L(t)$ such that it may be replaced by its average value and taken out of the integral. For a sharp delta function, on the other hand, the choice (25.91) is not the only possible one. Another option which is due to Itô reads

$$\Delta w = c(Q(t)) \int_t^{t+\Delta t} L(s) ds. \qquad (25.92)$$

It implies a Fokker–Planck equation of the form (25.89) but with (Risken, 1989), (van Kampen, 2001)

$$A(Q) = a(Q) \quad \text{and} \quad D(Q) = \frac{\Lambda}{2} [c(Q)]^2. \qquad (25.93)$$

Actually, as observed in (Bundschuh, 2004), a different type of equation is obtained if the integral is defined by extracting the value the c has at the end of the time interval. For $\Delta w = c(Q(t + \Delta t)) \int_t^{t+\Delta t} L(s) ds$ one gets a result which up to a factor of 2 is similar to the Stratonovich case, namely $A(Q) = a(Q) + \Lambda c(Q) \partial c(Q) / \partial Q$ and $D(Q) = (\Lambda/2) [c(Q)]^2$.

25.3.2 Properties of the general Fokker–Planck equation

As is easily seen the form (25.89) has the structure of the continuity equation

$$\frac{\partial}{\partial t} f(Q,t) + \frac{\partial}{\partial Q} j(Q,t) = 0 \qquad (25.94)$$

with the current density

$$j(Q,t) = A(Q,t) f(Q,t) - \frac{\partial}{\partial Q}\Big(D(Q,t) f(Q,t)\Big). \qquad (25.95)$$

From the previous discussion we know that the Fokker–Planck equation may be solved in terms of the propagator $K(Q,t|Q',t')$ which has to satisfy the initial condition (25.59). By way of the Chapman–Kolmogorov equation any other distribution may then serve as the initial condition for the $f(Q,t)$. If the transport coefficients are *in*dependent of time the conditional probability only depends on the time difference. As we recall from Section 25.2.2 such processes are called *homogeneous*. For eqn(25.89) this property may be specified as

$$\begin{aligned} A(Q,t) = A(Q), \quad D(Q,t) = D(Q) \quad &\text{(homogeneous process)} \\ \implies K(Q,t|Q',t') = K(Q,t-t'|Q',0) &= K(Q,0|Q',t'-t). \end{aligned} \qquad (25.96)$$

25.3.2.1 Boundary conditions
With respect to the coordinate Q the solutions $f(Q,t)$ must satisfy certain boundary conditions at the "surface S" of the region inside which they are to be calculated. The Gaussian solutions are commonly valid in the whole space and the boundary conditions simply means that the function has to vanish (sufficiently) fast at $\pm\infty$. However, there may be more subtle situations. For future reference it suffices to study two possibilities, which in the literature are referred to as "absorbing barrier" and "reflecting barrier". For one dimension the S reduces to the two ends of an interval, say $S = Q_1$ and Q_2 with $Q_1 < Q_2$.

Absorbing barrier: In this case the boundary condition reads

$$f(Q,t) = 0 \quad \text{for} \quad Q = Q_1, Q_2. \qquad (25.97)$$

It applies to a system where the "particles" are removed as soon as they reach the boundaries. In a sense this situation includes the conditions for the Gaussians at $Q_1 = Q_2 = \pm\infty$.

Reflecting barrier: For a situation in which the particles do not cross the boundaries the net flow to the outside of the interval must vanish. Formally this can be described by requiring

$$-j(Q = Q_1, t) = 0 = j(Q = Q_2, t). \qquad (25.98)$$

Evidently, both conditions could be mixed and in each case the barriers might be at infinity.

25.3.2.2 The backward equation
With respect to initial time t' and coordinate Q' the $K(Q,t|Q',t')$ satisfies the *backward or adjoint equation*

$$\frac{\partial}{\partial t'} K(Q,t|Q',t') = -\left\{ A(Q',t') \frac{\partial}{\partial Q'} + D(Q',t') \frac{\partial^2}{\partial Q'^2} \right\} K(Q,t|Q',t'). \tag{25.99}$$

The boundary conditions for the backward equation take on the forms

$$K(Q,t|Q',t') = 0 \quad \text{for} \quad Q' = Q_1, Q_2 \quad \text{absorbing b.}, \tag{25.100}$$

$$\frac{\partial}{\partial Q'} K(Q,t|Q',t') = 0 \quad \text{for} \quad Q' = Q_1, Q_2 \quad \text{reflecting b.} \tag{25.101}$$

25.4 The mean first passage time

In this section we want to present the concept of the mean first passage time (MFPT), exemplified for the one-dimensional case with the Fokker–Planck equation (25.89). Quite generally, the MFPT represents the average time a system spends within a certain interval (or region in the multidimensional case) once it is released at a certain point Q_0 initially, say at time $t_0 = 0$. Because of this condition it is to be calculated from the conditional probability $K(Q,t|Q_0,0)$. Various different situations may be treated in this way, which are discussed in appropriate textbooks, like (Risken, 1989), (van Kampen, 2001) and (Gardiner, 2002). Our presentation is oriented at the application just mentioned and largely follows the exposition given in (Gardiner, 2002).

Suppose the initial point Q_0 lies inside of some interval, say $Q_1 \leq Q_0 \leq Q_2$. The probability of finding at time t the particle still inside is given by

$$W(Q_0, t) = \int_{Q_1}^{Q_2} dQ \, K(Q,t|Q_0, 0). \tag{25.102}$$

Hence, the probability for the particle to leave the region during the time lapse $(t, t+dt)$ is determined by $-dW$ which may be written as

$$-dW(Q_0, t) \equiv -\frac{\partial}{\partial t} W(Q_0, t) dt = -\int_{Q_1}^{Q_2} dQ \, \frac{\partial}{\partial t} K(Q,t|Q_0, 0) \, dt. \tag{25.103}$$

The mean first passage time is thus given by

$$\tau_{\text{mfpt}}(Q_0) = -\int t \, dW(Q_0, t) = -\int_0^\infty t \frac{\partial}{\partial t} W(Q_0, t) \, dt. \tag{25.104}$$

By partial integration this expression turns into

$$\tau_{\text{mfpt}}(Q_0) = \int_0^\infty W(Q_0, t) \, dt \tag{25.105}$$

$$= \int_0^\infty \int_{Q_1}^{Q_2} K(Q,t|Q_0, 0) \, dQ \, dt. \tag{25.106}$$

Remember, however, that the solution for K to be used here must fulfill the right boundary condition. We will concentrate on the case of having a reflecting barrier at $Q = Q_1$ and an absorbing barrier at $Q = Q_2$, as defined in Section 25.1.2.

It is worthwhile to express the $\tau_{\mathrm{mfpt}}(Q_0)$ through an expression which involves the current. This is easily done through the continuity equation, which for the present case may be written as

$$\frac{\partial}{\partial t} K(Q,t\,|\,Q_0,0) + \frac{\partial}{\partial Q} j(Q,t\,|\,Q_0,0) = 0 \tag{25.107}$$

and, hence, implies

$$\frac{\partial}{\partial t} W(Q_0,t) = -\int_{Q_1}^{Q_2} dQ\, \frac{\partial}{\partial Q} j(Q,t\,|\,Q_0,0) = j(Q_1,t\,|\,Q_0,0) - j(Q_2,t\,|\,Q_0,0). \tag{25.108}$$

Thus the mean first passage time of (25.104) may be written as

$$\tau_{\mathrm{mfpt}}(Q_0) = \int_0^\infty dt\, t\, \Big(j(Q_2,t\,|\,Q_0,0) - j(Q_1,t\,|\,Q_0,0) \Big). \tag{25.109}$$

25.4.1 Differential equation for the MFPT

To get numerical values requires solving the transport equation (25.89) in one way or another. For a homogenous system, where the coefficients $A(Q)$ and $D(Q)$ are independent of time, it is possible to derive for $\tau_{\mathrm{mfpt}}(Q_0)$ an ordinary differential equation. We will assume such a case to be given and will sketch the derivation for a situation of a reflecting barrier on the left, say at Q_1, and an absorbing barrier at Q_2.

Take the basic formula (25.104) together with the definition of the probability $W(Q_0,t)$ of (25.102). Evidently the variation of the latter with Q_0 can be traced back to that of the $K(Q,t\,|\,Q_0,0)$ with respect to the initial coordinate Q_0. Because of (25.96) one has $K(Q,t\,|\,Q_0,0) = K(Q,0\,|\,Q_0,-t)$. Hence, by differentiating the $W(Q_0,t)$ of (25.102) with respect to time t one may involve the backward equation (25.99) for the integrand to get the following partial differential equation

$$\frac{\partial}{\partial t} K(Q,t\,|\,Q_0,0) = A(Q_0) \frac{\partial}{\partial Q_0} K(Q,t\,|\,Q_0,0) + D(Q_0) \frac{\partial^2}{\partial Q_0^2} K(Q,t\,|\,Q_0,t). \tag{25.110}$$

The minus sign on the right-hand side of (25.99) is cancelled because we have $t' = -t$. Integration over Q leads to the following equation for $W(Q_0,t)$:

$$\frac{\partial}{\partial t} W(Q_0,t) = A(Q_0) \frac{\partial}{\partial Q_0} W(Q_0,t)e + D(Q_0) \frac{\partial^2}{\partial Q_0^2} W(Q_0,t). \tag{25.111}$$

To specify its solution uniquely one needs initial conditions with respect to t and boundary conditions for the coordinate dependence. As $K(Q,t=0\,|\,Q_0,0) = \delta(Q-Q_0)$ the initial condition for $W(Q_0,t)$ turns out to be:

$$W(Q_0, t=0) = \begin{cases} 1 & \text{for } Q_1 \ll Q_0 \ll Q_2, \\ 0 & \text{elsewhere.} \end{cases} \qquad (25.112)$$

The boundary conditions follow from those for the backward equation discussed in Section 25.3.2.2. For a reflecting barrier at Q_1 and an absorbing barrier at Q_2 one gets from (25.101) and (25.100)

$$\frac{\partial}{\partial Q_0} W(Q_0, t)\Big|_{Q_0=Q_1} = 0 \qquad (25.113)$$

$$W(Q_0 = Q_2, t) = 0. \qquad (25.114)$$

To obtain the desired equation for $\tau_{\text{mfpt}}(Q_0)$ we may simply integrate (25.111) over times according to (25.105) and get the ordinary differential equation

$$-1 = A(Q_0) \frac{d\tau_{\text{mfpt}}(Q_0)}{dQ_0} + D(Q_0) \frac{d^2 \tau_{\text{mfpt}}(Q_0)}{dQ_0^2}. \qquad (25.115)$$

The left-hand side comes from

$$\int_0^\infty dt \frac{\partial}{\partial t} W(Q_0, t) = W(Q_0, \infty) - W(Q_0, 0) = -1, \qquad (25.116)$$

which in turn is a consequence of (25.112) and the same assumption which was behind the step from (25.104) to (25.105). For the situation we have in mind the boundary conditions for τ_{mfpt} are easily seen to be

$$\frac{d\tau_{\text{mfpt}}(Q_0)}{\partial Q_0}\Big|_{Q_0=Q_1} = 0 \qquad (25.117)$$

$$\tau_{\text{mfpt}}(Q_0 = Q_2) = 0. \qquad (25.118)$$

As one may verify by direct calculation the solution is given by

$$\tau_{\text{mfpt}}(Q_0) = 2 \int_{Q_0}^{Q_2} du \frac{1}{\Psi(u)} \int_{Q_1}^{u} dv \frac{\Psi(v)}{D(v)} \qquad (25.119)$$

where

$$\Psi(x) = \exp\left[\int_{Q_1}^{x} dv \frac{2A(y)}{D(y)}\right]. \qquad (25.120)$$

25.5 The multidimensional Kramers equation

It will turn out convenient to make use again of the summation convention and to distinguish further between covariant and contravariant quantities. They are defined with respect to a metric ds^2 fixed by the kinetic energy through the relation $2E_{\text{kin}} = ds^2/dt^2$ where $ds^2 = M_{\mu\nu} dQ^\mu dQ^\nu$ for $\mu, \nu = 1\ldots n$. As the inertial tensor $M_{\mu\nu}$ will be dependent on the coordinates one is dealing with a curvilinear system. With the contravariant inertial tensor $M^{\mu\nu}$ defined through $M_{\mu\eta}M^{\eta\nu} = \delta_\mu^\nu$ (with $\delta_\mu^\nu = 0$ for $\mu \neq \nu$ and 1 for $\mu = \nu$) one has $M_{\mu\nu}\dot{Q}^\mu\dot{Q}^\nu = M^{\mu\nu}P_\mu P_\nu$. Assuming a potential $V(Q) \equiv V(Q^1,..,Q^\mu,...Q^n)$ the Fokker–Planck equation may then be written as

$$\frac{\partial}{\partial t}f(Q,P,t) = \{\mathcal{H}(Q,P),\, f(Q,P,t)\} \\ + \gamma_{\mu\eta}(Q)\frac{\partial}{\partial P_\mu}\left(M^{\eta\nu}(Q)P_\nu + T(t)\frac{\partial}{\partial P_\eta}\right)f(Q,P,t). \quad (25.121)$$

The first line represents the classical Liouville equation with the Hamilton function $\mathcal{H}(Q,P) = \frac{1}{2}M^{\mu\nu}(Q)P_\mu P_\nu + V(Q)$ and the Poisson brackets defined as in (2.82) but with co- and contravariant vectors $\partial/\partial Q^\mu$ and $\partial/\partial P_\mu$. In writing these equations we have adopted the convention of leaving out indices of coordinates and momenta when they show up as arguments of functions. The diffusion coefficients are given by the classical Einstein relation

$$D_{P_\mu P_\nu} = \gamma_{\mu\nu}T \quad \text{with} \quad D^{Q^\mu Q^\nu} = D_{P_\mu}^{Q_\nu} = 0. \quad (25.122)$$

Besides the inertias the friction coefficients are also allowed to depend on the coordinates. The temperature, on the other hand, may vary in time but not in Q. The solutions of (25.121) should be normalized to unity like $\int d\Gamma f(Q,P,t) = 1$ with $d\Gamma = \prod_\mu^n (dP_\mu dQ^\mu)$. For constant temperature, and a stable potential, the stationary solutions are those of the canonical equilibrium, $f_{\text{eq}}(Q,P) = \exp(-\beta\mathcal{H})/Z$.

25.5.1 Gaussian solutions in curvilinear coordinates

We are now going to construct a Gaussian solution which is centered at the classical trajectory, defined through Newton's equations

$$P_\mu^c(t) = M_{\mu\nu}\frac{dQ_\nu^c}{dt} \qquad \frac{dP_\mu^c}{dt} = -\frac{\partial V(Q_\nu^c(t))}{\partial Q_\mu^c} - \gamma_{\mu\eta}M^{\eta\nu}P_\nu^c(t). \quad (25.123)$$

Using the shorthand notation

$$(\mathcal{X}_i) = \begin{pmatrix} P_\mu \\ Q_\mu \end{pmatrix} \quad \text{with} \quad \mu = 1\ldots n;\quad i = 1\ldots 2n$$

for the variables and the following ones for the first and second moments

$$\mathcal{X}_i^c(t) = \langle \mathcal{X}_i \rangle_t \equiv \int d\Gamma\, \mathcal{X}_i f_G(\mathcal{X},t)$$

$$\Sigma_{ij} = \langle (\mathcal{X}_i - \mathcal{X}_i^c(t))(\mathcal{X}_j - \mathcal{X}_j^c(t)) \rangle_t \equiv \int d\Gamma\, (\mathcal{X}_i - \mathcal{X}_i^c(t))(\mathcal{X}_j - \mathcal{X}_j^c(t))f_G(\mathcal{X},t),$$

the multidimensional Gaussians have the form

$$f_{\rm G}(Q,P,t) = (2\pi)^{-n/2} \left(\det \Sigma_{\mu\nu}\right)^{-1/2} \times$$
$$\exp\left\{-\frac{1}{2}\sum_{i,k}(\mathcal{X}_i - \mathcal{X}_i^c(t))(\mathcal{X}_k - \mathcal{X}_k^c(t))\left(\Sigma^{-1}\right)_{\mu\nu}\right\}. \tag{25.124}$$

This is a solution of (25.121) if and only if the fluctuations are sufficiently small, such that this Fokker–Planck equation can be linearized by applying an appropriate expansion around the $\mathcal{X}_i^c(t)$. The strategy for this procedure is as follows: One must keep only terms that lead to a new Fokker–Planck equation which is at most quadratic in the $\mathcal{X}_i - \mathcal{X}_i^c$ together with the partial derivatives $\partial f(\mathcal{X},t)/\partial \mathcal{X}_i$ (both are to be counted as being of the same order). For the conservative force this means to approximate

$$-\frac{\partial V(Q(t))}{\partial Q_\mu} \approx -\left.\frac{\partial V(Q(t))}{\partial Q_\mu}\right|_{Q=Q^c(t)} - \frac{1}{2}(Q_\eta - Q_\eta^c)\left.\frac{\partial^2 V(Q(t))}{\partial Q_\eta^2}\right|_{Q=Q^c(t)}.$$

Equivalently one may expand the potential inside the Poisson brackets, to second order. The same has to be done for the kinetic energy, but now to second order both with respect to coordinates *and* momenta:[58]

$$2E_{\rm kin} = P_\mu^c P_\nu^c M_c^{\mu\nu} + 2P_\mu^c\left(P_\nu - P_\nu^c\right)M_c^{\mu\nu} + \left(P_\mu - P_\mu^c\right)\left(P_\nu - P_\nu^c\right)M_c^{\mu\nu}$$
$$+ P_\mu^c P_\nu^c \left[(Q^\eta - Q_c^\eta)\frac{\partial M_c^{\mu\nu}}{\partial Q_c^\eta} + \frac{1}{2}(Q^\eta - Q_c^\eta)(Q^\xi - Q_c^\xi)\frac{\partial^2 M_c^{\mu\nu}}{\partial Q_c^\eta \partial Q_c^\xi}\right] \tag{25.125}$$
$$+ 2\left(P_\mu - P_\mu^c\right)P_\nu^c (Q^\eta - Q_c^\eta)\frac{\partial M_c^{\mu\nu}}{\partial Q_c^\eta}.$$

With the abbreviation

$$\begin{pmatrix} \Pi_{\mu\nu}(t) & \Psi_\mu^\nu(t) \\ \Psi_\mu^\nu(t) & \Xi^{\mu\nu}(t) \end{pmatrix} \equiv \begin{pmatrix} \Sigma_{P_\mu P_\nu} & \Sigma_{P_\mu}^{Q_\nu} \\ \Sigma_{P_\mu}^{Q_\nu} & \Sigma^{Q_\mu Q_\nu} \end{pmatrix}$$

for the second moments their set of equations become

$$\frac{d\Xi^{\alpha\beta}(t)}{dt} = M_c^{\mu\alpha}\Psi_\mu^\beta + M_c^{\mu\beta}\Psi_\mu^\alpha + P_\mu^c\left(\frac{\partial M_c^{\alpha\mu}}{\partial Q_c^\xi}\right)\Xi^{\xi\beta} + P_\mu^c\left(\frac{\partial M_c^{\beta\mu}}{\partial Q_c^\xi}\right)\Xi^{\xi\alpha}$$

$$\frac{d\Psi_\alpha^\beta(t)}{dt} = M_c^{\mu\beta}\Pi_{\mu\alpha} + P_\mu^c\left(\frac{\partial M_c^{\beta\mu}}{\partial Q_c^\xi}\right)\Xi_\alpha^\xi - P_\mu^c\left(\frac{\partial M_c^{\eta\mu}}{\partial Q_c^\alpha}\right)\Xi_\eta^\beta$$
$$- P_\mu^c P_\nu^c \left(\frac{\partial^2 M_c^{\beta\mu}}{\partial Q_c^\eta \partial Q_c^\alpha}\right)\Xi^{\eta\beta} - C_{\alpha\eta}^c \Xi^{\eta\beta} - \gamma_{\alpha\nu}^c M_c^{\eta\nu}\Psi_\eta^\beta - \gamma_{\alpha\nu}^c P_\eta^c\left(\frac{\partial M_c^{\eta\nu}}{\partial Q_c^\xi}\right)\Xi^{\xi\beta}$$

---

[58] In nuclear physics this technique was first applied in (Hofmann and Ngô, 1976), (Ngô and Hofmann, 1977). However, in eqn(14) of (Ngô and Hofmann, 1977) the following typing errors should be observed: There should be no factor 1/2 in the term with the second order fluctuations in the momenta and the very last term should be multiplied by 2.

$$\begin{aligned}\frac{d\Pi_{\alpha\beta}(t)}{dt} = &- P_\mu^c \left(\frac{\partial M_c^{\eta\mu}}{\partial Q_c^\alpha}\right)\Psi_{\beta\eta} - P_\mu^c \left(\frac{\partial M_c^{\eta\mu}}{\partial Q_c^\beta}\right)\Psi_{\alpha\eta} \\ &- P_\mu^c P_\nu^c \left(\frac{\partial^2 M_c^{\mu\nu}}{\partial Q_c^\eta \partial Q_c^\xi}\right)\left(\delta_\alpha^\xi \Psi_\beta^\eta + \delta_\beta^\xi \Psi_\alpha^\eta\right) \\ &- C_{\alpha\eta}^c \Psi_\beta^\eta - C_{\beta\eta}^c \Psi_\alpha^\eta - \gamma_{\alpha\nu}^c M_c^{\eta\nu}\Pi_{\eta\beta} - \gamma_{\beta\nu}^c M_c^{\eta\nu}\Pi_{\eta\alpha} \\ &- \gamma_{\alpha\nu}^c P_\eta^c \left(\frac{\partial M_c^{\eta\nu}}{\partial Q_c^\xi}\right)\Psi_\beta^\xi - \gamma_{\beta\nu}^c P_\eta^c \left(\frac{\partial M_c^{\eta\nu}}{\partial Q_c^\xi}\right)\Psi_\alpha^\xi + 2\gamma_{\alpha\beta}^c T(t).\end{aligned}$$

As in (Ngô and Hofmann, 1977) terms which would result from an expansion of the first term in the second line of (25.121) have been ignored, namely

$$\gamma_{\mu\eta}(Q)M^{\eta\nu}(Q)P_\nu = \gamma_{\mu\eta}^c M_c^{\eta\nu} P_\nu + P_\nu^c \left(Q^\xi - Q_c^\xi\right)\left(\frac{\partial M_c^{\mu\nu}}{\partial Q_c^\xi}\gamma_{\mu\eta}^c + M_{\mu\eta}^c \frac{\partial \gamma_c^{\mu\nu}}{\partial Q_c^\xi}\right). \quad (25.126)$$

Solutions of these equations may also be used to construct the propagators $K_G(Q, P, t; Q_0, P_0, t_0)$. One only has to make sure to start with zero initial conditions $\sum_{\mu\nu}(t_0) = 0$.

Finally, a note is in order on the normalization in curvilinear coordinates. The Gaussian (25.124) is normalized to unity with respect to the volume element $d\Gamma = \prod_j d\mathcal{X}_i$. Moreover, the integrals are understood to extend from $-\infty$ to ∞. In the case of curvilinear coordinates this may not make sense without further precautions. For instance, if some of the coordinates only extend over a finite range, it must be assumed that the distribution is centered sufficiently far away from the edge of the corresponding interval. Next we must care about possible transformations to a new system of coordinates $\mathcal{X}_i$ defined through $\mathcal{X}_j = \mathcal{X}_j(\mathcal{Y})$. If we still want to work with the same Gaussian structure this transformation must be considered to lowest order only, in the sense

$$\mathcal{X}_j - \mathcal{X}_j^c \approx (\mathcal{Y}_k - \mathcal{Y}_k^c)\frac{\partial \mathcal{X}_j^c}{\partial \mathcal{Y}_k^c}.$$

In other words, the transformation matrix has to be taken at its "classical" value, as determined by average motion. This ensures that the second moments do behave as tensors. Of course, the functional determinant which might appear in the volume element has to be treated in the same way. If these precautions are considered, the form (25.124) may be taken to apply in any coordinate system with the volume element being given as defined above.

25.5.2 Time dependence of first and second moments for the harmonic oscillator

In this section we shall present analytic solutions of the equations for the first and second moments (7.9) and (7.17)–(7.19). It is convenient to first renormalize coordinate and momentum in substituting

$$Q \longrightarrow Q\sqrt{|C|/\varpi} \quad \text{and} \quad P \longrightarrow P/\sqrt{\varpi M}. \quad (25.127)$$

Now both Q and P are of dimension $\sqrt{\hbar}$ and the new Hamiltonian becomes $\mathcal{H} = (\varpi/2)(P^2 + \mathrm{sgn}C\, q^2)$, where the stiffness may be positive or negative. The equations for the first moments become

$$\frac{dQ_c(t)}{dt} = \varpi P \qquad \frac{dP_c(t)}{dt} = -\mathrm{sgn}C\, \varpi Q_c(t) - 2\eta\, \frac{dQ_c(t)}{dt}\,. \tag{25.128}$$

Introducing the timescale τ_{kin} defined by $\tau_{\mathrm{kin}} = 1/(2\eta\varpi) = \hbar/\Gamma_{\mathrm{kin}}$ they may be combined to

$$\frac{d^2 Q_c(t)}{dt^2} + \frac{1}{\tau_{\mathrm{kin}}}\frac{dQ_c(t)}{dt} + \mathrm{sgn}C\, \varpi^2\, Q_c(t) = 0\,. \tag{25.129}$$

Its solution becomes

$$\begin{aligned}
Q_c(t) &= \frac{1}{2\mathcal{E}}\left(i\varpi P_0 - q_0\omega^-\right)\exp(-i\omega^+(t-t_0)) \\
&+ \frac{1}{2\mathcal{E}}\left(-i\varpi P_0 + q_0\omega^+\right)\exp(-i\omega^-(t-t_0))
\end{aligned} \tag{25.130}$$

with $\mathcal{E} = \varpi\sqrt{\mathrm{sgn}C - \eta^2}$ and $\omega^\pm = \pm\mathcal{E} - i\Gamma_{\mathrm{kin}}/2$. The substitution (25.127) changes the fluctuations to

$$\Sigma_{qq} \to \widetilde{\Sigma}_{qq} = \frac{|C|}{\varpi}\Sigma_{qq}\,,\quad \Sigma_{pp} \to \widetilde{\Sigma}_{pp} = \frac{1}{M\varpi}\Sigma_{pp}\quad \text{and}\quad \widetilde{\Sigma}_{qp} \equiv \Sigma_{qp}\,.$$

The $\widetilde{\Sigma}_{\mu\nu}$ are of dimension $\hbar$. The equations of motion now read

$$\frac{d}{dt}\widetilde{\Sigma}_{qq}(t) - 2\varpi\widetilde{\Sigma}_{qp}(t) = 0 \tag{25.131}$$

$$\frac{d}{dt}\widetilde{\Sigma}_{qp}(t) - \varpi\left(\widetilde{\Sigma}_{pp}(t) - \mathrm{sgn}C\,\widetilde{\Sigma}_{qq}(t)\right) + \frac{1}{\tau_{\mathrm{kin}}}\widetilde{\Sigma}_{qp}(t) = 2\widetilde{D}_{qp} \tag{25.132}$$

$$\frac{d}{dt}\widetilde{\Sigma}_{pp}(t) + 2\,\mathrm{sgn}C\,\varpi\widetilde{\Sigma}_{qp}(t) + \frac{2}{\tau_{\mathrm{kin}}}\widetilde{\Sigma}_{pp}(t) = 2\widetilde{D}_{pp} \tag{25.133}$$

with the diffusion coefficients given by

$$\widetilde{D}_{qq} = 0\,,\quad \widetilde{D}_{qp} = \varpi\left(\mathrm{sgn}C\,\widetilde{\Sigma}_{qq}^{eq} - \widetilde{\Sigma}_{pp}^{eq}\right)\quad \text{and}\quad \widetilde{D}_{pp} = \frac{1}{\tau_{\mathrm{kin}}}\widetilde{\Sigma}_{pp}^{eq}\,. \tag{25.134}$$

The solutions of (25.131) to (25.132) can be written as

$$\begin{pmatrix}\widetilde{\Sigma}_{pp}(t) \\ \widetilde{\Sigma}_{qq}(t) \\ \widetilde{\Sigma}_{qp}(t)\end{pmatrix} = \begin{pmatrix}\widetilde{\Sigma}_{pp}^{eq} \\ \widetilde{\Sigma}_{qq}^{eq} \\ 0\end{pmatrix} + A_1 \begin{pmatrix}-i\omega^- \\ i\frac{\varpi^2}{\omega^-} \\ \varpi\end{pmatrix}\exp(-2i\omega^-(t-t_0))$$

$$+ A_2 \begin{pmatrix}-i\omega^+ \\ i\frac{\varpi^2}{\omega^+} \\ \varpi\end{pmatrix}\exp(-2i\omega^+(t-t_0)) + A_3 \begin{pmatrix}2i\frac{\omega^-\omega^+}{\omega^-+\omega^+} \\ -2i\frac{\varpi^2}{\omega^-+\omega^+} \\ -\varpi\end{pmatrix}\exp(-i(\omega^-+\omega^+)(t-t_0))$$

with

$$A_1 = \frac{i\omega^-}{(\omega^+ - \omega^-)^2}\left[2i\frac{\omega^+}{\varpi}c - \left(\frac{\omega^+}{\varpi}\right)^2 b + a\right] \qquad (25.135)$$

$$A_2 = \frac{i\omega^+}{(\omega^+ - \omega^-)^2}\left[2i\frac{\omega^-}{\varpi}c - \left(\frac{\omega^-}{\varpi}\right)^2 b + a\right] \qquad (25.136)$$

$$A_3 = \frac{i(\omega^+ + \omega^-)}{(\omega^+ - \omega^-)^2}\left[i\frac{\omega^- + \omega^+}{\varpi}c - \frac{\omega^-\omega^+}{\varpi^2}b + a\right] \qquad (25.137)$$

where

$$\begin{pmatrix} a \\ b \\ c \end{pmatrix} = \begin{pmatrix} \widetilde{\Sigma}_{pp}(t_0) - \widetilde{\Sigma}_{pp}^{eq} \\ \widetilde{\Sigma}_{qq}(t_0) - \widetilde{\Sigma}_{qq}^{eq} \\ \widetilde{\Sigma}_{qp}(t_0) \end{pmatrix}. \qquad (25.138)$$

Here, the initial fluctuations appear which may be chosen according to the situation given. Finally, we may note that the vector multiplied by A_3 can be simplified further to

$$\begin{pmatrix} 2i\frac{\omega^-\omega^+}{\omega^-+\omega^+} \\ -2i\frac{\varpi^2}{\omega^-+\omega^+} \\ -\varpi \end{pmatrix} = \begin{pmatrix} \mathrm{sgn}C\frac{\varpi}{\eta} \\ \frac{\varpi}{\eta} \\ -\varpi \end{pmatrix}. \qquad (25.139)$$

25.6 Microscopic approach to transport problems

In a microscopic theory one starts from the full dynamics of a given system and derives effective equations of motion for the subsystem one is interested in. This can be done both on an entirely classical level as well as within the realm of quantum physics. In nuclear physics only the latter procedure is meaningful. Chapters 6 and 7 are devoted to this problem. There, the final aim is to derive an acceptable Fokker–Planck equation for collective motion. On the quantum level the corresponding equation is a generalized master equation for the collective density operator. A formal means to drive this type of equation is the projection technique of Nakajima–Zwanzig (Nakajima, 1958), (Zwanzig, 1960), for which reason we shall present here the essential elementary steps. A similar projection formalism has been developed by H. Mori (1965) to derive equations of motion for quantum observables, which may be considered the quantal analog of the Langevin equations (see e.g. (Brenig, 1989)).

25.6.1 The Nakajima–Zwanzig projection technique

Suppose we are given a Hamiltonian for the total system of all degrees of freedom of the form $\hat{\mathcal{H}} = \hat{H}_\mathrm{B} + \hat{H}_\mathrm{SB} + \hat{H}_\mathrm{S}$. Here, "S" denotes the (sub-)system we want to follow explicitly and "B" stands for the "heat bath". Averaging over the latter one may define the reduced density operator for "S" as $\hat{d}(t) = \operatorname{tr}_B \hat{W}(t)$, with $\hat{W}(t)$ being the solution of the von Neumann equation of the total system

$$\frac{d\hat{W}(t)}{dt} = -\frac{i}{\hbar}\left[\hat{\mathcal{H}}, \hat{W}(t)\right] \equiv -i\hat{L}\,\hat{W}(t). \tag{25.140}$$

Here, the notation of a Liouville operator has been introduced. Like the total Hamiltonian $\hat{\mathcal{H}}$ the Liouvillian $\hat{L}$, too, can be split into three parts:

$$\hat{L}\hat{O} \equiv \frac{1}{\hbar}\left[\hat{\mathcal{H}}, \hat{O}\right] = \frac{1}{\hbar}\left[\hat{H}_B + \hat{H}_{SB} + \hat{H}_S, \hat{O}\right] \equiv \left(\hat{L}_B + \hat{L}_{SB} + \hat{L}_S\right)\hat{O}. \tag{25.141}$$

To apply the Nakajima–Zwanzig projection technique one may introduce the projector $\hat{\mathcal{P}} = \hat{B}\operatorname{tr}_B$ (Haake, 1973). Here $\hat{B}$ is supposed to be some (Hermitian) operator in B-space normalized to unity as $\operatorname{tr}_B \hat{B} = 1$. Its precise form will be chosen later. Applying this $\hat{\mathcal{P}}$ to the total density $\hat{W}(t)$ one obtains: $\hat{\mathcal{P}}\hat{W}(t) = \hat{B}\,\hat{d}(t)$. After building the trace over the bath variables once more we are left with the reduced density operator for the system variables. Before we make use of such manipulations for eqn(25.140) let us first split $\hat{W}(t)$ into two parts, $\hat{W}(t) = \hat{\mathcal{P}}\hat{W}(t) + (1-\hat{\mathcal{P}})\hat{W}(t)$. To get individual equations for both let us multiply (25.140) from left first by $\hat{\mathcal{P}}$ and then by $(1-\hat{\mathcal{P}})$, which leads to

$$\hat{\mathcal{P}}\frac{d\hat{W}(t)}{dt} = -i\hat{\mathcal{P}}\hat{L}\left(\hat{\mathcal{P}}\hat{W}(t) + (1-\hat{\mathcal{P}})\hat{W}(t)\right), \tag{25.142}$$

$$(1-\hat{\mathcal{P}})\frac{d\hat{W}(t)}{dt} = -i(1-\hat{\mathcal{P}})\hat{L}\left(\hat{\mathcal{P}}\hat{W}(t) + (1-\hat{\mathcal{P}})\hat{W}(t)\right). \tag{25.143}$$

The latter equation can be understood as an inhomogeneous differential equation for $(1-\hat{\mathcal{P}})\hat{W}(t)$ with the inhomogeneity term being $-i(1-\hat{\mathcal{P}})\hat{L}\hat{\mathcal{P}}\hat{W}(t)$. Its formal solution will thus be

$$(1-\hat{\mathcal{P}})\hat{W}(t) = e^{-i(1-\hat{\mathcal{P}})\hat{L}(t-s)}(1-\hat{\mathcal{P}})\hat{W}(t_0) - i\int_{t_0}^{t} ds\, e^{-i(1-\hat{\mathcal{P}})\hat{L}(t-s)}(1-\hat{\mathcal{P}})\hat{L}\hat{\mathcal{P}}\hat{W}(s). \tag{25.144}$$

Here, every evolution operator acts on every operator on its right. After inserting this result into (25.142) and averaging over the bath coordinates one gets

$$\frac{d}{dt}\hat{d}(t) = -i\hat{L}_S^{(1)}\hat{d}(t) + \int_{t_0}^{t} ds\,\hat{\mathcal{K}}(t-s)\hat{d}(s) + \hat{I}(t) \tag{25.145}$$

for the reduced density $\hat{d}(t)$. Here, the Liouvillian $\hat{L}_S^{(1)}$, the kernel $\hat{\mathcal{K}}$ and the inhomogeneity $\hat{I}(t)$ are given by:

$$\hat{L}_S^{(1)} = \hat{L}_S + \operatorname{tr}_B \hat{L}_{SB}\hat{B} \tag{25.146}$$

$$\hat{\mathcal{K}}(t) = -\operatorname{tr}_B \hat{L}_{SB} e^{-i(1-\hat{\mathcal{P}})\hat{L}t}(1-\hat{\mathcal{P}})(\hat{L}_B + \hat{L}_{SB})\hat{B}$$

$$\hat{I}(t) = -i\operatorname{tr}_B \hat{L}_{SB} e^{-i(1-\hat{\mathcal{P}})\hat{L}(t-t_0)}(1-\hat{\mathcal{P}})\hat{W}(t=t_0).$$

In deriving these expressions the following relations were used:

$$\operatorname{tr}_B \hat{L}_B \ldots = \frac{1}{\hbar}\operatorname{tr}_B\left[\hat{H}_B, \ldots\right] = 0; \quad \operatorname{tr}_B(1-\hat{\mathcal{P}})\ldots = 0; \quad \operatorname{tr}_B \hat{L}_S \ldots = \hat{L}_S \operatorname{tr}_B \ldots .$$

No restriction has yet been imposed on the generalized master equation (25.145). Together with (25.143), or its formal solution (25.144), it is still equivalent to the original von Neumann equation. To make this equation (25.145) tractable one has to take advantage of further assumptions. Commonly, one treats the interaction to second order. For nuclear physics, however, where damping is known to be strong, such an assumption is not adequate. This problem is addressed in Chapter 7.

25.6.2 Perturbative approach for factorized coupling

In the following we want to simplify our formalism in a fourfold sense and assume that
(1) the initial density operator $\hat{W}(t = t_0)$ factorizes like $\hat{B}\hat{d}(t = t_0)$,
(2) $\hat{B} = \hat{\rho}(H_B, T)$ is that functional of $\hat{H}_B$ which represents thermal equilibrium,
(3) the Hamiltonian $\hat{\mathcal{H}}$ is at most quadratic in Q and P,
(4) the coupling is given by $\hat{H}_{SB} = \hat{q}\hat{F}(\hat{x}_i, \hat{p}_i)$ and treated to second order only.

As an immediate implication of the first assumption one has $(1-\hat{\mathcal{P}})\hat{W}(t = t_0) = 0$ which makes the inhomogeneous part $\hat{I}(t)$ of the master equation (25.145) vanish. Otherwise, this term would be non-zero for some finite time interval, which usually turns out to be of microscopic order and thus macroscopically small. Since furthermore $\hat{B} = \hat{\rho}(H_B, T)$ one has $\hat{L}_B\hat{B}\hat{d} = 0$, which greatly simplifies the perturbation expansion. It means that the projector $(1-\hat{\mathcal{P}})$ commutes with $(\hat{L}_S + \hat{L}_B)$, implying

$$e^{-i(1-\hat{\mathcal{P}})(\hat{L}_S+\hat{L}_B)t}(1-\hat{\mathcal{P}}) = e^{-i(\hat{L}_S+\hat{L}_B)t}(1-\hat{\mathcal{P}}).$$

To second order in L_{SB} one thus gets for $\hat{\mathcal{K}}$

$$\hat{\mathcal{K}}(t) = -\operatorname{tr}_B \hat{L}_{SB} e^{-i(\hat{L}_S+\hat{L}_B)t}(1-\hat{\mathcal{P}})\hat{L}_{SB}\hat{B}. \tag{25.147}$$

For the Nakajima–Zwanzig equation these simplifications lead to

$$\frac{d}{dt}\hat{d}(t) = -i\hat{L}_S^{(1)}\hat{d}(t) - \int_{t_0}^{t} ds \operatorname{tr}_B \hat{L}_{SB} e^{-i(\hat{L}_S+\hat{L}_B)(t-s)}(1-\hat{\mathcal{P}})\hat{L}_{SB}\hat{B}\hat{d}(s). \tag{25.148}$$

So far no assumption has been made about timescales being different for nucleonic or collective motion. Equation (25.148) still contains all kinds of memory effects and can thus be considered truly non-Markovian.

In accord with our assumptions the model Hamiltonian is of the form

$$\hat{H}_B + \hat{H}_{SB} + \hat{H}_S = \hat{H}(\hat{x}_i, \hat{p}_i,) + \hat{q}\hat{F}(\hat{x}_i, \hat{p}_i) + \left(\frac{1}{2m_0}\hat{P}^2 + \frac{m_0\omega_0^2}{2}\hat{q}^2\right). \tag{25.149}$$

To simplify matters let us assume further that $\langle \hat{F} \rangle_0 = \text{tr}_B \hat{F} \hat{\rho} = 0$. Then there will be no term of first order. It would be given by the second term on the right of (25.146), which can be rewritten as: $\hbar \text{tr}_B \hat{L}_{SB} \hat{\rho} \hat{d} = \text{tr}_B [\hat{q}\hat{F}, \hat{\rho}\hat{d}] = \hat{q}\hat{d}\,\text{tr}_B[\hat{F}, \hat{\rho}] + [\hat{q}, \hat{d}]\,\text{tr}_B \hat{F} \hat{\rho} = 0$. Next we look at the integral kernel of (25.148). For the same reasons as just explained, the term proportional to the projector $\hat{\mathcal{P}}$ vanishes. The remaining ones can be evaluated as follows:

$$\hbar^2 \text{tr}_B \hat{L}_{SB} e^{-i(\hat{L}_S + \hat{L}_B)(t-s)} \hat{L}_{SB} \hat{\rho} \hat{d}(s) = \text{tr}_B \left[\hat{q}\hat{F}, \left[\hat{q}^I(t-s)\hat{F}^I(t-s), \hat{\rho} e^{-i\hat{L}_S(t-s)} \hat{d}(s) \right] \right]$$

$$= \frac{1}{2} \left(\text{tr}_B\, \hat{\rho}\, \left[\hat{F},\, \hat{F}^I(s-t) \right] \right) \left[\hat{q},\, \left[\hat{q}^I(s-t),\, e^{-i\hat{L}_S(t-s)} \hat{d}(s) \right]_+ \right] + \qquad (25.150)$$

$$\frac{1}{2} \left(\text{tr}_B\, \hat{\rho}\, \left[\hat{F},\, \hat{F}^I(s-t) \right]_+ \right) \left[\hat{q},\, \left[\hat{q}^I(s-t),\, e^{-i\hat{L}_S(t-s)} \hat{d}(s) \right] \right].$$

Here, the time evolution of the operators is that of the interaction picture. For the collective coordinate it reads $\hat{q}^I(s-t) = \exp(-i\hat{L}_S(t-s))\hat{q}$. A similar expression holds true for the $\hat{F}$, with $\hat{L}_S$ replaced by $\hat{L}_B$, of course. In the two factors involving nucleonic operators we recognize the structures of the response and correlation functions $\tilde{\chi}(t-s)$ and $\tilde{\psi}(t-s)$. They are defined as in (23.45), (23.44) and (23.79), (23.78), respectively, with $A_\mu = F$. These terms take care of the one with $\hat{\mathcal{P}}$ in (25.147)). With the help of these functions the equation for the reduced density finally reads:

$$\frac{d}{dt}\hat{d}(t) = -\frac{i}{\hbar}\left[\frac{1}{2m_0}\hat{P}^2 + \frac{m_0 \omega_0^2}{2}\hat{q}^2,\, \hat{d}(t) \right]$$
$$+ \frac{i}{2\hbar} \int_{-\infty}^{+\infty} ds\, \tilde{\chi}(s) \left[\hat{q},\, \left[\left(e^{-i\hat{L}_S s}\hat{q}\right),\, (e^{-i\hat{L}_S s}\hat{d}(t-s)) \right]_+ \right] \qquad (25.151)$$
$$+ \frac{i}{2\hbar^2} \int_{-\infty}^{+\infty} ds\, \tilde{\psi}(s) \left[\hat{q},\, \left[\left(e^{-i\hat{L}_S s}\hat{q}\right),\, (e^{-i\hat{L}_S s}\hat{d}(t-s)) \right] \right]$$

We have put t_0 equal to $-\infty$ and have changed the integration variable from $s \to t-s$.

Notice that eqn(25.151) is still non-Markovian: In the integrals on the right-hand side the density operator appears at time $t-s$ rather than at time t itself. It is true that the evolution operator $e^{-i\hat{L}_S s}$ moves the density *forward* in s. But it is only the *unperturbed* evolution operator which appears here, whereas the argument in $\hat{d}(t-s)$ stands for full propagation. The latter is determined by the equation of motion itself including all induced forces. Therefore, the following relation is true only in a perturbative sense:

$$e^{-i\hat{L}_S s}\hat{d}(t-s) \approx \hat{d}(t). \qquad (25.152)$$

There is no apparent reason why it *must* be used in connection with eqn(25.151). However, as demonstrated in (Hofmann, 1997), this equation would otherwise be inconsistent, violating the fluctuation dissipation theorem, for instance. This inconsistency can be removed by requiring (25.152) to be fulfilled. The resulting equation,

$$\frac{d}{dt}\hat{d}(t) = -\frac{i}{\hbar}\left[\frac{1}{2m_0}\hat{P}^2 + \frac{m_0\omega_0^2}{2}\hat{q}^2, \hat{d}(t)\right] + \frac{i}{2\hbar}\int_{-\infty}^{+\infty} ds\, \tilde{\chi}(s)\left[\hat{q}, \left[\hat{q}^I(-s), \hat{d}(t)\right]_+\right]$$
$$+ \frac{i}{2\hbar^2}\int_{-\infty}^{+\infty} ds\, \tilde{\psi}(s)\left[\hat{q}, \left[\hat{q}^I(-s), \hat{d}(t)\right]\right]$$
(25.153)

then *is Markovian* but at the price of restricting all induced forces to second order perturbation theory. They can be evaluated using $\hat{q}^I(-s) = \hat{q}\cos\omega_0 s - \hat{P}\sin\omega_0 s/(m_0\omega_0)$ and by observing symmetry relations for the response and correlation functions. As the ω_0 is real these functions can easily be split into their real and imaginary parts (for details see Chapter 23). One obtains:

$$\int_{-\infty}^{+\infty} ds\, \tilde{\chi}(s)\hat{q}^I(-s) = \chi'(\omega_0)\,\hat{q} - \frac{\chi''(\omega_0)}{\omega_0}\frac{\hat{P}}{m_0} \qquad (25.154)$$

$$i\int_{-\infty}^{+\infty} ds\, \tilde{\psi}(s)\hat{q}^I(-s) = -\psi''(\omega_0)\,\hat{q} + \frac{\psi'(\omega_0)}{\omega_0}\frac{\hat{P}}{m_0}. \qquad (25.155)$$

The equation for the density operator is then found to be:

$$\frac{d}{dt}\hat{d}(t) = -\frac{i}{\hbar}\left[\frac{1}{2m_0}\hat{P}^2 + \frac{C}{2}\hat{q}^2, \hat{d}(t)\right]$$
$$-\frac{i}{\hbar}\gamma\left[\hat{q}, \frac{1}{2}\left[\frac{\hat{P}}{m_0}, \hat{d}(t)\right]_+\right] + \frac{1}{2\hbar^2}\left[\hat{q}, \left[\left(-D_{pp}\,\hat{q} + 2D_{qp}\frac{\hat{P}}{m_0}\right), \hat{d}(t)\right]\right].$$
(25.156)

The various coefficients appearing here are defined as

$$C \equiv C(\omega_0) = \left(m_0\omega_0^2 - \chi'(\omega_0)\right), \qquad (25.157)$$

$$\gamma \equiv \gamma(\omega_0) = \frac{\chi''(\omega_0)}{\omega_0}, \qquad (25.158)$$

$$2D_{qp}(\omega_0) = \frac{\psi'(\omega_0)}{2m_0\omega_0} \quad \text{and} \quad D_{pp}(\omega_0) = \frac{1}{2}\psi''(\omega_0) = \gamma(\omega_0)T^*(\omega_0). \quad (25.159)$$

with the "effective temperature" $T^*(\omega_0) = (\hbar\omega_0/2)\coth(\hbar\omega_0/2T)$. Their physical interpretation is easily recognized after employing a Wigner transformation from the density operator $\hat{d}(t)$ to the density distribution $d^W(Q, P, t)$. As outlined in Section 19.3 one may apply the correspondence rules (19.75) and (19.76)) to get

$$\frac{\partial}{\partial t}d(Q, P, t) = \left(-\frac{\partial}{\partial Q}\frac{P}{m_0} + \frac{\partial}{\partial P}CQ + \frac{\partial}{\partial P}\frac{P}{m_0}\gamma + 2D_{qp}\frac{\partial^2}{\partial Q\partial P} + D_{pp}\frac{\partial^2}{\partial P\partial P}\right)d(Q, P, t). \qquad (7.34)$$

Equation (7.34) is seen to have the form of Kramers' equation (25.16) with the exception of the appearance of a second diffusion coefficient D_{qp}; the factor of 2 is chosen to be in accord with the notation of the Kramers–Moyal expansion discussed in Section 25.2.3), see (25.58) in particular. Of course, the other diffusion coefficient D_{pp}, too, is influenced by quantum effects. Indeed, it is only in the high temperature limit that it turns into that used by Kramers,

$$D_{pp} = \gamma \frac{\hbar \omega_0}{2} \coth\left(\frac{\hbar \omega_0}{2T}\right) \longrightarrow \gamma T \quad \text{for} \quad \hbar \omega_0 \ll T. \qquad (25.160)$$

Notice that the condition here involves the *unperturbed* frequency ω_0. Indeed, because of the strict application of perturbation theory all transport coefficients for the induced forces depend only on this ω_0. The stiffness, for instance, simply experiences the renormalization from $m_0 \omega_0^2$ to the $C(\omega_0)$ one expects for the conservative force within linear response theory. In nuclear physics such features are not really acceptable. It is not only that one has to give up this simple minded perturbation theory, one should also aim at a level where these transport coefficients are determined in self-consistent fashion. In Chapter 7 this problem was described in greater detail and a solution was presented in terms of a modified approach in which the collective part of the time evolution is treated to all orders.

25.6.2.1 *Historical notes*
In the following we shall briefly discuss three cases in which such equations have been discussed in the literature in different contexts.

(1) Transport equations in quantum optics: In (1971) G.S. Agarwal derived an equation like (7.34) with the diffusion coefficient D_{pp} given in (25.159) and the D_{qp} put equal to zero. There, an oscillator bath was considered and the coupling operator $\hat{F}(\hat{x}_i, \hat{p}_i)$ was assumed to be a linear combination in the $\hat{x}_i$, $\hat{p}_i$. Agarwal did not only look at the equation for the Wigner transform of the density operator $\hat{d}(t)$. He also treated the case of a distribution corresponding to "normal ordering" as well as the one associated with the Sudarshan–Glauber representation. As demonstrated in eqn(2.11) of (Agarwal, 1971), the corresponding (c-number) transport equations differ only in the diffusion term. To get these other representations from (7.34) the D_{pp} simply has to be chosen according to

$$D_{pp} = \lambda \hbar \omega_0 + T^*(\omega_0) \quad \text{with} \quad \lambda = \begin{cases} -\frac{1}{2} & \text{Sudarshan–Glauber,} \\ 0 & \text{Wigner,} \\ \frac{1}{2} & \text{normal.} \end{cases} \qquad (25.161)$$

In the modern literature on quantum optics such "quantal master equations", as they are referred to at times, are discussed in various applications, see e.g. (Agarwal, 1970), (Weissbluth, 1989),(Gardiner, 1991), (Cohen-Tannouji et al., 1992) and (Scully and Zubairy, 1997). In (Gardiner, 1991) this phrase is even used for the case where the diffusion coefficient D_{pp} in (25.156) is calculated in the high temperature limit, where the noise is entirely classical in nature and for which the D_{qp} vanishes identically (remember the discussion in Chapter 7). It may also be said that the renormalization of the stiffness is often ignored, which in quantum optics is referred to as "ignoring the level shift". Such an approximation should go along with neglecting the cross term of the diffusion tensor. Indeed, in equilibrium the average unperturbed kinetic energy equals that of the potential, which according to (7.16) implies $D_{qp} = 0$.

2) Transport within linear response: In (Hofmann and Siemens, 1977) a transport equation like (25.153) was derived simply by using time-dependent perturbation theory of second order, without taking benefit of the Nakajima–Zwanzig technique. Actually, the integrals over time were calculated by expanding the $q^I(-s)$ into low powers of s. In this way transport coefficients appeared which in the notation of Section 6.5.2 are those of the zero frequency limit. Indeed, if in (25.158) one takes the limit $\omega_0 \to 0$ one immediately obtains the form $\gamma(0) = (\partial \chi''/\partial \omega)_{\omega=0}$ of (6.92). Likewise, the latter equation tells one that the diffusion coefficient $D_{pp}(\omega_0)$ of (25.159) becomes $\gamma(0)T$. The justification for these manipulations was seen in application to slow collective motion. As the latter was considered to be of large scale the transport coefficients depended on the collective variable and in the end arguments were given in favor of a global motion. These questions are discussed in great detail in Chapter 7 on the basis of applying local propagators to a Chapman–Kolmogorov equation. In (Hofmann et al., 1979) equation (25.153) itself was applied to treat charge equilibration in the entrance phase of a heavy-ion collision as a harmonic collective mode. As the latter must be considered a high frequency mode the assumption (25.160) could not be made. As in the cases mentioned for quantum optics the diffusion coefficient D_{qp} was put equal to zero.

3) Linearized transport equation within a Random Matrix approach: Section 23.4.4 examined how linear response may be formulated within a RMM. There in (23.110) response and correlation functions were given. They may be used in (25.156) to get a transport equation which in (Lutz and Weidenmüller, 1999) has been re-derived by different methods, but applying perturbation theory, too. A non-perturbative derivation of a quantal transport equation with a Random Matrix Model has been reported in (Bulgac et al., 1996) albeit only for the high temperature limit.

Exercises

1. (i) Starting from the Rayleigh equation derive the equations for the first and second moments shown in the text and convince yourself about the form of the diffusion coefficient.
   (ii) Find the relations to formulas (25.15) and (25.13).
   (iii) Show that the Gaussian (25.23) satisfies (25.20).

PART V

AUXILIARY INFORMATION

26

FORMAL MEANS

26.1 Gaussian integrals

Basic forms for one-dimensional Gaussian integrals are

$$I(a,b,c) = \int_{-\infty}^{\infty} du \, e^{-(au^2+2bu+c)} = \sqrt{\frac{\pi}{a}} \exp\left(\frac{b^2-ac}{a}\right) \quad (26.1)$$

$$\int_{-\infty}^{\infty} du \, u \, e^{-(au^2+2bu+c)} = \frac{-b}{a} I \qquad \int_{-\infty}^{\infty} du \, u^2 \, e^{-(au^2+2bu+c)} = \frac{a+2b^2}{2a^2} I \quad (26.2)$$

Integrals may involve the error function or its complement

$$\mathrm{erf}(\sqrt{az}) = \frac{2}{\sqrt{\pi}} \int_0^{\sqrt{az}} e^{-u^2} du = 1 - \mathrm{erfc}(\sqrt{az}) \quad (26.3)$$

like the double integral

$$J(a,b,c) = \int_{-\infty}^{\infty} ds \, e^{-(as^2+2bs)} \int_{-\infty}^{s} du \, e^{-cu^2} = \frac{\pi}{2\sqrt{ac}} e^{\frac{b^2}{a}} \left(1 - \mathrm{erf}\sqrt{\frac{b^2c}{a(a+c)}}\right) \quad (26.4)$$

For the nth moment:

$$J_n(a,b,c) = \int_{-\infty}^{\infty} ds \, s^n \, e^{-(as^2+2bs)} \int_{-\infty}^{s} du \, e^{-cu^2} \quad (26.5)$$

one easily derives the following recursion relation:

$$J_n(a,b,c) = -\frac{1}{2}\frac{d}{db} J_{n-1}(a,b,c) \quad (26.6)$$

Formula (26.1) allows one to generalize to the multidimensional case for which one may write (Negele and Orland, 1987)

$$\int dx_1 \ldots dx_M \, e^{-\frac{1}{2}x_l A_{lk} x_k + J_l x_l - \frac{1}{2} J_l A_{lk}^{-1} J_k} = \sqrt{\frac{(2\pi)^M}{\det A}}, \quad (26.7)$$

where summation over repeated indices is understood. It is assumed that A is a symmetric, Hermitian and positive definite matrix. The proof is readily performed by transforming first to a new system of variables $y_i = x_i - A_{ij}^{-1} J_j$ and then to a system $z_k = O_{ki}^{-1} y_i$ in which A is diagonal with eigenvalues a_m. Indeed, in this way one gets

$$\int dx_1 \ldots dx_M \, e^{-\frac{1}{2}x_l A_{lk} x_k + J_l x_l - \frac{1}{2} J_l A_{lk}^{-1} J_k} = \int dy_1 \ldots dy_M \, e^{-\frac{1}{2} y_l A_{lk} y_k}$$
$$= \prod_{m=1}^{M} \int dz_m e^{-\frac{1}{2} a_m z_m^2} = \prod_{m=1}^{M} \sqrt{\frac{2\pi}{a_m}} \equiv \sqrt{\frac{(2\pi)^M}{\det A}} \, . \tag{26.8}$$

26.2 Stationary phase and steepest decent

According to Stokes and Kelvin, for large γ the main contributions to the integral

$$f(\gamma) = \int_{t_1}^{t_2} g(t) e^{i\gamma h(t)} \, dt \qquad \gamma, \, g \text{ and } h \text{ real} \tag{26.9}$$

come from the neighborhoods of the

$$\text{stationary points } t_\nu \text{ with} \qquad h'(t_\nu) = \left.\frac{dh}{dt}\right|_{t=t_\nu} = 0 \, . \tag{26.10}$$

The integral may then be estimated as (see e.g. §2.9 of (Erdélyi, 1956))

$$f(\gamma) = \sum_\nu \sqrt{\frac{2\pi i}{\gamma h''(t_\nu)}} \, g(t_\nu) \, \exp\left(i\gamma h(t_\nu)\right) + \mathcal{O}(1/\gamma) \, . \tag{26.11}$$

Note, please, that this feature is intimately related to the generalized Gauss integral

$$\int_{-\infty}^{\infty} e^{i\gamma t^2} \, dt = \sqrt{\frac{\pi i}{\gamma}} + \mathcal{O}(1/\gamma) \, , \tag{26.12}$$

which is kind of a Fresnel integral.

The *method of steepest decent* or *saddle point approximation* allows one to get an asymptotic form for integrals of the type (26.9) even for complex $h(t)$ and $g(t)$ and integration along a contour in the complex t-plane. One gets

$$\begin{aligned} F(\gamma) &= \int_C dt \, g(t) \, \exp\left[\gamma h(t)\right] \\ &\simeq \sqrt{\frac{2\pi}{\gamma \, |\, h''(t_0) \,|}} \, g(t_0) \, \exp\left(\gamma h(t_0)\right) \qquad \text{for} \qquad |\gamma| \to \infty \, . \end{aligned} \tag{26.13}$$

It must be assumed that the contour can be deformed such that it passes through a stationary point at t_0 where

$$\left.\frac{dh}{dt}\right|_{t=t_0} = 0 \qquad \text{with} \qquad \text{Im}\left[\gamma h(t)\right] = \text{Im}\left[\gamma h(t_0)\right] \, . \tag{26.14}$$

The path must be chosen such that it concurs with that of *steepest descent* which is to say that path where $\text{Re}\left[\gamma h(t)\right]$ *decreases at maximum rate*. Care must be taken about the proper phase of the $t - t_0$ along this path; for details see e.g. (Cushing, 1975) or (Wyld, 1976).

26.3 The δ-function

The basic property of the δ-function is

$$\int_{-\infty}^{\infty} dx\, \delta(x-y) f(x) = f(y) \tag{26.15}$$

Dirac formula

$$\lim_{\epsilon \to 0} \frac{1}{z \pm i\epsilon} = \mathcal{P}\left(\frac{1}{z}\right) \mp i\pi \delta(z) \tag{26.16}$$

Representation through limiting processes

$$\begin{aligned}\delta(z) &= \frac{1}{\sqrt{\pi}} \lim_{\epsilon \to 0} \left[\frac{1}{\epsilon} \exp\left(-\frac{z^2}{\epsilon^2}\right)\right] \\ &= \frac{1}{\pi} \lim_{\epsilon \to 0} \left[\frac{\epsilon}{\epsilon^2 + z^2}\right]\end{aligned} \tag{26.17}$$

Fourier representations

$$\begin{aligned}\delta(z) &= \int_{-\infty}^{\infty} \frac{dk}{2\pi} e^{ikz} = \int_0^{\infty} \frac{dk}{2\pi} e^{ikz} + \int_0^{\infty} \frac{dk}{2\pi} e^{-ikz} \\ &\equiv -\frac{1}{2\pi i} \lim_{\epsilon \to 0} \left[\frac{1}{z + i\epsilon} - \frac{1}{z - i\epsilon}\right],\end{aligned} \tag{26.18}$$

where the second line is in accord with (26.23). In the spirit of (26.12) one may also accept the following representation of the δ function:

$$\delta(t - t_0) = \lim_{\gamma \to \infty} \sqrt{\frac{\gamma}{i\pi}}\, e^{i\gamma(t-t_0)^2}. \tag{26.19}$$

26.4 Fourier and Laplace transformations

For the Fourier transformation we use the following conventions,

$$f(\omega) = \int_{-\infty}^{\infty} dt\, e^{i\omega t} \tilde{f}(t) \qquad \tilde{f}(t) = \int_{-\infty}^{\infty} \frac{d\omega}{2\pi} e^{-i\omega t} f(\omega). \tag{26.20}$$

Hence the convolution theorem reads

$$\int_{-\infty}^{\infty} dx\, \tilde{f}(x)\, \tilde{g}(y - x) = \int_{-\infty}^{\infty} \frac{dk}{2\pi} e^{-iky} f(k)\, g(k), \tag{26.21}$$

with its inverse

$$\int_{-\infty}^{\infty} dy\, e^{iky} \tilde{f}(y)\, \tilde{g}(y) = \int_{-\infty}^{\infty} \frac{dK}{2\pi} f(K)\, g(k - K). \tag{26.22}$$

The Fourier transform of the Θ-function is given by

$$\int_{-\infty}^{\infty} dy\, e^{iky} \Theta(\pm y) = \frac{\pm i}{k \pm i\epsilon}. \tag{26.23}$$

The Laplace transform and its inverse are defined as

$$F(z) = \mathcal{L}[f(s)] \equiv \int_0^\infty ds\, e^{-zs} f(s) \tag{26.24}$$

$$f(s) = \mathcal{L}^{-1}[F(z)] \equiv \frac{1}{2\pi i} \int_{\eta-i\infty}^{\eta+i\infty} dz\, e^{sz} F(z), \tag{26.25}$$

where all singularities of $F(z)$ are supposed to lie to the left of the line $\mathrm{Re}\{z\} = \eta$.

26.5 Derivative of exponential operators

The derivative of the operator $\hat{k}(\beta; a) = \exp\left[-\beta\left(\hat{H} + a\hat{A}\right)\right]$ with respect to a is given by

$$\begin{aligned}\frac{\partial \hat{k}(\beta; a)}{\partial a} &\equiv \lim_{\delta a \to 0} \frac{1}{\delta a}\left(\hat{k}(\beta; a + \delta a) - \hat{k}(\beta; a)\right) \\ &= -\int_0^\beta \hat{k}(\beta - \lambda; a)\, \hat{A}\, \hat{k}(\lambda; a)\, d\lambda\,.\end{aligned} \tag{26.26}$$

The proof may be performed by exploiting the perturbation theory of Section 23.2 for the time evolution operator. One simply has to apply the "Wick-rotation" and identify $\hat{k}_{\mathrm{sc}} = \hat{\mathcal{U}}_{\mathrm{sc}}(t - t_0 = -i\hbar\beta)$ (see the exercise in Section 23.6.1). Notice that because of $\hat{k}(\lambda; a)\hat{k}(\beta - \lambda; a) = \hat{k}(\beta; a)$ and the invariance of the trace with respect to cyclic permutations, (26.26) implies

$$\frac{\partial}{\partial a}\mathrm{tr}\left(\hat{k}(\beta; a)\right) = -\beta\,\mathrm{tr}\,\hat{k}(\beta; a)\,A\,. \tag{26.27}$$

26.6 The Mori product

The Mori product of two operators is defined as[59]

$$\left(\hat{A}|\hat{B}\right) = \frac{1}{\beta}\int_0^\beta d\zeta\, \mathrm{tr}\,\rho(\beta)\hat{A}_{\mathrm{Im}}(\zeta)\hat{B} = \frac{1}{\beta}\int_0^\beta d\zeta\, \mathrm{tr}\,\rho(\beta)\hat{A}\hat{B}_{\mathrm{Im}}(-\zeta) \tag{26.28}$$

with $\rho(\beta) = \exp(-\beta\hat{H})/\mathrm{tr}\,\exp(-\beta\hat{H})$ and $\hat{A}_{\mathrm{Im}}(\zeta) = \exp(\zeta\hat{H})\hat{A}\exp(-\zeta\hat{H}) = \hat{A}(t = -i\hbar\zeta)$. The product (26.28) has the following properties

$$\left(\hat{A}|\hat{B}\right) = \left(\hat{B}|\hat{A}\right) \tag{26.29}$$

$$\left(\hat{A}|\hat{A}\right) > 0 \quad \text{if} \quad \hat{A} = \hat{A}^\dagger \tag{26.30}$$

$$\left(\hat{A}|\hat{B}\right) = \langle \hat{A}\hat{B}\rangle \quad \text{if either} \quad \left[\hat{A}, \hat{H}\right] = 0 \quad \text{or} \quad \left[\hat{B}, \hat{H}\right] = 0\,. \tag{26.31}$$

Whereas the third property is trivial, the second is easily seen from a spectral representation like that given in (23.154). The first property is proved by invoking the cyclic invariance of the trace as well as a transformation of the integration variable, $\zeta \to \beta - \zeta$. It implies that for Hermitian operators the Mori product is real: $\left(\hat{A}|\hat{B}\right)^* = \left(\hat{B}|\hat{A}\right) = \left(\hat{A}|\hat{B}\right)$.

---

[59] This definition follows that of (Kubo et al., 1991a), where $\left(\hat{A}|\hat{B}\right)$ (for which the symbol $\langle A; B\rangle$ is used) is called the *canonical correlation* of the two quantities $\hat{A}$ and $\hat{B}$.

26.7 Spin and isospin

The spin component $|s, s_z\rangle$ of the total wave function $\langle \mathbf{r}|\Psi\rangle = \psi(\mathbf{r})|s, s_z\rangle$ is an eigenstate of the square $\hat{s}^2$ of the spin operator $\hat{s}$ and its z-component $\hat{s}_z$:

$$\begin{aligned}\hat{s}^2|s, s_z\rangle &= s(s+1)\hbar^2|s, s_z\rangle & s &= 1/2 \\ \hat{s}_z|s, s_z\rangle &= s_z\hbar|s, s_z\rangle & s_z &= \pm 1/2.\end{aligned} \quad (26.32)$$

A convenient representation is defined by the three Pauli matrices σ

$$\mathbf{s} = \frac{1}{2}\boldsymbol{\sigma} \quad \sigma_x = \begin{pmatrix} 0 & 1 \\ 1 & 0 \end{pmatrix} \quad \sigma_y = \begin{pmatrix} 0 & -i \\ i & 0 \end{pmatrix} \quad \sigma_z = \begin{pmatrix} 1 & 0 \\ 0 & -1 \end{pmatrix}. \quad (26.33)$$

General spin states $|a\rangle = a_+|1/2, +1/2\rangle + a_-|1/2, -1/2\rangle$ can be expressed by two-dimensional spinors,

$$\chi_a = \begin{pmatrix} a_+ \\ a_- \end{pmatrix} \quad \text{with} \quad \chi_a^\dagger = (a_+^*, a_-^*), \quad (26.34)$$

with the orthonormality relation being $\langle a|b\rangle = \chi_b^\dagger \chi_a = b_+^* a_+ + b_-^* a_- = \delta_{ab}$.

The spin components fulfill the basic commutation rules for angular momenta: $[\hat{I}_x, \hat{I}_y] = i\hbar\hat{I}_z$ plus cyclic commutation. Any two such momenta, say $\hat{\mathbf{j}}_1$ and $\hat{\mathbf{j}}_2$, may be added to a third one, $\hat{\mathbf{J}}$, with eigenvectors $|JM\rangle$ obeying

$$\begin{aligned}\hat{J}^2|JM\rangle &= J(J+1)\hbar^2|JM\rangle \\ \hat{J}_z|JM\rangle &= M\hbar|JM\rangle \quad \text{with} \quad \hat{\mathbf{J}} = \hat{\mathbf{j}}_1 + \hat{\mathbf{j}}_2\end{aligned} \quad (26.35)$$

The quantum numbers J, M take on the following values

$$M = m_1 + m_2 \qquad |j_1 - j_2| \le J \le j_1 + j_2 \quad (26.36)$$

and the eigenfunctions can be constructed as

$$|JMj_1j_2\rangle = \sum_{m_1 m_2} |j_1 j_2 m_1 m_2\rangle \langle j_1 j_2 m_1 m_2 | JM j_1 j_2\rangle \quad (26.37)$$

Here, $|j_1 j_2 m_1 m_2\rangle = |j_1 m_1\rangle|j_2 m_2\rangle$ and the $\langle j_1 j_2 m_1 m_2 | IM j_1 j_2\rangle$ are Clebsch–Gordan coefficients. For the combination two spins one gets $\hat{\mathbf{S}} = \hat{\mathbf{s}}_1 + \hat{\mathbf{s}}_2$, leading to four states $|S, S_z\rangle$, the *singlet* state with $S = 0$

$$|0, 0\rangle = \frac{1}{\sqrt{2}} (|\uparrow\downarrow\rangle - |\downarrow\uparrow\rangle) \quad (26.38)$$

and the three *triplet* states with $S = 1$

$$|1, 1\rangle = |\uparrow\uparrow\rangle \quad |1, 0\rangle = \frac{1}{\sqrt{2}} (|\uparrow\downarrow\rangle + |\downarrow\uparrow\rangle) \quad |1, -1\rangle = |\downarrow\downarrow\rangle. \quad (26.39)$$

Please notice that the singlet state is *antisymmetric with respect to an interchange of the two particles* and that the triplet states are *symmetric*.

The operator
$$\hat{P}_\sigma = (1/2)\left(1 + \hat{\boldsymbol{\sigma}}_1 \cdot \hat{\boldsymbol{\sigma}}_2\right) \tag{26.40}$$
exchanges the spins of the two particles: Its eigenvectors are seen to be given by the singlet and triplet states, which means having $\hat{P}_\sigma|s\rangle = -|s\rangle$ and $\hat{P}_\sigma|t\rangle = +|t\rangle$. It allows one to define projectors onto these states, namely $\hat{\Pi}_t^\sigma = (1+\hat{P}_\sigma)/2$ and $\hat{\Pi}_s^\sigma = (1-\hat{P}_\sigma)/2$. For a formally elegant way to distinguish protons and neutrons one may exploit the concept of *isospin*. The vector operator $\hat{\mathbf{t}}$ is defined to fulfill the same commutation rules as a spin 1/2 particle and has the same representation. The states $|p\rangle$ and $|n\rangle$ may be introduced as the eigenstates of the z-component,

$$\hat{t}_z|p\rangle = +(1/2)|p\rangle \qquad \hat{t}_z|n\rangle = -(1/2)|n\rangle, \tag{26.41}$$

with eigenvalue 3/4 for $\hat{t}^2$. Either state can be changed into the other by introducing the common ladder operators,

$$\hat{t}_+|n\rangle \equiv \left(\hat{t}_x + i\hat{t}_y\right)|n\rangle = |p\rangle \quad \text{and} \quad \hat{t}_-|p\rangle \equiv \left(\hat{t}_x - i\hat{t}_y\right)|p\rangle = |n\rangle. \tag{26.42}$$

For a two particle system the total isospin $\hat{\mathbf{T}} = \hat{\mathbf{t}}_1 + \hat{\mathbf{t}}_2$, its eigenstates and eigenvalues can be calculated according to the rules of combining angular momenta. The two particle state may then be symmetric or antisymmetric with respect to an exchange of the partners, depending on the quantum number T.

Spin and isospin may be combined through a direct product $|m_s, m_\tau\rangle = |m_s\rangle|m_\tau\rangle$ of independent spinors, which may be represented by four-dimensional vectors, for which we use the shorthand notation $\chi(m_s, m_\tau)$

26.8 Second quantization for fermions

In the following we shall introduce the formalism of second quantization on a purely pragmatic level; a more profound introduction may be found in textbooks on many-body physics like (Negele and Orland, 1987). Suppose we are given a basis $|\alpha\rangle$ of single particle states, with the α meant to represent the set of quantum numbers necessary to specify the states completely. One introduces *annihilation and creation operators* $\hat{c}_\alpha$ and $\hat{c}_\alpha^\dagger \equiv (\hat{c}_\alpha)^\dagger$, respectively, together with a *vacuum state* $|\text{vac}\rangle$. They satisfy the relations
$$\hat{c}_\alpha|\text{vac}\rangle = 0 \quad \text{and} \quad \hat{c}_\alpha^\dagger|\text{vac}\rangle = |\alpha\rangle, \tag{26.43}$$
from which the notation becomes apparent. The many-body states $|\alpha_1, \alpha_2, \ldots \alpha_N\rangle$ of fermions have to be totally antisymmetric. This requirement is fulfilled for

$$|\alpha_1, \alpha_2, \ldots \alpha_N\rangle = \hat{c}_{\alpha_1}^\dagger \hat{c}_{\alpha_2}^\dagger \cdots \hat{c}_{\alpha_N}^\dagger |\text{vac}\rangle \tag{26.44}$$

if only the creation operators *anti-commute*

$$\hat{c}_\alpha^\dagger \hat{c}_\beta^\dagger + \hat{c}_\beta^\dagger \hat{c}_\alpha^\dagger \equiv \left[\hat{c}_\alpha^\dagger, \hat{c}_\beta^\dagger\right]_+ = 0, \tag{26.45}$$

which implies the same property for the annihilation operators. The $|\alpha_1, \alpha_2, \ldots \alpha_N\rangle$ represents a state of N independent particles. To go to a state of $N-1$ particles one should "annihilate" one. This is formally consistent with the basic meaning of $\hat{c}_\alpha$ if the anti-commutation relations (26.45) are accomplished by those involving both creation and annihilation operators, namely

$$\left[\hat{c}_\alpha^\dagger, \hat{c}_\beta\right]_+ = \delta_{\alpha\beta}. \tag{26.46}$$

Indeed, in this case one has

$$\hat{c}_\alpha |\alpha_1, \alpha_2, \ldots \alpha_N\rangle = \begin{cases} 0 & \text{if } \alpha \neq \alpha_k \\ (-)^{(k_0-1)} |\alpha_1, \ldots \alpha_{k_0-1}, \alpha_{k_0+1} \alpha_N\rangle & \text{if } \alpha \equiv \alpha_{k_0}. \end{cases} \tag{26.47}$$

The second line follows because exactly $(k_0 - 1)$ permutations are necessary to move the $\hat{c}_\alpha$ in front of the $\hat{c}_{\alpha_{k_0}}^\dagger$. The states $|\alpha_1, \alpha_2, \ldots \alpha_N\rangle$ are eigenstates of $\hat{n}_\alpha = \hat{c}_\alpha^\dagger \hat{c}_\alpha$ with eigenvalues 0 or 1, depending whether or not the single particle state $|\alpha\rangle$ is among the set $\alpha_1, \alpha_2, \ldots \alpha_N$ or not. Indeed, for a single particle state $|\beta\rangle$ one has $\hat{n}_\alpha |\beta\rangle = \delta_{\alpha\beta} |\beta\rangle$, with the generalization

$$\hat{n}_\alpha |\alpha_1, \alpha_2, \ldots \alpha_N\rangle = \left(\sum_k^N \delta_{\alpha\alpha_k}\right) |\alpha_1, \alpha_2, \ldots \alpha_N\rangle \tag{26.48}$$

to N-body states. These relations suggest the *number operator* $\hat{N}$ to be given as

$$\hat{N} = \sum_\alpha \hat{n}_\alpha = \sum_\alpha \hat{c}_\alpha^\dagger \hat{c}_\alpha. \tag{26.49}$$

Its eigenvalues N are identical to the number of particles in any given state.

Given a transformation to another (orthogonal) basis $|\tilde{\alpha}\rangle$ by $|\tilde{\alpha}\rangle = \sum_\alpha |\alpha\rangle\langle\alpha|\tilde{\alpha}\rangle$ the operators change like $\hat{c}_{\tilde{\alpha}}^\dagger = \sum_\alpha \langle\alpha|\tilde{\alpha}\rangle \hat{c}_\alpha^\dagger$ and $\hat{c}_{\tilde{\alpha}} = \sum_\alpha \langle\tilde{\alpha}|\alpha\rangle \hat{c}_\alpha$. In this way the fundamental anti-commutation relations are conserved.

Any one-body operator $\hat{F}$ can be expanded as

$$\hat{F} = \sum_{\alpha\beta} \langle\alpha|F|\beta\rangle \hat{c}_\alpha^\dagger \hat{c}_\beta \equiv \sum_{\alpha\beta} F_{\alpha\beta} \hat{c}_\alpha^\dagger \hat{c}_\beta. \tag{26.50}$$

This may be checked through the matrix representation. Indeed, it is easily seen that $\langle\alpha|\hat{F}|\beta\rangle = \langle\alpha|F|\beta\rangle$. A general two-body-operator can be written as

$$\hat{V} = \frac{1}{2} \sum_{\alpha\beta\gamma\delta} (\alpha\beta|V|\gamma\delta) \hat{c}_\alpha^\dagger \hat{c}_\beta^\dagger \hat{c}_\delta \hat{c}_\gamma, \tag{26.51}$$

where the two-body states are simple product states $|\gamma\delta) = |\gamma\rangle|\delta\rangle$ which are not antisymmetrized. For antisymmetrized matrix elements defined as

$$\{\alpha\beta|\overline{V}|\gamma\delta\} \equiv (\alpha\beta|V|\gamma\delta) - (\alpha\beta|V|\delta\gamma) \tag{26.52}$$

the expression (26.51) becomes

$$\hat{V} = \frac{1}{4} \sum_{\alpha\beta\gamma\delta} \{\alpha\beta|\overline{V}|\gamma\delta\} \hat{c}_\alpha^\dagger \hat{c}_\beta^\dagger \hat{c}_\delta \hat{c}_\gamma. \tag{26.53}$$

27

NATURAL UNITS IN NUCLEAR PHYSICS

In terms of the elementary units

$$[E] = \text{MeV} \quad [t] = 10^{-23} \text{ sec} \quad [L] = \text{fm } (Fermi) \tag{27.1}$$

one gets

$$e^2 = 1.44 \text{ MeV fm}, \quad \hbar = 65.82 \text{ MeV } [t], \quad c = 2.998 \frac{\text{fm}}{[t]} \tag{27.2}$$

for the elementary charge, Planck's constant and the velocity of light, respectively, and, hence,

$$\hbar c = 197.33 \text{ MeV fm} \quad \text{and} \quad \frac{\hbar c}{e^2} = 137.04. \tag{27.3}$$

The masses of neutron and proton are given by

$$m_n c^2 = 939.6 \text{ MeV} \quad m_p c^2 = 938.3 \text{ MeV} \tag{27.4}$$

such that the atomic mass unit $\quad 1 \text{ amu} \equiv C^{12}/12 \quad$ turns into

$$m = (931.5 \text{ MeV})/c^2 \equiv 103.5 \frac{\text{MeV}[t]^2}{\text{fm}^2}, \quad \text{such that} \quad \frac{\hbar^2}{2m} = 20.54 \text{ MeV fm}^2 \tag{27.5}$$

which corresponds to a Compton wavelength of $\hbar/mc = 0.21$ fm.

With the Boltzmann constant k_B put equal to unity, temperature is measured in MeV with the following relation to the kelvin:

$$1 \text{ K} \equiv 0.86171 \times 10^{-10} \text{ MeV}. \tag{27.6}$$

REFERENCES

Abe, Y. (2002). *European Physical Journal A - Hadrons and Nuclei*, **13**, 143–148.
Abe, Y., Ayik, S., Reinhard, P.-G., and Suraud, E. (1996). *Physics Reports*, **275**, 249.
Abe, Y., Bouriquet, B., Shen, C., and Kosenko, G. (2003). *Nuclear Physics A*, **722**, 241–247.
Abramowitz, M. and Stegun, I.A. (1968). *Handbook of Mathematical Functions*. Dover Publicications, New York.
Abrosimov, V.I. and Hofmann, H. (1994). *Nuclear Physics*, **A578**, 317.
Affleck, I. (1981). *Physical Review Letters*, **46**, 388.
Agarwal, G.S. (1970). *Quantum Optics*. Springer Tracts in Modern Physics, vol.70, Berlin.
Agarwal, G.S. (1971). *Physical Review A*, **4**, 739.
Agassi, D., Weidenmüller, H.A., and Mantzouranis, G. (1975). *Physics Reports*, **22**, 145–179.
Agassi, D., Ko, C.M., and Weidenmüller, H.A. (1977). *Annals of Physics*, **107**, 140–167.
Agrawal, B.K. and Ansari, A. (1998). *Nuclear Physics A*, **640**, 362–374.
Agrawal, B.K., Samaddar, S.K., Ansari, A., and De, J.N. (1999). *Physical Review C*, **59**, 3109–3115.
Akmal, A., Pandharipande, V.R., and Ravenhall, D.G. (1998). *Physical Review*, **C58**, 1804–1828.
Alhassid, Y. (1991). In *New trends in Nuclear Collective Motion* (ed. Y. Abe, H. Horiuchi, and K. Matsuyanagi). Springer, Heidelberg.
Alhassid, Y., Zingman, J., and Levit, S. (1987). *Nuclear Physics*, **A469**, 205.
Alhassid, Y., Bertsch, G.F., and Fang, L. (2003). *Physical Review C*, **68**, 044322.
Alhassid, Y., Bertsch, G.F., Fang, L., and Liu, S. (2005). *Physical Review C*, **72**, 064326.
Alkofer, R., Hofmann, H., and Siemens, P.J. (1988). *Nuclear Physics*, **A476**, 213.
Ankerhold, J., Grabert, H., and Ingold, G.-L. (1995). *Physical Review E*, **51**, 4267.
Ankerhold, J., Pechukas, P., and Grabert, H. (2001). *Physical Review Letters*, **87**, 0868021.
Aritomo, Y. and Ohta, M. (2005). *Nuclear Physics A*, **753**, 152–173.
Aritomo, Y. and Ohta, M. (2006). *Nuclear Physics A*, **764**, 149–159.
Aritomo, Y., Wada, T., Ohta, M., and Abe, Y. (1999). *Physical Review C*, **59**, 796–809.

Armbruster, P. (1999). *Reports on Progress in Physics*, **62**, 465–525.
Arve, P., Bertsch, G.F., Lauritzen, B., and Puddu, G. (1988). *Annals of Physics*, **183**, 309–319.
Ashcroft, N.W. and Mermin, N.D. (1976). *Solid State Physics*. Saunders College Publishers, Fort Worth.
Attias, H. and Alhassid, Y. (1997). *Nuclear Physics*, **A625**, 565.
Ayik, A. and Nörenberg, W. (1982). *Zeitschrift f. Physik A*, **309**, 121.
Back, B.B., Blumenthal, D.J., Davids, C.N., Henderson, D.J., Hermann, R., Hofman, D.J., Jiang, C.L., Penttilä, H.T., and Wuosmanaa, A.H. (1999). *Physical Review C*, **60**, 044602.
Balantekin, A.B. and Takigawa, N. (1998). *Rev. Mod. Phys*, **70**, 77.
Baldo, M. (1999). In *Nuclear Methods and the Nuclear Equation of State* (ed. M. Baldo). World Scientific, Singapore.
Baldo, M., Bombaci, I., and Burgio, G.F. (1997). *Astron. Astrophys.*, **328**, 274.
Balian, R. (1991). *From Microphysics to Macrophysics, Vol. I*. Springer, Berlin.
Balian, R. (1992). *From Microphysics to Macrophysics, Vol. II*. Springer, Berlin.
Balian, R. and Bloch, C. (1974). *Annals of Physics*, **85**, 514.
Balian, R. and Vénéroni, M. (1989). *Annals of Physics*, **195**, 324.
Baranger, H.U. and Stone, A.D. (1989). *Physical Review B*, **40**, 8169.
Bardeen, J., Cooper, L.N., and Schrieffer, J.R. (1957). *Physical Review*, **108**, 1175.
Barma, M. and Subrahmanyam, V. (1989). *J. Phys-Cond. Mat.*, **1**, 7681.
Barnett, A., Cohen, D., and Heller, E. (2001). *Journal of Physics A*, **34**, 413–438.
Barrett, B.R., Shlomo, S., and Weidenmüller, H.A. (1978). *Physical Review C*, **17**, 544–554.
Bartel, J., Quentin, P., Brack, M., Guet, C., and Hrakansson, H.B. (1982). *Nuclear Physics*, **A386**, 79.
Bauge, E., Delaroche, J.P., and Girod, M. (1998). *Physical Review C*, **58**, 1118.
Baym, G. (1969). *Lectures on Quantum Mechanics*. Benjamin/Cummings Publishing Company, Reading Massachussets.
Baym, G. and Pethick, C.J. (1991). *Landau Fermi Liquid Theory*. Wiley+Sons, New York.
Beane, S.R., Bedaque, P.F., Haxton, W.C., Phillips, D.R., and Savage, M.J. (2001). In *At the Frontier of Particle Physics-Handbook of QCD* (ed. M. Shifman). World Scientific, Singapore.
Behringer, H., Pleimling, M., and Hüller, A. (2005). *Journal of Physics A*, **38**, 973.
Benet, L., Rupp, T., and Weidenmüller, H.A. (2001). *Annals of Physics*, **292**, 67–94.
Berlanger, M., Grangé, P., Hofmann, H., Ngô, C., and Richert, J. (1978a). *Physical Review C*, **17**, 1495.
Berlanger, M., Ngô, C., Grangé, P., Richert, J., and Hofmann, H. (1978b). *Zeitschrift f. Physik A*, **286**, 207–214.

Berlanger, M., Ngô, C., Grangé, P., Richert, J., and Hofmann, H. (1978c). *Zeitschrift f. Physik A*, **284**, 61–64.

Berlanger, M., nd A. Gobbi, Hanappe, F., Lynen, U., Ngô, C., Olmi, A., Sann, H., Stelzer, H., Richel, H., and Rivet, M.F. (1979). *Zeitschrift f. Physik A*, **291**, 133.

Bertsch, G.F. (1978). *Zeitschrift f. Physik A*, **289**, 103.

Bertsch, G.F. and Broglia, R.A. (1994). *Oscillations in Finite Quantum Systems*. Cambridge University Press, Cambridge and New York.

Bertsch, G.F., Bortignon, P.F., and Broglia, R.A. (1983). *Review of Modern Physics*, **55**, 287.

Bethe, H.A. (1937). *Review of Modern Physics*, **9**, 69.

Bethe, H.A. (1940). *Physical Review*, **57**, 1125.

Bethe, H.A. (1971a). *Annual Review Nuclear Science*, **21**, 93.

Bethe, H.A. (1971b). In *Quanten and Felder*. F. Vieweg u. Sohn, Braunschweig.

Bethe, H.A. (1990). *Review of Modern Physics*, **62**, 801–866.

Bethe, H.A. and Brown, G.E. (1985). *Scientific American*, **May**, 40–48.

Bethe, H.A. and Goldstone, J. (1957). *Proc. Roy. Soc. (London)*, **A238**, 551.

Bhaduri, R.K. and Gupta, S. Das (1973). *Physics Letters B*, **47**, 129.

Bhaduri, R.K. and Ross, C.K. (1971). *Physical Review Letters*, **27**, 606.

Bhatt, K. H., Grangé, P., and Hiller, B. (1986). *Physical Review C*, **33**, 954.

Bjørnholm, S. (1994). In *Clusters of Atoms and Molecules* (ed. H. Haberland), Volume 52, Series in Chemical Physics. Springer, Berlin.

Blann, M. (1975). *Annual Review of Nuclear Science*, **25**, 123–166.

Blatt, J.M. and Weisskopf, V.F. (1952). *Theoretical Nuclear Physics*. Wiley, New York.

Bloch, C. (1954). *Physical Review*, **93**, 1094.

Bloch, C. (1968). In *Nuclear Physics* (ed. C. de Witt and V. Gillet), pp. 305–411. Gordon and Breach, New York.

Błocki, J., Boneh, Y., Nix, J.R., Randrup, J., Robel, M., Sierk, A.J., and Swiatecki, W.J. (1978). *Ann. Phys. (N.Y.)*, **113**, 330.

Blocki, J., Skalski, J., and Swiatecki, W.J. (1995). *Nuclear Physics*, **A594**, 137.

Bohigas, O. (1991). In *Chaos and Quantum Physics, Les Houches 1989* (ed. M.-J. Giannoni, A. Voros, and J. Zinn-Justin). North-Holland, Amsterdam.

Bohigas, O. and Leboeuf, P. (2002). *Physical Review Letters*, **88**, 092502.

Bohigas, O., Haq, R.U., and Pandey, A. (1983). In *Nuclear Data for Science and Technology* (ed. K. Böckhoff), p. 809. Reidel, Dordrecht.

Bohigas, O., Giannoni, M.-J., and Schmit, C. (1984). *Physical Review Letters*, **52**, 1.

Bohm, D. and Pines, D. (1953). *Physical Review*, **92**, 609–625.

Bohr, A. and Mottelson, B.R. (1958). *Kongel. Norske Vid. Selsk.*, **31(12)**, 1.

Bohr, A. and Mottelson, B.R. (1969). *Nuclear Structure, vol.I*. Benjamin, London.

Bohr, A. and Mottelson, B.R. (1975). *Nuclear Structure, vol.II*. Benjamin, London.

Bohr, N. (1936). *Nature*, **137**, 344.
Bohr, N. and Kalckar, F. (1937). *Mat. Fys. Medd. Dan. Vinsk. Selsk.*, **14**, no.10.
Bohr, N. and Wheeler, J.A. (1939). *Physical Review*, **56**, 426.
Bombaci, I. (1999). In *Nuclear Methods and the Nuclear Equation of State* (ed. M. Baldo). World Scientific, Singapore.
Bootermans, W. and Malfliet, R. (1990). *Physics Reports*, **198**, 115.
Bortignon, P.F., Bracco, A., and Broglia, R.A. (1998). *Giant Resonances: Nuclear Structure at finite Temperature*. Harwood Academic Publishers, Amsterdam.
Brack, M. (1992). In *Nuclear Structure Models* (ed. R. Bengtson, J. Draayer, and W. Nazarewicz), p. 165. World Scientific, Singapore.
Brack, M. (1993). *Review of Modern Physics*, **65**, 677.
Brack, M. (2001). In *Atomic Clusters and Nanoparticles, Les Houches 2000* (ed. C. G. et al.). Springer-Verlag, Berlin.
Brack, M. and Bhaduri, R.K. (1997). *Semiclassical Physics*. Addison-Wesley, Reading, MA.
Brack, M. and Quentin, P. (1974). In *Physics and Chemistry of Fission, Rochester 1973*, Volume I, Vienna, p. 231. IAEA.
Brack, M. and Quentin, P. (1981). *Nuclear Physics*, **A361**, 35–82.
Brack, M., Damgaard, J., Jensen, A.S., Pauli, H.C., Strutinsky, V.M., and Wong, C. Y. (1972). *Review of Modern Physics*, **44**, 320.
Braun, F. and von Delft, J. (1999). *Physical Review B*, **59**, 9527 – 9544.
Brenig, W. (1957). *Nuclear Physics*, **4**, 363.
Brenig, W. (1989). *Statistical Theory of heat, non-equlibrium phenomena*. Springer, Berlin.
Brenig, W. (1996). *Statistische Theorie der Wärme* (IV edn). Springer, Berlin.
Brockmann, R. and Machleidt, R (1984). *Physics Letters B*, **149**, 283.
Brockmann, R. and Machleidt, R (1999). In *Nuclear Methods and the Nuclear Equation of State* (ed. M. Baldo). World Scientific, Singapore.
Brody, T., Flores, J., French, J., Mello, P., Pandey, A., and Wong, S. (1981). *Review of Modern Physics*, **53**, 385–479.
Broglia, R.A., Dasso, C.H., and Winther, A. (1974). *Physics Letters B*, **53**, 301.
Brueckner, K.A. (1954). *Physical Review*, **97**, 508.
Bucurescu, D. and von Egidy, T. (2005). *Physical Review C*, **72**, 067304.
Bulgac, A., Dang, G. Do, and Kusnezov, D. (1996). *Physical Review E*, **54**, 3468.
Bulgac, A., Dang, G. Do, and Kusnezov, D. (2001). *Physica E*, **9**, 429–435.
Bunatian, G.G., Kolomietz, V.M., and Strutinsky, V.M. (1972). *Nuclear Physics*, **A188**, 225–258.
Bundschuh, R. (2004).
Bush, B.W., Bertsch, G.F., and Brown, B.A. (1992). *Physical Review C*, **45**, 1709.

Caldeira, A.O. and Leggett, A.J. (1983). *Annals of Physics*, **149**, 374.
Callan, C. and Coleman, S. (1977). *Physical Review D*, **16**, 1762.
Campi, X., Krivine, H., Plagnol, E., and Sator, N. (2005). *Physical Review C*, **71**, 041601.
Canosa, N., Rossignoli, R., and Ring, P. (1994). *Physical Review C*, **50**, 2850–2859.
Chabanat, E., Bonche, P., Kaensel, P., Meyer, J., and Schaeffer, R. (1997). *Nuclear Physics*, **A627**, 710–746.
Chaikin, P.M. and Lubensky, T.C. (1995). *Principles of Condensed Matter Physics*. Cambridge University Press, Cambridge and New York.
Charity, R.J. (1996). *Physical Review C*, **53**, 512–514.
Chomaz, Ph. and Gulminelli, F. (2005). *Nuclear Physics A*, **749**, 3c.
Coester, F., Cohen, S., Day, B.D., and Vincent, C.M. (1970). *Physical Review C*, **1**, 769.
Cohen, D. (2000). *Annals of Physics*, **283**, 175–231.
Cohen, D. (2003). *Physical Review B*, **68**, 155303.
Cohen, S. and Swiatecki, W.J. (1963). *Annals of Physics*, **22**, 406.
Cohen-Tannouji, C., Dupont-Roc, J., and Grynberg, G. (1992). *Atom-Photon Interactions; Basic Processes and Applications*. John-Wiley & Sons, New York.
Coleman, S. (1977). In *Erice Lectures 1977* (ed. A. Zuchichi). Plenum Press, London.
Courant, R. and Hilbert, D. (1953). *Methods of Mathematical Physics, Vol.1*. Interscience Publishers, New York.
Csatlós, M., Krasznahorkay, A., Thirolf, P.G., Habs, D., and et al. (2005). *Physics Letters B*, **615**, 175–185.
Curtin, W.A. and Ashcroft, N.W. (1985). *Physical Review A*, **32**, 2909.
Cushing, J.T. (1975). *Applied Analytical Mathematics for Physical Scientists*. Wiley, New York.
Danielewicz, P. (1984). *Annals of Physics*, **152**, 239.
Datta, S. (1995). *Electronic Transport in Mesoscopic Systems*. Cambridge University Press, Cambridge and New York.
Day, B.D. (1978). *Review of Modern Physics*, **50**, 495.
de Heer, W.A. (2000). In *Metal Clusters at Surfaces* (ed. K.-H. Meiwes-Broer). Springer, Berlin.
de Shalit, A. and Feshbach, H. (1974). *Theoretical Nuclear Physics*. Wiley, New York.
Denisov, V.Y. and Hofmann, S. (2000). *Physical Review C*, **61**, 034606.
des Cloizeaux, J. (1968). In *Many Body Physics* (ed. C. de Witt and R. Balian). Gordon and Breach, New York.
Deubler, H.H., Dietrich, K., and Hofmann, H. (1980). In *Heavy Ion Collisions, Vol.2* (ed. R. Bock). North Holland Publishing Company, Amsterdam.
Dickhoff, W.H. (1999). In *Nuclear Methods and the Nuclear Equation of State* (ed. M. Baldo). World Scientific, Singapore.
Dietrich, K., Pomorski, K., and Richert, J. (1995). *Zeitschrift f. Physik A*, **351**,

397.
Diószegi, I., Shaw, N.P., Mazumdar, I., Hatzikoutelis, A., and Paul, P. (2000). *Physical Review C*, **61**, 024613.
Dittrich, T., Hänggi, P., Ingold, G-L., Kramer, B., Schön, G., and Zwerger, W. (1998). *Quantum Transport and Dissipation*. Wiley-VCH, Weinheim.
Dorfman, J.R. (1999). *An Introduction to Chaos in Non-Equilibrium Statistical Mechanics*. Cambridge University Press, Cambridge.
Døssing, T. and Jensen, A.S. (1974). *Nuclear Physics*, **A222**, 493–511.
Dover, C.B., Lemmer, R.H., and Hahne, F.J.W. (1972). *Ann. Phys. (N.Y.)*, **70**, 458.
Duderstadt, J.J. and Martin, W.R. (1979). *Transport Theory*. Wiley+Sons, New York.
Erdélyi, A. (1956). *Asymptotic Expansions*. Dover Public., New York.
Ericson, T. (1960). *Advances in Physics*, **9**, 425.
Ericson, T. and Weise, W. (1988). *Pions and Nuclei*. Oxford Science Publication, Oxford.
Feldmeier, H. (1987). *Reports on Progress in Physics*, **50**, 915–994.
Feshbach, H. (1987). *Physics Today*, **40**(11), 9.
Feshbach, H. (1992). *Theoretical Nuclear Physics (Nuclear Reactions)*. Wiley, New York.
Feshbach, H., Porter, C.E., and Weisskopf, V.F. (1954). *Physical Review*, **96**, 448.
Feshbach, H., Kerman, A.K., and Lemmer, R.H. (1967). *Annals of Physics*, **41**, 230.
Feshbach, H., Kerman, A.K., and S.E.Koonin (1980). *Annals of Physics*, **125**, 429.
Fetter, A.L. and Walecka, J.D. (1971). *Quantum Theory of Many-Particle Systems*. McGraw-Hill, New York, New York.
Feynman, R.P. (1972). In *Nobel Lectures, Physics (1963-1970)* (ed. N. Foundation). Elsevier, London.
Feynman, R.P. (1988). *Statistical Mechanics*. Addison-Wesley, Reading, MA.
Feynman, R.P. and Hibbs, A.R. (1965). *Quantum Mechanics and Path Integrals*. Mc Graw Hill, New York.
Feynman, R.P. and Kleinert, H. (1986). *Physical Review A*, **34**, 5080.
Fick, E. and Sauermann, G. (1986). *Quantenmechanik dynamischer Prozesse, vol. IIa*. Harri Deutsch, Thun.
Forster, D. (1975). *Hydrodynamic Fluctuations, Broken Symmetry and Correlation Functions*. W.A. Benjamin, London.
Forster, J.S., Michtell, I.V., Andersen, J.U., Jensen, A.S., Laesgaard, E., Gibson, W.M., and Reicheldt, R. (1987). *Nuclear Physics*, **A464**, 497–524.
Frauendorf, S.G. and Guet, C. (2001). *Annual Review of Nuclear and Particle Science*, **51**, 219.
Friedman, B. and Pandharipande, V.R. (1981). *Nuclear Physics A*, **361**, 502–520.

Friedman, F.L. and Weisskopf, V.F. (1955). In *Niels Bohr and the Development of Physics* (ed. W. Pauli). Pergamon Press, London.
Friedman, W.A. (1980). *Physical Review C*, **22**, 22.
Friedman, W.A., Hussein, M.S., McVoy, K.W., and Mello, P.A. (1981). *Physics Reports*, **77**, 47.
Fritsch, S., Kaiser, N., and Weise, W. (2002). *Physics Letters B*, **545**, 73–81.
Fritsch, S., Kaiser, N., and Weise, W. (2005). *Nuclear Physics A*, **750**, 259.
Fröbrich, P. and Lipperheide, R. (1996). *Theory of Nuclear Reactions*. Clarendon Press, Oxford.
Fröbrich, P. and Tillack, G.-R. (1992). *Nuclear Physics*, **A540**, 353.
Fröbrich, P. and Xu, S. Y. (1988). *Nuclear Physics*, **A477**, 143.
Gadioli, E. and Hodgson, P.E. (1992). *Pre-Equilibrium Nuclear Reactions*. Clarendon Press, Oxford.
Gardiner, C.W. (1991). *Quantum Noise*. Springer, Series in Synergetics, Berlin.
Gardiner, C.W. (2002). *Handbook of Stochastic Methods*. Springer, Berlin.
Ghosh, S.K. and Jennings, B.K. (2001).
Gilbert, A. and Cameron, A.G.W. (1965). *Canadian Journal of Physics*, **43**, 1446.
Gleissl, P., Brack, M., Meyer, J., and Quentin, P. (1990). *Annals of Physics*, **197**, 205.
Goeppert-Mayer, M. and Jensen, J.H.D. (1955). *Elementary Theory of Nuclear Shell Structure*. Wiley, New York.
Gogny, D. (1975). *Nuclear Physics*, **A237**, 399.
Goldhaber, M. and Teller, E. (1948). *Physical Review*, **74**, 1046.
Gottfried, K. (1966). *Quantum Mechanics*. Benjamin, New York.
Grabert, H. and Weiss, U. (1984). *Physical Review Letters*, **53**, 1787.
Grabert, H., Weiss, U., and Hänggi, P. (1984a). *Physical Review Letters*, **52**, 2193.
Grabert, H., Weiss, U., and Talkner, P. (1984b). *Zeitschrift f. Physik*, **B 55**, 87.
Grabert, H., Schramm, P., and Ingold, G.-L. (1988). *Physics Reports*, **168**, 115.
Grangé, P., Li, J.-L., and Weidenmüller, H.A. (1983). *Physical Review C*, **27**, 2063.
Grègoire, C., Ngô, C., and Remaud, B. (1982). *Nuclear Physics*, **A383**, 392–420.
Griffin, J.J. (1966). *Physical Review Letters*, **17**, 478.
Griffin, J.J. (1967). *Physical Letters B*, **24**, 5.
Gross, D.H.E. (1975). *Nuclear Physics*, **A240**, 472–484.
Gross, D.H.E. (1979). *Zeitschrift f. Physik A*, **291**, 145–150.
Gross, D.H.E. (1980). In *Lecture Notes in Physics 117: Deep-Inelastic and Fusion Reactions with Heavy Ions* (ed. W. von Oertzen). Springer, Heidelberg.
Gross, D.H.E. (2001). *Microcanonical Thermodynamics: Phase Transitions in "Small" Systems*. Lecture Notes In Physics 66. World Scientific, Singapore.
Gross, D.H.E. and Kalinowski, H. (1978). *Physics Reports*, **45**, 175.

Guet, C., E.Strumberger, and Brack, M. (1988). *Physics Letters B*, **205**, 427.
Guhr, T., A.Müller-Groeling, and Weidenmüller, H.A. (1998). *Physics Report*, **299**, 190–425.
Gutzwiller, M.G. (1990). *Chaos in Classical and Quantum Mechanics.* Springer, Heidelberg.
Haake, F. (1973). *Statistical Treatment of Open Systems by Generalized Master Equations.* Springer Tracts in Modern Physics, vol.66, Berlin.
Haake, F. (1991). *Quantum Signatures of Chaos.* Springer, Heidelberg.
Haberland, H. (1999). *Nuclear Physics A*, **649**, 415–422.
Hänggi, P., Talkner, P., and Borkovec, M. (1990). *Review of Modern Physics*, **62**, 251–342.
Hasse, R.W. (1987). *Nuclear Physics*, **A467**, 407.
Hasse, R.W. and Myers, W.D. (1988). *Geometrical Relationships of Macroscopic Nuclear Physics.* Springer, Heidelberg.
Hauser, W. and Feshbach, H. (1952). *Physical Review*, **87**, 366.
Heiselberg, H. and Pandharipande, V.R. (2000). *Annual Review of Nuclear and Particle Science*, **50**, 481–524.
Henden, L., Bergholt, L., Guttormsen, M., Rekstad, J., and Tveter, T.S. (1995). *Nuclear Physics A*, **589**, 249.
Hertel, P. and Thirring, W. (1971). *Annals of Physics*, **63**, 520.
Hilscher, D. and Rossner, H. (1992). *Ann. Phys. (Paris)*, **17**, 471–552.
Hinde, D.J., Hilscher, D., and Rossner, H. (1989). *Nuclear Physics*, **A502**, 497c–514c.
Hodgson, P.E. (1994). *The Nucleon Optical Model.* World Scientific, Singapore.
Hofmann, H. (1976). *Physics Letters B*, **61**, 423.
Hofmann, H. (1977). *Fizika*, **9**, 441.
Hofmann, H. (1980). In *Lecture Notes in Physics 117: Deep-Inelastic and Fusion Reactions with Heavy Ions* (ed. W. von Oertzen). Springer, Heidelberg.
Hofmann, H. (1982). *Physics Letters B*, **119**, 7.
Hofmann, H. (1997). *Physics Reports*, **284**(4&5), 137–380.
Hofmann, H. and Ingold, G.-L. (1991). *Physics Letters B*, **264**, 253.
Hofmann, H. and Ivanyuk, F.A. (1993). *Z. Phys. A, Hadrons and Nuclei*, **344**, 285.
Hofmann, H. and Ivanyuk, F.A. (1999). *Physical Review Letters*, **82**, 4603.
Hofmann, H. and Ivanyuk, F.A. (2003). *Physical Review Letters*, **90**, 90.132701.
Hofmann, H. and Jensen, A.S. (1984). *Nuclear Physics*, **A428**, 1c–22c.
Hofmann, H. and Kiderlen, D. (1997). *Physical Review C*, **56**, 1025.
Hofmann, H. and Kiderlen, D. (1998). *Journal of Modern Physics E*, **7**, 243.
Hofmann, H. and Magner, A.G. (2003). *Physical Review C*, **68**, 014606.
Hofmann, H. and Ngô, C. (1976). *Physics Letters B*, **65**, 97.
Hofmann, H. and Nix, J.R. (1983). *Physics Letters B*, **122**, 117.
Hofmann, H. and Pomorski, K. (1991). *Physics Letters B*, **263**, 164.
Hofmann, H. and Siemens, P.J. (1975). *Physics Letters B*, **58**, 417.
Hofmann, H. and Siemens, P.J. (1976). *Nuclear Physics*, **A257**, 165.

Hofmann, H. and Siemens, P.J. (1977). *Nuclear Physics*, **A275**, 464.
Hofmann, H. and Sollacher, R. (1988). *Annals of Physics*, **184**, 62.
Hofmann, H., Grègoire, C., Lucas, R., and Ngô, C. (1979). *Z. Phys. A, Hadrons and Nuclei*, **293**, 229–240.
Hofmann, H., Samhammer, R., and Ockenfuß, G. (1989). *Nuclear Physics*, **A496**, 269.
Hofmann, H., Yamaji, S., and Jensen, A.S. (1992). *Physics Letters B*, **286**, 1.
Hofmann, H., Ingold, G.-L., and Thoma, M. (1993). *Physics Letters B*, **317**, 489.
Hofmann, H., Ivanyuk, F.A., and Yamaji, S. (1996). *Nuclear Physics A*, **598**, 187.
Hofmann, H., Ivanyuk, F.A., and Magner, A.G. (1998). *Acta Physica Pol. B*, **29**, 375.
Hofmann, H., Ivanyuk, F.A., Rummel, C., and Yamaji, S. (2001). *Physical Review C*, **64**, 054316.
Honerkamp, J. (1994.). *Stochastic Dynamical Systems*. VCH Publishers, New York.
Huizenga, J.R. and Moretto, L.G. (1972). *Annual Review of Nuclear Science*, **22**, 427–464.
Hurwitz, A. and Courant, R. (1964). *Funktionentheorie and elliptische Funktionen*. Springer, Berlin.
Ibach, H. and Lüth, H. (1995). *Festkörperphysik*. Springer, Berln.
Ignatyuk, A. V., Smirenkin, G.N., and Tishin, A.S. (1975). *Sov. J. of Nuclear Physics*, **21**, 255.
Ingold, G.-L. (1988). Ph.D. thesis, Universität Stuttgart.
Ivanyuk, F.A. (1989). *Z. Phys. A*, **334**, 69.
Ivanyuk, F.A. (2007). *submitted to Physical Review B*.
Ivanyuk, F.A. and Hofmann, H. (1999). *Nuclear Physics A*, **657**, 19.
Ivanyuk, F.A. and Strutinsky, V.M. (1979). *Z. Phys. A*, **293**, 337.
Ivanyuk, F.A. and Yamaji, S. (2001). *Nuclear Physics A*, **694**, 295.
Ivanyuk, F.A., Hofmann, H., Pashkevich, V.V., and Yamaji, S. (1997). *Physical Review C*, **55**, 1730.
Iwamoto, A., Möller, P., Nix, J.R., and Sagawa, H. (1996). *Nuclear Physics*, **A596**, 329–354.
Jackson, J.D. (1962). *Classical Electrodynamics*. John Wiley, New York.
Jarzynski, C. (1993). *Physical Review E*, **48**, 4340.
Jennings, B.K. (1973). *Nuclear Physics*, **A207**, 538.
Jennings, B.K. (1976). Ph.D. thesis, McMaster University, Hamilton, Ontario.
Jensen, A.S. and Damgaard, J. (1973). *Nuclear Physics*, **A203**, 578–596.
Jeukenne, J.P., Lejeune, A., and Mahaux, C. (1976). *Physics Reports*, **25**, 83.
Johansen, P.J., Siemens, P.J., Jensen, A.S., and Hofmann, H. (1977). *Nuclear Physics*, **A288**, 152.
Kadanoff, L.P. and Baym, G. (1962). *Nuclear Structure, vol.II*. Benjamin, London.

Kadanoff, L. P. and Martin, P. C. (1963). *Annals of Physics*, **24**, 419.
Kaiser, N., Fritsch, S., and Weise, W. (2002). *Nuclear Physics A*, **700**, 343.
Kaiser, N., Fritsch, S., and Weise, W. (2003). *Nuclear Physics A*, **724**, 47.
Kaneko, K. and Hasegawa, M. (2005). *Physical Review C*, **72**, 024307.
Kataria, S.K., Ramamurthy, V.S., and Kapoor, S.S. (1978). *Physical Review C*, **18**, 549.
Kawabata, A. and Kubo, R. (1966). *Journ.Phys.Soc.Japan*, **21**, 1765–1772.
Keldish, L.V. (1964). *Sov. Phys. JETP*, **20**, 1018.
Kelly, M.J. (1995). *Low Dimensional Semiconductors*. Clarendon Press, Oxford.
Kiderlen, D. (1994). Ph.D. thesis, Technische Universität München.
Kiderlen, D. and Hofmann, H. (1994). *Physics Letters B*, **332**, 8.
Kiderlen, D., Hofmann, H., and Ivanyuk, F.A. (1992). *Nuclear Physics*, **A550**, 473.
Kittel, C. (1988). *Physics Today*, **41**(5), 93.
Kleinert, H. (1990). *Pathintegrals in Quantum Mechanics, Statistics and Polymer Physics*. World Scientific, Singapore.
Klevansky, S.P. and Lemmer, R.H. (1983). *Physical Review C*, **28**, 1763.
Kolomietz, V.M., Sanzhur, A.I., Shlomo, S., and Firin, S.A. (2001). *Physical Review C*, **64**, 024315.
Koonin, S.E., Hatch, R.L., and Randrup, J. (1977). *Nuclear Physics*, **A283**, 87.
Koonin, S.E., Dean, D.J., and Langanke, K. (1997). *Physics Reports*, **278**, 1.
Kramers, H.A. (1940). *Physica*, **7**, 284.
Kreibig, U. and Vollmer, M. (1995). *Optical Properties of Metal Clusters*. Springer, Berlin.
Kubo, R. (1962). In *Fluctuation, Relaxation and Resonances in Magnetic Systems* (ed. D. ter Haar). Oliver and Boyd, London.
Kubo, R., Toda, M., and Hashitsume, N. (1991a). *Statistical Physics II: Non-Equilibrium Statistical Mechanics*. Springer, New York.
Kubo, R., Toda, M., and Saitô, N. (1991b). *Statistical Physics I: Equilibrium Statistical Mechanics*. Springer, New York.
Kudyaev, G.A., Yu.B.Ostapenko, and Smirenko, G.N. (1976). *Sov. J. Part. Nucl.*, **7**, 138.
Kuo, T.T.S., Shamanna, J., and Tzeng, Y. (1999). In *Nuclear Methods and the Nuclear Equation of State* (ed. M. Baldo). World Scientific, Singapore.
Küpper, W.A., Wegmann, G., and Hilf, E.R. (1974). *Annals of Physics*, **88**, 454–471.
Kusnezov, D. and Ormand, W.E. (2003). *Physical Review Letters*, **90**, 042501.
Kusnezov, D., Alhassid, Y., and Snover, K.A. (1998). *Physical Review Letters*, **81**, 542.
Kusnezov, D., Bulgac, A., and Dang, G. Do (2001). *Physica E*, **9**, 436–442.
Lamm, G. and Schulten, K. (1983). *Journal of Chemical Physics*, **78**, 2713.
Landau, L. (1946). *Journal of Physics (USSR)*, **10**, 25.
Landau, L.D. and Lifshitz, E.M. (1980). *Course of Theoretical Physics V:*

Statistical Physics. Pergamon, Oxford.
Landau, L.D. and Lifshitz, E.M. (1982). *Course of Theoretical Physics IV: Relativistic Quantum Theory.* Pergamon, Oxford.
Landau, L.D., Lifshitz, E.M., and Pitajewski, I.P. (1977). *Course of Theoretical Physics III: Quantum Mechanics.* Elsevier, Oxford.
Landau, L.D., Lifshitz, E.M., and Pitajewski, I.P. (1981). *Course of Theoretical Physics X: Physical Kinetics.* Pergamon, Oxford.
Landau, L.D., Lifshitz, E.M., and Pitajewski, I.P. (1984). *Course of Theoretical Physics VIII: Electrodynamics of Continuous Media.* Elsevier, Oxford.
Langer, J.S. (1967). *Annals of Physics*, **41**, 108.
Langer, J.S. (1969). *Annals of Physics*, **54**, 258–275.
Larkin, A.I. and Ochinnikov, Y.N. (1984). *Sov. Phys. JETP*, **59**, 420.
Lattimer, J.M. (1981). *Annual Review of Nuclear and Particle Science*, **31**, 337–74.
Lattimer, J.M. and Swesty, F.D. (1991). *Nuclear Physics A*, **535**, 331.
Lattimer, J.M., Pethick, C.J., Ravenhall, D.G., and Lamb, D.Q. (1985). *Nuclear Physics A*, **432**, 646.
Lauritzen, B., Døssing, T., and Broglia, R.A. (1986). *Nuclear Physics A*, **457**, 61–83.
Lavenda, B.H. (1991). *Statistical Physics, a Probabilistic Approach.* J. Wiley, New York.
Leboeuf, P. and Monastra, A.G. (2002). *Annals of Physics*, **297**, 127–156.
Leboeuf, P., Monastra, A.G., and Relaño, A. (2005). *Physical Review Letters*, **47**, 102502.
Lemmer, R.H. (1966). *Reports on Progress in Physics*, **29**, 131.
Lestone, J.P. (1993). *Physical Review Letters*, **70**, 2245.
Lighthill, M.J. (1962). *Introduction to Fourier Analysis and Generalized Functions.* Cambridge Monographs on Mechanics and Applied Mathematics. Cambridge University Press, Cambridge and New York.
Lindhard, J. (1974). *Mat. Fys. Medd. Dan. Vinsk. Selsk.*, **39**, no.1.
Lindhard, J. (1986). In *The Lesson of Quantum Theory* (ed. E. D. J. de Boer and O. Ulfbeck). Elsevier Sience, Amsterdam.
Liu, S. and Alhassid, Y. (2001). *Physical Review Letters*, **87**, 022501.
Lutz, E. and Weidenmüller, H.A. (1999). *Physica A*, **267**, 354.
Lutz, M.F.M, Friman, B., and Appel, C. (2000). *Physics Letters B*, **474**, 277.
Lyalin, A., Solovyov, A., and Greiner, W. (2002). *Physical Review A*, **65**, 043202.
Lynden-Bell, D. (1999). *Physica,* **A 263**, 293.
Mackey, M.C. (1991). *Time's Arrow: The Origins of Thermodynamic Behavior.* Springer, Heidelberg.
Magner, A.G., Vydrug-Vlasenko, S., and Hofmann, H. (1991). *Nuclear Physics A (1991)*, **524**, 31.
Magner, A. G., K. Arita, S.N. Fedotkin, and Matsuyanagi, K. (2002). *Progress of Theoretical Physics*, **108**, 853.

Mahan, G.D. (1983). *Many-Particle Physics*. Plenum Press, New York.
Mahaux, C. and Sartor, R. (1989). In *Nuclear Matter and Heavy Ion Collisions* (ed. M. Soyeur, H. Flocard, B. Tamain, and M. Porneuf). Plenum press, New York.
Mahaux, C. and Sartor, R. (1991). In *Advances in Nuclear Physics, Vol.20* (ed. J. Negele and E. Vogt). Plenum Press, New York.
Mahaux, C. and Weidenmüller, H.A. (1979). *Annual Review of Nuclear and Particle Science*, **79**, 1–31.
Mamdouh, A., Pearson, J.M., Rayet, M., and Tondeur, F. (1998). *Nuclear Physics*, **A644**, 389.
Mandelbrot, B.B. (1964). *Journal of Mathematical Physics*, **5**, 164.
Mandelbrot, B.B. (1989). *Physics Today*, **42**(1), 71.
McVoy, K.W. (1985). In *4th International Conference on Nuclear Reaction Mechanisms* (ed. E. Gadioli). Ricerca Scientifica ed Educazione Permanente, Milano.
McVoy, K.W. and Tsang, X.T. (1983). *Physics Reports*, **94**, 139.
Mehta, M.L. (2004). *Random Matrices*. Elsevier, Amsterdam.
Melby, E., Bergholt, L., Guttormsen, M., Hjorth-Jensen, M., Ingebretsen, F., Messelt, S., Rekstad, J., Schiller, A., Siem, S., and Ødegrard, S.W. (1999). *Physical Review Letters*, **83**, 3150–3153.
Messiah, A. (1965). *Quantum Mechanics, vol. I and II*. John Wiley, New York.
Mie, G. (1908). *Annalen der Physik (Leipzig)*, **25**, 377.
Möller, P. and Iwamoto, A. (2000). *Physical Review C*, **61**, 047602.
Möller, P., Nix, J.R., Myers, W.D., and Swiatecki, W.J. (1995). *Atomic Data and Nuclear Data Tables*, **59**, 185.
Möller, P., Nix, J.R., Armbruster, P., Hofmann, S., and Münzenberg, S. (1997). *Zeitschrift f. Physik*, **A 359**, 251–255.
Möller, P., Madland, D.G., Sierk, J., and Iwamoto, A. (2004a). *Nature*, **409**, 785–790.
Möller, P., Sierk, J., and Iwamoto, A. (2004b). *Physical Review Letters*, **92**, 072501.
Morel, P. and Nozières, P. (1962). *Physical Review*, **126**, 1909.
Moretto, L.G., K.X.Jing, Gatti, R., Wozniak, G.J., and Schmitt, R.P. (1995a). *Physical Review Letters*, **75**, 4186–4189.
Moretto, L.G., K.X.Jing, and Wozniak, G.J. (1995b). *Physical Review Letters*, **74**, 3557–3560.
Mori, H. (1965). *Progress of Theoretical Physics*, **33**, 423.
Morrison, I.A., Baumgarte, T.W., Shapiro, S.L., and Pandharipande, V.R. (2004). *Astrophysical Journal*, **617**, L135–L138.
Morrissey, D.J., Beneson, W., and Friedman, W.A. (1994). *Annual Review of Nuclear and Particle Science*, **44**, 27.
Moss, F. and McClintock, P.V.E. (1989). *Noise in Nonlinear Dynamical Systems*. Cambridge University Press, New York.
Mühlschlegel, B., Scalapino, D., and Denton, R. (1972). *Physical Review B*, **6**,

1767.
Muskens, O., Christofilos, D., Fatti, N. Del, and Vallée, F. (2006). *Journal of Optics A: Pure and Applied Optics*, **8**, S264–S272.
Myers, W.D. and Swiatecki, W.J. (1966). *Nuclear Physics*, **81**, 1.
Myers, W.D. and Swiatecki, W.J. (1969). *Annals of Physics*, **55**, 395.
Myers, W.D. and Swiatecki, W.J. (1996). *Nuclear Physics*, **A601**, 141–167.
Nakada, H. and Alhassid, Y. (1997). *Physical Review Letters*, **79**, 2939–2942.
Nakajima, S. (1958). *Progress of Theoretical Physics*, **20**, 948.
Negele, J.W. and Orland, H. (1987). *Quantum Many-Particle Systems*. Addison-Wesley, Reading, MA.
Nemes, M.C. and Weidenmüller, H.A. (1981a). *Physical Review C*, **24**, 450–457.
Nemes, M.C. and Weidenmüller, H.A. (1981b). *Physical Review C*, **24**, 944–953.
Newton, R.G. (1982). *Scattering Theory of Waves and Particles*. Springer, Berlin.
Ngô, C. (1986). *Progress in Particle and Nuclear Physics*, **16**, 139.
Ngô, C. and Hofmann, H. (1977). *Zeitschrift f. Physik*, **A 282**, 83.
Nilsson, S.G. (1955). *Mat. Fys. Medd. Dan. Vinsk. Selsk.*, **29**, no.16.
Nix, J.R., Sierk, A.J., Hofmann, H., Scheuter, F., and Vautherin, D. (1984). *Nuclear Physics*, **A424**, 239.
Nörenberg, W. (1974). *Physics Letters B*, **52**, 289.
Nörenberg, W. (1989). In *Heavy Ion Reaction Theory* (ed. W. Shen, J. Lin, and L. Ge), p. 1. World Scientific, Singapore.
Pandharipande, V.R. and Ravenhall, D.G. (1989). In *Nuclear Matter and Heavy Ion Collisions* (ed. M. Soyeur, H. Flocard, B. Tamain, and M. Porneuf). Plenum press, New York.
Pandharipande, V.R., Sick, I., and de Witt Huberts, P.K.A. (1997). *Review of Modern Physics*, **69**, 981.
Panofsky, W.K.H. and Phillips, M. (1962). *Classical Electricity and Magnetism*. Addison-Wesley, Reading, MA.
Pashkevich, V.V. (1971). *Nuclear Physics*, **A169**, 275.
Paul, P. and Thoennessen, M. (1994). *Annual Review of Nuclear and Particle Science*, **44**, 65–108.
Persson, P. and Aaberg, S. (1995). *Physical Review E*, **52**, 148–153.
Pines, D. and Nozières, P. (1966). *The Theory of Quantum Liquids*. W.A. Benjamin, Inc., New York.
Pluhař, Z. and Weidenmüller, H.A. (2002). *Annals of Physics*, **297**, 344–362.
Pomorski, K. and Dudek, J. (2003). *Physical Review C*, **67**, 044316.
Pomorski, K., Bartel, J., Richert, J., and Dietrich, K. (1996). *Nuclear Physics A*, **605**, 87.
Pomorski, K., Nerlo-Pomorska, B., Surowiec, A., and et al. (2000). *Nuclear Physics A*, **679**, 25.
Puddu, G., Bortignon, P.F., and Broglia, R.A. (1991). *Annals of Physics*, **206**, 409.
Ralph, D., Black, C., and Tinkham, M. (1995). *Physical Review Letters*, **74**,

3241–3244.
Ralph, D., Black, C., and Tinkham, M. (1996). *Physical Review Letters*, **76**, 688.
Ralph, D., Black, C., and Tinkham, M. (1997). *Physical Review Letters*, **78**, 4087.
Ramamurthy, V.S. and Kapoor, S.S. (1972). *Physics Letters B*, **42**, 399.
Ramamurthy, V.S., Kapoor, S.S., and Kataria, S.K. (1970). *Physical Review Letters*, **25**, 386.
Randrup, J. (1978). *Annals of Physics*, **112**, 356.
Randrup, J. (1979). *Nuclear Physics*, **A327**, 490–516.
Reid, R.V. (1968). *Annals of Physics*, **50**, 411.
Reif, F. (1984). *Fundamentals of Statistical and Thermal Physics*. Mc Graw Hill, New York.
Reimann, P., Schmid, G.J., and Hänggi, P. (1999). *Physical Review E*, **60**, R1.
Richter, K., Ullmo, D., and Jalabert, R.A. (1996). *Physics Reports*, **276**, 1–83.
Ring, P. and Schuck, P. (2000). *The Nuclear Many-Body problem*. Springer, Berlin.
Risken, H. (1989). *The Fokker–Planck Equation*. Springer, Berlin.
Rossignoli, R. and Canosa, N. (2001). *Physical Review B*, **63**, 134523.
Rossignoli, R., Canosa, N., and Ring, P. (1995). *Nuclear Physics A*, **591**, 15.
Rossignoli, R., Canosa, N., and Ring, P. (1998). *Physical Review Letters*, **80**, 1853–1856.
Rossignoli, R., Canosa, N., and Ring, P. (1999). *Annals of Physics*, **275**, 1.
Rummel, C. (2000).
Rummel, C. (2004). Ph.D. thesis, Technische Universität München.
Rummel, C. and Ankerhold, J. (2002). *Euro-Physics Journal*, **B 29**, 105.
Rummel, C. and Hofmann, H. (2001). *Physical Review E*, **64**, 066126.
Rummel, C. and Hofmann, H. (2003). *Nuclear Physics A*, **727**, 24.
Rummel, C. and Hofmann, H. (2005a). *Nuclear Physics A*, **756**, 136.
Rummel, C. and Hofmann, H. (2005b). *Nuclear Physics A*, **756**, 118.
Ruppin, R. and Yatom, H. (1976). *Phys. Stat. Sol. B*, **74**, 647.
Samhammer, R., Hofmann, H., and Yamaji, S. (1989). *Nuclear Physics*, **A503**, 404.
Sartor, R. (1999). In *Nuclear Methods and the Nuclear Equation of State* (ed. M. Baldo). World Scientific, Singapore.
Sato, K., Iwamoto, A., Harada, K., Yamaji, S., and Yoshida, S. (1978). *Zeitschrift f. Physik A*, **288**, 383.
Sato, K., Yamaji, S., Harada, K, and Yoshida, S. (1979). *Zeitschrift f. Physik A*, **290**, 149–156.
Schechter, M., Imry, Y., Levinson, Y., and von Delft, J. (2001). *Physical Review B*, **63**, 214518.
Schechter, M., von Delft, J., Imry, Y., and Levinson, Y. (2003). *Physical Review B*, **67**, 064506.
Scheuter, F. (1982). Ph.D. thesis, Technische Universität München.

Scheuter, F. and Hofmann, H. (1983). *Nuclear Physics*, **A394**, 477.
Scheuter, F., Grègoire, C., Hofmann, H., and Vautherin, J.R. Nix D. (1984). *Physics Letters B*, **149**, 303.
Schiller, A., Bjerve, A., Guttormsen, M., Hjorth-Jensen, M., Ingebretsen, F., Melby, E., Messelt, S., Rekstad, J., Siem, S., and S.W.Ødegrard (2001). *Physical Review C*, **63**, 021306(R).
Schmidt, K.E. and Pandharipande, V.R. (1979). *Physics Letters*, **87B**, 11.
Schmidt, M., Ellert, C., Kronmüller, W., and Haberland, H. (1999). *Physical Review B*, **59**, 10970–10979.
Schmidt, M., Kusche, R., Hippler, T., Donges, J., Kronmüller, W., von Issendorff, B., and Haberland, H. (2001). *Physical Review Letters*, **86**, 1191.
Schröder, W.U. and Huizenga, J.R. (1984). In *Treatise on Heavy-Ion Science, Vol.2: Fusion and Quasi-Fission Phenomena* (ed. D. Bromley). Plenum Press, New York.
Schulman, L.S. (1981). *Techniques and Applications of Path Integrals*. J. Wiley, New York.
Schwabl, F. (1995). *Quantum Mechanics*. Springer, Berlin.
Schwabl, F. (2000). *Statistische Mechanik*. Springer, Berlin.
Scully, M.O. and Zubairy, M.S. (1997). *Quantum Optics*. Cambridge University Press, New York.
Serot, B.D. and Walecka, J.D. (1986). *Adv. of Nuclear Physics*, **16**, 1.
Shapiro, S.L. and Teukolsky, S.A. (1983). *Black Holes, White Dwarfs and Neutron Stars, the Physics of Compact Objects*. John Wiley+Sons, New York.
Shen, C., Kosenko, G., and Abe, Y. (2002). *Physical Review C*, **66**, 061602.
Siemens, P.J. (1982). *Nuclear Physics*, **A387**, 247c.
Siemens, P.J. and Jensen, A.S. (1987). *Elements of Nuclei: Many-Body Physics with the Strong Interaction*. Addison and Wesley, New York.
Siemens, P.J., Jensen, A.S., and Hofmann, H. (1984). In *Nucleon-Nucleon Interaction and the Nuclear Many-Body Problem* (ed. S. Wu and T. Kuo), p. 231. World Scientific, Singapore.
Skyrme, T.H.R. (1956). *Phil. Mag.*, **1**, 1043.
Steinwedel, H. and Jensen, J.H.D. (1950). *Z. Naturforschung*, **5A**, 413.
Strutinsky, V.M. (1967). *Nuclear Physics*, **A95**, 420.
Strutinsky, V.M. (1973). *Physics Letters B*, **47**, 121.
Strutinsky, V.M. and Ivanyuk, F.A. (1975). *Nuclear Physics*, **A255**, 405.
Strutinsky, V.M. and Magner, A.G. (1976). *Sov. J. Part. Nucl.*, **7**, 138.
Strutinsky, V.M. and Magner, A.G. (1987). *Sov. J. Part. Nucl.*, **46**, 951–958.
Strutinsky, V.M., Ivanyuk, F.A., and Pashkevich, V.V. (1980). *Sov. J. Part. Nucl.*, **31**, 47.
Strutinsky, V. M., Magner, A. G., Ofengenden, S.R., and Døssing, T. (1977). *Z. Phys. A*, **283**, 269.
Süssmann, G. (1954). *Zeitschrift f. Physik*, **139**, 543.
Swiatecki, W.J. (1980). *Progress in Particle and Nuclear Physics*, **4**, 383.
Szilard, L. (1925). *Zeitschrift f. Physik*, **32**, 753.

ter Haar, B. and Malfliet, R. (1987). *Physics Reports*, **149**, 207.
Thirolf, P.G. and Habs, D. (2002). *Progress in Particle and Nuclear Physics*, **49**, 325.
Thirring, W., Narnhofer, H., and Posch, H.A. (2003). *Physical Review Letters*, **91**, 130601.
Thoennessen, M. (1996). *Nuclear Physics*, **A599**, 1c–16c.
Tishenko, V., Herbach, C.-M., Hilscher, D., Jahnke, U., Galin, J., Goldenbaum, F., Letourneau, A., and Schröder, W.-U. (2005). *Physical Review Letters*, **95**, 162701.
Tisza, L. (1963). *Review of Modern Physics*, **35**, 151.
Tisza, L. and Quay, P. (1963). *Ann. Phys. (N.Y.)*, **25**, 48.
Tõke, J. and Swiatecki, W.J. (1981). *Nuclear Physics*, **A372**, 141.
Tõke, J., Pieńkowski, L., Sobotka, L.G., Houk, M., and Schröder, W.U. (2005). *Physical Review C*, **72**, 031601(R).
Tolhoek, H.A. (1954). *Physica*, **21**, 1.
Tomonaga, S. (1955). *Progress of Theoretical Physics*, **13**, 467.
Trost, J. (1994).
Uehling, E.A. and Uhlenbeck, G.E. (1933). *Physical Review*, **43**, 552.
van Kampen, N.G. (2001). *Stochastic Processes in Physics and Chemistry. 2nd edition*. North-Holland, Amsterdam.
Vandenbosch, R. and Huizenga, J.R. (1973). *Nuclear Fission*. Academic Press, New York.
Vautherin, D. and Brink, D. (1972). *Physical Review C*, **5**, 626.
Vogel, K. and Risken, H. (1988). *Physical Review A*, **38**, 2409.
von Egidy, T. and Bucurescu, D. (2005). *Physical Review C*, **72**, 044311.
von Egidy, T., Behkami, A.N., and Schmidt, H.H. (1986). *Nuclear Physics A*, **454**, 109–127.
von Egidy, T., Schmidt, H.H., and Behkami, A.N. (1988). *Nuclear Physics A*, **481**, 189–206.
von Oppen, F. (1994). *Physical Review Letters*, **73**, 798–801.
Wambach, J. (1988). *Reports on Progress in Physics*, **51**, 989–1046.
Weber, F (1999). *Pulsars as Astrophysical Laboratories for Nuclear and Particle Physics*. IOP Publishing limited.
Weidenmüller, H.A. (1978*a*). *Journ.Phys.Soc.Japan*, **44**, 701.
Weidenmüller, H.A. (1978*b*, June). New York. Plenum Press.
Weidenmüller, H.A. (1980). *Progress in Particle and Nuclear Physics*, **3**, 49.
Weisbuch, C. and Vinter, B. (1991). *Quantum Semiconductors and Structures*. Academic Press, Boston.
Weise, W. (2005). *Nuclear Physics A*, **to appear**, nucl–th/0412075 (2004).
Weiss, U. (1993). *Quantum Dissipative Systems*. World Scientific, Singapore.
Weissbluth, M. (1989). *Photon-Atom Interactions*. Academic Press, Boston.
Weisskopf, V.F. and Ewing, P.H. (1940). *Physical Review*, **57**, 472.
Whittaker, E.T. and Watson, G.N. (1992). *A Course in Modern Analysis*. Cambridge University Press, New York.

Wigner, E.P. (1958). *Physics Report*, **67**, 325.
Wilets, L. (1964). *Theories of Nuclear Fission*. Clarendon Press, Oxford.
Wood, D.M. and Ashcroft, N.W. (1982). *Physical Review B*, **25**, 6255.
Woosley, S.E., Heger, A., and Weaver, T.A. (2002). *Review of Modern Physics*, **74**, 1015–1071.
Wyld, H.W. (1976). *Mathematics Methods for Physics*. Benjamin, London.
Yamaji, S. and Iwamoto, A. (1983). *Zeitschrift f. Physik A - Atoms and Nuclei*, **313**, 161–166.
Yamaji, S., Hofmann, H., and Samhammer, R. (1988). *Nuclear Physics*, **A475**, 487.
Yamaji, S., Jensen, A.S., and Hofmann, H. (1994). *Progress of Theoretical Physics*, **92**, 773.
Yamaji, S., Ivanyuk, F.A., and Hofmann, H. (1997). *Nuclear Physics A*, **612**, 1.
Yannouleas, C. and Broglia, R. A. (1992). *Annals of Physics*, **217**, 105–141.
Yuen, H.P. and Tombesi, P. (1986). *Optics Communications*, **59**, 155–159.
Yukawa, H. (1935). *Proc.Phys.Math.Soc.Japan*, **17**, 48.
Zagrebaev, V. and Greiner, W. (2005). *Journal of Physics G: Nuclear and Particle Physics*, **31**, 825–844.
Zakrzewski, J. and Delande, D. (1993). *Physical Review E*, **47**, 1650–1664.
Zelevinsky, V., Brown, B.A., Frazier, N., and Horoi, M. (1996). *Physics Report*, **276**, 85–176.
Zwanzig, R. (1960). *Journal of Chemical Physics*, **33**, 1338.

INDEX

action, 523, 524
  classical, 525–527
  Euclidean, 299, 533, 534
    effective, 545, 551
  fluctuation, 527
  minimal, 530
adiabatic limit, 310, 312
auto-correlation function, 561

ballistic transport, 344–345, 348
  conditions on mean free path, 344
barrier
  fission, see fission barrier
  fusion, 311
  motion across, 207–209, 320, 530
Bethe–Goldstone equation, 39–42
Bethe–Weizsäcker formula, 25, 26, 28, 105, 106, 111, 143, 228, 234
BHF, see Brueckner theory
binding energy, 24, 26, 28, 46, 52
  finite nuclei, 117, 228, 229, 234
  Jastrow, 50
  neutron, 15
  shell correction, 250, 255
Bohm–Pines method, 142, 187, 211, 212, 214, 304, 544
  Hamiltonian, 213, 217–219, 546
Bohr–Wheeler formula, 229–232, 234, 250, 287, 291, 292, 351
Boltzmann–Uehling–Uhlenbeck, see BUU
branching ratio, 228, 243, 400
Breit–Wigner, see resonance
Brownian
  motion, 534
  particle, 315, 554
Brueckner theory, 36, 46, 47
  BHF, 46, 47, 52, 65
  BHF at finite T, 48
BUU equation, 65, 151, 244, 431, 435, 436
  electronic transport, 343

Caldeira–Leggett model, 206, 211, 265, 299–301, 305, 306, 532
canonical ensemble, 51, 261, 278, 285, 438, 453, 466, 495
  isolated system, 157, 260, 454, 457
  level densities, 247, 248, 255

Cassini ovaloids, 75, 274
chaining hypothesis, 239
chaotic motion, 119, 450, 542
  Bohigas conjecture, 92
  complex systems, 92–93
Chapman–Kolmogorov equation, 190, 523, 562–566, 568, 572, 586
collective motion
  collisional damping, 129, 150–154, 178, 325, 333
  coupling constant, 132, 133, 139, 141, 212
    Bohr–Mottelson, 140
    diabatic, 172, 173
    isovector modes, 143
    quasi-static, 156, 157, 160, 168, 173
  irreversible behavior, 159, 163
  RPA, 136, 218
    finite T, 173
  secular equation, see secular equation
  strict Markov limit, 157, 160
    entropy production, 159
collective response, see response function
collective strength distribution, 166–171
collision term, 63–65, 102, 426, 434
  electronic transport, 342–344
  near equilibrium, see relaxation time approximation
  semi-classical approximation, 428–432
composite systems, 452
compound nucleus, 16, 23, 227, 311
  Bohr hypothesis, 225, 293
  decay, 225, 259
  formation cross section, 225–229, 387, 399, 400
conditional probability, 453, 562, 563, 572, 573
correlation function, 177, 178, 269, 271, 334, 335, 337, 357, 365, 489, 491–495
  at small ω, 158, 178, 180
  transport equation, 219, 220, 583, 584, 586
correspondence rules, 220, 416, 417, 584
critical temperature
  liquid–gas phase transition, 59
  pairing, 83, 185

unstable mode, 202, 253, 299, 305–307, 539
cross section
  absorption, 18, 351, 388, 395, 402, 403
  differential, 5, 65, 227, 376, 380
    heavy-ion collisions, 317–318, 335
  elastic, 19, 381, 390, 399, 402, 403, 405
  fluctuating, 397, 399–402
  inelastic, 381
  pre-equilibrium, 241
  reaction, 331, 387, 388, 402, 405
  scattering of light
    metal clusters, 353–354
    nuclear dipole, 330–331
  T-matrix, 383
  total, 18, 330, 353, 380, 390, 399, 402
crossover temperature, 299, 302
cumulant expansion, 559–560

Darwin–Fowler approach, 246, 253, 258, 457–460
DDD, see dissipative diabatic dynamics
degenerate model, 132, 137
density operator
  pure state, 407
  quasi-static, 155, 503, 513, 514
  statistical mixture, 407
detailed balance, 237, 241, 243, 401, 570
diabatic model, 172, 173
diabatic motion, 172, 173
dielectric function, 341, 350–355, 358, 359, 363, 520
diffusion coefficients
  coordinate dependent, 268
  equilibrium fluctuations, 197
  FDT, 198
  general FP-equations, 565, 568, 579, 585
  high T limit, 209, 284, 289, 322
  over-damped motion, 199
  quantal, 200–203, 207, 220, 264, 299, 321, 323–325, 585
    unstable modes, 200, 202
    weak damping, 201, 206, 209, 319, 586
  wall-window model, 316
dipole modes, 330
  large scale collective motion, 334, 335
  motional narrowing, 335
  sum rule, 331, 333
dispersion relation, see secular equation
dissipation, 177, 180, 189, 292, 311, 326, 335, 341, 348, 358, 512
  hydrodynamic, 180
  microscopic origin, 145–154, 365
  one-body, 313, 315, 344, 354

strength, 292, 294, 323
wall picture, 180, 348
dissipation rate, see energy transfer
dissipative diabatic dynamics, 174
doorway mechanism, 18, 242
doorway resonance, see resonance
Drude regularization, 195, 207, 321, 500
Drude–Lorentz model, 352, 354, 520

effective charge, 332
effective mass, 9, 37, 68, 152
  electron, 342, 344, 345, 352, 353
  Skyrme, 56
Einstein relation, 200, 206, 308, 315, 318, 325–327, 335, 557, 576
electric conductivity, 341, 352, 354, 358, 519
  quantization, 346–348
energy
  conservation, 516
    collective motion, 129, 161, 234
    collision term, 432
  energy fluctuation, 446, 457, 464
  energy transfer, 138, 377, 519
    irreversible, 354, 363
    rate, 363, 486, 520
    wall picture, 361–366
  entropy, 51, 59, 124, 154–161, 438, 444, 447, 448, 473, 515
    concavity, 445, 448
    independent particles, 424, 468, 470
    level density, see level density
    maximum, 448, 503, 569
    shell correction, 126, 472
    variation, 159, 439, 444, 449, 473, 503, 517
      in time, 510–512, 514
  equation of state, 52, 55, 469
    Brueckner theory, 60
    chiral perturbation theory, 34, 59
    in astrophysics, 60
    Jastrow, 60
    semi-phenomenological BHF, 47
    Skyrme–HF, 56
  equilibrium, see global, local equilibrium
  equilibrium fluctuations, 190, 194, 197, 266, 267, 419
    at instabilities, 200, 202
    oscillator, 195
    quantal, 199, 201, 321, 322, 492
    selfconsistency, 216
  equipartition theorem, 196, 204–206, 336, 418, 558
    quantal version, 195, 418
  ergodic
    behavior, 279, 508

hypothesis, nuclear reactions, 17, 395
  system, 156, 157, 160, 176
escape time, 288–291, 297, 298
evaporation rate, 229, 259
exciton model, 243
extensivity, 447, 455
  lack of, 464–466

FDT, see fluctuation dissipation theorem
Fermi gas, 7
  level density, 8
  total energy, 10
  two-body correlations, 11
Fermi liquid, 30, 61, 63, 343
  instabilities, 500
Fermi's golden rule, 376, 430, 482, 488, 569
first law of thermostatics, 439–440
fissility parameter, 27, 232, 234
fission barrier, 230, 232–234, 300, 311
fission rate
  Bohr–Wheeler, 229, 291
  Kramers, 283, 286, 292
  Langer, 285, 306
  over-damped motion, 287
  quantal correction, 299–302
  transition state, 229, 286
fluctuating force, 145, 193, 218–220, 269, 308, 309, 318, 329, 334, 335, 555, 560
fluctuation, 561, see quantum fluctuations
fluctuation dissipation theorem, 190, 334, 491–492
  diffusion coefficients, 198
  heavy-ion collisions, 319
  microcanonical for RMM, 494–496
  Nakajima–Zwanzig
    non-perturbative, 219
    perturbative, 583
  unstable modes, 499
Fokker–Planck equation, 262, 308, 524, 557, 568, 570–572
  backward equation, 573–575
  boundary conditions
    absorbing barrier, 572, 574, 575
    reflecting barrier, 572–575
  Chapman–Kolmogorov equation, 190
  for heavy-ion collisions, 317
  global, 262, 279
  Kramers–Moyal expansion, 270, 564
  linearized, 577
  microscopic derivation, 580
  pseudo-FPE, 198
  quantum effects, 262, 299
  stationary solutions, 264, 282, 284
    quantal oscillator, 203–206

  vs Langevin equation, 262, 268, 269, 308, 567, 571
  formation probability, 400–402
Fraunhofer diffraction, 381
free energy, 28, 82, 124, 157, 160, 169, 192, 274, 303, 440, 473, 535
  extremal property, 448
  free gas, 415
  imaginary part, 285, 286
  nuclear matter, 51, 57, 59
  probability distribution, 270
  shell correction, 126, 127, 182, 275
  SPA, 253, 303, 547
  stiffness, 157, 172, 173, 285, 304, 585
    negative, 282, 285, 305, 477, 539, 550, 579
friction, 183, 184, 209, 270, 292, 296, 299, 510
  coefficient, 162, 164, 176, 180, 183, 266, 272, 282, 316, 336, 344, 363, 367–369, 487
    Caldeira–Leggett, 265
    coordinate dependence, 273, 274
    variation with T, 275
  force, 145, 174, 177, 180, 309, 334, 365, 554
  surface, 314, 315, 321, 326
  tangential, 188
  wall-and-window, 312, 313, 326
  zero frequency limit, 177, 179, 362, 367
fusion
  formation probability, 321–325
  inferences on nuclear transport, 325

gamma down ($\Gamma^\downarrow$), gamma up ($\Gamma^\uparrow$), 101, 394
Gaussian
  process, 334, 555
  propagator, 192
generalized ensemble, 443
generalized forces, 515
global equilibrium, 82, 265, 266, 301, 347, 432, 435
grand canonical ensemble, 82, 133, 483
grand potential, 59, 84, 446
  independent particles, 466–470
  shell correction, 125, 249, 471
Green function, 348, 349, 426, 435, 539–541, 551
  temperature, see thermal
  thermal, 497, 548
Gutzwiller trace formula, 124, 249, 471, 472, 542

hard sphere gas, 32
Hartree–Fock, 67–70, 107

at finite T, 424–425
with density operators, 420–424
Hauser–Feshbach, *see* nuclear reaction
heat, 449
heat bath, 210, 254, 260, 267, 328, 348, 443, 454, 506, 532
heat capacity, *see* specific heat
heavy-ion collisions
  charge equilibration, 318
  deep inelastic, 311, 314, 318
  fusion, *see* fusion
  scenario, 312
Hubbard–Stratonovich transformation, 214, 304, 351, 543, 545

imaginary time propagation, *see* time evolution
incompressibility, 34, 57, 59, 169
inertia, 137, 175, 214
  coefficient, 162, 164, 183, 186, 187, 272
    coordinate dependence, 273
    sum rule limit, 138
  cranking model, 144, 158, 166, 186, 187, 211
  irrotational flow, 138, 184, 210
  zero frequency limit, 144, 166
interaction
  bare nucleon–nucleon, 4
  effective, 52
    Gogny, 69
    separable, 69
    Skyrme, 54
  nucleon–nucleus, 14
internal energy, 82, 157, 192, 438, 440
  Brueckner theory, 48
  conservative force, 159
  independent particles, 467
  Jastrow, 51
  shell correction, 126
  Skyrme, 57
  stiffness, 155
  Strutinsky smoothing, 122–124
inverse friction expansion, 267, 281, 288
irreversibility, 148, 316, 348, 450
irreversible behavior, 145, 149, 210, 279, 341, 348, 358, 362, 391, 450, 492
irreversible processes, 177, 448, 449
  linear, 507–517
isoscalar modes, 143, 166, 212, 544
isovector modes, 143, 319, 330

Jastrow method, 48, 52
jellium model, 350, 354

Kramers' rate formula, 182, 230, 281, 292, 297, 298
  quantum corrections, 299, 301
Kramers–Kronig relation, 97, 151, 170, 330, 484
Kramers–Moyal expansion, 221, 262, 263, 265, 267, 270, 315, 564, 585
Kubo formula, 348, 518

Landau damping, 145, 359
Landau–Vlasov equation, 63, 65, 151, 431
Landauer–Büttiker formulas, 348
Langevin equation, 193, 262, 267–269, 327, 336, 554, 570, 571
Legendre transformation, 440, 444, 447, 542
level density, 383, 395, 398, 399, 569
  angular momentum, 251–252
  Bethe formula, 247, 248, 251, 255
  empirical, 254–257
  energy distribution, 456
  many-body
    exact, 450, 452, 454, 455, 457, 466
    independent particles, 467
  parameter, 123, 231, 250, 253, 255, 323
  particle-hole, 239
  residual interactions, 252
  single particle, *see* single particle level density
  SMMC, 254
  spin-dependent, 253
  thermal entropy, 246–249, 251, 253, 456, 473
  with constraints, 460
LHA, *see* locally harmonic approximation
likelihood function, 462–463
Liouville
  equation, 62, 63, 197
  operator, 217, 512, 581
Lippmann–Schwinger equation, 374, 377
liquid drop model, 23, 25, 104, 109, 157, 314, 465
  and compound nucleus, 28
  deformed, 26
  finite range, 105, 117
  irrotational flow inertia, 166, 186, 308, 321
  macroscopic limit, 114
  static energy, 230, 234, 248, 250, 311, 351
  static stiffness, 157, 169, 183
local equilibrium, 82, 501
  collective variables, 154, 198, 281
  in space, 66, 186, 347, 435
locally harmonic approximation, 169, 190–194, 262, 279, 293

Index

level crossings, 275
  quantum effects, 206, 210, 211, 217, 299–302

macroscopic
  behavior, 23
  limit, 105, 114, 316, 321, 362, 456
    dynamics, 170, 180
    friction, 369
    level density, 247
    transport coefficients, 183–184, 366
  system, 79
Markov
  approximation, 156, 192, 326, 513
  limit, 193
master equation, 236, 568–570, 585
matrix
  G-, 42–44, 52, 56, 65, 436
    at small momenta, 52–54
  S-, 19, 348, 375, 380, 388, 395, 402, 403
  T-, 19, 42, 44, 64, 65, 226, 373–378, 392, 569
  angular momentum, 380, 389
  cross section, 383
  fluctuating part, 397
  reactions, 384–385
  resonances, 387, 389
Matsubara frequencies, 499, 500, 538, 546
maximum entropy principle, 187, 278, 443, 466
mean first passage time, 288, 294, 573, 574
mean free path, 20, 23, 28, 344, 346, 353, 442
metal cluster
  conductivity, 358–360
  optical properties, 351
    damping width, 354, 359
    dielectric function, 355–358
  shell effects, 350, 351
  superconductivity, 351
MFPT, *see* mean first passage time
microcanonical ensemble, 260, 437, 452, 457, 459
  compound nucleus, 17, 225, 229, 235, 400
  level density, 255, 256, 258, 259
  linear response, 495
  specific heat, 465
microreversibility, 64, 226, 378, 401
microscopic damping mechanism, 102, 551
Mori product, 511, 518, 592
multi-step compound reactions, 242
multipole vibrations, 140

Nakajima–Zwanzig

FDT, 220
  non-perturbative, 217–220, 315, 329
  perturbative, 582–585
  projection technique, 580–582
nearest level spacing, 86, 90, 276
neutron multiplicity, 292
noise
  additive, 568
  induced force, 267, 269, 568, 571
  internal, 269
  multiplicative, 268, 327, 335, 568
  white, 268
nuclear matter, 5, 9, 28, 31
  chiral dynamics, 34
  dilute or dense?, 35
  saturation, 24, 30, 32, 46, 47, 52
    density, 33
  three-body forces, 7
nuclear reaction
  Bohr hypothesis, 16, 395, 400, 403
  channel, 19, 226, 228
  compound, 15, 387
  Hauser–Feshbach theory, 227, 238, 241, 401–403
  potential scattering, 14, 15, 17, 373, 374, 376, 382
  pre-equilibrium, 18, 225, 235, 236
nuclear thermometry, 257–262

Onsager relations, 515
optical model, 18–22, 227, 397, 401–403, 436
  loss of probability, 20
  microscopic derivation, 390–391
  potential, 21

pairing, 69, 79–83, 108, 117
  field, 80, 83
  gap equation, 82–84
  gap parameter, 81, 83, 84, 351
  interaction, 70, 79, 252, 253, 351
  level density, 249, 255
  phase transition, 257
  rotations, 145, 185
  specific heat, 254, 256, 466
  transport coefficients, 181–183, 300, 301
  window, 79, 83
partition function, 438
  classical gas, 414
  composite systems, 453
  decay rate, 284–286, 301
  free energy, 440
  generalized ensemble, 444
  grand canonical, 446
  imaginary part, 204
    PSPA, 304

inverted oscillator, 204
Laplace transform, 458
level density, 460
likelihood function, 463
oscillator, 418
PSPA, 303
Wigner–Kirkwood, 114
path integral, 524
  continuous paths, 527
  Euclidean, 535
    metric, 533
  Jacobi determinant, 528
  semi-classical, 525
  short-time propagator, 523, 525
  statistical mechanics, 532–533
    classical, 535
    partition function, 535
    quantum corrections, 536
periodic orbit theory (POT), 105, 247, 366, 540, 542
perturbed-SPA, 203, 253, 254, 303–307, 548, 550, 551
  breakdown, 550
phase shift, 21, 379
piston model, 314
Poisson distribution, 86, 91, 93
Porter–Thomas distribution, 396
POT, see periodic orbit theory
potential energy, 44, 54, 424
  collective, 118, 171, 172, 182, 206, 214, 233, 292, 294, 319, 322, 327, 351, 537, 547
  collective motion
    free energy, 285
  per particle, 10, 31
  single particle, 31, 33, 345, 347, 545
pre-equilibrium
  cross section, 241
  time-dependent description, 236–239
pressure, 160, 447, 464, 473
  nuclear matter, 57, 60
prior density, 463, 474
process
  Markovian, 562, 563
  non-Markovian, 165, 174, 219, 305, 329, 582, 583
  stationary, 561, 562, 564
  stochastic, 277, 309, 335, 555, 560–570
propagator
  local, 196, 586
  path integral, see path integral
  short-time, 190, 564–568
PSPA, see perturbed-SPA

quadrupole vibrations, 165, 166, 186, 189, 356, 364, 367

quadrupole–quadrupole interaction, 70, 252, 253
quantum correction factor, 299, 301–303, 305, 537, 547, 551
  PSA, 549
quantum fluctuations
  equations of motion, 196–200
  equilibrium, see equilibrium fluctuations
  time dependence
    general solutions, 578–580
    stable modes, 206–209
    unstable modes, 207
quantum size effect, 354
quasi-static
  density operator, 516
  equilibrium, 286
  force, 155, 440
  picture, 154, 159, 184, 257, 270, 442, 516
  process, 159, 278, 449
quasiparticles, 61, 67, 81–83, 101, 183, 349

Random Matrix Model, 396, 449
  collective transport, 327, 586
  experimental information, 91
  Gaussian ensembles, 87–89
  level distribution, 276
  linear response, 493–496
    relaxation, 150
  pre-equlibrium, 242
randomization hypothesis, 316
reaction, see nuclear reaction
relativistic mean field theory, 61, 104
relaxation function, 156, 508, 509, 516
relaxation time, 342
  collective, 294, 298, 321
  intrinsic, 149, 153, 235, 312, 326, 336, 337
    DDD, 174
  kinetic energy, 319, 325
  Matthiesen's rule, 343
  mean free path, 344
  occupation numbers, 174
  RMM, 150
  self-energy, 66, 181, 343, 354, 433
relaxation time approximation, 66, 177, 180, 181, 273, 342–344, 434–435, 512
resonance, 396
  Δ, 7
  Breit–Wigner, 16, 101
  dipole, 337
  doorway, 392–395
  isolated, 385–388
  Mie, 353

neutron, 14, 16, 225, 255
  overlapping, 389, 398, 404
  potential scattering, 14, 16, 390
response function, 130–133, 146, 275, 277, 315, 352, 491, 492, 509
  1p-1h excitations, 133–134
  collective, 136–140, 329, see collective strength distribution
    qq-response, 161, 162
    quasi-static processes, 160–164
  complex frequencies, 496–500
  dipole, 333
  dissipative part, 146–149, 166, 173, 189, 272, 355, 363, 366, 476, 496, 499, 501
    collisional damping, 150
  energy dissipation, 486
  general definitions, 475, 482–484
  large frequencies, 485
  oscillator, 163, 164, 195, 475–478
  reactive part, 476, 488, 505, 506
  schematic model, 164
  spectral representations, 487–489
  symmetry relations, 484
  thermal Green function, 497
  time dependence, 149, 152
    RMM, 150
  unstable modes, 134, 499
rotational
  damping, 188
  invariance, 4, 91, 187, 188
  moment of inertia, 144, 185–187, 232, 251, 254
  motion, 185, 186
    at finite T, 185–189
    dissipative, 188
RPA, see collective motion

Sackur–Tetrode formula, 415
saddle point approximation, 282–285, 290, 292, 297, 298, 307, 539, 590
scattering theory, see nuclear reaction
second law of thermostatics, 159, 209, 280, 437, 449
secular equation, 132, 133, 140, 143, 149, 156, 158, 172, 189, 194, 200, 214, 215, 304
  unstable modes, 134
self-consistency condition, 137, 141, 216
self-energy, 66, 150, 153, 181, 435–436
semi-classical approximation, 120, 232, 252, 473
  collision term, 64
  electronic conductance, 342
  Green function, 541
  transition rate, 227

separation energy, see binding energy
shell effects, 110–113, 266, 307, 308, 312, 361, 465
  gross features, 113, 119, 120, 182, 272
  level density, 247, 249, 255
  thermostatic quantities, 124–127, 470–473
  variation with T, 124–127, 157, 167, 173, 184, 250, 272, 289
shell model, 15, 29, 107, 128, 147, 153, 275
  deformed, 30, 73, 107, 275, 350, 364
  independent particle model, 70
    validity, 103
  Monte Carlo, 254
  potential, 108, 113, 129, 144
  residual interactions, 85, 276
single particle level density, 83, 110, 124, 243, 247, 250, 540, 552
  angular momentum, 472
  coordinate dependence, 250
  ETF, 115, 543
  Fermi gas, 8
  fluctuating part, 110, 119
  Green function, 539–540
  gross shell behavior, 120
  oscillating, 124, 470
  POT, 542
  smooth, 110, 469
  Thomas–Fermi approximation, 115, 368, 369, 542
  Weyl formula, 370
single particle potential, 42
  Brueckner theory, 48
  Coulomb barrier, 73, 229
  folded Yukawa, 71, 76
  HF, 63, 67
  Nilsson model, 75, 112, 275
  Woods–Saxon, 22, 70, 75, 111, 152, 167, 275, 350
Slater determinant, 410, 419
SMMC, see shell model Monte Carlo
Smoluchowski equation, 267, 281, 289, 559, 570
Sommerfeld expansion, 123, 367, 469–470
SPA, see static path approximation
specific heat, 457, 504, 505
  at constant Q, 160
  empirical, 256
  energy fluctuation, 446, 455, 464
  negative, 464–466
  stability conditions, 464
  temperature fluctuations, 466
  variation with T, 253
spectral function, 99–103
  particles and holes, 100

quasiparticles and quasiholes, 100
spectroscopic factor, 95
spin-orbit
  density, 55, 67
  interaction, 15, 21, 22, 72, 76, 77, 141, 275
spreading width, 96, 101, 242, 395
static energy, 24, 104, 130, 172, 192, 216, 232, 310, 311
  diabatic model, 172
  finite nuclei, 115–118
  macroscopic limit, 183, 232
  stiffness, 27, 132, 162–168, 171, 189, 266, 304, 319
    T dependence, 169
    constrained energy, 143
    diabatic, 172, 176
    negative, 27, 143, 189, 200–202, 204, 206, 233
    nuclear matter, 57
    sum rule, 139
static path approximation, 253, 254, 303, 351
static response, 130, 132, 139, 164, 168, 172–174, 279, 505, 506, 549
stationary phase, 541, 543, 590
steepest descent, *see* saddle point approximation
stochastic differential equation, 268, 269
strength function, 95–99, 377, 405
  collective motion, 195, 334, 363, 492
  dielectric, 359
  periodic perturbation, 492
structure function, 450, 461, 463
Strutinsky
  energy theorem, 107, 108, 128
  method, procedure, 102, 104, 106, 108, 469
  smoothing, 105, 169, 172, 173, 362, 366
    occupation number, 113
    particle number, 114
sudden limit, 310, 312
sum rules, 137–139
  energy weighted sum, 138, 166, 213
  dielectric function, 352
susceptibility, 441
  adiabatic, 155, 157, 161, 507
  isolated, 506, 508
  isothermal, 157, 491, 505, 508

temperature
  effective, 204, 317, 418, 584
  inverse, 256, 438
  nuclear, 257, 258, 261
    possible justification, 260

uncertainty, 157, 247, 254, 256, 260, 261, 328, 457, 461–464, 495
  variation with Q, 160, 279, 442
thermostatic limit, 36, 258, 457, 464
Thomas–Fermi approximation, 369
  extended, 543
three-body force, 50, 54, 56, 57
time evolution
  imaginary time, 203, 299, 544, 545
  Liouville operator, 512, 581
  operator, 478, 480
  propagator, *see* propagator
time-dependent Hartree–Fock, 61
  extended, 426
  finite T, 425
time-ordering operator, 479, 497, 545, 548
transition rate, 259, 376, 481, 568
  collision term, 64, 65, 430
  compound nucleus, 225–228
  FDT, 493
  master equation, 568–570
  perturbation theory, 480–482
transition state theory, *see* fission rate
transmission coefficient, 19, 238, 240, 381, 401
transport coefficients, 201, 567, 568
  Q-dependence, 182, 194, 265, 311, 586
  T-dependence, 177–185
  finite frequency, 315
  heavy-ion collisions, 312, 316, 321, 325
  influence of rotation, 188
  large-scale motion, 271–275
  macroscopic, 322–324
  microscopic, 263, 294, 298, 301, 309, 321, 322, 324
  self-consistent, 162–165, 585
  zero frequency limit, 158, 162
tunneling, 306
  dissipative, 551

uncertainty relations
  Heisenberg, 32, 65, 195, 206
  in equilibrium, 419
  thermal, 459, 463–469
  time-energy, 16, 66
unstable modes, 207–209, 299, *see* critical temperature, *see* diffusion coefficients

variables
  collective, 124, 186, 190, 311, 336
    fluctuating, 211
    probability distribution, 269–270, *see* Fokker-Planck equation
  extensive, 258, 447

external vs internal, 131, 135, 271, 442, 443
intensive, 258, 447
viscosity, 28, 29, 286, 301
  limit of high, 283, 286, 287, 292
  limit of low, 286
  two-body, 28, 180, 292, 344
von Neumann eqn, 61, 217, 437, 512, 581
  Wigner transform, 197, 426–428

wall formula, 145, 180, 183, 184, 275, 316, 321, 362, 364, 365, 369
Weisskopf
  –Ewing relations, 400
  formula, 227, 259
Wigner
  distribution, 91
  function, 174, 301, 414
    equilibrium, 418
    time dependent, 190, 191
    wave packets, 415
  surmise, 91
  transform, 62, 195, 197, 203, 220, 412–416, 427, 584, 585
Wigner–Kirkwood expansion, 114, 124, 539
WKB approximation, 119, 531

yrast line, 187, 189

zero frequency limit, 158, 165, 273, 274, 315, 586
  conductivity, 519